UNIVERSITY OF STRATHCLYDE
30125 00581405 7

D1765405

ANDERSONIAN LIBRARY
WITHDRAWN
FROM
LIBRARY
STOCK
UNIVERSITY OF STRATHCLYDE

COMPUTATIONAL METHODS FOR ELECTROMAGNETICS

IEEE/OUP SERIES ON ELECTROMAGNETIC WAVE THEORY

The IEEE/OUP Series on Electromagnetic Wave Theory consists of new titles as well as reprintings and revisions of recognized classics that maintain long-term archival significance in electromagnetic waves and applications.

BOOKS IN THE IEEE/OUP SERIES ON ELECTROMAGNETIC WAVE THEORY

Chew, W. C., *Waves and Fields in Inhomogeneous Media*
Christopoulos, C., *The Transmission-Line Modeling Methods: TLM*
Clemmow, P. C., *The Plane Wave Spectrum Representation of Electromagnetic Fields*
Collin, R. E., *Field Theory of Guided Waves*, Second Edition
Dudley, D. G., *Mathematical Foundations for Electromagnetic Theory*
Elliot, R. S., *Electromagnetics: History, Theory, and Applications*
Felsen, L. B., and Marcuvitz, N., *Radiation and Scattering of Waves*
Harrington, R. F., *Field Computation by Moment Methods*
Jones, D. S., *Methods in Electromagnetic Wave Propagation*, Second Edition
Lindell, I. V., *Methods for Electromagnetic Field Analysis*
Tai, C. T., *Generalized Vector and Dyadic Analysis: Applied Mathematics in Field Theory*
Tai, C. T., *Dyadic Green Functions in Electromagnetic Theory*, Second Edition
Van Bladel, J., *Singular Electromagnetic Fields and Sources*
Wait, J., *Elecromagnetic Waves in Stratified Media*

COMPUTATIONAL METHODS FOR ELECTROMAGNETICS

IEEE PRESS Series on Electromagnetic Waves

Andrew F. Peterson
School of Electrical and Computer Engineering
Georgia Institute of Technology

Scott L. Ray
Modeling and Information Sciences Laboratory
Dow AgroSciences

Raj Mittra
Department of Electrical and Computer Engineering
Pennsylvania State University

IEEE Antennas & Propagation Society, *Sponsor*

The Institute of Electrical and Electronics Engineers, Inc., New York

Oxford University Press
Oxford, Tokyo, Melbourne

This book and other books may be purchased at a discount
from the publisher when ordered in bulk quantities. Contact:

IEEE Press Marketing
Attn: Special Sales
Piscataway, NJ 08855-1331
Fax: (732) 981-9334

For more information about IEEE PRESS products,
visit the IEEE Home Page: http://www.ieee.org/

Printed in the United States of America

10 9 8 7 6 5 4 3 2 1

ISBN 0-7803-1122-1
IEEE Order Number: PC5581

Library of Congress Cataloging-in-Publication Data

Peterson, Andrew F., 1960-
Computational methods for electromagnetics / Andrew F. Peterson, Scott L. Ray, Raj Mittra.
p. cm.
"IEEE Antennas & Propagation Society, sponsor."
"IEEE Press series on electromagnetic waves."
Includes bibliographic references and index.
ISBN 0-7803-1122-1
1. Electromagnetism. 2. Numerical analysis. I. Ray, Scott L., 1957- . II. Mittra, Raj. III. IEEE Antennas and Propagation Society. IV. Title.
QC760.P48 1997
621.3'01'5194--dc21 97-39612
CIP

IEEE Press
445 Hoes Lane, P.O. Box 1331
Piscataway, NJ 08855-1331

Cover Design: William T. Donnelly, *WT Design*

IEEE Antennas & Propagation Society, *Sponsor*
APS Liaison to IEEE Press, Robert Mailloux

Technical Reviewers

Andreas C. Cangellaris, University of Illinois
William A. Davis, Virginia Tech
Donald R. Wilton, University of Houston

Oxford University Press
Walton Street, Oxford OX2 6DP

Oxford New York
Athens Auckland Bangkok Bombay
Calcutta Cape Town Dar es Salaam Delhi
Florence Hong Kong Istanbul Karachi
Kuala Lumpur Madras Madrid Melbourne
Mexico City Nairobi Paris Singapore
Taipei Tokyo Toronto
and associated companies in
Berlin Ibadan

Contents

CHAPTER 2 INTEGRAL EQUATION METHODS FOR SCATTERING FROM INFINITE CYLINDERS 37

CHAPTER 3 DIFFERENTIAL EQUATION METHODS FOR SCATTERING FROM INFINITE CYLINDERS 95

Preface

A decade ago, when the task of developing this book was initiated, there were few available texts on computational techniques for electromagnetics. Although a large number have appeared since then, none attempt to treat both integral and differential equation formulations in a unified manner. The present text is intended to fill that gap and is designed for graduate-level classroom use or self-study. Its primary focus is open-region formulations, and while resonant cavity and antenna applications are touched on in places, the majority of the material is presented in the context of electromagnetic scattering. We have attempted to provide enough detail to enable a reader to implement the concepts in software. In addition to a few subroutines in Appendix C, a collection of related computer programs is available through the Internet. Earlier drafts of the material were tested in graduate courses taught at the University of Illinois and the Georgia Institute of Technology as well as in a number of continuing education courses. The authors sincerely appreciate the comments of former students, colleagues, and the dozen or more reviewers who offered critiques during the book's development.

Andrew F. Peterson
Scott L. Ray
Raj Mittra

Acknowledgments

I would like to thank Paul W. Klock for introducing me to computational electromagnetics and David R. Tanner for prodding me to seek a deeper understanding of almost every aspect of the discipline.

Andrew F. Peterson
Georgia Institute of Technology

1

Electromagnetic Theory

The success of electromagnetic analysis during the past century would not have been possible except for the existence of an accurate and complete theory. This chapter summarizes a number of concepts from electromagnetic field theory used in numerical formulations for scattering problems. Differential and integral equations that provide the basis for many of the computational techniques are introduced. In addition, expressions are developed for calculating the scattering cross section of a target. The presentation is intended as a review of these concepts rather than an introduction, and the reader is encouraged to study references [1–11] for an in-depth discussion of this material.

1.1 MAXWELL'S EQUATIONS

Consider a source-free region of space containing an inhomogeneity characterized by relative permittivity ε_r and permeability μ_r, both of which may be a function of position (Figure 1.1). If this region is illuminated by an electromagnetic field having time dependence $e^{j\omega t}$, the fields in the vicinity of the inhomogeneity must satisfy Maxwell's equations

$$\nabla \times \bar{E} = -j\omega\mu_0\mu_r\bar{H} \tag{1.1}$$

$$\nabla \times \bar{H} = j\omega\varepsilon_0\varepsilon_r\bar{E} \tag{1.2}$$

$$\nabla \cdot (\varepsilon_0\varepsilon_r\bar{E}) = 0 \tag{1.3}$$

$$\nabla \cdot (\mu_0\mu_r\bar{H}) = 0 \tag{1.4}$$

where $\bar{E}$ and $\bar{H}$ are the electric and magnetic fields, respectively. (More precisely, $\bar{E}$ and $\bar{H}$ are complex-valued phasors representing the vector amplitude and phase angle of the time-harmonic fields.) We have specialized these equations to a medium that is linear and isotropic.

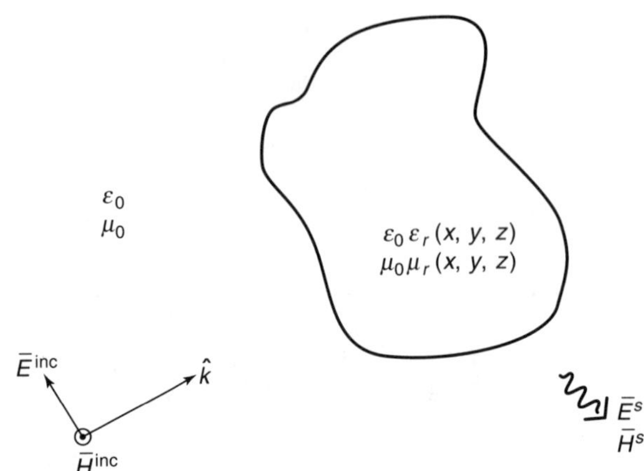

Figure 1.1 An inhomogeneity illuminated by an incident electromagnetic field.

The study of electromagnetics involves the application of Equations (1.1)–(1.4) to a specific geometry and their subsequent solution to determine the fields present within the inhomogeneity, the fields scattered in some direction by the presence of the inhomogeneity, or some similar observable quantity. The focus of this text will be the development of techniques for the numerical solution of Equations (1.1)–(1.4) or their equivalent.

Regions containing penetrable dielectric or magnetic material may be bounded by material having a very high conductivity, which is often approximated by infinite conductivity and termed a perfect electric conductor. Although ε_r and μ_r may in general be complex valued to represent conducting material, in the limit of infinite conductivity we denote the surface of a perfect electric conductor as a boundary of the problem domain. On such a boundary, the electric and magnetic field vectors satisfy the conditions

$$\hat{n} \times \bar{E} = 0 \tag{1.5}$$

$$\hat{n} \times \bar{H} = \bar{J}_s \tag{1.6}$$

$$\hat{n} \cdot \bar{E} = \frac{\rho_s}{\varepsilon_0 \varepsilon_r} \tag{1.7}$$

$$\hat{n} \cdot \bar{H} = 0 \tag{1.8}$$

where $\hat{n}$ is the normal vector to the surface that points into the problem domain (Figure 1.2), $\bar{J}_s$ is the surface current density, and ρ_s is the surface charge density.

Along an interface between two homogeneous regions specified by relative permittivity ε_r and permeability μ_r, appropriate continuity conditions involving the electric and magnetic fields are

$$\hat{n} \times (\bar{E}_1 - \bar{E}_2) = 0 \tag{1.9}$$

$$\hat{n} \times (\bar{H}_1 - \bar{H}_2) = 0 \tag{1.10}$$

$$\hat{n} \cdot (\varepsilon_{r1} \bar{E}_1 - \varepsilon_{r2} \bar{E}_2) = 0 \tag{1.11}$$

$$\hat{n} \cdot (\mu_{r1} \bar{H}_1 - \mu_{r2} \bar{H}_2) = 0 \tag{1.12}$$

where $\hat{n}$ is normal to the interface. Although the tangential components of the $\bar{E}$ and $\bar{H}$ fields are continuous, the normal components exhibit a jump discontinuity at material interfaces.

By combining Equations (1.1) and (1.2) and eliminating one of the fields, we obtain the "curl–curl" form of the vector Helmholtz equations

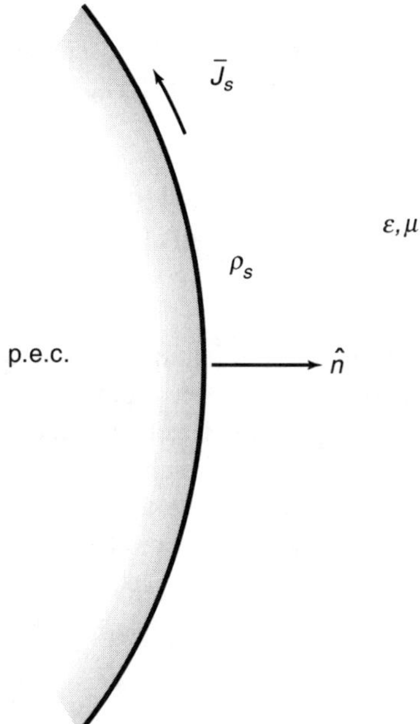

Figure 1.2 Electric current and charge density at the surface of a perfect conductor.

$$\nabla \times \left(\frac{1}{\mu_r} \nabla \times \bar{E} \right) - k^2 \varepsilon_r \bar{E} = 0 \tag{1.13}$$

$$\nabla \times \left(\frac{1}{\varepsilon_r} \nabla \times \bar{H} \right) - k^2 \mu_r \bar{H} = 0 \tag{1.14}$$

where $k^2 = \omega^2 \mu_0 \varepsilon_0$. The parameter k is known as the wavenumber of the medium (in this case, free space). We will consider several forms of these equations for application to two- and three-dimensional scattering problems.

Two-dimensional problems are those with invariance in the third dimension, such as an infinite cylindrical structure illuminated by a field that does not vary along the axis of the cylinder. Throughout this text, the term "cylinder" will be reserved for a structure whose geometry (ε_r, μ_r, and any conducting boundaries) does not vary with translation along z. If the cylinder axis lies along the $\hat{z}$ axis in a Cartesian coordinate system, it is usually convenient to separate the fields into transverse magnetic (TM) and transverse electric (TE) parts with respect to the variable z [2]. The $\hat{z}$-component of the magnetic field is absent in the TM case, while the $\hat{z}$-component of the electric field is absent in the TE case. Under the assumption of z-invariance, the $\hat{z}$-component of Equation (1.13) can be extracted and written in the form of a scalar Helmholtz equation

$$\nabla \cdot \left(\frac{1}{\mu_r} \nabla E_z \right) + k^2 \varepsilon_r E_z = 0 \tag{1.15}$$

In a two-dimensional problem, the TM part of the field can be found from the solution of Equation (1.15). Similarly, the TE part of the field can be found from

$$\nabla \cdot \left(\frac{1}{\varepsilon_r} \nabla H_z \right) + k^2 \mu_r H_z = 0 \tag{1.16}$$

Since the materials and the fields do not vary with z, all derivatives with respect to z vanish in the two-dimensional equations. After separate treatment, the TM and TE solutions can be superimposed to complete the analysis.

1.2 VOLUMETRIC EQUIVALENCE PRINCIPLE FOR PENETRABLE SCATTERERS [2, 7]

As alternatives to the differential equations presented above, integral equations are often chosen as the starting point for electromagnetic scattering analysis. To simplify the formulation of integral equations, it is convenient to convert the original scattering problem into an equivalent problem for which a formal solution may be directly written. One way of accomplishing this is to replace the inhomogeneous dielectric and magnetic material present in the problem by equivalent induced polarization currents and charges. Equations (1.1)–(1.4) can be rewritten to produce

$$\nabla \times \bar{E} = -j\omega\mu_0 \bar{H} - \bar{K} \tag{1.17}$$

$$\nabla \times \bar{H} = j\omega\varepsilon_0 \bar{E} + \bar{J} \tag{1.18}$$

$$\nabla \cdot (\varepsilon_0 \bar{E}) = \rho_e \tag{1.19}$$

$$\nabla \cdot (\mu_0 \bar{H}) = \rho_m \tag{1.20}$$

where

$$\bar{K} = j\omega\mu_0(\mu_r - 1)\bar{H} \tag{1.21}$$

$$\bar{J} = j\omega\varepsilon_0(\varepsilon_r - 1)\bar{E} \tag{1.22}$$

$$\rho_e = \varepsilon_0\varepsilon_r \bar{E} \cdot \nabla\left(\frac{1}{\varepsilon_r}\right) \tag{1.23}$$

$$\rho_m = \mu_0\mu_r \bar{H} \cdot \nabla\left(\frac{1}{\mu_r}\right) \tag{1.24}$$

Equations (1.17)–(1.20) describe the same fields as Equations (1.1)–(1.4) but appear to involve a homogeneous space characterized by permittivity ε_0 and permeability μ_0 instead of the original heterogeneous environment. The source terms compensate for the apparent difference between the two sets of equations, and we can think of these sources as replacing the dielectric or magnetic material explicit in (1.1)–(1.4). Since the two sets of equations are equivalent, we refer to the procedure of replacing the dielectric or magnetic material by induced sources as a *volumetric equivalence principle*. This type of equivalence principle will be used in the formulation of volume integral equations.

The sources of (1.21)–(1.24) radiate in free space. The task of finding electromagnetic fields in free space is much more straightforward than the original burden of solving Equations (1.1)–(1.4) directly in the inhomogeneous environment, and a solution will be presented in general form below (Section 1.4). However, at this point in the development the equivalent sources are unknowns to be determined, so the problem has not been solved by their introduction.

The expressions in Equations (1.23) and (1.24) require additional explanation. The quantities being differentiated, ε_r and μ_r, will be discontinuous at medium interfaces (including the scatterer surface). Therefore the derivatives in (1.23) and (1.24) must be interpreted in the context of generalized functions, that is, Dirac delta functions and their derivatives. Since ε_r and μ_r are constant throughout homogeneous regions of the original scatterer, Equations (1.23) and (1.24) show that there is no induced charge density in those regions. If ε_r and μ_r vary in a continuous manner and are differentiable in the classical sense, (1.23) and (1.24) produce the correct induced volume charge density. Furthermore, at an interface such as the scatterer surface, the Dirac delta function arising from the derivative signifies that there is actually a surface charge density rather than a volume charge density at that location. If (1.23) or (1.24) results in an induced surface charge density, the normal-field discontinuity produced in free space at that location can be shown to be identical to the discontinuity produced in the original scatterer by the material interface (Prob. P1.6). Therefore, when interpreted as generalized functions, the expressions in (1.23) and (1.24) conveniently account for all the possibilities. For additional information on generalized functions, the reader is referred to Chapter 1 of [12]. Throughout this text, the concepts of generalized functions are freely used wherever necessary. Problem P1.7 provides additional practice at manipulating delta functions and their derivatives.

Although we expressed $\bar{J}$ in terms of $\bar{E}$ and $\bar{K}$ in terms of $\bar{H}$ in Equations (1.21) and (1.22), we will have occasion to use the alternate forms

$$\bar{J} = \frac{\varepsilon_r - 1}{\varepsilon_r} \nabla \times \bar{H} \tag{1.25}$$

and

$$\bar{K} = -\frac{\mu_r - 1}{\mu_r} \nabla \times \bar{E} \tag{1.26}$$

These follow directly from Equations (1.21), (1.22), (1.1), and (1.2) and must also be interpreted as generalized functions.

1.3 GENERAL DESCRIPTION OF A SCATTERING PROBLEM [6, 7]

We are now in a position to describe one way of posing an electromagnetic scattering problem. Suppose the scatterer of Figure 1.1 is illuminated by a field produced by a primary source located somewhere outside the scatterer. We have shown that the inhomogeneous material can be replaced by equivalent induced sources radiating in free space. Consider splitting the fields into two parts, one associated with the primary source and another associated with the equivalent induced sources. The fields produced by the primary source in the absence of the scatterer will be denoted the *incident fields* $\bar{E}^{\text{inc}}$ and $\bar{H}^{\text{inc}}$. The secondary induced sources, which also radiate in free space, produce the *scattered fields* $\bar{E}^s$ and $\bar{H}^s$. The superposition of the incident and scattered fields yield the original fields in the presence of the scatterer. In other words, we can write

$$\bar{E} = \bar{E}^{\text{inc}} + \bar{E}^s \tag{1.27}$$

$$\bar{H} = \bar{H}^{\text{inc}} + \bar{H}^s \tag{1.28}$$

where the incident fields in the immediate vicinity of the scatterer (away from the primary source) satisfy the vector Helmholtz equations

$$\nabla^2 \bar{E}^{\text{inc}} + k^2 \bar{E}^{\text{inc}} = 0 \tag{1.29}$$

$$\nabla^2 \bar{H}^{\text{inc}} + k^2 \bar{H}^{\text{inc}} = 0 \tag{1.30}$$

and the scattered fields are solutions to the equations

$$\nabla^2 \bar{E}^s + k^2 \bar{E}^s = j\omega\mu_0 \bar{J} - \frac{\nabla\nabla \cdot \bar{J}}{j\omega\varepsilon_0} + \nabla \times \bar{K} \tag{1.31}$$

$$\nabla^2 \bar{H}^s + k^2 \bar{H}^s = -\nabla \times \bar{J} + j\omega\varepsilon_0 \bar{K} - \frac{\nabla\nabla \cdot \bar{K}}{j\omega\mu_0} \tag{1.32}$$

where $\bar{J}$ and $\bar{K}$ denote the equivalent sources defined in (1.21) and (1.22). Note that these sources are a function of the total fields $\bar{E}$ and $\bar{H}$. [As an alternate proof that the fields can be decomposed in this manner, combine Maxwell's equations for the incident field and primary source with Maxwell's equations for the scattered field and induced sources to obtain (1.1)–(1.4).]

In a source-free homogeneous medium, Equations (1.29) and (1.30) can be obtained from (1.13) and (1.14) using the vector Laplacian

$$\nabla^2 \bar{E} = \nabla(\nabla \cdot \bar{E}) - \nabla \times \nabla \times \bar{E} \tag{1.33}$$

and Maxwell's divergence equations. The derivation of Equations (1.31) and (1.32), which includes sources, is slightly more complicated and will be left as an exercise (Prob. P1.8).

Although the incident field may be arbitrary and may in fact be produced by sources immediately adjacent to the scatterer or within the scatterer, our primary interest is the case of an excitation produced by some source in the far zone. Often, we will consider the incident field to be a uniform plane wave.

Radiation conditions ensure that the fields satisfying Equations (1.31) and (1.32) propagate away from the scatterer. In a three-dimensional problem, where r is the conventional spherical coordinate variable, radiation conditions have the form

$$\lim_{r\to\infty} \hat{r} \times \nabla \times \bar{E}^s = jk\bar{E}^s \tag{1.34}$$

$$\lim_{r\to\infty} \hat{r} \times \nabla \times \bar{H}^s = jk\bar{H}^s \tag{1.35}$$

In the two-dimensional case, these simplify to a form of the Sommerfeld radiation conditions

$$\lim_{\rho\to\infty} \frac{\partial E_z^s}{\partial \rho} = -jkE_z^s \tag{1.36}$$

$$\lim_{\rho\to\infty} \frac{\partial H_z^s}{\partial \rho} = -jkH_z^s \tag{1.37}$$

for the TM and TE polarizations, respectively, where ρ is the radial variable in cylindrical coordinates.

1.4. SOURCE–FIELD RELATIONSHIPS IN HOMOGENEOUS SPACE [1–7]

There are a number of ways to approach the solution of the Helmholtz equations (1.31) and (1.32) in homogeneous, infinite space. The classical approach is to express the fields in terms of the magnetic vector potential $\bar{A}$ and the electric vector potential $\bar{F}$, according to

$$\bar{E}^s = \frac{\nabla\nabla \cdot \bar{A} + k^2 \bar{A}}{j\omega\varepsilon_0} - \nabla \times \bar{F} \tag{1.38}$$

$$\bar{H}^s = \nabla \times \bar{A} + \frac{\nabla\nabla \cdot \bar{F} + k^2 \bar{F}}{j\omega\mu_0} \tag{1.39}$$

By substitution into Maxwell's equations, it is easily demonstrated that the vector potentials satisfy

$$\nabla^2 \bar{A} + k^2 \bar{A} = -\bar{J} \tag{1.40}$$

$$\nabla^2 \bar{F} + k^2 \bar{F} = -\bar{K} \tag{1.41}$$

A solution to these equations satisfying the radiation condition can be concisely written in the form

$$\bar{A} = \bar{J} * G \tag{1.42}$$

$$\bar{F} = \bar{K} * G \tag{1.43}$$

where the scalar function G is the well-known three-dimensional Green's function

$$G = \frac{e^{-jk|\bar{r}|}}{4\pi|\bar{r}|} \tag{1.44}$$

and the asterisk ($*$) denotes three-dimensional convolution, that is,

$$\bar{A}(\bar{r}) = \iiint \bar{J}(\bar{r}') \frac{e^{-jk|\bar{r}-\bar{r}'|}}{4\pi|\bar{r}-\bar{r}'|} \, d\bar{r}' \tag{1.45}$$

The convolutional property of the solution is useful in a variety of ways and will be exploited in some of the numerical formulations to be presented in subsequent chapters.

In a two-dimensional problem, the integration over the third dimension only involves the Green's function and can be performed analytically. For generality, we first present the result [13, 3.876]

$$\int_{z=-\infty}^{\infty} \frac{e^{-jk\sqrt{p^2+z^2}}}{4\pi\sqrt{p^2+z^2}} e^{-j\gamma z} \, dz = \begin{cases} \dfrac{1}{4j} H_0^{(2)}\left(\rho\sqrt{k^2-\gamma^2}\right) & k^2 > \gamma^2 \\ \dfrac{1}{2\pi} K_0\left(\rho\sqrt{\gamma^2-k^2}\right) & \gamma^2 > k^2 \end{cases} \tag{1.46}$$

which may be useful if the geometry is z invariant but the excitation is not. In Equation (1.46), H_0 and K_0 are the zero-order Hankel and modified Bessel functions of the second kind, respectively. In the limiting case, as γ vanishes, we obtain the two-dimensional Green's function

$$G = \frac{1}{4j} H_0^{(2)}(k|\bar{\rho}|) \tag{1.47}$$

For two-dimensional problems, Equation (1.47) may be used within Equations (1.42) and (1.43) as two-dimensional convolutions.

To summarize, the above procedure requires that $\bar{A}$ and $\bar{F}$ be constructed by an integration of $\bar{J}$ and $\bar{K}$ according to (1.42) and (1.43). The electric and magnetic fields can then be produced by Equations (1.38) and (1.39), which involve differentiations. Unfortunately, the *integration-followed-by-differentiation* procedure dictated by these equations is not well suited for numerical implementation. The typical integrals arising from source–field relations can seldom be evaluated in closed form but usually must be evaluated at individual observation points by numerical quadrature algorithms. Both the accuracy and the efficiency of the computation will suffer if it is necessary to implement a subsequent derivative of the integral using finite-difference operations. In three dimensions, a finite-difference implementation of the second-order vector derivatives in (1.38) and (1.39) will require every vector component of each integral to be evaluated at a minimum of seven points around the desired location.

On the other hand, the free-space Green's function is easy to differentiate analytically. For observation points outside the source region, derivatives can be brought inside the integrals and carried out analytically, changing the procedure to one of *differentiation followed by integration*. The modified procedure eliminates the error introduced by the finite-difference operations and reduces the number of quadrature evaluations to one per integral. Because of the singularity of the Green's function in the region containing the sources, however, a direct interchange of integration and differentiation is not possible without violating Leibnitz's rule. Thus, unless an integral can be evaluated in closed form in the source region, an alternate approach may be necessary.

As an alternative to the pure vector potential source–field relationship, a mixed-potential formalism can be developed by seeking a solution of the form

$$\bar{E}^s = -j\omega\mu_0\bar{A} - \nabla\Phi_e - \nabla \times \bar{F} \tag{1.48}$$

$$\bar{H}^s = \nabla \times \bar{A} - j\omega\varepsilon_0\bar{F} - \nabla\Phi_m \tag{1.49}$$

where Φ_e and Φ_m are scalar potential functions. By carrying out a solution procedure similar to that employed above (Prob. P1.11), $\bar{A}$ and $\bar{F}$ can be shown to be the identical convolution expressions appearing in Equations (1.42) and (1.43). The scalar potentials are given by

$$\Phi_e = \frac{\rho_e}{\varepsilon_0} * G \tag{1.50}$$

$$\Phi_m = \frac{\rho_m}{\mu_0} * G \tag{1.51}$$

where the asterisk again denotes multidimensional convolution. Therefore, this particular choice of scalar and vector potentials results in a complete decoupling of the contribution to the field from the electric current density, magnetic current density, electric charge density, and magnetic charge density.

Once the scalar and vector potentials are determined by integration over the given sources, Equations (1.48) and (1.49) require only a single differentiation to obtain the electromagnetic fields. Because of the lower order derivative, the mixed-potential source–field representation is often used within numerical formulations in preference to Equations (1.38) and (1.39). For direct field calculations in the source region, this procedure is still an integration-followed-by-differentiation approach with the disadvantages noted above.

A third form of the source–field relationship can be developed using an analogy between Equations (1.40) and (1.41) and their general solutions (1.42) and (1.43) and the vector Helmholtz equations appearing in (1.31) and (1.32). Formally, we can write the solutions to Equations (1.31) and (1.32) directly as

$$\bar{E}^s = \left(-j\omega\mu_0\bar{J} + \frac{\nabla\nabla \cdot \bar{J}}{j\omega\varepsilon_0} - \nabla \times \bar{K}\right) * G \tag{1.52}$$

and

$$\bar{H}^s = \left(\nabla \times \bar{J} - j\omega\varepsilon_0\bar{K} + \frac{\nabla\nabla \cdot \bar{K}}{j\omega\mu_0}\right) * G \tag{1.53}$$

without the need of intermediate potential functions [14]. These equations can also be obtained from Equations (1.38) and (1.39) by employing the property that differentiation operators commute with the convolution operation. In contrast to Equations (1.38) and (1.39), which require an integration followed by a differentiation, Equations (1.52) and (1.53) require a differentiation followed by an integration. As noted above, it is often easier to differentiate a given expression in closed form than it is to obtain a closed-form expression for the relatively complex convolution integrals of (1.42) or (1.43). Thus, Equations (1.52) and (1.53) will often permit the closed-form evaluation of the first step of the process. As discussed in Section 1.2, these derivatives must be interpreted as generalized functions. Formal rules for manipulating Dirac delta functions and their derivatives must be followed to correctly carry out their evaluation.

It should be noted that (1.52) and (1.53) are not the equations that would result from a simple interchange of integration and differentiation in Equations (1.38) and (1.39). In fact, carrying out such an interchange leads to a fourth type of source–field relationship, obtained in the form of a convolution between the sources $\bar{J}$ and $\bar{K}$ and the so-called dyadic Green's functions [2, 4, 9]. The expressions can be written

$$\bar{E}^s = \bar{J} * \bar{\bar{G}}_{ej} + \bar{K} * \bar{\bar{G}}_{ek} \tag{1.54}$$

$$\bar{H}^s = \bar{J} * \bar{\bar{G}}_{mj} + \bar{K} * \bar{\bar{G}}_{mk} \tag{1.55}$$

where $\bar{\bar{G}}_{ej}$ and $\bar{\bar{G}}_{ek}$ are the dyadic Green's functions for the electric field, symbolically denoted

$$\bar{\bar{G}}_{ej} = \frac{1}{j\omega\varepsilon_0}(\nabla\nabla + k^2\bar{\bar{I}})G \tag{1.56}$$

$$\bar{\bar{G}}_{ek} = -\nabla \times (\bar{\bar{I}}G) \tag{1.57}$$

and $\bar{\bar{G}}_{mj}$ and $\bar{\bar{G}}_{mk}$ are the dyadic Green's functions for the magnetic field,

$$\bar{\bar{G}}_{mj} = \nabla \times (\bar{\bar{I}}G) \tag{1.58}$$

$$\bar{\bar{G}}_{mk} = \frac{1}{j\omega\mu_0}(\nabla\nabla + k^2\bar{\bar{I}})G \tag{1.59}$$

As a consequence of the singularity in the Green's function, Leibnitz's rule is violated by this interchange of integration and differentiation when the source and observation regions overlap. In this situation, formal classical integration is not sufficient to evaluate the integrals required in (1.54) and (1.55). The evaluation of these integrals using "regularization" procedures is possible [15–17] but is beyond the scope of this text.

Although we will not have occasion to use the dyadic representation, the first three source–field relationships will be used throughout the text when formulating numerical schemes for solving electromagnetic scattering problems. Occasionally, it will be possible to find closed-form expressions for the vector potential functions, which readily permit their differentiation according to Equations (1.38) and (1.39). Often, the mixed-potential representation of Equations (1.48) and (1.49) will be preferred because of the lower order derivative appearing in front of the scalar potential terms. The mixed-potential representation is also preferred if the charge densities are defined separately from the current densities, as occurs when (1.23) and (1.24) are used to define the charge densities within a volumetric formulation. Equations (1.52) and (1.53) require no differentiation of the integral, which may make them preferable for situations where the integration over the sources must be performed numerically. Of course, if applied correctly, all of the above source–field relationships are equivalent and produce identical results. As will be demonstrated in chapters to follow, however, one approach is usually easier to implement than the others within a specific numerical treatment.

1.5 DUALITY RELATIONSHIPS [2]

Because the equations describing the electric and magnetic fields and the associated sources exhibit almost perfect symmetry, any relationship between the fields and sources can be used to directly arrive at a dual relationship describing the complementary fields and sources. The idea of duality is a useful aid to generating new formulas or just remembering some of the existing ones. Table 1.1 summarizes the duality relationships.

TABLE 1.1 Principle of Duality. The equations describing electromagnetic fields remain valid if all the quantities in the left column are replaced by those in the right.

$\bar{E}$	$\bar{H}$
$\bar{H}$	$-\bar{E}$
$\bar{J}$	$\bar{K}$
$\bar{K}$	$-\bar{J}$
ρ_e	ρ_m
ρ_m	$-\rho_e$
ε	μ
μ	ε
$\bar{A}$	$\bar{F}$
$\bar{F}$	$-\bar{A}$

1.6 SURFACE EQUIVALENCE PRINCIPLE [2, 7, 8]

Equations (1.17)–(1.24) demonstrate that mathematical volume sources can be used to replace dielectric and magnetic materials. Equivalent sources distributed on a surface are of similar utility. To illustrate the surface equivalence principle, consider the hypothetical situation posed in Figure 1.3. Figure 1.3 shows two regions of space separated by a mathematical surface S. Region 1 is homogeneous with ε_1 and μ_1, whereas region 2 contains inhomogeneities that may include perfectly conducting materials. A source $(\bar{J}_2, \bar{K}_2)$ located in region 2 and radiating in the presence of the inhomogeneities produces fields $\bar{E}_2$ and $\bar{H}_2$ throughout region 1. We also postulate a second source $(\bar{J}_1, \bar{K}_1)$ located in region 1 but radiating fields $\bar{E}_1$ and $\bar{H}_1$ in a homogeneous space having constitutive parameters ε_1 and μ_1. The fields of both sources satisfy the radiation condition on the boundary at infinity (S_∞).

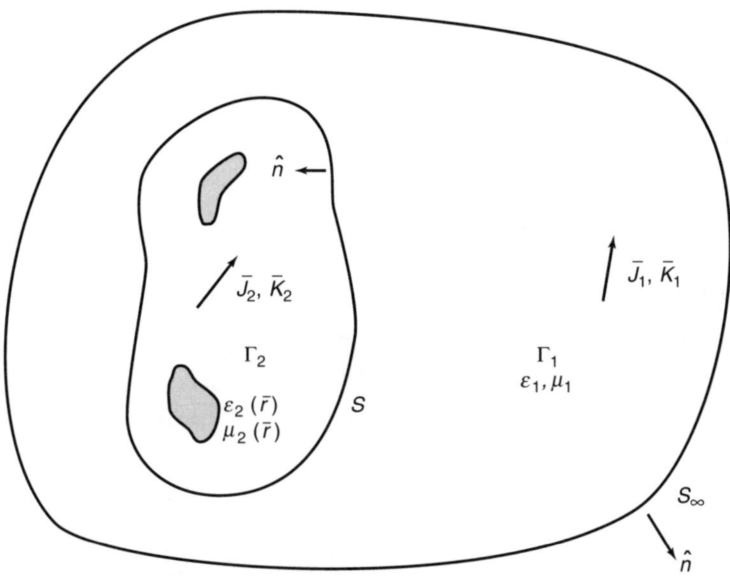

Figure 1.3 Two regions separated by a surface S.

Throughout region 1, Maxwell's curl equations can be written

$$\nabla \times \bar{E}_1 = -j\omega\mu_1 \bar{H}_1 - \bar{K}_1 \tag{1.60}$$

$$\nabla \times \bar{H}_1 = j\omega\varepsilon_1 \bar{E}_1 + \bar{J}_1 \tag{1.61}$$

$$\nabla \times \bar{E}_2 = -j\omega\mu_1 \bar{H}_2 \tag{1.62}$$

$$\nabla \times \bar{H}_2 = j\omega\varepsilon_1 \bar{E}_2 \tag{1.63}$$

Therefore, in region 1 we can construct the following equations:

$$\bar{H}_2 \cdot \nabla \times \bar{E}_1 = -j\omega\mu_1 \bar{H}_2 \cdot \bar{H}_1 - \bar{H}_2 \cdot \bar{K}_1 \tag{1.64}$$

$$\bar{E}_2 \cdot \nabla \times \bar{H}_1 = j\omega\varepsilon_1 \bar{E}_2 \cdot \bar{E}_1 + \bar{E}_2 \cdot \bar{J}_1 \tag{1.65}$$

$$\bar{H}_1 \cdot \nabla \times \bar{E}_2 = -j\omega\mu_1 \bar{H}_1 \cdot \bar{H}_2 \tag{1.66}$$

$$\bar{E}_1 \cdot \nabla \times \bar{H}_2 = j\omega\varepsilon_1 \bar{E}_1 \cdot \bar{E}_2 \tag{1.67}$$

By combining these equations, we obtain

$$\bar{H}_2 \cdot \nabla \times \bar{E}_1 - \bar{E}_1 \cdot \nabla \times \bar{H}_2 + \bar{E}_2 \cdot \nabla \times \bar{H}_1 - \bar{H}_1 \cdot \nabla \times \bar{E}_2 = \bar{E}_2 \cdot \bar{J}_1 - \bar{H}_2 \cdot \bar{K}_1 \tag{1.68}$$

which is equivalent to

$$\nabla \cdot (\bar{E}_1 \times \bar{H}_2 - \bar{E}_2 \times \bar{H}_1) = \bar{E}_2 \cdot \bar{J}_1 - \bar{H}_2 \cdot \bar{K}_1 \tag{1.69}$$

Equation (1.69) is a form of the *Lorentz reciprocity theorem* [2]. Integrating both sides of Equation (1.69) over region 1 and applying the divergence theorem

$$\iiint_{\Gamma_1} \nabla \cdot \bar{Q}\, dv = \iint_S \bar{Q} \cdot \hat{n}\, dS + \iint_{S_\infty} \bar{Q} \cdot \hat{n}\, dS \tag{1.70}$$

where $\hat{n}$ is the normal vector on the surface pointing out of region 1, produce

$$\iint_S (\bar{E}_1 \times \bar{H}_2 - \bar{E}_2 \times \bar{H}_1) \cdot \hat{n}\, dS = \iiint_{\Gamma_1} (\bar{E}_2 \cdot \bar{J}_1 - \bar{H}_2 \cdot \bar{K}_1)\, dv \tag{1.71}$$

(The integral over the surface at infinity vanishes as a consequence of the radiation condition.) Vector identities dictate that

$$\bar{E}_1 \times \bar{H}_2 \cdot \hat{n} = -\bar{E}_1 \cdot (\hat{n} \times \bar{H}_2) \tag{1.72}$$

and

$$\bar{E}_2 \times \bar{H}_1 \cdot \hat{n} = -\bar{H}_1 \cdot (\bar{E}_2 \times \hat{n}) \tag{1.73}$$

Therefore, Equation (1.71) can be rewritten as

$$\iint_S [\bar{E}_1 \cdot (-\hat{n} \times \bar{H}_2) - \bar{H}_1 \cdot (-\bar{E}_2 \times \hat{n})] \, dS = \iiint_{\Gamma_1} (\bar{E}_2 \cdot \bar{J}_1 - \bar{H}_2 \cdot \bar{K}_1) \, dv \tag{1.74}$$

Equation (1.74) is a generalized statement of reciprocity.

Now, let us suppose that the sources in region 1 are

$$\bar{J}_1 = \hat{u}\delta(\bar{r} - \bar{r}') \tag{1.75}$$

and

$$\bar{K}_1 = 0 \tag{1.76}$$

where $\bar{r}$ denotes the source point in region 1 and $\bar{r}'$ represents the integration variable in (1.74). For these sources, Equation (1.74) can be written as

$$\hat{u} \cdot \bar{E}_2|_{\bar{r}} = \iint_S [\bar{E}_1 \cdot (-\hat{n} \times \bar{H}_2) - \bar{H}_1 \cdot (-\bar{E}_2 \times \hat{n})] \, dS' \tag{1.77}$$

where $\bar{E}_1$ and $\bar{H}_1$ are the fields produced at location $\bar{r}'$ in an infinite homogeneous space by sources $\bar{J}_1$ and $\bar{K}_1$ located at $\bar{r}$. These fields can be expressed in terms of the first source–field relationship derived in Section 1.4, to obtain

$$\bar{E}_1(\bar{r}') = \frac{\nabla'\nabla' \cdot + k^2}{j\omega\varepsilon_1} \left(\hat{u} \frac{e^{-jk|\bar{r}-\bar{r}'|}}{4\pi|\bar{r}-\bar{r}'|} \right) \tag{1.78}$$

$$\bar{H}_1(\bar{r}') = \nabla' \times \left(\hat{u} \frac{e^{-jk|\bar{r}-\bar{r}'|}}{4\pi|\bar{r}-\bar{r}'|} \right) \tag{1.79}$$

where $k = \omega(\mu_1\varepsilon_1)^{1/2}$. Note that the derivatives are taken with respect to the primed coordinates. Because of the symmetry of the Green's function, however, it is easily shown that

$$\nabla'\nabla' \cdot \left(\hat{u} \frac{e^{-jk|\bar{r}-\bar{r}'|}}{4\pi|\bar{r}-\bar{r}'|} \right) = \nabla\nabla \cdot \left(\hat{u} \frac{e^{-jk|\bar{r}-\bar{r}'|}}{4\pi|\bar{r}-\bar{r}'|} \right) \tag{1.80}$$

and

$$\nabla' \times \left(\hat{u} \frac{e^{-jk|\bar{r}-\bar{r}'|}}{4\pi|\bar{r}-\bar{r}'|} \right) = -\nabla \times \left(\hat{u} \frac{e^{-jk|\bar{r}-\bar{r}'|}}{4\pi|\bar{r}-\bar{r}'|} \right) \tag{1.81}$$

Therefore, (1.77) becomes

$$\hat{u} \cdot \bar{E}_2|_{\bar{r}} = \iint_S \left[\frac{\nabla\nabla \cdot + k^2}{j\omega\varepsilon_1} \left(\hat{u} \frac{e^{-jk|\bar{r}-\bar{r}'|}}{4\pi|\bar{r}-\bar{r}'|} \right) \cdot (-\hat{n} \times \bar{H}_2) + \nabla \times \left(\hat{u} \frac{e^{-jk|\bar{r}-\bar{r}'|}}{4\pi|\bar{r}-\bar{r}'|} \right) \cdot (-\bar{E}_2 \times \hat{n}) \right] dS' \tag{1.82}$$

The integration is to be performed in primed coordinates over the surface S. Note that the derivatives appearing in Equation (1.82) are now taken with respect to unprimed coordinates, while $\bar{E}_2$ and $\bar{H}_2$ functions of primed variables. Therefore, the first term in (1.82) can be modified using

$$(-\hat{n} \times \bar{H}_2) \cdot \nabla\nabla \cdot \left(\hat{u} \frac{e^{-jk|\bar{r}-\bar{r}'|}}{4\pi |\bar{r} - \bar{r}'|} \right)$$

$$= \sum_{i=1}^{3} \hat{x}_i \cdot (-\hat{n} \times \bar{H}_2) \frac{\partial}{\partial x_i} \frac{\partial}{\partial u} \left(\frac{e^{-jk|\bar{r}-\bar{r}'|}}{4\pi |\bar{r} - \bar{r}'|} \right)$$

$$= \frac{\partial}{\partial u} \sum_{i=1}^{3} \frac{\partial}{\partial x_i} \left(\hat{x}_i \cdot (-\hat{n} \times \bar{H}_2) \frac{e^{-jk|\bar{r}-\bar{r}'|}}{4\pi |\bar{r} - \bar{r}'|} \right)$$

$$= \hat{u} \cdot \nabla\nabla \cdot \left((-\hat{n} \times \bar{H}_2) \frac{e^{-jk|\bar{r}-\bar{r}'|}}{4\pi |\bar{r} - \bar{r}'|} \right) \tag{1.83}$$

where $\{x_i\}$ denote the three Cartesian variables and u is a variable defined along $\hat{u}$. Furthermore, the second term in (1.82) can be converted using vector identities and the fact that $\hat{u}$ is constant to produce

$$\nabla \times \left(\hat{u} \frac{e^{-jk|\bar{r}-\bar{r}'|}}{4\pi |\bar{r} - \bar{r}'|} \right) \cdot (-\bar{E}_2 \times \hat{n})$$

$$= \nabla \left(\frac{e^{-jk|\bar{r}-\bar{r}'|}}{4\pi |\bar{r} - \bar{r}'|} \right) \times \hat{u} \cdot (-\bar{E}_2 \times \hat{n})$$

$$= -\hat{u} \cdot \nabla \left(\frac{e^{-jk|\bar{r}-\bar{r}'|}}{4\pi |\bar{r} - \bar{r}'|} \right) \times (-\bar{E}_2 \times \hat{n})$$

$$= -\hat{u} \cdot \nabla \times \left((-\bar{E}_2 \times \hat{n}) \frac{e^{-jk|\bar{r}-\bar{r}'|}}{4\pi |\bar{r} - \bar{r}'|} \right) \tag{1.84}$$

After these results are substituted into (1.82), the derivatives (taken with respect to unprimed variables) can be moved outside the integrals to produce

$$\hat{u} \cdot \bar{E}_2|_{\bar{r}} = \hat{u} \cdot \frac{\nabla\nabla \cdot +k^2}{j\omega\varepsilon_1} \iint_S (-\hat{n} \times \bar{H}_2) \frac{e^{-jk|\bar{r}-\bar{r}'|}}{4\pi |\bar{r} - \bar{r}'|} \, dS'$$

$$-\hat{u} \cdot \nabla \times \iint_S (-\bar{E}_2 \times \hat{n}) \frac{e^{-jk|\bar{r}-\bar{r}'|}}{4\pi |\bar{r} - \bar{r}'|} \, dS' \tag{1.85}$$

Equation (1.85) is a statement that the field produced by $(\bar{J}_2, \bar{K}_2)$ at some location outside of region 2 can be expressed in the form of an integration over tangential fields on the surface of region 2. In fact, by comparing Equation (1.85) with Equation (1.38), it is immediately apparent that the field is equivalent to that produced by surface current densities

$$\bar{J}_s = -\hat{n} \times \bar{H}_2 \tag{1.86}$$

and

$$\bar{K}_s = -\bar{E}_2 \times \hat{n} \tag{1.87}$$

located on the surface S and radiating in a homogeneous space having constitutive parameters ε_1 and μ_1. This property is a fundamental theorem of electromagnetics generally known as *Huygens' surface equivalence principle*. [Note that the normal vector $\hat{n}$ in Equations (1.86) and (1.87) points into region 2, i.e., into the closed surface S. This is actually the opposite of our usual convention, which requires that the normal vector point out of the region containing the sources. In the following discussion, we revert back to the usual convention.]

We will now state the surface equivalence principle in a slightly different form. Consider Figure 1.4, which shows a source in region 1 radiating in the presence of inhomogeneities located in region 2. Fields produced in region 1 are denoted $\bar{E}_1$ and $\bar{H}_1$; those produced in region 2 are denoted $\bar{E}_2$ and $\bar{H}_2$. Now, consider equivalent sources $\bar{J}_s$ and $\bar{K}_s$ located on the mathematical surface S and satisfying

$$\bar{J}_s = \hat{n} \times \bar{H}_1 \tag{1.88}$$

$$\bar{K}_s = \bar{E}_1 \times \hat{n} \tag{1.89}$$

where $\hat{n}$ is the outward normal vector. According to the equivalence principle, the combination of the original source and the equivalent sources produces fields $\bar{E}_1$ and $\bar{H}_1$ in region 1 identical to that of the original problem illustrated in Figure 1.4. The fields in region 2 are not identical to those of the original problem; in fact, null fields are produced throughout region 2 by the combination of the original and equivalent sources. (This result is sometimes known as the *extinction theorem*.) The modified problem is illustrated in Figure 1.5.

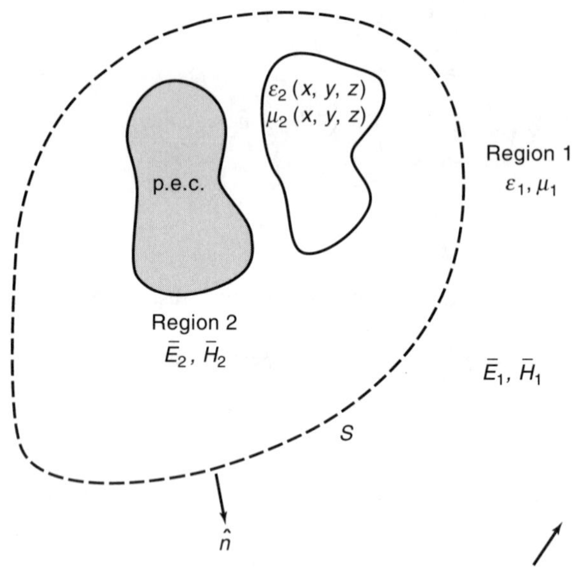

Figure 1.4 Electromagnetic source in region 1 radiating in the presence of inhomogeneities located in region 2. The surface S separates the two regions.

Since the fields throughout region 2 of the modified problem vanish, any inhomogeneities present in region 2 may be replaced at will without affecting the fields in region 1. Figure 1.6 shows one possibility, that of removing all the inhomogeneities from region 2 and

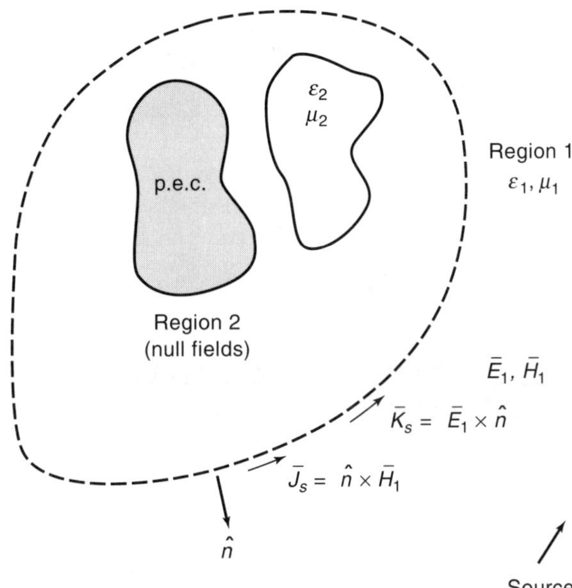

Figure 1.5 Intermediate step in the construction of the equivalent exterior problem associated with Figure 1.4. Sources $\bar{J}_s$ and $\bar{K}_s$ are introduced on the surface S and replicate the original fields in region 1. Null fields are produced throughout region 2.

leaving a homogeneous medium with the same constitutive parameters as region 1. This is often the approach followed in practice, since it effectively replaces the original problem involving complicated inhomogeneous media with a problem involving only sources radiating in homogeneous space. The effect is that the original scattering geometry has been replaced by an *equivalent exterior problem.* As was the case in the volume equivalence principle described previously, we have not solved the electromagnetic field problem by the introduction of equivalent sources. In fact, $\bar{J}_s$ and $\bar{K}_s$ are unknowns that remain to be determined. However, we have converted the problem from one requiring the solution of

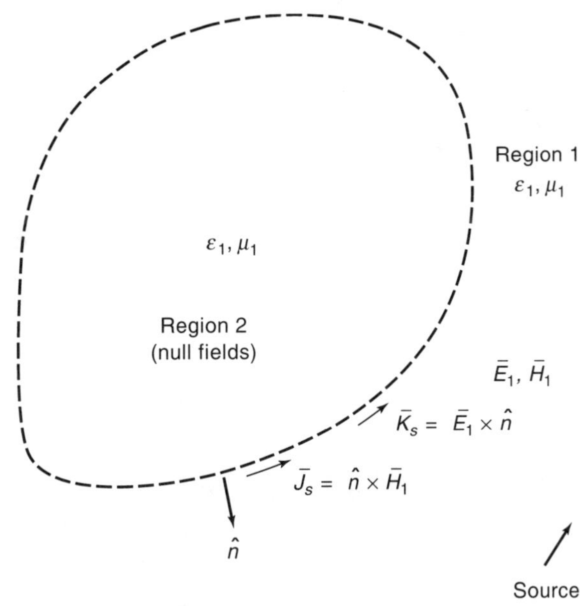

Figure 1.6 Equivalent exterior problem constructed from Figure 1.4. The inhomogeneities within region 2 have been replaced by a homogeneous medium identical to that of region 1. The combination of the original source and the equivalent sources $\bar{J}_s$ and $\bar{K}_s$ produce the original fields throughout region 1 and null fields throughout region 2.

Equations (1.1)–(1.4) to one requiring the solution of Equations (1.17)–(1.20), with equivalent surface currents and charges defined respectively by Equations (1.88) and (1.89) and the continuity equations

$$\rho_e = \frac{-1}{j\omega}\nabla_s \cdot \bar{J}_s \tag{1.90}$$

$$\rho_m = \frac{-1}{j\omega}\nabla_s \cdot \bar{K}_s \tag{1.91}$$

where ∇_s is the surface divergence operator.

1.7. SURFACE INTEGRAL EQUATIONS FOR PERFECTLY CONDUCTING SCATTERERS

Figure 1.7 shows a scatterer of perfect electric conducting (p.e.c.) material illuminated by a source. Consider a mathematical surface enclosing the scatterer, over which equivalent sources $\bar{J}_s$ and $\bar{K}_s$ are defined according to Equations (1.88) and (1.89). If the mathematical surface is permitted to shrink until it coincides with the surface of the perfect conductor, Equation (1.5) dictates that the tangential electric field must vanish on the surface. It follows that equivalent sources

$$\bar{J}_s = \hat{n} \times \bar{H} \tag{1.92}$$

$$\bar{K}_s = 0 \tag{1.93}$$

located on the surface of the p.e.c. scatterer will produce the correct scattered fields in the exterior region. The equivalent problem is depicted in Figure 1.8.

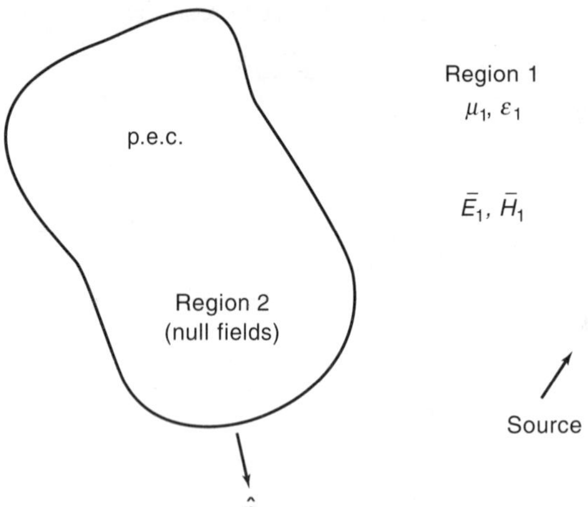

Figure 1.7 Original problem involving a p.e.c. scatterer.

We now combine the surface equivalence principle, the source–field relationships, and the boundary conditions discussed previously in order to formulate integral equations for the unknown equivalent sources. Assuming that the incident fields $\bar{E}^{\text{inc}}$ and $\bar{H}^{\text{inc}}$ are specified, the source–field relationships from Equations (1.38) and (1.39) may be combined

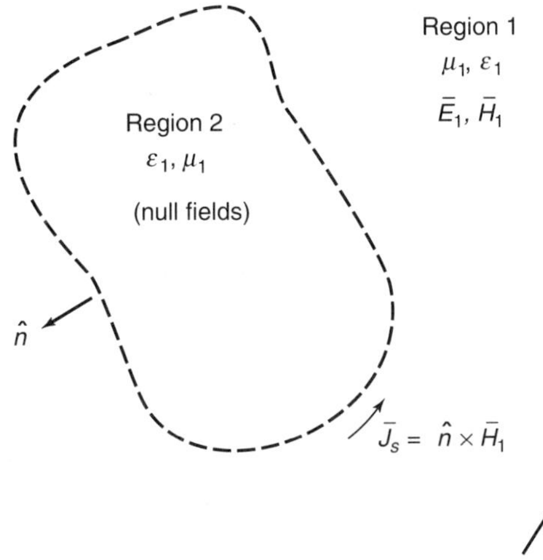

Figure 1.8 Equivalent exterior problem associated with Figure 1.7. An equivalent source $\bar{J}_s$ is introduced along the location of the conducting surface, and the conductor is replaced by a homogeneous medium with the same constitutive parameters as the exterior region.

with the equivalent sources from (1.92) and (1.93) to produce

$$\bar{E}^{\text{inc}}(\bar{r}) = \bar{E}(\bar{r}) - \frac{\nabla\nabla \cdot \bar{A} + k^2 \bar{A}}{j\omega\varepsilon_0} \tag{1.94}$$

$$\bar{H}^{\text{inc}}(\bar{r}) = \bar{H}(\bar{r}) - \nabla \times \bar{A} \tag{1.95}$$

At present, these relationships simply define the fields $\bar{E}$ and $\bar{H}$ produced by the excitation in the presence of the p.e.c. scatterer and hold throughout the exterior region (in this case they hold throughout the interior region also, since the fields within a p.e.c. vanish). However, if the boundary condition of Equation (1.5) is imposed on the surface of the scatterer, Equation (1.94) becomes

$$\hat{n} \times \bar{E}^{\text{inc}} = -\hat{n} \times \left\{ \frac{\nabla\nabla \cdot \bar{A} + k^2 \bar{A}}{j\omega\varepsilon_0} \right\}_S \tag{1.96}$$

which is an integro-differential equation for the unknown equivalent surface current density $\bar{J}_s$. Equation (1.96) holds only for points on the surface S of the scatterer and is one form of the *electric field integral equation* (EFIE). If Equation (1.6) is combined with Equation (1.95), we obtain the *magnetic field integral equation* (MFIE)

$$\hat{n} \times \bar{H}^{\text{inc}} = \bar{J}_s - \{\hat{n} \times \nabla \times \bar{A}\}_{S^+} \tag{1.97}$$

Equation (1.97) is also an integro-differential equation for the unknown surface current $\bar{J}_s$ and is enforced an infinitesimal distance outside the scatterer surface (S^+). It is common practice to refer to these equations as integral equations rather than integro-differential equations. Note that any of the source–field relationships presented in Section 1.4 could be employed as alternatives, producing equivalent equations.

In principle, either of Equations (1.96) or (1.97) can be solved to produce the unknown equivalent source $\bar{J}_s$. Once $\bar{J}_s$ is determined, the electric and magnetic fields everywhere in space may be found from the source–field relationships presented previously, superimposing the incident field with the scattered fields produced by $\bar{J}_s$. In deriving the EFIE and MFIE, we imposed only one of the conditions (1.5) and (1.6). Because of this, there are scatterers for which the solution of these equations is not unique. The uniqueness issue will be discussed in Chapter 6.

Suppose that, instead of a closed body, we wish to treat scattering from an infinitesimally thin open p.e.c. shell, strip, or plate (Figure 1.9). The surface equivalence principle can be applied in the same fashion as in the case of a solid scatterer. However, if the surface S collapses to the scatterer surface, the equivalent current densities on either side of the scatterer become superimposed at the location of the thin shell. The equations are unable to distinguish between the two equivalent sources, and we are forced to work with a single equivalent source that represents the sum of the sources on either side. Since the boundary condition of (1.5) remains valid for infinitesimally thin p.e.c. structures, however, an EFIE of the form of Equation (1.96) can be used to treat this type of scattering problem.

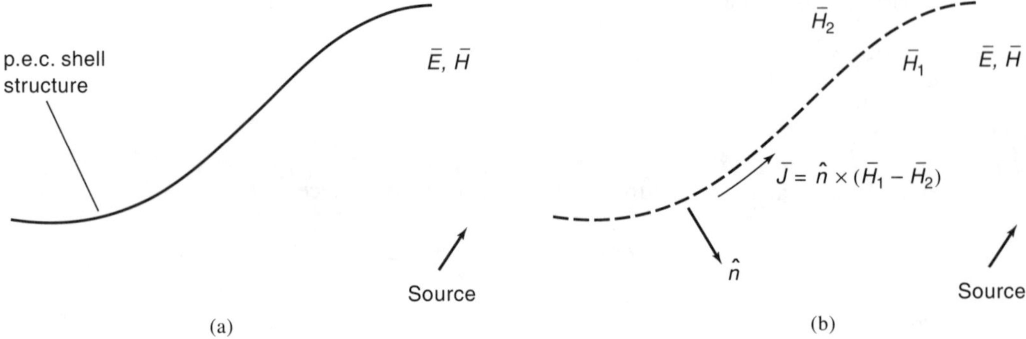

Figure 1.9 (a) Original problem involving an infinitesimally thin p.e.c. scatterer; (b) Equivalent problem. The p.e.c. material is replaced by a single source $\bar{J}_s$ that represents the superposition of the electric currents on both surfaces of the thin scatterer.

Although the EFIE can be employed to model a thin shell, the MFIE of Equation (1.97) is based on a boundary condition that is not valid for extremely thin geometries. Equation (1.6) is actually a special case of the general boundary condition

$$\hat{n} \times (\bar{H}_1 - \bar{H}_2) = \bar{J}_s \tag{1.98}$$

For closed bodies or other situations where the magnetic field vanishes on one side of the surface, Equation (1.98) reduces to (1.6). For infinitesimally thin structures with nonzero fields on both sides of the surface, however, Equation (1.6) is not equivalent to (1.98) and does not actually constitute a boundary condition. Consequently, the MFIE of (1.97) is restricted to closed bodies and cannot be used to describe scattering from an infinitesimally thin p.e.c. structure.

1.8. VOLUME INTEGRAL EQUATIONS FOR PENETRABLE SCATTERERS

The volumetric equivalence principle from Section 1.2 can be used to construct integro-differential equations describing the interaction of electromagnetic fields with penetrable scatterers. Combining the sources from Equations (1.21)–(1.24) with the source–field relationships from Equations (1.48) and (1.49), we obtain

$$\bar{E}^{\text{inc}}(\bar{r}) = \bar{E}(\bar{r}) + jk\eta\bar{A} + \nabla\Phi_e + \nabla \times \bar{F} \tag{1.99}$$

$$\bar{H}^{\text{inc}}(\bar{r}) = \bar{H}(\bar{r}) - \nabla \times \bar{A} + j\frac{k}{\eta}\bar{F} + \nabla\Phi_m \tag{1.100}$$

where $\eta = (\mu_0/\varepsilon_0)^{1/2}$. (Throughout the text, we will frequently interchange $\omega\mu_0 = k\eta$ and $\omega\varepsilon_0 = k/\eta$.) Equations (1.99) and (1.100) can be thought of as an EFIE and MFIE, respectively, although it is noteworthy that there are some differences between these equations and those of the previous section describing the p.e.c. scatterer. Specifically, these are volume equations that hold everywhere throughout the penetrable scatterer rather than just on the scatterer surface. Instead of designating the equivalent sources as the primary unknowns to be determined, it is usually more convenient to pose the problem directly in terms of the internal $\bar{E}$ or $\bar{H}$ fields. By using Equations (1.22), (1.23), and (1.26) with the EFIE, all equivalent sources can be defined as functions of $\bar{E}$. Similarly, with the MFIE it is convenient to employ Equations (1.21), (1.24), and (1.25) to define all quantities in terms of $\bar{H}$.

In the special case in which the body in question is composed entirely of dielectric material, terms involving equivalent magnetic currents and charges drop out, leaving

$$\bar{E}^{\text{inc}}(\bar{r}) = \bar{E}(\bar{r}) + jk\eta\bar{A} + \nabla\Phi_e \tag{1.101}$$

$$\bar{H}^{\text{inc}}(\bar{r}) = \bar{H}(\bar{r}) - \nabla \times \bar{A} \tag{1.102}$$

If the scatterer is composed entirely of magnetic material, terms involving electric current and charge density vanish, producing

$$\bar{E}^{\text{inc}}(\bar{r}) = \bar{E}(\bar{r}) + \nabla \times \bar{F} \tag{1.103}$$

$$\bar{H}^{\text{inc}}(\bar{r}) = \bar{H}(\bar{r}) + j\frac{k}{\eta}\bar{F} + \nabla\Phi_m \tag{1.104}$$

When simultaneously treating dielectric and magnetic materials with (1.99) and (1.100), fewer unknowns are required if the internal $\bar{E}$ or $\bar{H}$ field instead of the equivalent sources is designated the primary unknown.

These volume integral equations are suitable for the analysis of inhomogeneous material. If the penetrable scatterer under consideration is homogeneous with constant μ_r and ε_r, the problem can be formulated with either volume integral equations or surface integral equations. Surface integral equations are usually more manageable for numerical solution since the unknowns to be determined are confined to the scatterer surface rather than distributed throughout the scatterer volume.

1.9. SURFACE INTEGRAL EQUATIONS FOR HOMOGENEOUS SCATTERERS

Figure 1.10 depicts a homogeneous, penetrable body illuminated by an incident electromagnetic field. Region 1 is free space and region 2 is characterized by a constant μ_r and ε_r. The terms $\bar{E}_1$ and $\bar{H}_1$ denote the fields in region 1, and $\bar{E}_2$ and $\bar{H}_2$ denote the fields throughout region 2. Using the surface equivalence principle, we wish to define equivalent sources on the scatterer surface that replicate the original fields in both regions.

The equivalent exterior problem, as shown in Figure 1.11, is constructed in a manner identical to the general situation presented in Section 1.6. Equivalent sources $\bar{J}_1$ and $\bar{K}_1$ have been placed on a surface coinciding with the original scatterer. These sources have been defined so that

$$\bar{J}_1 = \hat{n} \times \bar{H}_1 \tag{1.105}$$

$$\bar{K}_1 = \bar{E}_1 \times \hat{n} \tag{1.106}$$

where $\hat{n}$ is the outward normal vector at points on the surface. These sources, radiating in conjunction with the original source, replicate the original fields $\bar{E}_1$ and $\bar{H}_1$ throughout region 1. Null fields are produced in region 2, allowing us to replace medium 2 with free space without changing the fields in region 1. Thus, the exterior part of the original problem of Figure 1.10 is equivalent to the problem of Figure 1.11, which only involves sources radiating in free space.

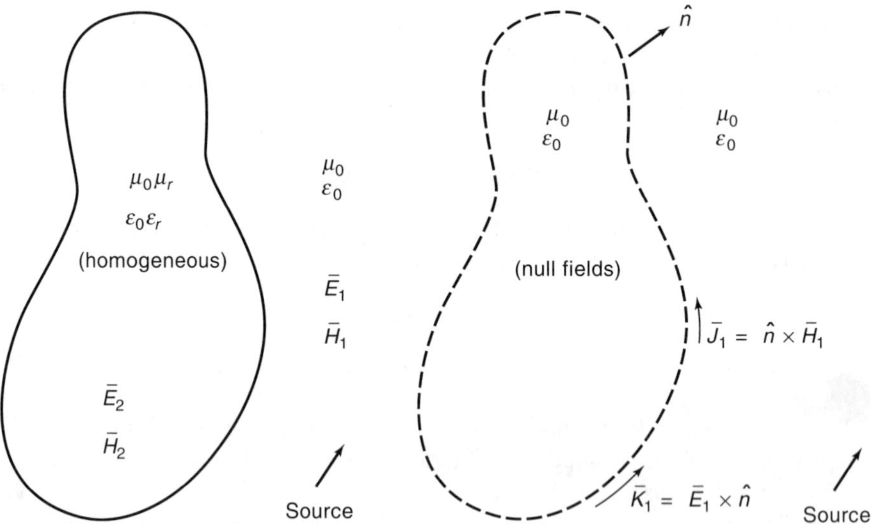

Figure 1.10 Original problem, involving a homogeneous body.

Figure 1.11 Equivalent exterior problem associated with Figure 1.10.

To describe the interior problem, a second equivalence relationship must be constructed. The equivalent interior problem is depicted in Figure 1.12. Sources $\bar{J}_2$ and $\bar{K}_2$ are defined according to

$$\bar{J}_2 = (-\hat{n}) \times \bar{H}_2 \tag{1.107}$$

$$\bar{K}_2 = \bar{E}_2 \times (-\hat{n}) \tag{1.108}$$

where $\hat{n}$ is still the normal vector pointing into region 1 (out of the scatterer). Radiating *in the absence* of the original source, these equivalent sources replicate the original fields throughout region 2 and produce null fields throughout the entirety of region 1. Since the region 1 fields vanish, we are free to insert material having $\mu = \mu_2$ and $\varepsilon = \varepsilon_2$ throughout region 1 in order to convert the problem to one involving infinite, homogeneous space. Thus, the equivalent interior problem involves sources $\bar{J}_2$ and $\bar{K}_2$ radiating in homogeneous space characterized by permittivity ε_2 and permeability μ_2.

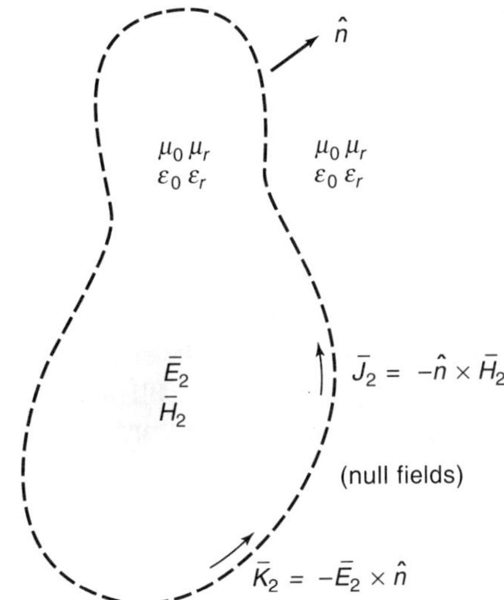

Figure 1.12 Equivalent interior problem associated with Figure 1.10.

The continuity of the tangential $\bar{E}$ and $\bar{H}$ fields at the dielectric interface dictates that

$$\bar{J}_1 = -\bar{J}_2 \tag{1.109}$$

and

$$\bar{K}_1 = -\bar{K}_2 \tag{1.110}$$

Therefore, it suffices to work with $\bar{J}_1$ and $\bar{K}_1$ as the primary unknowns to be determined. Since we have two equivalent problems and two unknown sources, we must employ a system of two coupled equations. The source–field relationships from Equations (1.38) and (1.39) can be combined with conditions (1.105)–(1.108) to produce the coupled EFIEs

$$\hat{n} \times \bar{E}^{\text{inc}} = -\bar{K}_1 - \hat{n} \times \left\{ \frac{\eta_1}{jk_1}(\nabla\nabla \cdot \bar{A}_1 + k_1^2 \bar{A}_1) - \nabla \times \bar{F}_1 \right\}_{S^+} \tag{1.111}$$

$$0 = \bar{K}_1 - \hat{n} \times \left\{ \frac{\eta_2}{jk_2}(\nabla\nabla \cdot \bar{A}_2 + k_2^2 \bar{A}_2) - \nabla \times \bar{F}_2 \right\}_{S^-} \tag{1.112}$$

where

$$\bar{A}_1 = \bar{J}_1 * \frac{e^{-jk_1 r}}{4\pi r} \tag{1.113}$$

$$\bar{F}_1 = \bar{K}_1 * \frac{e^{-jk_1 r}}{4\pi r} \tag{1.114}$$

$$\bar{A}_2 = \bar{J}_1 * \frac{e^{-jk_2 r}}{4\pi r} \tag{1.115}$$

$$\bar{F}_2 = \bar{K}_1 * \frac{e^{-jk_2 r}}{4\pi r} \tag{1.116}$$

$k_1 = \omega(\mu_1\varepsilon_1)^{1/2}$, $k_2 = \omega(\mu_2\varepsilon_2)^{1/2}$, $\eta_1 = (\mu_1/\varepsilon_1)^{1/2}$, and $\eta_2 = (\mu_2/\varepsilon_2)^{1/2}$ (the subscript on $\bar{A}$ and $\bar{F}$ indicates the medium into which the sources $\bar{J}_1$ and $\bar{K}_1$ radiate). Equation (1.111) is evaluated an infinitesimal distance *outside* the scatterer surface (S^+), while Equation

(1.112) is evaluated an infinitesimal distance *inside* the surface (S^-). As an alternative, coupled MFIEs are obtained as

$$\hat{n} \times \bar{H}^{\text{inc}} = \bar{J}_1 - \hat{n} \times \left\{ \nabla \times \bar{A}_1 + \frac{\nabla\nabla \cdot \bar{F}_1 + k_1^2 \bar{F}_1}{jk_1\eta_1} \right\}_{S^+} \tag{1.117}$$

$$0 = -\bar{J}_1 - \hat{n} \times \left\{ \nabla \times \bar{A}_2 + \frac{\nabla\nabla \cdot \bar{F}_2 + k_2^2 \bar{F}_2}{jk_2\eta_2} \right\}_{S^-} \tag{1.118}$$

Either of these formulations could be used to represent homogeneous scatterers. The extension of these equations to treat layered homogeneous regions will be left as an exercise (Prob. P1.20). In common with the EFIE and MFIE for p.e.c. scatterers, the coupled surface integral equations do not always guarantee a unique solution (see Chapter 6).

1.10. SURFACE INTEGRAL EQUATION FOR AN APERTURE IN A CONDUCTING PLANE

Consider the problem of scattering from an aperture in an infinite p.e.c. plane. Figures 1.13 and 1.14 show the geometry. The source is located in region 1. We will develop an integral equation formulation based on two equivalent problems, representing regions 1 and 2, respectively.

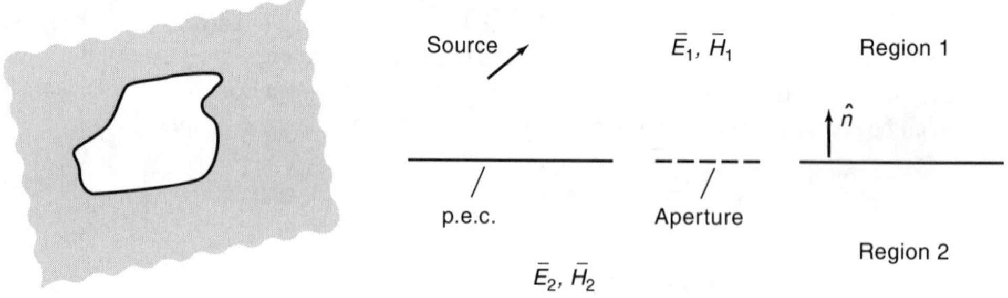

Figure 1.13 Aperture in an infinite p.e.c. plane.

Figure 1.14 Side view of an aperture in an infinite p.e.c. plane.

An equivalent representation for region 1 can be constructed by placing a mathematical surface S on the region 1 side of the conducting plane and introducing equivalent sources

$$\bar{J}_1 = \hat{n} \times \bar{H}_1 \tag{1.119}$$

$$\bar{K}_1 = \bar{E}_1 \times \hat{n} \tag{1.120}$$

on S. These sources, radiating together with the original source, will replicate the original fields in region 1 and create null fields throughout region 2. Although $\bar{J}_1$ is nonzero over the entire surface, the tangential electric field vanishes on the p.e.c. part of the plane, and therefore the magnetic source $\bar{K}_1$ is nonzero only over the aperture. The fact that $\bar{K}_1$ is confined to the aperture motivates its use as the primary unknown within an integral equation formulation.

Since the fields in region 2 of the equivalent problem vanish, we are free to modify the material present without changing the fields in region 1. Previous examples have employed this property in order to remove p.e.c. material. In this situation, suppose instead that we introduce additional p.e.c. material to completely close the aperture and create a uniform p.e.c. plane. The equivalent problem now involves the original source and the sources $\bar{J}_1$ and $\bar{K}_1$ radiating in front of the infinite conducting plane. As a second step, the method of images [2, 3, 7] can be employed to remove the p.e.c. plane. The image of a magnetic current over a perfect electric conductor is the mirror image; the image of an electric source is the negative mirror image. For tangential sources immediately adjacent to the plane, the image of the electric source cancels the original, while that of the magnetic source adds to the original. Thus, application of image theory eliminates the p.e.c. plane and the electric source $\bar{J}_1$, leaving only an equivalent magnetic source $2\bar{K}_1$ located in the original aperture. Consequently, the superposition of the original source, the image of the original source, and an equivalent source $2\bar{K}_1$ radiating in free space replicate the original fields in region 1 (Figure 1.15). These sources produce nonzero fields in region 2 as well, but these fields differ from the original fields in region 2 and the equivalence only holds for region 1.

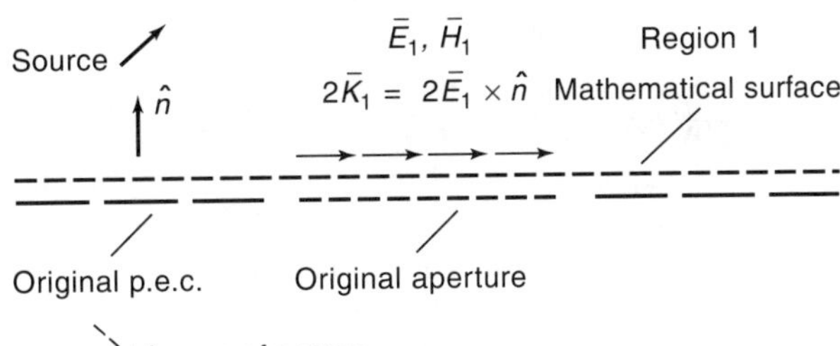

Figure 1.15 Equivalent problem for region 1 associated with Figure 1.14.

An equivalent problem for region 2 can be constructed in a similar manner and is depicted in Figure 1.16. A mathematical surface is introduced on the region 2 side of the plane, containing equivalent sources

$$\bar{J}_2 = (-\hat{n}) \times \bar{H}_2 \tag{1.121}$$

$$\bar{K}_2 = \bar{E}_2 \times (-\hat{n}) \tag{1.122}$$

where $\hat{n}$ is still the normal vector pointing into region 1. These sources replicate the original fields in region 2 and produce null fields throughout region 1. Again, the aperture can be closed with p.e.c. material, and image theory can be employed to reduce the problem to one involving just an equivalent magnetic source $2\bar{K}_2$ radiating in free space. This source is confined to the original aperture and, radiating in the absence of any other source, replicates the original fields throughout region 2. (The source also produces nonzero fields throughout region 1 that differ from the original fields in region 1.)

Original p.e.c.
Original aperture
$2\bar{K}_2 = -2\bar{K}_1$
$\bar{E}_2, \bar{H}_2$
Region 2

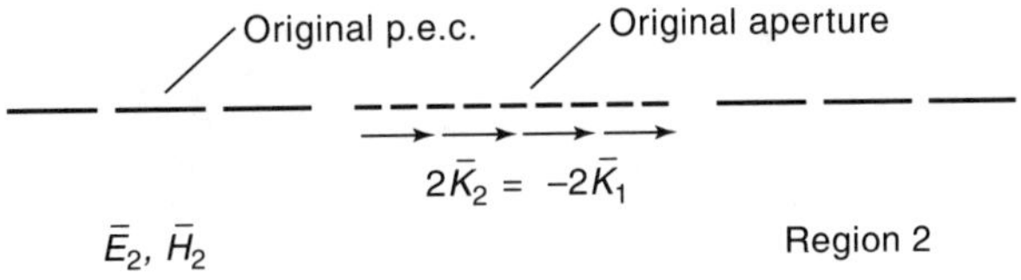

Figure 1.16 Equivalent problem for region 2 associated with Figure 1.14.

Because of the continuity of the original tangential electric field through the aperture, $\bar{K}_1 = -\bar{K}_2$, and it suffices to work with $\bar{K}_1$ as the primary unknown. Relationships similar to Equations (1.117) and (1.118) can be written involving the tangential magnetic field on either side of the aperture. Combining these two expressions to eliminate the aperture $\bar{H}$-field yields the integral equation

$$\hat{n} \times \bar{H}^{\text{inc}} = -4\hat{n} \times \left\{ \frac{\nabla\nabla \cdot \bar{F} + k^2 \bar{F}}{j\omega\mu_0} \right\}_S \tag{1.123}$$

where $\bar{H}^{\text{inc}}$ is the field produced by the original source *and its image* and $\bar{F}$ is the electric vector potential produced by $\bar{K}_1$ radiating in free space. The term $\bar{H}^{\text{inc}}$ can also be thought of as the field produced by the original source *radiating in the presence of the infinite p.e.c. plane* (aperture closed). Equation (1.123) holds in the original aperture and can be solved in principle to find $\bar{K}_1$.

1.11. SCATTERING CROSS SECTION CALCULATION FOR TWO-DIMENSIONAL PROBLEMS

A useful characterization of the scattering properties of an electromagnetic target is given by the *bistatic scattering cross section.* This quantity is an equivalent area proportional to the apparent size of the target in a particular direction (with the apparent size determined by the amount of power scattered in that direction in response to an excitation that may be incident from some other direction). More precisely, it is the area that, if multiplied by the power flux density of the incident field, would yield sufficient power to produce by isotropic radiation the same intensity in a given direction as that actually produced by the scatterer. In a two-dimensional problem, the scattering cross section (sometimes known as the "echo width") can be defined in a similar fashion as an equivalent width proportional to the apparent size of the scatterer in a particular direction.

Consider the two-dimensional situation involving a TM plane wave of the form

$$E_z^{\text{inc}}(x, y) = e^{-jk(x\cos\phi^{\text{inc}} + y\sin\phi^{\text{inc}})} \tag{1.124}$$

impinging upon an infinite, cylindrical geometry. The only electric field component present in the problem is E_z. The two-dimensional bistatic scattering cross section can be expressed as

$$\sigma_{\text{TM}}(\phi, \phi^{\text{inc}}) = \lim_{\rho\to\infty} 2\pi\rho \frac{|E_z^s(\rho, \phi)|^2}{|E_z^{\text{inc}}(0, 0)|^2} \tag{1.125}$$

where (ρ, ϕ) are ordinary polar coordinates. The scattered electric field can be found from Equation (1.38), which simplifies for the TM polarization to

$$E_z^s(x, y) = -jk\eta A_z - \frac{\partial F_y}{\partial x} + \frac{\partial F_x}{\partial y} \tag{1.126}$$

where

$$A_z(x, y) = \iint J_z(x', y') \frac{1}{4j} H_0^{(2)}(kR)\, dx'\, dy' \tag{1.127}$$

$$\bar{F}(x, y) = \iint \bar{K}(x', y') \frac{1}{4j} H_0^{(2)}(kR)\, dx'\, dy' \tag{1.128}$$

and

$$R = \sqrt{(x - x')^2 + (y - y')^2} \tag{1.129}$$

Since the observation point (x, y) is in the far field, it is convenient to work in cylindrical coordinates, that is,

$$R = \sqrt{(\rho \cos\phi - x')^2 + (\rho \sin\phi - y')^2} \tag{1.130}$$

which can be rearranged and written as

$$R = \rho \sqrt{1 - \frac{2}{\rho}(x' \cos\phi + y' \sin\phi) + \frac{(x')^2 + (y')^2}{\rho^2}} \tag{1.131}$$

As $\rho \to \infty$, the third expression under the radical is negligible compared to the others and may be omitted. The approximation

$$\sqrt{1 + \alpha} \cong 1 + \frac{1}{2}\alpha \tag{1.132}$$

can be used for small α to simplify Equation (1.131) to the "far-field" form

$$R \cong \rho - x' \cos\phi - y' \sin\phi \quad \text{as } \rho \to \infty \tag{1.133}$$

This result can be obtained from a purely geometrical argument, as illustrated in Figure 1.17.

To further simplify the calculation, the large-argument asymptotic form of the Hankel function

$$H_0^{(2)}(\alpha) \approx \sqrt{\frac{2j}{\pi\alpha}} e^{-j\alpha} \quad \text{as } \alpha \to \infty \tag{1.134}$$

may be employed. Substituting (1.133) and (1.134) into the previous expressions, we obtain

$$A_z(\rho, \phi) = \frac{1}{4j} \sqrt{\frac{2j}{\pi k \rho}} e^{-jk\rho} \iint J_z(x', y') e^{jk(x' \cos\phi + y' \sin\phi)} \, dx' \, dy' \tag{1.135}$$

$$\frac{\partial F_y}{\partial x} - \frac{\partial F_x}{\partial y} = \frac{-k}{4} \sqrt{\frac{2j}{\pi k \rho}} e^{-jk\rho} \iint (K_y \cos\phi - K_x \sin\phi) e^{jk(x' \cos\phi + y' \sin\phi)} \, dx' \, dy' \tag{1.136}$$

which may be combined with Equation (1.126) to produce

$$\sigma_{\text{TM}}(\phi, \phi^{\text{inc}}) = \frac{k}{4} \left| \iint (\eta J_z + K_x \sin\phi - K_y \cos\phi) e^{jk(x' \cos\phi + y' \sin\phi)} \, dx' \, dy' \right|^2 \tag{1.137}$$

As shown for emphasis, the scattering cross section is a function of the direction of the incident field and the far-zone observation angle. Although Equation (1.137) contains a double integral over volume sources defined throughout the scatterer, the integral collapses in an obvious way to a surface integral if the equivalent sources are confined to surfaces. For instance, in the special case of a p.e.c. cylinder represented by electric sources, the scattering cross section is given by

$$\sigma_{\text{TM}}(\phi, \phi^{\text{inc}}) = \frac{k\eta^2}{4} \left| \int J_z(t') e^{jk(x(t') \cos\phi + y(t') \sin\phi)} \, dt' \right|^2 \tag{1.138}$$

where t is a parametric variable defined along the contour of the cylinder surface.

In a two-dimensional TE problem, the magnetic field has only a $\hat{z}$-component. For a plane-wave excitation of the form

$$H_z^{\text{inc}}(x, y) = e^{-jk(x \cos\phi^{\text{inc}} + y \sin\phi^{\text{inc}})} \tag{1.139}$$

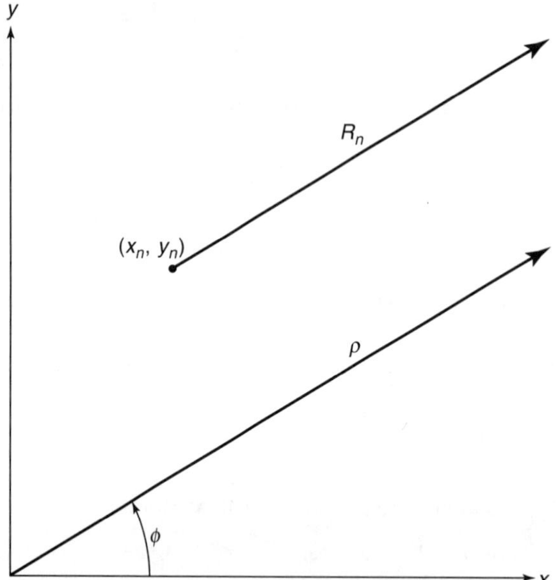

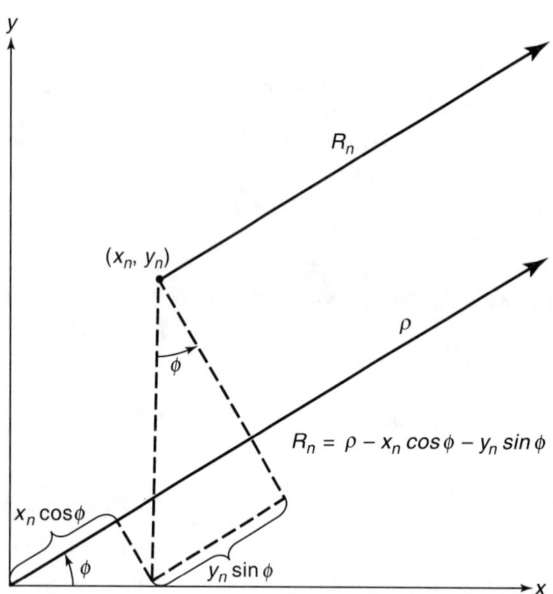

Figure 1.17 Relative path lengths for the far-field approximation.

the bistatic scattering cross section may be determined from

$$\sigma_{\text{TE}}(\phi, \phi^{\text{inc}}) = \lim_{\rho \to \infty} 2\pi\rho \frac{|H_z^s(\rho, \phi)|^2}{|H_z^{\text{inc}}(0, 0)|^2} \tag{1.140}$$

The scattered magnetic field can be found from Equation (1.39), which simplifies in the TE case to

$$H_z^s(x, y) = \frac{\partial A_y}{\partial x} - \frac{\partial A_x}{\partial y} - j\frac{k}{\eta} F_z \tag{1.141}$$

The far-zone approximations from Equations (1.133) and (1.134) may be used to obtain

$$\frac{\partial A_y}{\partial x} - \frac{\partial A_x}{\partial y} = \frac{-k}{4}\sqrt{\frac{2j}{\pi k\rho}}e^{-jk\rho} \iint (J_y \cos\phi - J_x \sin\phi) e^{jk(x'\cos\phi + y'\sin\phi)}\, dx'\, dy' \tag{1.142}$$

$$F_z(\rho, \phi) = \frac{1}{4j}\sqrt{\frac{2j}{\pi k\rho}}e^{-jk\rho} \iint K_z(x', y') e^{jk(x'\cos\phi + y'\sin\phi)}\, dx'\, dy' \tag{1.143}$$

Therefore, the scattering cross section is given by

$$\sigma_{\mathrm{TE}}(\phi, \phi^{\mathrm{inc}}) = \frac{k}{4}\left|\iint \left(J_x \sin\phi - J_y \cos\phi - \frac{K_z}{\eta}\right) e^{jk(x'\cos\phi + y'\sin\phi)}\, dx'\, dy'\right|^2 \tag{1.144}$$

The general expressions for σ_{TM} and σ_{TE} will be specialized to a variety of specific examples in the chapters to follow.

1.12. SCATTERING CROSS SECTION CALCULATION FOR THREE-DIMENSIONAL PROBLEMS

For a three-dimensional geometry where all components of the electric and magnetic field are present, the bistatic scattering cross section can be expressed for plane-wave incidence as

$$\sigma(\theta, \phi, \theta^{\mathrm{inc}}, \phi^{\mathrm{inc}}) = \lim_{r\to\infty} 4\pi r^2 \frac{\left|\bar{E}^s(\theta, \phi)\right|^2}{\left|\bar{E}^{\mathrm{inc}}(0, 0)\right|^2} \tag{1.145}$$

In the far zone, the scattered electric field has the form

$$\bar{E}^s \cong \hat{\theta} E_\theta^s + \hat{\phi} E_\phi^s \tag{1.146}$$

Since Equation (1.145) involves the expression

$$|\bar{E}^s|^2 \cong \left|E_\theta^s\right|^2 + \left|E_\phi^s\right|^2 \tag{1.147}$$

it is sufficient to compute the θ- and ϕ-components separately. In the far zone, these can be obtained from

$$E_\theta^s \cong -jk\eta A_\theta + \frac{\partial F_\phi}{\partial r} \tag{1.148}$$

$$E_\phi^s \cong -jk\eta A_\phi + \frac{\partial F_\theta}{\partial r} \tag{1.149}$$

where the potential functions are defined in Equations (1.42) and (1.43). Because the sources of the scattered field are often described in Cartesian coordinates, it may be necessary to transform to the spherical system using

$$A_\theta = \cos\theta \cos\phi A_x + \cos\theta \sin\phi A_y - \sin\theta A_z \tag{1.150}$$

$$A_\phi = -\sin\phi A_x + \cos\phi A_y \tag{1.151}$$

For three-dimensional analysis, the argument of the Green's function within the vector potentials is given by

$$R = \sqrt{(x - x')^2 + (y - y')^2 + (z - z')^2} \tag{1.152}$$

After converting the observation point (x, y, z) to spherical coordinates and grouping terms according to powers of r, we obtain

$$R = r\sqrt{1 - \frac{2}{r}(x' \sin\theta \cos\phi + y' \sin\theta \sin\phi + z' \cos\theta) + \frac{(x')^2 + (y')^2 + (z')^2}{r^2}} \tag{1.153}$$

Approximations similar to those employed to derive Equation (1.133) produce the far-field expression

$$R \cong r - x' \sin\theta \cos\phi - y' \sin\theta \sin\phi - z' \cos\theta \tag{1.154}$$

It follows that far-field forms of the vector potential functions are

$$\bar{A}(r, \theta, \phi) = \frac{e^{-jkr}}{4\pi r} \iiint \bar{J}(x', y', z') e^{jk(x' \sin\theta \cos\phi + y' \sin\theta \sin\phi + z' \cos\theta)}\, dx'\, dy'\, dz' \tag{1.155}$$

and

$$\bar{F}(r, \theta, \phi) = \frac{e^{-jkr}}{4\pi r} \iiint \bar{K}(x', y', z') e^{jk(x' \sin\theta \cos\phi + y' \sin\theta \sin\phi + z' \cos\theta)}\, dx'\, dy'\, dz' \tag{1.156}$$

Consequently, the scattering cross section can be written as

$$\sigma(\theta, \phi, \theta^{\text{inc}}, \phi^{\text{inc}}) = \sigma_\theta(\theta, \phi) + \sigma_\phi(\theta, \phi) \tag{1.157}$$

where

$$\sigma_\theta(\theta, \phi) = \frac{k^2}{4\pi} \left| \iiint (\eta J_x \cos\theta \cos\phi + \eta J_y \cos\theta \sin\phi - \eta J_z \sin\theta \right.$$
$$\left. - K_x \sin\phi + K_y \cos\phi) e^{jk(x' \sin\theta \cos\phi + y' \sin\theta \sin\phi + z' \cos\theta)}\, dx'\, dy'\, dz' \right|^2 \tag{1.158}$$

$$\sigma_\phi(\theta, \phi) = \frac{k^2}{4\pi} \left| \iiint (-\eta J_x \sin\phi + \eta J_y \cos\phi + K_x \cos\theta \cos\phi \right.$$
$$\left. + K_y \cos\theta \sin\phi) - K_z \sin\theta) e^{jk(x' \sin\theta \cos\phi + y' \sin\theta \sin\phi + z' \cos\theta)}\, dx'\, dy'\, dz' \right|^2 \tag{1.159}$$

and where we have assumed that the magnitude of the incident electric field is unity. Equations (1.158) and (1.159) are written in terms of triple integrals over volume sources; as in the two-dimensional case the integrals collapse to surface integrals in an obvious way if the equivalent sources are confined to surfaces.

As shown for emphasis in Equation (1.157), σ is a function of the direction of the incident field and the far-zone observation angle. The bistatic scattering cross section is also a function of the polarization of the incident wave. To explicitly characterize the scatterer as a function of polarization, the scattering cross section data can be obtained for two orthogonal polarizations and arranged in the form of a scattering matrix such as

$$\Sigma = \begin{bmatrix} \sigma_{\theta\theta} & \sigma_{\theta\phi} \\ \sigma_{\phi\theta} & \sigma_{\phi\phi} \end{bmatrix} \tag{1.160}$$

The entries of this 2×2 matrix remain a function of the direction of the incident field and the far-zone observation angle.

1.13 APPLICATION TO ANTENNA ANALYSIS

Although preceding sections of this chapter have dealt exclusively with scattering problems, it is important to note that most antenna radiation problems can be analyzed using identical techniques. To illustrate the connection between scattering formulations and antenna analysis in detail, consider a monopole antenna of radius a and height $L/2$ radiating over a p.e.c. ground plane (Figure 1.18). The monopole is coincident with the z-axis of a cylindrical coordinate system and is fed by a coaxial transmission line with outer radius b. The ground plane is located at $z = 0$, and for simplicity we assume that the aperture of the transmission line contains only a transverse electromagnetic mode with electric field distribution

$$\bar{E}(\rho, \phi) = \hat{\rho}\frac{E_0}{\rho \ln(b/a)} \tag{1.161}$$

Following the procedure discussed in Section 1.10, we introduce an equivalent magnetic current density defined at the transmission line aperture. The method of images is used to remove the ground plane, leaving a dipole antenna of length L (Figure 1.19) illuminated by a "magnetic frill" source that can be expressed by the volume current density

$$\bar{K}(\rho, \phi, z) = -\hat{\phi}\frac{2E_0}{\rho \ln(b/a)} p(\rho; a, b)\delta(z) \tag{1.162}$$

where the pulse function

$$p(\rho; a, b) = \begin{cases} 1 & a < \rho < b \\ 0 & \text{otherwise} \end{cases} \tag{1.163}$$

serves as a window to identify the original aperture location. The equivalent dipole produces the same fields in the upper half space as the original monopole.

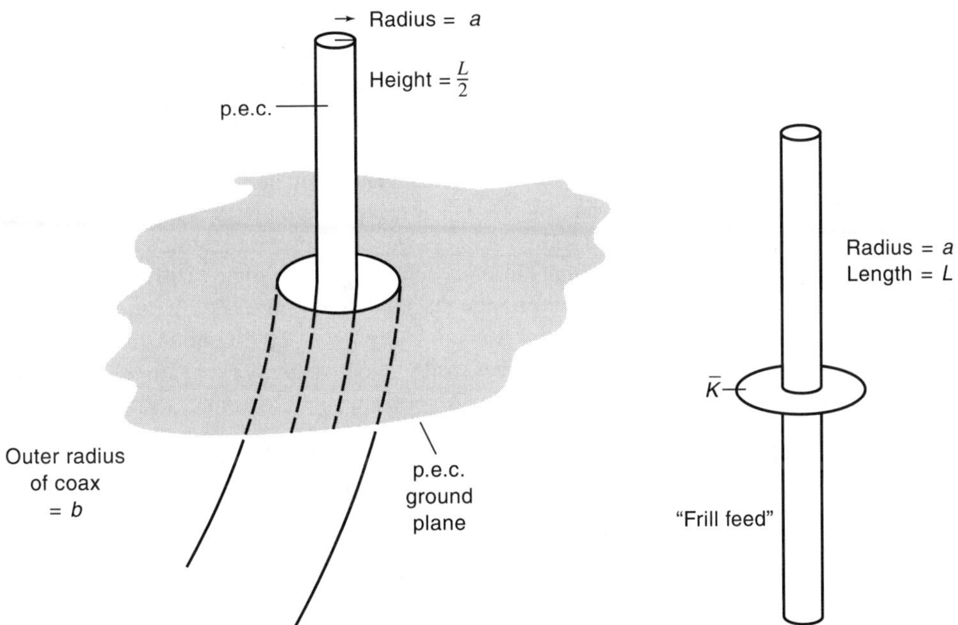

Figure 1.18 Monopole antenna radiating over a p.e.c. ground plane.

Figure 1.19 Equivalent problem of a dipole fed by a magnetic frill source.

At this point in the development, the problem is that of a p.e.c. scatterer (the dipole) illuminated by an incident field (produced by the magnetic frill source). The electric field produced by a frill source is the topic of Prob. P1.17, and the z-component of the incident field produced on the surface of the monopole is

$$E_z^{\text{inc}}(a, \phi, z) = \frac{2E_0}{\ln(b/a)} \int_{\phi'=0}^{2\pi} \left(\frac{e^{-jkR_1}}{4\pi R_1} - \frac{e^{-jkR_2}}{4\pi R_2} \right) d\phi' \tag{1.164}$$

where R_1 and R_2 are defined in Prob. P1.17. Thus, the antenna currents can be determined from a surface integral equation, as discussed in Section 1.7. (A specific EFIE for the dipole antenna will be presented in Chapter 8.)

In general, the primary difference between antenna analysis and the scattering formulations considered earlier is that the primary source in an antenna geometry is located immediately adjacent to the scatterer, rather than an infinite distance away. In fact, antennas actually function by scattering the energy emitted by the primary feed. For instance, the arms of the dipole in Figure 1.19 scatter (or focus) the energy radiated by the magnetic frill source in order to produce the characteristic dipole radiation pattern. Yagi antennas contain parasitic elements that clearly act as scatterers to enhance the radiation pattern. Reflector antennas focus the energy from a primary feed horn to achieve a narrow radiation beam. Almost all types of antennas can be thought of as scatterers and posed in terms of differential equation or integral equation formulations.

1.14 SUMMARY

Chapter 1 has reviewed concepts from electromagnetic theory that play an important role in the numerical procedures of interest. Of particular importance are the source–field relations summarized in Section 1.4 and the equivalence principles presented in Sections 1.2 and 1.6. These ideas are central to the formulation of integral equations for scatterers or antennas.

Integral equations for conducting bodies, penetrable bodies, and aperture problems have been developed in Sections 1.7–1.10. These equations will be specialized to a variety of situations in the chapters to follow. For instance, Chapter 2 considers integral equation formulations for two-dimensional (infinite cylinder) geometries. Differential equations have also been presented and will provide the foundation for alternate numerical solution methods. Chapter 3 presents several ways of using the scalar Helmholtz equation to treat two-dimensional open-region geometries. Subsequent chapters extend these procedures to three dimensions and to a variety of other situations.

REFERENCES

[1] J. A. Stratton, *Electromagnetic Theory*, New York: McGraw-Hill, 1941.

[2] R. F. Harrington, *Time-Harmonic Electromagnetic Fields*, New York: McGraw-Hill, 1961.

[3] E. C. Jordan and K. G. Balmain, *Electromagnetic Waves and Radiating Systems*, Englewood Cliffs, NJ: Prentice-Hall, 1968.

[4] J. Van Bladel, *Electromagnetic Fields*, New York: Hemisphere, 1985.

[5] J. A. Kong, *Electromagnetic Wave Theory*, New York: Wiley, 1986.

[6] A. J. Poggio and E. K. Miller, “Integral equation solutions of three-dimensional scattering problems,” in *Computer Techniques for Electromagnetics*, ed. R. Mittra, New York: Hemisphere, 1987.

[7] C. A. Balanis, *Advanced Engineering Electromagnetics*, New York: Wiley, 1989.

[8] K. M. Chen, "A mathematical formulation of the equivalence principle," *IEEE Trans. Microwave Theory Tech.*, vol. 37, pp. 1576–1581, 1989.

[9] W. C. Chew, *Waves and Fields in Inhomogeneous Media*, New York: Van Nostrand Reinhold, 1990.

[10] R. E. Collin, *Field Theory of Guided Waves*, New York: IEEE Press, 1991.

[11] A. Ishimaru, *Electromagnetic Wave Propagation, Radiation, and Scattering*, Englewood Cliffs, NJ: Prentice-Hall, 1991.

[12] J. Van Bladel, *Singular Electromagnetic Fields and Sources*, Oxford: Oxford University Press, 1991.

[13] I. S. Gradshteyn and I. M. Ryzhik, *Table of Integrals, Series, and Products*, New York: Academic, 1980.

[14] D. F. Hanson and P. E. Mayes, "The current source-function technique solution of electromagnetic scattering from a half-plane," *Radio Science*, vol. 13, pp. 49–58, 1978.

[15] A. D. Yaghjian, "Electric dyadic Green's functions in the source region," *Proc. IEEE*, vol. 68, pp. 248–263, 1980.

[16] S. W. Lee, J. Boersma, C. L. Law, and G. A. Deschamps, "Singularity in Green's function and its numerical evaluation," *IEEE Trans. Antennas Propagat.*, vol. AP-28, pp. 311–317, 1980.

[17] D. B. Miron, "The singular integral problem in surfaces," *IEEE Trans. Antennas Propagat.*, vol. AP-31, pp. 507–509, May 1983.

PROBLEMS

P1.1 (a) Derive Equations (1.13) and (1.14).

(b) In a homogeneous region with $\varepsilon_r = 1$ and $\mu_r = 1$, containing electric and magnetic sources $\bar{J}$ and $\bar{K}$, respectively, Equations (1.1) and (1.2) become

$$\nabla \times \bar{E} = -j\omega\mu_0 \bar{H} - \bar{K} \qquad \nabla \times \bar{H} = j\omega\varepsilon_0 \bar{E} + \bar{J}$$

Under these conditions, show that the equivalent curl–curl equation for $\bar{E}$ can be obtained as

$$\nabla \times \nabla \times \bar{E} - k^2 \bar{E} = -j\omega\mu_0 \bar{J} - \nabla \times \bar{K}$$

Find the corresponding equation for $\bar{H}$.

P1.2 Repeat Prob. P1.1(a) for the case where ε_r and μ_r are replaced by tensors to represent anisotropic material.

P1.3 Under the assumption that the fields in some region do not depend on z, show (using Maxwell's equations) that the field components E_z, H_x, and H_y (the TM part) are completely independent from H_z, E_x, and E_y (the TE part).

P1.4 (a) Under the assumption that the z-dependence of the electric field is $e^{-j\gamma z}$, show that the z-component of Equation (1.13) can be expressed as

$$\nabla_t \cdot \left(\frac{1}{\mu_r} \nabla_t E_z\right) + k^2 \varepsilon_r E_z = \frac{1}{\mu_r} \gamma^2 E_z - j\gamma \nabla_t \left(\frac{1}{\mu_r \varepsilon_r}\right) \cdot \varepsilon_r \bar{E}_t$$

where ∇_t is the transverse part of the operator, for instance,

$$\nabla_t E_z = \hat{x} \frac{\partial E_z}{\partial x} + \hat{y} \frac{\partial E_z}{\partial y}$$

(b) Using duality (Table 1.1), write down the analogous equation involving the magnetic field $\bar{H}$.

(c) In certain cases, the z-component of the vector Helmholtz equation can be used instead of the complete vector Helmholtz equation to produce a solution. In general, this occurs whenever the z-components of the field decouple from the transverse components, so that the equations in (a) and (b) coincide with the scalar Helmholtz equations in (1.15) and (1.16). Identify two situations where the equations obtained in (a) and (b) constitute a sufficient description of the fields E_z and H_z. (*Hint:* Obviously, the case $\gamma = 0$ is one answer. What is the other situation?)

(d) What is the physical interpretation of zero γ? Of nonzero γ?

P1.5 (TM–TE Decomposition) In a homogeneous region with fields having z-dependence $e^{-j\gamma z}$, show that the transverse-to-z field components can be expressed as a function of the z-components according to

$$\bar{E}_t = \frac{1}{k^2 - \gamma^2}\left(-j\gamma\nabla_t E_z + j\omega\mu_0\hat{z} \times \nabla_t H_z\right)$$

$$\bar{H}_t = \frac{1}{k^2 - \gamma^2}\left(-j\omega\varepsilon_0\hat{z} \times \nabla_t E_z - j\gamma\nabla_t H_z\right)$$

P1.6 (a) Assuming that ε_r is sufficiently differentiable, derive Equation (1.23) using the identity

$$\nabla\left(\frac{1}{\varepsilon_r}\right) = -\left(\frac{1}{\varepsilon_r}\right)^2 \nabla\varepsilon_r$$

(b) At a jump discontinuity in the permittivity, the normal component of $\bar{E}$ must behave according to Equation (1.11), that is,

$$(1) \qquad 0 = \varepsilon_0\varepsilon_1 E_1^{\text{nor}} - \varepsilon_0\varepsilon_2 E_2^{\text{nor}}$$

In the equivalent problem constructed by replacing the dielectric material with induced sources $\bar{J}$ and ρ_e, the proper behavior at the location of the original dielectric interface is

$$(2) \qquad \rho_{\text{es}} = \varepsilon_0 E_1^{\text{nor}} - \varepsilon_0 E_2^{\text{nor}}$$

where ρ_{es} represents a surface charge density, and the normal direction points from region 2 into region 1. Demonstrate that Equation (1.23) is consistent with (1) and (2) and therefore produces the proper surface charge density at an interface when the permittivity has a jump discontinuity. Because of the generalized function interpretation of (1.23), the volume charge density in that equation will appear as a Dirac delta function in the three-dimensional space in order to represent a surface charge density.

P1.7 (a) Consider the subsectional "triangle" function

$$t(x) = \begin{cases} \dfrac{x + \Delta}{\Delta} & -\Delta < x < 0 \\ \dfrac{\Delta - x}{\Delta} & 0 < x < \Delta \\ 0 & \text{otherwise} \end{cases}$$

Show that the second derivative with respect to x is

$$t''(x) = \frac{d^2 t}{dx^2} = \frac{1}{\Delta}\delta(x + \Delta) - \frac{2}{\Delta}\delta(x) + \frac{1}{\Delta}\delta(x - \Delta)$$

Sketch t, t', and t''.

(b) A "sinusoidal triangle" function can be defined

$$s(x) = \begin{cases} \dfrac{\sin(kx + k\Delta)}{\sin(k\Delta)} & -\Delta < x < 0 \\ \dfrac{\sin(k\Delta - kx)}{\sin(k\Delta)} & 0 < x < \Delta \\ 0 & \text{otherwise} \end{cases}$$

Show that

$$\frac{d^2 s}{dx^2} + k^2 s(x)$$
$$= \frac{k}{\sin(k\Delta)}\delta(x+\Delta) - \frac{2k\cos(k\Delta)}{\sin(k\Delta)}\delta(x) + \frac{k}{\sin(k\Delta)}\delta(x-\Delta)$$

P1.8 (a) In a homogeneous medium, use Maxwell's equations

$$\nabla \times \bar{E} = -j\omega\mu_0 \bar{H} - \bar{K} \qquad \nabla \times \bar{H} = j\omega\varepsilon_0 \bar{E} + \bar{J}$$
$$\nabla \cdot (\varepsilon_0 \bar{E}) = \rho_e \qquad \nabla \cdot (\mu_0 \bar{H}) = \rho_m$$

to derive the equations of continuity

$$\nabla \cdot \bar{J} = -j\omega\rho_e \qquad \nabla \cdot \bar{K} = -j\omega\rho_m$$

(b) Using the preceding results and the vector Laplacian

$$\nabla^2 \bar{A} = \nabla(\nabla \cdot \bar{A}) - \nabla \times \nabla \times \bar{A}$$

derive Equations (1.31) and (1.32).

P1.9 Show that the three-dimensional Green's function

$$G = \frac{e^{-jkr}}{4\pi r}$$

satisfies the scalar Helmholtz equation

$$\nabla^2 G + k^2 G = -\delta(r)$$

(*Hint:* The calculation for $r \neq 0$ is straightforward. In the vicinity of $r = 0$, integrate the equation throughout a sphere of radius r, and use the divergence theorem to obtain

$$\iiint_V \nabla \cdot \nabla G \, dv = \iint_S \nabla G \cdot \hat{r} \, ds = -1$$

as $r \to 0$.)

P1.10 A subsectional "pulse" function can be defined as

$$p\left(x; -\frac{\Delta}{2}, \frac{\Delta}{2}\right) = \begin{cases} 1 & -\dfrac{\Delta}{2} < x < \dfrac{\Delta}{2} \\ 0 & \text{otherwise} \end{cases}$$

Show that the convolution

$$p\left(x; -\frac{\Delta}{2}, \frac{\Delta}{2}\right) * p\left(x; -\frac{\Delta}{2}, \frac{\Delta}{2}\right)$$
$$= \int_{x'=-\Delta/2}^{\Delta/2} p\left(x - x'; -\frac{\Delta}{2}, \frac{\Delta}{2}\right) dx'$$
$$= \Delta t(x)$$

where $t(x)$ is the subsectional triangle function defined in Prob. P1.7.

P1.11 The potential functions in Equations (1.48) and (1.49) satisfy the Lorentz gauge conditions

$$\nabla \cdot \bar{A} = -j\omega\varepsilon_0 \Phi_e \qquad \nabla \cdot \bar{F} = -j\omega\mu_0 \Phi_m$$

Using these conditions, substitute (1.48) and (1.49) into Maxwell's equations to show that $\bar{A}$, $\bar{F}$, Φ_e, and Φ_m can be decoupled from one another to produce

$$\nabla^2 \bar{A} + k^2 \bar{A} = -\bar{J} \qquad \nabla^2 \bar{F} + k^2 \bar{F} = -\bar{K}$$

$$\nabla^2 \Phi_e + k^2 \Phi_e = -\frac{\rho_e}{\varepsilon_0} \qquad \nabla^2 \Phi_m + k^2 \Phi_m = -\frac{\rho_m}{\mu_0}$$

P1.12 Using properties of the Fourier transform integral (see, e.g., Chapter 7), demonstrate that differentiation and convolution operations commute. Specifically, for two differentiable functions $a(x)$ and $b(x)$, show that

$$\frac{d}{dx}[a(x) * b(x)] = \frac{da}{dx} * b(x) = a(x) * \frac{db}{dx}$$

This concept provides an alternative way of deriving Equations (1.52) and (1.53).

P1.13 A rectangular waveguide of dimension $a \times b$ radiates through an aperture in an infinite p.e.c. ground plane into the half-space $z > 0$. Assume that the only fields present in the aperture are those associated with the TE_{10} mode.

(a) Identify equivalent surface currents located at $z = 0^+$ that, radiating in the presence of an infinite ground plane (no aperture) at $z = 0$, reproduce the fields in the region $z > 0$. (Give an explicit expression for $\bar{J}_s$ and $\bar{K}_s$ in the location of the original aperture, and comment on their values away from the aperture.)

(b) Use the method of images to remove the p.e.c. ground plane and provide expressions for the equivalent surface currents that, radiating in free space, reproduce the $z > 0$ fields. (*Hint:* The image of a magnetic current over a p.e.c. is the mirror image, while the image of an electric source is the negative mirror image.)

P1.14 Repeat P1.13 for the fields of the TE_{11} mode in the aperture of a circular waveguide having radius a, radiating in the presence of an infinite p.e.c. ground plane.

P1.15 Section 1.4 presents several alternative expressions for the fields produced by sources. In particular, given an equivalent source density $\bar{J}$ radiating in free space, the following are just three of the possible formulas:

$$(a) \qquad \bar{E}^s = \frac{\nabla\nabla \cdot + k^2}{j\omega\varepsilon} \bar{A}$$

$$(b) \qquad \bar{E}^s = -j\omega\mu\bar{A} - \nabla\Phi_e$$

$$(c) \qquad \bar{E}^s = \left(\frac{\nabla\nabla \cdot + k^2}{j\omega\varepsilon} \bar{J}\right) * G$$

where $\bar{A}$, Φ_e, and G are defined in Section 1.4. For the case in which $\bar{J}$ is a sphere of uniform current density, that is,

$$\bar{J}(r, \theta, \phi) = \hat{x}\, p(r; 0, a) = \hat{x} \begin{cases} 1 & r < a \\ 0 & \text{otherwise} \end{cases}$$

the electric field at the center of the sphere has been obtained using dyadic Green's functions (D. E. Livesay and K. M. Chen, *IEEE Trans. Microwave Theory Tech.*, vol. MTT-22, Dec. 1974) in the form

$$\bar{E}^s(0, 0, 0) = \hat{x} \frac{-1}{j\omega\varepsilon}[1 - \tfrac{2}{3} e^{-jka}(1 + jka)]$$

(a) Derive this result using Equation (c) above. [*Hint:* You should obtain

$$\nabla \cdot \bar{J} = -J_r(\theta, \phi)\delta(r-a) = -\sin\theta\cos\phi\delta(r-a)$$
$$\hat{x} \cdot (\nabla\nabla \cdot \bar{J}) = -\delta'(r-a)\sin^2\theta\cos^2\phi - \tfrac{\delta(r-a)}{r}(\cos^2\theta\cos^2\phi + \sin^2\phi)$$

as intermediate results.]

(b) Why is it difficult to derive the result using Equations (a) or (b)? Explain.

P1.16 The current density along a thin linear center-fed dipole antenna of length 2Δ is often approximated by a sinusoidal function with support confined to the dipole axis. If written in terms of a volume current, this function has the form

$$\bar{J}(\rho, z) = \hat{z} I_0 s(z)\delta(\rho)$$

where s is the sinusoidal triangle function defined in Prob. P1.7. Following a procedure similar to that employed in Prob. P1.15, find the z-component of the electric field at a general location (ρ, ϕ, z) produced by this current density in free space.

P1.17 The magnetic current density

$$\bar{K}(\rho, z) = \hat{\phi}\frac{-1}{\rho\ln(b/a)} p(\rho; a, b)\delta(z)$$

obtained from the aperture transverse electromagnetic field of an open-ended coaxial cable is often used as a "magnetic frill" feed model for a dipole antenna. Using Equation (1.52), show that the z-component of the electric field produced by this magnetic current density radiating in free space can be expressed as

$$E_z(\rho, \phi, z) = \frac{1}{\ln(b/a)} \int_{\phi'=0}^{2\pi} \left(\frac{e^{-jkR_1}}{4\pi R_1} - \frac{e^{-jkR_2}}{4\pi R_2} \right) d\phi'$$

where

$$R_1 = \sqrt{z^2 + \rho^2 + a^2 - 2\rho a\cos\phi'}$$

and

$$R_2 = \sqrt{z^2 + \rho^2 + b^2 - 2\rho b\cos\phi'}$$

P1.18 The surface integral equations for p.e.c. scatterers, (1.96) and (1.97), are readily specialized to the two-dimensional case using

$$\bar{A}(t) = \int_S \bar{J}(t')\frac{1}{4j}H_0^{(2)}\left(k\sqrt{[x(t)-x(t')]^2 + [y(t)-y(t')]^2}\right) dt'$$

(a) For the TM polarization, identify the components of $\bar{E}$, $\bar{H}$, and $\bar{J}$ present and write down a scalar form of the EFIE.

(b) Repeat part (a) for the TE case, producing a scalar form of the MFIE.

P1.19 Equations (1.96) and (1.97) are obtained using the tangential-field boundary conditions $\hat{n} \times \bar{E} = 0$ and $\hat{n} \times \bar{H} = \bar{J}_s$ on the scatterer surface. Similar relationships obtained from the normal-field boundary conditions $\hat{n} \cdot \bar{E} = \rho_s/\varepsilon_0$ and $\hat{n} \cdot \bar{H} = 0$ can be expressed as

$$\hat{n} \cdot \bar{E}^{\text{inc}} = \frac{\rho_s}{\varepsilon_0} - \hat{n} \cdot \left\{ \frac{\nabla\nabla \cdot + k^2}{j\omega\varepsilon_0}\bar{A} \right\}_{S+}$$

$$\hat{n} \cdot \bar{H}^{\text{inc}} = -\hat{n} \cdot \left\{ \nabla \times \bar{A} \right\}_{S+}$$

Are these valid equations? Can they be used instead of (1.96) and/or (1.97)? Discuss their possible utility for both two-dimensional and three-dimensional problems.

P1.20 Figure 1.20 depicts a layered dielectric scatterer illuminated by an incident wave. We wish to extend the procedure of Section 1.9 in order to produce a system of coupled surface integral equations describing this problem.

(a) By introducing equivalent sources $(\bar{J}_1, \bar{K}_1)$ on S_1 and $(\bar{J}_2, \bar{K}_2)$ on S_2, identify three equivalent problems involving sources radiating in homogeneous space that reproduce the original fields in region A, region B, and the exterior of the scatterer.

(b) For each equivalent problem in (a), construct a surface EFIE. Develop a notation that clearly indicates the medium employed within each equation, the surface over which each integral is evaluated, and the surface on which each equation is enforced. (*Hint:* Use four surfaces: S_1^+, S_1^-, S_2^+, and S_2^-.)

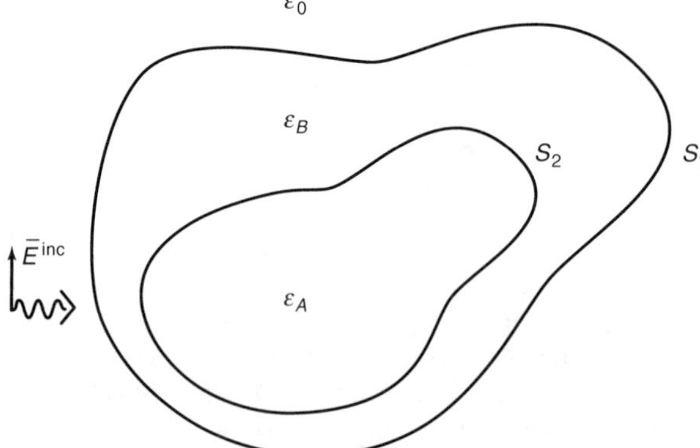

Figure 1.20 Layered dielectric scatterer (for Prob. P1.20).

P1.21 (a) Specialize the two-dimensional form of Equations (1.38) and (1.39) in order to obtain formulas valid in the far field of the sources $\bar{J}$ and $\bar{K}$. Retain only the dominant-order terms as $\rho \to \infty$.

(b) Repeat part (a) for the three-dimensional situation.

P1.22 A wave having the form of Equation (1.124) produces scattered fields $E_z^s(\phi)$ and $H_\phi^s(\phi)$ on a circular boundary of radius a enclosing a two-dimensional obstacle. Express the two-dimensional scattering cross section $\sigma_{TM}(\phi)$ in terms of an integral over E_z^s and H_ϕ^s on the circular contour.

P1.23 Repeat Prob. P1.22 for the three-dimensional case by obtaining an expression for the scattering cross section in the form of an integral over tangential electric fields on the surface of a sphere of radius a.

P1.24 Exterior to the circular contour defined in Prob. P1.22, E_z^s can be written as a Fourier series of the form

$$E_z^s(\rho, \phi) = \sum_{n=-\infty}^{\infty} j^{-n} A_n H_n^{(2)}(k\rho) e^{jn\phi}$$

where

$$A_n = \frac{1}{2\pi j^{-n} H_n^{(2)}(ka)} \int_{\phi'=0}^{2\pi} E_z^s(a, \phi') e^{-jn\phi'}\, d\phi'$$

and where H_n denotes the nth-order Hankel function. Using the asymptotic approximation

$$H_n^{(2)}(kp) \approx \sqrt{\frac{2j}{\pi k\rho}}\, j^n e^{-jkp}$$

to simplify your result, find an expression for the two-dimensional scattering cross section σ_{TM} as a function of the coefficients $\{A_n\}$.

2

Integral Equation Methods for Scattering from Infinite Cylinders

We begin our investigation of numerical techniques by considering two-dimensional scatterers illuminated by normally incident plane waves. This chapter considers formulations based on the electric and magnetic field integral equations (EFIE and MFIE, respectively). Although the problems considered and methods employed are relatively simple, they illustrate the approach to be followed in more complicated formulations. In addition, a goal of this chapter is the presentation of sufficient detail to enable the reader to implement these techniques in software. To demonstrate possible approaches, sample FORTRAN programs are described in the Appendices.

2.1 TM-WAVE SCATTERING FROM CONDUCTING CYLINDERS: EFIE DISCRETIZED WITH PULSE BASIS AND DELTA TESTING FUNCTIONS [1]

The cross section of an infinite, perfectly conducting (p.e.c.) cylinder is depicted in Figure 2.1. Using the surface equivalence principle from Section 1.6, the p.e.c. material may be replaced by equivalent electric currents radiating in free space. For a normally incident TM wave, the field components present are E_z, H_x, and H_y and the only current component present is J_z. Since $\nabla \cdot \bar{J} = 0$ for the TM polarization, which in turn implies that $\nabla \cdot \bar{A} = 0$, the EFIE from (1.96) can be specialized to

$$E_z^{\text{inc}}(t) = jk\eta A_z(t) \tag{2.1}$$

where

$$A_z(t) = \int J_z(t') \frac{1}{4j} H_0^{(2)}(kR)\, dt' \tag{2.2}$$

$$R = \sqrt{[x(t) - x(t')]^2 + [y(t) - y(t')]^2} \tag{2.3}$$

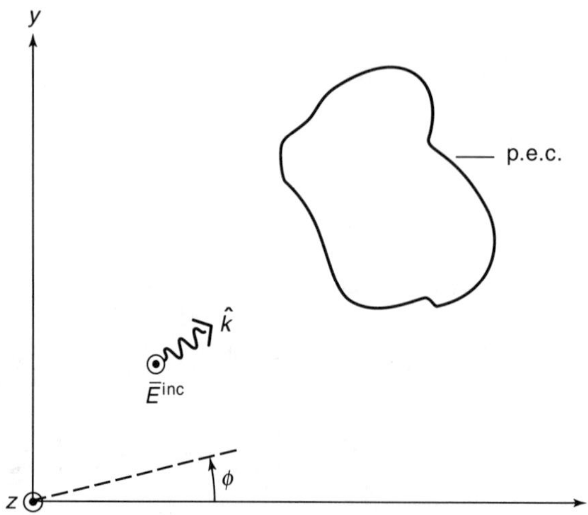

Figure 2.1 A p.e.c. cylinder illuminated by an incident TM wave.

and t is a parametric variable denoting the position around the contour of the cylinder surface. Note that Equation (2.1) is only valid at points on the surface of the original cylinder.

To construct an approximate solution for J_z, we divide the cylinder contour into cells as illustrated in Figure 2.2. As a function of position around the contour, the equivalent current density can be approximated by the superposition of subsectional pulse basis functions

$$p_n(t) = \begin{cases} 1 & \text{if } t \in \text{cell } n \\ 0 & \text{otherwise} \end{cases} \tag{2.4}$$

so that

$$J_z(t) \cong \sum_{n=1}^{N} j_n p_n(t) \tag{2.5}$$

Substituting (2.5) into Equation (2.1) produces

$$E_z^{\text{inc}}(t) \cong jk\eta \sum_{n=1}^{N} j_n \int_{\text{cell } n} \frac{1}{4j} H_0^{(2)}(kR)\, dt' \tag{2.6}$$

This substitution replaces the original problem of finding $J_z(t)$ with that of finding N unknown coefficients $\{j_n\}$. The N linearly independent equations may be obtained by enforcing Equation (2.6) at the centers of each of the N cells, to produce the $N \times N$ system

$$\begin{bmatrix} E_z^{\text{inc}}(t_1) \\ E_z^{\text{inc}}(t_2) \\ \cdot \\ \cdot \\ \cdot \\ E_z^{\text{inc}}(t_N) \end{bmatrix} = \begin{bmatrix} Z_{11} & Z_{12} & \cdots & Z_{1N} \\ Z_{21} & Z_{22} & & Z_{2N} \\ \cdot & & & \\ \cdot & & & \\ \cdot & & & \\ Z_{N1} & Z_{N2} & \cdots & Z_{NN} \end{bmatrix} \begin{bmatrix} j_1 \\ j_2 \\ \cdot \\ \cdot \\ \cdot \\ j_N \end{bmatrix} \tag{2.7}$$

We say that Equation (2.1) has been *discretized* to form the matrix equation (2.7). The specific procedure we followed to convert the continuous integral equation to a discrete matrix equation is one form of a general approach known as the method of moments [1].

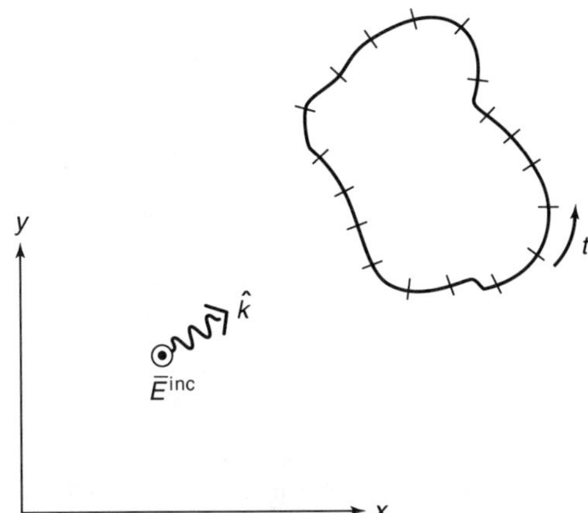

Figure 2.2 Cylinder contour divided into cells.

This discretization procedure will be used throughout the text and will be discussed in more generality in Chapter 5. Once Equation (2.7) is constructed, it can be solved by standard algorithms (Chapter 4).

The $N \times N$ matrix in Equation (2.7) is often called the *moment method impedance matrix* since its entries represent the mutual impedance between different cells in the model. For this example, these entries are given by

$$Z_{mn} = \frac{k\eta}{4} \int_{\text{cell } n} H_0^{(2)}(kR_m)\, dt' \tag{2.8}$$

where

$$R_m = \sqrt{[x_m - x(t')]^2 + [y_m - y(t')]^2} \tag{2.9}$$

and (x_m, y_m) represents the phase center of the mth strip in the model. Although the integral in Equation (2.8) cannot be evaluated exactly, if the cells are small compared to the wavelength, a possible approximation is [1]

$$Z_{mn} \cong \frac{k\eta}{4} w_n H_0^{(2)}(kR_{mn}) \qquad m \neq n \tag{2.10}$$

where

$$R_{mn} = \sqrt{(x_m - x_n)^2 + (y_m - y_n)^2} \tag{2.11}$$

The diagonal elements of the impedance matrix cannot be approximated using Equation (2.10) because the Hankel function is singular (infinite) for $R_{mn} = 0$. In addition, it is usually important to accurately evaluate the diagonal impedance matrix entries, since they are relatively large and tend to give a greater contribution to the solution of the system. For small arguments, the Hankel function can be replaced by a power series expansion [2]

$$H_0^{(2)}(x) \approx \left(1 - \frac{x^2}{4}\right) - j\left\{\frac{2}{\pi}\ln\left(\frac{\gamma x}{2}\right) + \left[\frac{1}{2\pi} - \frac{1}{2\pi}\ln\left(\frac{\gamma x}{2}\right)\right]x^2\right\} + O(x^4) \tag{2.12}$$

where

$$\gamma = 1.781072418\ldots \tag{2.13}$$

Assuming that the total curvature of each cell is small enough so that each cell may be considered flat, the dominant terms in Equation (2.12) can be retained to produce

$$\int_{\text{cell } m} H_0^{(2)}(kR_m)\,dt' \cong 2\int_0^{w_m/2}\left[1 - j\frac{2}{\pi}\ln\left(\frac{\gamma k u}{2}\right)\right]du$$
$$= w_m - j\frac{2}{\pi}w_m\left[\ln\left(\frac{\gamma k w_m}{4}\right) - 1\right] \tag{2.14}$$

It follows that

$$Z_{mm} \cong \frac{k\eta w_m}{4}\left\{1 - j\frac{2}{\pi}\left[\ln\left(\frac{\gamma k w_m}{4}\right) - 1\right]\right\} \tag{2.15}$$

The above approach will produce an approximate solution for the equivalent current density J_z. Once J_z is determined, secondary quantities of interest such as the bistatic scattering cross section $\sigma_{\text{TM}}(\phi)$ can be computed. The calculation of scattering cross section for a two-dimensional cylinder geometry has been discussed in Chapter 1. If the incident field is of the form

$$E_z^{\text{inc}}(x, y) = e^{-jk(x\cos\phi^{\text{inc}} + y\sin\phi^{\text{inc}})} \tag{2.16}$$

Equation (1.138) can be specialized to this example to produce

$$\sigma_{\text{TM}}(\phi) \cong \frac{k\eta^2}{4}\left|\sum_{n=1}^{N} j_n w_n e^{jk(x_n\cos\phi + y_n\sin\phi)}\right|^2 \tag{2.17}$$

In order to implement the above procedure in software, two key issues must be addressed. The first concerns the data structure used to describe the scatterer model within the computer program. This information must unambiguously specify the model yet should be concise and relatively easy to access when required. The second issue concerns the particular algorithm used to evaluate the matrix elements. One possible approach for evaluating Equation (2.8) is suggested in the preceding discussion and leads to the approximate expressions in Equations (2.10) and (2.15). Employing these approximations for the matrix entries not only eliminates the need for some form of numerical quadrature but also greatly simplifies the modeling task. Specifically, the cylinder contour can be completely described by the coordinates of the phase center (x_n, y_n) of each cell and the width w_n of each cell. No additional information is required to construct Equation (2.7). It is interesting that no information about the orientation of each cell is incorporated into this model. Clearly, because of the approximations employed, there is a *resolution limit* inherent in the modeling. As a consequence, different physical scatterers may produce identical computer models. It is important not to expect accuracy beyond the resolution available for a given model.

Successful implementation of the numerical procedure hinges on the use of a meaningful scatterer model. Proper modeling is something of an art. The cell sizes in use must always be small compared to the wavelength λ_0, and a suggested maximum size is $\lambda_0/10$. Note that the approximation employed in Equation (2.14) for the diagonal matrix elements is based on a flat-cell assumption, which also may limit the cell size. The procedure based on Equations (2.10) and (2.15) has been found to give the best accuracy when the phase centers (x_n, y_n) of each cell are located on the surface of the original cylinder. In other words, rather than model the original contour with a polygon having the edges of each cell located on the original contour, it appears best to locate the cell centers on the original scatterer contour. In addition, the total circumference of the cylinder model should be scaled to equal that of the original scatterer contour. This scaling can be performed by modifying the widths $\{w_n\}$ without changing the locations (x_n, y_n). Despite the fact that the resulting scatterer model is slightly nonphysical, such a scaling appears to produce a more accurate current density.

For ease of use, it is convenient to organize general-purpose computational software packages into several distinct stages. The first stage, known as *preprocessing*, encompasses the generation of a suitable scatterer model. Depending on the specific geometry under consideration, this task may be highly automated (simple geometries) or may require many man-hours of human input and verification (complex geometries). As an aid to visualization, preprocessing often involves a color graphic display of the scatterer model. The end result of the preprocessing stage is a data file containing the model in a form that is directly compatible with the main analysis routines. The second stage of the process, *analysis*, primarily consists of the creation and solution of a matrix equation such as (2.7). The analysis stage is computationally intensive but requires little human input or effort. Specialized calculations such as the scattering cross section computation may also be performed in the analysis stage. The output from the analysis stage consists of all numerical results arranged in data files in some fashion that may not be particularly convenient for interpretation. These results are collected and reorganized in the *postprocessing stage*. Postprocessing may include the generation of graphs, contour plots, color pictures, and even color videos in order to permit the user to visualize and interpret the numerical results.

The preprocessing and postprocessing stages are generic to a wide variety of engineering applications and can be adapted from software packages that have no direct link to electromagnetic modeling. The analysis stage, however, must be developed for the specific electromagnetic scattering problem of interest. Therefore, we will emphasize the analysis stage of the numerical process throughout this text.

It is appropriate to consider the sources of error in the above analysis, and we propose the following classifications:

Modeling errors: errors introduced by replacing the actual geometry by the perfectly conducting, infinite cylinder; the precise location of the cell phase centers (x_n, y_n); and the flat-strip model of the smooth-cylinder contour.

Discretization errors: errors due to the replacement of $J_z(t)$ by the pulse expansion and the approximate enforcement of the integral equation only at the cell centers (an approach known as *point matching* [1]).

Approximations: errors introduced by the approximations employed to simplify the formulation, such as the "single-point" evaluation of the integral suggested in Equation (2.10).

Numerical errors: round-off errors such as those occurring in the calculation of the Bessel functions and the solution of the matrix equation.

Each type of error may dominate in a specific situation.

Relatively simple procedures were used to obtain the approximate expressions in (2.10) and (2.15) for the matrix elements. These approximations appear sensible, in the sense that as the cell sizes are reduced, the accuracy of the approximation improves. Note that the approximation used for the current density should also improve as the cell sizes are reduced, and thus we might expect the solution to converge to some limit as the model in question is refined (Prob. P2.2). These observations seem consistent with the general goal of this type of numerical analysis, which is to construct a solution in such a way that the accuracy improves as the cell sizes are reduced. However, since the cell sizes will never shrink to zero in practice, it may be necessary to improve the approximations in order to obtain good accuracy for a finite value of N. One way of accomplishing this is to accurately evaluate the matrix entries. The entries nearest the main diagonal often represent interactions between cells in close proximity, where (2.10) can be a poor approximation.

If a more accurate evaluation of Equation (2.8) is desired, numerical quadrature may be employed to evaluate the integrals to a somewhat arbitrary degree of accuracy. The only difficulty arises for the diagonal elements, which contain a singularity in the integrand. To evaluate the diagonal entries, the integral can be written [3]

$$\int_a^b H_0^{(2)}(kx)\,dx = \int_a^b J_0(kx)\,dx - j\int_a^b \left[Y_0(kx) - \frac{2}{\pi}\ln\left(\frac{\gamma kx}{2}\right)\right] dx \\ -j\frac{2}{\pi}\int_a^b \ln\left(\frac{\gamma kx}{2}\right) dx \tag{2.18}$$

where we have separated the real and imaginary parts of the expression and added and subtracted an asymptotic form of the singular Neumann function Y_0 for small arguments. Since the singularities in the second integrand cancel, the first two integrals can be computed by numerical quadrature. The final integral in Equation (2.18) contains a singularity but is easily evaluated analytically. A brief review of numerical quadrature is provided in Appendix A. As mentioned there, singular integrands can sometimes be treated using a quadrature rule that specifically incorporates the singularity. FORTRAN subroutines that compute Equation (2.8) by numerical quadrature are provided for illustration in Appendix C. As an alternative to quadrature, the series for the Hankel function in (2.12) may be integrated on a term-by-term basis.

To illustrate the accuracy of the single-point approximation employed in Equation (2.10), Table 2.1 presents numerical values for the off-diagonal matrix elements produced by a source cell with width equal to $0.1\lambda_0$, where λ_0 is the free-space wavelength. (In all the examples discussed throughout this text, we will use units of free-space wavelength for the scatterer geometry. In these examples, the wavenumber k equals 2π.) The "exact" results are obtained by numerical quadrature. For these data, the observation point is in the plane containing the source strip, thus maximizing the error in the single-point approximation. In this case, the error stabilizes at about 2% when the observer is more than $0.2\lambda_0$ away from the center of the source cell. The error in the plane of the source cell remains at 2% even

TABLE 2.1 Comparison of Exact and Single-Point Evaluation of Integral Used for Off-diagonal Matrix Entries

	$\int_{-w_n/2}^{w_n/2} \frac{1}{4j} H_0^{(2)}(k\|x-x'\|)\,dx'$		$\frac{w_n}{4j} H_0^{(2)}(kx)$		
$x\ (\lambda_0)$	Magnitude	Angle (deg)	Magnitude	Angle (deg)	Percent Error
0.1	0.023662	–71.333	0.023605	–73.161	3
0.2	0.017147	–111.385	0.017348	–112.190	2
0.3	0.014131	–149.030	0.014334	–149.551	2
0.4	0.012289	174.063	0.012478	173.678	2
0.5	0.011017	137.490	0.011191	137.184	2
0.6	0.010071	101.095	0.010232	100.841	2
0.7	0.009331	64.807	0.009483	64.590	2
0.8	0.008734	28.588	0.008876	28.398	2
0.9	0.008238	–7.584	0.008373	–7.752	2

Note: The source cell has width equal to $w_n = 0.1$; $k = 2\pi$.

in the far field. This error is primarily due to neglecting the partial interference that arises from the difference of up to $0.1\lambda_0$ in path length across the source cell. Alternative ways of approximating Equation (2.8) can be developed that improve on the simple evaluation suggested above by building in the proper interference effects (Prob. P2.3).

An indication of the accuracy of the overall approach is illustrated by Tables 2.2 and 2.3. Values of the current density and scattering cross section found for different models of a circular cylinder illuminated by a TM plane wave are compared to the exact analytical solution. In each case, the models consist of N equal-sized cells with phase centers (x_n, y_n) located on the circular contour. Cell sizes were chosen so that the total circumference used in the model equals the original circumference of the cylinder $(1\lambda_0)$. Since we are using units of free-space wavelengths, the two-dimensional scattering cross section is also in wavelengths and is presented in decibels ($\sigma_{\text{dB}} = 10\log_{10}\sigma$). The matrix equation is

TABLE 2.2 Current Density J_z Induced on Perfectly Conducting Cylinder Having $1\lambda_0$ Circumference

N	Magnitude	Phase (deg)	Percent Error
	a. For $\phi = \pi$		
8	0.006391	41.261	2.9
16	0.006302	41.134	1.7
32	0.006271	40.792	0.9
64	0.006254	40.567	0.4
128	0.006245	40.451	0.2
Exact	0.006237	40.335	

N	Magnitude	Phase (deg)
	b. For $\phi = \pi/2$	
8	0.002983	−38.969
16	0.003020	−39.651
32	0.003009	−39.455
64	0.003001	−39.304
128	0.002997	−39.223
Exact	0.002993	−39.140

N	Magnitude	Phase (deg)
	c. For $\phi = 0$	
8	0.000845	154.158
16	0.000784	152.361
32	0.000773	152.678
64	0.000766	152.995
128	0.000763	153.171
Exact	0.000760	153.351

Note: Comparison of numerical (EFIE) and exact results as a function of the number of basis functions. The wave is incident from the $\phi = \pi$direction.

TABLE 2.3 Bistatic Scattering Cross Section σ_{TM} of Perfectly Conducting Cylinder Having $1\lambda_0$ Circumference

N	$\phi = 0$	$\phi = \pi/2$	$\phi = \pi$
8	2.8339	−1.8739	−2.0299
16	2.8195	−1.8746	−2.0821
32	2.7947	−1.8783	−2.0959
64	2.7820	−1.8797	−2.1045
128	2.7755	−1.8805	−2.1087
Exact	2.7689	−1.8812	−2.1129

Note: In decibels free-space wavelength. Comparison of numerical (EFIE) and exact results for three observation angles as a function of the number of basis functions. The wave is incident from the $\phi = \pi$ direction.

based on the approximate formulas of Equations (2.10) and (2.15). The numerical results appear to be converging to the exact solution as the model is refined, and it appears that accuracy of 1% in the current density is possible even in the shadow region of the cylinder, provided that the cell densities in use exceed 32 cells/λ_0. Because of the smooth nature of the scatterer, however, the circular cylinder example should be considered a "best case" unlikely to be representative of more complicated geometries.

As an additional illustration, Figure 2.3 compares two numerical solutions for the current density induced on one-half of a $10\lambda_0 \times 3\lambda_0$ ogival cylinder. The 80-unknown

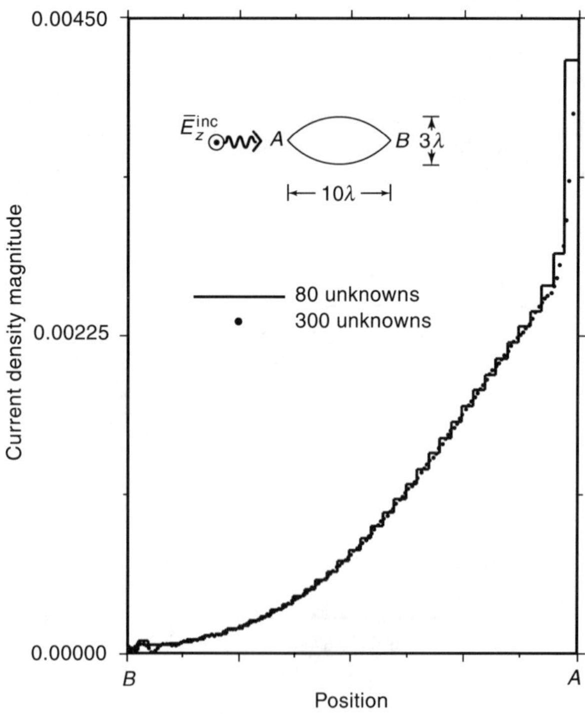

Figure 2.3 The TM electric current density (EFIE result) induced on a $10\lambda \times 3\lambda$ ogival cylinder.

result employed cells with $w_n \cong 0.265\lambda_0$, larger than the recommended guideline of $0.1\lambda_0$. Despite the large cells, the current distribution obtained with 80 unknowns exhibits good agreement with the result obtained using 300 unknowns ($w_n \cong 0.07\lambda_0$).

2.2 TE-WAVE SCATTERING FROM CONDUCTING CYLINDERS: MFIE DISCRETIZED WITH PULSE BASIS AND DELTA TESTING FUNCTIONS [1]

The scattering of TE fields incident upon an infinite, perfectly conducting cylinder (Figure 2.4) can also be analyzed using an integral equation formulation. The surface equivalence principle can be used to replace the perfectly conducting material by equivalent electric currents radiating in free space. For normally incident TE illumination, the field components present are H_z, E_x, and E_y and the equivalent current density has only a transverse component. Provided that the cylinder is a solid body and not a thin shell, a MFIE is convenient since only a single component of $\bar{H}$ is present. If specialized to the TE polarization, the MFIE can be expressed as

$$H_z^{\text{inc}}(t) = -J_t(t) - \left\{\hat{z} \cdot \nabla \times \bar{A}\right\}_{S^+} = -J_t(t) - \left\{\frac{\partial A_y}{\partial x} - \frac{\partial A_x}{\partial y}\right\}_{S^+} \tag{2.19}$$

where

$$\bar{A}(t) = \int \hat{t}(t') J_t(t') \frac{1}{4j} H_0^{(2)}(kR)\, dt' \tag{2.20}$$

and

$$R = \sqrt{[x(t) - x(t')]^2 + [y(t) - y(t')]^2} \tag{2.21}$$

In Equation (2.19), the subscript S^+ is a reminder that the function in brackets is to be evaluated an infinitesimal distance outside the surface of the cylinder. The unit tangent vector along the surface can be expressed in terms of the orientation parameter Ω defined in Figure 2.5 as

$$\hat{t}(t) = \hat{x} \cos \Omega(t) + \hat{y} \sin \Omega(t) \tag{2.22}$$

where t denotes the location around the contour.

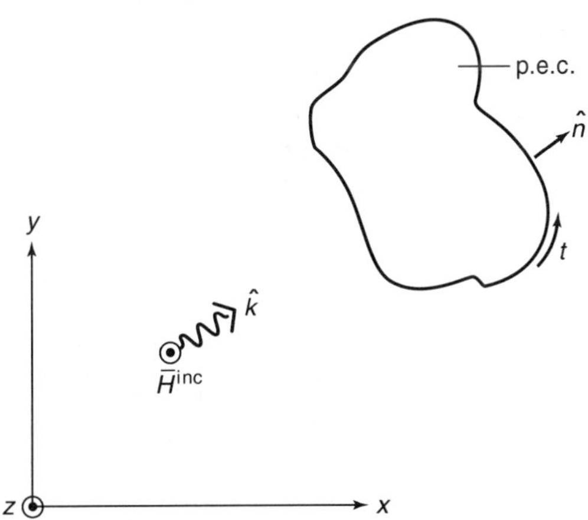

Figure 2.4 A p.e.c. cylinder illuminated by an incident TE wave.

Following the procedure of Section 2.1, the cylinder contour is divided into N cells. The equivalent current density may be approximated by the superposition of subsectional pulse functions

$$p_n(t) = \begin{cases} 1 & \text{if } t \in \text{cell } n \\ 0 & \text{otherwise} \end{cases} \tag{2.23}$$

to produce

$$J_t(t) \cong \sum_{n=1}^{N} j_n p_n(t) \tag{2.24}$$

If (2.24) is substituted into (2.19), the resulting equation can be enforced at the centers of each of the N cells in the cylinder model (point matching) to produce the $N \times N$ system

$$\begin{bmatrix} H_z^{\text{inc}}(t_1) \\ H_z^{\text{inc}}(t_2) \\ \cdot \\ \cdot \\ \cdot \\ H_z^{\text{inc}}(t_N) \end{bmatrix} = \begin{bmatrix} Z_{11} & Z_{12} & \cdots & Z_{1N} \\ Z_{21} & Z_{22} & & Z_{2N} \\ \cdot & \cdot & & \\ \cdot & \cdot & & \\ \cdot & \cdot & & \\ Z_{N1} & Z_{N2} & \cdots & Z_{NN} \end{bmatrix} \begin{bmatrix} j_1 \\ j_2 \\ \cdot \\ \cdot \\ \cdot \\ j_N \end{bmatrix} \tag{2.25}$$

The off-diagonal matrix elements are given by

$$Z_{mn} = \frac{k}{4j} \int_{\text{cell } n} \left(\sin\Omega(t') \frac{x_m - x(t')}{R_m} - \cos\Omega(t') \frac{y_m - y(t')}{R_m} \right) H_1^{(2)}(kR_m)\, dt' \tag{2.26}$$

where

$$R_m = \sqrt{[x_m - x(t')]^2 + [y_m - y(t')]^2} \tag{2.27}$$

The diagonal matrix entries can be obtained by constructing the limit

$$\begin{aligned} Z_{mm} = -1 + \lim_{x \to x_m, y \to y_m} \frac{k}{4j} \int_{\text{cell } m} &\left(\sin\Omega(t') \frac{x - x(t')}{R} \right. \\ &\left. - \cos\Omega(t') \frac{y - y(t')}{R} \right) H_1^{(2)}(kR)\, dt' \end{aligned} \tag{2.28}$$

as the observation point (x, y) approaches (x_m, y_m) from the exterior of S. The limiting procedure is easily carried out for flat cells (Prob. P2.8) and produces $Z_{mm} = -\frac{1}{2}$ regardless of the cell size. In fact, the sole contribution to the integral arises from the immediate neighborhood of the singularity of the Green's function, since electric currents on a flat plane produce no tangential magnetic field elsewhere on the same plane. Therefore, even if the cells of the cylinder model are slightly curved, the diagonal matrix elements are approximately given by

$$Z_{mm} \cong -\tfrac{1}{2} \tag{2.29}$$

Although Equation (2.26) cannot be evaluated exactly, if the the cell sizes in the model are sufficiently small compared to the wavelength, a single-point evaluation similar to that employed for the TM example of Section 2.1 is possible and produces [1]

$$Z_{mn} \cong \frac{k w_n}{4j} \left(\sin\Omega_n \frac{x_m - x_n}{R_{mn}} - \cos\Omega_n \frac{y_m - y_n}{R_{mn}} \right) H_1^{(2)}(kR_{mn}) \tag{2.30}$$

where

$$R_{mn} = \sqrt{(x_m - x_n)^2 + (y_m - y_n)^2} \tag{2.31}$$

As will be shown below, Equation (2.30) can be a poor approximation to Equation (2.26) for closely spaced cells near corners. For better accuracy, it is a straightforward matter to evaluate Equation (2.26) over flat cells by numerical quadrature.

The simplest cylinder model consists of a flat-strip representation of the original contour, and the associated data structure incorporates the phase centers (x_n, y_n), the widths w_n, and the orientation angles Ω_n of each cell (Figure 2.5). This information allows an evaluation of the off-diagonal matrix entries using either the approximate formula (2.30) or a more exact numerical quadrature. Because the cell orientation is needed for the TE matrix entries, the modeling requirements are slightly more complicated than the TM EFIE example discussed in Section 2.1. (Since the TM currents flow in the $\hat{z}$ direction, the cell orientation in the transverse plane is not critical as long as the cell is small with respect to the wavelength. For the TE polarization, the fields are maximized if the observer is in a direction perpendicular to the source cell; the fields are zero if the observer is in the plane of the cell. Thus, the cell orientation is a critical parameter.) It may be necessary to employ a large number of flat cells to represent a complex scatterer, and cylinder models that include the effect of cell curvature could be developed. Of course, a more sophisticated evaluation of (2.26) and (2.28) would be required in conjunction with better models.

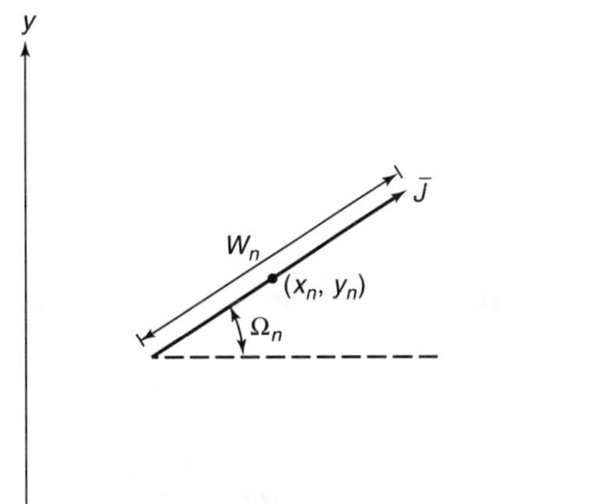

Figure 2.5 Parameters describing a p.e.c. cell for the TE polarization.

Once Equation (2.25) is solved for the coefficients $\{j_n\}$, secondary quantities such as the bistatic scattering cross section may be computed. Details of the calculation are provided in Chapter 1. For the TE polarization, the bistatic scattering cross section can be approximated according to

$$\sigma_{\text{TE}}(\phi) \cong \frac{k}{4} \left| \sum_{n=1}^{N} j_n w_n \sin(\Omega_n - \phi) e^{jk(x_n \cos\phi + y_n \sin\phi)} \right|^2 \tag{2.32}$$

The overall accuracy of the approach is illustrated by Table 2.4, which compares the scattering cross section of a circular cylinder computed using the approximate formulas of Equations (2.29) and (2.30) with the exact solution for several different models employing equal-sized cells. In common with the EFIE formulation, the MFIE results are obtained using scatterer models that have the same circumference as the desired cylinder and phase

TABLE 2.4 Two-dimensional Bistatic Scattering Cross Section as a Function of Observation Angle ϕ and Number of Cells

ϕ (deg)	$N = 36$	$N = 72$	$N = 100$	Exact
0	10.93	10.82	10.79	10.70
30	3.68	3.86	3.90	4.00
60	1.82	2.01	2.06	2.17
90	0.92	0.72	0.68	0.53
120	3.23	2.97	2.91	2.74
150	3.96	4.04	4.06	4.09
180	3.31	3.40	3.43	3.47

Note: In decibels free-space wavelength. Numerical (MFIE) results are compared with the exact solution. The scatterer is a circular cylinder with circumference of 5 λ_0.

centers (x_n, y_n) located on the surface of the original cylinder. For the same $5\lambda_0$ circular cylinder, Figure 2.6 shows the surface currents obtained from a discretization with $N = 36$. Despite the fact that the cell sizes are somewhat larger than the $\lambda_0/10$ "rule of thumb," the accuracy is good.

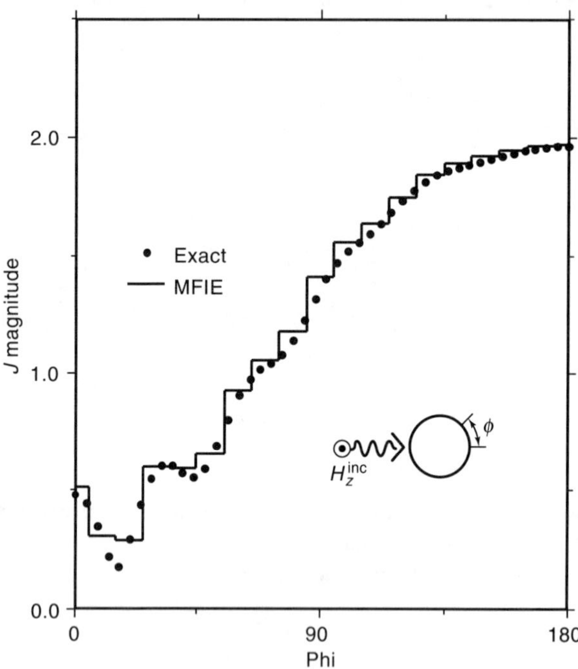

Figure 2.6 Comparison of the numerical and exact results for the TE electric current density induced by a uniform plane wave on a p.e.c. circular cylinder with 5λ circumference. The numerical (MFIE) result was obtained using 36 basis functions.

In order to judge the effect of the single-point approximation proposed in Equation (2.30), Table 2.5 compares numerical solutions for the surface current magnitude on a cylinder with circumference equal to $1\lambda_0$. These results were obtained using Equation (2.30), an accurate evaluation of (2.26) by quadrature, and the exact analytical solution. Clearly, the overall accuracy is improved by a more accurate evaluation of the matrix elements. In fact, Equation (2.30) is not a good approximation to (2.26) for closely spaced

TABLE 2.5 Magnitude of Surface Current Density J_t Induced on a Circular, Perfectly Conducting Cylinder with Circumference $1\lambda_0$ by an Incident TE Plane Wave Having Unity Magnitude H_z^{inc}

ϕ (deg)	Equation (2.30)	Equation (2.26)	Exact
0	0.9742	0.9321	0.8882
36	0.7405	0.7267	0.7050
72	0.9363	0.9378	0.8914
108	1.517	1.478	1.410
144	1.778	1.708	1.656
180	1.829	1.749	1.707

Note: Results are reported as a function of observation angle ϕ for a 10-cell model treated with the approximate formula (2.30) and an accurate evaluation of (2.26) by numerical quadrature. The exact analytical solution is also shown for comparison.

cells and is particularly poor in the vicinity of corners. To illustrate, Table 2.6 presents numerical data comparing scattered magnetic fields obtained using (2.26) and an accurate evaluation of (2.30) for an observer in the near zone of a source cell. Near the source, the error exceeds 8%. In fact, by comparing these data with Table 2.1, we conclude that the error in the single-point approximation is at least twice as bad for the scattered magnetic field calculation as it was for the scattered electric field calculation considered in Section 2.1. The additional error can be attributed to the presence of a differential operator in the source–field relationship for the TE case, which tends to amplify the error in the approximation. Because the magnetic field vanishes in the plane of the source strip, there is little interaction between adjacent cells for very smooth geometries, and the error will be minimal. However, the error is likely to be large for scatterers with corners.

TABLE 2.6 Numerical Values of Scattered Magnetic Field

x, y (λ_0)	Using (2.26)		Using (2.30)		Percent Error
0.05	−0.1975	$+j0.0240$	−0.1807	$+j0.0241$	8.4
0.075	−0.1345	$+j0.0349$	−0.1279	$+j0.0350$	4.7
0.1	−0.1012	$+j0.0445$	−0.0982	$+j0.0446$	2.7
0.125	−0.0782	$+j0.0524$	−0.0766	$+j0.0526$	1.7
0.15	−0.0593	$+j0.0583$	−0.0584	$+j0.0588$	1.2
0.175	−0.0425	$+j0.0625$	−0.0418	$+j0.0628$	0.9
0.2	−0.0269	$+j0.0643$	−0.0264	$+j0.0645$	0.8
10.0	−0.00927	$+j0.00104$	−0.00934	$+j0.00105$	0.8

Note: As obtained from Equations (2.26) and (2.30) for an observer located along the line $x = y$ with respect to a unit source located at the origin with $w_n = 0.1$ and $\Omega_n = 0$. The wavenumber is $k = 2\pi$.

2.3. LIMITATIONS OF PULSE BASIS/DELTA TESTING DISCRETIZATIONS

The previous two examples used pulse basis functions and a procedure known as *point matching* to convert integral equations into matrix form. The discretizations appeared to produce stable, converging solutions as the number of expansion functions was increased. However, we sometimes encounter difficulties if we attempt to use a similar procedure to discretize the TE EFIE.

The source of the difficulty lies in the behavior of the fields produced by the pulse expansion functions. For the TE polarization, the $\hat{x}$-component of the electric field produced by an electric current density J_x can be found from a specialization of Equation (1.52) to obtain

$$E_x(x, y) = \frac{1}{j\omega\varepsilon_0}\left(\frac{\partial^2}{\partial x^2} + k^2\right) J_x * \frac{1}{4j} H_0^{(2)}(k\rho) \tag{2.33}$$

where the asterisk denotes two-dimensional convolution. An infinite strip of constant current density is depicted in Figure 2.7 and can be expressed mathematically as

$$J_x(x, y) = p(x; x_0, x_1)\delta(y) = \begin{cases} \delta(y) & x_0 < x < x_1 \\ 0 & \text{otherwise} \end{cases} \tag{2.34}$$

The source function can be obtained as

$$\left(\frac{\partial^2}{\partial x^2} + k^2\right) J_x = [k^2 p(x; x_0, x_1) + \delta'(x - x_0) - \delta'(x - x_1)]\delta(y) \tag{2.35}$$

Therefore, the convolution of Equation (2.35) with the Green's function produces

$$\begin{aligned} E_x(x, y) = &\frac{-k^2}{4\omega\varepsilon_0} \int_{x_0}^{x_1} H_0^{(2)}(kR)\, dx' \\ &+ \frac{1}{4\omega\varepsilon_0}\left(\frac{\partial}{\partial x'} H_0^{(2)}(kR)\,|_{x'=x_0} - \frac{\partial}{\partial x'} H_0^{(2)}(kR)\,|_{x'=x_1}\right) \end{aligned} \tag{2.36}$$

where

$$R = \sqrt{(x - x')^2 + y^2} \tag{2.37}$$

Equation (2.36) reduces to

$$\begin{aligned} E_x(x, y) = &\frac{-k^2}{4\omega\varepsilon_0} \int_{x_0}^{x_1} H_0^{(2)}(kR)\, dx' \\ &- \frac{k}{4\omega\varepsilon_0}\left(H_1^{(2)}(kR_0)\frac{x - x_0}{R_0} - H_1^{(2)}(kR_1)\frac{x - x_1}{R_1}\right) \end{aligned} \tag{2.38}$$

where

$$R_0 = \sqrt{(x - x_0)^2 + y^2} \tag{2.39}$$

$$R_1 = \sqrt{(x - x_1)^2 + y^2} \tag{2.40}$$

The terms R_0 and R_1 represent the distance from the observer to the leading and trailing edges of the current source.

The potential difficulty lies in the character of the Hankel function $H_1(\alpha)$ for small arguments. The leading behavior for small α is

$$H_1^{(2)}(\alpha) \approx \frac{j2}{\pi\alpha} \quad \text{as } \alpha \to 0 \tag{2.41}$$

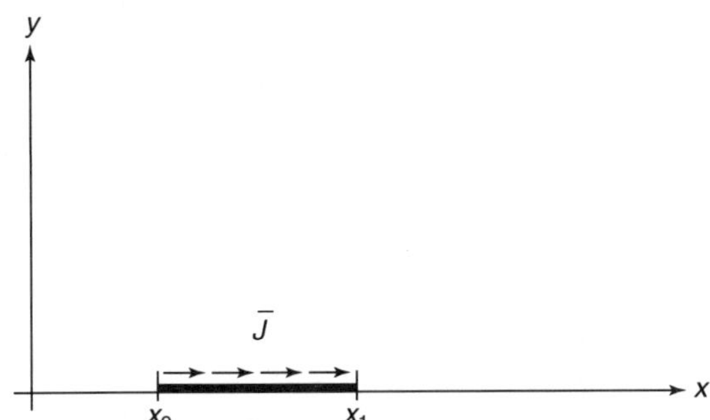

Figure 2.7 Geometry of a source consisting of a transverse electric current density.

which indicates that the electric field produced by the constant pulse of current density $J_x(x)$ is singular (infinite) at the edges of the source strip. The singular behavior can be related to the presence of line charges associated with the discontinuity in current density at the strip ends.

Now, consider the effect of using pulse basis functions to discretize the TE EFIE for a closed, perfectly conducting cylinder. For these basis functions, the moment method process requires that the cylinder geometry be represented by a finite number of cells over which the current density is assumed constant. Although we expect the true current density to be a continuous function along the cylinder perimeter, the discontinuous representation leads to the presence of fictitious line charges at the junctions between each cell in the model [4]. These line charges arise because the continuity equation (1.90) is implicitly built into the form of the EFIE operator used in Equation (2.33). These fictitious line charges give rise to infinite tangential electric fields at the cell junctions. If point matching is used to enforce the integral equation in the center of each cell, the matrix entries will be finite and a numerical solution can be found. However, because of the crude representation of the charge density, the accuracy is usually poor. Furthermore, as additional expansion functions are used in an attempt to get better results, the fictitious singularities at the cell edges may tend to dominate the near-zone scattered field and prevent the p.e.c. boundary condition $\bar{E}_{\text{tan}} = 0$ from being satisfied near the cell junctions. Although acceptable results can sometimes be obtained, there is no reason to expect numerical solutions produced by a pulse basis/point-matching discretization of the TE EFIE to improve as more and more expansion functions are employed.

The preceding example of Section 2.2 used pulse functions and point matching with the MFIE. Although the transverse electric field is singular at cell edges, the magnetic field H_z produced by a pulse basis function is finite along the source strip. Thus, the field singularities associated with the TE EFIE are not present in a TE MFIE formulation. As observed, the discretization of the MFIE appears to produce stable, converging solutions.

The example from Section 2.1 using the EFIE for TM-wave scattering also appeared to produce accurate solutions. In fact, the electric field produced by a $\hat{z}$-component of electric current density is finite and continuous throughout the source region. Because the fields and geometry are invariant with respect to z, the divergence of $\bar{J}$ vanishes in the TM case. Thus, no fictitious line charges or field singularities are introduced from the use of pulse expansion functions with the EFIE for the TM polarization.

To discretize the EFIE for TE excitation and avoid the difficulties associated with fictitious line charges, we should employ smoother basis functions to eliminate the abrupt discontinuities in the current density. Subsectional triangle basis functions are illustrated in the following section. An alternate possibility is to generalize the point-matching procedure used to enforce the integral equation. Instead of enforcing the equation at discrete points, it is possible to multiply both sides by a weighting or "testing" function and integrate over the cylinder surface. The equation is enforced in the sense of a weighted average, which can help compensate for discontinuities or singularities in the fields. To preserve the stability of the discretization, N linearly independent testing functions must be employed with N basis functions. The point-matching procedure is equivalent to the use of Dirac delta testing functions. The use of pulse testing functions will be described in the following section in the context of discretizing the TE EFIE.

In summary, there are a variety of cases where a pulse basis function/Dirac delta testing function discretization of an integral equation will produce accurate, converging numerical solutions. In other cases, however, the use of pulse functions introduces fictitious charge densities and undesired singularities in the fields. As a general rule, the proper choice of basis and testing functions must take into account the specific integro-differential operator to be discretized. This topic will be explored in Chapter 5. For the present discussion, it suffices to say that the basis and testing functions must contain enough smoothness to compensate for derivatives in the original equation. If we think of operators with weakly singular kernels (such as the TM EFIE) as the "baseline," then the combination of pulse basis functions and Dirac delta testing functions provides the minimum degree of smoothness because the "baseline" operator can be discretized without the presence of singularities. However, pulse basis and point-matching discretizations are only well suited for use with equations that impose no additional derivatives within the operator. The TE EFIE involves two additional derivative operators, and thus two additional degrees of differentiability are required in the basis and testing functions in order to produce a robust discretization.

2.4 TE-WAVE SCATTERING FROM PERFECTLY CONDUCTING STRIPS OR CYLINDERS: EFIE DISCRETIZED WITH TRIANGLE BASIS AND PULSE TESTING FUNCTIONS [4]

In Section 2.2, the MFIE was used to treat closed-body, perfectly conducting scatterers for the TE polarization. A disadvantage of the MFIE approach is that it cannot be used to treat infinitesimally thin structures (or any geometry with thickness much less than the wavelength) such as strips, plates, or scatterers with fins. The EFIE can be used for thin structures, but its implementation is slightly more complicated than the MFIE because of the differential operators appearing in the equation.

As discussed in the preceding section, it is advisable to use basis and testing functions having additional degrees of differentiability to compensate for the additional derivatives present in the TE EFIE. We will consider the use of subsectional triangle basis functions with pulse testing functions (Figure 2.8), which together provide two degrees of differentiability beyond that of the pulse/Dirac delta combination. For the TE polarization, the surface equivalence principle dictates that only a transverse component of equivalent surface current density $\bar{J}$ is necessary to model the conducting material. Because we intend to use pulse testing functions, it will be particularly convenient to use the mixed-potential form of the EFIE, which for the TE case can be written as

$$\hat{t} \cdot \bar{E}^{\text{inc}} = jk\eta\hat{t} \cdot \bar{A} + \hat{t} \cdot \nabla\Phi_e \tag{2.42}$$

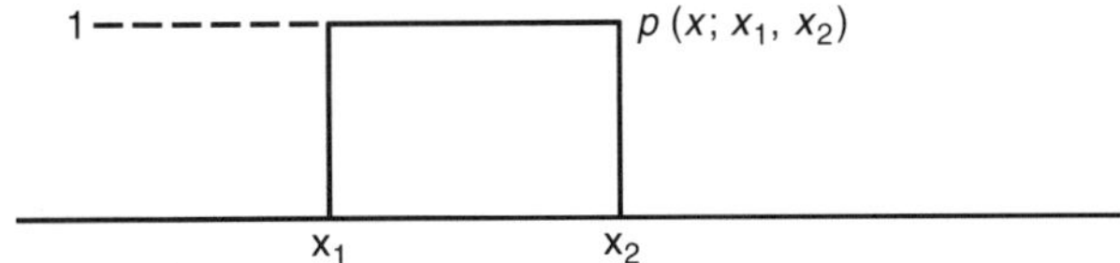

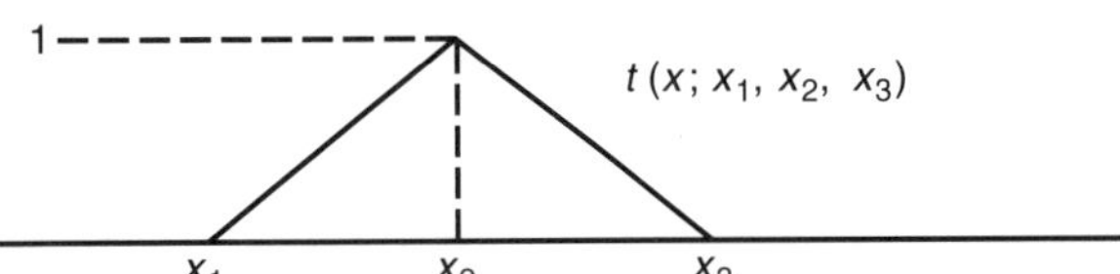

Figure 2.8 Definition of pulse basis function $p(x; x_1, x_2)$ and triangle basis function $t(x; x_1, x_2, x_3)$.

where $\bar{A}$ is the magnetic vector potential

$$\bar{A}(t) = \int \hat{t}(t')J_t(t')\frac{1}{4j}H_0^{(2)}(kR)\,dt' \tag{2.43}$$

Φ_e is the electric scalar potential

$$\Phi_e(t) = \int \frac{\rho_e(t')}{\varepsilon_o}\frac{1}{4j}H_0^{(2)}(kR)\,dt' \tag{2.44}$$

and, in practice, the continuity equation

$$-j\omega\rho_e = \nabla_s \cdot \bar{J} = \frac{\partial J_t}{\partial t} \tag{2.45}$$

is used to define the surface charge density ρ_e. As in the previous examples, Equation (2.42) is only valid for points on the surface of the original cylinder. In terms of the parametric variable t defined in Figure 2.4,

$$R = \sqrt{[x(t) - x(t')]^2 + [y(t) - y(t')]^2} \tag{2.46}$$

The contour of the cylinder can be modeled by the superposition of M flat strips, as depicted in Figure 2.9. Previous examples used pulse basis functions for the current density and assigned one basis function per strip in this type of model. In this example, we consider triangle basis functions that overlap two adjacent strips. The current density can be written in terms of the expansion functions and the coordinates shown in Figure 2.9 as

$$J_t(t) \cong \sum_{n=1}^{N} j_n t(t; t_{n-1}, t_n, t_{n+1}) \tag{2.47}$$

In the case of a closed cylinder, the basis functions overlap continuously around the contour, and there will be M basis functions for an M-cell model ($N = M$). In the case of an open structure, the basis function adjacent to an edge is centered between the first and second strip pair and forces the current density to equal zero on the ends of the structure. This is consistent with the physical behavior that the current flowing into the strip edge must vanish. (Recall from Section 1.7 that the equivalent current density actually represents the superposition of the currents on both sides of the strip, and therefore J_t can not flow around the edge.) Thus, there will be $M - 1$ basis functions for an open structure modeled with M cells ($N = M - 1$).

A consequence of this type of basis function is that a minimum of 2 cells (strips) are required to model any structure, no matter how small that scatterer may be. In common

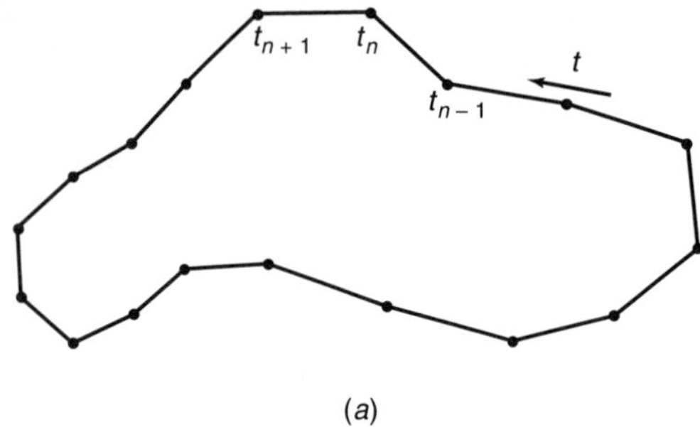

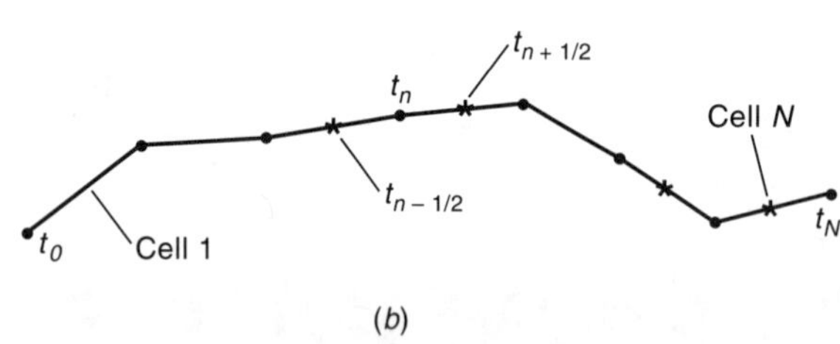

Figure 2.9 (a) Flat-strip model of a closed-cylinder cross section. (b) Flat-strip model of the cross section of an open cylindrical structure.

with previous formulations, cell sizes must be much smaller than the wavelength in order to accurately represent the current density, and a minimum of 10 cells per wavelength is usually recommended.

The surface charge density consistent with a triangle basis function for J_t can be expressed as a combination of two pulse basis functions, each with support over one of the two original cells straddled by the subsectional triangle function. This combination of pulse functions is sometimes denoted a *pulse doublet*. The superposition of each doublet of charge density produces a pulse expansion, consistent with Equation (2.45), that can be written as

$$\frac{\rho_e(t)}{\varepsilon_0} \cong \frac{-\eta}{jk} \sum_{n=1}^{N} j_n \left(\frac{1}{t_n - t_{n-1}} p(t; t_{n-1}, t_n) - \frac{1}{t_{n+1} - t_n} p(t; t_n, t_{n+1}) \right) \tag{2.48}$$

Equations (2.47) and (2.48) can be substituted into the vector and scalar potential functions, respectively.

The integral equation (2.42) can be enforced by "testing" the equation with pulse functions whose domain begins in the center of one strip and extends to the center of an adjacent strip. The process, a generalization of the point-matching approach employed in Sections 2.1 and 2.2 to approximately enforce the equation, involves multiplying both sides of the integral equation with a testing function and integrating over the scatterer surface (see Chapter 5 and [1]). For notational convenience, consider the numbering scheme proposed in Figure 2.9b. The testing function spanning strips m and $m+1$ is a pulse function defined as

$$T_m(t) = p(t; t_{m-1/2}, t_{m+1/2}) \tag{2.49}$$

The choice of pulse testing functions permits the analytical treatment of the gradient operator appearing in the mixed-potential form of the EFIE according to the idea

$$\int p(t; a, b)\hat{t}(t) \cdot \nabla F \, dt = \int_a^b \frac{dF}{dt} dt = F(b) - F(a) \tag{2.50}$$

Therefore, the pulse testing function absorbs the derivative present in the gradient operator.

After completing the discretization, the system of equations can be expressed as an $N \times N$ matrix of the form

$$\begin{bmatrix} e_1 \\ e_2 \\ \cdot \\ \cdot \\ \cdot \\ e_N \end{bmatrix} = \begin{bmatrix} Z_{11} & Z_{12} & \cdots & Z_{1N} \\ Z_{21} & Z_{22} & & Z_{2N} \\ \cdot & \cdot & & \\ \cdot & \cdot & & \\ \cdot & \cdot & & \\ Z_{N1} & Z_{N2} & \cdots & Z_{NN} \end{bmatrix} \begin{bmatrix} j_1 \\ j_2 \\ \cdot \\ \cdot \\ \cdot \\ j_N \end{bmatrix} \tag{2.51}$$

where

$$e_m = \int_{t_{m-1/2}}^{t_{m+1/2}} \hat{t}(t) \cdot \bar{E}^{\text{inc}}(t)\, dt \tag{2.52}$$

$$\begin{aligned} Z_{mn} = \frac{k\eta}{4} \cdot \int_{t_{m-1/2}}^{t_{m+1/2}} \hat{t}(t) \cdot \int_{t_{n-1}}^{t_{n+1}} \hat{t}(t')t(t'; t_{n-1}, t_n, t_{n+1}) H_0^{(2)}(kR)\, dt'\, dt \\ + \frac{\eta}{4k} \left(\frac{1}{t_n - t_{n-1}} \int_{t_{n-1}}^{t_n} H_0^{(2)}(kR_2)\, dt' - \frac{1}{t_{n+1} - t_n} \int_{t_n}^{t_{n+1}} H_0^{(2)}(kR_2)\, dt' \right. \\ \left. - \frac{1}{t_n - t_{n-1}} \int_{t_{n-1}}^{t_n} H_0^{(2)}(kR_1)\, dt' + \frac{1}{t_{n+1} - t_n} \int_{t_n}^{t_{n+1}} H_0^{(2)}(kR_1)\, dt' \right) \end{aligned} \tag{2.53}$$

$$R_1 = \sqrt{[x(t_{m-1/2}) - x(t')]^2 + [y(t_{m-1/2} - y(t')]^2} \tag{2.54}$$

and

$$R_2 = \sqrt{[x(t_{m+1/2} - x(t')]^2 + [y(t_{m+1/2}) - y(t')]^2} \tag{2.55}$$

The matrix entries of (2.53) can be evaluated to any necessary degree of accuracy by numerical quadrature. However, the two-dimensional quadrature required for the double integral may be time consuming, and it is worthwhile to consider a more efficient alternative. One possible approximation is given by

$$\begin{aligned} \int_{t_{m-1/2}}^{t_{m+1/2}} \hat{t}(t) \cdot \int_{t_{n-1}}^{t_{n+1}} \hat{t}(t')t(t'; t_{n-1}, t_n, t_{n+1}) H_0^{(2)}(kR)\, dt'\, dt \\ \cong [(t_m - t_{m-1/2})\hat{t}(t_{m-1/2}) + (t_{m+1/2} - t_m)\hat{t}(t_{m+1/2})] \\ \cdot \left(\hat{t}(t_{n-1/2}) \int_{t_{n-1/2}}^{t_n} H_0^{(2)}(k\tilde{R})\, dt' + \hat{t}(t_{n+1/2}) \int_{t_n}^{t_{n+1/2}} H_0^{(2)}(k\tilde{R})\, dt' \right) \end{aligned} \tag{2.56}$$

where

$$\tilde{R} = \sqrt{[x(t_m) - x(t')]^2 + [y(t_m) - y(t')]^2} \tag{2.57}$$

This approximation reduces all the necessary integrals to the form

$$\int H_0^{(2)}(kR)\, dt' \tag{2.58}$$

where R is defined in Equation (2.46). Equation (2.58) is the same integral encountered in the TM EFIE example, and its evaluation has been discussed in Section 2.1.

There are several good reasons for considering the approximation introduced in Equation (2.56). The approximation is sensible and improves as the cell sizes are reduced. One practical advantage of using the approximation is that it permits all of the integrals required for the matrix entry calculation to be evaluated by a common subroutine. An additional advantage is that it eliminates the need for time-consuming two-dimensional quadrature. In justification for the approximation, it is worth noting that as the cell sizes are made smaller and smaller compared to the wavelength, the matrix element Z_{mn} is dominated by the contribution from the scalar potential part (which is not affected by the approximation). For small cell sizes, the vector potential part is of secondary importance. (The decoupling of the scalar and vector potentials leads to numerical difficulties in the case of electrically small scatterers. These will be examined in the context of three-dimensional bodies in Chapter 10.)

The scattering cross section can be found once the coefficients $\{j_n\}$ are obtained from the solution of Equation (2.51). An exact formula for the bistatic cross section that incorporates the flat-strip model and triangular basis functions is

$$\sigma_{\mathrm{TE}}(\phi) = \frac{k}{4}\left|\sum_{n=1}^{N}\int_{t_{n-1}}^{t_{n+1}} [\hat{y}\cdot\hat{t}(t')\cos\phi - \hat{x}\cdot\hat{t}(t')\sin\phi] j_n \times t(t'; t_{n-1}, t_n, t_{n+1}) e^{jk[x(t')\cos\phi + y(t')\sin\phi]}\, dt'\right|^2 \tag{2.59}$$

These integrals can either be evaluated in closed form or approximated for simplicity.

The accuracy of this moment method approach is illustrated by Table 2.7, which shows the scattering cross section of a circular cylinder produced by the above method compared to the exact eigenfunction solution. The numerical results appear to improve as the cell density is increased. As an illustration of a different geometry and one that cannot be treated using an MFIE, Figure 2.10 shows the current density induced on a flat strip of width $3\lambda_0$, obtained using a 10-cell/λ_0 and a 50-cell/λ_0 discretization.

TABLE 2.7 Bistatic Scattering Cross Section as a Function of Observation Angle ϕ and Number of Cells

ϕ (deg)	$N = 10$	$N = 20$	$N = 40$	Exact
0	–6.83	–6.05	–5.88	–5.82
30	–8.78	–8.02	–7.84	–7.78
60	–10.94	–10.11	–9.90	–9.84
90	–6.72	–6.10	–5.95	–5.90
120	–3.94	–3.59	–3.51	–3.49
150	–2.90	–2.77	–2.75	–2.74
180	–2.67	–2.63	–2.64	–2.64

Note: In decibels free-space wavelength. The scatterer is a circular cylinder with circumference of $1\lambda_0$ illuminated by a TE plane wave propagating in the $\phi = 0$ direction. Triangle basis functions and pulse testing functions were used to discretize the EFIE.

It is worthwhile to investigate the relative accuracy of the MFIE formulation described in Section 2.2 and this EFIE approach. Table 2.8 compares the current density on a circular cylinder obtained using these two formulations. In addition, results are shown for an EFIE discretized with pulse basis functions and Dirac delta testing functions (the approach recommended *against* in Section 2.3) and an MFIE discretized with triangle basis functions and

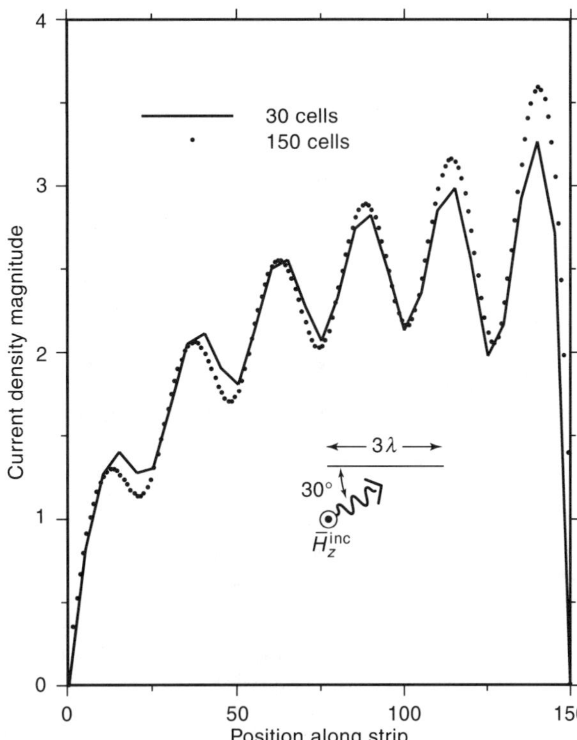

Figure 2.10 Two numerical (EFIE) results for the TE surface current density along a flat strip of 3λ width. The first result is obtained with 10 cells/λ; the second with 50 cells/λ.

Dirac delta functions (Prob. P2.11). The cylinder has $1\lambda_0$ circumference and is represented with 40 equal-sized cells. By comparing these results with the exact analytical solution, it is apparent that the pulse/delta MFIE formulation and the triangle/pulse EFIE formulation exhibit similar accuracy and are substantially more accurate than the pulse/delta EFIE formulation. However, neither is as accurate for this problem as the triangle/delta MFIE approach suggested in Prob. P2.11.

A computer program to implement the TE EFIE approach described above will require a somewhat more complicated cylinder model than those employed in Sections 2.1 and 2.2. Because of the need for additional generality, including the possibility of junctions between multiple cells in the model, we consider the following scheme. First, we number the endpoints (x_n, y_n) of each strip used to represent the cylinder contour. (We often refer to the endpoints as *nodes* of the model.) Second, we number the cells themselves. Because of the manner in which the basis functions straddle two adjacent cells, it is convenient to construct a table linking the endpoints at which basis functions are centered to the adjacent cells. A second pointer array links each cell with the associated nodes. To illustrate this modeling scheme, Figure 2.11 shows a sample cylinder model and Table 2.9 contains the pointer arrays describing the *connectivity* between cells and nodes.

Scatterer models having junctions between strips or between strips and solid cylinders can also be treated using this EFIE formulation. Figure 2.11 depicts a hypothetical model containing several strip cells. Consider the cells numbered 4, 5, and 6, which form a junction at node 2. (Furthermore, suppose that these cells are located in open space rather than as a part of a closed scatterer.) It appears possible to represent the current flow through the junction of these three cells by superimposing three triangle basis functions (straddling cells 4 and 5, 5 and 6, and 4 and 6, respectively). However, only two of the basis functions

TABLE 2.8 Comparison of Current Density Induced on a Circular Cylinder with Circumference $1\lambda_0$ by TE Plane Wave Propagating in $\phi = 0$ Direction

	EFIE		MFIE		
ϕ (deg)	Pulse/ Delta	Triangle/ Pulse	Pulse/ Delta	Triangle/ Delta	Exact
			Magnitude		
0	0.8000	0.8875	0.8907	0.8891	0.8882
45	0.6386	0.6754	0.6733	0.6729	0.6722
90	1.1829	1.1741	1.1751	1.1708	1.1713
135	1.6052	1.6216	1.6232	1.6201	1.6199
180	1.6800	1.7084	1.7094	1.7076	1.7071
			Phase (deg)		
0	68.69	66.89	66.29	66.66	66.56
45	119.06	113.83	113.41	113.57	113.56
90	−163.66	−164.88	−164.88	−164.80	−164.82
135	−125.65	−125.86	−125.83	−125.82	−125.84
180	−110.65	−110.85	−110.77	−110.82	−110.83

Note: An EFIE result obtained with pulse basis functions and Dirac delta testing functions is compared with the EFIE formulation of this section (triangle basis functions and pulse testing functions), the MFIE formulation of Section 2.2, another MFIE formulation (Prob. P2.11), and the exact solution. Each result was obtained using 40 equal-sized cells to represent the cylinder.

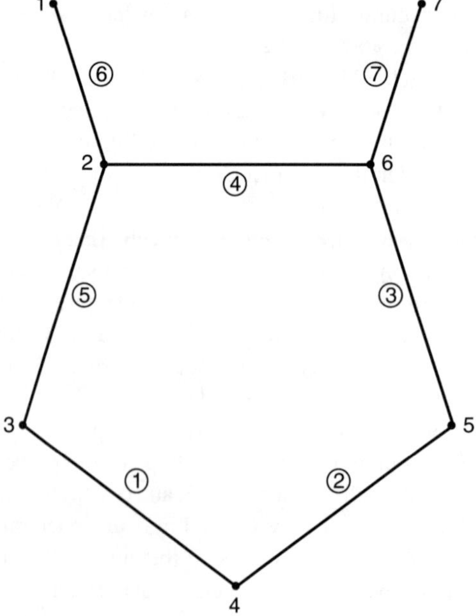

Figure 2.11 Sample cylinder cross section depicting the indexing of nodes and cells used in the pointer arrays of Table 2.9.

are linearly independent. Therefore, it is necessary to arbitrarily discard one of the three functions, leaving two at the node (x_2, y_2). If more than three cells join at a node, a similar procedure is followed using one fewer basis function than cells. With the exception of this adjustment, the numerical formulation proceeds as described above. Because of the way triangle basis functions straddle adjacent cells, Kirchhoff's current law is automatically satisfied at the junction.

TABLE 2.9 Pointer Arrays Describing Cylinder Model in Figure 2.11

List of node locations	
x_1	y_1
x_2	y_2
•	•
•	•
•	•
x_n	y_n

Cell Index	Node 1	Node 2
Pointer from cell indices to node indices		
1	4	3
2	5	4
3	6	5
4	2	6
5	3	2
6	2	1
7	6	7

Node Index	Cell 1	Cell 2
Pointer from nodes where a basis function is located to adjacent cells		
2	5	6
2	5	4
6	4	7
6	4	3
5	3	2
4	2	1
3	1	5

2.5. TM-WAVE SCATTERING FROM INHOMOGENEOUS DIELECTRIC CYLINDERS: VOLUME EFIE DISCRETIZED WITH PULSE BASIS AND DELTA TESTING FUNCTIONS [5]

The preceding sections have examined surface integral equation formulations for scattering from conducting cylinders. An inhomogeneous dielectric cylinder characterized by a complex relative permittivity $\varepsilon_r(x, y)$ can be analyzed using a volume integral equation. The field components excited by a normally incident TM wave are E_z, H_x, and H_y. Following the volume equivalence principle discussed in Section 1.2, the dielectric material may be replaced by equivalent polarization currents

$$\bar{J}(x, y) = \hat{z} j\omega\varepsilon_0[\varepsilon_r(x, y) - 1]E_z(x, y) \tag{2.60}$$

radiating in free space. For the TM polarization, the EFIE appearing in Equation (1.101) can be specialized to

$$E_z^{\text{inc}}(x, y) = \frac{J_z}{j\omega\varepsilon_0(\varepsilon_r - 1)} + j\omega\mu_0 A_z \tag{2.61}$$

where

$$A_z(x, y) = \iint J_z(x', y') \frac{1}{4j} H_0^{(2)}(kR)\, dx'\, dy' \tag{2.62}$$

and

$$R = \sqrt{(x - x')^2 + (y - y')^2} \tag{2.63}$$

The specific form of Equation (2.61) is slightly different than Equation (1.101); here we have chosen J_z as the primary unknown instead of E_z.

The cylinder cross section can be divided into cells, as illustrated in Figure 2.12. If the unknown polarization current density is approximated by the superposition of subsectional pulse basis functions, defined in two-dimensional space as

$$p_n(x, y) = \begin{cases} 1 & \text{if } (x, y) \in \text{cell } n \\ 0 & \text{otherwise} \end{cases} \tag{2.64}$$

the current density can be expressed as

$$J_z(x, y) \cong \sum_{n=1}^{N} j_n p_n(x, y) \tag{2.65}$$

Equation (2.61) reduces to

$$E_z^{\text{inc}}(x, y) \cong \sum_{n=1}^{N} j_n \left(\frac{\eta p_n(x, y)}{jk[\varepsilon_r(x, y) - 1]} + jk\eta \iint_{\text{cell } n} \frac{1}{4j} H_0^{(2)}(kR)\, dx'\, dy' \right) \tag{2.66}$$

Enforcing Equation (2.66) at the centers of each of the N cells produces an $N \times N$ system

$$\begin{bmatrix} E_z^{\text{inc}}(x_1, y_1) \\ E_z^{\text{inc}}(x_2, y_2) \\ \cdot \\ \cdot \\ \cdot \\ E_z^{\text{inc}}(x_N, y_N) \end{bmatrix} = \begin{bmatrix} Z_{11} & Z_{12} & \cdots & Z_{1N} \\ Z_{21} & Z_{22} & & Z_{2N} \\ \cdot & \cdot & & \\ \cdot & \cdot & & \\ \cdot & \cdot & & \\ Z_{N1} & Z_{N2} & \cdots & Z_{NN} \end{bmatrix} \begin{bmatrix} j_1 \\ j_2 \\ \cdot \\ \cdot \\ \cdot \\ j_N \end{bmatrix} \tag{2.67}$$

whose entries are given by

$$Z_{mn} = \frac{k\eta}{4} \iint_{\text{cell } n} H_0^{(2)}(kR_m)\, dx'\, dy' \qquad m \neq n \tag{2.68}$$

and

$$Z_{mm} = \frac{\eta}{jk(\varepsilon_{rm} - 1)} + \frac{k\eta}{4} \iint_{\text{cell } m} H_0^{(2)}(kR_m)\, dx'\, dy' \tag{2.69}$$

where

$$R_m = \sqrt{(x_m - x')^2 + (y_m - y')^2} \tag{2.70}$$

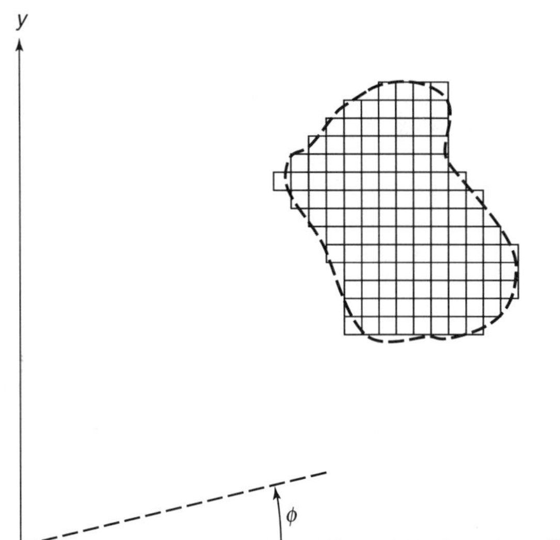

Figure 2.12 Cross section of a dielectric cylinder illustrating the manner in which the region is divided into square cells.

In Equation (2.69), ε_{rn} is the average complex relative permittivity of the nth cell in the cylinder model.

Although the integrals in (2.68) and (2.69) must in general be evaluated by numerical quadrature, they can be evaluated analytically if the cell shapes are approximated by circles of the same area using [6, 6.684]

$$\int_{\phi'=0}^{2\pi}\int_{\rho'=0}^{a} H_0^{(2)}(kR)\rho'\,d\rho'\,d\phi' = \begin{cases} \dfrac{2\pi a}{k}J_0(k\rho)H_1^{(2)}(ka) - \dfrac{j4}{k^2} & \rho < a \\ \dfrac{2\pi a}{k}J_1(ka)H_0^{(2)}(k\rho) & \rho > a \end{cases} \tag{2.71}$$

where (ρ, ϕ) represent conventional cylindrical coordinates, a denotes the radius of the equivalent circle, J_0 and J_1 are the Bessel functions of order zero and one, respectively, and H_1 is the Hankel function of order one. Using the circular-cell approximation [5], we obtain

$$Z_{mn} = \frac{\eta\pi a_n}{2}J_1(ka_n)H_0^{(2)}(kR_{mn}) \qquad m \neq n \tag{2.72}$$

where

$$R_{mn} = \sqrt{(x_m - x_n)^2 + (y_m - y_n)^2} \tag{2.73}$$

for the off-diagonal entries and

$$Z_{mm} = \frac{\eta\pi a_m}{2}H_1^{(2)}(ka_m) - \frac{j\eta\varepsilon_{rm}}{k(\varepsilon_{rm} - 1)} \tag{2.74}$$

for the diagonal entries. As a consequence of using J_z as the primary unknown instead of E_z, ε_{rm} only appears in the diagonal matrix entries.

The solution of matrix equation (2.67) yields the coefficients $\{j_n\}$. Once the current density is obtained, other quantities such as the bistatic scattering cross section can be computed. Equation (1.137) can be specialized to this example to produce the scattering cross section

$$\sigma_{\text{TM}}(\phi) \cong \frac{k\eta^2}{4}\left|\sum_{n=1}^{N} j_n \frac{2\pi a_n}{k}J_1(ka_n)e^{jk(x_n\cos\phi + y_n\sin\phi)}\right|^2 \tag{2.75}$$

Assuming that we employ the approximate expressions based on Equation (2.71), the data structure used to store the cylinder model reduces to the phase center (x_n, y_n), the equivalent radius a_n, and the average complex relative permittivity ε_{rn} for each cell in the model. Thus, the modeling is no more complicated than the approaches discussed in Sections 2.1 and 2.2.

Principal sources of error associated with the above procedure include the approximation of the polarization current by pulse basis functions (discretization error), the implicit assumption that $\varepsilon_r(x, y)$ can be accurately modeled as a constant per cell, and the replacement of the original cylinder geometry by a superposition of cells that are approximately circular in shape (modeling errors). As in the previous examples of this chapter, each cell in the model must be relatively small in terms of the free-space wavelength. However, for this example the cells also must be small in terms of the wavelength in the dielectric medium. If λ_0 is the free-space wavelength, we define the wavelength in the dielectric material as

$$\lambda_d = \frac{1}{\sqrt{|\varepsilon_r|}}\lambda_0 \tag{2.76}$$

In accordance with our usual "rule of thumb," we recommend a minimum of 100 cells/λ_d^2 of cross-sectional area.

Some indication of the accuracy of the overall moment method approach can be found by considering circular, homogeneous cylinders. Several models constructed with equal-sized square cells are shown in Figure 2.13. Tables 2.10 and 2.11 show the scattering cross section and internal E_z-field for a circular cylinder with circumference of $0.5137\lambda_0$ and relative permittivity $\varepsilon_r = 10$. The numerical results are compared to the exact solutions. In these examples, the cross-sectional area of the models is scaled in order to equal that of the desired circular geometries. Figure 2.14 shows a comparison of internal E_z-fields for a cylinder with $1.0\lambda_0$ circumference and relative permittivity $\varepsilon_r = 2.56$. Overall, the accuracy of the numerical result appears excellent.

TABLE 2.10 Scattering Cross Section Obtained for Homogeneous Circular Dielectric Cylinder Having $\varepsilon_r = 10$ and $0.5137\lambda_0$ Circumference

N	σ_{TM} (dB λ_0)
21	−1.8484
61	−1.8469
101	−1.8442
Exact	−1.8426

Note: Comparison of numerical (EFIE) and exact results as a function of the number N of basis functions for $\phi = \pi$.

TABLE 2.11 Electric Field E_z Induced at Center of Homogeneous Circular Dielectric Cylinder Having $\varepsilon_r = 10$ and Circumference $0.5137\lambda_0$ by Plane Wave with Unit Magnitude E_z^{inc}

N	Magnitude	Phase
21	0.770	−94.75
61	0.779	−94.66
101	0.779	−94.75
Exact	0.780	−94.82

Note: Comparison of numerical (EFIE) and exact results as a function of the number N of basis functions.

Because exact results are available only for circular geometries having relatively simple permittivity profiles, it is desirable to have independent methods for estimating the accuracy of a given moment method solution. Since the boundary condition embodied in the integral equation is enforced only at the N match points, one way of studying the

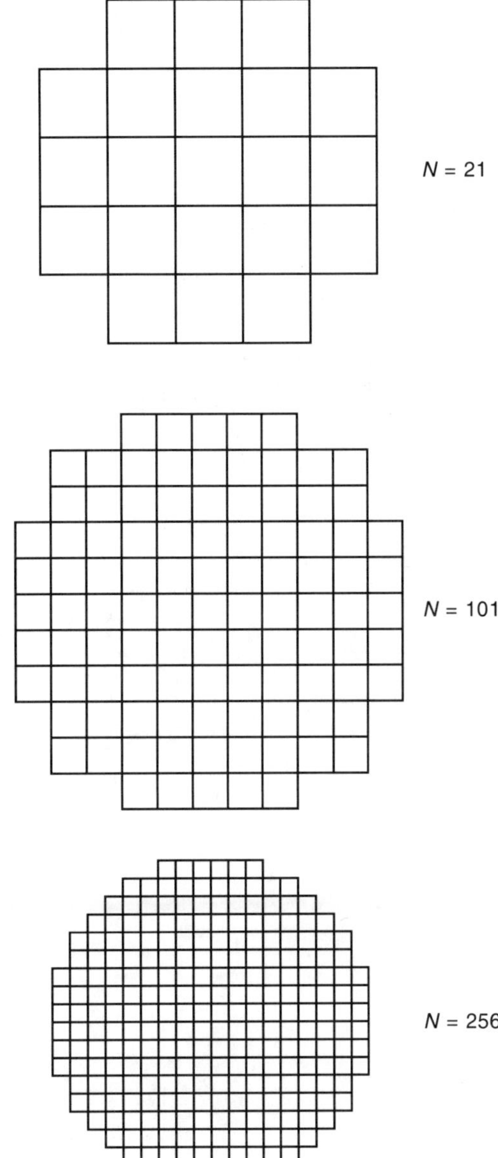

Figure 2.13 Three models of circular dielectric cylinders.

accuracy is to compute the electric field at other points within the cylinder. The boundary condition embodied in the relationship

$$E_z^{\text{inc}} = E_z - E_z^s \tag{2.77}$$

will be satisfied exactly only by the true solution, and thus we do not expect perfect agreement from our approximate result. The amount of deviation is indicative of the overall accuracy, however, and should provide useful guidance to aid an experienced user in estimating the accuracy of a given numerical result. Figure 2.15 shows the boundary condition error in magnitude for the example used in Figure 2.14.

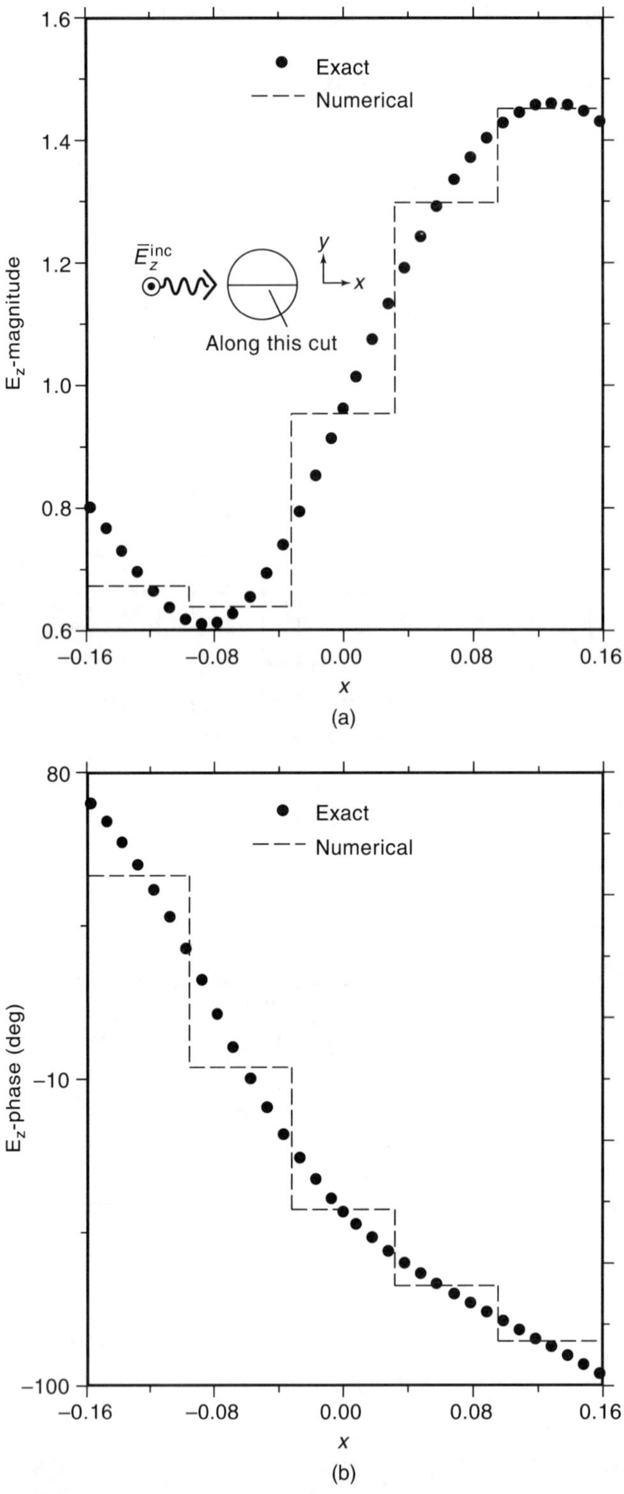

Figure 2.14 Comparison of the numerical and exact results for the TM electric field within a dielectric cylinder with circumference $1.0\lambda_0$ and relative permittivity $\varepsilon_r = 2.56$. The numerical results were obtained using 21 equal-sized cells (the first model in Figure 2.13) and the EFIE: (a) field magnitude; (b) field phase.

If greater accuracy for a given number of cells is desired, the above approach could be improved by incorporating a more flexible modeling scheme. In fact, it may be difficult to model highly heterogeneous permittivity profiles with "roughly circular" cells. Triangular or polygonal cells provide much greater flexibility at the cost of necessitating two-dimensional numerical integration to evaluate Equations (2.68) and (2.69) [7]. Additional accuracy could also be obtained by the use of smoother basis and testing functions.

Although we presented results for homogeneous cylinders to illustrate accuracy, the volume integral equation technique is intended for heterogeneous scatterers. For homogeneous geometries, the surface integral equation formulations described in Section 2.8 will usually be more computationally efficient alternatives.

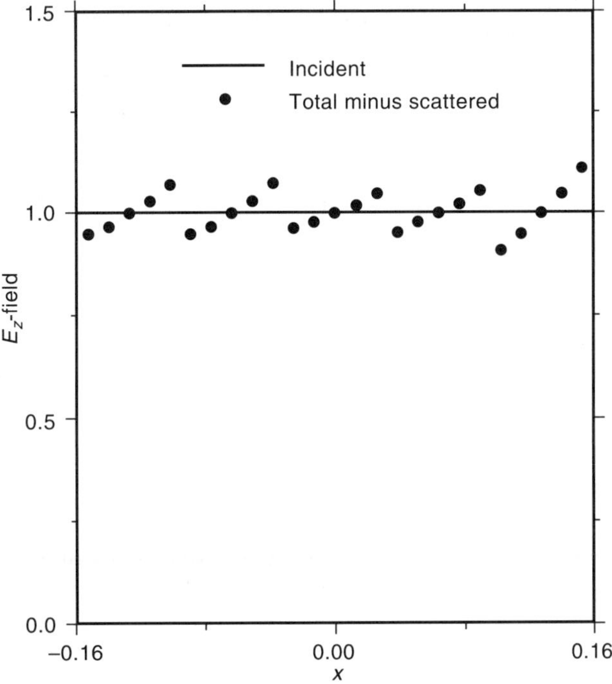

Figure 2.15 Error in satisfying the electric field boundary condition for the TM example depicted in Figure 2.14.

2.6. TE-WAVE SCATTERING FROM DIELECTRIC CYLINDERS: VOLUME EFIE DISCRETIZED WITH PULSE BASIS AND DELTA TESTING FUNCTIONS [8]

An inhomogeneous dielectric cylinder illuminated by a normally incident TE wave can be treated in a manner analogous to that of the previous example, although we will see that the accuracy obtained with a similar pulse basis discretization can be significantly worse than that observed in the TM case. For the TE polarization, the field components present are H_z, E_x, and E_y. The dielectric material may be replaced by equivalent electric polarization current densities

$$J_x(x, y) = j\omega\varepsilon_0[\varepsilon_r(x, y) - 1]E_x(x, y) \tag{2.78}$$

and

$$J_y(x, y) = j\omega\varepsilon_0[\varepsilon_r(x, y) - 1]E_y(x, y) \tag{2.79}$$

An electric field integral equation equivalent to (1.101) can be expressed in the form [8]

$$\bar{E}^{\text{inc}}(x, y) = \bar{E}(x, y) - \frac{\nabla \times (\nabla \times \bar{A}) - \bar{J}}{j\omega\varepsilon_0} \tag{2.80}$$

Equation (2.80) can be separated into its components to produce the coupled system

$$E_x^{\text{inc}} = \frac{J_x \varepsilon_r}{j\omega\varepsilon_0(\varepsilon_r - 1)} - \frac{1}{j\omega\varepsilon_0}\left(\frac{\partial^2 A_y}{\partial x\,\partial y} - \frac{\partial^2 A_x}{\partial y^2}\right) \tag{2.81}$$

$$E_y^{\text{inc}} = \frac{J_y \varepsilon_r}{j\omega\varepsilon_0(\varepsilon_r - 1)} - \frac{1}{j\omega\varepsilon_0}\left(\frac{\partial^2 A_x}{\partial x\,\partial y} - \frac{\partial^2 A_y}{\partial x^2}\right) \tag{2.82}$$

where

$$\bar{A}(x, y) = \iint \bar{J}(x', y') \frac{1}{4j} H_0^{(2)}(kR)\, dx'\, dy' \tag{2.83}$$

and

$$R = \sqrt{(x - x')^2 + (y - y')^2} \tag{2.84}$$

If the cylinder cross section is divided into cells as depicted in Figure 2.12, and the current density is approximated by a superposition of pulse basis functions

$$\bar{J}(x, y) \cong \sum_{n=1}^{N} (\hat{x} j_{xn} + \hat{y} j_{yn}) p_n(x, y) \tag{2.85}$$

where

$$p_n(x, y) = \begin{cases} 1 & \text{if } (x, y) \in \text{cell } n \\ 0 & \text{otherwise} \end{cases} \tag{2.86}$$

Equations (2.81) and (2.82) can be enforced at the centers of each of the N cells to yield a $2N \times 2N$ system of the form

$$\begin{bmatrix} E_x^{\text{inc}}(x_1, y_1) \\ E_x^{\text{inc}}(x_2, y_2) \\ \cdot \\ \cdot \\ \cdot \\ E_x^{\text{inc}}(x_N, y_N) \\ E_y^{\text{inc}}(x_1, y_1) \\ E_y^{\text{inc}}(x_2, y_2) \\ \cdot \\ \cdot \\ \cdot \\ E_y^{\text{inc}}(x_N, y_N) \end{bmatrix} = \begin{bmatrix} A_{11} & A_{12} & \cdots & A_{1N} & B_{11} & B_{12} & \cdots & B_{1N} \\ A_{21} & A_{22} & \cdots & A_{2N} & B_{21} & B_{22} & \cdots & B_{2N} \\ \cdot & & & \cdot & \cdot & & & \cdot \\ \cdot & & & \cdot & \cdot & & & \cdot \\ \cdot & & & \cdot & \cdot & & & \cdot \\ A_{N1} & A_{N2} & \cdots & A_{NN} & B_{N1} & B_{N2} & \cdots & B_{NN} \\ C_{11} & C_{12} & \cdots & C_{1N} & D_{11} & D_{12} & \cdots & D_{1N} \\ C_{21} & C_{22} & \cdots & C_{2N} & D_{21} & D_{22} & \cdots & D_{2N} \\ \cdot & & & \cdot & \cdot & & & \cdot \\ \cdot & & & \cdot & \cdot & & & \cdot \\ \cdot & & & \cdot & \cdot & & & \cdot \\ C_{N1} & C_{N2} & \cdots & C_{NN} & D_{N1} & D_{N2} & \cdots & D_{NN} \end{bmatrix} \begin{bmatrix} j_{x1} \\ j_{x2} \\ \cdot \\ \cdot \\ \cdot \\ j_{xN} \\ j_{y1} \\ j_{y2} \\ \cdot \\ \cdot \\ \cdot \\ j_{yN} \end{bmatrix} \tag{2.87}$$

Each of the four blocks of this system is an $N \times N$ submatrix, with the specific entries given in terms of the integral

$$I_n(x, y) = \iint_{\text{cell } n} H_0^{(2)}(kR)\, dx'\, dy' \tag{2.88}$$

as

$$A_{mn} = \frac{-\eta}{4k} \frac{\partial^2 I_n}{\partial y^2}\bigg|_{x=x_m, y=y_m} \qquad m \neq n \tag{2.89}$$

$$A_{mn} = \frac{\eta \varepsilon_{rm}}{jk(\varepsilon_{rm} - 1)} - \frac{\eta}{4k} \frac{\partial^2 I_m}{\partial y^2}\bigg|_{x=x_m, y=y_m} \tag{2.90}$$

$$B_{mn} = C_{mn} = \frac{\eta}{4k} \frac{\partial^2 I_n}{\partial x\, \partial y}\bigg|_{x=x_m, y=y_m} \tag{2.91}$$

$$D_{mn} = \frac{-\eta}{4k} \frac{\partial^2 I_n}{\partial x^2}\bigg|_{x=x_m, y=y_m} \qquad m \neq n \tag{2.92}$$

$$D_{mm} = \frac{\eta \varepsilon_{rm}}{jk(\varepsilon_{rm} - 1)} - \frac{\eta}{4k} \frac{\partial^2 I_m}{\partial x^2}\bigg|_{x=x_m, y=y_m} \tag{2.93}$$

where ε_{rn} is the average complex-valued relative permittivity of cell n. If the cell shapes are approximated by circular cross sections, Equation (2.71) can be used to provide a closed-form evaluation of I_n and its derivatives [8]. We leave the detailed evaluation of the matrix entries as an exercise for the reader (Prob. P2.25).

If the circular-cell approximation suggested in the previous example (Section 2.5) is used for the TE case as well, the cylinder model necessary for computer implementation is limited to the phase centers (x_n, y_n), the equivalent radii a_n, and the average relative permittivities of each cell. The cells in question must be small relative to the wavelength in the dielectric medium, as defined by Equation (2.76), and a reasonable minimum density is 100 cells/λ_d^2. Because there are two unknown coefficients per cell, this translates into 200 unknowns/λ_d^2 of cross-sectional area.

After Equation (2.87) is solved for the coefficients $\{j_{xn}\}$ and $\{j_{yn}\}$ of the current distribution, secondary quantities such as the scattering cross section can be computed. Assuming a plane-wave incident field

$$H_z^{\text{inc}}(x, y) = e^{-jk(x \cos\phi^{\text{inc}} + y \sin\phi^{\text{inc}})} \tag{2.94}$$

the bistatic scattering cross section is given by

$$\sigma_{\text{TE}}(\phi) \cong \frac{k}{4} \left| \sum_{n=1}^{N} \frac{2\pi a_n}{k} J_1(ka_n) \left(j_{yn} \cos\phi - j_{xn} \sin\phi\right) e^{jk(x_n \cos\phi + y_n \sin\phi)} \right|^2 \tag{2.95}$$

The accuracy of the approach can be quite good for scatterers with permittivity not drastically different from that of the surrounding medium. Figure 2.16 shows the electric field within a circular, homogeneous cylinder having $\varepsilon_r = 2.56 - j2.56$ and radius λ_0/π. These results were obtained using the square-cell model shown in Figure 2.13. The cross-sectional area of the model is scaled to equal that of the circular cylinder.

It was previously suggested that one way of estimating the error in an approximate solution is to check the degree to which the boundary condition embodied in the EFIE is satisfied. This has been carried out for an example illustrated in Figure 2.17. From a comparison with similar results presented for the TM polarization (Figure 2.15), we see a larger variation across each cell in the TE case. This might be expected due to the derivatives

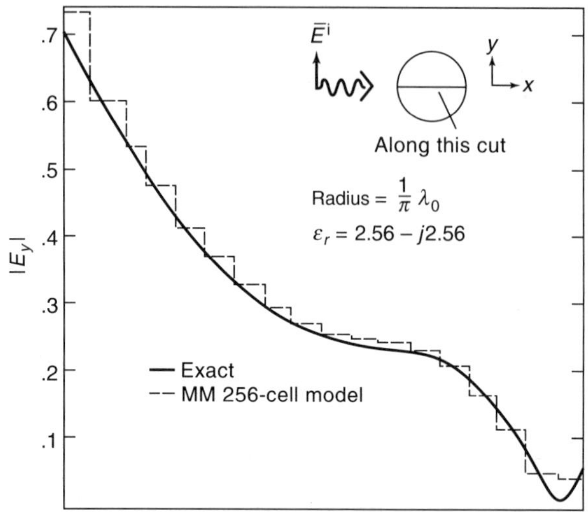

Figure 2.16 Comparison of the numerical (EFIE) and exact results for the TE electric field magnitude within a dielectric cylinder with circumference $2.0\lambda_0$ and relative permittivity $\varepsilon_r = 2.56 - j2.56$. The numerical results were obtained using 256 equal-sized cells (the third model in Figure 2.13), equivalent to a cell density of 222 cells per square dielectric wavelength.

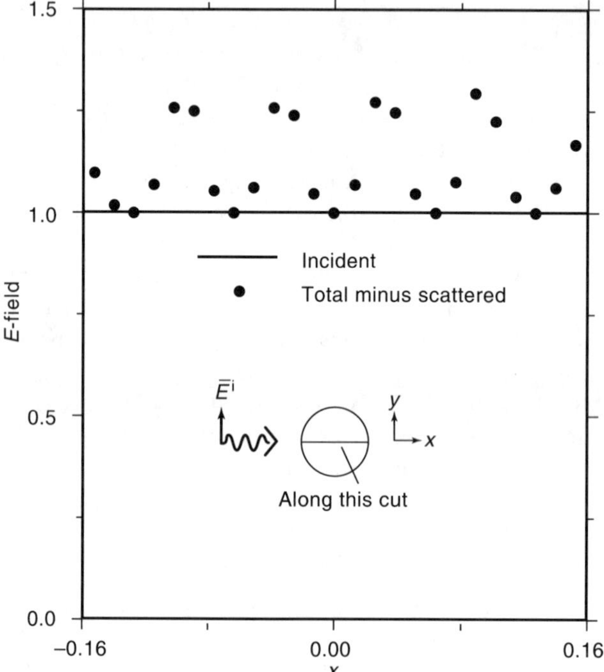

Figure 2.17 Error in satisfying the electric field boundary condition for the TE polarization for a dielectric cylinder with circumference $1.0\lambda_0$ and relative permittivity $\varepsilon_r = 2.56$. The numerical (EFIE) results were obtained using the 21 equal-sized cell model depicted in Figure 2.13.

in the integral equation, which tend to amplify the error arising from approximations. Because of the additional derivatives in the TE EFIE, we expect to require smaller cell sizes to achieve the same accuracy as was possible for the TM example.

Unfortunately, smaller cell sizes are not enough to ensure accurate results, especially for larger values of ε_r. For example, Figure 2.18 shows the magnitude and phase of the

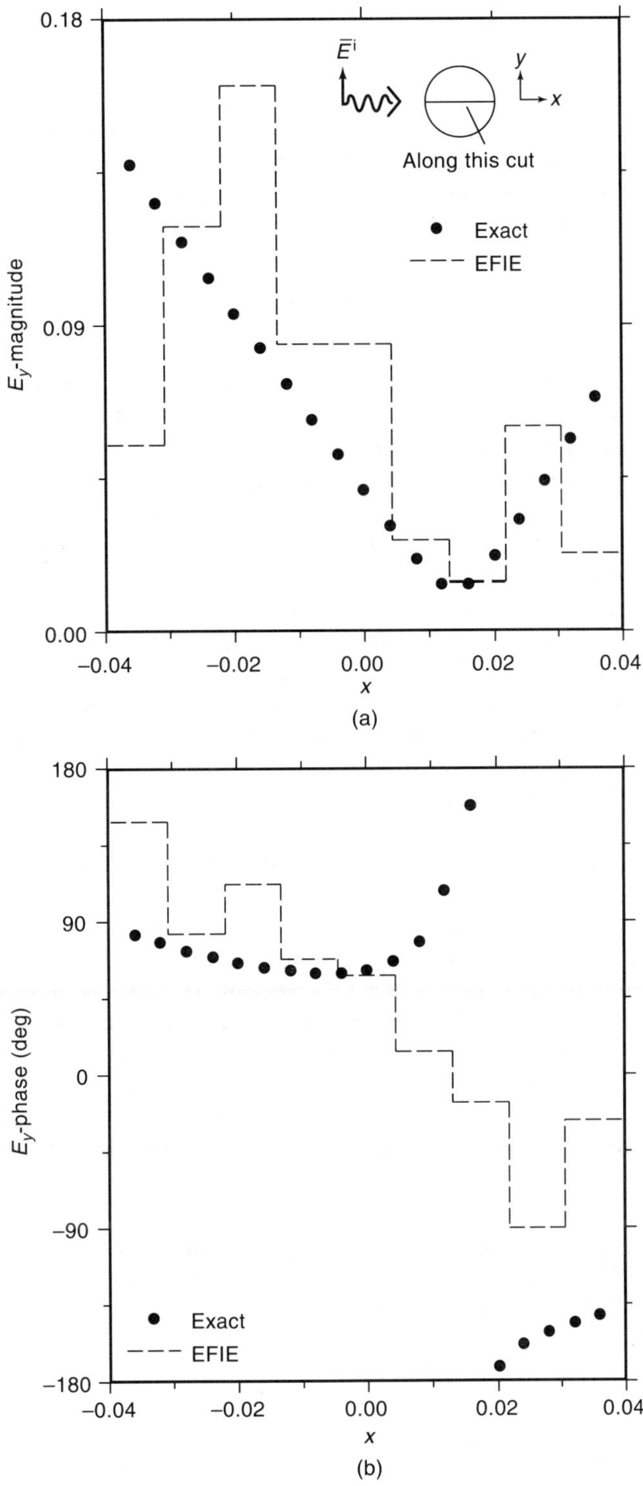

Figure 2.18 Comparison of the numerical (EFIE) and exact results for the TE electric field within a dielectric cylinder with circumference $0.248\lambda_0$ and relative permittivity $\epsilon_r = 2 - j50$. The numerical results were obtained using 61 equal-sized cells, for a density of 250 cells per square dielectric wavelength: (a) field magnitude; (b) field phase.

total electric field within a circular, homogeneous dielectric cylinder having $\varepsilon_r = 2 - j50$. The cylinder has a circumference of $0.248\lambda_0$, and is modeled with 61 square cells. Clearly, the accuracy is poor, despite the fact that the cell density is 250 cells/λ_d^2. In addition, the results are very sensitive to changes in the cylinder model and get worse instead of better in this case as larger numbers of cells are used. A similar behavior is observed whenever the magnitude of ε_r is greater than 10 and can probably be found for all values of ε_r.

To understand the failure of the approach for large values of ε_r, recall the previous discussion regarding the limitations associated with pulse basis functions and point matching for discretizing integro-differential equations. For the TE polarization, charge density is induced only at interfaces between homogeneous regions (or in regions with varying permittivity, which our model excludes from consideration). However, a fictitious surface charge density is introduced between each cell in the model through the pulse basis functions and the specific source–field relationship embodied in Equation (2.80). One effect of this charge is to produce fictitious discontinuities in the normal component of the electric field at each cell boundary. A second effect seems to be stronger fields in the vicinity of cell boundaries, as is apparent from Figure 2.17. As the cell sizes within the model decrease (as they must to maintain a fixed cell density relative to λ_d for increasing ε_r), the detrimental effects become more pronounced. In addition, the approximation employed to evaluate the matrix elements in closed form involved replacing the true cell shapes by circles, which changes the location of the fictitious charge layers. Hagmann and Levin have identified this approximation as a source of additional error [9]. Borup, Sullivan, and Gandhi have also investigated this formulation and report that significant errors are introduced due to the staircase approximation to the true cylinder surface necessitated by the square cell or circular cell modeling scheme [10]. Apparently, the local electric fields are a strong function of the precise location of surface charge density, and staircase boundaries are not adequate for placing the charge density in the proper location.

It is noteworthy that the TM approach presented in Section 2.5 (which involves no charge density since the divergence of the current density is zero) does not exhibit the instability or inaccuracy of the TE formulation. Better accuracy for the TE case might be obtained by employing smoother basis and testing functions in order to eliminate the fictitious charges. A triangular-cell discretization of the EFIE might also be in order to eliminate the staircase approximation of the true cylinder boundaries [7]. A third possibility is to use a volumetric MFIE formulation, which may prove less sensitive to proper charge modeling. The following section considers a combination of these three ideas.

2.7. TE-WAVE SCATTERING FROM INHOMOGENEOUS DIELECTRIC CYLINDERS: VOLUME MFIE DISCRETIZED WITH LINEAR PYRAMID BASIS AND DELTA TESTING FUNCTIONS [11]

The previous example employed the EFIE within a numerical formulation for TE-wave scattering from inhomogeneous dielectric cylinders. The MFIE can also be used to treat the TE problem and has several advantages over the specific EFIE approach discussed in the previous section.

For normally incident TE excitation, only a $\hat{z}$-component of the magnetic field is present. The dielectric material may be replaced by equivalent polarization currents according to Equation (1.25), which in the two-dimensional situation specializes to

$$\bar{J}(x, y) = \frac{\varepsilon_r(x, y) - 1}{\varepsilon_r(x, y)} \nabla \times (\hat{z} H_z) = \frac{\varepsilon_r(x, y) - 1}{\varepsilon_r(x, y)} \left(\hat{x} \frac{\partial H_z}{\partial y} - \hat{y} \frac{\partial H_z}{\partial x} \right) \tag{2.96}$$

Using the source–field relationship presented in Equation (1.53), the MFIE may be expressed in the form

$$H_z^{\text{inc}}(x, y) = H_z(x, y) - \iint (\hat{z} \cdot \nabla' \times \bar{J}) \frac{1}{4j} H_0^{(2)}(kR)\, dx'\, dy' \tag{2.97}$$

where

$$R = \sqrt{(x - x')^2 + (y - y')^2} \tag{2.98}$$

In general, the current density $\bar{J}$ will exhibit an abrupt discontinuity at the cylinder edge. The domain of the integral appearing in Equation (2.97) must include the Dirac delta function arising from this discontinuity.

In a departure from the square-cell modeling scheme followed in the previous two sections, we consider a cylinder model comprised of triangular cells (Figure 2.19). Each cell in the model is assumed to have constant permittivity ε_{rn}. Throughout the interior of the scatterer, the magnetic field is a continuous function and may be expanded in a basis consisting of a superposition of *linear pyramid functions* centered at each node of the triangular-cell model, so that

$$H_z(x, y) \cong \sum_{n=1}^{N} h_n P_n(x, y) \tag{2.99}$$

Figure 2.20 illustrates this type of expansion function. Each basis function straddles the triangular cells grouped around a given vertex and vanishes at every other node in the model. The expansion in (2.99) is a piecewise-linear interpolation between each node of the model, and the coefficient h_n represents the value of H_z at (x_n, y_n). Within a cell,

$$H_z(x, y) = c_1 + c_2 x + c_3 y \tag{2.100}$$

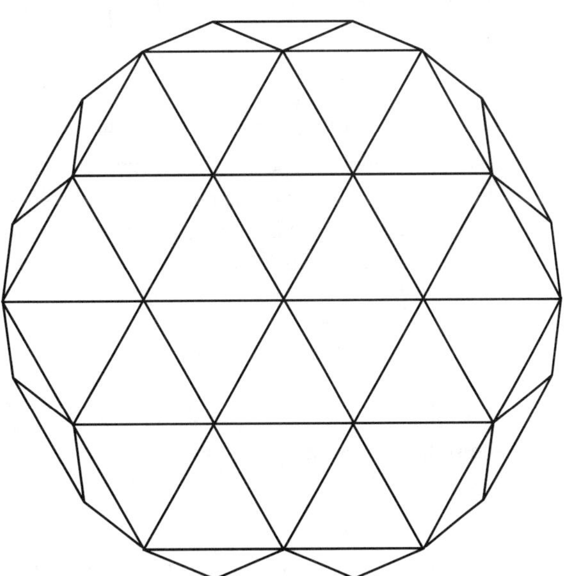

Figure 2.19 Triangular-cell model representing the cross section of a circular cylinder. The model involves 31 nodes, 42 cells, and 72 edges.

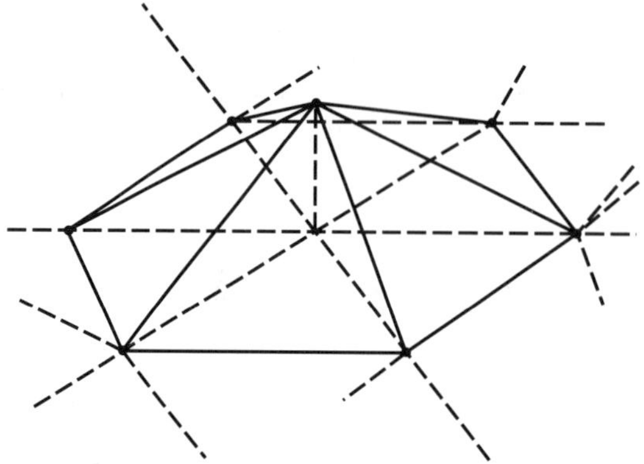

Figure 2.20 Linear pyramid basis function. The function has unit value at one node of a triangular-cell model and is zero at all other nodes.

where c_1, c_2, and c_3 are constants that can be easily determined from the values of H_z at the cell corners (Prob. P2.26). It follows from Equation (2.96) that the current density will be constant within each cell. In terms of the coefficients $\{h_n\}$ of the three basis functions overlapping a particular cell, the current density is given by

$$J_x = \frac{\varepsilon_{rn} - 1}{\varepsilon_{rn}} \frac{h_1(x_3 - x_2) + h_2(x_1 - x_3) + h_3(x_2 - x_1)}{y_1(x_3 - x_2) + y_2(x_1 - x_3) + y_3(x_2 - x_1)} \tag{2.101}$$

and

$$J_y = \frac{\varepsilon_{rn} - 1}{\varepsilon_{rn}} \frac{h_1(y_3 - y_2) + h_2(y_1 - y_3) + h_3(y_2 - y_1)}{y_1(x_3 - x_2) + y_2(x_1 - x_3) + y_3(x_2 - x_1)} \tag{2.102}$$

where h_n is the magnetic field at corner (x_n, y_n). Since the scattered magnetic field produced by these equivalent sources is a continuous function across cell boundaries, it suffices to point match the MFIE at the nodes of the model. This provides one equation for each basis function, which can be arranged in matrix form as

$$\begin{bmatrix} H_z^{\text{inc}}(x_1, y_1) \\ H_z^{\text{inc}}(x_2, y_2) \\ \cdot \\ \cdot \\ \cdot \\ H_z^{\text{inc}}(x_N, y_N) \end{bmatrix} = \begin{bmatrix} Z_{11} & Z_{12} & \cdots & Z_{1N} \\ Z_{21} & Z_{22} & \cdots & Z_{2N} \\ \cdot & \cdot & & \cdot \\ \cdot & \cdot & \ddots & \cdot \\ \cdot & \cdot & & \cdot \\ Z_{N1} & \cdots & \cdots & Z_{NN} \end{bmatrix} \begin{bmatrix} h_1 \\ h_2 \\ \cdot \\ \cdot \\ \cdot \\ h_N \end{bmatrix} \tag{2.103}$$

As a consequence of the piecewise-constant representation employed for the current density, the curl operator appearing in Equation (2.97) produces a Dirac delta function over each cell edge. In other words, the two-dimensional integration required in (2.97) collapses to a one-dimensional integral over each cell edge in the model. An explicit expression for Z_{mn} in terms of the coefficients $\{h_n\}$ is rather complicated and not needed to implement the scheme. It is more convenient to consider each edge of the model as a source and obtain an expression for the contribution from that edge to the entries of (2.103).

Figure 2.21 depicts a hypothetical edge located between cells 1 and 2 and a local coordinate system (n, t) associated with the normal and tangential directions. At this edge, the source function can be written

$$\hat{z} \cdot \nabla \times \bar{J} = \frac{\partial J_t}{\partial_n} = (J_{t2} - J_{t1})\,\delta(n) \tag{2.104}$$

where $\delta(n)$ denotes a Dirac delta function having support at the edge and J_{ti} is the component of equivalent current density in cell i tangential to the edge. Equations (2.101) and (2.102) provide the functional dependence of the current density on the coefficients $\{h_n\}$, the node locations $\{x_n, y_n\}$, and the cell permittivity ε_{rn}. For a particular edge, these equations can be combined to determine the four weighting factors associated with the coefficients. We leave this calculation as an exercise (Prob. P2.27). The contribution from a given edge in the model to row m of the $N \times N$ matrix of Equation (2.103) can be expressed as

$$(\alpha_1 h_1 + \alpha_2 h_2 + \alpha_3 h_3 + \alpha_4 h_4) \int \frac{1}{4j} H_0^{(2)}(kR)\, dt' \tag{2.105}$$

where $\{\alpha_n\}$ are the weighting coefficients and R is the distance from the source point on the edge to node (x_m, y_m). The equivalent source at each edge contributes to four locations in each row of the matrix (three locations if the edge is located on the outer surface of the cylinder). The evaluation of the integral over the Hankel function has been discussed in Section 2.1.

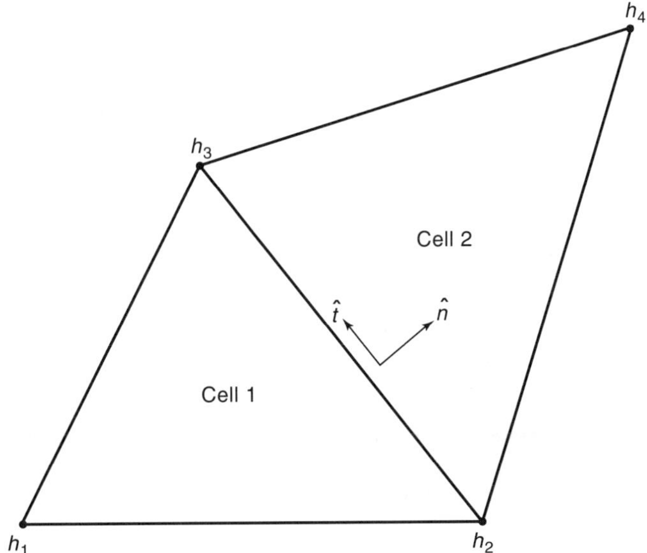

Figure 2.21 Local geometry of a source edge between two triangular cells, with the indices associated with the magnetic field indicated at the nodes.

The complete matrix can be constructed by systematically scanning through each edge in the model and computing the contribution of Equation (2.104) to each row of the matrix. Once the coefficients $\{h_n\}$ are determined from the solution of the matrix equation, the interior electric fields can be found from (2.101), (2.102), and

$$\bar{E} = \frac{-j\eta}{k(\varepsilon_r - 1)} \bar{J} \tag{2.106}$$

Assuming that the incident magnetic field has unity magnitude, the bistatic scattering cross section is given by

$$\sigma_{\text{TE}}(\phi) = \frac{1}{4k} \left| \iint (\hat{z} \cdot \nabla' \times \bar{J}) e^{jk(x' \cos\phi + y' \sin\phi)}\, dx'\, dy' \right|^2 \tag{2.107}$$

The integration in (2.107) collapses to the cell edges and can be approximated by a summation for convenience.

To illustrate the accuracy of this procedure, Figure 2.22 shows a plot of the internal H_z-field within a circular, homogeneous cylinder having circumference of $1\lambda_0$ and $\varepsilon_r = 2.56$. Several different triangular-cell models were employed, with each scaled to have the same cross-sectional area as the desired circular cylinder. The numerical results appear to be converging toward the exact solution.

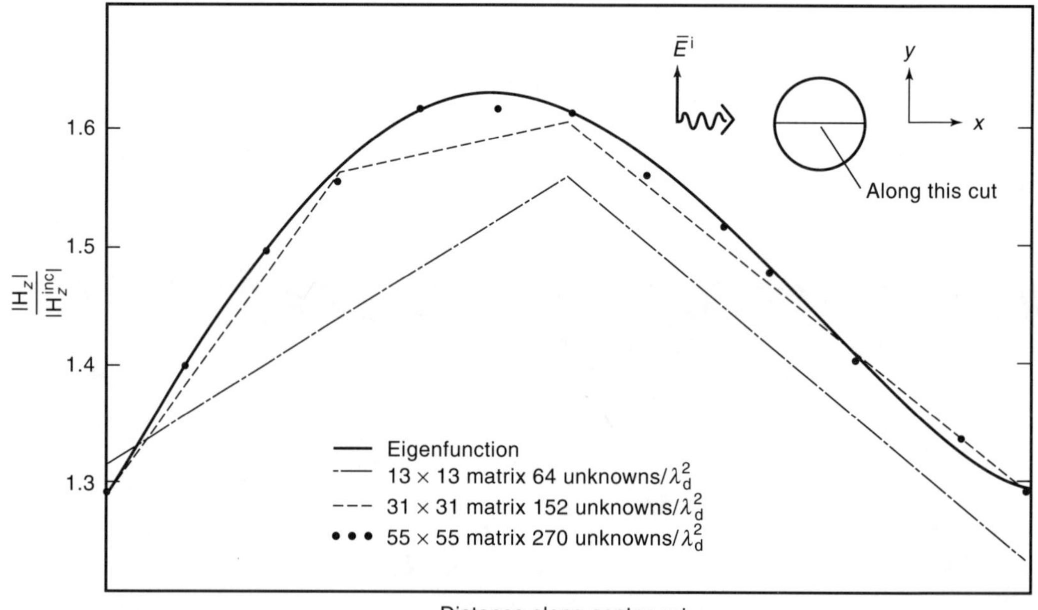

Figure 2.22 The TE magnetic field within a circular, homogeneous dielectric cylinder with circumference $1.0\lambda_0$ and relative permittivity $\varepsilon_r = 2.56$ (exact and MFIE results). After [11]. ©1988 IEEE.

It is often desired to compute the electric field within dielectric materials, and Equation (2.106) can be employed to generate the electric field as a secondary calculation once the numerical solution for H_z is obtained. Figure 2.23 shows the electric field within a layered dielectric cylinder having outer radius $0.15\lambda_0$, core relative permittivity $\varepsilon_r = 10 - j5$, and cladding relative permittivity $\varepsilon_r = 6$. The numerical result was obtained using a cylinder model with 196 triangular cells and 115 nodes. Agreement between the numerical result and the exact solution is excellent.

To provide a comparison between this formulation and the TE EFIE approach of the previous section, Table 2.12 shows the scattering cross section produced by both approaches for a circular cylinder with $\varepsilon_r = 2.56 - j2.56$ and circumference $2.0\lambda_0$. The results are presented as a function of the required number of unknowns. For this case and for a variety of other examples [11], the MFIE result is more accurate for a given number of unknowns than the pulse basis/delta testing EFIE result. Much of the improvement in accuracy can likely be attributed to the triangular-cell model employed with the MFIE formulation, which allows a much better approximation of the circular cross section than the square-cell model used with the EFIE.

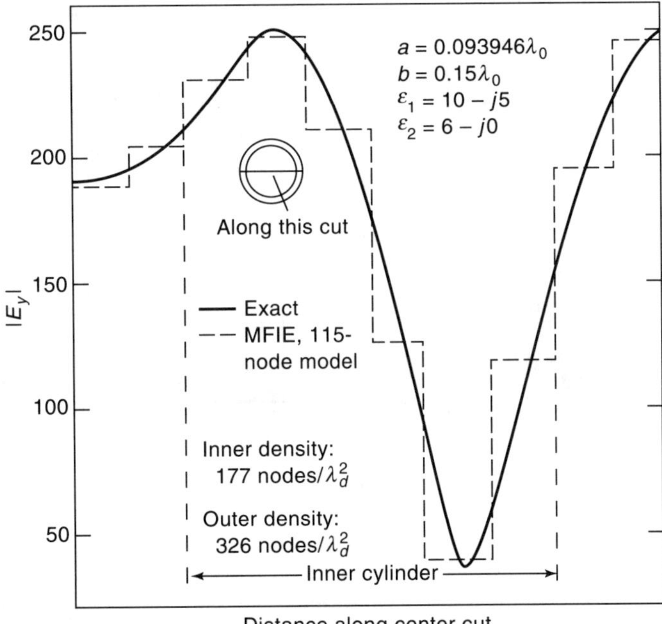

Figure 2.23 The TE electric field within a layered dielecric cylinder (exact and MFIE results). After [11]. ©1998 IEEE.

TABLE 2.12 Comparison of TE Bistatic Scattering Cross Section[a] for Dielectric Cylinder with $\varepsilon_r = 2.56 - j2.56$ and Circumference $2.0\lambda_0$

	EFIE		MFIE		
ϕ (deg)	42 Unknowns	512 Unknowns	31 Unknowns	55 Unknowns	Exact
0	6.19	5.76	5.54	5.57	5.58
30	3.02	2.51	2.53	2.45	2.36
60	–6.79	–6.24	–5.54	–5.74	–5.94
90	–7.76	–7.91	–7.32	–7.66	–7.94
120	–14.19	–15.01	–13.08	–14.26	–15.12
150	–9.79	–9.48	–8.84	–8.89	–8.85
180	–6.98	–7.20	–6.15	–6.35	–6.45

[a] In decibels free-space wavelength.

Note: Numerical results obtained from the EFIE of Section 2.6 (square-cell models) are compared to results from the MFIE of Section 2.7 (triangular-cell models) and the exact solution. The matrix order is provided with each result.

The instabilities mentioned in connection with the TE EFIE do not seem to arise with the MFIE approach. As an example, Figure 2.24 shows the internal electric field produced by the MFIE formulation for the lossy dielectric cylinder of radius $0.248\lambda_0$ and $\varepsilon_r = 2-j50$. Clearly, the MFIE result is superior to the EFIE result shown in Figure 2.18. The additional accuracy obtained from the MFIE formulation appears to be the result of several factors. The particular scheme employed to discretize the MFIE in this section employs a continuous

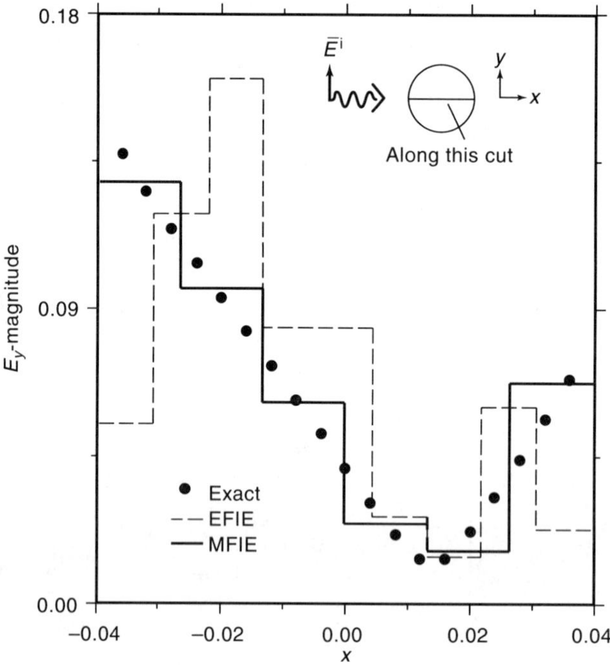

Figure 2.24 The TE electric field produced by the volume MFIE for the identical scattering problem presented in Figure 2.18. The MFIE result used an 84-cell model resulting in a matrix of order 55.

representation for the primary unknown (H_z) and in the process eliminates fictitious charge densities (Prob. P2.26). In addition, more accurate modeling is possible because of the flexibility associated with triangular cells. Finally, the TE MFIE operator itself seems to be more stable for numerical discretization than the TE EFIE operator.

The price to be paid for the flexibility of triangular cells is the additional complexity associated with the cylinder model. The cylinder model used with this MFIE formulation consists of a list of nodes $\{x_n, y_n\}$, a list of the relative permittivities of each cell, an array to act as a pointer from the cell indices to the three associated node indices, a second array to act as a pointer from the edge indices to the adjacent cell indices, and for convenience a third pointer array to link edge indices with the nodes defining the edge. This model exhibits a significant increase in complexity as compared with the simple models required with the EFIE approaches of Sections 2.5 and 2.6.

2.8. SCATTERING FROM HOMOGENEOUS DIELECTRIC CYLINDERS: SURFACE INTEGRAL EQUATIONS DISCRETIZED WITH PULSE BASIS AND DELTA TESTING FUNCTIONS [12]

A volume discretization is necessary to properly model a heterogeneous scatterer. However, homogeneous or layered geometries can be treated with surface integral equation formulations. Since the basis and testing functions are confined to the surfaces, rather than distributed throughout the volume of the scatterer, far fewer unknowns arise with surface integral equations as the electrical size of the scatterer increases. Therefore, they are usually considered a much more efficient alternative than the volume equations discussed previously, provided the geometry in question consists of relatively few homogeneous regions.

A homogeneous cylinder characterized by permittivity ε_d and permeability μ_d is depicted in Figure 2.25. Suppose this cylinder is illuminated by a TM wave. According to the surface equivalence principle, equivalent sources J_z and K_t defined on the cylinder surface S are sufficient to represent either the exterior or interior problem. Coupled EFIEs were developed in Section 1.9 and can be specialized to the TM polarization to produce

$$E_z^{\text{inc}}(t) = K_t(t) + jk_0\eta_0 A_z^{(0)} + \left\{ \frac{\partial F_y^{(0)}}{\partial x} - \frac{\partial F_x^{(0)}}{\partial y} \right\}_{S^+} \tag{2.108}$$

$$0 = -K_t(t) + jk_d\eta_d A_z^{(d)} + \left\{ \frac{\partial F_y^{(d)}}{\partial x} - \frac{\partial F_x^{(d)}}{\partial y} \right\}_{S^-} \tag{2.109}$$

where t is a parametric variable describing the cylinder surface,

$$A_z^{(i)} = \int J_z(t') \frac{1}{4j} H_0^{(2)}(k_i R)\, dt' \tag{2.110}$$

$$\bar{F}_t^{(i)} = \int \hat{t}(t') K_t(t') \frac{1}{4j} H_0^{(2)}(k_i R)\, dt' \tag{2.111}$$

$$R = \sqrt{[x(t) - x(t')]^2 + [y(t) - y(t')]^2} \tag{2.112}$$

and $\hat{t}$ is the unit vector tangent to the cylinder contour (Figure 2.26). The wavenumbers of the exterior medium and the dielectric region are denoted k_0 and k_d, respectively; the intrinsic impedances of the exterior medium and the dielectric region are denoted η_0 and η_d, respectively. The expressions in brackets in (2.108) and (2.109) differ depending on which side of the surface the observer is located. Equation (2.108) should be evaluated with the observer an infinitesimal distance *outside* S, while (2.109) should be evaluated with the observer an infinitesimal distance *inside* S.

In common with previous surface integral equation formulations, we assume that the cylinder model is represented by a superposition of flat strips, as illustrated in Figure 2.25. If pulse basis functions are used to represent the unknowns J_z and K_t and Equations (2.108) and (2.109) are enforced in the center of each of the cells in the model, the result is a matrix equation having a 2×2 block structure

$$\begin{bmatrix} \mathbf{E} \\ \mathbf{0} \end{bmatrix} = \begin{bmatrix} \mathbf{A} & \mathbf{B} \\ \mathbf{C} & \mathbf{D} \end{bmatrix} \begin{bmatrix} \mathbf{j} \\ \mathbf{k} \end{bmatrix} \tag{2.113}$$

where each entry in Equation (2.113) is an $N \times N$ matrix having elements

$$A_{mn} = \frac{k_0\eta_0}{4} \int_{\text{cell } n} H_0^{(2)}(k_0 R)\, dt' \tag{2.114}$$

$$B_{mm} = \tfrac{1}{2} \tag{2.115}$$

$$B_{mn} = \frac{k_0}{4j} \int_{\text{cell } n} \left(\cos\phi_n \frac{\Delta x}{R_m} + \sin\phi_n \frac{\Delta y}{R_m} \right) H_1^{(2)}(k_0 R_m)\, dt' \qquad m \neq n \tag{2.116}$$

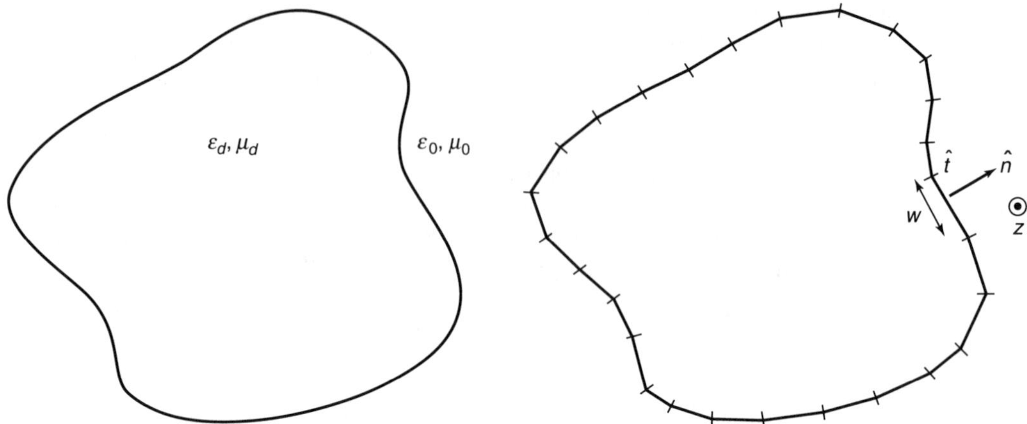

Figure 2.25 Flat-strip model of the surface of a homogeneous dielectric cylinder.

$$C_{mn} = \frac{k_d \eta_d}{4} \int_{\text{cell } n} H_0^{(2)}(k_d R)\, dt' \tag{2.117}$$

$$D_{mm} = -\tfrac{1}{2} \tag{2.118}$$

and

$$D_{mn} = \frac{k_d}{4j} \int_{\text{cell } n} \left(\cos\phi_n \frac{\Delta x}{R_m} + \sin\phi_n \frac{\Delta y}{R_m} \right) H_1^{(2)}(k_d R_m)\, dt' \qquad m \neq n \tag{2.119}$$

where

$$\Delta x = x_m - x(t') \tag{2.120}$$

$$\Delta y = y_m - y(t') \tag{2.121}$$

$$R_m = \sqrt{(\Delta x)^2 + (\Delta y)^2} \tag{2.122}$$

and ϕ_n is the polar angle defining the outward normal vector to the nth strip in the model, as illustrated in Figure 2.26. Expressions for B_{mm} and D_{mm} were obtained by a limiting procedure similar to that discussed in Section 2.2 and Prob. P2.8.

The data structure representing the cylinder model and the specific approach used to evaluate the expressions in (2.114)–(2.119) are generally identical to those used in previous surface integral equation formulations (Sections 2.1–2.2). The evaluation is slightly more complicated if the cylinder is lossy, in which case k_d is complex valued and the computations require a subroutine for Hankel functions of complex argument. Once these equivalent electric and magnetic sources are obtained by the solution of (2.113), the bistatic scattering cross section can be determined from

$$\sigma_{\text{TM}}(\phi) \cong \frac{k_0}{4} \left| \sum_{n=1}^{N} [\eta_0 j_n - k_n \cos(\phi - \phi_n)] w_n e^{jk_0(x_n \cos\phi + y_n \sin\phi)} \right|^2 \tag{2.123}$$

Fields within the dielectric cylinder (Prob. P2.30) can be obtained using source–field relationships that take the interior medium into account in conjunction with the sources J_z and K_t for the equivalent interior problem (which differ from the exterior sources obtained from the matrix solution by a 180° phase shift).

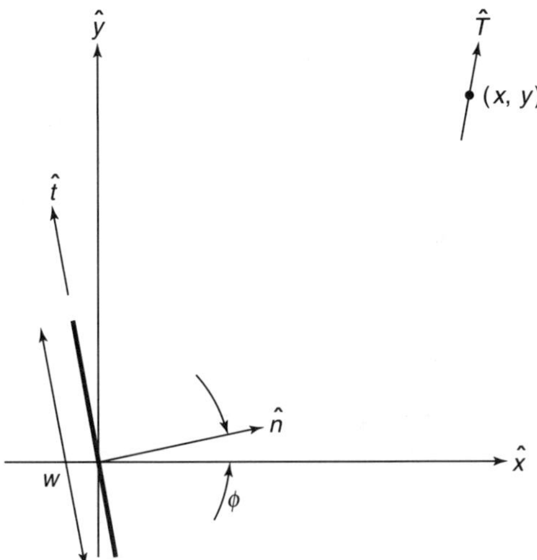

Figure 2.26 Geometry of a source cell, showing normal and tangent vectors.

Coupled MFIEs describing the TE case can be formulated and discretized in a similar manner, and we leave this exercise to the reader (Prob. P2.31). We note in passing that, as an alternative formulation, coupled magnetic field equations can be developed for the TM polarization or coupled electric field equations can be used for the TE case.

We will illustrate the accuracy of the moment method formulation for the TE polarization based on coupled magnetic field equations (Prob. P2.31). Table 2.13 shows the magnitude and phase of the H_z-field within a circular, homogeneous cylinder having circumference of $1\lambda_0$ and $\varepsilon_r = 2.56$. These field values were computed as a secondary calculation after J_z and K_t were obtained from a 10-cell model and a 40-cell model (using equal-sized cells) of the circular contour. (This geometry was previously analyzed using the volume MFIE formulation of Section 2.7, with the results appearing in Figure 2.22.)

TABLE 2.13 H_z-Field Along $y = 0$ Cut through Circular, Homogeneous Cylinder with Circumference $1\lambda_0$ and Relative Permittivity $\varepsilon_r = 2.56$ in Response to TE Plane Wave Propagating in $\phi = 0$ Direction

	H_z-Field			H_z-Field		
$x(\lambda_0)$	$N = 10$	$N = 40$	Exact	$N = 10$	$N = 40$	Exact
	Magnitude			Phase (deg)		
–0.14	1.334	1.359	1.359	–130.51	–133.78	–134.05
–0.10	1.502	1.507	1.509	–150.71	–151.31	–151.57
–0.06	1.587	1.600	1.605	–166.63	–167.12	–167.37
–0.02	1.600	1.623	1.630	177.79	177.49	177.29
0.02	1.547	1.578	1.587	161.36	161.41	161.31
0.06	1.453	1.486	1.496	143.08	143.66	143.69
0.10	1.356	1.384	1.392	122.23	123.57	123.71
0.14	1.291	1.309	1.314	96.03	101.28	101.60

Note: A coupled MFIE formulation was used with N pulse basis functions and Dirac delta testing functions located on the surface of the dielectric cylinder; the fields within the cylinder were obtained as a secondary calculation. The exact solution is shown for comparison.

The numerical results produced by the surface integral equation exhibit good agreement with the exact solution and appear to improve as the number of basis functions is increased.

As a second example, the scattering cross section of a circular, homogeneous cylinder having circumference of $0.248\lambda_0$ and $\varepsilon_r = 2 - j50$ is displayed in Table 2.14, where it is compared with data obtained from the volume EFIE formulation from Section 2.6 and the volume MFIE formulation from Section 2.7. Although the EFIE result is questionable (see Figure 2.18 and related discussion in Section 2.6), the volume MFIE and the surface MFIE results exhibit good agreement with the exact solution.

TABLE 2.14 TE Bistatic Scattering Cross Section[a] of a Circular Cylinder with Circumference $0.248\lambda_0$ and $\varepsilon_r = 2 - j50$ for Three Different Formulations

	Volume		Surface MFIE		
ϕ (deg)	EFIE 61 Cells, 122 Unknowns	MFIE 84 Cells, 55 Unknowns	10 Cells, 20 Unknowns	40 Cells, 80 Unknowns	Exact
0	−21.08	−22.37	−21.52	−22.24	−22.32
30	−22.29	−23.69	−22.81	−23.56	−23.64
60	−26.81	−28.70	−27.64	−28.55	−28.66
90	−41.31	−37.86	−38.41	−37.93	−37.70
120	−27.95	−26.78	−26.70	−26.74	−26.69
150	−23.13	−22.66	−22.39	−22.60	−22.58
180	−21.87	−21.53	−21.22	−21.47	−21.46

[a] In decibels free-space wavelength.

Note: Results from the volume EFIE approach from Section 2.6, the volume MFIE approach from Section 2.7, and the surface MFIE approach from this section are compared. The exact solution is shown for comparison.

Using the general principles discussed in Chapter 1, the surface integral equation formulation may be extended to treat scatterers made of several homogeneous regions. This topic is left as an exercise (Prob. P2.33).

2.9. INTEGRAL EQUATIONS FOR TWO-DIMENSIONAL SCATTERERS HAVING AN IMPEDANCE SURFACE

If the relative permittivity or permeability of a scatterer is large, an approximate impedance boundary condition (IBC) can sometimes be used to simplify the problem formulation beyond that of the coupled surface integral equations discussed in Section 2.8. If an IBC is valid, even inhomogeneous scatterers may be treated using a single surface integral equation instead of the volume formulations discussed previously.

Consider the scatterer illustrated in Figure 2.27. As discussed in Section 1.9, an equivalent exterior problem is obtained through the use of sources $\bar{J}$ and $\bar{K}$ defined on the surface of the original scatterer according to

$$\bar{J} = \hat{n} \times \bar{H} \tag{2.124}$$

$$\bar{K} = \bar{E} \times \hat{n} \tag{2.125}$$

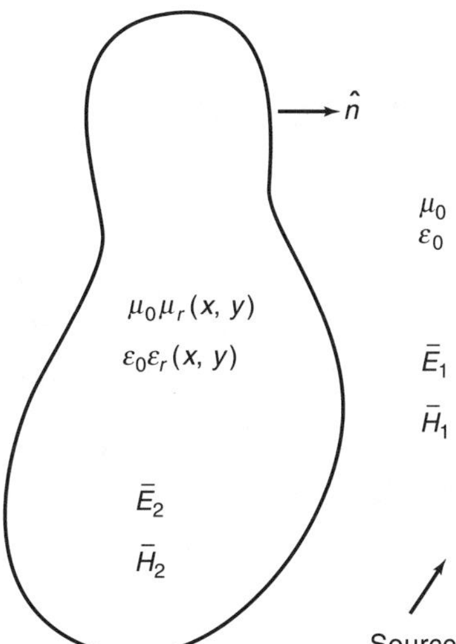

Figure 2.27 Cross section of a general cylindrical scatterer.

and radiating in an infinite space having the same parameters as the exterior medium. If the scatterer material is such that a ray field impinging at any angle to the surface is transmitted at an internal angle very close to normal, the equivalent sources of (2.124) and (2.125) may satisfy the IBC [13]

$$\bar{K}(t) = \eta_s(t)\bar{J}(t) \times \hat{n}(t) \tag{2.126}$$

where $\hat{n}$ is the outward normal vector to the surface and, in general, the surface impedance η_s of the scatterer may vary with location around the contour. In addition to the constraint mentioned above on the transmission direction of a ray field, in order for the IBC to be valid, sufficient loss must be present in the scatterer to ensure that the fields decay within the cylinder and do not emerge from the opposite side [14, 15].

Integral equations may be formulated in the usual manner to relate the incident electric or magnetic fields to the equivalent sources. In previous examples involving homogeneous dielectric bodies, the fact that both $\bar{J}$ and $\bar{K}$ were unknowns prompted the use of surface integral equations for both interior and exterior regions. If the impedance condition of Equation (2.126) is valid, it eliminates one of the two unknowns and the need to consider the interior region when formulating equations for penetrable bodies. As compared to a rigorous surface integral equation formulation, the IBC approach reduces the matrix order by a factor of 2. Since an impedance condition eliminates the need to find a Green's function for the interior region, its use permits a surface integral treatment of inhomogeneous bodies.

Consider a TM wave incident on a scatterer for which an impedance boundary condition is valid. Equivalent sources have components J_z and K_t, and the IBC reduces to

$$K_t(t) = E_z(t) = \eta_s(t)H_t(t) = \eta_s(t)J_z(t) \tag{2.127}$$

An electric field equation can be written entirely in terms of the equivalent electric current density J_z as

$$E_z^{\text{inc}}(t) = \eta_s(t)J_z(t) + jk\eta A_z + \left\{ \frac{\partial F_y}{\partial x} - \frac{\partial F_x}{\partial y} \right\}_{S^+} \tag{2.128}$$

where

$$A_z(t) = \int J_z(t')\frac{1}{4j}H_0^{(2)}(kR)\,dt' \tag{2.129}$$

$$\bar{F}_t(t) = \int \hat{t}(t')\eta_s(t')J_z(t')\frac{1}{4j}H_0^{(2)}(kR)\,dt' \tag{2.130}$$

and

$$R = \sqrt{[x(t) - x(t')]^2 + [y(t) - y(t')]^2} \tag{2.131}$$

The expression in brackets is to be evaluated an infinitesimal distance outside the surface.

For the case of a normally incident TE wave impinging on a scatterer that can be represented using an IBC, equivalent sources have components J_t and K_z. The impedance condition is given by

$$K_z(t) = -E_t(t) = \eta_s(t)H_z(t) = -\eta_s(t)J_t(t) \tag{2.132}$$

A TE EFIE can be written entirely in terms of the equivalent electric current density J_t according to

$$E_t^{\text{inc}}(t) = \eta_s(t)J_t(t) - \left\{\hat{t}\cdot\frac{\nabla\nabla\cdot + k^2}{j\omega\varepsilon}\bar{A}_t\right\} + \{\hat{t}\cdot\nabla\times\bar{F}_z\}_{S^+} \tag{2.133}$$

where

$$\bar{A}_t(t) = \int \hat{t}(t')J_t(t')\frac{1}{4j}H_0^{(2)}(kR)\,dt' \tag{2.134}$$

$$F_z(t) = -\int \eta_s(t')J_t(t')\frac{1}{4j}H_0^{(2)}(kR)\,dt' \tag{2.135}$$

and R is defined in Equation (2.131). As an alternative, a TE MFIE can be written in terms of J_t as

$$H_z^{\text{inc}}(t) = -J_t(t) - \{\hat{z}\cdot\nabla\times\bar{A}_t\}_{S^+} + j\omega\varepsilon F_z \tag{2.136}$$

In either of these equations, the subscript S^+ indicates that the bracketed term is to be evaluated an infinitesimal distance outside the scatterer surface.

For either polarization, the discretization of these integral equations into matrix form can be carried out following a procedure similar to that of previous examples. Issues such as the model used to represent the cylinder surface, the basis and testing functions, and the algorithm employed to evaluate the required integrals have been discussed in Sections 2.1, 2.2, and 2.4. Thus, we omit further details of the implementation.

To illustrate the use of an impedance boundary condition, consider the triangular dielectric cylinder depicted in Figure 2.28. We will limit ourselves to the TE polarization. This cylinder has perimeter equal to $3.24\lambda_0$ and relative permittivity $\varepsilon_r = 20 - j50$. Although the material appears to be compatible with an impedance boundary condition, the small size of the scatterer and the presence of sharp corners may invalidate the IBC approximation.

In order to investigate the validity of the IBC, we first treat this scatterer using the rigorous surface integral equation formulation from Section 2.8 (Prob. P2.31) for a 55-cell model and pulse basis functions for both J_t and K_z. Once the coefficients of the basis functions have been determined by a solution of the matrix equation, the ratio of K_z to J_t can be examined in order to justify the impedance concept. Tables 2.15 and 2.16 show this ratio for the equivalent currents on the top and back sides of the triangular cylinder. The

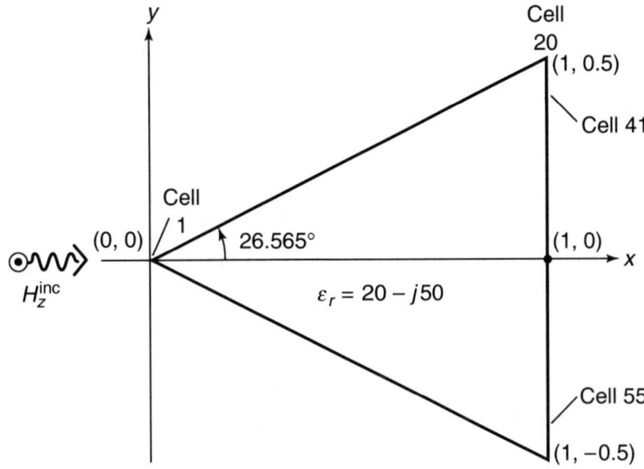

Figure 2.28 Cross section of a triangular dielectric cylinder. The coordinate locations are in units of free-space wavelength.

TABLE 2.15 Ratio of $-K_z$ to J_t Along Lit Face of Homogeneous, Triangular Dielectric Cylinder with $\varepsilon_r = 20 - j50$ Produced by Pulse Basis/Delta Testing Discretization of Coupled MFIE from Section 2.8

Cell Number	Magnitude	Phase (deg)
1	31.839	69.32
2	57.125	46.37
3	52.003	38.51
4	49.546	39.97
5	49.926	40.82
6	50.118	40.67
7	50.064	40.59
8	50.037	40.59
9	50.042	40.59
10	50.050	40.58
11	50.061	40.58
12	50.070	40.60
13	50.068	40.62
14	50.052	40.64
15	50.015	40.65
16	50.014	40.60
17	50.062	40.77
18	49.299	40.60
19	51.040	37.85
20	52.751	47.53

Note: The ratio should be compared with the intrinsic impedance of the dielectric material, which in this case has a magnitude equal to 51.34 Ω and a phase angle of 34.1°.

TABLE 2.16 Ratio of $-K_z$ to J_t Along Back Wall of Homogeneous, Triangular Dielectric Cylinder with $\varepsilon_r = 20 - j50$ Produced by Pulse Basis/Delta Testing Discretization of Coupled MFIE from Section 2.8

Cell Number	Magnitude	Phase (deg)
41	2.873	116.83
42	52.832	52.14
43	49.384	40.65
44	46.074	41.97
45	46.717	42.23
46	47.343	42.62
47	47.402	42.63
48	47.404	42.64
49	47.402	42.63
50	47.344	42.62
51	46.717	42.23
52	46.074	41.97
53	49.384	40.65
54	52.832	52.14
55	2.873	116.83

Source: The ratio should be compared with the intrinsic impedance of the dielectric material, which in this case has a magnitude equal to 51.34 Ω and a phase angle of 34.1°.

numerical values in these tables support the use of an IBC but suggest a surface impedance of approximately $38 + j33\Omega$ for the lit face and $35 + j32\Omega$ for the back face of the cylinder. The intrinsic impedance of this material is $42.51 + j28.78\Omega$, which differs from these values by 12 and 16%, respectively. Despite the difference, the scattering cross section obtained from a TE MFIE code incorporating the IBC with $\eta_s = 42.51 + j28.78\Omega$ exhibits reasonable agreement with the result from the rigorous coupled MFIE formulation (Table 2.17). Both

TABLE 2.17 Bistatic Scattering Cross Section σ_{TE} of Triangular Dielectric Cylinder Having $\varepsilon_r = 20 - j50$

ϕ	MFIE with IBC	Coupled MFIE
0	7.20	7.13
30	3.39	3.34
60	2.35	2.99
90	–5.14	–4.70
120	–15.56	–15.83
150	–11.46	–11.49
180	–9.25	–9.54

Note: In decibels free-space wavelength. Comparison of numerical results obtained using an (IBC) of $\eta_s = 42.51 + j28.78$ within an MFIE formulation and a rigorous coupled surface MFIE formulation. Results are based on the identical 55-cell model.

procedures use pulse basis functions and Dirac delta testing functions and a 55-cell model of the cylinder contour.

2.10. SUMMARY

Throughout this chapter, procedures were presented for the numerical solution of a variety of integral equations representing two-dimensional scattering geometries. Both surface integral equations and volume integral equations were considered, since the discretization process is similar in both cases. In an attempt to maintain an introductory level of presentation, the methods discussed are among the simplest that appear to produce accurate, converging results. We also discussed some procedures that do not always work well in an attempt to understand why these techniques fail. Throughout, we tried to provide enough detail in order to facilitate a computer implementation. Appendices C and D describe sample FORTRAN codes that demonstrate these techniques.

We have deferred a number of interesting topics to later chapters. Chapter 4 addresses common matrix solution algorithms and considers the effect of the stability of the matrix equation. Chapter 5 presents a theoretical examination of the discretization process and attempts to address the question of convergence. Alternative surface integral equation formulations are considered in Chapter 6, where difficulties that arise in several special situations are investigated. Chapter 7 presents an extension of the present formulations to geometries with one-dimensional periodicity. Chapter 8 considers the case of waves incident on infinite cylinders from oblique angles with respect to the cylinder axis, which is a building block needed to treat a finite source radiating in the presence of an infinite cylinder. We have mentioned that smoother basis and testing functions should provide improved accuracy, and a variety of examples of alternative basis and testing functions are presented in Chapter 9. Chapter 9 also introduces the use of curved-cell models. Finally, the development of integral equation formulations for three-dimensional scatterers is treated in Chapter 10.

Integral equations encompass only one of several different ways of posing an electromagnetic scattering problem. In the following chapter, alternative formulations are considered that employ a direct discretization of the scalar Helmholtz equations for two-dimensional scattering. These procedures may be more efficient for heterogeneous scatterers than the volume integral equation formulations discussed in the present chapter.

REFERENCES

[1] R. F. Harrington, *Field Computation by Moment Methods*, Malabar, FL: Krieger, 1982.

[2] M. Abramowitz and I. A. Stegun, *Handbook of Mathematical Functions*, New York: Dover, 1965.

[3] D. R. Wilton and C. M. Butler, "Effective methods for solving integral and integro-differential equations," *Electromagnetics*, vol. 1, pp. 289–308, 1981.

[4] A. W. Glisson and D. R. Wilton, "Simple and efficient methods for problems of electromagnetic radiation and scattering from surfaces," *IEEE Trans. Antennas Propagat.*, vol. AP-28, pp. 593–603, Sept. 1980.

[5] J. H. Richmond, "Scattering by a dielectric cylinder of arbitrary cross section shape," *IEEE Trans. Antennas Propagat.*, vol. AP-13, pp. 334–341, May 1965.

[6] I. S. Gradshteyn and I. M. Ryzhik, *Table of Integrals, Series, and Products*, New York: Academic, 1980.

[7] O. M. Al-Bundak, "Electromagnetic scattering of arbitrarily shaped inhomogeneous cylinders whose cross-sections are modeled by triangular patches," M.S. thesis, University of Mississippi, University, MS, 1983.

[8] J. H. Richmond, "TE-wave scattering by a dielectric cylinder of arbitrary cross section shape," *IEEE Trans. Antennas Propagat.*, vol. AP-14, pp. 460–464, July 1966.

[9] M. J. Hagmann and R. L. Levin, "Criteria for accurate usage of block models," *J. Microwave Power*, vol. 22, pp. 19–27, Jan. 1987.

[10] D. T. Borup, D. M. Sullivan, and O. P. Gandhi, "Comparison of the FFT conjugate gradient method and the finite-difference time domain method for the 2-D absorption problem," *IEEE Trans. Microwave Theory Tech.*, vol. MTT-35, pp. 383–395, Apr. 1987.

[11] A. F. Peterson and P. W. Klock, "An improved MFIE formulation for TE-wave scattering from lossy, inhomogeneous dielectric cylinders," *IEEE Trans. Antennas Propagat.*, vol. AP-36, pp. 45–49, Jan. 1988.

[12] T. K. Wu and L. L. Tsai, "Scattering by arbitrarily cross sectioned layered lossy dielectric cylinders," *IEEE Trans. Antennas Propagat.*, vol. AP-25, pp. 518–524, July 1977.

[13] K. M. Mitzner, "An integral equation approach to scattering from a body of finite conductivity," *Radio Science*, vol. 2, pp. 1459–1470, Dec. 1967.

[14] D. S. Wang, "Limits and validity of the impedance boundary condition," *IEEE Trans. Antennas Propagat.*, vol. AP-35, pp. 453–457, Apr. 1987.

[15] S. W. Lee and W. Gee, "How good is the impedance boundary condition?" *IEEE Trans. Antennas Propagat.*, vol. AP-35, pp. 1313–1315, Nov. 1987.

PROBLEMS

P2.1 (a) Consider the matrix operator arising in the TM EFIE formulation of Section 2.1. Under what conditions is the matrix in (2.7) symmetric ($Z_{mn} = Z_{nm}$)? Propose a simple way of modifying the formulation to always ensure symmetry.

(b) Other types of matrix symmetry appear in special cases. A matrix with $Z_{mn} = Z_{m-n}$ is a Toeplitz matrix, which appears in the form

$$\begin{bmatrix} z_0 & z_1 & z_2 & \\ z_1 & z_0 & z_1 & \cdots \\ z_2 & z_1 & z_0 & \\ & \cdot & & \\ & \cdot & & \\ & \cdot & & \end{bmatrix}$$

in the symmetric case. Identify two different scatterer geometries that, when used with equal-sized cells, cause Equation (2.7) to exhibit the Toeplitz structure.

P2.2 The worst-case error in Equation (2.10) occurs when the source and observation cells are closely spaced. To study this error, consider the formulation of Section 2.1 applied to TM scattering from a flat p.e.c. strip. If the cell sizes are restricted to width w, the matrix entry immediately adjacent to the main diagonal is given by

$$Z_{12} = \frac{k\eta}{4} \int_{w/2}^{3w/2} H_0^{(2)}(kx)\, dx$$

According to Equation (2.10), this expression may be approximated by

$$Z_{12}^{\text{app}} = \frac{k\eta w}{4} H_0^{(2)}(kw)$$

Using

$$H_0^{(2)}(kx) \cong 1 - j\frac{2}{\pi} \ln\left(\frac{\gamma kx}{2}\right) \qquad \text{as } kx \to 0$$

show that

$$\frac{Z_{12} - Z_{12}^{\text{app}}}{Z_{12}} \cong \frac{-j(2/\pi)\ln(3\sqrt{3}/2e)}{1 - j(2/\pi)\ln[(\gamma kw/2)(3\sqrt{3}/2e)]}$$

as $w \to 0$. What does this result suggest about the approximation in (2.10)?

P2.3 (a) Data from Table 2.1 suggest that the approximation of Equation (2.10) produces an error in the plane of the source cell that is independent of the distance to the observer for large separations. It is conjectured that this error is due to the phase cancellation ignored by the single-point approximation. Since $R \cong \rho - x' \cos\phi - y' \sin\phi$ as $\rho \to \infty$, integrate over the asymptotic form

$$H_0^{(2)}(kR) \approx \sqrt{\frac{2j}{\pi k\rho}} e^{-jkR} \qquad \text{as } \rho \to \infty$$

in order to derive the "improved" approximation

$$\int_{-w/2}^{w/2} \frac{1}{4j} H_0^{(2)}(k|x - x'|)\, dx' \cong \frac{1}{4j} H_0^{(2)}(kx) \frac{2}{k} \sin\left(\frac{kw}{2}\right)$$

(b) Using numerical data from Table 2.1, tabulate the error in this improved expression for $w = 0.1$ and $k = 2\pi$.

(c) Generalize this approximation to obtain a formula valid for all orientations of the source strip with respect to the observation point.

P2.4 Using the computer program described in Appendix A for TM scattering from p.e.c. cylinders, find the surface currents and scattering cross section for circular cylinders of circumference $1.0\lambda_0$, $2.0\lambda_0$, and $2.405\lambda_0$. If possible, compare your numerical results with exact solutions. Do the results seem to converge in every case with an increasing number of cells in the models? Based on your observations, propose guidelines for minimum cell densities.

P2.5 The formulation for TM scattering from p.e.c. cylinders can be improved by replacing Equation (2.10) with an evaluation of (2.8) by numerical quadrature. Assume that we retain the flat-cell model, the pulse basis functions, and the Dirac delta testing functions but extend the original approach to include the orientation of cells. Provide a detailed expression for the integrand as a function of the integration variable t' along the source cell. Discuss the impact of this change on the required scatterer model and the computational overhead.

P2.6 At the tip of a p.e.c. wedge with interior angle Ω (Figure 2.29), the TM current density is known to behave according to

$$J_z \approx \rho^{-(\pi-\Omega)/(2\pi-\Omega)} \quad \text{as } \rho \to 0$$

if ρ is the distance from the tip (N. Morita, N. Kumagai, and J. R. Mautz, *Integral Equation Methods for Electromagnetics*, Boston: Artech House, 1990). Basis functions used to represent J_z at the tip can be modified to incorporate this singularity for $0 \le \Omega < \pi$. Propose at least one way of modifying the piecewise-constant basis functions used in Section 2.1 to obtain a new function with a singularity of the proper order at one edge of a cell.

P2.7 Data in Table 2.2 suggest that the error in the TM current density is proportional to $O(\Delta)$ as $\Delta \to 0$, where Δ is the cell size. Given such knowledge, two successive results can be used to extrapolate to a more accurate value. For instance, if J_N is the numerical value obtained for the current density using N cells of uniform width Δ_N, then the dominant error in the current density is roughly $K\Delta_N$, and we can write

$$J_\infty \cong J_N + K\Delta_N$$

where J_∞ is the exact result. It follows that by doubling the number of cells we obtain a new result J_{2N} satisfying

$$J_\infty \cong J_{2N} + K(\Delta_N/2)$$

These two equations can be combined to eliminate K and produce

$$J_\infty \cong 2J_{2N} - J_N$$

which should be a more accurate result than either J_N or J_{2N}. This procedure is a special case of *Richardson extrapolation* (K. E. Atkinson, *An Introduction to Numerical Analysis*, New York: Wiley, 1989).

Use the values in Table 2.2 for $N = 32$ and $N = 64$ to extrapolate to a more accurate result for the current density. What is the percentage error in the extrapolated value in each case? Is the extrapolation process more efficient than using additional unknowns?

P2.8 (a) Show that the H_z-field due to a current density

$$\bar{J}_t(t) = \hat{y}p(y; -a, a)\delta(x)$$

radiating in free space is given at a point $(x, 0)$ by the expression

$$H_z^s(x, 0) = \frac{\partial A_y}{\partial x} = -\frac{k}{4j}\int_{y'=-a}^{a} \frac{x}{R} H_1^{(2)}(kR)\, dy'$$

where

$$R = \sqrt{x^2 + (y')^2}$$

(b) Using the approximation

$$H_1^{(2)}(kR) \cong \frac{kR}{2} + j\frac{2}{\pi kR} \quad \text{as } kR \to 0$$

evaluate the integral to show that

$$\lim_{x \to 0} H_z^s(x, 0) = \begin{cases} -\frac{1}{2} & x > 0 \\ +\frac{1}{2} & x < 0 \end{cases}$$

This completes the derivation of Equation (2.29).

(c) Although this derivation has been carried out for the two-dimensional case, the result can be extended to produce the scattered magnetic field an infinitesimal distance from an electric source on a smooth surface in three dimensions. Explain the equivalence.

P2.9 Following the basic procedure presented for the TE polarization in Section 2.2, develop a tangential-field MFIE formulation for TM scattering from closed, perfectly conducting cylinders.

(a) Identify the field components and current components present, and specialize Equation (1.97) to the two-dimensional TM case.
(b) Introduce pulse basis functions and Dirac delta testing functions in order to discretize the MFIE into matrix form. Provide expressions for the diagonal and off-diagonal matrix entries (the off-diagonal entries may be left in the form of an integral, but use a limiting process analogous to that of Prob. P2.8 to obtain the diagonal entries) as well as the entries of the excitation vector.
(c) Describe the minimum data structure required to represent the cylinder geometry.

P2.10 Generalize Prob. P2.8 in order to obtain the scattered H_z-field at the tip of a wedge produced by a uniform transverse current density J_t, as depicted in Figure 2.29. Show that H_z^s is given by

$$\lim_{x \to 0} H_z^s(x, 0) = -\frac{\Omega}{2\pi}$$

[*Hint:* You should obtain

$$H_z^s(x, 0) = -\frac{x \sin(\Omega/2)}{\pi} \int_{t'=0}^{w} \frac{dt'}{x^2 + 2xt' \cos(\Omega/2) + (t')^2}$$

as an intermediate result.]

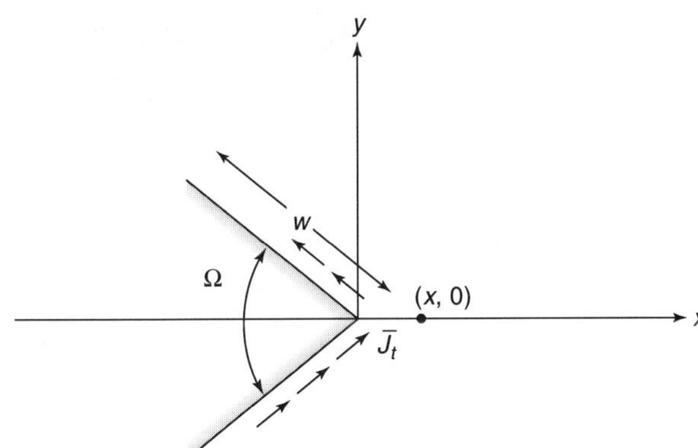

Figure 2.29 Geometry of a p.e.c. wedge (for Prob. P2.6).

P2.11 Using the result of Prob. P2.10, develop a TE MFIE formulation for closed p.e.c. scatterers based on subsectional triangle basis functions for J_t and Dirac delta testing functions. The basis and testing functions are to be centered at the edges of flat cells comprising the model, so that the triangle functions overlap two adjacent cells. Provide detailed expressions for the matrix entries (diagonal and off-diagonal cases) under the assumption that numerical quadrature will be used to evaluate the off-diagonal entries. Describe the minimum data structure necessary to represent the cylinder geometry.

P2.12 The MFIE of (2.19) is enforced an infinitesimal distance *outside* the scatterer surface. Using concepts from Sections 1.6 and 1.7 and the result of Prob. P2.8, develop an equivalent TE MFIE expressed an infinitesimal distance *inside* the surface. Evaluate the diagonal matrix entries that would arise from a numerical discretization analogous to that used in Section 2.2. Explain any differences between the original MFIE and this equation.

P2.13 Develop a TE implementation of the normal-field MFIE introduced in Prob. P1.19 by treating the transverse current density as the primary unknown and employing pulse basis functions and Dirac delta testing functions defined on flat cells. Provide expressions for the diagonal and off-diagonal matrix entries and the entries of the excitation vector. Compare the normal-field formulation to the approach discussed in Section 2.2. Can you identify any advantages or disadvantages?

P2.14 Equation (2.33) was obtained by specializing (1.52) to produce the x-component of the electric field due to a two-dimensional x-directed current density. Generalize this result to obtain an expression for the y-component of the electric field produced by the current in Equation (2.34). By combining your result with (2.38), provide the complete expression for the electric field due to the current density.

P2.15 Develop a FORTRAN subroutine to compute the scattering cross section according to Equation (2.59) for the TE EFIE approach of Section 2.4. Assume that the input data consists of a cylinder model (Figure 2.11 and Table 2.9) and the coefficients $\{j_n\}$ determined by the prior solution of Equation (2.51).

P2.16 The TE EFIE formulation of Section 2.4 can be extended to treat junctions between strips and closed cylinders. For a junction between three cells, develop a scheme for placing basis and testing functions that satisfies Kirchhoff's current law at the junction. How does the TE situation differ from the TM approach presented in Section 2.1?

P2.17 Using the subroutine (Appendix D) for the entries of Equation (2.51), generate a FORTRAN program to implement the TE EFIE formulation for p.e.c. strips and cylinders discussed in Section 2.4. You may assume that a suitable cylinder model has been placed in an input data file in the form described in Table 2.9 (a list of nodes, a pointer from cell indices to adjacent node indices, and a second pointer from basis function locations to the adjacent cell indices). You may also neglect the treatment of junctions between multiple strips. Develop sections of code to read the cylinder model from the data file, fill the matrix equation, create the excitation vector for an incident field of the form

$$H_z^{\mathrm{inc}}(x, y) = e^{-jk(x\cos\theta + y\sin\theta)}$$

solve the matrix equation, and write the coefficients of the basis functions to an output file. Test your program by comparing results for circular cylinders with exact solutions. Study the convergence of the numerical solutions as a function of the number of cells used in the model (i.e., try 5 cells/λ, 10 cells/λ, etc).

P2.18 Study the accuracy of the approximation employed in Equation (2.56) for the special case when the basis and testing functions are located on a common plane by developing software that provides an accurate evaluation of the double integral required in (2.53). Tabulate the error in the approximation as a function of displacement between the source cell and the observation cell for cell sizes of $w = 0.1\lambda_0$ and $w = 0.05\lambda_0$.

P2.19 Verify the integral presented in Equation (2.71).

P2.20 An analytical result is obtained in Equation (2.72) for the off-diagonal matrix entries associated with the TM EFIE formulation of Section 2.5. Compare this expression with the obvious single-point approximation to the integral, that is,

$$Z_{mn} \cong \frac{k\eta}{4}\pi a_n^2 H_0^{(2)}(kR_{mn})$$

Would an improved single-point approximation similar to that introduced in Prob. P2.3 be likely to enhance the accuracy of (2.72)?

P2.21 The volume integral formulation of Section 2.5 for TM scattering from dielectric cylinders employed J_z instead of E_z as the primary unknown to be determined. Formulate a matrix equation similar to (2.67) based on E_z as the primary unknown. Compare the two approaches and identify the differences. Can you think of an advantage of one formulation over the other?

P2.22 Reformulate the EFIE discretization presented for TM-wave scattering from dielectric cylinders in Section 2.5 in order to replace the square cells with triangular cells.

(a) Provide integral expressions for the diagonal and off-diagonal matrix entries obtained using pulse basis functions and Dirac delta testing functions (assume that a quadrature algorithm will be employed to evaluate integrals over triangular domains).

(b) Develop a procedure for evaluating the diagonal matrix entries, which are complicated by the singularity in the Hankel function.

(c) Develop a data structure to represent the triangular-cell scatterer geometry. How does the minimum model compare with that required in Section 2.5?

P2.23 The problem of a TE wave incident on an inhomogeneous magnetic (permeable) cylinder is the electromagnetic "dual" to that of a TM wave incident on a dielectric cylinder (duality is reviewed in Section 1.5). Based on the EFIE formulation of Section 2.5, develop an MFIE formulation for the TE magnetic cylinder problem, assuming that the discretization employs pulse basis functions and Dirac delta testing functions on square cells. Treat the magnetic polarization current density

$$K_z(x, y) = j\omega\mu_0[\mu_r(x, y) - 1]H_z(x, y)$$

as the primary unknown. Provide expressions for the matrix entries using approximations analogous to those employed in Section 2.5.

P2.24 By combining the essential features of the formulations described in Sections 2.1 and 2.5, develop a procedure for TM scattering from composite cylinders containing p.e.c. and dielectric materials. Describe the matrix equation and provide expressions for the matrix entries. What features of this approach make the matrix entries so easy to obtain?

P2.25 By employing the "circular-cell" approximation introduced in Equation (2.71), evaluate Equations (2.89)–(2.93) to obtain analytical expressions.

P2.26 (a) Derive Equations (2.101) and (2.102).

(b) With the help of Maxwell's equations, show that the use of a continuous expansion for H_z in Section 2.7 guarantees the proper continuity of the normal component of the electric field at a cell interface and therefore eliminates fictitious charge density.

P2.27 An edge located between two cells is depicted in Figure 2.21. If the tangent vector along the edge is defined in terms of the angle Ω by

$$\hat{t}(t) = \hat{x}\cos\Omega(t) + \hat{y}\sin\Omega(t)$$

develop explicit expressions for the parameters $\{\alpha_n\}$ appearing in Equation (2.105) as a function of Ω, the relative permittivity ε_r of the two adjacent cells, and the coordinates (x_n, y_n) of the surrounding nodes.

P2.28 Extend Prob. P2.22 to produce a volume EFIE formulation for TM scattering from dielectric cylinders based on a linear pyramid representation of the internal E_z-field. Provide expressions for the matrix entries obtained using pyramid basis functions and Dirac delta testing functions centered at the corners of triangular cells. Develop a procedure for evaluating the diagonal matrix entries in order to treat the singularity when the testing function location coincides with a source cell. Propose a data structure to represent the triangular-cell model.

P2.29 By combining the TM EFIE formulation from Prob. P2.28 with the electromagnetic dual of the MFIE formulation described in Section 2.7, develop a triangular-cell EFIE formulation for TM scattering from cylinders having dielectric and magnetic inhomogeneities. In this formulation, both electric polarization currents and magnetic polarization currents can be simultaneously defined in terms of the E_z-field using

$$J_z(x, y) = j\omega\varepsilon_0[\varepsilon_r(x, y) - 1]E_z(x, y)$$

$$\bar{K}(x, y) = -\frac{\mu_r(x, y) - 1}{\mu_r(x, y)}\nabla \times \bar{E}_z$$

Provide expressions for the matrix entries arising from a linear pyramid representation for E_z and point matching at the corners of triangular cells.

P2.30 The integral formulation of Section 2.8 produces a numerical solution for the tangential fields on the surface of a homogeneous cylinder. The internal fields can be found from a secondary calculation by treating the surface fields as equivalent sources. For the TM polarization, provide expressions for the internal E_z-field and the internal H_x- and H_y-fields at an arbitrary point within a homogeneous cylinder of permittivity ε_d and permeability μ_d as a function of the tangential fields $E_z(t)$ and $H_t(t)$ on the cylinder surface.

P2.31 Using electromagnetic duality (Section 1.5), formulate an MFIE approach for TE scattering from a homogeneous dielectric cylinder that parallels the TM formulation discussed in Section 2.8. Provide all the matrix entries.

P2.32 A volume discretization requires unknowns distributed throughout the interior of the geometry under consideration, whereas surface integral equations confine the unknowns to the scatterer surface. For a homogeneous dielectric cylinder having radius $a = 50\lambda_0$ and relative permittivity $\varepsilon_r = 10$, estimate the number of unknowns required with the volume integral formulation of Section 2.5 and the surface integral formulation of Section 2.8. Assume that a density of 100 cells/λ_d^2 is required for the volume approach and 10 cells/λ_d is required for the surface approach.

P2.33 Specialize the coupled EFIE formulation developed in Prob. P1.20 for a layered dielectric scatterer to the two-dimensional TM case. Using pulse basis functions and Dirac delta testing functions, convert the resulting equations into matrix form. Show that the matrix structure has the form

$$\begin{bmatrix} X & X & & \\ X & X & X & X \\ X & X & X & X \\ & & X & X \end{bmatrix}\begin{bmatrix} J_1 \\ K_1 \\ J_2 \\ K_2 \end{bmatrix} = \begin{bmatrix} E^{\text{inc}}_{S_s^+} \\ 0 \\ 0 \\ 0 \end{bmatrix}$$

Extend this result to the case of a core dielectric region surrounded by two layers of homogeneous material. In that case, you should obtain a matrix having the structure

$$\begin{bmatrix} X & X & & & & \\ X & X & X & X & & \\ X & X & X & X & & \\ & & X & X & X & X \\ & & X & X & X & X \\ & & & & X & X \end{bmatrix}\begin{bmatrix} J_1 \\ K_1 \\ J_2 \\ K_2 \\ J_3 \\ K_3 \end{bmatrix} = \begin{bmatrix} E^{\text{inc}}_{S_1^+} \\ 0 \\ 0 \\ 0 \\ 0 \\ 0 \end{bmatrix}$$

Does the structure present in these systems suggest a recursive elimination procedure?

Outline such a process, and comment on the savings possible for a scatterer containing a large number of thin layers of homogeneous material.

P2.34 Develop expressions for the matrix entries associated with a discretization of Equation (2.128) employing pulse basis functions and Dirac delta testing functions.

P2.35 Discuss the discretization of Equation (2.133). What basis and testing functions should be used to compensate for the additional derivatives?

P2.36 A thin absorber known as a "resistive card" can be modeled as a shell of lossy dielectric material. If the shell has thickness Δ, where Δ is small compared with the wavelength, the relationship

$$\bar{J} = j\omega\varepsilon_0(\varepsilon_r - 1)\bar{E}$$

between the volume current density and the electric field can be used to obtain the approximate boundary condition

$$\bar{E}_{\tan} = \frac{\Delta}{j\omega\varepsilon_0(\varepsilon_r - 1)}\bar{J}_{\tan} = R_s \bar{J}_{\tan}$$

where $\bar{J}_{\tan}$ is an equivalent surface current density and R_s is the resistance of the card. Observe that a p.e.c. surface may be modeled by $R_s = 0$. Reformulate the TM EFIE formulation of Section 2.1 in order to incorporate this boundary condition. What changes are necessary in the matrix entries?

P2.37 Incorporate the resistive boundary condition from Prob. P2.36 into the TE EFIE formulation presented in Section 2.4. Describe any alterations in the entries of the moment method impedance matrix.

3

Differential Equation Methods for Scattering from Infinite Cylinders

Our investigation of numerical techniques for the analysis of infinite cylinders continues with formulations involving the two-dimensional scalar Helmholtz equations in the frequency domain. The integral equation approaches presented in Chapter 2 confined the unknown fields or currents to the surface or interior volume of the scatterer under consideration. In contrast, the computational domain associated with differential equation methods often includes an additional region of space outside the scatterer. To ensure that the scattered fields represent outward-propagating solutions, this region must be terminated with a radiation boundary condition. Exact and approximate radiation boundary conditions will be investigated. In addition, we consider two distinctly different ways of combining them with the Helmholtz equation to treat two-dimensional scattering.

Since different equivalent sources arise for dielectric, magnetic, and perfectly conducting materials, Chapter 2 presented separate integral equations for each type of scatterer. We adopt a more general approach in the present chapter and develop formulations capable of simultaneously treating scatterers containing any isotropic material. In fact, the procedure for treating highly heterogeneous materials with these techniques is identical to that used for homogeneous regions.

Because we strive for an introductory level of presentation, the discussion is limited to discretizations based on triangular-cell models of the cylinder cross section and piecewise-linear basis and testing functions. Extensions to other basis functions and more general cell shapes are deferred until Chapter 9. A sample computer program is described in Appendix D to illustrate the software implementation.

3.1 WEAK FORMS OF THE SCALAR HELMHOLTZ EQUATIONS

Consider a cylindrical scatterer characterized by a relative permittivity ε_r and permeability μ_r that may vary with position (Figure 3.1). The electromagnetic fields produced in the

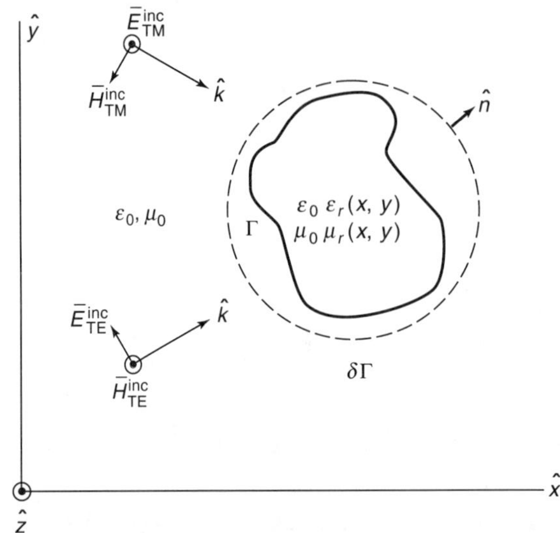

Figure 3.1 Inhomogeneity illuminated by an incident electromagnetic field. A mathematical boundary $\partial\Gamma$ encloses the scatterer.

vicinity of the cylinder by a TM illumination can be determined from the scalar Helmholtz equation

$$\nabla \cdot \left(\frac{1}{\mu_r} \nabla E_z \right) + k^2 \varepsilon_r E_z = 0 \tag{3.1}$$

The analogous equation for the TE case involves the magnetic field and is

$$\nabla \cdot \left(\frac{1}{\varepsilon_r} \nabla H_z \right) + k^2 \mu_r H_z = 0 \tag{3.2}$$

Equations (3.1) and (3.2) hold throughout a source-free region Γ containing the scatterer and must be augmented with boundary conditions on the surface $\partial\Gamma$ illustrated in Figure 3.1. The boundary conditions on $\partial\Gamma$ serve to couple the incident field into the equation and to ensure that the scattered field represents an outward-propagating solution. In principle, $\partial\Gamma$ may recede to infinity. In practice, $\partial\Gamma$ is located as close as possible to the scatterer surface in order to limit the number of required unknowns.

Equations (3.1) and (3.2) illustrate the "strong" form of the Helmholtz equations. In the strong form, the unknown appears within a second-order differential operator. To make these equations more amenable to numerical solution, they can be converted into their so-called "weak" form by multiplying both sides with a testing function $T(x, y)$ and performing an integration over Γ [1]. The vector identity

$$\nabla \cdot \{T\bar{A}\} = T\nabla \cdot \bar{A} + \nabla T \cdot \bar{A} \tag{3.3}$$

and the divergence theorem

$$\iint_\Gamma \nabla \cdot \bar{A} \, dx \, dy = \int_{\partial\Gamma} \bar{A} \cdot \hat{n} \, dt \tag{3.4}$$

may be used to cast (3.1) and (3.2) into

$$\iint_\Gamma \left(\frac{1}{\mu_r} \nabla T \cdot \nabla E_z - k^2 \varepsilon_r T E_z \right) dx \, dy = \int_{\partial\Gamma} \frac{1}{\mu_r} T \frac{\partial E_z}{\partial n} \, dt \tag{3.5}$$

$$\iint_{\Gamma} \left(\frac{1}{\varepsilon_r} \nabla T \cdot \nabla H_z - k^2 \mu_r T H_z \right) dx\, dy = \int_{\partial\Gamma} \frac{1}{\varepsilon_r} T \frac{\partial H_z}{\partial n} dt \tag{3.6}$$

Equations (3.5) and (3.6) are known as weak equations because the order of differentiation of the unknown fields is less than that of the original Helmholtz equations. By relaxing the strict differentiability requirements of the original equations, the weak equations permit the use of a wider variety of basis functions to represent the fields. In addition, medium discontinuities that would give rise to Dirac delta function terms in Equations (3.1) and (3.2) pose no such difficulty in Equations (3.5) and (3.6).

In deriving (3.5) and (3.6), we have tacitly assumed that the fields E_z and H_z are continuous functions. Later, when we introduce a subsectional basis representation for the fields, this assumption must be maintained in order to ensure the boundedness of ∇E_z and ∇H_z at cell interfaces. If the field representation is not assumed to be continuous, additional contour integrals arise between cells and can be included on the right-hand sides of (3.5) and (3.6). Our development will ignore these integrals and recapture them if needed (they will not be) via a Dirac delta dependence in ∇E_z and ∇H_z.

We will generally locate the surface $\partial\Gamma$ so that it completely encloses all inhomogeneities, leaving a buffer layer around the scatterer. Consequently, we will assume that $\mu_r = 1$ and $\varepsilon_r = 1$ in the integrals over $\partial\Gamma$ appearing in (3.5) and (3.6) and omit the constitutive parameters from the boundary integrals in many of the examples to follow.

The introduction of a testing function to construct a weak equation may seem very different from the previous use of testing functions to approximately enforce integral equations (Chapter 2). In fact, the two procedures are completely analogous. In the formulation of Section 2.4, for instance, the integral equation was cast into a weak form by shifting a derivative onto the testing function. An apparent distinction arises because, in Chapter 2, testing functions were introduced as the final step of the discretization process. By employing testing functions as the first step of the process with differential equations, our attention is directed toward the boundary integrals over $\partial\Gamma$ in (3.5) and (3.6). These integrals provide a convenient way of incorporating boundary conditions and therefore play a prominent role in the formulation.

Boundary conditions may take several forms. The classical literature on boundary-value problems usually emphasizes conditions of the Dirichlet and Neumann variety. A Dirichlet type of boundary condition can be expressed as

$$E_z = f(t) \quad \text{on } \partial\Gamma \tag{3.7}$$

where f is a given function. If f is specified on the entire surface $\partial\Gamma$, Equation (3.1) or (3.5) uniquely describes E_z everywhere throughout Γ. A Neumann boundary condition prescribes the normal derivative of the unknown field on $\partial\Gamma$ and can be written

$$\frac{\partial E_z}{\partial n} = g(t) \quad \text{on } \partial\Gamma \tag{3.8}$$

where g is a given function. If g is specified on the entire boundary, Equation (3.1) or (3.5) again uniquely describes E_z throughout Γ. As a third possibility, a Dirichlet condition may be imposed on part of the boundary while a Neumann condition is imposed on the remaining part.

The given information in most electromagnetic scattering problems is limited to the target geometry and the incident field. The total field and its normal derivative are not initially known on a boundary surrounding the scatterer, and consequently Dirichlet and Neumann boundary conditions are not directly specified. Despite this, it is still possible to pose the scattering problem in terms of indirect Dirichlet or Neumann conditions, and a formulation incorporating this idea will be described in Section 3.12.

Although the nature of scattering problems prevents the direct specification of Dirichlet or Neumann conditions on $\partial\Gamma$, boundary conditions can assume other forms. Two examples that often arise in electromagnetics are the impedance boundary conditions (Section 2.9) and the classical radiation boundary conditions (Section 1.3). These conditions have the form

$$\frac{\partial E_z}{\partial n} = h(t)E_z(t) \quad \text{on } \partial\Gamma \tag{3.9}$$

and relate the field and its normal derivative along $\partial\Gamma$ in such a way as to ensure unique solutions. If the incident field is known, an alternative statement of the scattering problem is to solve (3.5) or (3.6) to find the scattered field throughout Γ, subject to a radiation boundary condition (RBC) on $\partial\Gamma$ that forces the scattered field to propagate away from the region. In practice, this approach requires an RBC that can be applied close to the scatterer, in contrast to the Sommerfeld radiation condition from (1.36), which is only valid as $\rho \to \infty$. Several "near-field" radiation conditions will be developed in the following sections. These conditions are always more complicated than the simple Sommerfeld RBC, and for arbitrary boundary shapes they are only available in the form of integral equations. In addition, exact near-field RBCs are global conditions and couple the fields around the entire radiation boundary. In an attempt to facilitate an efficient numerical implementation, we will also consider one type of approximate RBC that is local in nature.

Generally, the most convenient way of incorporating boundary conditions such as (3.8) or (3.9) is by direct substitution into the surface integrals appearing on the right-hand sides of Equations (3.5) and (3.6). The resulting solution will satisfy the boundary condition in a weak sense, meaning that the degree of satisfaction is not exact for a finite number of basis functions. A Dirichlet condition, however, requires a substantially different implementation. The unknown function must be constructed from a basis that exactly satisfies the Dirichlet boundary condition (i.e., in the strong sense). The details of these two implementations will be discussed in the following section.

3.2 INCORPORATION OF PERFECTLY CONDUCTING BOUNDARIES

An obvious use of Dirichlet or Neumann boundary conditions is to incorporate materials within the scatterer geometry. Consider Figure 3.2, which depicts a p.e.c. region imbedded in a general scatterer. We redefine the interior volume Γ to exclude the conducting region and denote the surface of the conducting region $\partial\Gamma_c$ in order to distinguish it from the exterior surface $\partial\Gamma$. Equation (3.5) must be generalized to include the additional surface around the imbedded conductor and takes the modified form

$$\iint_\Gamma \left(\frac{1}{\mu_r}\nabla T \cdot \nabla E_z - k^2\varepsilon_r T E_z\right) dx\, dy = \int_{\partial\Gamma} T\frac{\partial E_z}{\partial n}\, dt + \int_{\partial\Gamma_c} \frac{1}{\mu_r} T\frac{\partial E_z}{\partial n}\, dt \tag{3.10}$$

where n is a local coordinate defined along the normal vector $\hat{n}$ pointing out of Γ. The TM

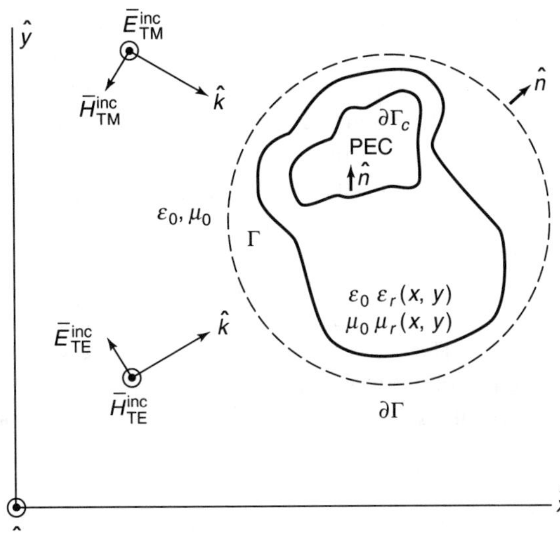

Figure 3.2 Cross section of the cylinder geometry showing the imbedded p.e.c. region whose surface is denoted $\partial\Gamma_c$.

fields must satisfy the Dirichlet boundary condition

$$E_z = 0 \quad \text{on } \partial\Gamma_c \tag{3.11}$$

The boundary condition in (3.11) can be enforced by building that behavior into the representation for E_z, or equivalently by zeroing the coefficient of any basis function that contributes to a nonzero E_z along $\partial\Gamma_c$. An alternative possibility is not to assign basis functions to $\partial\Gamma_c$ in the first place. In either case, there will be no unknowns associated with basis functions on $\partial\Gamma_c$, and the basis representation for E_z will exactly satisfy (3.11). Since there are no unknowns assigned to $\partial\Gamma_c$, there will be no need for testing functions along this boundary either. In practice, this implies that the integral over $\partial\Gamma_c$ in (3.10) does not contribute to the finite-element system and can be ignored. Thus, Equation (3.5) suffices, with the only modification necessary to incorporate (3.11) being to force the basis representation for E_z to vanish along the p.e.c. boundary.

For the TE polarization, the magnetic field at a perfectly conducting surface satisfies the Neumann boundary condition

$$\frac{\partial H_z}{\partial n} = 0 \quad \text{on } \partial\Gamma_c \tag{3.12}$$

Equation (3.6) can be modified to include the presence of the additional surface, yielding an equation for H_z of the form

$$\iint_\Gamma \left(\frac{1}{\varepsilon_r} \nabla T \cdot \nabla H_z - k^2 \mu_r T H_z \right) dx\, dy = \int_{\partial\Gamma} T \frac{\partial H_z}{\partial n}\, dt + \int_{\partial\Gamma_c} \frac{1}{\varepsilon_r} T \frac{\partial H_z}{\partial n}\, dt \tag{3.13}$$

To impose the Neumann boundary condition, Equation (3.12) is substituted into the boundary integral over $\partial\Gamma_c$ in (3.13). The integral vanishes, and (3.13) reverts back to the original form (3.6). In this case, values of H_z on $\partial\Gamma_c$ remain as unknowns to be determined, and consequently basis functions must be assigned to represent the nonzero field on this boundary. What has changed as a result of the Neumann condition? Although the weak equation remains unchanged, the scatterer geometry now excludes the region enclosed by $\partial\Gamma_c$ from the problem domain. By not imposing any explicit condition on the fields adjacent to $\partial\Gamma_c$, the procedure automatically enforces (3.12) in a weak sense.

The Dirichlet boundary condition of Equation (3.11) is often denoted as an *essential* boundary condition, while the Neumann condition of Equation (3.12) is termed a *natural* boundary condition. The distinction between these is discussed by Strang and Fix [2]. The essential condition has to be explicitly built into the expansion functions in order to be satisfied by the solution, even in the limit of an infinite number of unknowns. In contrast, the natural condition will be satisfied in the limiting case, even if not satisfied by the individual expansion functions. For a finite number of basis functions, the natural boundary condition will only be satisfied in an approximate sense.

The Dirichlet and Neumann boundary conditions permit the incorporation of perfectly conducting material. To complete the formulation of the scattering problem, a radiation condition must be imposed along the outer boundary $\partial\Gamma$.

3.3 EXACT NEAR-ZONE RADIATION CONDITION ON A CIRCULAR BOUNDARY [3, 4]

If unconstrained by an appropriate boundary condition on $\partial\Gamma$, the scalar Helmholtz equation generally admits two solutions, one propagating away from the scatterer and one propagating toward the scatterer. To ensure a unique solution, some type of RBC must be imposed along $\partial\Gamma$. This RBC must ensure that the scattered field is propagating away from the cylinder and at the same time incorporate the incident field into the problem description.

Provided that the shape of $\partial\Gamma$ is constrained to be circular, an exact radiation boundary condition can be obtained from an exterior eigenfunction expansion. Assuming that we know the scattered electric field E_z^s on a circular boundary of radius $\rho = a$, the cylindrical eigenfunctions may be used to express the scattered fields exterior to the boundary as [5]

$$E_z^s(\rho, \phi) = \sum_{n=-\infty}^{\infty} e_n H_n^{(2)}(k\rho) e^{jn\phi} \qquad \rho \geq a \tag{3.14}$$

where

$$e_n = \frac{1}{2\pi H_n^{(2)}(ka)} \int_0^{2\pi} E_z^s(a, \phi') e^{-jn\phi'} \, d\phi' \tag{3.15}$$

From this expansion, we construct

$$\frac{\partial E_z^s}{\partial \rho} = \sum_{n=-\infty}^{\infty} e_n k H_n^{(2)\prime}(k\rho) e^{jn\phi} \tag{3.16}$$

where the prime in (3.16) denotes differentiation with respect to the argument of the Hankel function. By substituting (3.15) into (3.16) and interchanging the order of summation and integration, we obtain

$$\left.\frac{\partial E_z^s}{\partial \rho}\right|_{\rho=a} = \int_0^{2\pi} E_z^s(a, \phi') \left[\frac{k}{2\pi} \sum_{n=-\infty}^{\infty} \frac{H_n^{(2)\prime}(ka)}{H_n^{(2)}(ka)} e^{jn(\phi-\phi')} \right] d\phi' \tag{3.17}$$

Although the form of Equation (3.17) is somewhat different from that of (3.9), it is an exact radiation boundary condition valid for finite ρ.

To facilitate a formulation in which the total electric field is the primary unknown, instead of the scattered field, Equation (3.17) can be combined with a similar expansion of the incident field. A plane-wave incident field can be expressed as [5]

$$E_z^{\text{inc}}(\rho, \phi) = \sum_{n=-\infty}^{\infty} e_n^{\text{inc}} J_n(k\rho) e^{jn\phi} \tag{3.18}$$

where

$$e_n^{\text{inc}} = \frac{1}{2\pi J_n(ka)} \int_0^{2\pi} E_z^{\text{inc}}(a, \phi') e^{-jn\phi'} \, d\phi' \tag{3.19}$$

By a similar procedure, we construct

$$\frac{\partial E_z^{\text{inc}}}{\partial \rho} = \sum_{n=-\infty}^{\infty} e_n^{\text{inc}} k J_n'(k\rho) e^{jn\phi} \tag{3.20}$$

and combine (3.19) and (3.20) to obtain

$$\left.\frac{\partial E_z^{\text{inc}}}{\partial \rho}\right|_{\rho=a} = \int_0^{2\pi} E_z^{\text{inc}}(a, \phi') \left(\frac{k}{2\pi} \sum_{n=-\infty}^{\infty} \frac{J_n'(ka)}{J_n(ka)} e^{jn(\phi-\phi')} \right) d\phi' \tag{3.21}$$

Equations (3.17) and (3.21) and the Wronskian relationship [5]

$$J_n'(ka) H_n^{(2)}(ka) - J_n(ka) H_n^{(2)\prime}(ka) = \frac{j2}{\pi ka} \tag{3.22}$$

produce the alternative radiation boundary condition

$$\begin{aligned} \left.\frac{\partial E_z}{\partial \rho}\right|_{\rho=a} = & \int_0^{2\pi} E_z^{\text{inc}}(a, \phi') \left(\frac{j}{\pi^2 a} \sum_{n=-\infty}^{\infty} \frac{1}{J_n(ka) H_n^{(2)}(ka)} e^{jn(\phi-\phi')} \right) d\phi' \\ & + \int_0^{2\pi} E_z(a, \phi') \left(\frac{k}{2\pi} \sum_{n=-\infty}^{\infty} \frac{H_n^{(2)\prime}(ka)}{H_n^{(2)}(ka)} e^{jn(\phi-\phi')} \right) d\phi' \end{aligned} \tag{3.23}$$

This condition provides an exact relationship between the total and incident fields on a circular boundary of radius $\rho = a$. The same RBC describes the TE polarization, provided that E_z is replaced with H_z in Equation (3.23).

The second summation appearing in Equation (3.23) is divergent, since

$$\frac{H_n^{(2)\prime}(ka)}{H_n^{(2)}(ka)} \approx -\frac{|n|}{ka} \quad \text{as } |n| \to \infty \text{ with } ka \text{ fixed} \tag{3.24}$$

The divergent summation is a consequence of interchanging the integration and summation when constructing (3.17) and reflects the fact that the bracketed term in (3.17) behaves as a type of Green's function with a singularity at $\phi = \phi'$. Any computational difficulty, however, can be avoided by integrating term by term. The decay in the Fourier spectrum of $E_z(a, \phi)$ as $n \to \infty$ will ensure the convergence of the net result.

3.4 OUTWARD-LOOKING FORMULATION COMBINING THE SCALAR HELMHOLTZ EQUATION WITH THE EXACT RADIATION BOUNDARY CONDITION

The radiation boundary condition of Equation (3.23) can be substituted into Equation (3.5) to produce

$$\iint_\Gamma \left(\frac{1}{\mu_r}\nabla T \cdot \nabla E_z - k^2 \varepsilon_r T E_z\right) dx\, dy + \int_0^{2\pi} T(a, \phi) \int_0^{2\pi} E_z(a, \phi') G_2(\phi - \phi')\, d\phi'\, a\, d\phi = \int_0^{2\pi} T(a, \phi) \int_0^{2\pi} E_z^{\text{inc}}(a, \phi') G_1(\phi - \phi')\, d\phi'\, a\, d\phi \tag{3.25}$$

where

$$G_1(\phi) = \frac{j}{\pi^2 a} \sum_{n=-\infty}^{\infty} \frac{1}{J_n(ka) H_n^{(2)}(ka)} e^{jn\phi} \tag{3.26}$$

and

$$G_2(\phi) = -\frac{k}{2\pi} \sum_{n=-\infty}^{\infty} \frac{H_n^{(2)\prime}(ka)}{H_n^{(2)}(ka)} e^{jn\phi} \tag{3.27}$$

Equation (3.25) provides a complete description of the TM scattering problem in terms of the interior E_z-field as the primary unknown to be determined. Because the equation is expressed explicitly in terms of interior field quantities (as if we stand inside the scatterer and look out), we denote this type of approach as an *outward-looking* formulation [6]. (This terminology will be clarified later in the chapter, when we introduce an alternative formulation we denote *inward looking*.)

Since the scatterer is inhomogeneous and arbitrary in shape, analytical solutions to (3.25) are usually not possible. Instead, we seek an approximate solution using basis and testing functions. Although the procedure to be followed is identical to the *method of moments* introduced in Chapter 2, it is popular usage to call it the *finite-element method* [1, 2, 7]. (Historically, the use of basis and testing functions to discretize integral equations of electromagnetics is most often named the "method of moments"; the same process applied to differential equations is usually known as the "weighted-residual" method or the "finite-element" method. Purists prefer to reserve the term finite-element method for variational methods, that is, explicitly minimizing a quadratic functional. However, for all aspects considered in this text the difference is only one of terminology.)

In order to discretize Equation (3.25) into matrix form, consider the division of the region Γ into triangular cells of constant permittivity and permeability, as depicted in Figure 3.3. We refer to the collection of cells as a *finite-element mesh* and introduce subsectional basis functions defined on the triangular-cell domain. The weak equation (3.25) involves the derivative ∇E_z, which will be unbounded unless E_z is continuous, and thus we exclude piecewise-constant basis functions from consideration. The simplest basis functions that provide a continuous representation for E_z are the linear "pyramid" functions introduced in Section 2.7, which are depicted in Figure 3.4. Each basis function has unity amplitude

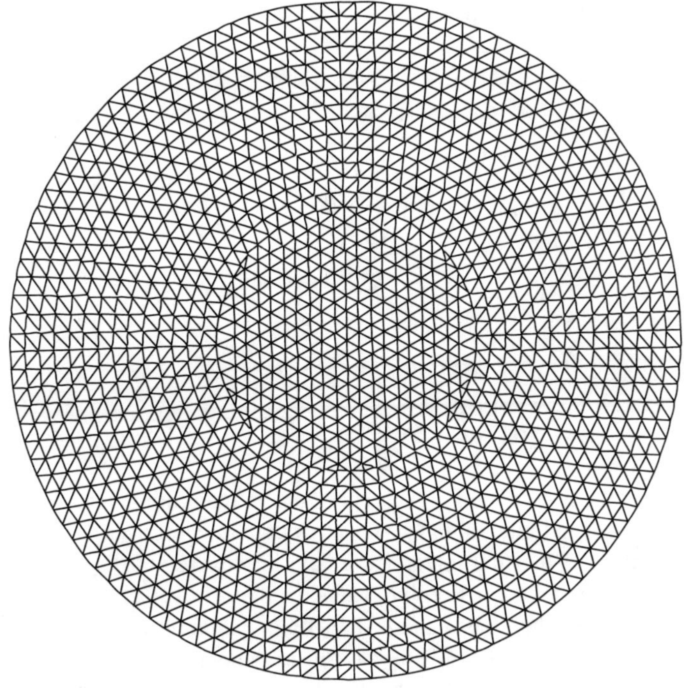

Figure 3.3 Triangular-cell mesh used to represent the cylinder cross section.

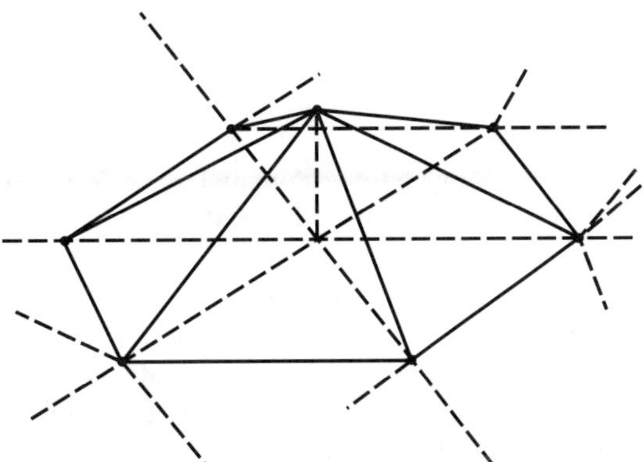

Figure 3.4 Linear pyramid basis function employed with the first-order finite-element representation.

at one node and vanishes at all other nodes in the mesh. Three basis functions overlap each triangular cell to provide a continuous piecewise-linear representation. If the mesh contains N nodes, the field representation can be expressed as

$$E_z(x, y) \cong \sum_{n=1}^{N} e_n B_n(x, y) \tag{3.28}$$

where $B_n(x, y)$ denotes a pyramid basis function centered at node n and e_n represents the interpolated value of E_z at that node. If (3.28) is substituted into Equation (3.25), N linearly independent equations can be generated using the same functions as testing functions:

$$T(x, y) = B_m(x, y) \qquad m = 1, 2, \ldots, N \tag{3.29}$$

The resulting matrix equation has the form

$$\mathbf{A}\mathbf{e} = \mathbf{b} \tag{3.30}$$

with entries given by

$$\begin{aligned} A_{mn} = & \iint_{\Gamma} \left(\frac{1}{\mu_r} \nabla B_m \cdot \nabla B_n - k^2 \varepsilon_r B_m B_n \right) dx\, dy \\ & + \int_0^{2\pi} B_m(\phi) \int_0^{2\pi} B_n(\phi') G_2(\phi - \phi')\, d\phi'\, a\, d\phi \end{aligned} \tag{3.31}$$

and

$$b_m = \int_0^{2\pi} B_m(\phi) \int_0^{2\pi} E_z^{\text{inc}}(a, \phi') G_1(\phi - \phi')\, d\phi'\, a\, d\phi \tag{3.32}$$

Note that the first term in (3.31) represents an integration over the interior cross section Γ, while the second integral in (3.31) is a double integral over the boundary $\partial\Gamma$.

Consider the first integral over the interior region Γ appearing in Equation (3.31). Because each of the functions $\{B_n\}$ is nonzero only over a few cells of the mesh, this integral is zero except when basis and testing functions are centered at the same or immediately adjacent nodes. As a result of this "nearest-neighbor" or *local* interaction, many of the entries A_{mn} will be zero. We say that **A** is a *sparse* matrix. The storage and computational requirements associated with a sparse-matrix equation can be significantly less than those of a full matrix, and several algorithms for the solution of sparse systems are discussed in Chapter 4.

As a general rule, the discretization of a differential operator usually produces a sparse matrix, in contrast to the full matrices produced by the integral operators considered in Chapter 2. Consequently, if compared with volume integral equation techniques (which require a computational domain Γ of similar overall size), differential equation formulations such as (3.25) are potentially more efficient and should enable the treatment of electrically larger regions.

The second term in (3.31) involves an integration over the radiation boundary condition imposed on $\partial\Gamma$. Unfortunately, this RBC is a *global* condition and couples information around the entire boundary. As a result, the part of the matrix with rows and columns associated with nodes on the boundary will be fully populated. For illustration, Figure 3.5 shows a hypothetical mesh and the resulting sparsity pattern in the **A** matrix. The fill-in due to the global RBC is cause for concern, since it can significantly degrade the expected efficiency of the differential equation formulation. This difficulty has motivated the development of *local* boundary conditions that alleviate the fill-in problem, such as the Bayliss–Turkel boundary condition to be considered in Section 3.8.

We have now described the formulation in general terms but still need to develop a specific implementation for the entries of the matrix equation (3.30). Consider the boundary integral in (3.31),

$$I_{mn} = a \int_0^{2\pi} B_m(\phi) \int_0^{2\pi} B_n(\phi') G_2(\phi - \phi')\, d\phi'\, d\phi \tag{3.33}$$

Assuming that the triangular cells are small enough to closely approximate the curvature

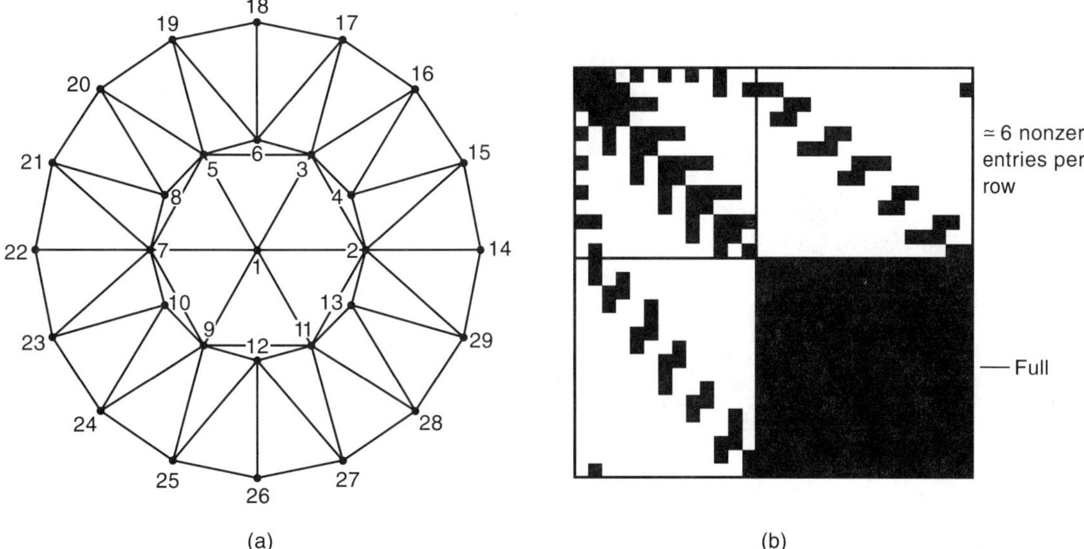

Figure 3.5 Triangular-cell mesh and the associated matrix sparsity pattern obtained with linear pyramid functions and the exact RBC.

of the circular boundary $\partial\Gamma$, B_m may be expressed along the boundary as

$$B_m(\phi) = \begin{cases} \dfrac{\phi - \phi_{m-1}}{\phi_m - \phi_{m-1}} & \phi_{m-1} < \phi < \phi_m \\[2ex] \dfrac{\phi_{m+1} - \phi}{\phi_{m+1} - \phi_m} & \phi_m < \phi < \phi_{m+1} \\ 0 & \text{otherwise} \end{cases} \tag{3.34}$$

where for convenience we denote the ϕ-coordinates of the three nodes associated with the mth basis function as ϕ_{m-1}, ϕ_m, and ϕ_{m+1}. After interchanging the order of integration and summation, we obtain

$$I_{mn} = -\frac{ka}{2\pi} \sum_{i=-\infty}^{\infty} \frac{H_i^{(2)\prime}(ka)}{H_i^{(2)}(ka)} \int_0^{2\pi} B_m(\phi) e^{ji\phi}\, d\phi \int_0^{2\pi} B_n(\phi') e^{-ji\phi'}\, d\phi' \tag{3.35}$$

A closed-form evaluation of the inner integrals is readily obtained in the form

$$\int_0^{2\pi} B_m(\phi) e^{ji\phi} d\phi = \frac{e^{ji\phi_m}}{i^2} \left(\frac{1 - e^{-ji(\phi_m - \phi_{m-1})}}{\phi_m - \phi_{m-1}} + \frac{1 - e^{ji(\phi_{m+1} - \phi_m)}}{\phi_{m+1} - \phi_m} \right) \tag{3.36}$$

Observe that the $O(i^{-2})$ decay in each integral is sufficient to ensure the convergence of the summation in (3.35), despite the $O(i)$ growth in the ratio of the Hankel functions.

The entries of the excitation vector from Equation (3.32) will be considered for a uniform plane-wave incident field of the form

$$E_z^{\text{inc}} = e^{-jk(x\cos\theta + y\sin\theta)} = \sum_{i=-\infty}^{\infty} j^{-i} J_i(k\rho) e^{ji(\phi-\theta)} \tag{3.37}$$

After interchanging the order of integration and summation, the required integral simplifies to

$$b_m = \frac{j2}{\pi} \sum_{i=-\infty}^{\infty} \frac{1}{H_i^{(2)}(ka)} j^{-i} e^{-ji\theta} \int_0^{2\pi} B_m(\phi) e^{ji\phi} \, d\phi \tag{3.38}$$

which can also be evaluated using (3.36). Each entry of the excitation vector not associated with a testing function on the outer boundary is identically zero.

The remaining integrals involve the interior interaction of the basis and testing functions throughout Γ. These integrals consist of polynomial functions and can be evaluated exactly in closed form. We will consider their evaluation shortly (Section 3.7). First, however, let us explore a specific example to illustrate the solution process.

3.5 EXAMPLE: TM-WAVE SCATTERING FROM A DIELECTRIC CYLINDER

As a problem to illustrate the finite-element procedure, consider a TM plane wave incident on a circular dielectric cylinder and the task of determining the fields within the cylinder and the two-dimensional scattering cross section. The cylinder geometry will be represented by a triangular-cell mesh, and the interior E_z-field will be represented with piecewise-linear basis functions. For ease of implementation, let us assume that the circular outer boundary $\partial\Gamma$ is located one layer of cells away from the cylinder surface. Figure 3.5 shows a coarse finite-element mesh that could be used to describe the geometry, containing 40 cells and 29 nodes. The entire data structure describing this mesh would likely consist of (1) the number of cells, number of nodes, and number of nodes located on the outer boundary $\partial\Gamma$; (2) a list of the x, y-coordinates of each node; (3) a list giving the indices of the three nodes associated with each of the 40 cells (the *connectivity* array); (4) a list of the constitutive parameters of each cell; and (5) a list giving the indices of all the nodes located along the outer boundary, perhaps in order of increasing cylindrical angle ϕ (or possibly a boundary connectivity array listing the two node indices associated with each boundary edge).

To construct the global finite-element system $\mathbf{Ae} = \mathbf{b}$, the matrix entries given in Equations (3.31) and (3.32) must be computed for all possible combinations of basis and testing functions. The boundary integrals have been evaluated in (3.35)–(3.38), and the indices of the boundary basis and testing functions can be determined using the boundary pointer array in the mesh data structure. In other words, if nodes m and n are both located on the radiation boundary, there will be a contribution from I_{mn} in (3.35) to entry A_{mn}, and a contribution to both b_m and b_n from (3.38). The angles defining the limits of these integrals can be determined from the x, y-coordinates tabulated in the mesh data structure. If either index m or n refers to an interior node, there will be no boundary contribution to that location in $\mathbf{A}$.

The remaining integral in (3.31) represents the interaction of the basis and testing functions throughout the interior Γ. The interior interaction embodied in a single entry of $\mathbf{A}$ involves integrals over a variable number of triangular cells. For example, there are six cells surrounding node 1 in Figure 3.5, implying that the evaluation of entry A_{11} involves integrals over six different cells. However, the data structure describing the finite-element mesh does not provide a direct way of determining the indices of these six cells, or even an easy way of determining that there are precisely six cells adjacent to node 1. The situation may even be worse for an entry such as A_{13}, since the program would have to determine if any cells are shared by the basis and testing functions at nodes 1 and 3 (two are) and which cell indices were involved. In general, to evaluate entry A_{mn} directly, the computer program would require access to the number of cells touching each node, the indices of these cells, and a list of the other nodes associated with each of these cells. Fortunately, there is an easier (although conceptually indirect) way of computing the entries of **A**.

To minimize the amount of data needed to define the mesh structure, the entries of **A** can be evaluated on a cell-by-cell basis instead of a node-by-node approach. For any cell in the mesh, the connectivity array in the mesh data structure can be used to identify the x, y-coordinates of the vertices of that cell, after which the nine integrals arising from the possible combinations of the three testing functions and the three basis functions can be evaluated. Once these integrals have been evaluated, they make up the 3×3 *element matrix* associated with that cell. The nine numbers in the element matrix can be added to the appropriate locations in **A**, using the connectivity array to identify the global indices (i.e., the row and column locations in **A**) of the three nodes. The cell-by-cell approach permits the matrix entries to be constructed without knowing how many cells are shared by two adjacent basis functions. In addition, the element matrix entries depend only on the cell coordinates and can be conveniently evaluated in a standard subroutine.

For illustration, a hypothetical portion of the connectivity array associated with the mesh of Figure 3.5 is displayed in Table 3.1. Observe from the table that cell 3 is located between nodes 7, 1, and 5. The integrals over cell 3 can be evaluated and stored temporarily in the 3×3 element matrix. (The closed-form evaluation of the necessary integrals will be discussed in Section 3.7.) Element matrix entry E_{11} actually represents the portion of A_{77} that involves an integration over cell 3, while entry E_{12} represents the portion of A_{71}, and so on. Once the nine entries of the element matrix are evaluated in a standard subroutine, the connectivity array can be used to transfer them into the appropriate location within **A**.

TABLE 3.1 Portion of Connectivity Array Containing Relative Locations of Nodes and Cells in the Mesh Shown in Figure 3.5

Cell Number	Node 1	Node 2	Node 3
1	1	2	3
2	1	3	5
3	7	1	5
⋮	⋮	⋮	⋮
25	7	8	21
26	7	21	22
27	7	22	23
28	7	10	23

Note: Each row "points" to the indices of the three nodes associated with that cell.

As the matrix entries are evaluated, the sparse global matrix **A** is constructed and stored in some manner designed to minimize memory requirements and yet remain compatible with a suitable matrix solution algorithm. We defer a discussion of sparse matrix organization and solution to Chapter 4.

After $\mathbf{Ae} = \mathbf{b}$ has been solved to produce the coefficients of E_z at each node of the mesh, secondary calculations such as the bistatic scattering cross section $\sigma(\phi)$ can be performed. There are several different ways of obtaining the scattering cross section, including the direct integration over equivalent electric and magnetic sources distributed throughout the penetrable scatterer or located on its surface, as discussed in Section 1.11 and Chapter 2. However, since the boundary $\partial\Gamma$ is circular in this case, a simpler alternative is an eigenfunction expansion of the exterior fields. In the TM case, the total fields external to $\partial\Gamma$ have the form [5]

$$E_z(\rho, \phi) = \sum_{n=-\infty}^{\infty} j^{-n}[J_n(k\rho)e^{-jn\theta} + \alpha_n H_n^{(2)}(k\rho)]e^{jn\phi} \tag{3.39}$$

where θ defines the polar angle into which the incident plane wave propagates. The coefficients can be found from the decomposition of the boundary E_z-field according to

$$\alpha_n = \frac{j^n(1/2\pi)\displaystyle\int_0^{2\pi} E_z(a, \phi)e^{-jn\phi}\, d\phi - J_n(ka)e^{-jn\theta}}{H_n^{(2)}(ka)} \tag{3.40}$$

where a is the radius of the outer boundary. The scattering cross section is given by

$$\sigma_{\text{TM}}(\phi) = \frac{4}{k}\left|\sum_{n=-\infty}^{\infty} \alpha_n e^{jn\phi}\right|^2 \tag{3.41}$$

Similar expressions involving the H_z-field can be used to find the scattering cross section for the TE polarization. It should be noted that the eigenfunction expansion procedure seems more sensitive to error in the field values than the direct volume integration method of computing σ discussed in Section 1.11 [8]. Although apparently more accurate, the direct integration method is more difficult to implement because it requires additional pointer arrays to describe the cell-to-edge connectivity and involves volumetric integrations over Γ rather than integrals along the boundary $\partial\Gamma$.

To illustrate the accuracy and convergence of the numerical results produced by the formulation, consider a circular dielectric cylinder with one wavelength circumference ($ka = 1$) and relative permittivity $\varepsilon_r = 3$. Figures 3.5 and 3.6 illustrate typical triangular-cell models containing 29 and 51 nodes, respectively. A plot of the total E_z-field produced along a cut through the center of the dielectric cylinder by a TM wave incident in the $\hat{x}$ direction is shown in Figure 3.7. The numerical results obtained with linear basis and testing functions exhibit good agreement with the exact solution and improve as the cell density is increased. These numerical results were obtained with a 51-node model having $\partial\Gamma$ located at $\rho = 0.22\lambda_0$ and 83- and 117-node models having the outer boundary located at $\rho = 0.20\lambda_0$ (λ_0 is the free-space wavelength). Numerical values for the magnitude and phase of the total E_z-field at the cylinder center are provided in Table 3.2 for these results and results from a fourth cylinder model employing 29 nodes with the RBC imposed at $\rho = 0.25\lambda_0$ and a fifth model having 255 nodes with $\partial\Gamma$ located at $\rho = 0.20\lambda_0$. For each model, Table 3.2 shows the density of nodes within the scatterer and the largest cell edge.

Table 3.2 also presents the percentage error associated with each solution. A study of this error suggests that it decreases roughly in proportion to $O(\Delta^2)$ as $\Delta \to 0$, where Δ is

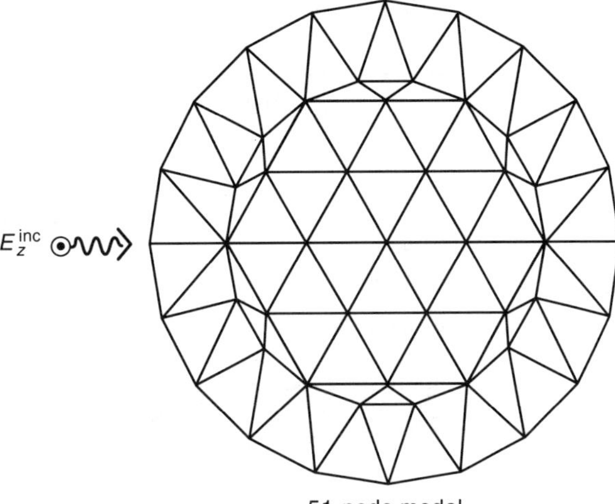

Figure 3.6 A 51-node triangular-cell model.

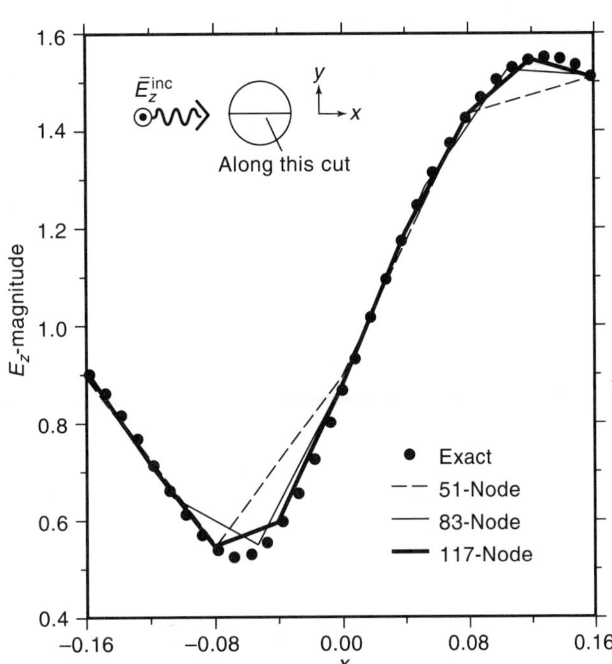

Figure 3.7 The TM electric field within a homogeneous dielectric cylinder with circumference $1.0\lambda_0$ and relative permittivity $\varepsilon_r = 3$. The exact solution is compared with several first-order finite-element results.

the largest edge in the finite-element mesh. This behavior is consistent with the theoretical interpolation error associated with piecewise-linear basis functions (Chapter 5).

Table 3.3 presents scattering cross section data for the cylinder having $ka = 1$ and $\varepsilon_r = 3$, obtained from a decomposition of the boundary fields according to (3.39)–(3.41). The numerical data appear to be converging toward the exact solution as the cell density is increased. These models have a density of unknowns in excess of 100 nodes/λ_d^2 and produce numerical results for scattering cross section exhibiting an error of less than 1 dB.

TABLE 3.2 Numerical Values of E_z-Field Produced at Center of Circular Dielectric Cylinder with $\varepsilon_r = 3$ and $ka = 1$

Model	Average Node Density (nodes/λ_d^2)	Longest Edge in Mesh (λ_0)	Magnitude of E_Z	Phase of E_Z (deg)	Percent Error
29-Node	54	0.163	0.992	−50.63	14.9
51-Node	130	0.093	0.915	−55.20	4.1
83-Node	230	0.057	0.903	−56.44	1.8
117-Node	356	0.054	0.898	−56.76	1.1
255-Node	683	0.032	0.895	−57.09	0.5
Exact	—	—	0.892	−57.36	—

Note: Results from five triangular-cell models are compared with the exact solution.

TABLE 3.3 Two-Dimensional Bistatic Scattering Cross Section[a] σ_{TM} for Circular Dielectric Cylinder with $ka = 1$ and $\varepsilon_r = 3$

ϕ (deg)	51-Node	83-Node	117-Node	255-Node	Exact
0	−0.24	0.09	0.17	0.24	0.289
30	−0.65	−0.37	−0.30	−0.23	−0.192
60	−1.87	−1.69	−1.64	−1.60	−1.567
90	−3.71	−3.62	−3.60	−3.57	−3.550
120	−5.56	−5.51	−5.51	−5.49	−5.469
150	−6.68	−6.62	−6.60	−6.58	−6.554
180	−6.97	−6.90	−6.88	−6.85	−6.819

[a] In decibels free-space wavelength.

Note: Numerical results obtained using four cylinder models are compared with the exact solution.

3.6 SCATTERING FROM CYLINDERS CONTAINING CONDUCTORS

To illustrate the treatment of conducting material imbedded in the region Γ, we consider the problem of a circular conducting cylinder of radius $0.25\lambda_0$ coated with a homogeneous layer of dielectric having $\varepsilon_r = 4$ and outer radius $0.30\lambda_0$. Two triangular-cell models will be employed having 22 and 44 nodes on the conductor and a total of 106 and 200 nodes, respectively. Both models contain three layers of cells around the conductor, two in the dielectric coating and a third outside the coating and terminating at $\rho = 0.33\lambda_0$.

For the TM polarization, the matrix **A** can be constructed as outlined in the preceding section, except that special treatment is required to account for the p.e.c. boundary. As discussed in Section 3.2, there is no need to assign unknowns for E_z along the conductor boundary $\partial\Gamma_c$, and there will be no basis or testing functions (and no rows or columns in the system matrix) associated with nodes located on $\partial\Gamma_c$. Consequently, the order of the global system will be smaller than the number of nodes in the mesh. (The matrix orders associated with the two models of the coated cylinder discussed above are 84 and 156, respectively.) Since the nodes do not exhibit a one-to-one relationship with rows and columns of the system matrix, a more complex numbering scheme must be used to associate node indices with row and column indices. Perhaps the simplest scheme is to order the node indices so that any nodes on p.e.c. boundaries appear at the end of the list, leaving a one-to-one correspondence between the remaining nodes and the row and column indices. For each triangular cell bordering $\partial\Gamma_c$, the nine-element matrix entries are again obtained in the usual manner. However, only entries associated with interior nodes are transferred to the global system. In order to facilitate the identification and exclusion of the appropriate element matrix entries, an additional pointer array is required to identify nodes on $\partial\Gamma_c$.

A more general approach for the TM case would instead employ a pointer assigning a row and column index to each node at which an unknown is located and simultaneously identifying nodes on conducting boundaries. A third approach is to introduce "dummy" unknowns associated with p.e.c. boundary nodes in order to preserve the one-to-one relationship between the node indices and the row and column indices at the price of a slightly larger matrix order. Since the coefficients are known a priori, the rows associated with boundary nodes must be replaced with rows from an identity matrix, and the associated entries of the right-hand side are set to zero in order to force those coefficients to vanish after matrix solution. Since these rows are treated specially, the additional pointer is still required to identify boundary nodes. Thus, there does not appear to be a computational advantage to the use of dummy unknowns.

Table 3.4 presents the TM scattering cross section σ_{TM} for the coated cylinder geometry. Similar data for the TE polarization (σ_{TE}) is presented in Table 3.5. The TE formulation is obtained by exchanging ε_r and μ_r and replacing E_z with H_z in Equation (3.25). For the TE polarization, the H_z-field values on $\partial\Gamma_c$ remain unknowns to be determined, and all nine entries of each element matrix contribute to **A**. The order of the system matrix is equal to the number of nodes in the mesh, and no additional pointers are needed to identify p.e.c. boundary nodes since they are not treated differently from the interior nodes. In essence, by placing a "hole" somewhere in the mesh and imposing no special boundary condition around it, the TE formulation will automatically treat that region as a perfect conductor.

TABLE 3.4 Bistatic Scattering Cross Section[a] σ_{TM} for Circular Conducting Cylinder with Radius $0.25\lambda_0$ Coated with Dielectric Layer Having $\varepsilon_r = 4$ and Outer Radius $0.3\lambda_0$

ϕ (deg)	106-Node	200-Node	Exact
0	5.45	5.41	5.372
30	3.79	3.77	3.748
60	0.05	0.05	0.043
90	−1.15	−1.19	−1.209
120	−0.73	−0.74	−0.749
150	−0.64	−0.64	−0.634
180	−0.63	−0.63	−0.627

[a] In decibels free-space wavelength.

Note: Numerical results obtained using two cylinder models are compared with the exact solution.

TABLE 3.5 Bistatic Scattering Cross Section[a] σ_{TE} for Circular Conducting Cylinder with Radius $0.25\lambda_0$ Coated with Dielectric Layer Having $\varepsilon_r = 4$ and Outer Radius $0.3\lambda_0$

ϕ(deg)	106-Node	200-Node	Exact
0	3.99	4.17	4.231
30	–0.11	0.00	0.034
60	–1.96	–1.91	–1.916
90	0.95	1.01	1.026
120	–1.83	–1.87	–1.864
150	–1.89	–1.86	–1.867
180	–0.19	–0.10	–0.107

[a] In decibels free-space wavelength.

Note: Numerical results obtained using two cylinder models are compared with the exact solution.

For the TE polarization, the current density J_t on the imbedded conductor can be obtained directly from the coefficients for H_z at nodes bordering the conductor. Figure 3.8 compares the surface current magnitude obtained using the 106-node model with the exact solution. The agreement is excellent. For the TM polarization, the surface current J_z cannot be obtained directly from the coefficients for E_z but can be estimated by differentiating the E_z-field to obtain H_t. The process of differentiating reduces the accuracy of J_z compared with E_z. An alternative approach for calculating J_z is explored in Prob. P3.11.

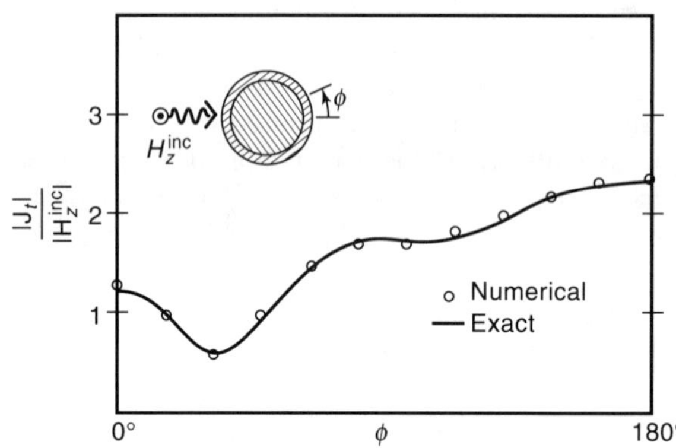

Figure 3.8 Comparison of the exact and numerical TE electric current density induced on the conducting surface of a circular p.e.c. cylinder imbedded within a circular dielectric shell by a uniform plane wave. The geometry is identical to that used in Tables 3.4 and 3.5.

3.7 EVALUATION OF VOLUMETRIC INTEGRALS FOR THE MATRIX ENTRIES [7]

The volume integral terms appearing in Equation (3.31) are common to scalar finite-element discretizations of the Laplace, Poisson, and Helmholtz equations and have been discussed in a variety of texts. In this section, we consider the evaluation of

$$A_{pq}^{(1)} = \iint_{\Gamma} \nabla B_p \cdot \nabla B_q \, dx \, dy \tag{3.42}$$

and

$$A_{pq}^{(2)} = \iint_{\Gamma} B_p B_q \, dx \, dy \tag{3.43}$$

where $\{B_q\}$ are linear pyramid functions (Figure 3.4) and the domain of integration is a single triangular cell. As discussed in preceding sections, the matrix **A** can be constructed indirectly by scanning through the mesh and evaluating the necessary integrals on a cell-by-cell basis. We will assume that the constitutive parameters ε_r and μ_r are constant within each cell of the mesh and therefore omit them from the integrals over a single cell.

Because the integrand consists of polynomials, the integrations can be evaluated in closed form. The evaluation over triangular cells is described in detail by Silvester and Ferrari [7]. It is convenient to carry out the analysis in terms of *simplex* coordinates $\{L_1, L_2, L_3\}$. These coordinates specify the position of a point within a triangle by giving the relative perpendicular distance measured from each side to the point, with the distance expressed as a fraction of the triangle altitude (Figure 3.9). Lines of constant L_i are parallel to side i of the triangle. (Each simplex coordinate is also the ratio of the area of the respective interior triangle formed by the point to the area of the entire triangle, hence the alternate name *local-area coordinates*.) The simplex coordinates are related to Cartesian coordinates by

$$x = L_1 x_1 + L_2 x_2 + L_3 x_3 \tag{3.44}$$

$$y = L_1 y_1 + L_2 y_2 + L_3 y_3 \tag{3.45}$$

where (x_i, y_i) are the coordinates of the ith vertex (see Figure 3.9). Since only two of the area coordinates can be independent, the additional condition

$$L_1 + L_2 + L_3 = 1 \tag{3.46}$$

is applied to yield

$$\begin{bmatrix} x \\ y \\ 1 \end{bmatrix} = \begin{bmatrix} x_1 & x_2 & x_3 \\ y_1 & y_2 & y_3 \\ 1 & 1 & 1 \end{bmatrix} \begin{bmatrix} L_1 \\ L_2 \\ L_3 \end{bmatrix} \tag{3.47}$$

Equation (3.47) can be inverted to produce

$$L_i = \frac{1}{2A}(a_i + b_i x + c_i y) \tag{3.48}$$

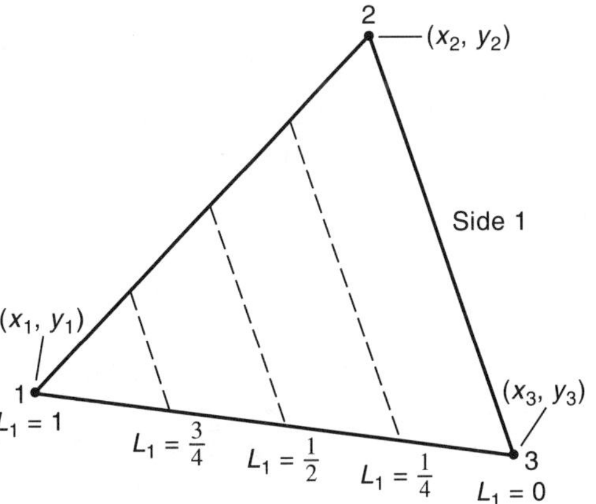

Figure 3.9 Local-area coordinate L_1 at different points throughout a triangular element.

where

$$a_i = x_{i+1}y_{i+2} - x_{i+2}y_{i+1} \tag{3.49}$$

$$b_i = y_{i+1} - y_{i-1} \tag{3.50}$$

$$c_i = x_{i-1} - x_{i+1} \tag{3.51}$$

and A represents the area of the triangle, that is,

$$A = \left|\tfrac{1}{2}(a_1 + a_2 + a_3)\right| = \left|\tfrac{1}{2}(b_{i+1}c_{i-1} - b_{i-1}c_{i+1})\right| \tag{3.52}$$

The index i in Equations (3.49)–(3.52) assumes values 1, 2, and 3 cyclically, so that if $i = 3$, then $i + 1 = 1$.

Consider the evaluation of

$$A_{pq}^{(1)} = \iint_{\Gamma} \nabla B_p \cdot \nabla B_q \, dx \, dy \tag{3.53}$$

over one triangular cell Γ_t, where indices p and q range from 1 to 3 and represent the local index of the basis or testing function. (In other words, we seek to evaluate the 3×3 element matrix. These entries will subsequently be transferred into appropriate rows and columns of the global system using the connectivity matrix to map local indices to global indices.) Within the cell, linear pyramid basis functions are identical to the simplex coordinates themselves and can be expressed as

$$B_1(L_1, L_2, L_3) = L_1 \tag{3.54}$$

$$B_2(L_1, L_2, L_3) = L_2 \tag{3.55}$$

$$B_3(L_1, L_2, L_3) = L_3 \tag{3.56}$$

Since

$$\nabla L_i = \frac{1}{2A}(\hat{x}b_i + \hat{y}c_i) \tag{3.57}$$

the integrand of Equation (3.53) is constant and we immediately obtain

$$A_{pq}^{(1)} = \frac{b_p b_q + c_p c_q}{4A} \tag{3.58}$$

as an expression for the entries of the 3×3 element matrix.

The element matrix for the integral

$$A_{pq}^{(2)} = \iint_{\Gamma} B_p B_q \, dx \, dy \tag{3.59}$$

can be evaluated after transforming the integral from Cartesian coordinates to the simplex coordinates $\{L_1, L_2, L_3 = 1 - L_1 - L_2\}$ using the Jacobian relationship

$$dx \, dy = dL_1 \, dL_2 \frac{\partial(x, y)}{\partial(L_1, L_2)} = 2A \, dL_1 \, dL_2 \tag{3.60}$$

For the basis functions defined in (3.54)–(3.56), the integral

$$A_{pq}^{(2)} = 2A \iint_{\Gamma_t} B_p B_q \, dL_1 \, dL_2 \tag{3.61}$$

is a special case of the general formula [7]

$$I = \iint_{\Gamma_t} L_1^a L_2^b L_3^c \, dL_1 \, dL_2 = \frac{a!b!c!}{(a + b + c + 2)!} \tag{3.62}$$

where a, b, and c represent integer powers. Therefore,

$$A_{pq}^{(2)} = \begin{cases} \frac{1}{6}A & p = q \\ \frac{1}{12}A & p \neq q \end{cases} \tag{3.63}$$

Thus, the entries of both element matrices are simple functions of the triangular-cell geometry.

In summary, the volumetric integrals within the matrix **A** can be constructed on a cell-by-cell basis using element matrix entries from the simple closed-form expressions given in (3.58) and (3.63). These entries are subsequently multiplied with the constitutive parameters and added to appropriate locations within **A** using the connectivity matrix to identify the global row and column indices. Appendix D describes a computer program incorporating this feature.

3.8 LOCAL RADIATION BOUNDARY CONDITIONS ON A CIRCULAR SURFACE: THE BAYLISS–TURKEL CONDITIONS [9]

The Helmholtz equation admits solutions for the scattered field that represent both outgoing and incoming waves. In classical electromagnetic analysis, a form of the Sommerfeld radiation condition

$$\lim_{\rho\to\infty} \frac{\partial E_z^s}{\partial \rho} = -jkE_z^s \tag{3.64}$$

is imposed on the boundary at infinity to suppress the inward-propagating part of the solution. Equation (3.64) is a *local* condition, since the derivative of the scattered field at a point on the boundary depends only on the field at that same point. Unfortunately, the Sommerfeld condition must be applied an infinite distance from the scatterer and is therefore not practical for near-field use. Although the exact RBC from Equation (3.17) is applicable for finite ρ, it is a *global* condition that couples field information around the entire boundary $\partial\Gamma$. In fact, any exact RBC valid for near-field use must be global. In an attempt to avoid the fill-in associated with exact global radiation conditions, several approximate RBCs have been developed that can be localized in a similar manner as the Sommerfeld condition. One family of near-field RBCs has been proposed by Bayliss and Turkel [9], and we review the derivation of those conditions in this section.

To derive a radiation or "absorbing" boundary condition that can be imposed in the near zone of a scatterer, note that an outward-propagating two-dimensional field may be expressed in the asymptotic form, valid for large ρ,

$$E_z^s(\rho, \phi) \approx \frac{e^{-jk\rho}}{\sqrt{\rho}} \left(E_0(\phi) + \frac{E_1(\phi)}{\rho} + \frac{E_2(\phi)}{\rho^2} + \frac{E_3(\phi)}{\rho^3} + \cdots \right) \tag{3.65}$$

or, in a more compact notation,

$$E_z^s \approx \frac{e^{-jk\rho}}{\sqrt{\rho}} \sum_{n=0}^{\infty} \frac{E_n(\phi)}{\rho^n} \tag{3.66}$$

As $\rho \to \infty$, Equation (3.65) satisfies the Sommerfeld condition. However, imposing the

Sommerfeld condition at finite ρ only forces the resulting solution to agree with the first term of (3.65), as can be demonstrated by writing

$$\frac{\partial E_z^s}{\partial \rho} \approx \frac{e^{-jk\rho}}{\sqrt{\rho}} \sum_{n=0}^{\infty} \left[-jk\frac{E_n(\phi)}{\rho^n} - \left(n + \frac{1}{2}\right) \frac{E_n(\phi)}{\rho^{n+1}} \right] \tag{3.67}$$

and combining Equations (3.66) and (3.67) to arrive at

$$\frac{\partial E_z^s}{\partial \rho} + jkE_z^s \approx \frac{e^{-jk\rho}}{\sqrt{\rho}} \sum_{n=0}^{\infty} \left[-\left(n + \frac{1}{2}\right) \frac{E_n(\phi)}{\rho^{n+1}} \right] \approx O(\rho^{-3/2}) \tag{3.68}$$

The Sommerfeld condition produces a spurious reflection of asymptotic order $O(\rho^{-3/2})$. In fact, since the residual produced by an outward-propagating wave is $O(\rho^{-3/2})$ while that produced by an inward-propagating wave is $O(\rho^{-1/2})$, the Sommerfeld radiation condition only provides an $O(\rho)$ degree of discrimination between incoming and outgoing waves. Although sufficient if applied at infinity, this RBC is not usually adequate for near-field use.

Bayliss and Turkel have derived a family of higher order radiation conditions that force the scattered field to agree with any number of terms from Equation (3.65) and provide a much greater degree of discrimination between inward-propagating and outward-propagating waves [9]. Their approach is based on the observation that the leading-order term in the residual on the right-hand side of (3.68) is

$$\frac{e^{-jk\rho}}{\sqrt{\rho}} \left(-\frac{1}{2} \frac{E_0(\phi)}{\rho} \right) \tag{3.69}$$

Since this quantity happens to equal the first term of (3.65) divided by -2ρ, we are motivated to inspect

$$\frac{\partial E_z^s}{\partial \rho} + jkE_z^s + \frac{E_z^s}{2\rho} \approx \frac{e^{-jk\rho}}{\sqrt{\rho}} \sum_{n=1}^{\infty} \left(-n \frac{E_n(\phi)}{\rho^{n+1}} \right) \approx O(\rho^{-5/2}) \tag{3.70}$$

In the asymptotic sense, an improved radiation condition can be obtained by equating the left-hand side of (3.70) with zero, to obtain the "first-order" RBC

$$\left(jk + \frac{\partial}{\partial \rho} + \frac{1}{2\rho} \right) E_z^s = 0 \tag{3.71}$$

Imposing (3.71) forces the solution to agree with the first two terms of the asymptotic expansion given in Equation (3.65), which should be an improvement over the Sommerfeld condition for finite ρ. Equation (3.71) provides an $O(\rho^2)$ degree of discrimination between the inward-propagating and outward-propagating waves.

The leading-order residual in (3.70) is

$$\frac{e^{-jk\rho}}{\sqrt{\rho}} \left(-\frac{E_1(\phi)}{\rho^2} \right) \tag{3.72}$$

Since (3.72) is not directly proportional to the leading-order term in the series for the scattered field, we are unable to continue combining simple multiples of the field to obtain a reduction in the residual. However, observe that

$$\begin{aligned}\frac{\partial}{\partial \rho}\left(\frac{\partial E_z^s}{\partial \rho} + jkE_z^s + \frac{E_z^s}{2\rho}\right) \\ \approx \frac{e^{-jk\rho}}{\sqrt{\rho}} \sum_{n=1}^{\infty}\left(jkn\frac{E_n(\phi)}{\rho^{n+1}} + \frac{n}{2}\frac{E_n(\phi)}{\rho^{n+2}} + n(n+1)\frac{E_n(\phi)}{\rho^{n+2}}\right)\end{aligned} \tag{3.73}$$

The leading-order term of the residual series in (3.73) is

$$\frac{e^{-jk\rho}}{\sqrt{\rho}}\left(jk\frac{E_1(\phi)}{\rho^2}\right) \tag{3.74}$$

This immediately motivates the inspection of

$$\begin{aligned}\left(\frac{\partial}{\partial \rho} + jk\right)\left(\frac{\partial E_z^s}{\partial \rho} + jkE_z^s + \frac{E_z^s}{2\rho}\right) &\approx \frac{e^{-jk\rho}}{\sqrt{\rho}} \sum_{n=1}^{\infty}\left[n\left(n + \frac{3}{2}\right)\frac{E_n(\phi)}{\rho^{n+2}}\right] \\ &\approx O(\rho^{-7/2})\end{aligned} \tag{3.75}$$

which has leading-order residual

$$\frac{e^{-jk\rho}}{\sqrt{\rho}}\left(\frac{5}{2}\frac{E_1(\phi)}{\rho^3}\right) \tag{3.76}$$

Continuing this procedure, we consider

$$\begin{aligned}\left(\frac{\partial}{\partial \rho} + jk + \frac{5}{2\rho}\right)\left(\frac{\partial E_z^s}{\partial \rho} + jkE_z^s + \frac{E_z^s}{2\rho}\right) &\approx \frac{e^{-jk\rho}}{\sqrt{\rho}} \sum_{n=1}^{\infty}\left(n(n-1)\frac{E_n(\phi)}{\rho^{n+2}}\right) \\ &\approx O(\rho^{-9/2})\end{aligned} \tag{3.77}$$

From the size of the residual in (3.77), we conclude that imposing the "second-order" RBC

$$\left(jk + \frac{\partial}{\partial \rho} + \frac{5}{2\rho}\right)\left(jk + \frac{\partial}{\partial \rho} + \frac{1}{2\rho}\right)E_z^s = 0 \tag{3.78}$$

forces the numerical solution to agree with the first four terms of (3.65). Equation (3.78) provides an $O(\rho^4)$ degree of discrimination between the inward-propagating and outward-propagating waves and suppresses incoming waves down to $O(\rho^{-9/2})$.

The procedure outlined above can be continued, leading to the general Nth order condition [9]

$$\prod_{n=1}^{N}\left(jk + \frac{\partial}{\partial \rho} + \frac{4n-3}{2\rho}\right)E_z^s = 0 \tag{3.79}$$

The Nth-order Bayliss–Turkel RBC forces the solution to agree with the first $2N$ terms of the asymptotic expansion given in (3.65) and eliminates spurious reflections from the artificial boundary down to terms of order $O(\rho^{-2N-1/2})$.

We will be primarily interested in the second-order Bayliss–Turkel condition appearing in Equation (3.78). Using the scalar Helmholtz equation in cylindrical coordinates, this condition can be rewritten in the form

$$\frac{\partial E_z^s}{\partial \rho} = \alpha(\rho) E_z^s + \beta(\rho) \frac{\partial^2 E_z^s}{\partial \phi^2} \tag{3.80}$$

where

$$\alpha(\rho) = \frac{-jk - 3/2\rho + j3/8k\rho^2}{1 - j/k\rho} \tag{3.81}$$

and

$$\beta(\rho) = \frac{-j/2k\rho^2}{1 - j/k\rho} \tag{3.82}$$

Because this boundary condition involves only second-order tangential derivatives, it is local in nature and does not couple the fields around the entire boundary. In fact, since it has an order of differentiation identical with that of the Helmholtz equation, it should lead to a comparable degree of sparsity after discretization. If used with linear pyramid expansion and testing functions, (3.80) will only couple information between adjacent cells around the periphery of Γ and introduce no additional fill-in beyond that already present due to internal interactions.

Except in the limit as $\rho \to \infty$, Equation (3.80) is approximate. The solution produced using the second-order RBC can be written in the vicinity of the boundary as

$$\begin{aligned} E_z^s = &\frac{e^{-jk\rho}}{\sqrt{\rho}} \left(E_0(\phi) + \frac{E_1(\phi)}{\rho} + \frac{E_2(\phi)}{\rho^2} + \frac{E_3(\phi)}{\rho^3} \right) \\ &+ \frac{e^{-jk\rho}}{\sqrt{\rho}} \left(\frac{A_4(\phi)}{\rho^4} + \cdots \right) + \frac{e^{+jk\rho}}{\sqrt{\rho}} \left(\frac{B_4(\phi)}{\rho^4} + \cdots \right) \end{aligned} \tag{3.83}$$

where $\{E_n\}$ are the "correct" outward-propagating terms produced by the Helmholtz equation and $\{A_n\}$ and $\{B_n\}$ denote terms that have been corrupted by the approximate RBC. Clearly, the Bayliss–Turkel condition will produce an accurate result only if the second and third summations in (3.83) are small compared with the first. This is easily accomplished in principle by making ρ large, but to minimize the number of unknowns, it is desirable to place $\partial\Gamma$ relatively close to the scatterer surface. As we shall demonstrate, accurate results can sometimes be obtained with boundaries almost circumscribing the scatterer. For these examples, the convergence rate of (3.65) appears to be very fast, so that terms such as A_4 and B_4 are negligible even for small ρ. However, in other situations the convergence rate of (3.65) may be quite slow. Canning has demonstrated the slow convergence of (3.65) for scattered waves emanating from locations other than the origin of the coordinate system [10]. In other words, it appears that the Bayliss–Turkel RBC will readily absorb a wave that is normally incident to the boundary but not a wave that approaches $\partial\Gamma$ at angles close to grazing. Consequently, if the geometry of interest happens to contain strong scattering centers located far from the origin, it may not be possible to locate $\partial\Gamma$ immediately adjacent to the scatterer. Numerical experimentation can provide useful feedback concerning the sensitivity of the results to boundary location.

An alternate approach for systematically studying the accuracy of the local RBC employs cylindrical harmonics [11]. An outward-propagating harmonic has the form

$$E_z^s(\rho, \phi) = E_0 H_n^{(2)}(k\rho) e^{jn\phi} \tag{3.84}$$

and will satisfy any exact radiation condition. For this nth harmonic, the second-order Bayliss–Turkel condition can be interpreted as an attempt to approximate the exact relationship

$$\frac{\partial E_z^s}{\partial \rho} = \frac{kH_n^{(2)\prime}(k\rho)}{H_n^{(2)}(k\rho)} E_z^s \tag{3.85}$$

with the approximate relationship

$$\frac{\partial E_z^s}{\partial \rho} \cong [\alpha(\rho) - n^2 \beta(\rho)] E_z^s \tag{3.86}$$

Equations (3.85) and (3.86) can be compared as a function of the radius of the outer boundary to determine the range of validity of the local RBC. Figure 3.10 presents a comparison for a boundary of radius $k\rho = 51$. Although the lower order harmonics satisfy the Bayliss–Turkel condition very well, the higher order harmonics do not. The agreement degenerates in the vicinity of $n = k\rho$, which for a given harmonic is the approximate region where the predominantly evanescent character ($k\rho < n$) of the Hankel functions changes to a propagating behavior ($k\rho > n$). The Bayliss–Turkel condition was based on the propagating form presented in Equation (3.65), and it is not surprising that the second-order RBC will fail for an evanescent field. In fact, the Nth-order RBC would fail in an evanescent region even for arbitrarily large N. Consequently, for the Bayliss–Turkel RBC to work well, it appears that $\partial\Gamma$ must always be located beyond the evanescent region of the scatterer.

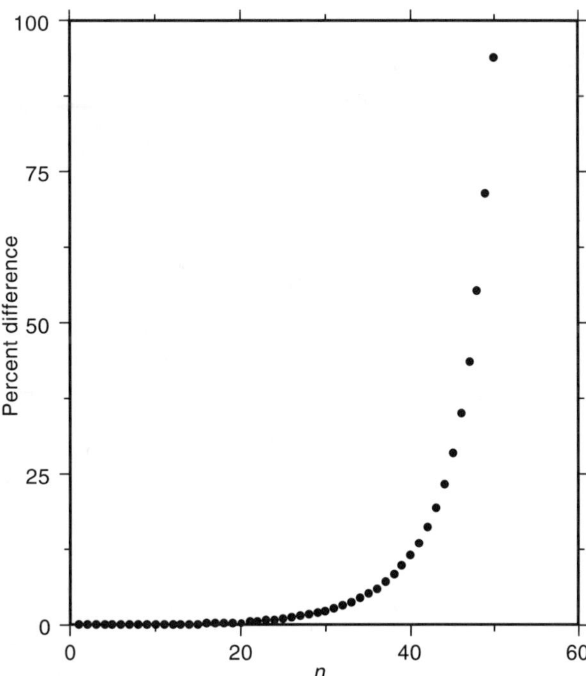

Figure 3.10 Comparison of Equations (3.85) and (3.86) as a function of the harmonic index n for $k\rho = 51$.

3.9 OUTWARD-LOOKING FORMULATION COMBINING THE SCALAR HELMHOLTZ EQUATION AND THE SECOND-ORDER BAYLISS–TURKEL RBC [12]

To illustrate the use of the local RBC, we return to the problem of TM-wave scattering from an inhomogeneous cylinder. Equation (3.80) can be substituted into the boundary integral of (3.5) and terms involving the incident field added in order to preserve the total field as the primary unknown. After integration by parts is employed to eliminate the second-order differential operator, the equation has the form

$$\iint_\Gamma \left(\frac{1}{\mu_r} \nabla T \cdot \nabla E_z - k^2 \varepsilon_r T E_z \right) dx\, dy - \int_{\partial\Gamma} \left(\alpha T E_z - \beta \frac{\partial T}{\partial \phi} \frac{\partial E_z}{\partial \phi} \right) \rho\, d\phi$$
$$= \int_{\partial\Gamma} T \left(\frac{\partial E_z^{\text{inc}}}{\partial \rho} - \alpha E_z^{\text{inc}} - \beta \frac{\partial^2 E_z^{\text{inc}}}{\partial \phi^2} \right) \rho\, d\phi \tag{3.87}$$

Equation (3.87) is an outward-looking formulation that can be used as an alternative to (3.25). Although the equation is approximate because of the second-order Bayliss–Turkel RBC, this formulation preserves the sparsity inherent in the differential equation and usually requires much less storage and computation than (3.25).

The finite-element discretization of Equation (3.87) proceeds in a manner similar to that carried out in Section 3.3 and produces a matrix equation

$$\mathbf{A}\mathbf{e} = \mathbf{b} \tag{3.88}$$

with entries given by

$$A_{mn} = \iint_\Gamma \left(\frac{1}{\mu_r} \nabla B_m \cdot \nabla B_n - k^2 \varepsilon_r B_m B_n \right) dx\, dy$$
$$- \int_{\partial\Gamma} \left(\alpha B_m B_n - \beta \frac{\partial B_m}{\partial \phi} \frac{\partial B_n}{\partial \phi} \right) \rho\, d\phi \tag{3.89}$$

and

$$b_m = \int_{\partial\Gamma} B_m \left(\frac{\partial E_z^{\text{inc}}}{\partial \rho} - \alpha E_z^{\text{inc}} - \beta \frac{\partial^2 E_z^{\text{inc}}}{\partial \phi^2} \right) \rho\, d\phi \tag{3.90}$$

where α and β are defined in Equations (3.81) and (3.82).

The evaluation of the volume integral terms in (3.89) for linear pyramid expansion and testing functions has already been discussed in Section 3.7. The additional calculations arising from boundary integrals

$$A_{pq}^{(3)} = \int_{\partial\Gamma} B_p B_q\, dt \tag{3.91}$$

and

$$A_{pq}^{(4)} = \int_{\partial\Gamma} \frac{\partial B_p}{\partial t} \frac{\partial B_q}{\partial t}\, dt \tag{3.92}$$

can be arranged in the form of 2×2 element matrices. For piecewise-linear basis functions spanning an interval of length w on the boundary, these integrals are easily evaluated to produce

$$A_{pq}^{(3)} = \begin{cases} \frac{1}{3}w & p = q \\ \frac{1}{6}w & \text{otherwise} \end{cases} \tag{3.93}$$

and

$$A_{pq}^{(4)} = \begin{cases} \dfrac{1}{w} & p = q \\ -\dfrac{1}{w} & \text{otherwise} \end{cases} \tag{3.94}$$

Equation (3.90) can be determined for a plane-wave incident field of the form

$$E_z^{\text{inc}} = e^{-jk(x\cos\theta + y\sin\theta)} \tag{3.95}$$

by incorporating the intermediate result

$$\begin{aligned} &\frac{\partial E_z^{\text{inc}}}{\partial \rho} - \alpha E_z^{\text{inc}} - \beta \frac{\partial^2 E_z^{\text{inc}}}{\partial \phi^2} \\ &\quad = -[\alpha + (1 + \beta\rho) jk\cos(\theta - \phi) - \beta k^2 \rho^2 \sin^2(\theta - \phi)] e^{-jk\rho\cos(\theta-\phi)} \end{aligned} \tag{3.96}$$

and evaluating the remaining integral.

If linear pyramid functions are employed for basis and testing, **A** involves only nearest-neighbor interactions and has a high degree of sparsity. Figure 3.11 presents the sparsity pattern for the cylinder model previously illustrated in Figure 3.5, assuming linear basis and testing functions. The right-hand side of (3.90) is also sparse, as the nonzero entries are restricted to rows associated with nodes on the boundary.

To illustrate the validity of the formulation incorporating the second-order Bayliss–Turkel RBC, we return to the coated conducting cylinder previously considered in Section

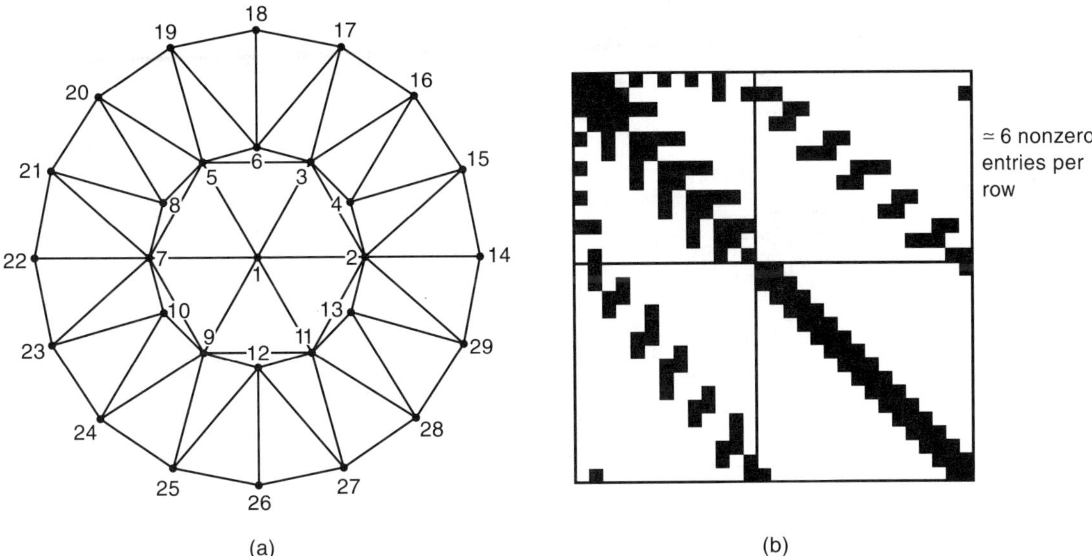

Figure 3.11 Triangular-cell mesh and the associated matrix sparsity pattern obtained with linear pyramid functions and the second-order Bayliss–Turkel RBC.

3.6. The circular conductor has radius $\rho = 0.25\lambda_0$ and is surrounded by a dielectric cladding with $\varepsilon_r = 4$ and outer radius $\rho = 0.30\lambda_0$. Tables 3.6 and 3.7 present the two-dimensional scattering cross section as a function of the location of the outer boundary $\partial\Gamma$ for triangular-cell models having 44 nodes on the conductor and two layers of cells in the cladding. These results are compared with the numerical solutions from Tables 3.4 and 3.5, which were obtained using the "exact" eigenfunction RBC. For this example, the TM results are in good agreement with the exact solutions even when $\partial\Gamma$ is located at $\rho = 0.33\lambda_0$ and do not seem very sensitive to the placement of the boundary. The TE results exhibit a larger amount of error for boundary locations close to the scatterer but consistently improve as the boundary radius increases.

TABLE 3.6 Bistatic Scattering Cross Section[a] σ_{TM} for Circular Conducting Cylinder with Radius $0.25\lambda_0$ Coated with Dielectric Layer Having $\varepsilon_r = 4$ and Outer Radius $0.3\lambda_0$

ϕ (deg)	$\rho = 0.33\,\lambda_0$	$\rho = 0.37\,\lambda_0$	$\rho = 0.41\,\lambda_0$	Exact RBC
0	5.42	5.42	5.40	5.41
30	3.76	3.76	3.76	3.76
60	0.05	0.08	0.10	0.05
90	−1.15	−1.15	−1.15	−1.19
120	−0.78	−0.79	−0.78	−0.74
150	−0.69	−0.68	−0.66	−0.64
180	−0.67	−0.66	−0.65	−0.63

[a] In decibels free-space wavelength.

Note: Numerical results obtained using the Bayliss–Turkel RBC located at three different radii are compared with the results obtained using the exact RBC imposed at $\rho = 0.33\,\lambda_0$.

TABLE 3.7 Bistatic Scattering Cross Section[a] σ_{TE} for Circular Conducting Cylinder with Radius $0.25\lambda_0$ Coated with Dielectric Layer Having $\varepsilon_r = 4$ and Outer Radius $0.3\lambda_0$

ϕ (deg)	$\rho = 0.33\,\lambda_0$	$\rho = 0.37\,\lambda_0$	$\rho = 0.41\,\lambda_0$	Exact RBC
0	4.43	4.32	4.23	4.17
30	0.12	0.03	−0.04	0.00
60	−1.37	−1.63	−1.76	−1.91
90	1.00	0.94	0.93	1.01
120	−2.51	−2.24	−2.09	−1.87
150	−1.64	−1.71	−1.77	−1.86
180	0.35	0.07	−0.08	−0.10

[a] In decibels free-space wavelength.

Note: Numerical results obtained using the Bayliss–Turkel RBC located at three different radii are compared with the results obtained using the exact RBC imposed at $\rho = 0.33\,\lambda_0$.

As a second example, consider a TE wave incident upon a hollow circular dielectric shell having $\varepsilon_r = 4$, inner radius $0.25\lambda_0$, and outer radius $0.30\lambda_0$. Figure 3.12 shows the

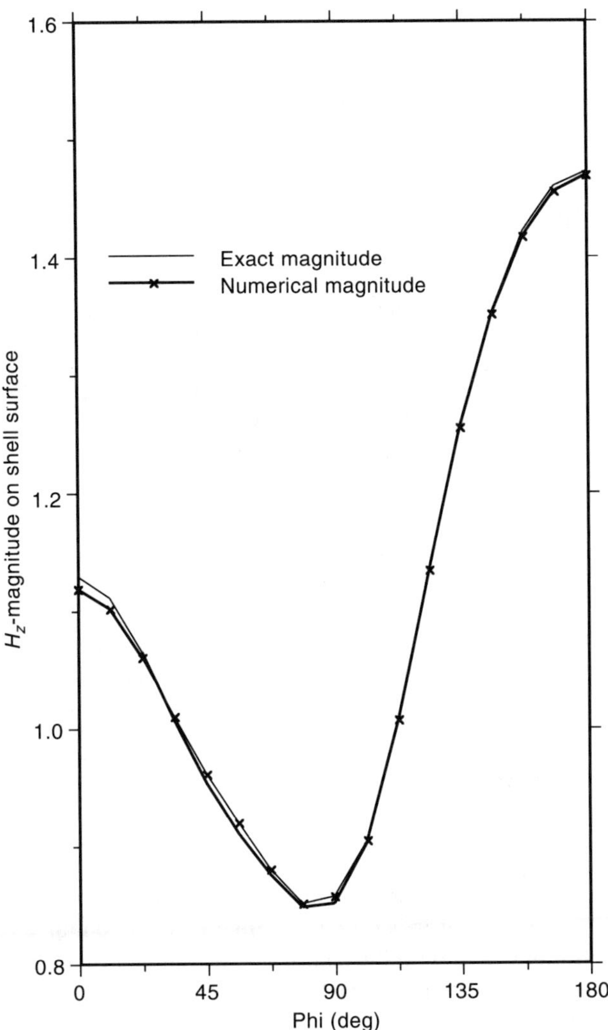

Figure 3.12 The TE magnetic field induced by a uniform plane wave along the surface of a hollow dielectric shell with $\varepsilon_r = 4$, $r_{\text{inner}} = 0.25\lambda_0$, and $r_{\text{outer}} = 0.30\lambda_0$.

magnitude of H_z around the outer surface of the cylinder. The numerical result exhibits excellent agreement with the exact analytical solution. The numerical solution was obtained from a 237-node model that employed a single layer of 62 triangular cells to represent the shell. The region inside the shell is discretized into 138 cells, and three exterior layers containing a total of 228 cells are used to place $\partial\Gamma$ at $\rho = 0.42\lambda_0$. The largest cell edge in the entire model has length $0.074\lambda_0$.

Figure 3.13 presents a comparison of numerical solutions for the magnetic field produced on the surface of a homogeneous triangular cylinder with $\varepsilon_r = 5$ by an incident TE plane wave. The result from the differential equation formulation incorporating the second-order Bayliss–Turkel RBC is compared with the numerical solution produced by the volume MFIE formulation of Section 2.7 applied to the identical triangular-cell model.

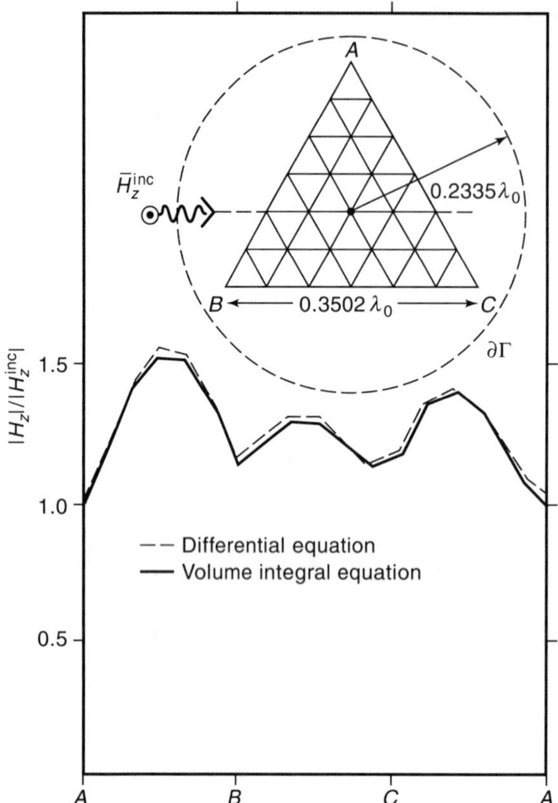

Figure 3.13 The TE magnetic field induced by a uniform plane wave along the surface of a triangular dielectric cylinder with $\varepsilon_r = 5$. The radiation boundary is located at radius $0.2335\lambda_0$. The first-order finite-element result is compared to that produced by the volume MFIE of Section 2.7. After [19]. ©1989 VSP BV.

The numerical results are almost indistinguishable, even though the Bayliss–Turkel condition is applied on a boundary $\partial\Gamma$ that almost circumscribes the scatterer. Since the outer boundary is circular, it is necessary to employ additional unknowns with the differential equation formulation to fill the computational domain Γ. However, as compared with the volume integral equation approach, the additional unknowns are offset by a reduction in the overall storage due to the sparsity of the matrix **A** in (3.88) and a significant reduction in the computational effort required to fill the matrix.

Although the Bayliss–Turkel RBC is approximate, the numerical solutions produced by the differential equation formulation appear reasonably accurate as long as several layers of cells are used to separate the boundary $\partial\Gamma$ from the scatterer surface. While the smattering of results presented in this section does not constitute a comprehensive validation, the high degree of sparsity in the matrix **A** and the resulting computational efficiency provide strong motivation for the general-purpose use of a local RBC. Additional numerical results illustrating the Bayliss–Turkel RBC will be presented in Section 3.12. A computer program implementing this formulation is described in Appendix D.

Like the eigenfunction RBC developed in Section 3.3, the Bayliss–Turkel condition requires a circular outer boundary $\partial\Gamma$. In an attempt to reduce the overall size of the computational domain, recent work has investigated similar RBCs that can be imposed on general boundary shapes [13, 14]. These conditions are also local and seem to introduce the same level of error as the Bayliss–Turkel RBC. In the following section, we consider more accurate global RBCs that can be imposed on general boundaries.

3.10 EXACT NEAR-ZONE RADIATION BOUNDARY CONDITIONS FOR SURFACES OF GENERAL SHAPE

The formulations presented in Sections 3.4 and 3.9 are restricted to cylinder models that terminate on circular boundaries. While it is always possible in principle to surround a scatterer with a region of free space, an unacceptable number of additional unknowns may be required to extend the computational domain around an elongated geometry (such as an airfoil) out to a circle. In other situations, such as a scatterer in the near field of an antenna, the primary source may be located so close to the scatterer surface that it may not be possible to incorporate a circular boundary. For these situations, alternative radiation boundary conditions can be obtained from surface integral equations. Although these are exact prior to discretization, they suffer from the fact that they are also global conditions and create fill-in beyond that normally associated with the discrete form of a differential operator. However, since they can circumscribe the scatterer without introducing substantial error, these RBCs may sometimes be able to reduce the number of unknowns required with the previous formulations enough to compensate for the additional fill-in.

Radiation boundary conditions can be developed using the surface equivalence principle from Section 1.6. Refer once more to the cylindrical geometry of Figure 3.1, which depicts a scatterer surrounded by a surface $\partial\Gamma$. In the exterior region the total fields can be produced by the superposition of the original "incident" fields and "scattered" fields produced by equivalent secondary sources located on the surface $\partial\Gamma$ and radiating in free space. For the TM polarization, these equivalent sources are

$$J_z = \hat{z} \cdot \hat{n} \times (\hat{t} H_t) = H_t = \frac{1}{jk\eta} \frac{\partial E_z}{\partial n} \quad \text{on } \partial\Gamma \tag{3.97}$$

$$K_t = \hat{t} \cdot (\hat{z} E_z) \times \hat{n} = E_z \quad \text{on } \partial\Gamma \tag{3.98}$$

For the TE polarization, equivalent sources for the scattered fields are

$$J_t = \hat{t} \cdot \hat{n} \times (\hat{z} H_z) = -H_z \quad \text{on } \partial\Gamma \tag{3.99}$$

$$K_z = \hat{z} \cdot (\hat{t} E_t) \times \hat{n} = -E_t = -\frac{j\eta}{k} \frac{\partial H_z}{\partial n} \quad \text{on } \partial\Gamma \tag{3.100}$$

The unit vector $\hat{n}$ is the outward normal vector to the surface, and $\hat{t}$ is the tangent vector defined so that $\hat{n} \times \hat{t} = \hat{z}$. The parameters η and k denote the intrinsic impedance and wavenumber, respectively, of the exterior medium. The equivalent exterior problem is depicted in Figure 3.14.

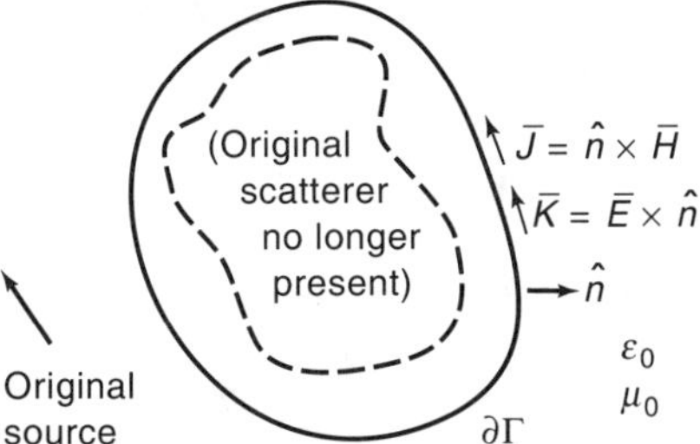

Figure 3.14 Equivalent exterior problem showing equivalent sources located on the radiation boundary $\partial\Gamma$.

The introduction of secondary sources $\bar{J}$ and $\bar{K}$ is merely an intermediate step in the derivation of integral equations. It was shown in Chapters 1 and 2 that the secondary sources must satisfy the surface EFIE and MFIE (for the TM and TE polarizations, respectively) given by

$$E_z^{\rm inc}(t) = K_t(t) + \hat{z} \cdot \nabla \times \int_{\partial\Gamma} \bar{K}_t(t') \frac{1}{4j} H_0^{(2)}(kR)\, dt' + jk\eta \int_{\partial\Gamma} J_z(t') \frac{1}{4j} H_0^{(2)}(kR)\, dt' \tag{3.101}$$

$$H_z^{\rm inc}(t) = -J_t(t) - \hat{z} \cdot \nabla \times \int_{\partial\Gamma} \bar{J}_t(t') \frac{1}{4j} H_0^{(2)}(kR)\, dt' + j\frac{k}{\eta} \int_{\partial\Gamma} K_z(t') \frac{1}{4j} H_0^{(2)}(kR)\, dt' \tag{3.102}$$

where

$$R = \sqrt{[x(t) - x(t')]^2 + [y(t) - y(t')]^2} \tag{3.103}$$

Neither integral equation can uniquely specify both secondary sources, since they incorporate no information describing the interior medium. However, they do provide a direct linear relationship between the equivalent sources and the incident field on $\partial\Gamma$, and this relationship is equivalent to a radiation boundary condition.

A discretization is necessary in order to obtain an explicit radiation condition from the integral equations. Consider the TM polarization and the use of linear pyramid basis functions to represent the interior E_z-field. Since the transverse magnetic field obtained from E_z would be piecewise constant, the boundary discretization compatible with these interior functions would involve piecewise-linear basis functions for K_t and piecewise-constant basis functions for J_z. Simple testing functions, such as Dirac delta functions located in the center of each cell edge around the periphery, can be used to convert Equation (3.101) into the matrix equation

$$\mathbf{e}^{\rm inc} = \mathbf{L}\mathbf{k}_t + \mathbf{M}\mathbf{j}_z \tag{3.104}$$

where $\mathbf{L}$ and $\mathbf{M}$ are square matrix operators and $\mathbf{e}^{\rm inc}$, $\mathbf{k}_t$, and $\mathbf{j}_z$ are column vectors. The details of this type of discretization have been described in Chapter 2. Symbolically, Equation (3.104) is equivalent to

$$\mathbf{j}_z = \mathbf{M}^{-1}\mathbf{e}^{\rm inc} - \mathbf{M}^{-1}\mathbf{L}\mathbf{k}_t \tag{3.105}$$

which, from an inspection of (3.97) and (3.98), is clearly a linear relationship between the normal derivative of E_z, E_z itself, and $E_z^{\rm inc}$ on $\partial\Gamma$. Except for the discretization, Equation (3.105) has the same form as the exact RBC presented in Equation (3.23). It is somewhat more general than (3.23), however, since the shape of the boundary is not restricted. [To be precise, the equation relates coefficients of basis functions to sampled values of the incident field. However, as long as the discretization of (3.101) is compatible with that of the Helmholtz equation, the distinction is of no consequence.]

To illustrate the implementation of the integral equation RBC, consider the weak scalar Helmholtz equation in (3.5), which can be rewritten in the form

$$\iint_{\Gamma} \left(\frac{1}{\mu_r} \nabla T \cdot \nabla E_z - k^2 \varepsilon_r T E_z \right) dx\, dy = j\omega\mu_0 \int_{\partial\Gamma} T J_z\, dt \tag{3.106}$$

where J_z is the equivalent surface current density defined in (3.97). The interior field can be expanded in basis functions

$$E_z(x, y) \cong \sum_{n=1}^{N_{\rm int}} e_{zn}^{\rm int} B_n(x, y) + \sum_{n=N_{\rm int}+1}^{N_{\rm int}+N_{\rm bound}} e_{zn}^{\rm bound} B_n(x, y) \tag{3.107}$$

where $N_{\rm int}$ is the number of interior unknowns and $N_{\rm bound}$ is the number of unknowns located on the radiation boundary $\partial\Gamma$. The equivalent surface current J_z can be expressed as

$$J_z(t) \cong \sum_{n=1}^{N_{\text{bound}}} j_{zn}\tilde{B}_n(t) \tag{3.108}$$

where t is a parametric variable along $\partial\Gamma$ and $\tilde{B}$ may be a different basis function than B. Since $E_z = K_t$ by (3.98), the magnetic surface current density is automatically given by

$$K_t(t) \cong \sum_{n=N_{\text{int}}+1}^{N_{\text{int}}+N_{\text{bound}}} e_{zn}^{\text{bound}} B_n(t) \tag{3.109}$$

where $B_n(t)$ is the projection of $B_n(x, y)$ onto the boundary. Therefore, prior to imposing a radiation condition, the discretized weak equation can be written as

$$\begin{bmatrix} \mathbf{I} & \mathbf{I}_b^T & \mathbf{0} \\ \mathbf{I}_b & \mathbf{E} & \mathbf{J} \end{bmatrix} \begin{bmatrix} \mathbf{e}_z^{\text{int}} \\ \mathbf{e}_z^{\text{bound}} \\ \mathbf{j}_z \end{bmatrix} = \begin{bmatrix} \mathbf{0} \\ \mathbf{0} \end{bmatrix} \tag{3.110}$$

where the entries of $\mathbf{I}$, $\mathbf{I}_b$, and $\mathbf{E}$ represent interior interactions and have the common form

$$I_{mn} = \iint_{\Gamma} \frac{1}{\mu_r} \nabla B_m \cdot \nabla B_n - k^2 \varepsilon_r B_m B_n \tag{3.111}$$

and the entries of $\mathbf{J}$ can be expressed as

$$J_{mn} = -j\omega\mu_0 \int_{\partial\Gamma} B_m \tilde{B}_n \tag{3.112}$$

Finally, the RBC of (3.105) can be substituted for $\mathbf{j}_z$ to produce

$$\begin{bmatrix} \mathbf{I} & \mathbf{I}_b^T \\ \mathbf{I}_b & \mathbf{E} - \mathbf{J}\mathbf{M}^{-1}\mathbf{L} \end{bmatrix} \begin{bmatrix} \mathbf{e}_z^{\text{int}} \\ \mathbf{e}_z^{\text{bound}} \end{bmatrix} \begin{bmatrix} \mathbf{0} \\ -\mathbf{J}\mathbf{M}^{-1}\mathbf{e}^{\text{inc}} \end{bmatrix} \tag{3.113}$$

Equation (3.113) constitutes an outward-looking formulation that can be solved for the coefficients of the E_z-field throughout the computational domain. The blocks of the matrix containing $\mathbf{I}$, $\mathbf{I}_b$, and $\mathbf{I}_b^T$ are sparse, while the block $\mathbf{E} - \mathbf{J}\mathbf{M}^{-1}\mathbf{L}$ (the submatrix associated with nodes on $\partial\Gamma$) is fully populated. The basic structure of the global finite-element system is similar to that shown in Figure 3.5.

In this formulation, the number of unknown coefficients is equal to the number of unconstrained nodes in the mesh (for the TM case, all nodes not located on conducting boundaries). Because the surface integral equation is an "exact" RBC prior to discretization, the error arising within this procedure should be comparable to that of the formulation discussed in Section 3.4. Applications of this type of approach are described in the literature [15, 16], and we refer the reader to these references for examples.

Two principal drawbacks associated with the preceding formulation are the full submatrices arising from the global nature of the RBC and the need to construct the inverse of $\mathbf{M}$ as the first step of the procedure (at the least, $\mathbf{M}^{-1}$ must be constructed implicitly in terms of an $\mathbf{LU}$ factorization; see Chapter 4). The initial inversion of $\mathbf{M}$ can be avoided by using Equation (3.104) instead of (3.105), at the expense of additional unknowns representing J_z on $\partial\Gamma$. The latter approach is equivalent to the simultaneous solution of (3.104) and (3.110) and constitutes an alternative formulation [17]. However, because (3.104) contains two full matrices, and since the combination of (3.104) and (3.110) will produce a larger order matrix than (3.113) because of the additional unknowns, the alternative formulation does not appear to offer a savings in either storage or computation when compared with (3.113).

3.11 CONNECTION BETWEEN THE SURFACE INTEGRAL AND EIGENFUNCTION RBCS

The exact RBC developed in Section 3.3 can also be obtained from a surface integral equation, and we illustrate the alternative derivation in order to conceptually connect the integral equation and eigenfunction expansion ideas. As discussed in the preceding section, (3.105) constitutes an integral equation RBC but involves a numerical inversion to obtain $\mathbf{M}^{-1}$. If the boundary $\partial\Gamma$ is circular, however, an exact radiation condition may be developed without the need to numerically invert the integral operator. Consider the TM polarization and Equation (3.101). Assuming that $\partial\Gamma$ is circular with radius a, that the unknown functions are expanded in cylindrical harmonics according to

$$J_z(\phi) \cong \sum_{n=-M}^{M} j_n e^{jn\phi} \tag{3.114}$$

$$K_t(\phi) \cong \sum_{n=-M}^{M} k_n e^{jn\phi} \tag{3.115}$$

and that Equation (3.101) is enforced on the circular boundary by integrating over ϕ with testing functions $e^{-jn\phi}$, the result is an $N \times 2N$ matrix equation (where $N = 2M + 1$) of the form

$$\begin{bmatrix} \alpha_{-M} & 0 & \cdots & 0 & \beta_{-M} & 0 & \cdots & 0 \\ 0 & \alpha_{1-M} & & 0 & 0 & \beta_{1-M} & & 0 \\ \cdot & \cdot & & \cdot & \cdot & \cdot & & \cdot \\ \cdot & \cdot & & \cdot & \cdot & \cdot & & \cdot \\ \cdot & \cdot & & \cdot & \cdot & \cdot & & \cdot \\ 0 & 0 & \cdots & \alpha_M & 0 & 0 & \cdots & \beta_M \end{bmatrix} \begin{bmatrix} j_{-M} \\ \cdot \\ \cdot \\ \cdot \\ j_M \\ k_{-M} \\ \cdot \\ \cdot \\ \cdot \\ k_M \end{bmatrix} = \begin{bmatrix} e_{-M} \\ \cdot \\ \cdot \\ \cdot \\ e_M \end{bmatrix} \tag{3.116}$$

By carrying out the usual integrations over the basis and testing functions (Prob. P3.21), the diagonal elements of Equation (3.116) are found to be

$$\alpha_n = \tfrac{1}{2}(\eta\pi ka) J_n(ka) H_n^{(2)}(ka) \tag{3.117}$$

$$\beta_n = \tfrac{1}{2}(j\pi ka) J_n(ka) H_n^{(2)\prime}(ka) \tag{3.118}$$

and the right-hand side is given by

$$e_n = \frac{1}{2\pi} \int_0^{2\pi} E_z^{\text{inc}}(\phi) e^{-jn\phi}\, d\phi \tag{3.119}$$

The prime in Equation (3.118) denotes differentiation with respect to the argument of the Hankel function.

A similar development follows for the TE polarization, incorporating the expansions

$$J_t(\phi) \cong \sum_{n=-M}^{M} j_n e^{jn\phi} \tag{3.120}$$

$$K_z(\phi) \cong \sum_{n=-M}^{M} k_n e^{jn\phi} \tag{3.121}$$

to discretize Equation (3.102). Enforcing the equation with testing functions $e^{-jn\phi}$ produces the system

$$\begin{bmatrix} \alpha_{-M} & 0 & \cdots & 0 & \beta_{-M} & 0 & \cdots & 0 \\ 0 & \alpha_{1-M} & & 0 & 0 & \beta_{1-M} & & 0 \\ \cdot & \cdot & & \cdot & \cdot & \cdot & & \cdot \\ \cdot & \cdot & & \cdot & \cdot & \cdot & & \cdot \\ \cdot & \cdot & & \cdot & \cdot & \cdot & & \cdot \\ 0 & 0 & \cdots & \alpha_M & 0 & 0 & \cdots & \beta_M \end{bmatrix} \begin{bmatrix} j_{-M} \\ \cdot \\ \cdot \\ \cdot \\ j_M \\ k_{-M} \\ \cdot \\ \cdot \\ \cdot \\ k_M \end{bmatrix} = \begin{bmatrix} h_{-M} \\ \cdot \\ \cdot \\ \cdot \\ h_M \end{bmatrix} \tag{3.122}$$

where

$$\alpha_n = -\frac{1}{2}(j\pi ka) J_n(ka) H_n^{(2)\prime}(ka) \tag{3.123}$$

$$\beta_n = \frac{\pi ka}{2\eta} J_n(ka) H_n^{(2)}(ka) \tag{3.124}$$

and

$$h_n = \frac{1}{2\pi} \int_0^{2\pi} H_z^{\text{inc}}(\phi) e^{-jn\phi} \, d\phi \tag{3.125}$$

In contrast to the fully populated matrices encountered in Chapter 2, Equations (3.116) and (3.122) contain nonzero entries only on the main and minor diagonals. The diagonal nature of these matrices is a consequence of employing the eigenfunctions of the integral operators as basis and testing functions, which is usually only practical for boundaries $\partial\Gamma$ that are separable (i.e., circular or elliptical). For other surface shapes or other basis or testing functions, the systems would generally be fully populated [18].

Since the matrices are diagonal, the inversion necessary to cast these equations into the form of a radiation boundary condition is easily carried out analytically. Consider the TM polarization and one of the individual equations from (3.116), which can be written

$$\alpha_n j_n + \beta_n k_n = e_n \tag{3.126}$$

where α_n and β_n have been defined in (3.117) and (3.118),

$$j_n = \frac{1}{2\pi} \int_0^{2\pi} J_z(\phi') e^{-jn\phi'} \, d\phi' \tag{3.127}$$

$$k_n = \frac{1}{2\pi} \int_0^{2\pi} K_t(\phi') e^{-jn\phi'} \, d\phi' \tag{3.128}$$

and e_n is defined in (3.119). Solving Equation (3.126) for j_n, we rewrite (3.114) in the form

$$J_z(\phi) = \sum_{n=-\infty}^{\infty} \frac{e_n - \beta_n k_n}{\alpha_n} e^{jn\phi} \tag{3.129}$$

By replacing the Fourier coefficients e_n and k_n by their explicit integral representation and exchanging the order of integration and summation, we obtain the RBC previously given in Equation (3.23). A similar procedure can be carried out for the TE polarization.

The equivalence between the surface integral equation RBC and the RBC obtained in Section 3.3 is a simple consequence of the inherent connection between the cylindrical eigenfunction expansion and the integral equation representations. Although the former is restricted in practice to situations where the boundaries conform to separable surfaces, the two descriptions are conceptually equivalent.

3.12 INWARD-LOOKING DIFFERENTIAL EQUATION FORMULATION: THE UNIMOMENT METHOD

Preceding sections have presented *outward-looking* formulations for combining the scalar Helmholtz equations with radiation boundary conditions. These approaches employ an RBC to augment the Helmholtz equation, producing a formulation in which the primary unknown is the E_z- or H_z-field throughout Γ. We now consider a complementary approach where the equation representing the interior problem is used to constrain equivalent sources located on $\partial\Gamma$ and representing the exterior problem. Since the primary unknowns in this formulation are coefficients of equivalent sources on the outer boundary, as would be "seen" by an observer standing outside the region Γ and looking in, we denote this as an *inward-looking* formulation [6].

The starting point in this inward-looking formulation is the relationship embodied in Equations (3.116) and (3.122). Recall that these equations provide a description of the exterior scattering problem in terms of $\bar{J}$ and $\bar{K}$ located on the boundary but contain no information about the interior medium. What is needed is an additional equation relating $\bar{J}$ and $\bar{K}$ to the interior problem. In fact, such a relationship is provided by the scalar Helmholtz equations in (3.1) and (3.2) or their weak forms in (3.5) and (3.6). Consider the TM polarization. Knowledge of K_t on the boundary is equivalent to a Dirichlet boundary condition for E_z. Given K_t on the entire boundary, Equation (3.5) uniquely describes the interior E_z-field. After E_z is determined, the associated J_z could be constructed from the normal derivative of the E_z-field on the surface. Similarly, if J_z is prescribed on the entire boundary, Equation (3.5) can be solved subject to the equivalent Neumann boundary condition to produce a corresponding K_t. Thus, Equation (3.5) can be used to provide a linear relationship between J_z and K_t, which can be combined with Equation (3.116) to determine the secondary sources.

To construct a secondary $N \times 2N$ system providing the linear relationship between the unknowns J_z and K_t, consider the repeated finite-element solution of Equation (3.5) subject to the Neumann boundary condition

$$\frac{\partial E_z}{\partial \rho} = jk\eta e^{jm\phi} \quad \text{on } \partial\Gamma \tag{3.130}$$

which is equivalent to

$$J_z = e^{jm\phi} \tag{3.131}$$

In common with the preceding formulations, we divide the region Γ into triangular cells of constant permittivity and permeability, as depicted in Figure 3.3, and represent E_z by subsectional linear interpolation ("pyramid") functions

$$E_z(x, y) \cong \sum_{q=1}^{N} e_q B_q(x, y) \tag{3.132}$$

where $B_q(x, y)$ denotes a basis function centered at node q. If this expression is substituted into Equation (3.5), N linearly independent equations can be generated using

$$T(x, y) = B_p(x, y) \qquad p = 1, 2, \ldots, N \tag{3.133}$$

The resulting matrix equation has the form

$$\mathbf{Ae} = \mathbf{b} \tag{3.134}$$

with entries given by

$$A_{pq} = \iint_{\Gamma} \left(\frac{1}{\mu_r} \nabla B_p \cdot \nabla B_q - k^2 \varepsilon_r B_p B_q \right) dx\, dy \tag{3.135}$$

and

$$b_p = jk\eta \int_{\partial\Gamma} B_p e^{jm\phi} \rho \, d\phi \tag{3.136}$$

Because the interaction between the basis and testing functions is entirely local, $\mathbf{A}$ is a sparse matrix. Since the boundary contributions are also local, the sparsity pattern in $\mathbf{A}$ is comparable to that illustrated in Figure 3.11 (i.e., that associated with the outward-looking formulation employing the Bayliss–Turkel radiation condition).

After Equation (3.134) is solved to produce values of E_z throughout Γ, the fields on the boundary $\partial\Gamma$ can be decomposed into the expansion functions $e^{jn\phi}$ to produce

$$E_z(\phi) \cong \sum_{n=-M}^{M} \gamma_{n,m} e^{jn\phi} \cong K_t(\phi) \tag{3.137}$$

where

$$\gamma_{n,m} = \frac{1}{2\pi} \int_0^{2\pi} E_z(\phi) e^{-jn\phi} \, d\phi \tag{3.138}$$

This process of solving the matrix equation and decomposing the boundary fields must be repeated for each harmonic retained in the expansion of Equations (3.114) and (3.115). The linear relationship between the coefficients of J_z and K_t follows from the fact that an arbitrary Neumann condition

$$J_z(\phi) \cong \sum_{m=-M}^{M} j_m e^{jm\phi} \tag{3.139}$$

would produce the boundary expansion

$$K_t(\phi) \cong \sum_{n=-M}^{M} \sum_{m=-M}^{M} \gamma_{n,m} j_m e^{jn\phi} \tag{3.140}$$

Equating this expansion with (3.115), that is,

$$K_t(\phi) \cong \sum_{n=-M}^{M} k_n e^{jn\phi} \tag{3.141}$$

yields the $N \times 2N$ system

$$\begin{bmatrix} \gamma_{-M,-M} & \gamma_{-M,1-M} & \cdots & \gamma_{-M,M} & -1 & 0 & \cdots & 0 \\ \gamma_{1-M,-M} & \gamma_{1-M,1-M} & \cdots & \gamma_{1-M,M} & 0 & -1 & \cdots & 0 \\ \cdot & \cdot & \ddots & \cdot & \cdot & \cdot & \ddots & \cdot \\ \cdot & \cdot & & \cdot & \cdot & \cdot & & \cdot \\ \gamma_{M,-M} & \gamma_{M,1-M} & \cdots & \gamma_{M,M} & 0 & 0 & \cdots & -1 \end{bmatrix} \begin{bmatrix} j_{-M} \\ \cdot \\ \cdot \\ j_M \\ k_{-M} \\ \cdot \\ \cdot \\ k_M \end{bmatrix} = \begin{bmatrix} 0 \\ 0 \\ \cdot \\ \cdot \\ 0 \end{bmatrix} \tag{3.142}$$

Equation (3.142) is a matrix description of the relationship between E_z and its normal derivative on the interior side of the boundary $\partial\Gamma$ and takes into account the effects of the inhomogeneous material in the region Γ. The relationship of the boundary fields to the incident field is provided by Equation (3.116). Together, Equations (3.116) and (3.142) comprise a $2N \times 2N$ matrix description of the scattering problem. Once this matrix is constructed for a given cylinder geometry, any number of different incident fields may be treated by solving the system for additional right-hand sides. Since three of the four blocks of the $2N \times 2N$ system are diagonal, it can be reduced to a fully populated $N \times N$ matrix for computational purposes.

The TE polarization can be treated in a similar fashion using Equation (3.6) and working with the magnetic field H_z instead of the electric field [19]. (As an alternative to the above scheme, applied Dirichlet boundary conditions can be used with either the TM or TE polarization instead of Neumann conditions. The general procedure remains the same. However, if Dirichlet conditions are employed, the approximate differentiation process necessary to construct the other source may introduce additional error into the results.)

In summary, this inward-looking formulation requires the solution of the sparse finite-element system representing the interior problem for the $2M - 1$ right-hand sides required to construct (3.142). The $2N \times 2N$ system obtained from Equations (3.116) and (3.142) can be solved to produce the equivalent sources $\bar{J}$ and $\bar{K}$. These sources can be used directly to produce the far fields. If the interior field distribution is required, the finite-element system must be solved one additional time using either $\bar{J}$ or $\bar{K}$ as an applied boundary condition.

The accuracy of the inward-looking formulation is comparable to that of the outward-looking formulations of the preceding sections. As an example, consider a layered dielectric cylinder having a core region with $\varepsilon_r = 10 - j5$ and radius $\rho = 0.0939\lambda_0$ surrounded by a cladding with $\varepsilon_r = 6$ and $\rho = 0.15\lambda_0$. Several numerical results for the internal electric field induced by an incident TE wave are compared with exact solutions in Figure 3.15. For this example, H_z is represented by piecewise-linear functions. The transverse electric field is obtained by differentiation and is therefore piecewise constant. The inward-looking differential equation result is virtually identical with the MFIE result (Section 2.7) obtained from the same 75-node model, and both exhibit excellent agreement with the exact solution. (The model employed 42 cells to represent the core region, 38 to represent the cladding,

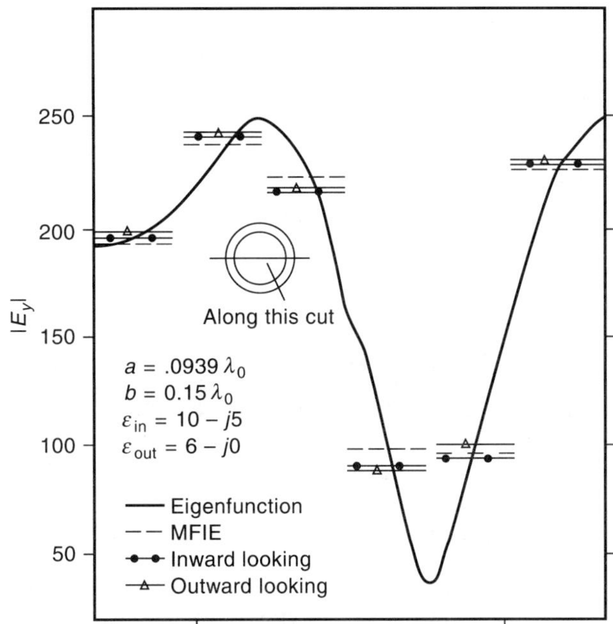

Figure 3.15 The TE electric field within a layered dielectric cylinder. Results from the inward-looking finite-element approach, the outward-looking approach using the second-order Bayliss–Turkel RBC, and the volume MFIE formulation of Section 2.7 are compared with the exact solution. The outward-looking result incorporating the Bayliss–Turkel RBC is obtained with the radiation boundary located at radius $0.41\lambda_0$. After [19]. ©1989 VSP BV.

and an additional layer of 44 cells outside the cylinder to place the outer boundary at $\rho = 0.23\lambda_0$.) The result from the outward-looking formulation using the second-order Bayliss–Turkel radiation boundary condition is also shown and is almost identical with the other numerical solutions. (Three layers of cells outside the scatterer were used in order to locate the Bayliss–Turkel boundary at $\rho = 0.41\lambda_0$.) Similar results from the numerical methods are shown in Figure 3.16 for a cylinder model with a greater density of cells. In Figure 3.16, the outer boundary on which the approximate Bayliss–Turkel condition is applied is brought in closer to the surface of the dielectric scatterer, resulting in some additional error in that result. Table 3.8 compares the scattering cross section data produced by the inward-looking formulation for these two models with the exact solution. An eigenfunction expansion procedure similar to that described in Equations (3.39)–(3.41) was used to calculate σ_{TE}. The numerical results appear to be converging toward the exact solution as the model is refined.

As an additional example, consider a lossy circular dielectric cylinder of radius $0.0179\lambda_0$ and relative permittivity $\varepsilon_r = 75 - j300$ illuminated by a TE wave. Numerical and exact results for this scatterer are presented in Figure 3.17. Although the relative permittivity is quite large, results suggest that the inward-looking formulation of this section, the outward-looking formulation employing the second-order Bayliss–Turkel RBC, and the volume MFIE formulation of Section 2.7 all produce stable, accurate solutions. In fact, the results from the inward-looking and outward-looking formulations are essentially identical.

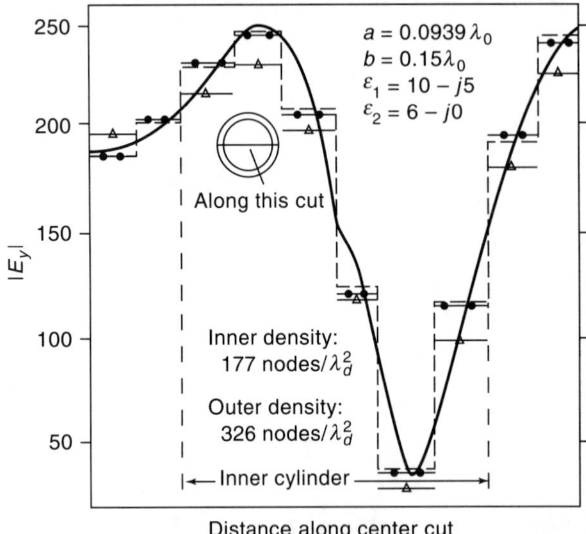

Figure 3.16 The TE electric field produced within the layered dielectric cylinder of Figure 3.15 for a finer triangular-cell model. The outward-looking result incorporating the Bayliss–Turkel RBC is obtained with the radiation boundary located at radius $0.25\lambda_0$: (—) exact; (- - -) MFIE (Section 2.7); (•) inward-looking PDE (Section 3.12); (Δ) outward-looking PDE with Bayliss–Turkel RBC (Section 3.9) applied at $\rho = 0.25\lambda_0$. After [19]. ©1989 VSP BV.

TABLE 3.8 Bistatic Scattering Cross Section[a] σ_{TE} for a Circular Dielectric Cylinder with Radius $0.0939\lambda_0$ and $\varepsilon_r = 10 - j5$ Coated with Dielectric Layer Having $\varepsilon_r = 6$ and Outer Radius $0.15\lambda_0$

ϕ (deg)	75-Node	151-Node	Exact
0	0.83	1.08	1.102
30	–0.22	0.02	0.044
60	–3.21	–3.05	–3.059
90	–6.60	–6.74	–6.814
120	–6.77	–6.79	–6.801
150	–5.51	–5.24	–5.157
180	–5.02	–4.65	–4.544

[a] In decibels free-space wavelength.

Note: Numerical results obtained with the two cylinder models used in Figures 3.15 and 3.16 are compared with the exact solution.

There are several advantages to the inward-looking formulation compared with the outward-looking formulation of Section 3.4. Because of the global RBC, the finite-element system of Equation (3.30) is not as sparse as that of Equation (3.134). In addition, if the scatterer is lossless, (3.134) involves a real-symmetric matrix as opposed to the complex-valued system of (3.30). Consequently, the inward-looking formulation will generally require less storage than (3.30). Although the outward-looking formulation employing the Bayliss–Turkel condition (Section 3.9) has sparsity that is comparable to this inward-looking approach, that scheme employs an approximate radiation condition that may introduce additional error into the solution or involve additional unknowns because of the need to locate the radiation boundary relatively far from the scatterer surface. The matrix arising in that formulation is also complex valued.

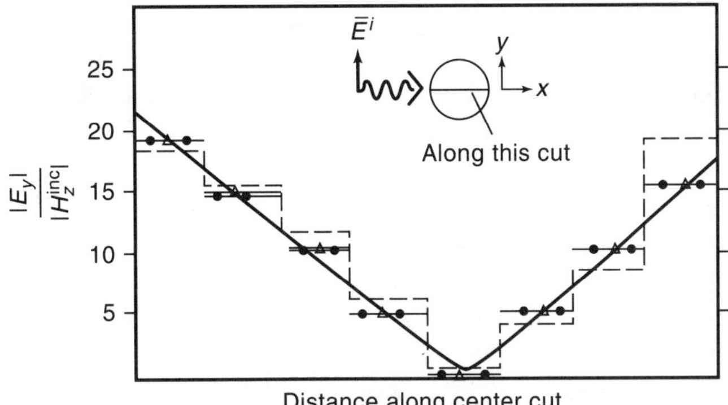

Figure 3.17 The TE electric field within a homogeneous dielectric cylinder with radius $0.0179\lambda_0$ and $\varepsilon_r = 75 - j300$. Results from the inward-looking finite-element approach, the outward-looking approach using the second-order Bayliss–Turkel RBC, and the volume MFIE formulation of Section 2.7 are compared with the exact solution: (—) exact; (- - -) MFIE; (•) inward looking; (Δ) outward looking. The outward-looking result incorporating the Bayliss–Turkel RBC is obtained with the radiation boundary located at radius $0.032\lambda_0$. After [19]. ©1989 VSP BV.

There are also disadvantages to the inward-looking approach. Because of the need to solve (3.134) repeatedly and decompose the boundary fields according to Equation (3.137), there is additional computational overhead not encountered in the outward-looking approaches. Another drawback to the inward-looking formulation is that the interior problem is being modeled as a closed region. Even if some loss is present, the system of equations may be nearly singular if the surface $\partial\Gamma$ happens to coincide with the surface of a resonant cavity. (This difficulty is explored in Chapter 6.) For electrically large domains, the cavity resonances may be difficult to avoid. In the outward-looking formulations, however, radiation loss eliminates the resonances and ensures that the system of equations is nonsingular.

Although presented from a different point of view, the formulation described in this section is equivalent to the *unimoment method* discussed in the literature [20, 21]. It is also possible to generalize this approach in order to employ an arbitrarily shaped boundary by using a surface integral equation to relate $\bar{J}$ and $\bar{K}$ on $\partial\Gamma$ to the incident field through the exterior medium. The discretization of the integral equation produces a fully populated matrix equation that takes the place of Equation (3.116). The interior problem is then solved as described above using the basis functions employed with the surface integral equation as the applied excitation instead of the exponential functions of (3.114) and (3.115). The result is a second fully populated system replacing Equation (3.134). Additional implementation details are available in several recent articles [22–24].

3.13 SUMMARY

This chapter has introduced several differential equation formulations for electromagnetic scattering from two-dimensional targets. The computational domain associated with these approaches must be truncated with a radiation boundary condition, and the primary focus of this chapter has been to investigate several near-field RBCs and explore various ways of incorporating them into the formulation. The inward-looking formulation of Section 3.12 requires that a large finite-element system be solved in order to construct a smaller fully populated matrix equation representing the scatterer. In contrast, the outward-looking approaches developed in the previous sections of this chapter permit the treatment of the scattering problem with a single finite-element system. The sparse nature of the finite-element system is degraded by the fill-in resulting from the global nature of any exact near-field radiation condition. Approximate local radiation conditions, such as the Bayliss–Turkel RBC, alleviate the fill-in problem but may introduce some inaccuracy.

Details have been provided to illustrate the finite-element discretization of the scalar Helmholtz equation and the associated boundary integrals using first-order linear interpolation functions on triangular cells. An analysis of discretization error (carried out in Chapter 5) suggests that linear basis functions limit the accuracy for electrically large regions. To produce greater accuracy, higher order interpolation functions and cells having an arbitrary shape will be considered in Chapter 9.

As compared to the integral equation formulations considered in Chapter 2, the primary computational advantage of the differential equation approaches is the resulting matrix sparsity. Algorithms that exploit this sparsity during the solution of the system of equations are the focus of Chapter 4.

REFERENCES

[1] J. N. Reddy, *An Introduction to the Finite Element Method*, New York: McGraw-Hill, 1984.

[2] G. Strang and G. J. Fix, *An Analysis of the Finite Element Method*, Englewood Cliffs, NJ: Prentice-Hall, 1973.

[3] R. B. Wu and C. H. Chen, "Variational reaction formulation of scattering problem for anisotropic dielectric cylinders," *IEEE Trans. Antennas Propagat.*, vol. 34, pp. 640–645, May 1986.

[4] L. W. Pearson, R. A. Whitaker, and L. J. Bahrmasel, "An exact radiation boundary condition for the finite-element solution of electromagnetic scattering on an open domain," *IEEE Trans. Magnet.*, vol. 25, pp. 3046–3048, July 1989.

[5] R. F. Harrington, *Time-Harmonic Electromagnetic Fields*, New York: McGraw-Hill, 1961.

[6] L. W. Pearson, A. F. Peterson, L. J. Bahrmasel, and R. A. Whitaker, "Inward-looking and outward-looking formulations for scattering from penetrable objects," *IEEE Trans. Antennas Propagat.*, vol. 40, pp. 714–720, June 1992.

[7] P. P. Silvester and R. L. Ferrari, *Finite Elements for Electrical Engineers*, Cambridge: Cambridge University Press, 1983.

[8] A. F. Peterson and S. P. Castillo, "Differential equation methods for electromagnetic scattering from inhomogeneous cylinders," in *Radar Cross Sections of Complex Objects*, ed. W. R. Stone, New York: IEEE Press, 1990.

[9] A. Bayliss and E. Turkel, "Radiation boundary conditions for wave-like equations," *Comm. Pure Appl. Math.*, vol. 33, pp. 707–725, 1980.

[10] F. X. Canning, "On the application of some radiation boundary conditions," *IEEE Trans. Antennas Propagat.*, vol. 38, pp. 740–745, May 1990.

[11] R. Mittra and O. Ramahi, "Absorbing boundary conditions for the direct solution of partial differential equations arising in electromagnetic scattering problems," in *Finite Element and Finite Difference Methods in Electromagnetic Scattering*, ed. M. A. Morgan, New York: Elsevier, 1990.

[12] A. F. Peterson and S. P. Castillo, "A frequency-domain differential equation formulation for electromagnetic scattering from inhomogeneous cylinders," *IEEE Trans. Antennas Propagat.*, vol. AP-37, pp. 601–607, May 1989.

[13] G. A. Kriegsmann, A. Taflove, and K. R. Umashankar, "A new formulation of electromagnetic wave scattering using an on-surface radiation boundary condition approach," *IEEE Trans. Antennas Propagat.*, vol. 35, pp. 153–161, Feb. 1987.

[14] B. Lichtenberg, K. J. Webb, D. B. Meade, and A. F. Peterson, "Comparison of two-dimensional conformable local radiation boundary conditions," *Electromagnetics*, vol. 16, pp. 359–384, 1996.

[15] B. H. McDonald and A. Wexler, "Finite element solution of unbounded field problems," *IEEE Trans. Microwave Theory Tech.*, vol. MTT-20, pp. 841–847, Dec. 1972.

[16] S. P. Marin, "Computing scattering amplitudes for arbitrary cylinders under incident plane waves," *IEEE Trans. Antennas Propagat.*, vol. AP-30, pp. 1045–1049, Nov. 1982.

[17] Z. Gong and A. W. Glisson, "A hybrid approach for the solution of electromagnetic scattering problems involving two-dimensional inhomogeneous dielectric cylinders," *IEEE Trans. Antennas Propagat.*, vol. 38, pp. 60–68, Jan. 1990.

[18] R. F. Harrington, *Field Computation by Moment Methods*, Malabar, FL: Krieger, 1982.

[19] A. F. Peterson, "A comparison of integral, differential, and hybrid methods for TE-wave scattering from inhomogeneous dielectric cylinders," *J. Electromagnetic Waves Appl.*, vol. 3, pp. 87–106, 1989.

[20] K. K. Mei, "Unimoment method for solving antenna and scattering problems," *IEEE Trans. Antennas Propagat.*, vol. AP-22, pp. 760–766, Nov. 1974.

[21] S. K. Chang and K. K. Mei, "Application of the unimoment method to electromagnetic scattering by dielectric cylinders," *IEEE Trans. Antennas Propagat.*, vol. AP-24, pp. 35–42, Jan. 1976.

[22] J. M. Jin and V. V. Liepa, "A note on hybrid finite element method for solving scattering problems," *IEEE Trans. Antennas Propagat.*, vol. AP-36, pp. 1486–1490, Oct. 1988.

[23] X. Yuan, D. R. Lynch, and J. W. Strohbehn, "Coupling of finite element and moment methods for electromagnetic scattering from inhomogeneous objects," *IEEE Trans. Antennas Propagat.*, vol. AP-38, pp. 386–393, March 1990.

[24] A. C. Cangellaris and R. Lee, "The bymoment method for two-dimensional electromagnetic scattering," *IEEE Trans. Antennas Propagat.*, vol. AP-38, pp. 1429–1437, Sept. 1990.

PROBLEMS

P3.1 Beginning with Maxwell's equations, generalize Equation (3.1) to include sources J_z and K_t located in the region Γ.

P3.2 Finite-element formulations are often based on direct variational methods, that is, finding the stationary point of a functional. Consider the functional

$$F(\Psi) = \frac{1}{2} \iint_{\Gamma} \left(\frac{1}{\mu_r} \nabla\Psi \cdot \nabla\Psi - k^2 \varepsilon_r \Psi^2 \right) dx\, dy$$

and a trial solution $\Psi = E_z + \varepsilon\Phi$. The function E_z satisfies the scalar Helmholtz equation

$$\nabla \cdot \left(\frac{1}{\mu_r} \nabla E_z \right) + k^2 \varepsilon_r E_z = 0$$

subject to specified Dirichlet boundary conditions along some portion of the boundary $\partial\Gamma$ and homogeneous Neumann boundary conditions along the rest of $\partial\Gamma$. The function $\Phi(x, y)$ vanishes on the portion of $\partial\Gamma$ over which Dirichlet conditions are specified and ε is a scalar parameter.

By direct substitution and the use of Equations (3.3) and (3.4), show that

$$F(\Psi) - F(E_z) = O(\varepsilon^2) \quad \text{as } \varepsilon \to 0$$

and therefore the functional F has a stationary point at $\Psi = E_z$.

P3.3 (a) Problem P3.2 demonstrated that the quadratic functional

$$F(E_z) = \frac{1}{2} \iint_{\Gamma} \left(\frac{1}{\mu_r} \nabla E_z \cdot \nabla E_z - k^2 \varepsilon_r (E_z)^2 \right) dx\, dy$$

is stationary about the solution to the scalar Helmholtz equation

$$\nabla \cdot \left(\frac{1}{\mu_r} \nabla E_z \right) + k^2 \varepsilon_r E_z = 0$$

for the case of Dirichlet boundary conditions or homogeneous Neumann boundary conditions on $\partial\Gamma$. Assume that the unknown field is expanded in basis functions

$$E_z(x, y) \cong \sum_{n=1}^{N} e_n B_n(x, y) + \sum_{n=N+1}^{M} e_n B_n(x, y)$$

where $\{e_n\}, n = 1, 2, \ldots, N$ are *unknown* coefficients and $\{e_n\}, n = N + 1, N + 2, \ldots, M$ are *known* coefficients chosen to satisfy the Dirichlet boundary conditions on $\partial\Gamma$. By differentiating the functional with respect to each of the first N coefficients $\{e_n\}$, obtain a matrix equation for $\{e_n\}$.

(b) How does this equation compare to that arising from the weak equation (3.5) if the same functions are used as basis functions and testing functions to discretize the weak equation?

(c) Consider the incorporation of nonhomogeneous Neumann boundary conditions. Assume that the expansion has the form

$$E_z(x, y) \cong \sum_{n=1}^{N} e_n B_n(x, y) + \sum_{n=N+1}^{M} e_n B_n(x, y) + \sum_{n=M+1}^{P} e_n B_n(x, y)$$

where $\{e_n\}, n = 1, 2, \ldots, N$ are *unknown* coefficients associated with interior bases, $\{e_n\}, n = N+1, N+2, \ldots, M$ are *known* coefficients associated with Dirichlet boundary conditions on part of $\partial\Gamma$, say $\partial\Gamma_D$, and $\{e_n\}, n = M+1, M+2, \ldots, P$ are *unknown* coefficients associated with the bases on the part of the boundary $\partial\Gamma_N$ where the Neumann boundary conditions

$$\frac{\partial E_z}{\partial n} = f$$

are to be applied. Generalize the functional in order to include a boundary integral [similar to that appearing in Equation (3.5)] in the corresponding matrix equation representing the contribution from the Neumann condition.

(d) Generalize the functional to include an electric current source $J_z(x, y)$ within the region Γ.

P3.4 By a procedure similar to that used in Prob. P3.2, show that the functional

$$F(\bar{A}) = \frac{1}{2}\iint_{\Gamma}\left(\frac{1}{\mu_r}\nabla\times\bar{A}\cdot\nabla\times\bar{A} + k^2\varepsilon_r\bar{A}\cdot\bar{A}\right) dx\, dy + \int_{\partial\Gamma}\frac{1}{\mu_r}\bar{A}\cdot\bar{h}\, dt$$

is stationary about the solution $\bar{A} = \bar{E}$, where $\bar{E}$ satisfies the two-dimensional vector Helmholtz equation

$$\nabla\times\left(\frac{1}{\mu_r}\nabla\times\bar{E}\right) - k^2\varepsilon_r\bar{E} = 0$$

subject to either Dirichlet boundary conditions of the form

$$\hat{n}\times\bar{E} = \bar{g} \quad \text{on } \partial\Gamma$$

or Neumann boundary conditions of the form

$$\hat{n}\times(\nabla\times\bar{E}) = \bar{h} \quad \text{on } \partial\Gamma$$

where $\bar{g}$ or $\bar{h}$ is specified. In other words, seek a solution of the form

$$\bar{A} = \bar{E} + \varepsilon\bar{\Phi}$$

where $\bar{\Phi}(x, y)$ satisfies $\hat{n}\times\bar{\Phi} = 0$ on the portion of $\partial\Gamma$ where Dirichlet conditions are imposed, and show that $F(\bar{A}) - F(\bar{E}) = O(\varepsilon^2)$ as $\varepsilon \to 0$.

P3.5 Using Maxwell's equations, show that the boundary condition $\hat{n}\times\bar{E} = 0$ is equivalent to

$$\frac{\partial H_z}{\partial n} = 0$$

for the TE polarization. In the general three-dimensional case, what would be an equivalent boundary condition applied to $\bar{H}$ on a p.e.c. surface?

P3.6 Demonstrate that Equation (3.17) is equivalent to the Sommerfeld RBC in the limit as $a \to \infty$.

P3.7 Suppose that the imbedded p.e.c. boundary considered in Section 3.2 is replaced by a surface $\partial\Gamma_c$ over which, for the TM polarization, an impedance boundary condition

$$E_z = \eta_s H_t$$

holds. Describe the modifications necessary in order to incorporate the impedance boundary condition into the formulation described in Section 3.4.

P3.8 (a) Equation (3.25) is written with the "total" field as the primary unknown. However, the scalar Helmholtz equation can be manipulated into a slightly different "weak" equation, written for the TE case as

$$\iint_{\Gamma}\left(\frac{1}{\varepsilon_r}\nabla T\cdot\nabla H_z^s - k^2 T H_z^s\right) dx\, dy - \int_{\partial\Gamma} T\frac{\partial H_z^s}{\partial n} dt$$
$$= \iint_{\Gamma} T\left[\nabla\cdot\left(\frac{1}{\varepsilon_r}\nabla H_z^{\text{inc}}\right) + k^2 H_z^{\text{inc}}\right] dx\, dy$$

in which the "scattered" field constitutes the primary unknown. Using the TE form of Equation (3.17) as a radiation boundary condition, develop the complete scattered field equation describing the H_z-field within an inhomogeneous cylinder.

(b) What are the essential differences between the total and scattered field formulations? Is the matrix operator different? Does the similarity of the matrix suggest any conclusion regarding the relative accuracy of either approach?

(c) Describe the implementation of boundary conditions on imbedded perfect electric conductors or impedance surfaces within the scattered f ield formulation. Are these conditions more or less convenient than with a total field formulation?

P3.9 The error associated with linear interpolation on triangles is known to behave as $O(\Delta^2)$ as $\Delta \to 0$, where Δ is the longest edge in the mesh. From the numerical data for E_z presented in Table 3.2, construct a graph of the error in the field at the center of the cylinder as a function of Δ. Assume that the error is of order $O(\Delta^q)$ as $\Delta \to 0$ and find the real-valued exponent q that best fits the data.

P3.10 Suppose the scattering cross section σ_{TM} is to be computed by integrating over equivalent volumetric polarization currents, as described in Section 1.11. Describe the manner in which the polarization currents would be obtained from the coefficients of E_z for the TM polarization. What additional pointer arrays are necessary to carry out the calculations? (*Hint:* Review the volume MFIE formulation described in Section 2.7.)

P3.11 Maxwell's equations dictate that

$$\frac{1}{\mu_r}\frac{\partial E_z}{\partial n} = j\omega\mu_0 J_z$$

at the surface $\partial\Gamma_c$ of a perfect conductor. Consequently, for testing functions T located on $\partial\Gamma_c$, the weak equation in (3.5) can be rewritten in the form

$$\int_{\partial\Gamma_c} T J_z \, dt = \frac{1}{j\omega\mu_0} \iint_{\Gamma} \left(\frac{1}{\mu_r} \nabla T \cdot \nabla E_z - k^2 \varepsilon_r T E_z \right) dx\, dy$$

Consider the use of this equation as a way of computing J_z at the surface of a conducting region imbedded in the volume Γ after the E_z-field has been calculated by the solution of Equation (3.30). Assuming that J_z is a piecewise-linear function along the perimeter of the p.e.c. region, develop a procedure for obtaining the coefficients as a secondary calculation. Discuss the advantages and disadvantages of this approach compared with the direct calculation

$$J_z = \frac{1}{j\omega\mu_0\mu_r}\frac{\partial E_z}{\partial n}$$

P3.12 Derive Equation (3.48) by carrying out the inversion of (3.47).

P3.13 Using the scalar Helmholtz equation, derive Equation (3.80) from (3.78).

P3.14 The third-order Bayliss–Turkel RBC can be obtained from Equation (3.79). Express the third-order condition in the form

$$\frac{\partial E_z^s}{\partial \rho} = \cdots$$

Consider the numerical implementation of this RBC within a finite-element discretization employing linear interpolation functions. Is there an obvious difficulty associated with its use?

P3.15 The Bayliss–Turkel type of local absorbing boundary condition is just one of numerous RBCs that have been developed. Engquist and Majda have derived boundary conditions based on pseudo-differential operator theory (B. Engquist and A. Majda, "Absorbing boundary conditions for the numerical simulation of waves," *Math. Comp.*, vol. 31, pp. 629–651, 1977). If applied to a circular boundary, their second-order condition has the form

$$\frac{\partial E_z^s}{\partial \rho} = \gamma(\rho) E_z^s + \delta(\rho) \frac{\partial^2 E_z^s}{\partial \phi^2}$$

where

$$\gamma(\rho) = \left(-jk - \frac{1}{2\rho} \right)$$

$$\delta(\rho) = \left(\frac{-j}{2k\rho^2} + \frac{1}{2k^2\rho^3} \right)$$

The error associated with this RBC can be studied for cylindrical harmonics following a procedure similar to that described in Equations (3.85) and (3.86), in order to obtain

$$\text{Error}(n, \rho) = \left| \frac{k H_n^{(2)\prime}(k\rho) - (\gamma - n^2\delta) H_n^{(2)}(k\rho)}{k H_n^{(2)\prime}(k\rho)} \right|$$

Compare the accuracy of this RBC to that of the second-order Bayliss–Turkel condition by computing the error associated with each RBC for several values of n and ρ. For a circular boundary, which appears to be more accurate?

P3.16 Verify the boundary integral element matrices in (3.93) and (3.94).

P3.17 Rewrite Equation (3.87) in order to produce a scattered field formulation, as described in Prob. P3.8.

P3.18 In Prob. P1.22, the two-dimensional scattering cross section σ_{TM} was expressed as an integral over E_z and H_ϕ on a circular boundary. In general, the error associated with approximating H_ϕ on the boundary can be significant and thus usually prevents us from employing the approach of Prob. P1.22 with the outward-looking formulations of Chapter 3. However, there is a way in which this might be accomplished.

(a) By combining the second-order Bayliss–Turkel RBC with the result of Prob. P1.22, show that

$$\sigma_{\text{TM}}(\phi) \cong \frac{k}{4} \left| \int_{\phi'=0}^{2\pi} \left\{ \left[\frac{\alpha}{jk} - \cos(\phi - \phi') \right] E_z^s(\phi') + \frac{\beta}{jk} \frac{\partial^2 E_z^s}{\partial (\phi')^2} \right\} \times e^{jka\cos(\phi-\phi')} a \, d\phi' \right|^2$$

Discuss the implementation of this expression, assuming that E_z is represented by linear interpolation functions.

(b) Compare the resulting expression with that obtained from the eigenfunction expansion described in Equations (3.39)–(3.41) by applying it to individual harmonics $E_z^s = E_n e^{jn\phi}$ in order to quantitatively characterize the approximation error.

P3.19 By a coordinate transformation, the second-order Bayliss–Turkel RBC from Equation (3.80) can be expressed directly in terms of the normal and tangential variables along a particular boundary. By transforming the cylindrical coordinate system (ρ, ϕ, z) into the right-handed system (n, t, z) local to some region along the boundary, where the polar angle from $\hat{\rho}$ to $\hat{n}$ is θ, show that the RBC can be written as

$$\frac{\partial E_z^s}{\partial n} = \frac{\alpha - k^2 \sin^2\theta}{\cos\theta} E_z^s + \tan\theta \frac{\partial E_z^s}{\partial t} + 2\sin\theta \frac{\partial^2 E_z^s}{\partial n \partial t} + \frac{\beta\rho^2\cos^2\theta - \sin^2\theta}{\cos\theta} \frac{\partial^2 E_z^s}{\partial t^2}$$

Identify the two factors that complicate the direct implementation of this RBC. Can you suggest a way of replacing the mixed-derivative term with a reasonable approximation?

P3.20 Develop explicit expressions for the entries of the matrices **L**, **M**, and $\mathbf{e}^{\text{inc}}$ in Equation (3.104), assuming piecewise-linear basis functions are used to represent K_t, piecewise-constant basis functions are used for J_z, and Dirac delta testing functions are located in the center of each cell edge around the boundary.

P3.21 The *addition theorem* for Hankel functions is given by

$$H_0^{(2)}(kR) = \begin{cases} \displaystyle\sum_{n=-\infty}^{\infty} H_n^{(2)}(k\rho')J_n(k\rho)e^{jn(\phi-\phi')} & \rho < \rho' \\ \displaystyle\sum_{n=-\infty}^{\infty} J_n(k\rho')H_n^{(2)}(k\rho)e^{jn(\phi-\phi')} & \rho > \rho' \end{cases}$$

where R is defined in Equation (3.103).

(a) Use the addition theorem in combination with Equations (3.101), (3.114), and (3.115) to derive Equations (3.117) and (3.118).

(b) Repeat your derivation for the TE case to obtain (3.123) and (3.124).

P3.22 Recast the TM inward-looking formulation described in Section 3.12 for the TE polarization.

P3.23 Extend the inward-looking formulation of Section 3.12 in order to use a surface integral equation such as (3.101) on a boundary of arbitrary shape.

4

Algorithms for the Solution of Linear Systems of Equations

The procedures of the previous two chapters produce a linear system of equations that must be solved to determine the unknown coefficients of the basis functions. In this chapter we discuss some of the algorithms in widespread use for solving matrix equations. Initially, we review direct methods for the solution of general full- and sparse-matrix equations. The error introduced by finite precision arithmetic is explored. One iterative algorithm, the conjugate gradient method, is considered in detail. The conjugate gradient–fast Fourier transform implementation is discussed for the solution of certain specialized integral equations. Finally, we briefly introduce the fast multipole method, which offers similar computational advantages for the iterative solution of general problems.

4.1 NAIVE GAUSSIAN ELIMINATION

Consider the general matrix equation $\mathbf{Ax} = \mathbf{b}$, or

$$\begin{bmatrix} a_{11} & a_{12} & a_{13} & \cdots & a_{1N} \\ a_{21} & a_{22} & a_{23} & \ddots & \ddots \\ a_{31} & a_{32} & a_{33} & & \\ \cdot & \ddots & \ddots & & \\ \cdot & & & & \\ \cdot & & & & \\ a_{N1} & \cdots & \cdots & \cdots & a_{NN} \end{bmatrix} \begin{bmatrix} x_1 \\ x_2 \\ x_3 \\ \cdot \\ \cdot \\ \cdot \\ x_N \end{bmatrix} = \begin{bmatrix} b_1 \\ b_2 \\ b_3 \\ \cdot \\ \cdot \\ \cdot \\ b_N \end{bmatrix} \tag{4.1}$$

A solution to this system of equations may be found using one of the various forms of Gaussian elimination. The procedure we describe requires the row equations of the system to be systematically replaced with linear combinations of other rows of the matrix in order to recast Equation (4.1) into the form of an upper triangular system:

$$\begin{bmatrix} 1 & u_{12} & u_{13} & \cdots & u_{1N} \\ 0 & 1 & u_{23} & \ddots & \ddots \\ 0 & 0 & 1 & & \\ \cdot & \cdot & & & \\ \cdot & \cdot & & & \\ \cdot & \cdot & & & \\ 0 & 0 & \cdots & \cdots & 1 \end{bmatrix} \begin{bmatrix} x_1 \\ x_2 \\ x_3 \\ \cdot \\ \cdot \\ \cdot \\ x_N \end{bmatrix} = \begin{bmatrix} c_1 \\ c_2 \\ c_3 \\ \cdot \\ \cdot \\ \cdot \\ c_N \end{bmatrix} \tag{4.2}$$

Once the system of equations is converted to the form of Equation (4.2), the solution can be obtained by back substitution, following the recipe

$$x_N = c_N \tag{4.3}$$

$$x_m = c_m - \sum_{i=m+1}^{N} u_{mi} x_i \qquad m = N-1,\ N-2, \ldots, 1 \tag{4.4}$$

To systematically convert Equation (4.1) into the form of (4.2), consider the first row of the system

$$a_{11}x_1 + a_{12}x_2 + \cdots + a_{1N}x_N = b_1 \tag{4.5}$$

This equation may be divided by a_{11} to produce a new first equation

$$x_1 + u_{12}x_2 + u_{13}x_3 + \cdots + u_{1N}x_N = c_1 \tag{4.6}$$

The new equation is to be employed in "sweeping out" the first column of the matrix. The operation proceeds as follows. After multiplication by a_{21}, the new equation can be subtracted from the second row of the system, altering the matrix equation to the equivalent form

$$\begin{bmatrix} 1 & u_{12} & u_{13} & \cdots & u_{1N} \\ 0 & a_{22} - a_{21}u_{12} & a_{23} - a_{21}u_{13} & \cdots & a_{2N} - a_{21}u_{1N} \\ a_{31} & a_{32} & a_{33} & \cdots & a_{3N} \\ \cdot & \cdot & \cdot & & \cdot \\ \cdot & \cdot & \cdot & \ddots & \cdot \\ \cdot & \cdot & \cdot & & \cdot \\ a_{N1} & a_{N2} & a_{N3} & \cdots & a_{NN} \end{bmatrix} \begin{bmatrix} x_1 \\ x_2 \\ x_3 \\ \cdot \\ \cdot \\ \cdot \\ x_N \end{bmatrix} = \begin{bmatrix} c_1 \\ b_2 - a_{21}c_1 \\ b_3 \\ \cdot \\ \cdot \\ \cdot \\ b_N \end{bmatrix} \tag{4.7}$$

We have "eliminated" the (2, 1) entry from the matrix. This procedure is repeated until the first column of the matrix has been completely swept out, leaving an equivalent linear system

$$\begin{bmatrix} 1 & u_{12} & u_{13} & \cdots & u_{1N} \\ 0 & a_{22} - a_{21}u_{12} & a_{23} - a_{21}u_{13} & \cdots & a_{2N} - a_{21}u_{1N} \\ 0 & a_{32} - a_{31}u_{12} & a_{33} - a_{31}u_{13} & \cdots & a_{3N} - a_{31}u_{1N} \\ \cdot & \cdot & \cdot & & \cdot \\ \cdot & \cdot & \cdot & \ddots & \cdot \\ \cdot & \cdot & \cdot & & \cdot \\ 0 & a_{N2} - a_{N1}u_{12} & a_{N3} - a_{N1}u_{13} & \cdots & a_{NN} - a_{N1}u_{1N} \end{bmatrix} \begin{bmatrix} x_1 \\ x_2 \\ x_3 \\ \cdot \\ \cdot \\ \cdot \\ x_N \end{bmatrix} = \begin{bmatrix} c_1 \\ b_2 - a_{21}c_1 \\ b_3 - a_{31}c_1 \\ \cdot \\ \cdot \\ \cdot \\ b_N - a_{N1}c_1 \end{bmatrix} \tag{4.8}$$

For notational convenience, we divide the second equation by $a_{22} - a_{21}u_{12}$ to obtain

$$\begin{bmatrix} 1 & u_{12} & u_{13} & \cdots & u_{1N} \\ 0 & 1 & u_{23} & \cdots & u_{2N} \\ 0 & a_{32} - a_{31}u_{12} & a_{33} - a_{31}u_{13} & \cdots & a_{3N} - a_{31}u_{1N} \\ \cdot & \cdot & \cdot & & \cdot \\ \cdot & \cdot & \cdot & \ddots & \cdot \\ \cdot & \cdot & \cdot & & \cdot \\ 0 & a_{N2} - a_{N1}u_{12} & a_{N3} - a_{N1}u_{13} & \cdots & a_{NN} - a_{N1}u_{1N} \end{bmatrix} \begin{bmatrix} x_1 \\ x_2 \\ x_3 \\ \cdot \\ \cdot \\ \cdot \\ x_N \end{bmatrix} = \begin{bmatrix} c_1 \\ c_2 \\ b_3 - a_{31}c_1 \\ \cdot \\ \cdot \\ \cdot \\ b_N - a_{N1}c_1 \end{bmatrix} \tag{4.9}$$

The new second equation is used to sweep out the second column of the matrix. We continue this procedure throughout the remaining columns of the matrix, finally producing an upper triangular system having the form of Equation (4.2).

Numerous variations exist on the basic algorithm outlined in the preceding paragraph, and the reader is referred to the literature where Gaussian elimination is studied in detail [1–9]. In practice, the procedure discussed above has a drawback: The operations performed on the right-hand side **b** must be remembered if additional systems involving the same matrix **A** are to be treated. (This need arises in the solution of scattering problems involving more than one incident field; refer to Chapters 2 and 3.) This difficulty can be remedied by observing that Gaussian elimination is equivalent to factorizing the matrix **A** into the product of two matrices **L** and **U** according to

$$\begin{bmatrix} l_{11} & 0 & 0 & \cdots & 0 \\ l_{21} & l_{22} & 0 & \cdots & 0 \\ l_{31} & l_{32} & l_{33} & \cdots & 0 \\ \cdot & \cdot & \cdot & & \cdot \\ \cdot & \cdot & \cdot & \ddots & \cdot \\ \cdot & \cdot & \cdot & & \cdot \\ l_{N1} & l_{N2} & l_{N3} & \cdots & l_{NN} \end{bmatrix} \begin{bmatrix} 1 & u_{12} & u_{13} & \cdots & u_{1N} \\ 0 & 1 & u_{23} & \ddots & \ddots \\ 0 & 0 & 1 & & \\ \cdot & \cdot & & & \\ \cdot & \cdot & & & \\ \cdot & \cdot & & & \\ 0 & 0 & \cdots & \cdots & 1 \end{bmatrix} \begin{bmatrix} x_1 \\ x_2 \\ x_3 \\ \cdot \\ \cdot \\ \cdot \\ x_N \end{bmatrix} = \begin{bmatrix} b_1 \\ b_2 \\ b_3 \\ \cdot \\ \cdot \\ \cdot \\ b_N \end{bmatrix} \tag{4.10}$$

The two triangular systems can be obtained with the same number of arithmetic operations as the procedure described above and stored in place of the original matrix as they are constructed. Once **L** and **U** are obtained, the vector $\mathbf{c} = \mathbf{L}^{-1}\mathbf{b}$ can be computed for each right-hand side **b** by forward substitution, yielding an equivalent system having the form of Equation (4.2).

The factorization of **A** into the LU form of Equation (4.10) requires approximately $N^3/3$ multiplications and additions. After factorization, the additional forward and backward substitution involves N^2 multiplications and additions for each right-hand side. (The identical effort would be required to produce each solution using the inverse $\mathbf{A}^{-1}$, whose computation requires many more operations than the factorization. Consequently, the factorization can be thought of as an implicit inverse matrix.) For large N the factorization accounts for the bulk of the computational effort. In the context of the electromagnetic scattering applications discussed in Chapters 2 and 3, only a single factorization is needed to characterize a scatterer illuminated by a number of different incident fields.

There are numerous forms of LU factorization described in the literature, including the Gauss, Crout, Doolittle, and bifactorization algorithms [1–9]. These procedures require a similar number of arithmetic operations and differ little from each other in overall performance on traditional serial-architecture computers. However, these algorithms differ substantially in the order in which matrix entries are accessed from computer memory and in the number of times certain entries are altered. Consequently, one algorithm may give substantially better performance than another on a particular machine, especially in a distributed-memory environment. The recently introduced LAPACK library [10] for linear equations and eigenvalue problems was designed for efficient data handling on modern high-performance computers.

4.2 PIVOTING

An obvious difficulty arises in the implementation of the elimination algorithm if the ith diagonal entry is zero at the start of sweeping out the ith column. The algorithm will fail attempting to divide by zero. This potential problem is readily circumvented by the use of pivoting. Partial row pivoting is the exchange of two rows of the system in order to replace one diagonal entry with a "stronger entry" (i.e., one that is greater in magnitude than the original entry). Row pivoting requires the algorithm to search for the best pivot among the ith column of the rows appearing below row i in the matrix. In the simplest scheme, the row having the strongest entry in column i is exchanged with the original row i before sweeping out column i of the matrix. This procedure, if repeated for each column of the matrix, ensures that the elimination algorithm will never require a division by zero unless the original system of equations is singular.

Although any form of pivoting will prevent a catastrophic division by zero, alternative pivoting strategies are sometimes sought to minimize the growth of round-off errors. Even if all the pivots are nonzero, large errors can arise during elimination in finite machine precision from the combined effects of division by small pivots followed by the required subtractions. A variety of different pivoting schemes are in use. For instance, scaling can be used to alter the manner in which the best pivots are selected [4, 5]. Column pivoting is possible, as is complete pivoting (complete pivoting selects the best pivot available in any of the remaining columns of the remaining rows of the system) [3, 4]. It is noteworthy that the additional computation associated with complete pivoting is usually not offset by improved accuracy over partial pivoting. General-purpose LU factorization codes, such as those of the LINPACK [6] or LAPACK [10] libraries, always provide some form of pivoting to ensure stability.

By rearranging operations, pivoting can provide stability in order to control unnecessary numerical error during elimination in finite precision. However, it cannot compensate for some of the errors that may arise. In order to gain an appreciation for the fundamental limits associated with the solution of linear systems, the following section discusses the propagation of error in more detail.

4.3 CONDITION NUMBERS AND ERROR PROPAGATION IN THE SOLUTION OF LINEAR SYSTEMS

Numerical errors arise during factorization due to the finite precision of the machine and the limited accuracy of the various quantities. The *matrix condition number* is a useful concept for treating error propagation. The condition number of a general, Nth-order complex-valued matrix **A** can be defined as

$$\kappa(\mathbf{A}) = \sqrt{\frac{\lambda_{\max}}{\lambda_{\min}}} \tag{4.11}$$

where $\lambda_{\min}$ and $\lambda_{\max}$ denote the smallest and largest eigenvalues, respectively, of the matrix $\mathbf{A}^\dagger\mathbf{A}$, where $\mathbf{A}^\dagger$ is the transpose conjugate of **A**. A more general definition of the condition number is given by

$$\kappa(\mathbf{A}) = \|\mathbf{A}\|\,\|\mathbf{A}^{-1}\| \tag{4.12}$$

where the matrix norm is

$$\|\mathbf{A}\| = \sup_{x \neq 0} \frac{\|\mathbf{Ax}\|}{\|\mathbf{x}\|} \tag{4.13}$$

and a variety of vector norms are possible. Use of the Euclidean 2-norm

$$\|\mathbf{x}\| = \sqrt{\mathbf{x}^\dagger \mathbf{x}} \tag{4.14}$$

is common, as is the use of the infinity norm. The actual condition number varies with the definition of the norm.

The condition number is a measure of the stability of the linear system of equations. A singular matrix, containing at least one degenerate row, has at least one zero eigenvalue and an infinite condition number. The identity matrix, with every eigenvalue equal to 1, has a condition number of 1. If the condition number is much greater than unity, we say that the matrix is *ill-conditioned*. The greater the condition number of a linear system, the more sensitive the equations to slight perturbations and the more numerical error likely to appear in the solution.

Since the matrix equations generated using the procedures of Chapters 2 and 3 can occasionally be poorly conditioned, we want to consider the effects of ill-conditioning in more detail. Because of numerical errors, the computed solution of $\mathbf{Ax} = \mathbf{b}$ will differ from the desired **x** by an amount $\Delta\mathbf{x}$. Two distinct sources of solution error are (1) error due to approximations in the construction of the entries of **A** and **b** or the truncation of the entries of **A** or **b** due to finite machine precision and (2) additional round-off and truncation error introduced by the Gaussian elimination process.

The effect of errors in the entries of **A** or **b** can be estimated by considering the perturbed equation

$$(\mathbf{A} + \Delta\mathbf{A})(\mathbf{x} + \Delta\mathbf{x}) = (\mathbf{b} + \Delta\mathbf{b}) \tag{4.15}$$

To estimate bounds on the error $\Delta\mathbf{x}$, consider Equation (4.15) under the condition when $\Delta\mathbf{A} = \mathbf{0}$. The Schwartz inequality can be applied to $\mathbf{Ax} = \mathbf{b}$ to produce

$$\|\mathbf{b}\| \leq \|\mathbf{A}\|\,\|\mathbf{x}\| \tag{4.16}$$

The combination of Equation (4.15) with $\Delta\mathbf{A} = \mathbf{0}$ and the equation $\mathbf{Ax} = \mathbf{b}$ yields the inequality

$$\|\Delta\mathbf{x}\| \leq \|\mathbf{A}^{-1}\|\,\|\Delta\mathbf{b}\| \tag{4.17}$$

The preceding two inequalities can be combined to produce

$$\frac{\|\Delta\mathbf{x}\|}{\|\mathbf{x}\|} \leq \kappa(\mathbf{A}) \frac{\|\Delta\mathbf{b}\|}{\|\mathbf{b}\|} \tag{4.18}$$

Similar inequalities constructed under the assumption that $\Delta\mathbf{b} = \mathbf{0}$ can be combined to yield

$$\frac{\|\Delta\mathbf{x}\|}{\|\mathbf{x}+\Delta\mathbf{x}\|} \leq \kappa(\mathbf{A})\frac{\|\Delta\mathbf{A}\|}{\|\mathbf{A}\|} \tag{4.19}$$

Equations (4.18) and (4.19) provide upper bounds on the error in the solution $\mathbf{x}$ due to errors in the entries of $\mathbf{A}$ or $\mathbf{b}$. Note the presence of the condition number $\kappa(\mathbf{A})$ in these inequalities. Since the condition number is always greater than 1 in practice, the solution error $\Delta\mathbf{x}$ can be significantly larger than the errors in the individual matrix entries. In a sense, the error originally present in the matrix entries can be amplified in proportion to the condition number.

The effect of machine precision on the entries of $\mathbf{A}$ and $\mathbf{b}$ can be estimated using Equations (4.18) and (4.19) with $\Delta\mathbf{A}$ and $\Delta\mathbf{b}$ set to the value associated with the machine word length. For instance, on a machine with 32-bit (24-bit mantissa, 8-bit exponent) binary words, this truncation error ε is on the order of 2^{-24}. Of course, unless the entries of $\mathbf{A}$ and $\mathbf{b}$ are accurate to this order, the additional error due to machine truncation will be negligible compared to the uncertainty already present.

In summary, the finite accuracy of the entries in the matrix and the right-hand side can produce an error $\Delta\mathbf{x}$ in the solution that is proportional to the error in $\mathbf{A}$ and $\mathbf{b}$ amplified by the condition number $\kappa(\mathbf{A})$. Since this error is present in the original system, it cannot be reduced by subsequent operations. It can only be reduced by a more accurate evaluation of the matrix entries and, if necessary, by increased machine precision. In addition, this error is completely independent of additional errors introduced by the solution process.

The additional numerical error introduced during the factorization procedure was originally analyzed by Wilkinson [11] and is discussed in several recent references [3, 4, 7, 12]. Wilkinson's backward-error analysis postulated that the actual solution $(\mathbf{x} + \Delta\mathbf{x})$ exactly satisfies a perturbed equation

$$(\mathbf{A} + \Delta\mathbf{A})(\mathbf{x} + \Delta\mathbf{x}) = \mathbf{b} \tag{4.20}$$

A bound on the equivalent error $\Delta\mathbf{A}$ can be obtained in terms of the infinity norm and has the form [3, 4, 7–10, 12]

$$\frac{\|\Delta\mathbf{A}\|}{\|\mathbf{A}\|} \leq \alpha\rho\varepsilon + O(\varepsilon^2) \tag{4.21}$$

where ε is the machine unit round-off [approximately 2^{-24} on a 32-bit machine employing the Institute of Electrical and Electronics Engineers (IEEE) arithmetic], α is approximately N^3 (where N is the order of $\mathbf{A}$), and ρ is Wilkinson's growth factor. The growth factor is proportional to the largest number arising throughout the elimination process and is difficult to estimate. However, for general matrices factorized with partial pivoting, the growth factor is bounded by

$$\rho < 2^{N-1} \tag{4.22}$$

Combining these estimates with Equation (4.19) produces the approximate solution error bound

$$\frac{\|\Delta\mathbf{x}\|}{\|\mathbf{x}+\Delta\mathbf{x}\|} \leq \kappa(\mathbf{A})N^3 2^{N-1}\varepsilon \tag{4.23}$$

It is necessary to employ a variety of assumptions in order to obtain a simple expression for the error bound in Equation (4.21). These include the assumption that $N\varepsilon < 0.1$, the assumption that every entry of the matrix is nonzero, and the worst-case assumption that each matrix element contributes to the error [11, 12]. The resulting expression in Equation (4.23) is a very loose bound but does indicate that the expected error is proportional to the condition number and can be improved by increasing the machine precision. Unfortunately, the bound is almost useless for estimating the error as a function of N, since authorities appear to agree (!) that in practice the dependence on N^3 is extremely pessimistic and that the growth factor is almost never as bad as 2^{N-1} for actual applications (although matrices can be constructed for which ρ does grow as 2^{N-1}) [3, 4]. Because the bound in (4.23) is so loose, the additional error introduced by the elimination process is usually estimated as being comparable to the error introduced by the truncation of the matrix entries. This error estimate is embodied in the often-quoted rule of thumb: *Gaussian elimination produces a solution that has about*

$$t \log_{10} \beta - \log_{10} \kappa(\mathbf{A}) \tag{4.24}$$

correct decimal places, where β is the base ($\beta = 2$ for binary operations) and t is the bit length of the mantissa ($t = 24$ for a typical 32-bit binary word). This is equivalent to saying that the solution will lose about $\log_{10}\kappa$ digits of accuracy during elimination.

In summary, there are two independent sources of error in the solution of a matrix equation. The first is introduced by the inaccuracy of the matrix entries, the second by the elimination process. Although the latter could easily dominate for large-order systems, it is difficult to estimate and can only be controlled by increasing the precision. Practical experience suggests that the primary source of error is likely to be due to inaccuracies in the entries of $\mathbf{A}$ and $\mathbf{b}$, assuming that the entries are far less accurate than machine precision would allow. Since these inaccuracies may produce an error $\kappa(\mathbf{A})$ times as large in the solution, it is important to ensure the accurate evaluation of the matrix entries to more digits than $\log_{10}\kappa$. However, the matrix entries arising from the integral equation formulations of Chapter 2 were seldom available in exact, closed-form expressions. In practice, these entries are usually computed numerically or approximated by some convenient expression. (The finite-element modeling described in Chapter 3 permitted an accurate closed-form evaluation of some of the matrix entries.) In addition, both the integral and differential equation formulations required scatterer models that were, at best, only approximations to the desired geometry. These inherent approximations place a fundamental limit on the accuracy possible in entries of the matrix equation. In order to monitor the likely presence of error in the numerical solution, the condition number is usually estimated during elimination. For example, the LINPACK routine CGECO performs an LU factorization and estimates the condition number of a complex matrix [6]. Based upon the known accuracy of the matrix entries, the condition number, and the machine precision, a decision can be made as to the likely accuracy of the numerical result.

4.4 CHOLESKY DECOMPOSITION FOR COMPLEX-SYMMETRIC SYSTEMS

Often, the matrix equations arising from the formulations of Chapter 2 and 3 are complex symmetric in nature. For symmetric systems, an alternative factorization known as Cholesky decomposition can exploit symmetry to reduce the required computer storage and the required computation by a factor of 2 over that required by general-purpose LU factorization routines.

Naive Cholesky decomposition seeks a factorization in the form

$$\mathbf{A} = \mathbf{L}\mathbf{L}^T \tag{4.25}$$

where $\mathbf{L}^T$ is the transpose of the matrix $\mathbf{L}$, which is lower triangular. (Note that the transpose is different from the transpose conjugate.) It is readily verified that such a factorization is only possible if $\mathbf{A}$ is symmetric, that is,

$$\mathbf{A} = \mathbf{A}^T \tag{4.26}$$

The entries of $\mathbf{L}$ may be found by expanding (4.25), producing the formulas

$$l_{11} = \sqrt{a_{11}} \tag{4.27}$$

$$l_{n1} = \frac{a_{n1}}{l_{11}} \qquad n = 2, 3, \ldots, N \tag{4.28}$$

$$l_{22} = \sqrt{a_{22} - (l_{21})^2} \tag{4.29}$$

$$l_{n2} = \frac{a_{n2} - l_{21}l_{n1}}{l_{22}} \qquad n = 3, 4, \ldots, N \tag{4.30}$$

and so on. Since the operations are performed with complex arithmetic, the square root of a number having a negative real part is not a difficulty.

Although it is generally not possible to incorporate complete or partial pivoting into the Cholesky factorization procedure, alternative pivoting schemes have been proposed [13–16]. One approach, the diagonal pivoting method, is incorporated into the LINPACK library [16].

The advantage of the Cholesky algorithm is that the factorization requires approximately $N^3/6$ multiplications and additions, half of that necessary for LU factorization. Because of symmetry, it is only necessary to compute half of the matrix $\mathbf{A}$. The lower triangular matrix $\mathbf{L}$ is stored in the same space originally used by $\mathbf{A}$. As an additional hedge against rounding errors, the summations appearing in Equations (4.29), (4.30), and the rest that follow can be accumulated in double precision.

4.5 REORDERING ALGORITHMS FOR SPARSE SYSTEMS OF EQUATIONS

The differential equation formulations discussed in Chapter 3 produce matrix equations that are *sparse* (most of the entries are zero). It is obviously not economical to employ full-storage Gaussian elimination with sparse systems, and we turn our attention to special-purpose methods that exploit the matrix sparsity. Unfortunately, the LU factorization is usually not as sparse as the original system. *Fill-in* occurs whenever entries that were zero in the original matrix are replaced by nonzero entries in the factorization. However, a property of Gaussian elimination ensures that, in the absence of pivoting, all entries located to the left of the first nonzero entry in a given row of the matrix remain

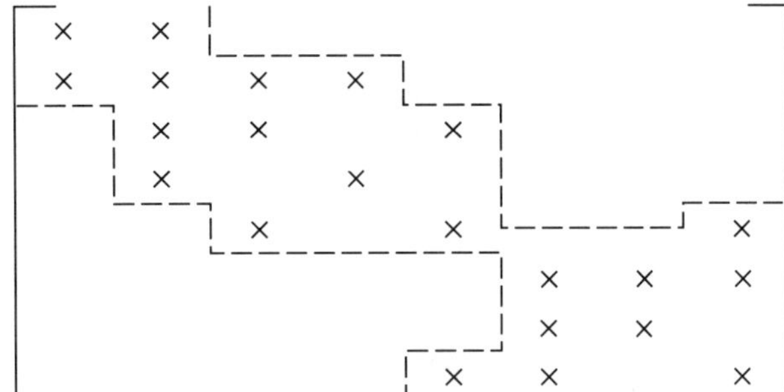

Figure 4.1 Envelope of a matrix.

zero after factorization [2, 7, 17, 18]. This property can be exploited to provide a simple way of reducing the necessary storage to the *envelope* containing the nonzero entries (Figure 4.1).

One way of attempting to optimize the storage and computation associated with factorization algorithms is to minimize the size of the envelope. The actual sparsity pattern depends on the ordering of the matrix rows and columns, which in turn depends on the numbering scheme used within the finite-element mesh. Thus, it is possible in principle to optimize the numbering scheme used to construct the scatterer model in order to enhance the efficiency of the matrix solution algorithm. A variety of automatic reordering algorithms have been proposed for this purpose.

Chapter 3 considered the first-order finite-element discretization of the scalar Helmholtz equation. A two-dimensional model and the associated matrix sparsity pattern are depicted in Figure 3.11. The specific sparsity pattern is a direct result of the scheme used to assign numbers to different nodes of the triangular-cell mesh representing the cylinder cross section. Because adjacent nodes in Figure 3.11 are occasionally assigned widely differing numbers, there are nonzero entries of the matrix very far from the main diagonal.

Figure 4.2 depicts an identical model employing a numbering scheme that attempts to minimize the effective *bandwidth* of the matrix (the maximum deviation from the diagonal). Note that nodes have been numbered in an attempt to avoid large differences between the indices of adjacent nodes. Figure 4.3 shows the resulting sparsity pattern. In this case, the automatic node-reordering algorithm of Puttonen [19] was employed to produce the numbering scheme of Figures 4.2 and 4.3. Algorithms that attempt to minimize the bandwidth of the matrix are generally used in connection with a solver that stores only a banded portion of the matrix.

Of the numerous automatic reordering algorithms that have been proposed, we will examine only the Cuthill–McKee algorithm [2, 7, 17, 20–23]. Although the Cuthill–McKee algorithm will generally not produce a sparsity pattern with optimum bandwidth, variations on it are widely used in connection with solvers that store only the envelope of the matrix. We will describe the algorithm by example while using it to optimize the ordering of the mesh depicted in Figure 3.11.

The starting point in the Cuthill–McKee algorithm is to determine the *degree* of each node in the mesh. For first-order interpolation functions, the degree of a node is the number of immediately adjacent nodes in the mesh. Table 4.1 lists the degree of each node in the mesh of Figure 3.11.

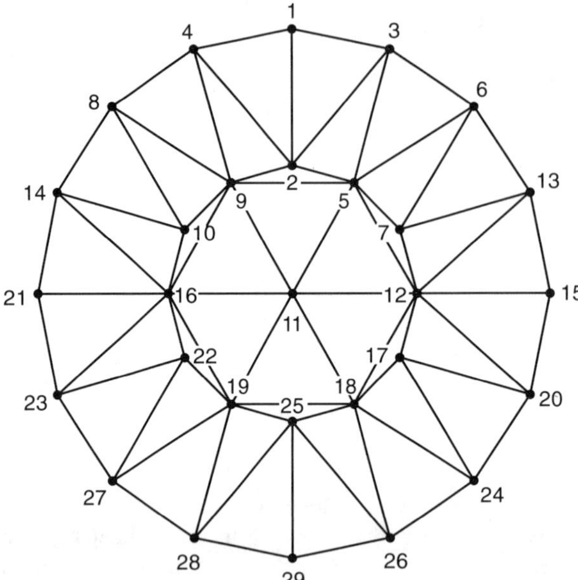

Figure 4.2 Triangular-cell model showing a node-indexing scheme that attempts to minimize the bandwidth of the associated finite-element matrix.

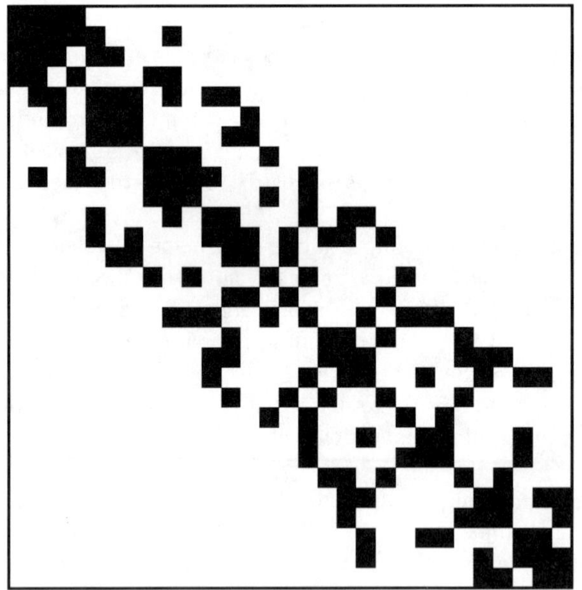

Figure 4.3 Matrix sparsity pattern produced by the node-numbering scheme of Figure 4.2 for a first-order finite-element representation.

The procedure begins with the identification of the node with the smallest degree of connectivity (nodes numbered 14, 18, 22, or 26 in Table 4.1 are each connected to three other nodes). We select node 14 arbitrarily and relabel it as the new node 1. The second step of the procedure is to consider all the nodes connected to the "new" node 1 and order these so that they are arranged from smallest to largest degree of connectivity. Nodes 2, 15,

TABLE 4.1 Degree of Connectivity of Each Node in Mesh of Figure 3.11 Assuming First-Order Interpolation Functions

Node Number	Connectivity (deg)	Node Number	Connectivity (deg)
1	6	16	4
2	8	17	4
3	7	18	3
4	4	19	4
5	7	20	4
6	5	21	4
7	8	22	3
8	4	23	4
9	7	24	4
10	4	25	4
11	7	26	3
12	5	27	4
13	4	28	4
14	3	29	4
15	4		

and 29 are connected to the new node 1, and we renumber these according to the following permutation list:

Old Number	New Number
14	1
15	2
29	3
2	4

(Since nodes 15 and 29 have the same degree, we arbitrarily place the lowest index first.) We next consider each of the remaining nodes connected to any on the new list, adding them to the new list in the order of increasing degree. In this case, nodes 4 and 16 are connected to the original node 15 (the new node 2); nodes 13 and 28 are connected to the original node 29 (the new node 3); and nodes 1, 3, and 11 are connected to the original node 2 (new node 4). After adding these in order of lowest degree, we obtain the following:

Old Number	New Number
4	5
16	6
13	7
28	8
1	9
3	10
11	11

We now continue this procedure, identifying any nodes connected to new nodes 5, 6, and so on, and adding them to the list in the order of increasing degree. There are no nodes connected to new node 5 that have not already been added to the permutation list. New node 6 is connected to node 17, so we renumber it as "new" node 12. We add nodes 27, 5, 7, 9, 6, and 12 in a similar fashion. Continuing through the entire list, we finally obtain the following:

Old Number	New Number
17	12
27	13
5	14
9	15
7	16
6	17
12	18
18	19
26	20
8	21
19	22
20	23
10	24
24	25
25	26
22	27
21	28
23	29

(If at some point in the procedure we find that we have exhausted the nodes connected to those on the permutation list, we begin anew by identifying the remaining node with the smallest connectivity. This would only happen if the model consisted of several isolated parts.) After permutation according to the Cuthill–McKee procedure, the matrix sparsity pattern for this example is displayed in Figure 4.4.

In several places throughout the above procedure, an arbitrary choice was made about the order of nodes. Consequently, there may be numerous different permutation lists produced by this procedure, and there is no guarantee that the specific permutation we obtained is the best that can be found (see Prob. P4.6). In addition, it is important to note that the Cuthill–McKee (CM) procedure is a relatively simple algorithm that will not, in general, produce the best possible numbering. In fact, the numbering employed to generate Figure 4.2 appears to be slightly better than this permutation (Figure 4.4).

George observed that the Cuthill–McKee algorithm could be improved for the purpose of minimizing the envelope of a sparse matrix by reversing the order of the Cuthill–McKee numbering [2, 7]. This produces the common implementation of the procedure, known as the *reverse Cuthill–McKee* (RCM) algorithm. Figure 4.5 shows the sparsity pattern and envelope obtained using the reverse permutation of the preceding example. Table 4.2 summarizes the relative efficiency of the four numbering schemes presented in this section.

Alternative reordering algorithms are available that may be more effective than the RCM approach [7, 17–26]. These procedures are beyond the scope of the present text.

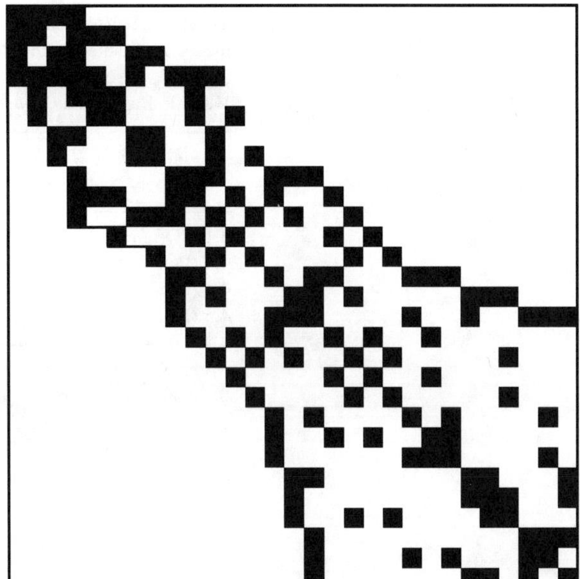

Figure 4.4 Matrix sparsity pattern produced by the Cuthill–McKee algorithm.

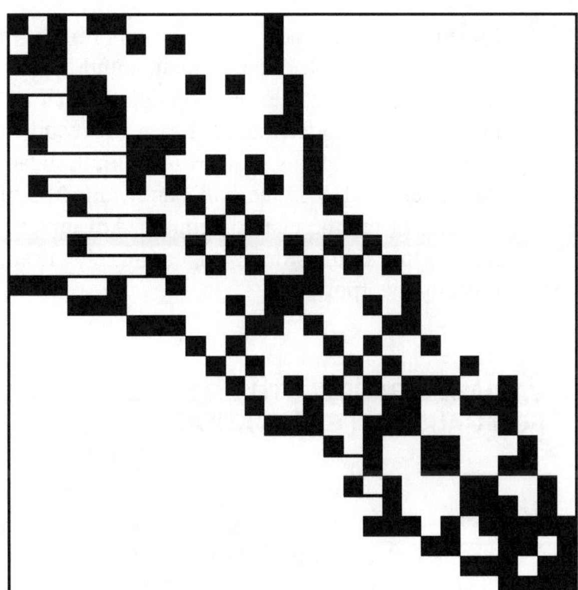

Figure 4.5 Matrix sparsity pattern produced by the reverse Cuthill–McKee algorithm.

Although our example involved the nodes of a two-dimensional mesh, reordering algorithms are readily applied to order the indices associated with cells, edges, or any other quantity of interest. Commercial preprocessors used for general mesh generation often incorporate some type of automatic reordering algorithm [27].

TABLE 4.2 Comparison of Storage Requirements Associated with Matrices of Figures 3.11, 4.3, 4.4, and 4.5 for Fixed-Bandwidth and Variable-Bandwidth (Envelope) Schemes

Figure Number	Banded	Envelope
3.11 (Original)	839	631
4.3 (Puttonen)	461	337
4.4 (CM)	601	403
4.5 (RCM)	601	335

Note: Numbers reflect only the storage of the matrix entries and do not include additional overhead or pointer arrays. Full-matrix storage for this example requires 841 locations, although there are only 165 nonzero entries in the original matrix.

4.6 BANDED STORAGE FOR GAUSSIAN ELIMINATION

If the numbering scheme effectively optimizes the bandwidth of the system (as perhaps the case with Figure 4.3), an efficient solution may be obtained by storing only a fixed band of the matrix entries located around the main diagonal. The LU factorization algorithms incorporating banded matrix storage are readily available in libraries such as LINPACK [6]. Factorization requires about $NB^2/2$ multiplications and additions, where N is the order of the system and B is the half-bandwidth (the maximum distance that a nonzero entry is displaced from the main diagonal). Advantages of banded-storage elimination algorithms include the simple storage allocation scheme and the fact that a modified form of pivoting is easily incorporated [2].

4.7 VARIABLE-BANDWIDTH OR ENVELOPE STORAGE FOR GAUSSIAN ELIMINATION

An improvement in efficiency over the banded matrix solvers can usually be realized through the use of factorization algorithms that store only the envelope of the matrix (Figure 4.1). Since they do not require a fixed band, these algorithms permit more flexibility in the numbering scheme used within the scatterer model. As illustrated by the comparisons presented in Table 4.2, envelope schemes require less storage than banded solvers. The improvement in efficiency comes at a cost that includes (1) the increased complexity of the storage scheme, which must use an additional array as a pointer to specify the shape of the envelope, and (2) the fact that it is generally more difficult to incorporate pivoting into the factorization algorithm.

In practice, the matrix envelope is stored in three parts: the main diagonal of the matrix (the *diagonal*), the matrix entries located below the main diagonal (the *lower enve-*

lope), and the matrix entries located above the main diagonal (the *upper envelope*). (If the matrix is symmetric and Cholesky decomposition is to be employed, the upper envelope is unnecessary.) Each of these three parts can be stored in a one-dimensional array. A pointer of some sort is necessary to specify the location of the original (i, j)th matrix entry in the one-dimensional arrays. The applications considered in this text always involve a matrix sparsity pattern that is symmetric across the main diagonal, which simplifies the required bookkeeping (i.e., only one pointer is necessary to specify the shape of the upper and lower envelopes, even if the matrix entries are not symmetric). The pointer array used to identify the location of the (i, j) matrix entry in the one-dimensional arrays must be constructed prior to the "matrix fill" procedure and may be generated as part of the scatterer model in anticipation of a variable-bandwidth factorization. (To construct the pointer, it is only necessary to identify the locations of nonzero entries in the matrix. Since these are completely determined by the numbers assigned to the nodes of the mesh, the pointer is easily constructed once the mesh is specified.)

A simple storage scheme is proposed by George and Liu [17]. For the lower envelope, a pointer identifies the location in the single dimensional array of the first entry in each row of the matrix. The number of entries in a given row can be found from the difference between successive indices within the pointer. As an example, consider the matrix depicted in Figure 4.6, which has a lower envelope containing eight entries. The pointer $P = [1, 1, 1, 2, 5, 7, 9]$ identifies the location in the envelope of the first entry in each row of the $N \times N$ matrix. One additional entry at the end of the pointer acts as a flag to the end

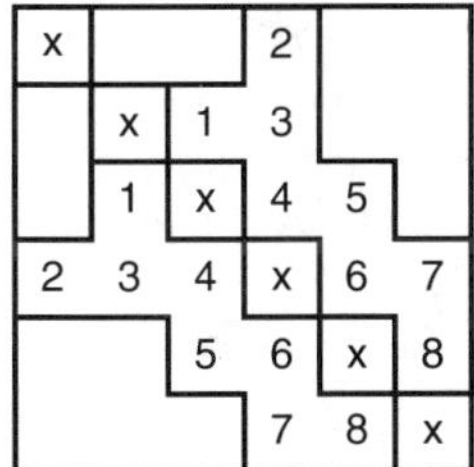

Row	Entries	Starting entry	Pointer
1	0	—	1
2	0	—	1
3	1	1	1
4	3	2	2
5	2	5	5
6	2	7	7
			9

Figure 4.6 Illustration of a scheme for storing the lower envelope of a sparse matrix in a one-dimensional array.

of the envelope. Assuming that $i > j$, the (i, j)th entry of the matrix is located at the $P[i+1] - i + j$ location in the lower envelope array. For instance, the (4, 2) entry of the matrix is stored at the third position in the envelope (location $P[5] - 4 + 2 = 3$).

Factorization algorithms employing variable-bandwidth storage have been presented in FORTRAN or pseudocode by Jennings [2], George and Liu [17], and Hoole [23] for the case of positive-definite, real symmetric sparse matrices. It is possible to modify these algorithms to perform the naive LU factorization of complex matrices. Unfortunately, it is not easy to incorporate pivoting into the factorization process (since the interchange of rows or columns necessarily alters the matrix envelope).

4.8 SPARSE MATRIX METHODS EMPLOYING DYNAMIC STORAGE ALLOCATION

Banded or envelope solution schemes have fixed or *static* storage requirements that are established prior to the start of factorization. These algorithms are relatively unsophisticated and, since many zeros may be stored within the envelope, are often far from optimum for large-order systems. To avoid storing unnecessary zeros, more sophisticated factorization algorithms allocate storage *dynamically*. Examples of these algorithms include the Yale Sparse Matrix Package (YSMP) [28], MA28 and ME28 from the Harwell library [7, 29, 30], and Y12M [31, 32].

A matrix can be stored in packed form in a variety of ways, including (1) a linked list of the nonzero entries and (2) a one-dimensional array with pointers describing the row and column locations of each entry. These storage structures are flexible enough to incorporate any new entries that arise during factorization at the end of the list. For user convenience, dynamic factorization algorithms usually accept the original matrix entries in any order and sort them prior to factorization. Additional sorting, or "garbage collecting," may be performed throughout factorization in order to free workspace [32]. Since the storage structure can incorporate fill-in in any region of the matrix, pivoting can be included in the algorithm. However, in contrast to the full-storage case, pivots are primarily chosen in an attempt to minimize fill-in or minimize the number of arithmetic operations. By ensuring that pivots are always larger than some predetermined value, sufficient numerical stability can be provided as well. The details of the available algorithms are beyond the scope of the present discussion, and readers are encouraged to consult the texts by Duff, Erisman, and Reid [7], George and Liu [17], and Pissanetzky [18] for comprehensive overviews of these procedures.

We illustrate the performance of one algorithm, a complex-valued adaptation of Y12M, when used to solve the sparse system arising from a discretization of the scalar Helmholtz equation augmented with the Bayliss–Turkel radiation condition (Section 3.9). Piecewise-linear interpolation functions are used, as described in Chapter 3. No node re-ordering is applied prior to factorization. As an indication of the storage efficiency, Figure 4.7 shows the number of nonzero entries in the packed matrix before and after factorization by Y12M. For these two-dimensional finite-element problems, Figure 4.7 suggests that fill-in during factorization amplifies the storage requirements by a factor of 5. For large systems, this factor translates into a storage requirement of roughly 30 entries per row of the matrix, a smaller number than usually possible with banded or envelope storage.

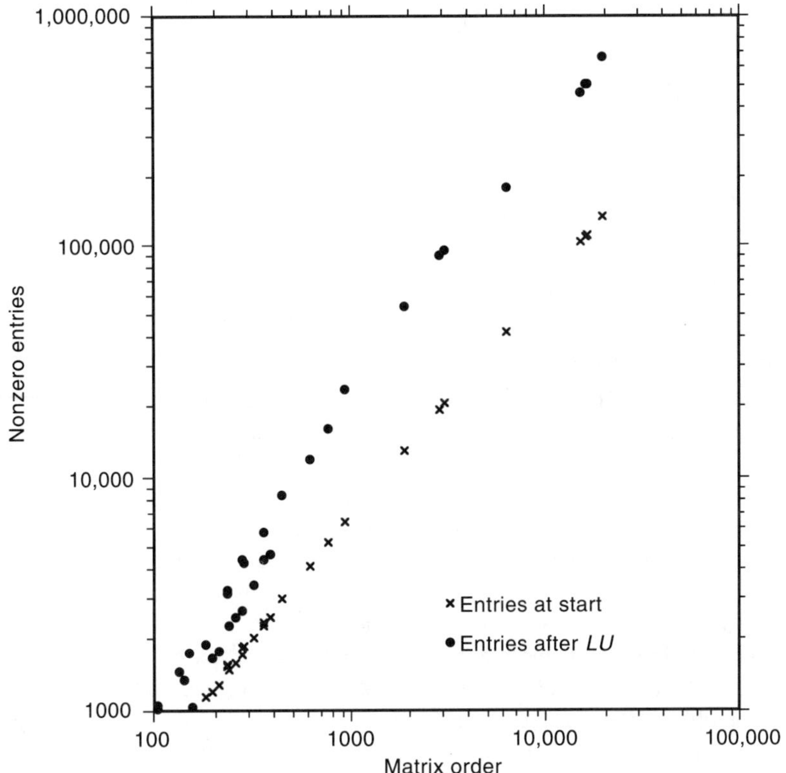

Figure 4.7 Performance of the sparse LU factorization code Y12M when used to solve a number of first-order finite-element systems. To illustrate fill-in, the graph shows the number of entries before and after factorization as a function of matrix order.

4.9 FRONTAL ALGORITHM FOR GAUSSIAN ELIMINATION

The frontal solution method is an alternative direct method pioneered by Irons [33] as a way to reduce the amount of fast-access memory required to solve large systems. It is of interest for use with machines having a limited amount of fast-access memory and in modified form for certain parallel-architecture computers. To illustrate the frontal algorithm, consider the finite-element mesh shown in Figure 4.8. Figure 4.8 depicts eight triangular elements $a, b, \ldots, h$. Nine nodes in the mesh are identified by number. The initial step of the frontal method requires the ordering of the nodes to determine the order by which equations are eliminated from the system. In step 1 of the procedure, we assemble element a and activate the equations associated with nodes 1, 2, and 3. Step 2 involves the recognition that element a is the only cell that contributes to equation 1 of the global system. Thus, equation 1 is complete and Gaussian elimination can be performed to eliminate the corresponding unknown from the system. At this point, the reduced equation is written to secondary storage (i.e., a hard disk, solid-state disk, etc.) to release primary computer memory for future use.

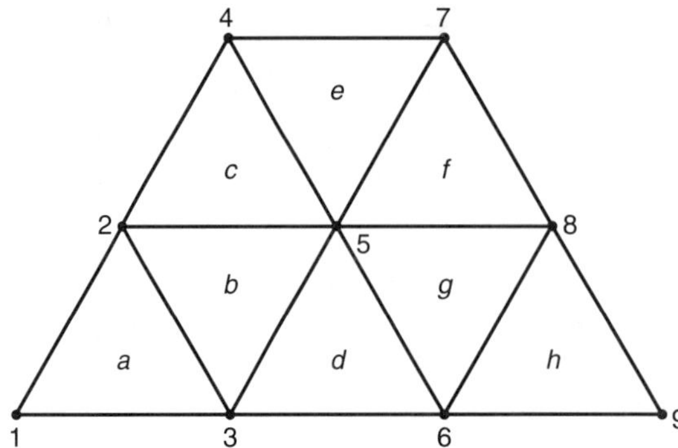

Figure 4.8 Hypothetical triangular-cell mesh.

Step 3 of the frontal procedure is to assemble element b into the "front" containing the two equations surviving from step 2 and activate the additional equation corresponding to node 5. Step 4 is to assemble element c into the front and activate the equation associated with node 4. Step 5 is to recognize that equation 2 is now complete and can be eliminated from those equations remaining in primary memory (i.e., transferred to disk). Step 6 of the algorithm assembles element d and activates the equation for node 6. Step 7 is to eliminate equation 3 from the system. In step 8, element e is assembled and the equation at node 7 activated. Step 8 involves the elimination of the equation corresponding to node 4.

At this point, the algorithm has activated seven equations and eliminated four of these from primary computer memory. The process continues with the remaining cells, assembling f, g, and h until the frontal algorithm has reduced the entire system by Gaussian elimination. To obtain the solution, the final step employs back substitution in reverse order through the equations. For this simple example, only four equations are held in primary storage at any given time, compared to nine equations that would have to be retained if the complete system was treated in a conventional manner. As the mesh grows in size, the savings become more apparent. Memory savings will be improved as the aspect ratio of the mesh becomes more extreme. For example, a mesh that is 400 elements long and 10 elements wide yields an aspect ratio of 40. If the elements are sequenced across the width of the mesh, the maximum "frontwidth" at any time will be 22 nodes. On the other hand, if elements are improperly sequenced along the length of the mesh, the frontwidth will exceed 800 nodes. Reordering algorithms can be adapted to minimize frontwidth and can greatly improve the efficiency of this type of solution.

The drawback to the frontal solver is the speed of solution compared to in-core factorization algorithms. Since the frontal scheme requires transfers to secondary memory, its relative efficiency is machine dependent. However, the idea behind the frontal algorithm can also be applied to decompose problem domains for distributed-memory parallel-architecture computers.

4.10 ITERATIVE METHODS FOR MATRIX SOLUTION

The direct methods of the preceding sections are generally the preferred avenue of approach for treating general linear systems of equations. However, it is often difficult to fully exploit sparsity or special structure with direct methods. Furthermore, direct methods for sparse systems are difficult to adapt to parallel-architecture computers. A wide variety of iterative methods have been developed [34, 35], but few are applicable to the general complex-valued matrix equations arising in electromagnetic radiation and scattering formulations. One approach that is applicable to general complex systems is the conjugate gradient algorithm [36, 37], and we consider this technique in detail in the following sections. For problems formulated in terms of integral equations, a common iterative implementation is the conjugate gradient–fast Fourier transform (CG–FFT) procedure. The CG–FFT implementation has found widespread use for treating systems with slightly perturbed Toeplitz symmetries. For problems lacking Toeplitz symmetries, the CG algorithm can be used with the fast multipole procedure (Section 4.13) to realize an efficiency similar to that of the CG–FFT approach.

Various extensions of the CG algorithm appear to offer better efficiency for poorly conditioned systems and may be more effective for treating the sparse systems arising from differential equation formulations than the CG algorithm we describe below. We refer the reader to the literature for a discussion of alternative approaches such as the biconjugate gradient procedure [38–40], the generalized minimal-residual method (GMRES) [40–42], and the quasi-minimal-residual (QMR) algorithm [40, 42–44].

4.11 THE CONJUGATE GRADIENT ALGORITHM FOR GENERAL LINEAR SYSTEMS [45]

There are several different ways in which the CG algorithm can be developed. For instance, we can view the algorithm as a procedure for the minimization of an error functional or as a process for constructing an orthogonal expansion of the solution. In actuality, these two ideas are linked together, that is, each functional is associated with a specific orthogonal expansion. The CG algorithm reduces to the process of generating the orthogonal vectors and finding the proper coefficients to represent the desired solution.

Consider the nonsingular matrix equation

$$\mathbf{A}\mathbf{x} = \mathbf{b} \tag{4.31}$$

where $\mathbf{A}$ denotes an $N \times N$ matrix, $\mathbf{x}$ is the unknown $N \times 1$ column vector to be determined, and $\mathbf{b}$ is a given $N \times 1$ column vector that we usually denote as the "right-hand side." It is necessary to define an inner product, and we employ the conventional Euclidean scalar product

$$\langle \mathbf{x}, \mathbf{y} \rangle = \mathbf{x}^\dagger \mathbf{y} \tag{4.32}$$

and its associated norm

$$\|\mathbf{x}\| = \sqrt{\langle \mathbf{x}, \mathbf{x} \rangle} \tag{4.33}$$

where the dagger denotes the transpose-conjugate matrix.

All iterative algorithms for solving $\mathbf{A}\mathbf{x} = \mathbf{b}$ seek an estimate of the solution in the form

$$\mathbf{x}_n = \mathbf{x}_{n-1} + \alpha_n \mathbf{p}_n \tag{4.34}$$

where $\mathbf{x}_{n-1}$ is a previous estimate of the solution, $\mathbf{p}_n$ is a "direction" vector ($\mathbf{p}_n$ determines the direction in the N-dimensional space in which the algorithm moves to correct the estimate of $\mathbf{x}$), and α_n is a scalar coefficient that determines how far the algorithm moves in the $\mathbf{p}_n$ direction. Although all iterative methods are similar in that they follow the form of Equation (4.34), they differ in the procedure by which they generate α_n and $\mathbf{p}_n$. Nondivergence can be guaranteed by selecting α_n in order to minimize an error functional. The CG algorithm to be presented is based on the error functional

$$E_n(\mathbf{x}_n) = \|\mathbf{A}\mathbf{x}_n - \mathbf{b}\|^2 \tag{4.35}$$

Note that other functionals have been used and give rise to related members of the family of CG algorithms. The coefficient α_n from (4.34) that minimizes the functional is given by (Prob. P4.7)

$$\alpha_n = \frac{-\langle \mathbf{A}\mathbf{p}_n, \mathbf{r}_{n-1}\rangle}{\|\mathbf{A}\mathbf{p}_n\|^2} \tag{4.36}$$

where for convenience we define the residual vector

$$\mathbf{r}_n = \mathbf{A}\mathbf{x}_n - \mathbf{b} \tag{4.37}$$

This process has the geometric interpretation of minimizing E_n along the line in N-dimensional space defined by $\mathbf{p}_n$.

It is reasonable to expect that an improved algorithm would seek the minimum of E_n in a plane spanned by two direction vectors. For example, consider a solution estimate of the form

$$\mathbf{x}_n = \mathbf{x}_{n-1} + \alpha_n(\mathbf{p}_n + \beta_n \mathbf{q}_n) \tag{4.38}$$

where the direction vectors $\mathbf{p}_n$ and $\mathbf{q}_n$ span a plane in N-dimensional space and the scalar coefficients α_n and β_n are to be obtained in order to simultaneously minimize the error functional E_n. Carrying out the simultaneous minimization, we find that α_n is given by Equation (4.36) with $\mathbf{p}_n$ replaced by $\mathbf{p}_n + \beta_n\mathbf{q}_n$ and β_n is given by

$$\beta_n = \frac{\langle \mathbf{A}\mathbf{q}_n, \mathbf{r}_{n-1}\rangle \|\mathbf{A}\mathbf{p}_n\|^2 - \langle \mathbf{A}\mathbf{p}_n, \mathbf{r}_{n-1}\rangle\langle \mathbf{A}\mathbf{q}_n, \mathbf{A}\mathbf{p}_n\rangle}{\langle \mathbf{A}\mathbf{p}_n, \mathbf{r}_{n-1}\rangle \|\mathbf{A}\mathbf{q}_n\|^2 - \langle \mathbf{A}\mathbf{q}_n, \mathbf{r}_{n-1}\rangle\langle \mathbf{A}\mathbf{p}_n, \mathbf{A}\mathbf{q}_n\rangle} \tag{4.39}$$

Suppose that $\mathbf{p}_n$ and $\mathbf{q}_n$ are arbitrary direction vectors but that $\mathbf{q}_n$ has been previously used in the iterative procedure, so that

$$\mathbf{x}_{n-1} = \mathbf{x}_{n-2} + \alpha_{n-1}\mathbf{q}_n \tag{4.40}$$

where α_{n-1} was previously found to minimize the error functional along the line defined by $\mathbf{q}_n$, that is,

$$\alpha_{n-1} = \frac{-\langle \mathbf{A}\mathbf{q}_n, \mathbf{r}_{n-2}\rangle}{\|\mathbf{A}\mathbf{q}_n\|^2} \tag{4.41}$$

It immediately follows that

$$\langle \mathbf{A}\mathbf{q}_n, \mathbf{r}_{n-1}\rangle = \langle \mathbf{A}\mathbf{q}_n, \mathbf{r}_{n-2} + \alpha_{n-1}\mathbf{A}\mathbf{q}_n\rangle = 0 \tag{4.42}$$

$$\beta_n = \frac{-\langle \mathbf{A}\mathbf{q}_n, \mathbf{A}\mathbf{p}_n\rangle}{\|\mathbf{A}\mathbf{q}_n\|^2} \tag{4.43}$$

and

$$\langle \mathbf{A}(\mathbf{p}_n + \beta_n\mathbf{q}_n), \mathbf{A}\mathbf{q}_n\rangle = 0 \tag{4.44}$$

Therefore, if the solution is already optimized along the $\mathbf{q}_n$ direction, the best solution in the plane spanned by $\mathbf{p}_n$ and $\mathbf{q}_n$ is obtained in a direction orthogonal to $\mathbf{q}_n$ in the sense of Equation (4.44).

This analysis shows that the process of selecting direction vectors and coefficients to minimize the error functional E_n is optimized when vectors satisfying the orthogonality condition

$$\langle \mathbf{A}\mathbf{p}_i, \mathbf{A}\mathbf{p}_j \rangle = 0 \qquad i \neq j \tag{4.45}$$

are used. If an arbitrary set of direction vectors is employed, the process of minimizing E_n will adjust their coefficients in order to generate a sequence satisfying this generalized orthogonality. Furthermore, if direction vectors satisfying (4.45) are used, there is no advantage in simultaneously minimizing the error functional along more than one direction. Vectors that satisfy (4.45) are said to be mutually conjugate with respect to the operator $\mathbf{A}^\dagger\mathbf{A}$, where $\mathbf{A}^\dagger$ is the adjoint with respect to the inner product, that is,

$$\langle \mathbf{A}^\dagger \mathbf{x}, \mathbf{y} \rangle = \langle \mathbf{x}, \mathbf{A}\mathbf{y} \rangle \tag{4.46}$$

In accordance with our definition for the inner product, the matrix $\mathbf{A}^\dagger$ is the transpose conjugate of $\mathbf{A}$.

Suppose that a set of direction vectors satisfying the orthogonality condition of Equation (4.45) is readily available. Since $\mathbf{A}$ is nonsingular, these vectors are linearly independent and span the N-dimensional space. The solution can be expressed in the form

$$\mathbf{x} = \mathbf{x}_0 + \alpha_1 \mathbf{p}_1 + \alpha_2 \mathbf{p}_2 + \cdots + \alpha_N \mathbf{p}_N \tag{4.47}$$

where, for generality, the arbitrary vector $\mathbf{x}_0$ can be thought of as an initial estimate, or "guess," for the solution $\mathbf{x}$. Because of the orthogonality of (4.45), each coefficient can be found independently according to

$$\alpha_n = \frac{-\langle \mathbf{A}\mathbf{p}_n, \mathbf{r}_0 \rangle}{\|\mathbf{A}\mathbf{p}_n\|^2} \tag{4.48}$$

where $\mathbf{r}_0 = \mathbf{A}\mathbf{x}_0 - \mathbf{b}$.

From the above relationships, it is apparent that

$$\mathbf{r}_n = \mathbf{r}_0 + \alpha_1 \mathbf{A}\mathbf{p}_1 + \alpha_2 \mathbf{A}\mathbf{p}_2 + \cdots + \alpha_N \mathbf{A}\mathbf{p}_N \tag{4.49}$$

and recursive relationships are given as

$$\mathbf{r}_n = \mathbf{r}_{n-1} + \alpha_n \mathbf{A}\mathbf{p}_n \tag{4.50}$$

$$\mathbf{x}_n = \mathbf{x}_{n-1} + \alpha_n \mathbf{p}_n \tag{4.51}$$

and

$$\|\mathbf{r}_n\|^2 = \|\mathbf{r}_{n-1}\|^2 - |\alpha_n|^2 \|\mathbf{A}\mathbf{p}_n\|^2 \tag{4.52}$$

From Equations (4.45), (4.48), and (4.49), we can readily deduce that

$$\langle \mathbf{A}\mathbf{p}_m, \mathbf{r}_n \rangle = \begin{cases} \langle \mathbf{A}\mathbf{p}_m, \mathbf{r}_0 \rangle & n < m \\ 0 & n \geq m \end{cases} \tag{4.53}$$

It follows that (4.36) and (4.48) are equivalent, indicating that the error minimization process and the orthogonal expansion procedure yield the same results.

The process of expanding a solution in terms of mutually conjugate direction vectors is known as the *conjugate direction method*, after Hestenes and Stiefel [36]. The conjugate direction method does not specify the means for generating a mutually conjugate sequence, however. The CG method is a conjugate direction method that includes a recursive procedure for generating the **p**-vectors. The CG algorithm begins with the choice

$$\mathbf{p}_1 = -\mathbf{A}^\dagger \mathbf{r}_0 \tag{4.54}$$

which is proportional to the gradient of the functional E_n at $\mathbf{x} = \mathbf{x}_0$ (Prob. P4.10). Subsequent functions are found from

$$\mathbf{p}_{n+1} = -\mathbf{A}^\dagger \mathbf{r}_n + \beta_n \mathbf{p}_n \tag{4.55}$$

where the scalar coefficient β_n is chosen to ensure

$$\langle \mathbf{A}^\dagger \mathbf{A} \mathbf{p}_n, \mathbf{p}_{n+1} \rangle = 0 \tag{4.56}$$

We will demonstrate that enforcing Equation (4.56) is sufficient to ensure that the **p**-vectors form a mutually conjugate set satisfying (4.45). To illustrate, we first present several relationships involving the vectors generated within the CG algorithm.

From Equation (4.55), we write

$$\langle \mathbf{A}^\dagger \mathbf{r}_m, \mathbf{p}_{n+1} \rangle = -\langle \mathbf{A}^\dagger \mathbf{r}_m, \mathbf{A}^\dagger \mathbf{r}_n \rangle + \beta_n \langle \mathbf{A}^\dagger \mathbf{r}_m, \mathbf{p}_n \rangle \tag{4.57}$$

According to Equation (4.53), the first and last inner product in (4.57) vanish for $m > n$, leaving

$$\langle \mathbf{A}^\dagger \mathbf{r}_n, \mathbf{A}^\dagger \mathbf{r}_m \rangle = 0 \qquad m \neq n \tag{4.58}$$

which is the orthogonality associated with the residual vectors. Equations (4.53) and (4.55) can be combined to produce

$$\langle \mathbf{p}_{n+1}, \mathbf{A}^\dagger \mathbf{r}_n \rangle = -\langle \mathbf{A}^\dagger \mathbf{r}_n, \mathbf{A}^\dagger \mathbf{r}_n \rangle = -\|\mathbf{A}^\dagger \mathbf{r}_n\|^2 \tag{4.59}$$

Therefore, α_n can be expressed in the alternate form

$$\alpha_n = \frac{\|\mathbf{A}^\dagger \mathbf{r}_{n-1}\|^2}{\|\mathbf{A}\mathbf{p}_n\|^2} \tag{4.60}$$

From Equation (4.50),

$$\mathbf{A}^\dagger \mathbf{r}_n = \mathbf{A}^\dagger \mathbf{r}_{n-1} + \alpha_n \mathbf{A}^\dagger \mathbf{A} \mathbf{p}_n \tag{4.61}$$

Because of the orthogonality expressed in Equation (4.58), an inner product between $\mathbf{A}^\dagger \mathbf{r}_m$ and Equation (4.61) leads to the result

$$\langle \mathbf{A}^\dagger \mathbf{r}_m, \mathbf{A}^\dagger \mathbf{A} \mathbf{p}_n \rangle = \begin{cases} \dfrac{\|\mathbf{A}^\dagger \mathbf{r}_m\|^2}{\alpha_n} & m = n \\ \dfrac{-\|\mathbf{A}^\dagger \mathbf{r}_m\|^2}{\alpha_n} & m = n - 1 \\ 0 & \text{otherwise} \end{cases} \tag{4.62}$$

Using Equation (4.62) with $m = n$, we find the value of β_n from Equations (4.55) and (4.56) to be

$$\beta_n = \frac{\|\mathbf{A}^\dagger \mathbf{r}_n\|^2}{\|\mathbf{A}^\dagger \mathbf{r}_{n-1}\|^2} \tag{4.63}$$

To see that this formula for β_n guarantees the proper orthogonality between vectors when (4.45) is not explicitly enforced, consider the following. During the first iteration, Equations (4.36) and (4.50) are imposed, so that

$$\langle \mathbf{p}_1, \mathbf{A}^\dagger \mathbf{r}_1 \rangle = 0 \tag{4.64}$$

From Equation (4.54), this is equivalent to

$$\langle \mathbf{A}^\dagger \mathbf{r}_1, \mathbf{A}^\dagger \mathbf{r}_0 \rangle = 0 \tag{4.65}$$

Because of Equation (4.65), the expression for β_1 from (4.63) is sufficient to ensure that

$$\langle \mathbf{p}_1, \mathbf{A}^\dagger \mathbf{A} \mathbf{p}_2 \rangle = 0 \tag{4.66}$$

On the second iteration, Equations (4.50), (4.64), and (4.66) guarantee that

$$\langle \mathbf{p}_1, \mathbf{A}^\dagger \mathbf{r}_2 \rangle = -\langle \mathbf{A}^\dagger \mathbf{r}_0, \mathbf{A}^\dagger \mathbf{r}_2 \rangle = 0 \tag{4.67}$$

Taking an inner product of Equation (4.55) (with $n = 1$) and $\mathbf{A}^\dagger \mathbf{r}_2$, we find that

$$\langle \mathbf{A}^\dagger \mathbf{r}_2, \mathbf{A}^\dagger \mathbf{r}_1 \rangle = 0 \tag{4.68}$$

Therefore, the value of β_2 from (4.63) is sufficient to ensure that

$$\langle \mathbf{p}_2, \mathbf{A}^\dagger \mathbf{A} \mathbf{p}_3 \rangle = 0 \tag{4.69}$$

What remains is the validity of

$$\langle \mathbf{p}_1, \mathbf{A}^\dagger \mathbf{A} \mathbf{p}_3 \rangle = 0 \tag{4.70}$$

From Equations (4.55) and (4.63), $\mathbf{p}_n$ can be written as

$$\mathbf{p}_n = -\|\mathbf{A}^\dagger \mathbf{r}_{n-1}\|^2 \sum_{i=0}^{n-1} \frac{\mathbf{A}^\dagger \mathbf{r}_i}{\|\mathbf{A}^\dagger \mathbf{r}_i\|^2} \tag{4.71}$$

Using Equation (4.71) with $n = 3$ we obtain

$$\langle \mathbf{p}_3, \mathbf{A}^\dagger \mathbf{A} \mathbf{p}_1 \rangle = -\|\mathbf{A}^\dagger \mathbf{r}_2\|^2 \sum_{i=0}^{2} \frac{\langle \mathbf{A}^\dagger \mathbf{r}_i, \mathbf{A}^\dagger \mathbf{A} \mathbf{p}_1 \rangle}{\|\mathbf{A}^\dagger \mathbf{r}_i\|^2} \tag{4.72}$$

But, by the relationship established in (4.62), which is valid for these values of n and m as established in Equations (4.65), (4.67), and (4.68), the above reduces to

$$\langle \mathbf{p}_3, \mathbf{A}^\dagger \mathbf{A} \mathbf{p}_1 \rangle = -\|\mathbf{A}^\dagger \mathbf{r}_2\|^2 \left(\frac{-1}{\alpha_1} + \frac{1}{\alpha_1} \right) = 0 \tag{4.73}$$

Thus, in an inductive fashion we see that the direction vectors generated by the above procedure satisfy the assumed orthogonality properties of the conjugate direction method.

The CG algorithm is summarized in Table 4.3. In the computer science literature, this particular form of the CG algorithm is referred to as the "conjugate gradient method applied to the normal equations." The conventional CG algorithm discussed in many texts is restricted to the special case of a Hermitian positive-definite matrix $\mathbf{A}$. To extend the algorithm to arbitrary linear systems, the matrix equation is premultiplied by $\mathbf{A}^\dagger$ to produce the *normal equations* $\mathbf{A}^\dagger \mathbf{A} \mathbf{x} = \mathbf{A}^\dagger \mathbf{b}$. Note that it is unnecessary to compute the product $\mathbf{A}^\dagger \mathbf{A}$. By using a different error functional or a different definition of the inner product, a variety of related CG algorithms can be constructed [46].

For an arbitrary nonsingular matrix $\mathbf{A}$, the CG algorithm outlined above produces a solution in at most N iteration steps (assuming infinite-precision arithmetic). This is a direct

TABLE 4.3 Conjugate Gradient Algorithm

Initial steps:

Guess $\mathbf{x}_0$

$\mathbf{r}_0 = \mathbf{A}\mathbf{x}_0 - \mathbf{b}$

$\mathbf{p}_1 = -\mathbf{A}^\dagger \mathbf{r}_0$

Iterate ($n = 1, 2, \ldots$):

$$\alpha_n = -\frac{\langle \mathbf{A}\mathbf{p}_n, \mathbf{r}_{n-1} \rangle}{\|\mathbf{A}\mathbf{p}_n\|^2} = \frac{\|\mathbf{A}^\dagger \mathbf{r}_{n-1}\|^2}{\|\mathbf{A}\mathbf{p}_n\|^2}$$

$\mathbf{x}_n = \mathbf{x}_{n-1} + \alpha_n \mathbf{p}_n$

$\mathbf{r}_n = \mathbf{A}\mathbf{x}_n - \mathbf{b} = \mathbf{r}_{n-1} + \alpha_n \mathbf{A}\mathbf{p}_n$

$$\beta_n = \frac{\|\mathbf{A}^\dagger \mathbf{r}_n\|^2}{\|\mathbf{A}^\dagger \mathbf{r}_{n-1}\|^2}$$

$\mathbf{p}_{n+1} = -\mathbf{A}^\dagger \mathbf{r}_n + \beta_n \mathbf{p}_n$

Terminate when a norm of $\mathbf{r}_n$ falls below some predetermined value; see cautionary note following Equation (4.76).

consequence of the fact that N **p**-vectors span the solution space. Finite-step termination is a significant advantage of the CG method over other iterative algorithms. In addition, the CG algorithm produces solution estimates that satisfy (Prob. P4.11)

$$\|\mathbf{x} - \mathbf{x}_n\| \le \|\mathbf{x} - \mathbf{x}_m\| \qquad n > m \tag{4.74}$$

In words, the error in $\mathbf{x}_n$ decreases monotonically as the algorithm progresses. Consequently, it will usually be possible to terminate the algorithm prior to the Nth iteration step.

For the purpose of terminating the CG algorithm, it is necessary to estimate the accuracy of x_n at each iteration step. Since the solution **x** is not known, the error vector

$$\mathbf{e}_n = \mathbf{x} - \mathbf{x}_n \tag{4.75}$$

is not directly computable. Instead, it is convenient to compute the residual norm

$$N_n = \frac{\|\mathbf{r}_n\|}{\|\mathbf{b}\|} = \frac{\|\mathbf{A}\mathbf{x}_n - \mathbf{b}\|}{\|\mathbf{b}\|} \tag{4.76}$$

As illustrated by Equation (4.52), the residual norm decreases monotonically (a direct consequence of minimizing E_n at each iteration step). The CG algorithm can be terminated when the residual norm decreases to some predetermined value. As long as **A** is fairly well conditioned, $N_n < 10^{-4}$ suggests that several decimal places of accuracy are obtained in $\mathbf{x}_n$. However, note that the residual norm only provides an indirect bound on the error, according to

$$\frac{\|\mathbf{e}_n\|}{\|\mathbf{e}_0\|} \le \kappa(\mathbf{A}) \frac{\|\mathbf{r}_n\|}{\|\mathbf{r}_0\|} \tag{4.77}$$

where $\kappa(\mathbf{A})$ is the condition number of the matrix **A**. If the matrix **A** becomes ill-conditioned, $\kappa(\mathbf{A})$ will grow large and the residual norm N_n may be a poor indication of the accuracy of $\mathbf{x}_n$.

The convergence rate of the CG algorithm depends on specific properties of the matrix equation under consideration. An upper bound on the accuracy of the solution estimate has been obtained in terms of the condition number of **A** in the form [47]

$$\frac{\|\mathbf{e}_n\|}{\|\mathbf{e}_0\|} \leq 2\left[\frac{\kappa(\mathbf{A}) - 1}{\kappa(\mathbf{A} + 1}\right]^n \tag{4.78}$$

According to Equation (4.78), if the matrix is poorly conditioned, the convergence may be quite slow. As a general observation, it has been reported that roughly $P\kappa(\mathbf{A})$ iterations are required to produce accuracy to P decimal places [47]. When applied to integral equation formulations of electromagnetics, the observed convergence rate of the algorithm usually approximates a linear function of the decimal places of accuracy [48]. As an illustration, Figure 4.9 shows the performance of the CG algorithm applied to several examples of perfectly conducting strips or cylinders illuminated by a TM wave, following the approach discussed in Section 2.1.

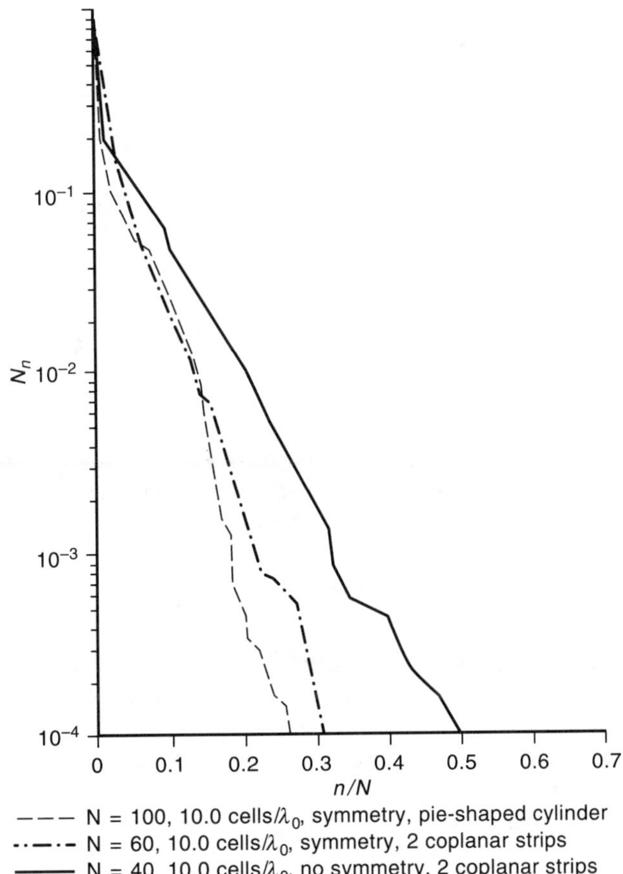

Figure 4.9 Plot of the residual norm versus normalized iteration step for the conjugate gradient algorithm. The matrix equations represent the TM EFIE formulation for p.e.c. cylinders (Section 2.1). After [48]. ©1986 IEEE.

A precise characterization of the convergence behavior of the CG algorithm can be obtained in terms of the matrix **A** and the excitation **b** [49–51]. Observe that the residual after the first iteration step can be written as

$$\mathbf{r}_1 = [\mathbf{I} - \alpha_1 \mathbf{A}\mathbf{A}^\dagger]\mathbf{r}_0 \tag{4.79}$$

After the second iteration step, the residual can be expressed as

$$\mathbf{r}_2 = [\mathbf{I} - (\alpha_1 + \alpha_2 + \alpha_2\beta_1)\mathbf{A}\mathbf{A}^\dagger + \alpha_1\alpha_2(\mathbf{A}\mathbf{A}^\dagger)^2]\mathbf{r}_0 \tag{4.80}$$

In general, the residual at the nth iteration step can be written as

$$\mathbf{r}_n = R_n(\mathbf{A}\mathbf{A}^\dagger)\mathbf{r}_0 \tag{4.81}$$

where $R_n(\mathbf{A}\mathbf{A}^\dagger)$ is a polynomial of order n in the matrix $\mathbf{A}\mathbf{A}^\dagger$, that is,

$$R_n(\mathbf{A}\mathbf{A}^\dagger) = \sum_{k=0}^{n} \xi_{nk}(\mathbf{A}\mathbf{A}^\dagger)^k \tag{4.82}$$

Since the coefficients $\{\xi_{nk}\}$ in Equation (4.82) are combinations of the previous scalars α and β from (4.60) and (4.63), they are real valued. Here, R_n is known as the *residual polynomial* of order n.

The residual polynomial can be related to the error norm produced by the CG algorithm at the nth iteration step. Let $\{\lambda_i\}$ denote the eigenvalues of the matrix $\mathbf{A}\mathbf{A}^\dagger$ (these are also the eigenvalues of the matrix $\mathbf{A}^\dagger\mathbf{A}$) and let $\{\mathbf{u}_i\}$ denote the orthonormal eigenvectors of $\mathbf{A}\mathbf{A}^\dagger$. The initial residual vector can be decomposed into an eigenvector expansion according to

$$\mathbf{r}_0 = \sum_{i=1}^{N} \langle \mathbf{u}_i, \mathbf{r}_0 \rangle \mathbf{u}_i \tag{4.83}$$

Inserting this expansion into Equation (4.81) and replacing $(\mathbf{A}\mathbf{A}^\dagger)^k \mathbf{u}_i$ by $(\lambda_i)^k \mathbf{u}_i$ produce

$$\mathbf{r}_n = \sum_{i=1}^{N} \langle \mathbf{u}_i, \mathbf{r}_0 \rangle R_n(\lambda_i) \mathbf{u}_i \tag{4.84}$$

Finally, using the orthonormal property of the eigenvectors, the error functional being minimized by CG at the nth step can be written

$$E_n = \langle \mathbf{r}_n, \mathbf{r}_n \rangle = \sum_{i=1}^{N} |\langle \mathbf{u}_i, \mathbf{r}_0 \rangle|^2 \, [R_n(\lambda_i)]^2 \tag{4.85}$$

Since the coefficients $\langle \mathbf{u}_i, \mathbf{r}_0 \rangle$ are fixed by the excitation $\mathbf{b}$ and the initial estimate $\mathbf{x}_0$, the reduction in E_n as the algorithm progresses is entirely due to changes in the residual polynomial R_n.

Equation (4.85) can be used to draw a variety of conclusions about the CG algorithm. First, note that at some point in the iteration process the algorithm will terminate with $E_n = 0$. At this point, the algorithm will have generated a residual polynomial $R_n(\lambda)$ having zeros at each of the eigenvalues λ_i (Figure 4.10). Since the algorithm can place one additional zero in this polynomial at each iteration step and will place these zeros in order to minimize E_n, it follows that the CG algorithm will require at most M steps to converge, where M is the number of independent eigenvalues of $\mathbf{A}\mathbf{A}^\dagger$. In addition, if eigenvalues are repeated or clustered together in groups, the algorithm may only need to place one zero somewhere within the cluster in order to significantly reduce the error as measured by E_n.

The terms in Equation (4.85) are weighted by the coefficients of the eigenvector decomposition of the initial residual. If some of these coefficients vanish or are very

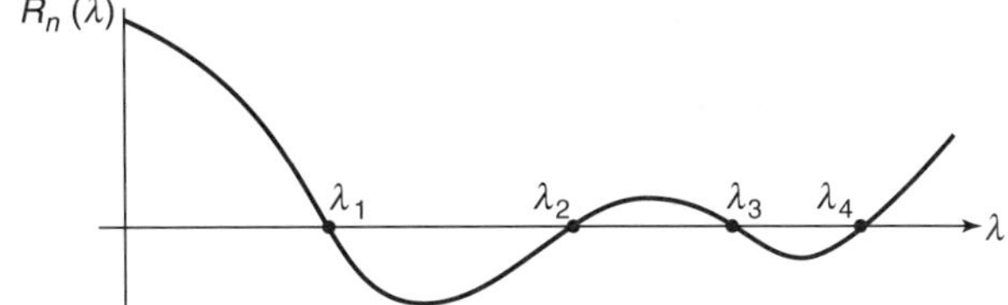

Figure 4.10 Residual polynomial after the CG algorithm has converged.

small, the algorithm will not need to place a zero of the polynomial at the corresponding eigenvalues. This suggests that fewer than M iteration steps may be required. In fact, if the initial residual can be represented by exactly Q eigenvectors, the CG algorithm will converge in exactly Q iterations!

The effect of the initial estimate x_0 on the convergence behavior is to alter the initial residual and therefore the coefficients in (4.85). The choice of $\mathbf{x}_0 = 0$ as an initial estimate of the solution produces an initial residual norm of $N_0 = 1$ in accordance with Equation (4.76). Intuitively, a "good" initial guess would appear be one that reduces N_0 from unity to some smaller value ($N_0 = 10^{-2}$?). However, it would be counterproductive to employ an initial estimate of the solution that excites more eigenvectors in the decomposition of (4.83) than would be excited by $\mathbf{x}_0 = 0$, because more iteration steps would be required despite a smaller initial residual. In other words, a good initial guess would attempt to minimize the number of nonzero coefficients in (4.83), rather than the initial residual norm. Because of the difficulty of ensuring this property, the zero estimate for $\mathbf{x}_0$ is often employed in practice and is sometimes called an "optimal" starting value. (Of course, the "best" starting value would be the solution $\mathbf{x}$.)

Equation (4.85) also sheds light on the observation that the convergence rate of the CG algorithm is typically slower for poorly conditioned systems. At termination, the residual polynomial $R_n(\lambda)$ must vanish at each important eigenvalue. However, the residual polynomial always has unity value at the origin ($\lambda = 0$). For a poorly conditioned matrix where the eigenvalue spectrum of $\mathbf{AA}^\dagger$ is widely spread along the positive real axis, more degrees in the polynomial are necessary in order to initially reduce E_n, even in infinite-precision arithmetic. In finite-precision arithmetic, increased round-off errors during the computation of the matrix–vector operations exacerbate the generation of the polynomial and negate the finite step termination property of the CG algorithm. If the equation is very badly conditioned, convergence may slow to the point of stagnation. Even if the algorithm does appear to converge, because of (4.77), the residual norm usually employed to estimate the accuracy of the solution will not be reliable if $\mathbf{A}$ is poorly conditioned.

The observed performance of the CG algorithm for electromagnetic applications supports the preceding conclusions [51]. As an example, consider the EFIE formulation of Section 2.1 for TM scattering by a perfectly conducting circular cylinder of one wavelength circumference. Table 4.4 presents the residual norm versus the iteration step for a number of different discretizations. As the discretization is refined, the error E_n observed after each iteration step appears to stabilize at certain values. For this example there are clearly only five important eigenvalues in the spectrum of the matrix operator, and refining the discretization beyond 10 cells/λ does not significantly change the important part of the spectrum. (The eigenvalues of the matrix operator are necessarily related to those of the continuous integral operator; this relationship is explored in Chapter 5.)

For EM scattering problems formulated in terms of integral equations and involving uniform plane-wave excitations, numerical experiments suggest that there are typically

TABLE 4.4 Values of Residual Norm Produced by CG Versus Iteration Step for Four Different Discretizations of Same Integral Equation

	N_n			
n	$N = 4, \dfrac{4.0 \text{ cells}}{\lambda_0}$	$N = 8, \dfrac{8.0 \text{ cells}}{\lambda_0}$	$N = 16, \dfrac{16.0 \text{ cells}}{\lambda_0}$	$N = 32, \dfrac{32.0 \text{ cells}}{\lambda_0}$
0	1.0	1.0	1.0	1.0
1	0.359	0.366	0.361	0.358
2	0.100	0.114	0.115	0.115
3	8.9×10^{-10}	0.0142	0.0161	0.0161
4	—	0.000706	0.00128	0.00132
5	—	2.2×10^{-7}	6.9×10^{-5}	8.0×10^{-5}

on the order of $N/3$ important eigenvalues in the spectrum of $\mathbf{AA}^{\dagger}$ when a discretization involving about 10 subsectional cells per wavelength is employed [48, 51]. This suggests that the algorithm will require roughly $N/3$ iterations to produce a solution. For differential equation formulations such as those discussed in Chapter 3, the convergence rate is usually much slower [39].

In infinite-precision arithmetic, the CG algorithm produces a solution in at most N iteration steps. Unfortunately, for a general linear system, CG requires approximately six times the number of operations as LU factorization to attain N complete steps. Thus, to be competitive with direct methods, the CG algorithm would have to converge to necessary accuracy in fewer than $N/6$ iteration steps. As noted above, seldom is the observed convergence sufficiently rapid for general linear systems. Consequently, the CG algorithm is usually reserved for treating matrix equations having sparsity or special structure that cannot easily be exploited using direct methods of solution. Because the integral equations of electromagnetics involve convolutional kernels, the matrix equations of interest often possess discrete-convolutional symmetries. An implementation of the CG method for treating these systems is discussed in the following section.

4.12 THE CONJUGATE GRADIENT–FAST FOURIER TRANSFORM PROCEDURE [37, 52–59]

Direct methods for general matrix solution require the full $N \times N$ matrix to be stored in computer memory, placing a bottleneck on the solution process for large systems. If sufficient structure or sparsity is present in the equation of interest, iterative methods offer the possibility of avoiding this storage bottleneck. Iterative algorithms only require an implicit matrix operator (a subroutine that when given a column vector returns the product of the $N \times N$ system matrix with the column vector) and can easily exploit any type of matrix structure.

Electromagnetics problems posed in terms of integral equations with convolutional kernels can sometimes be discretized to yield matrices having discrete-convolutional symmetries. A general discrete convolution is an operation of the form [60, 61]

$$e_m = \sum_{n=0}^{N-1} j_n g_{m-n} \tag{4.86}$$

where e, j, and g denote sequences of numbers. Equation (4.86) is equivalent to the matrix equation

$$\begin{bmatrix} g_0 & g_{-1} & g_{-2} & \cdots & g_{1-N} \\ g_1 & g_0 & g_{-1} & \ddots & \\ g_2 & g_1 & g_0 & \ddots & \\ \cdot & & & & \\ \cdot & & & & \\ \cdot & & & & \\ g_{N-1} & \cdots & \cdots & \cdots & g_0 \end{bmatrix} \begin{bmatrix} j_0 \\ j_1 \\ \cdot \\ \cdot \\ \cdot \\ j_{N-1} \end{bmatrix} = \begin{bmatrix} e_0 \\ e_1 \\ \cdot \\ \cdot \\ \cdot \\ e_{N-1} \end{bmatrix} \tag{4.87}$$

The $N \times N$ matrix depicted in Equation (4.87) is a general *Toeplitz* matrix. All of the elements of this matrix are described by the $2N - 1$ entries of the first row and column. If the elements of the sequence g repeat with period N, so that

$$g_{n-N} = g_n \qquad n = 1, 2, \ldots, N-1 \tag{4.88}$$

the operation is known as a *circular* discrete convolution and the $N \times N$ matrix in Equation (4.87) is *circulant.* Otherwise, the operation is a *linear* discrete convolution. Note that any linear discrete convolution of length N can be embedded into a circular discrete convolution of length $2N - 1$. This can be accomplished by extending the original sequence g to repeat with period $2N - 1$, zero padding the sequence j to length $2N - 1$, and changing the upper limit of the summation in Equation (4.86) to $2N - 2$.

The FFT algorithm is an efficient way of implementing the discrete Fourier transform [60, 61]

$$\tilde{g}_n = \sum_{k=0}^{N-1} g_k e^{-j\frac{2\pi nk}{N}} \qquad n = 0, 1, \ldots, N-1 \tag{4.89}$$

The inverse discrete Fourier transform is defined

$$g_k = \frac{1}{N} \sum_{n=0}^{N-1} \tilde{g}_n e^{(j\frac{2\pi nk}{N})} \qquad k = 0, 1, \ldots, N-1 \tag{4.90}$$

For notational purposes, we use

$$\tilde{g} = \mathrm{FFT}_N(g) \tag{4.91}$$

$$g = \mathrm{FFT}_N^{-1}(\tilde{g}) \tag{4.92}$$

to denote the discrete Fourier transform pair for a sequence of length N. The discrete convolution theorem states that if Equation (4.86) is a circular discrete convolution of length N, it is equivalent to

$$\tilde{e}_n = \tilde{j}_n \tilde{g}_n \qquad n = 0, 1, \ldots, N-1 \tag{4.93}$$

If Equation (4.86) is a linear discrete convolution, the equivalence holds if the linear convolution is embedded in a circular convolution of length $2N - 1$ as described above.

To summarize, the discrete convolution operation of Equation (4.86) is equivalent to a Toeplitz matrix multiplication. Furthermore, either can be implemented using the FFT and inverse FFT algorithm according to the discrete convolution theorem [60, 61]

$$e = \mathrm{FFT}_N^{-1}\left[\mathrm{FFT}_N(j)\mathrm{FFT}_N(g)\right] \tag{4.94}$$

If the discrete convolution is of the linear type, the FFTs must be of length $2N - 1$ rather than length N.

The above conclusions are easily generalized to two or three dimensions. A two-dimensional discrete convolution is an operation of the form

$$e_{pq} = \sum_{n=0}^{N-1} \sum_{m=0}^{M-1} j_{nm} g_{p-n,q-m} \qquad \left\{ \begin{array}{c} p \\ q \end{array} \right\} = 0, 1, \ldots, N-1 \tag{4.95}$$

This equation is equivalent to the matrix operation

$$\begin{bmatrix} \mathbf{G}_0 & \mathbf{G}_{-1} & \cdots & \mathbf{G}_{1-N} \\ \mathbf{G}_1 & \mathbf{G}_0 & \ddots & \\ \cdot & & & \\ \cdot & & & \\ \cdot & & & \\ \mathbf{G}_{N-1} & \cdots & \cdots & \mathbf{G}_0 \end{bmatrix} \begin{bmatrix} \mathbf{J}_0 \\ \mathbf{J}_1 \\ \cdot \\ \cdot \\ \cdot \\ \mathbf{J}_{N-1} \end{bmatrix} = \begin{bmatrix} \mathbf{E}_0 \\ \mathbf{E}_1 \\ \cdot \\ \cdot \\ \cdot \\ \mathbf{E}_{N-1} \end{bmatrix} \tag{4.96}$$

where each element of the $N \times N$ *block Toeplitz* matrix of (4.96) is itself a Toeplitz matrix of the form depicted in (4.87). The relationship established in Equation (4.94) can be extended to multiple dimensions in an obvious manner.

To illustrate the appearance of discrete convolutional structure in electromagnetics equations, consider the scattering of a TM wave by an inhomogeneous dielectric cylinder, following the formulation of Section 2.5. Suppose we initially restrict our attention to the model appearing in Figure 4.11, which consists of one row of equal-sized square cells. Each cell in the model may represent a region of different permittivity, and the relative permittivity of the nth cell is denoted ε_n. Following the procedure of Section 2.5, we obtain the discrete system

$$e_m = \frac{\eta j_m}{jk(\varepsilon_m - 1)} + \sum_{n=0}^{N-1} j_n g_{m-n} \tag{4.97}$$

where

$$g_{m-n} = \frac{k\eta}{4} \iint_{\text{cell } n} H_0^{(2)} \left(k\sqrt{(x_m - x')^2 + (y_m - y')^2} \right) dx'dy' \tag{4.98}$$

This system can be written in matrix form as

$$\begin{bmatrix} g_0 + \chi_0 & g_{-1} & g_{-2} & \cdots & g_{1-N} \\ g_1 & g_0 + \chi_1 & g_{-1} & \ddots & \\ g_2 & g_1 & g_0 + \chi_2 & \ddots & \\ \cdot & & & & \\ \cdot & & & & \\ \cdot & & & & \\ g_{N-1} & \cdots & \cdots & \cdots & g_0 + \chi_{N-1} \end{bmatrix} \begin{bmatrix} j_0 \\ j_1 \\ \cdot \\ \cdot \\ \cdot \\ j_{N-1} \end{bmatrix} = \begin{bmatrix} e_0 \\ e_1 \\ \cdot \\ \cdot \\ \cdot \\ e_{N-1} \end{bmatrix} \tag{4.99}$$

where

$$\chi_n = \frac{\eta}{jk(\varepsilon_n - 1)} \tag{4.100}$$

The presence of the relative permittivity on the main diagonal perturbs the $N \times N$ matrix from a purely Toeplitz form. Now consider a more general geometry modeled by the lattice

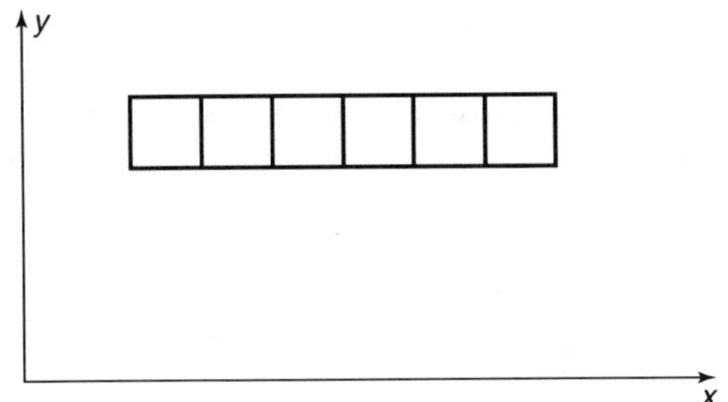

Figure 4.11 One-dimensional lattice of uniform cells.

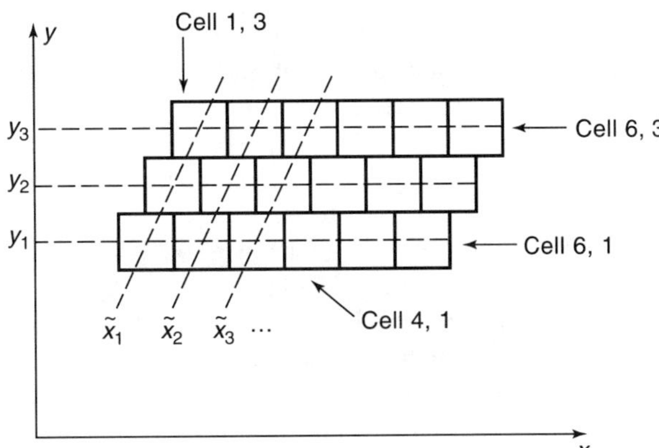

Figure 4.12 Skewed two-dimensional lattice of uniform cells showing the indexing scheme used to build translational symmetries into the matrix equation.

of cells depicted in Figure 4.12. In this case, a two-dimensional numbering scheme is employed for the cells. For this geometry, the system of equations can be written as

$$e_{pq} = \frac{\eta j_{pq}}{jk(\varepsilon_{pq} - 1)} + \sum_{m=0}^{M-1} \sum_{n=0}^{N-1} j_{mn} g_{p-m,q-n} \tag{4.101}$$

where

$$g_{p-m,q-n} = \frac{k\eta}{4} \iint_{\text{cell}\, m,n} H_0^{(2)} \left(k\sqrt{(x_p - x')^2 + (y_q - y')^2} \right) dx'\, dy' \tag{4.102}$$

The corresponding matrix has the perturbed block-Toeplitz form

$$\begin{bmatrix} \mathbf{G}_0 + \mathbf{X}_0 & \mathbf{G}_{-1} & \cdots & \mathbf{G}_{1-N} \\ \mathbf{G}_1 & \mathbf{G}_0 + \mathbf{X}_1 & \ddots & \\ \cdot & & & \\ \cdot & & & \\ \cdot & & & \\ \mathbf{G}_{N-1} & \cdots & \cdots & \mathbf{G}_0 + \mathbf{X}_{N-1} \end{bmatrix} \begin{bmatrix} \mathbf{J}_0 \\ \mathbf{J}_1 \\ \cdot \\ \cdot \\ \cdot \\ \mathbf{J}_{N-1} \end{bmatrix} = \begin{bmatrix} \mathbf{E}_0 \\ \mathbf{E}_1 \\ \cdot \\ \cdot \\ \cdot \\ \mathbf{E}_{N-1} \end{bmatrix} \tag{4.103}$$

where $\mathbf{X}_n$ denotes a diagonal matrix containing the perturbation due to the presence of the relative permittivity. Each matrix $\mathbf{G}_n$ is a Toeplitz matrix having the form of Equation (4.87). (The entries on the main diagonal are the only perturbation from a true block-Toeplitz system comprised of Toeplitz blocks.) In addition, for this example $\mathbf{G}_{-n} = \mathbf{G}_n$.

Because of the perturbation along the main diagonal, these systems are not suited for treatment with conventional Toeplitz or block-Toeplitz routines. However, they do contain a significant degree of structure. Iterative methods offer a way of exploiting this symmetry to reduce the storage to the level required by a conventional Toeplitz algorithm. For example, during each iteration step of the CG method, the only computations involving the matrix $\mathbf{A}$ are the product of $\mathbf{A}$ with a given column vector and the product of $\mathbf{A}^{\dagger}$ with a given column vector. The subroutines that perform the necessary matrix–vector multiplications can incorporate any sparsity or structure to minimize storage and computation costs. In addition, the specific type of matrix structure need not affect the organization of the part of the computer program that involves the main body of the iterative algorithm. The CG driver routine can be thought of as a "black box" similar in form to library routines that perform Gaussian elimination.

For the treatment of Equation (4.103), the CG algorithm can be implemented in two different ways. In the first approach, the matrix is stored in reduced form (one row and the main diagonal) and the required matrix–vector multiplications are performed directly. This implementation requires a minimum amount of storage but $O(N^2)$ operations to implement the matrix–vector multiplications. Since the discrete convolutions can be performed in $O(N \log N)$ operations using FFT algorithms, an alternative approach is to employ a multidimensional FFT to perform the discrete convolutions according to Equation (4.94). This latter approach is known as the CG–FFT implementation. Because of the zero padding required with linear discrete convolutions, use of the FFT involves a trade-off between storage and computational costs. For the two-dimensional scatterer depicted in Figure 4.12, use of the FFT roughly quadruples the required storage.

When the FFT is used to perform the required convolution between two sequences, the arrays must include terms corresponding to every possible location throughout the lattice. For scatterer geometries that do not completely fill the lattice shape, "dummy cells" are usually employed to complete the lattice. After each convolution is performed, locations in the resulting array that correspond to dummy cells are set to zero so that the iterative algorithm does not see additional unknowns at these cells.

The CG–FFT procedure has been a common use of the CG algorithm for treating integral equation formulations in electromagnetics. Unfortunately, to preserve the convolutional symmetry in the matrix, the scatterer geometry must be restricted to a relatively simple shape such as a flat plate, a surface of constant curvature, or a penetrable dielectric body discretized with uniform cells. The restrictions on the scatterer geometry necessary to preserve the convolutional symmetries limit the generality of the approach and prevent practical scatterers such as airplanes or bent wires from being analyzed this way. For electrically large geometries not imposing any type of structure on the $N \times N$ matrix, the fast multipole method introduced in the following section may offer an efficiency similar to that of the CG–FFT approach.

Iterative algorithms also suffer in comparison with direct methods of solution when it is necessary to treat multiple excitations. For example, Table 4.5 illustrates computation times associated with the analysis of flat-plate scattering using the CG–FFT procedure and LU factorization [62]. The CG–FFT procedure is more efficient for the solution of a large plate illuminated by a single incident field. However, when treating 91 different incident fields, the relative efficiency of LU factorization (which only requires an additional forward and back substitution for each additional excitation vector) is superior. Although some progress has been made in improving the performance of the CG algorithm for multiple

TABLE 4.5 Comparison of CPU Time (CRAY XMP/24) Required to Solve the Matrix Equation Representing EFIE for Scattering from Square Conducting Plate of Side Dimension a Using CG–FFT Algorithm and LU Factorization with Forward and Back Substitution

	CG–FFT Solution				LU Factorization	
a (λ)	Unknowns	FFT Size	Average Number of Iterations	Average CPU Time for CG–FFT (sec)	Average CPU Time for 1 RHS (sec)	CPU Time for 91 RHS (sec)
1.000	176	16	42	5.56	1.71	5.17
2.000	736	32	43	16.26	65.54	138.21
3.125	3,008	64	58	68.50	—	—
5.078	12,160	128	78	616.52	—	—

Note: Both procedures were used to treat 91 different excitations or "right-hand sides" (RHS). The averages were obtained by averaging over the 91 different excitations. After [62].

excitations [63], it is doubtful that iterative approaches will ever prove comparable to direct methods in this regard.

4.13 FAST MATRIX–VECTOR MULTIPLICATION: AN INTRODUCTION TO THE FAST MULTIPOLE METHOD

The choice of direct versus iterative methods is usually made based on the ultimate computational efficiency of the algorithm. Direct methods for dense matrices require $O(N^3)$ operations, while iterative methods require $O(PQ)$ operations, where P is the number of iterations and Q is the operation count per iteration. For a CG–FFT implementation, where Q may grow as slowly as $N \log N$, an overall computational complexity of $O(PN \log N)$ may be obtained. The CG–FFT approach is attractive when N is large, and the $O(N^3)$ operation count for dense matrices is prohibitive.

Unfortunately, the CG–FFT approach is restricted to translationally invariant geometries, which seriously limits its application to practical structures. An alternative scheme, known as the *fast multipole method*, offers the possibility of achieving $O(PN \log N)$ operations for arbitrarily shaped scatterers [64–68]. Variations on this approach appear to offer the most efficient possibilities yet proposed for the accurate numerical analysis of electrically large geometries, where N may be far greater than 10^4.

The central ideas behind the fast multipole method will be illustrated by considering a much simpler approximate scheme proposed by Lu and Chew [69]. Consider a TM-to-z p.e.c. scatterer, and assume that the pulse basis–point matching discretization of Section 2.1 is used. Suppose that the scatterer perimeter is divided into N cells and grouped into p segments of roughly equal size and numbers of unknowns. The segments will be indexed $i = 1, 2, \ldots, p$. There will be N/p cells per segment, which can be indexed $n = 0, 1, \ldots, N/p - 1$. One cell per segment will be centered at a local origin (x_{i0}, y_{i0}), while the other cell centroids will be denoted (x_{in}, y_{in}). Figure 4.13 illustrates two such segments.

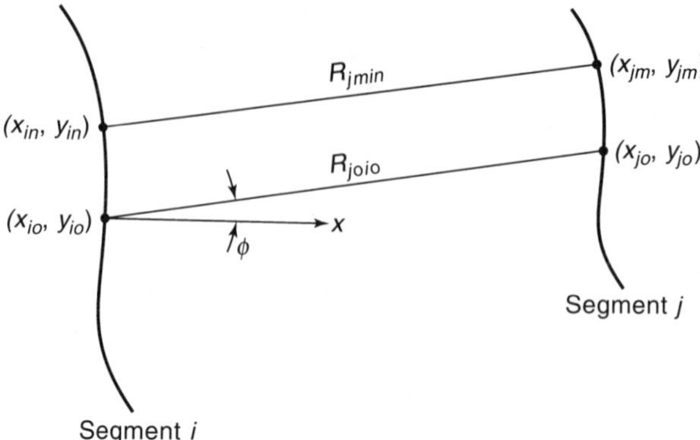

Figure 4.13 Two segments in the far zone of each other, showing the parameter R_{jmin}.

Regardless of the iterative algorithm employed, each iteration step requires a matrix–vector multiplication operation equivalent to finding the scattered E_z-field at the center of each cell given the N coefficients of the current density, $\{j_{in}\}$. To improve the efficiency of this step without sacrificing accuracy, the interactions between segments can be classified as either *near zone* or *far zone*, depending on the proximity of the segments. The near-zone interactions will be carried out by an explicit matrix–vector multiplication involving the conventional matrix entries, while the far-zone interactions will be replaced by a more efficient calculation. Let us assume that immediately adjacent segments are in the near zone of each other and all others fall in the far zone. Thus, the interactions between a given segment and its two adjacent neighbors are near zone and require $(3N/p)^2$ multiplications and additions. For all p segments, the computational complexity of the near-zone interactions grows as $O(N^2/p)$.

The far-zone interactions can be simplified using a far-field approximation [69]. Consider the calculation of the field at (x_{jm}, y_{jm}) due to sources on segment i,

$$E_z^s(x_{jm}, y_{jm}) = -\frac{\omega\mu}{4} \sum_{n=0}^{N/p-1} j_n w_n H_0^{(2)}(kR_{jmin}) \tag{4.104}$$

where R_{jmin} is defined in Figure 4.13 and we have incorporated the "single-point" approximation of the integral as used in Section 2.1. Segments i and j are in the far zone of each other, and we consider the far-field approximation (similar to that illustrated in Figure 1.17)

$$R_{jmin} \cong R_{j0i0} + R_{jm} - R_{in} \tag{4.105}$$

where

$$R_{j0i0} = \sqrt{(x_{j0} - x_{i0})^2 + (y_{j0} - y_{i0})^2} \tag{4.106}$$

$$R_{jm} = (x_{jm} - x_{j0}) \cos\phi + (y_{jm} - y_{j0}) \sin\phi \tag{4.107}$$

and

$$R_{in} = (x_{in} - x_{i0}) \cos\phi + (y_{in} - y_{i0}) \sin\phi \tag{4.108}$$

The angle ϕ denotes the orientation of R_{j0i0} with respect to the x-axis (Figure 4.13). The asymptotic form of the Hankel function for large arguments

$$H_0^{(2)}(k\rho) \approx \sqrt{\frac{2j}{\pi k\rho}} e^{-jk\rho} \tag{4.109}$$

motivates the approximation

$$H_0^{(2)}(kR_{jmin}) \cong H_0^{(2)}(kR_{j0i0})e^{-jkR_{jm}}e^{+jkR_{in}} \tag{4.110}$$

and we replace Equation (4.104) with

$$E_z^s(x_{jm}, y_{jm}) \cong -\frac{\omega\mu}{4} H_0^{(2)}(kR_{j0i0})e^{-jkR_{jm}} \sum_{n=0}^{N/p-1} j_n w_n e^{+jkR_{in}} \tag{4.111}$$

Using Equation (4.111), all interactions between the cells in segments i and j can be obtained from a single summation over the coefficients $\{j_{in}\}$ and one Hankel function calculation. This procedure involves $O(N/p)$ operations for a single segment pair. Since there are $p(p-3)$ combinations of far-zone segments in all, the entire far-zone computation grows with complexity $O(Np)$.

By combining the operation counts for the near-zone and the far-zone parts of the computation, we obtain a total that grows as

$$O\left(K_1\frac{N^2}{p} + K_2 Np\right) \tag{4.112}$$

Clearly, the growth with respect to N is minimized if the number of segments is proportional to the square root of N, producing an overall complexity for the matrix–vector multiplication process of $O(N^{1.5})$. For large N, this is superior to the $O(N^2)$ complexity of explicit matrix–vector multiplication.

To illustrate the relative efficiency of the procedure, suppose that the scatterer has perimeter dimension of 50,000 wavelengths, and a total of one million unknowns are employed. The perimeter is divided into 1000 segments. In this case, both the near-zone and far-zone interactions involve $O(10^9)$ operations, compared with $O(10^{12})$ for explicit matrix multiplication. Thus, the procedure requires approximately 1/1000 of the computation associated with a conventional implementation.

It is instructive to separate the far-zone calculation of Equation (4.111) into three parts, as suggested by Lu and Chew [69]. First, the sources on segment i can be *aggregated* together via the summation

$$S_i \cong \sum_{n=0}^{N/p-1} j_n w_n e^{+jkR_{in}} \tag{4.113}$$

The second step, that of *translation*, uses the Hankel function

$$E_z^s(x_{j0}, y_{j0}) \cong -\frac{\omega\mu}{4} H_0^{(2)}(kR_{j0i0}) S_i \tag{4.114}$$

to shift the field to the center of segment j. Finally, the scattered field is *disaggregated* throughout segment j via a multiplication with the phase correction

$$E_z^s(x_{jm}, y_{jm}) \cong e^{-jkR_{jm}} E_z^s(x_{j0}, y_{j0}) \tag{4.115}$$

These three distinct steps are analogous to those employed in the fast multipole method, which differs from the above primarily in the fact that it does not require a far-field approximation.

In the fast multipole method (FMM) of Rokhlin and his colleagues [64–67], the aggregation step replaces the field calculation in Equation (4.104) by a multipole expansion of the form

$$E_z^s(x_{jm}, y_{jm}) = \sum_q \alpha_q H_q^{(2)}(kR_{jmi0})e^{jq\phi} \tag{4.116}$$

whose coefficients can be obtained from (4.104) using the addition theorem (Prob. P3.21). Equation (4.116) is an expansion around a local origin on segment i. The translation step of the process is carried out using diagonalized translation operators associated with the multipole expansion to produce a similar expansion about a local origin on segment j. Finally, a disaggregation step is applied to obtain the scattered fields at individual cells [65, 66]. Unlike the simpler approach outlined above, Rokhlin's FMM does not involve a far-field approximation and can be carried out to a prescribed floating-point accuracy. The computational complexity per iteration step grows as $O(N^{1.5})$, and the storage requirements are similar. The efficiency improvement is obtained by replacing the individual interactions between cells in two far-zone segments with a single aggregated interaction.

A further reduction in computational complexity can be obtained by recursively nesting the segmentation concept developed above, by subdividing the segments into smaller segments, and by applying the same procedure to the internal interactions within each segment. In the limiting case, the multilevel algorithm involves $O(N \log N)$ operations per iteration [66]. Details of the FMM procedure for two- and three-dimensional geometries may be found in a number of recent articles [64–69]. (In the two-dimensional case, the FMM involves the multipole expansion described above. In contrast, the proposed three-dimensional FMM implementation employs a superposition of plane waves in geometrical directions adequate to reproduce the angular dependence of the original field, rather than an explicit multipole expansion [67].)

The FMM, and several other recently proposed algorithms similar to the FMM in many respects [70–73], permit a computational burden on the order of that obtained using the CG–FFT solution procedure but without the limitation of a translationally invariant scatterer geometry. These approaches are geared for electrically large structures, where direct methods of solution are prohibitive.

4.14 PRECONDITIONING STRATEGIES FOR ITERATIVE ALGORITHMS

The convergence rate of iterative algorithms, and the CG method in particular, is highly dependent on the eigenvalue spectrum of the matrix operator under consideration. Consequently, a scaling or transformation that converts the original system of equations into one with a more favorable eigenvalue spectrum might significantly improve the rate of convergence. This procedure is known as *preconditioning*. Although preconditioning strategies have been attempted for integral equation formulations [74–75], experience suggests that these formulations usually produce systems that are fairly well conditioned. However, the matrices arising from differential equation formulations (Chapter 3) are not always as well conditioned in general, and it is reasonable to expect that the performance of the CG algorithm applied to these systems can be improved by preconditioning.

A number of systematic approaches for sparse-matrix preconditioning have been developed, and an overview is presented by Evans [76]. The essential idea is to convert the system

$$\mathbf{A}\mathbf{x} = \mathbf{b} \tag{4.117}$$

into an equivalent equation

$$\mathbf{M}^{-1}\mathbf{A}\mathbf{x} = \mathbf{M}^{-1}\mathbf{b} \tag{4.118}$$

where $\mathbf{M}^{-1}$ approximates $\mathbf{A}^{-1}$. For example, $\mathbf{M}^{-1}$ may be found by *incomplete LU factorization.* A complete LU factorization of $\mathbf{A}$ might require many times the storage originally required by $\mathbf{A}$ (Figure 4.7), exceeding available memory. By ignoring some or all of the fill-in, an approximate or incomplete factorization can be found with storage comparable to that of the original matrix. A number of other preconditioning schemes are possible [40, 42, 76]. As a general rule, the product of the matrix $\mathbf{M}^{-1}\mathbf{A}$ and a column vector can be constructed by repeated matrix–vector multiplications, so that it is not necessary to explicitly construct the product $\mathbf{M}^{-1}\mathbf{A}$.

If applied to Equation (4.118), the CG algorithm exhibits convergence behavior that depends on the operator $\mathbf{M}^{-1}\mathbf{A}$, which should prove much better conditioned than $\mathbf{A}$. Since the iterative algorithm constructs a solution to (4.118), the final solution will be correct regardless of the specific choice for $\mathbf{M}^{-1}$. In other words, there is no approximation inherent in preconditioning. However, preconditioning schemes can significantly increase the storage requirements and the computation required per iteration. Thus, they are usually reserved for problems where the convergence is prohibitively slow.

A simple preconditioning strategy that often proves worthwhile is to scale the system of equations so that the entries along the main diagonal are all equal. This procedure can compensate for orders-of-magnitude differences in scale that sometimes arise when several different equations are coupled together within a hybrid formulation. However, simple scaling is unlikely to have a significant impact on a system based on an inherently ill-conditioned underlying formulation.

4.15 SUMMARY

The matrix solution process is usually the largest computational task associated with the numerical treatment of an electromagnetic scattering problem. It is therefore important to employ an efficient, robust algorithm. This chapter has reviewed some of the techniques in use for full- and sparse-matrix solution. In addition, the relevant issue of matrix stability has been discussed. A variety of excellent software libraries have been developed for the purpose of matrix solution, and their use is generally recommended whenever possible.

The relative efficiency of direct and iterative methods depends on a number of factors. As a rule, direct methods tend to be more efficient for systems that can be stored entirely in the fast-access memory of the computer. This is especially true if it is necessary to solve one equation for many excitations. For moderately sized problems, iterative methods might be more efficient if the convergence rate is fast, but the convergence rate is difficult to determine in practice except by trial and error. Iterative methods are generally superior for specialized situations where significant matrix structure exists that cannot be easily exploited by available direct methods or for electrically large problems that cannot be treated with direct methods.

REFERENCES

[1] G. Strang, *Linear Algebra and Its Applications*, San Diego: Harcourt Brace Javanovich, 1988.

[2] A. Jennings, *Matrix Computation for Engineers and Scientists*, New York: Wiley, 1977.

[3] G. H. Golub and C. F. Van Loan, *Matrix Computations*, Baltimore: Johns Hopkins University Press, 1983.

[4] K. E. Atkinson, *An Introduction to Numerical Analysis*, New York: Wiley, 1989.

[5] W. Cheney and D. Kincaid, *Numerical Mathematics and Computing*, Monterey: Brooks/Cole, 1980.

[6] J. Dongarra, J. R. Bunch, C. B. Moler, and G. W. Stewart, *LINPACK User's Guide*, Philadelphia: SIAM, 1979.

[7] I. S. Duff, A. M. Erisman, and J. K. Reid, *Direct Methods for Sparse Matrices*, Oxford: Clarendon, 1986.

[8] W. H. Press, B. P. Flannery, S. A. Teukolsky, and W. T. Vetterling, *Numerical Recipes*, Cambridge: Cambridge University Press, 1986.

[9] D. Kahaner, C. Moler, and S. Nash, *Numerical Methods and Software*, Englewood Cliffs, NJ: Prentice-Hall, 1989.

[10] E. Anderson, Z. Bai, C. Bischof, J. Demmel, J. Dongarra, J. Du Croz, A. Greenbaum, S. Hammarling, A. McKenney, S. Ostrouchov, and D. Sorensen, *LAPACK User's Guide*, 2nd ed., Philadelphia: SIAM, 1995.

[11] J. H. Wilkinson, "Error analysis of direct methods of matrix inversion," *J. ACM*, vol. 8, pp. 281–330, July 1961.

[12] J. K. Reid, "A note on the stability of Gaussian elimination," *J. Inst. Math. Applicat.*, vol. 8, pp. 374–375, 1971.

[13] J. R. Bunch, "Partial pivoting strategies for symmetric matrices," *SIAM J. Num. Anal.*, vol. 11, pp. 521–528, 1974.

[14] J. R. Bunch, L. Kaufman, and B. N. Parlett, "Decomposition of a symmetric matrix," *Num. Math.*, vol. 27, pp. 95–110, 1976.

[15] V. Barwell and A. George, "A comparison of algorithms for solving symmetric indefinite systems of linear equations," *ACM Trans. Math. Software*, vol. 2, pp. 242–251, 1976.

[16] F. X. Canning, "Direct solution of the EFIE with half the computation," *IEEE Trans. Antennas Propagat.*, vol. 39, pp. 118–119, Jan. 1991.

[17] A. George and J. W. Liu, *Computer Solution of Large Sparse Positive Definite Systems*, Englewood Cliffs, NJ: Prentice-Hall, 1981.

[18] S. Pissanetzky, *Sparse Matrix Technology*, New York: Academic, 1984.

[19] J. Puttonen, "Simple and effective bandwidth reduction algorithm," *Int. J. Num. Methods Eng.*, vol. 19, pp. 1139–1152, 1983.

[20] E. Cuthill and J. McKee, "Reducing the bandwidth of sparse symmetric matrices," *Proc. 24th Nat. Conf. Assoc. Comput. Mach.*, ACM Publ., pp. 157–172, 1969.

[21] E. Cuthill, "Several strategies for reducing the bandwidth of matrices," in *Sparse Matrices and Their Applications*, eds. D. J. Rose and R. A. Willoughby, New York: Plenum, 1972.

[22] H. R. Schwartz, *Finite Element Methods*, New York: Academic, 1988.

[23] S. R. H. Hoole, *Computer-Aided Analysis and Design of Electromagnetic Devices*, New York: Elsevier, 1989.

[24] H. R. Grooms, "Algorithm for matrix bandwidth reduction," *ASCE J. Struct. Div.*, vol. 98, pp. 203–214, 1972.

[25] R. J. Colins, "Bandwidth reduction by automatic renumbering," *Int. J. Num. Methods Eng.*, vol. 6, pp. 345–356, 1973.

[26] N. E. Gibbs, W. G. Poole, and P. K. Stockmeyer, "An algorithm for reducing the bandwidth and profile of a sparse matrix," *SIAM J. Num. Anal.*, vol. 13, April 1976.

[27] *PATRAN User's Guide*, Santa Ana, CA: PDA Engineering, 1984.

[28] S. C. Eisenstat, M C. Gursky, M. H. Schultz, and A. H. Sherman, "The Yale sparse matrix package II: The non-symmetric codes," Report 114, Department of Computer Science, Yale University, 1977.

[29] I. S. Duff, "MA28—A set of FORTRAN subroutines for sparse unsymmetric linear equations," Harwell Report AERE R-8730, Her Majesty's Stationery Office, London, 1977.

[30] I. S. Duff, "ME28: A sparse unsymmetric linear equation solver for complex equations," *ACM Trans. Math. Software*, vol. 7, pp. 505–511, Dec. 1981.

[31] Z. Zlatev, J. Wasniewski, and K. Schaumburg, *Y12M—Solution of Large and Sparse Systems of Linear Algebraic Equations* (Lecture Notes in Computer Science 121), Berlin: Springer-Verlag, 1981.

[32] O. Osterby and Z. Zlatev, *Direct Methods for Sparse Matrices* (Lecture Notes in Computer Science 157), Berlin: Springer-Verlag, 1983.

[33] B. M. Irons, "A frontal solution program for finite element analysis," *Int. J. Num. Methods Eng.*, vol 2, pp. 5–32, 1970.

[34] R. S. Varga, *Matrix Iterative Analysis*, Englewood Cliffs, NJ: Prentice-Hall, 1962.

[35] D. M. Young, *Iterative Solution of Large Linear Systems*, New York: Academic, 1971.

[36] M. R. Hestenes and E. Stiefel, "Methods of conjugate gradients for solving linear systems," *J. Res. Nat. Bur. Stand.*, vol. 49, pp. 409–435, 1952.

[37] T. K. Sarkar, ed., *Application of Conjugate Gradient Method to Electromagnetics and Signal Analysis*, New York: Elsevier, 1991.

[38] D. A. H. Jacobs, "The exploitation of sparsity by iterative methods," in *Sparse Matrices and Their Uses*, ed. I. S. Duff, Berlin: Springer-Verlag, 1981.

[39] C. F. Smith, A. F. Peterson, and R. Mittra, "The biconjugate gradient method for electromagnetic scattering," *IEEE Trans. Antennas Propagat.*, vol. AP-38, pp. 938–940, June 1990.

[40] R. Barret, M. Berry, T. F. Chan, J. Demmel, J. Donato, J. Dongarra, V. Eijkhout, R. Pozo, C. Romaine, and H. van der Vorst, *Templates for the Solution of Linear Systems: Building Blocks for Iterative Methods*, Philadelphia: SIAM, 1994.

[41] Y. Saad and M. H. Schultz, "GMRES: A generalized minimal residual algorithm for solving non-symmetric linear systems," *SIAM J. Sci. Stat. Comp.*, vol. 7, pp. 856–869, July 1986.

[42] Y. Saad, *Iterative Methods for Sparse Linear Systems*, Boston: PWS Publishing, 1996.

[43] R. Freund and N. Nachtigal, "An implementation of the look-ahead Lanczos algorithm for non-Hermitian matrices," RIACS Tech. Reports 90.45 and 90.46 (Parts I and II), Nov. 1990.

[44] N. Nachtigal, "A look-ahead variant of the Lanczos algorithm and its application to the quasi-minimal residual method for non-Hermitian linear systems," Ph.D. dissertation, Department of Mathematics, Massachusetts Institute of Technology, Cambridge, MA, Aug. 1991.

[45] A. F. Peterson, S. L. Ray, C. H. Chan, and R. Mittra, "Numerical implementations of the conjugate gradient method and the CG-FFT for electromagnetic scattering, in *Application of Conjugate Gradient Method to Electromagnetics and Signal Analysis*, ed. T. K. Sarkar, New York: Elsevier, 1991.

[46] S. F. Ashby, T. A. Manteuffel, and P. E. Saylor, "A taxonomy for conjugate gradient methods," Dept. of Computer Science Report UIUC-DCS-R-88-1414, UILU-ENG-88-1719, University of Illinois, Urbana, IL, April 1988.

[47] J. Stoer, "Solution of large linear systems of equations by conjugate gradient type methods," in *Mathematical Programming: The State of the Art*, eds. A. Bachem, M. Grötschel, and B. Korte, New York: Springer-Verlag, 1983.

[48] A. F. Peterson and R. Mittra, "Convergence of the conjugate gradient method when applied to matrix equations representing electromagnetic scattering problems," *IEEE Trans. Antennas Propagat.*, vol. AP-34, pp. 1447–1454, Dec. 1986.

[49] E. L. Stiefel, "Kernel polynomials in linear algebra and their numerical applications," *Nat. Bur. Stand. Appl. Math. Ser.*, vol. 49, pp. 1–22, 1958.

[50] A. Jennings, "Influence of the eigenvalue spectrum on the convergence rate of the conjugate gradient method," *J. Inst. Math. Appl.*, vol. 20, pp. 61–72, 1977.

[51] A. F. Peterson, C. F. Smith, and R. Mittra, "Eigenvalues of the moment method matrix and their effect on the convergence of the conjugate gradient method," *IEEE Trans. Antennas Propagat.*, vol. AP-36, pp. 1177–1179, Aug. 1988.

[52] P. M. van den Berg, "Iterative computational techniques in scattering based upon the integrated square error criterion," *IEEE Trans. Antennas Propagat.*, vol. AP-32, pp. 1063–1071, Oct. 1984.

[53] D. T. Borup and O. P. Gandhi, "Calculation of high resolution SAR distribution in biological bodies using the FFT algorithm and conjugate gradient method," *IEEE Trans. Microwave Theory Tech.*, vol. MTT-33, pp. 417–419, May 1985.

[54] A. F. Peterson and R. Mittra, "Method of conjugate gradients for the numerical solution of large body electromagnetic scattering problems," *J. Opt. Soc. Am. A*, vol. 2, pp. 971–977, June 1985.

[55] L. W. Pearson, "A technique for organizing large moment calculations for use with iterative solution methods," *IEEE Trans. Antennas Propagat.*, vol. AP-33, pp. 1031–1033, Sept. 1985.

[56] A. F. Peterson, "An analysis of the spectral iterative technique for electromagnetic scattering from individual and periodic structures," *Electromagnetics*, vol. 6, pp. 255–276, 1986.

[57] A. F. Peterson and R. Mittra, "Iterative-based computational methods for electromagnetic scattering from individual or periodic structures," *IEEE J. Oceanic Engineering*, Special Issue on Scattering, vol. OE-12, pp. 458–465, April 1987.

[58] C. C. Su, "Calculation of electromagnetic scattering from a dielectric cylinder using the conjugate gradient method and FFT," *IEEE Trans. Antennas Propagat.*, vol. AP-35, pp. 1418–1425, Dec. 1987.

[59] T. J. Peters and J. L. Volakis, "Application of a conjugate gradient FFT method to scattering from thin material plates," *IEEE Trans. Antennas Propagat.*, vol. AP-36, pp. 518–526, April 1988.

[60] E. O. Brigham, *The Fast Fourier Transform*, Englewood Cliffs, NJ: Prentice-Hall, 1974.

[61] A. V. Oppenheim and R. W. Schafer, *Digital Signal Processing*, Englewood Cliffs, NJ: Prentice-Hall, 1975.

[62] C. H. Chan, "Investigation of iterative and spectral Galerkin techniques for solving electromagnetic boundary value problems," Ph.D. dissertation, University of Illinois, Urbana, IL, 1987.

[63] C. F. Smith, A. F. Peterson, and R. Mittra, "A conjugate gradient algorithm for the treatment of multiple incident electromagnetic fields," *IEEE Trans. Antennas Propagat.*, vol. AP-37, pp. 1490–1493, Nov. 1989.

[64] V. Rokhlin, "Rapid solution of integral equations of classical potential theory," *J. Computat. Phys.*, vol. 60, pp. 187–207, 1985.

[65] V. Rokhlin, "Rapid solution of integral equations of scattering theory in two dimensions," *J. Computat. Phys.*, vol. 86, pp. 414–439, 1990.

[66] N. Engheta, W. D. Murphy, V. Rokhlin, and M. S. Vassiliou, "The fast multipole method (FMM) for electromagnetic scattering problems," *IEEE Trans. Antennas Propagat.*, vol. 40, pp. 634–641, June 1992.

[67] R. Coifman, V. Rokhlin, and S. Wandzura, "The fast multipole method for the wave equation: A pedestrian prescription," *IEEE Antennas Propagat. Mag.*, vol. 35, pp. 7–12, June 1993.

[68] C. C. Lu and W. C. Chew, "Fast algorithm for solving hybrid integral equations," *IEE Proc. Part H*, vol. 140, no. 6, pp. 455–460, Dec. 1993.

[69] C. C. Lu and W. C. Chew, "Fast far field approximation for calculating the RCS of large objects," *Proceedings of the Eleventh Annual Review of Progress in Applied Computational Electromagnetics*, The Applied Computational Electromagnetic Society, Monterey, CA, pp. 576–583, March 1995.

[70] C. R. Anderson, "An implementation of the fast multipole method without multipoles," *SIAM J. Sci. Stat. Comput.*, vol. 13, no. 4, pp. 923–947, July 1992.

[71] W. C. Chew, "Fast algorithms for wave scattering developed at the University of Illinois' Electromagnetics Laboratory," *IEEE Antennas Propagat. Mag.*, vol. 35, pp. 22–32, Aug. 1993.

[72] E. Bleszynski, M. Bleszynski, and T. Jaroszewicz, "AIM: Adaptive integral method for solving large-scale electromagnetic scattering and radiation problems," *Radio Science*, vol. 31, pp. 1225–1251, Sept./Oct. 1996.

[73] E. Michielssen and A. Boag, "A multilevel matrix decomposition algorithm for analyzing scattering from large structures," *IEEE Trans. Antennas Propagat.*, vol. 44, pp. 1086–1093, Aug. 1996.

[74] A. Kas and E. L. Yip, "Preconditioned conjugate gradient methods for solving electromagnetic problems," *IEEE Trans. Antennas Propagat.*, vol. AP-35, pp. 147–152, Feb. 1987.

[75] C. F. Smith, "The performance of preconditioned iterative methods in computational electromagnetics," Ph.D. dissertation, University of Illinois, Urbana, IL, 1987.

[76] D. J. Evans, ed., *Preconditioning Methods: Analysis and Application*, New York: Gordon & Breach, 1983.

PROBLEMS

P4.1 For the LU factorization procedure described in Section 4.1, calculate the number of additions, multiplications, and divisions required to treat an $N \times N$ system.

P4.2 Find the condition number of the matrix

$$\mathbf{A} = \begin{bmatrix} 1 & 10{,}000 \\ 0 & 2 \end{bmatrix}$$

using the definition in (4.11). How does the condition number $\kappa(\mathbf{A})$ compare with the ratio of the largest to smallest eigenvalues of $\mathbf{A}$?

P4.3 The matrix

$$\mathbf{B} = \begin{bmatrix} 1 & 0 \\ 0 & 10^{10} \end{bmatrix}$$

is not normally considered ill conditioned, despite the fact that $\kappa(\mathbf{B}) = 10^{10}$. Why not? What is the essential difference between this matrix and the matrix $\mathbf{A}$ presented in Prob. P4.2?

P4.4 By following a procedure similar to that outlined in Equations (4.15)–(4.18), derive (4.19).

P4.5 The relation

$$\|(\mathbf{A} + \Delta\mathbf{A})^{-1}\| \le \frac{\|\mathbf{A}^{-1}\|}{1 - \|\mathbf{A}^{-1}\|\|\Delta\mathbf{A}\|}$$

is valid when [4, p. 493]

$$\|\Delta\mathbf{A}\| < \frac{1}{\|\mathbf{A}^{-1}\|}$$

Using (4.15) and this relation, derive the inequality

$$\frac{\|\Delta\mathbf{x}\|}{\|\mathbf{x}\|} \le \frac{\kappa(\mathbf{A})}{1 - \kappa(\mathbf{A})\dfrac{\|\Delta\mathbf{A}\|}{\|\mathbf{A}\|}} \left\{ \frac{\|\Delta\mathbf{A}\|}{\|\mathbf{A}\|} + \frac{\|\Delta\mathbf{b}\|}{\|\mathbf{b}\|} \right\}$$

This result is more general than (4.18) or (4.19).

P4.6 Repeat the node-numbering exercise of Section 4.5, but begin the Cuthill–McKee procedure using node 18 in Figure 3.11 instead of node 14. Compare the resulting bandwidth and envelope size with the numbers in Table 4.2. Is this a better ordering?

P4.7 Verify that (4.36) minimizes the error functional $E_n(\mathbf{x}_n)$ in (4.35). *Hint:* Since $\mathbf{x}_n = \mathbf{x}_{n-1} + \alpha_n \mathbf{p}_n$, E_n can be expanded as

$$E_n(\mathbf{x}_n) = \langle \mathbf{r}_{n-1}, \mathbf{r}_{n-1} \rangle + \alpha_n^\dagger \langle \mathbf{A}\mathbf{p}_n, \mathbf{r}_{n-1} \rangle + \alpha_n \langle \mathbf{r}_{n-1}, \mathbf{A}\mathbf{p}_n \rangle + \alpha_n \alpha_n^\dagger \langle \mathbf{A}\mathbf{p}_n, \mathbf{A}\mathbf{p}_n \rangle$$

where $\alpha_n^\dagger$ is the complex conjugate of α_n and the inner product is defined in (4.32). Because α_n may be complex valued, it is necessary to consider both

$$\frac{\partial E_n}{\partial \alpha_n} = 0 \quad \text{and} \quad \frac{\partial E_n}{\partial \alpha_n^\dagger} = 0$$

and show that these two conditions are satisfied by α_n in (4.36).

P4.8 There are a number of alternative choices that could be used for the error functional E_n on which a CG algorithm is based. For each of the following, determine the value of α_n that minimizes the functional $E_n(\mathbf{x}_n)$.

(a) $E^1(\mathbf{x}_n) = \langle \mathbf{x}_n - \mathbf{x}, \mathbf{x}_n - \mathbf{x} \rangle$
(b) $E^2(\mathbf{x}_n) = \langle \mathbf{A}\mathbf{x}_n - \mathbf{b}, \mathbf{x}_n - \mathbf{x} \rangle$
(c) $E^3(\mathbf{x}_n) = \langle \frac{1}{2}\mathbf{A}\mathbf{x}_n - \mathbf{b}, \mathbf{x}_n \rangle$

Use the definition in (4.32) for the inner product. For part (a), assume $\mathbf{A}$ is a general matrix operator. For parts (b) and (c), assume that $\mathbf{A}$ is self-adjoint, that is, $\mathbf{A} = \mathbf{A}^\dagger$.

P4.9 For the case where $\mathbf{A}$ is a real, symmetric 3×3 matrix, demonstrate that the gradient of the functional

$$f(\mathbf{x}) = \frac{1}{2}\mathbf{x}^T\mathbf{A}\mathbf{x} - \mathbf{b}^T\mathbf{x}$$

where $\mathbf{x}^T$ is the transpose of $\mathbf{x}$, is given by $\nabla \mathbf{f} = \mathbf{A}\mathbf{x} - \mathbf{b}$.

P4.10 For the case where $\mathbf{A}$ is a real, nonsymmetric 3×3 matrix, demonstrate that the gradient of the functional

$$g(\mathbf{x}) = \langle \mathbf{A}\mathbf{x} - \mathbf{b}, \mathbf{A}\mathbf{x} - \mathbf{b} \rangle$$

is given by $\nabla g = 2\mathbf{A}^T(\mathbf{A}\mathbf{x} - \mathbf{b})$, where $\mathbf{A}^T$ is the transpose of $\mathbf{A}$.

P4.11 Prove Equation (4.74). *Hint:* One approach is to use (4.71) to show that

$$\langle \mathbf{p}_i, \mathbf{p}_j \rangle \geq 0$$

Then, use

$$\mathbf{x}_n - \mathbf{x}_m = \sum_{i=m+1}^{n} \alpha_i \mathbf{p}_i$$

to obtain the inequality

$$\langle \mathbf{x}_n - \mathbf{x}_m, \mathbf{x}_N - \mathbf{x}_n \rangle \geq 0 \qquad N > n > m$$

The identity

$$\|\mathbf{x} - \mathbf{x}_m\|^2 = \|\mathbf{x}_n - \mathbf{x}_m\|^2 + \|\mathbf{x} - \mathbf{x}_n\|^2 + 2\,\mathrm{Re}\left(\langle \mathbf{x}_n - \mathbf{x}_m, \mathbf{x} - \mathbf{x}_n \rangle\right)$$

may also be helpful.

P4.12 An alternative conjugate gradient algorithm for solving $\mathbf{A}\mathbf{x} = \mathbf{b}$ may be based on the error functional $E_n(\mathbf{x}_n) = \langle \mathbf{x}_n - \mathbf{x}, \mathbf{x}_n - \mathbf{x} \rangle$, where the inner product is defined in (4.32) and the solution is expanded in direction vectors $\{q_n\}$ according to $\mathbf{x}_n = \mathbf{x}_{n-1} + \alpha_n \mathbf{q}_n$. Using $q_1 = -\mathbf{A}^\dagger \mathbf{r}_0$ and $\mathbf{q}_{n+1} = -\mathbf{A}^\dagger \mathbf{r}_n + \beta_n \mathbf{q}_n$, derive this CG algorithm by following a development similar to that used in Section 4.11. *Hint:* The appropriate direction vectors satisfy $\langle \mathbf{q}_i, \mathbf{q}_j \rangle = 0, i \neq j$, and the residual vectors satisfy $\langle \mathbf{r}_i, \mathbf{r}_j \rangle = 0, i \neq j$. The coefficients α_n and β_n can be expressed as

$$\alpha_n = \frac{\|\mathbf{r}_{n-1}\|^2}{\|\mathbf{q}_n\|^2}$$

$$\beta_n = \frac{\|\mathbf{r}_n\|^2}{\|\mathbf{r}_{n-1}\|^2}$$

In the CG algorithm presented in Section 4.11, the error vectors $\mathbf{e}_n = \mathbf{x} - \mathbf{x}_n$ and the residual vectors $\mathbf{r}_n = \mathbf{A}\mathbf{x}_n - \mathbf{b}$ decrease monotonically. Is the same true for this CG algorithm?

P4.13 Consider the problem of TM scattering from a dielectric cylinder using the volume integral equation formulation discussed in Sections 2.5 and 4.12. Suppose that the cylinder has cross-sectional dimension $10\lambda \times 10\lambda$, relative permittivity $\varepsilon = 4$, and a uniform lattice of square cells employed with a cell density of 100 cells/λ_d^2. Compare the efficiency of a CG–FFT solution to a CG solution in which the FFT is not employed. For both methods, estimate the required computation per iteration and the required storage.

P4.14 Repeat Prob. P4.13 for a cubical dielectric scatterer of side dimension 10λ and $\varepsilon_r = 4$ assuming that an analogous volume integral equation formulation is discretized using pulse basis functions, point matching, and a uniform lattice of cubical cells with a cell density of 1000 cells/λ_d^3. (Note that three components of the electric field are involved and that the matrix operator will therefore have a 3×3 block structure, with each block being a block-Toeplitz matrix. Each entry of the block-Toeplitz matrix is itself block-Toeplitz, with each entry of that matrix being Toeplitz.) Compare the efficiency of a CG–FFT solution to a CG solution in which the FFT is not employed by estimating the operations required per iteration for each approach. How much storage is required if (a) FFTs are used with minimum amount of zero padding, (b) no FFTs are used but all the Toeplitz symmetries are exploited to minimize storage, and (c) the entire matrix is stored?

5

The Discretization Process: Basis/Testing Functions and Convergence

Chapters 2 and 3 presented a variety of examples involving the discretization of continuous equations into matrix form. The specific discretization places a limit on the accuracy of a numerical result for a fixed number of basis and testing functions and determines whether or not the numerical result will converge to the exact solution as the number of basis and testing functions is increased. These are fundamental issues, and they require an examination from a perspective that is more theoretical than operational. To accomplish this, we employ some of the tools of functional analysis, the branch of applied mathematics said to be concerned with providing "solutions to equations, often by means of convergent sequences of approximation" [1]. This chapter also introduces a variety of subsectional basis functions and explores the role that the basis and testing functions play in the solution accuracy.

5.1 INNER PRODUCT SPACES

We desire to solve an equation of the form $Lf = g$, where L is a continuous linear operator such as the integral operators of Chapter 2 or the differential operators of Chapter 3. The function f is the unknown to be determined, and g represents a known excitation. If a unique solution exists, it is given by $f = L^{-1}g$, where L^{-1} is the inverse operator. In practice, we are usually not able to determine L^{-1} and resort to numerical solutions.

The linear operator L maps functions in its domain (such as the unknown f) to functions in its range (such as the excitation g). As a general rule, the domain and range are different linear spaces. For instance, in the case where L is a differential operator, the domain of L will generally include boundary conditions not imposed on functions in the range.

It is convenient to introduce the notion of an inner product, which is a scalar quantity denoted $\langle a, b\rangle$ satisfying the following properties:

$$\langle a, b\rangle = \langle b, a\rangle^\dagger \tag{5.1}$$

$$\langle \alpha a, \beta b + c\rangle = \alpha^\dagger \beta \langle a, b\rangle + \alpha^\dagger \langle a, c\rangle \tag{5.2}$$

$$\begin{aligned} \langle a, a\rangle &> 0 \quad \text{if } a \neq 0 \\ \langle a, a\rangle &= 0 \quad \text{if } a = 0 \end{aligned} \tag{5.3}$$

where a, b, and c are functions and α and β are scalars. Complex conjugation is denoted using a dagger (†). Any inner product satisfying these properties can be used to define a natural norm

$$\|a\| = \sqrt{\langle a, a\rangle} \tag{5.4}$$

and the associated metric

$$d(a, b) = \|a - b\| \tag{5.5}$$

The metric provides us with the notion of "distance" between two functions.

Two functions a and b in an inner product space are said to be *orthogonal* if

$$\langle a, b\rangle = 0 \tag{5.6}$$

In a similar fashion, functions $\{B_n\}$ in an inner product space form an orthogonal set if

$$\langle B_m, B_n\rangle = 0 \qquad m \neq n \tag{5.7}$$

The set $\{B_n\}$ is said to be *complete* if the zero function is the only function in the inner product space orthogonal to each member of the set. A set $\{B_n\}$ that is both complete and orthogonal is said to be a basis and can be used to represent any function f in the inner product space in the sense that

$$\|f - \sum_n \alpha_n B_n\| = 0 \tag{5.8}$$

where the $\{a_n\}$ are scalar coefficients uniquely determined by [1, 2]

$$\alpha_n = \frac{\langle B_n, f\rangle}{\langle B_n, B_n\rangle} \tag{5.9}$$

In practice, we are forced to project the functions of interest onto a finite-dimensional subspace of the original inner product space. In the subspace, the basis is truncated to the form $\{B_1, B_2, \ldots, B_N\}$, and the representation is given by

$$f \cong f^N = \sum_{n=1}^{N} \alpha_n B_n \tag{5.10}$$

The scalar coefficients $\{\alpha_1, \alpha_2, \ldots, \alpha_N\}$ are selected to minimize the distance between the function f and the representation f^N. The error

$$d(f, f^N) = \|f - f^N\| \tag{5.11}$$

is minimized when the coefficients are chosen to make the error orthogonal to the N-dimensional basis, that is,

$$\langle B_n, f - f^N\rangle = 0 \qquad n = 1, 2, \ldots, N \tag{5.12}$$

This is known as an orthogonal projection. Because of the orthogonality of the basis functions, the coefficients are the same in the subspace as in the original inner product space. Therefore, the orthogonal projection (the "best" representation as measured by the metric) is realized using coefficients from Equation (5.9).

Now, consider the equation $Lf = g$. We seek a representation of the solution f in an N-dimensional subspace of the original domain of L, and Equation (5.10) provides the general form. The best approximation is obtained when the coefficients from Equation (5.9) are employed. Unfortunately, since f is not known, the coefficients $\{\alpha_n\}$ cannot be determined directly from (5.9).

On the other hand, quantities defined on the range of the linear operator L are known and might be more convenient to work with. If the set $\{T_n\}$ forms a basis (complete and orthogonal) for the range space of the operator L, any function in the range may be represented in the N-dimensional subspace spanned by $\{T_1, T_2, \ldots, T_N\}$ according to

$$g \cong g^N = \sum_{m=1}^{N} \beta_m T_m \tag{5.13}$$

The projection that minimizes the error $d(g, g^N)$ employs coefficients

$$\beta_m = \frac{\langle T_m, g\rangle}{\langle T_m, T_m\rangle} \tag{5.14}$$

If g is a known function, the coefficients $\{\beta_m\}$ are readily determined. In a similar fashion, the function LB_n can be represented by

$$LB_n \cong \sum_{m=1}^{N} l_{mn} T_m \tag{5.15}$$

where the coefficients $\{l_{mn}\}$ that minimize the error

$$\left\| LB_n - \sum_{m=1}^{N} l_{mn} T_m \right\| \tag{5.16}$$

are given by

$$l_{mn} = \frac{\langle T_m, LB_n\rangle}{\langle T_m, T_m\rangle} \tag{5.17}$$

The coefficients of Equation (5.17) achieve an orthogonal projection in the range of the operator and therefore provide the best approximation as measured by the metric.

Returning to the approximate solution of the equation $Lf = g$, we represent the unknown solution f in the form of Equation (5.10), where $\{\alpha_n\}$ are unknowns to be determined. This representation produces a function on the range space having the form

$$Lf^N = \sum_{n=1}^{N} \alpha_n LB_n \tag{5.18}$$

Projecting this function on the N-dimensional subspace spanned by the set $\{T_1, T_2, \ldots, T_N\}$ yields

$$Lf^N \cong \sum_{m=1}^{N} \sum_{n=1}^{N} l_{mn} \alpha_n T_m \tag{5.19}$$

where the coefficients $\{l_{mn}\}$ are obtained from Equation (5.17). Equating this representation for Lf^N with the representation from (5.13) for g^N produces the discrete system of equations

$$\sum_{n=1}^{N} l_{mn}\alpha_n = \beta_m \qquad m = 1, 2, \ldots, N \tag{5.20}$$

This system is an $N \times N$ matrix equation that can be solved for the coefficients $\{\alpha_n\}$.

Although the above procedure provides a way of obtaining f^N, the coefficients $\{\alpha_1, \alpha_2, \ldots, \alpha_N\}$ obtained from the solution of Equation (5.20) are generally *not* the same as those specified in Equation (5.9). In other words, despite the fact that the projections in the range space are orthogonal, (5.20) does not ensure an orthogonal projection in the domain space and will usually not produce the best approximation as measured by the metric $d(f, f^N)$ (see Prob. P5.3). We momentarily defer a discussion of the consequences of this fact.

5.2 THE METHOD OF MOMENTS [3]

The preceding discussion indicates that an approximate solution of the linear equation $Lf = g$ may be obtained in the form

$$f \cong \sum_{n=1}^{N} \alpha_n B_n \tag{5.21}$$

where the functions $\{B_n\}$ are known basis functions defined on the domain of L and the scalars $\{\alpha_n\}$ are unknown coefficients to be determined. From an operational standpoint, Equation (5.21) is substituted into $Lf = g$, and a system of linear equations is obtained by forcing the residual

$$L\left(\sum_{n=1}^{N} \alpha_n B_n\right) - g = \sum_{n=1}^{N} \alpha_n L B_n - g \tag{5.22}$$

to be orthogonal to a set of testing functions $\{T_1, T_2, \ldots, T_N\}$. This produces the matrix equation $\mathbf{L}\alpha = \beta$ having entries

$$l_{mn} = \langle T_m, LB_n\rangle \tag{5.23}$$

and

$$\beta_m = \langle T_m, g\rangle \tag{5.24}$$

The matrix equation $\mathbf{L}\alpha = \beta$ is formally identical to Equation (5.20), except for the normalization. Provided the matrix $\mathbf{L}$ is nonsingular, the unknown coefficients can be found using standard matrix solution algorithms (Chapter 4).

Since the system of equations is obtained by forcing the residuals to be orthogonal to the testing functions, this procedure is often given the name *weighted-residual method* [4, 5]. In electromagnetics, it is also known as the *method of moments* [3, 6–8]. The roots of this procedure originate in the methods of Rayleigh, Ritz, and Galerkin developed near the turn of the century [4, 9]. All discretization procedures can be placed on a direct correspondence with this approach, at least approximately. The *finite-element method* is equivalent, although it is often presented in the context of minimizing a quadratic functional [10, 11]. The classical *finite-difference method* can also be interpreted in the context of basis and testing functions (Prob. P5.5).

The discretization of a continuous equation by the method of moments necessarily involves the projection of the continuous linear operator onto finite-dimensional subspaces defined by the basis and testing functions. Although the method-of-moments system $\mathbf{L}\alpha = \beta$ is formally equivalent to Equation (5.20), the process outlined in (5.21)–(5.24) can be applied to produce an approximate solution regardless of whether or not the functions $\{B_n\}$ and $\{T_m\}$ form complete, orthogonal sets. As illustrated in the following section, the basis and testing functions used in practice are often not orthogonal sets. (We continue to denote these as "basis functions" despite the fact that they do not satisfy the definition of a basis in the strict sense.) If the $\{T_m\}$ form an orthogonal set, the projection of the range space onto the testing functions is orthogonal and therefore a best approximation. Unfortunately, even if the basis functions are orthogonal, the projection of the domain space onto the basis functions is not guaranteed to be orthogonal. This makes it difficult to make firm statements about the convergence of the numerical approximation in Equation (5.21) to the exact solution as $N \to \infty$. In any case, since N is necessarily finite for numerical calculations, the result obtained from (5.21) is always approximate.

The choice of basis and testing functions is the principal issue arising within a method-of-moments implementation. As discussed by Harrington [3], practical factors affecting the selection of basis functions include the desired accuracy of the approximate solution, the relative complexity of the resulting matrix entries, and computational constraints that place an upper limit on the matrix size. While one would expect that the desired goal would always be to obtain the best accuracy with the fewest basis functions, the need to adapt the approach (i.e., the finished computer program) to a wide variety of different problems may motivate a less than optimal formulation. The basis and testing functions should be linearly independent and able to accurately approximate the f and Lf, respectively. (Actually, the basis functions do not have to be linearly independent, as long as the set $\{LB_n\}$ is linearly independent [12].) It cannot be overemphasized that for good results the basis and testing functions must be chosen with the particular operator L in mind. Although the domain of L may be restricted to functions that satisfy certain differentiability requirements, in many cases the differentiability, or "smoothness," requirements of the basis functions can be shifted to the testing functions. We have already seen this property illustrated in specific examples throughout Chapters 2 and 3 and consider it in more detail in Section 5.6.

The applications of interest range from those involving rather simple geometries to those involving structures of arbitrary shape and composition and include unknown quantities that may be scalar or vector functions of two or three variables. Because of this generality, functions that are commonly employed as a basis in other applications (such as the exponential or trigonometric functions) may not be well suited for many of the electromagnetic problems we encounter. From the examples presented in Chapters 2 and 3, we have seen that it is often more convenient to employ subsectional functions on irregular domains. Subsectional functions differ from the classical basis introduced in Section 5.1 in two respects. First, subsectional functions usually do not form orthogonal sets. Second, increasing the order N of the approximation usually alters every element of the set $\{B_1, B_2, \ldots, B_N\}$, rather than just adding an additional function. However, subsectional functions are quite flexible in that they can be easily adapted to arbitrary domains. The following section illustrates some of the common subsectional basis and testing functions in widespread use for discretizing electromagnetics equations.

5.3 EXAMPLES OF SUBSECTIONAL BASIS FUNCTIONS

The examples considered in Chapters 2 and 3 illustrate the use of the simplest subsectional basis and testing functions. In this section, several families of subsectional functions are introduced for one-dimensional scalar quantities. The generalization to the multidimensional, vector case is deferred until Chapter 9.

For single-dimension scalar quantities, the simplest basis functions in use are illustrated in Figure 5.1. These include the Dirac delta function

$$B_0(x) = \delta(x - x_0) \tag{5.25}$$

the pulse, or piecewise-constant, function

$$B_1(x) = p(x; x_1, x_2) = \begin{cases} 1 & x_1 < x < x_2 \\ 0 & \text{otherwise} \end{cases} \tag{5.26}$$

and the subsectional triangle function

$$B_2(x) = t(x; x_3, x_4, x_5) = \begin{cases} \dfrac{x - x_3}{x_4 - x_3} & x_3 < x < x_4 \\[2ex] \dfrac{x_5 - x}{x_5 - x_4} & x_4 < x < x_5 \end{cases} \tag{5.27}$$

On uniform intervals, the subsectional pulse and triangle functions are actually the simplest members of a family of spline functions generated by the convolution

$$\begin{aligned} B_n(x) &= B_{n-1}(x) * \frac{1}{\Delta} p\left(x; -\frac{\Delta}{2}, \frac{\Delta}{2}\right) \\ &= \frac{1}{\Delta} \int_{-\Delta/2}^{\Delta/2} B_{n-1}(x - x')\, dx' \end{aligned} \tag{5.28}$$

The next member of this family is the quadratic spline

$$B_3(x) = q\left(x; -\frac{3\Delta}{2}, -\frac{\Delta}{2}, \frac{\Delta}{2}, \frac{3\Delta}{2}\right) = \begin{cases} 0 & x < -\dfrac{3\Delta}{2} \\[2ex] \dfrac{9}{8} + \dfrac{3x}{2\Delta} + \dfrac{x^2}{2\Delta^2} & -\dfrac{3\Delta}{2} < x < -\dfrac{\Delta}{2} \\[2ex] \dfrac{3}{4} - \dfrac{x^2}{\Delta^2} & -\dfrac{\Delta}{2} < x < \dfrac{\Delta}{2} \\[2ex] \dfrac{9}{8} - \dfrac{3x}{2\Delta} + \dfrac{x^2}{2\Delta^2} & \dfrac{\Delta}{2} < x < \dfrac{3\Delta}{2} \\[2ex] 0 & x > \dfrac{3\Delta}{2} \end{cases} \tag{5.29}$$

which is illustrated in Figure 5.2. Higher order functions can be constructed from the recursive formula in (5.28). The next such spline, denoted B_4, is a cubic function.

Suppose the domain of interest is divided into intervals, or *cells*, along the x-axis. The pulse function has support confined to a single cell and is orthogonal to pulse functions located at other cells. An expansion in pulse functions produces a piecewise-constant representation. The triangle function overlaps two adjacent cells and shares each cell with

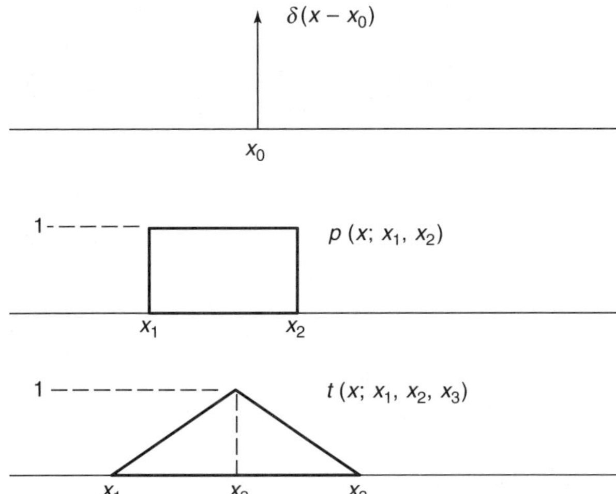

Figure 5.1 Definition of basis and testing functions.

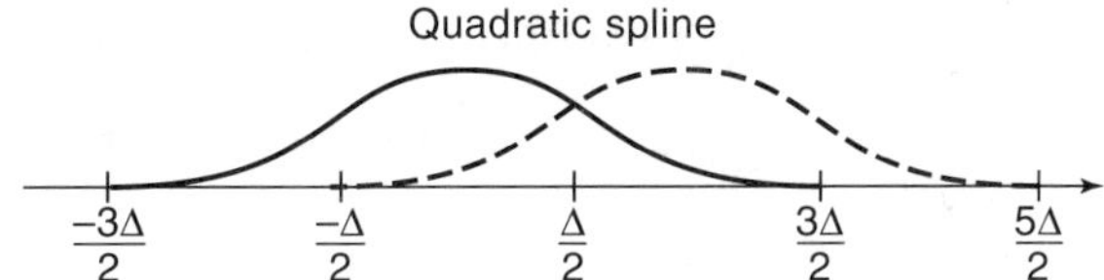

Figure 5.2 Quadratic spline functions.

adjacent triangle functions. By continuously superimposing triangle functions along the entire domain of interest, a global piecewise-linear approximation is achieved as depicted in Figure 5.3. The quadratic spline overlaps three cells and shares each cell with two other quadratic splines. A piecewise-quadratic representation throughout the domain can be obtained by a superposition of shifted quadratic splines (Figure 5.2). Triangle functions ensure continuity of the function they represent; quadratic splines provide continuity of the function and its first derivative. It is noteworthy that neither the triangle functions nor the quadratic splines form orthogonal sets.

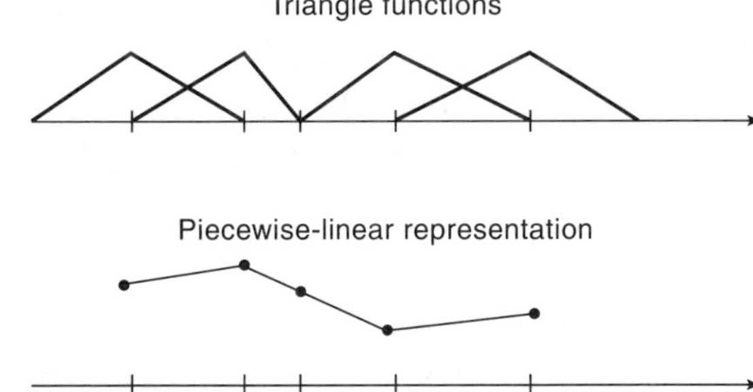

Figure 5.3 Linear interpolation using triangle functions.

Another important family of subsectional functions is obtained from the Lagrangian interpolation polynomials. These polynomials interpolate between function values at a finite number of locations throughout the domain of interest. The first-order Lagrangian functions are piecewise linear and interpolate between the function at endpoints of the domain. These functions are identical to the subsectional triangle functions defined in Equation (5.27). The second-order Lagrangian functions consist of three quadratic polynomials that interpolate between two endpoints and one interior point in the interval of interest. On the interval spanning $[-1, 1]$ and employing the origin as the interior point, the three quadratic functions have the form

$$\phi_1(x) = \tfrac{1}{2}x(x-1) \tag{5.30}$$

$$\phi_2(x) = 1 - x^2 \tag{5.31}$$

and

$$\phi_3(x) = \tfrac{1}{2}x(x+1) \tag{5.32}$$

These are illustrated in Figure 5.4. All three functions span the domain $[-1, 1]$; the first equals one at $x = -1$ and vanishes at $x = 0$ and $x = 1$, the second has unity value at $x = 0$ and equals zero at the endpoints, and the third is zero at $x = -1$ and $x = 0$ and equals one at $x = 1$. The superposition of these functions, appropriately weighted, provides a quadratic representation of the function over the interval.

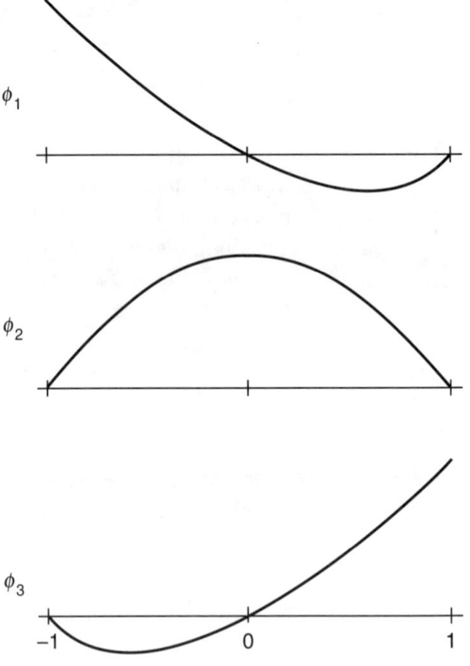

Figure 5.4 Quadratic Lagrangian functions.

In the general case, higher order Lagrangian functions can be defined so that the jth function of order $N-1$ on a general point set within the interval $[x_1, x_N]$ is given by the formula

$$\phi_j(x) = \frac{(x-x_1)(x-x_2)\cdots(x-x_{j-1})(x-x_{j+1})\cdots(x-x_N)}{(x_j-x_1)(x_j-x_2)\cdots(x_j-x_{j-1})(x_j-x_{j+1})\cdots(x_j-x_N)} \tag{5.33}$$

where $j = 1, 2, \ldots, N$ and

$$x_1 < x_2 < \cdots < x_N \tag{5.34}$$

There are N polynomials of order $N - 1$ spanning the interval.

The Lagrangian polynomials could be used as an entire-domain basis spanning the global region of interest. In practice, however, higher order polynomial interpolation is often unstable [13]. Consequently, the traditional approach is to divide the global domain of interest into cells and employ a local expansion in Lagrangian polynomials of some fixed order within each cell. To improve the representation, polynomials of the original degree are employed on a finer subdivision of cells. In other words, the functions are used as a subsectional basis.

The two Lagrangian functions that interpolate (from each side) to a point located at a boundary between two cells can be thought of as one continuous function spanning two cells (Figure 5.5). Thus, the actual basis set consists of (a) functions spanning two cells that interpolate at cell boundaries and (b) functions spanning only one cell that interpolate to an interior point. It is also obvious from Figure 5.5 that the quadratic Lagrangian functions maintain continuity across cell boundaries but do not ensure continuity of any derivatives. This illustrates a significant difference between the quadratic Lagrangian functions and the quadratic spline functions defined in Equation (5.29). Another difference is that the splines are not interpolatory for orders higher than linear.

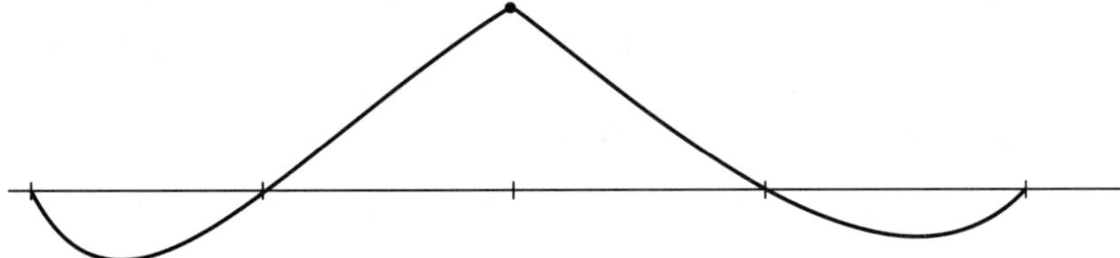

Figure 5.5 Global quadratic Lagrangian function interpolatory at a node between adjacent cells.

If it is necessary to ensure continuity of derivatives across cell boundaries, a set of functions known as Hermitian interpolates can be used. The lowest order Hermitian set consists of four cubic polynomials that interpolate between the function values at the endpoints of the interval. On the domain $[-1, 1]$, these four functions can be expressed as

$$\Psi_1(x) = \tfrac{1}{4}(1 - x)^2(2 + x) \tag{5.35}$$

$$\Psi_2(x) = \tfrac{1}{4}(1 + x)^2(2 - x) \tag{5.36}$$

$$\Psi_3(x) = \tfrac{1}{4}(1 - x^2)(1 - x) \tag{5.37}$$

$$\Psi_4(x) = \tfrac{1}{4}(-1 + x^2)(1 + x) \tag{5.38}$$

The four Hermitian functions are used in the same manner as the Lagrangian polynomials. Functions Ψ_1 and Ψ_2 interpolate the values at $x = -1$ and $x = 1$, respectively; these functions have vanishing first derivative at both endpoints. The functions Ψ_3 and Ψ_4 are zero at both endpoints but have unity first derivative at $x = -1$ and $x = 1$, respectively. Thus Ψ_1 and Ψ_2 maintain continuity of the function while Ψ_3 and Ψ_4 maintain continuity of the first derivative across cell boundaries. The representation requires the superposition of all four functions. Each basis function can be thought of as spanning the two cells adjacent to their interpolation point.

A variety of other ad hoc subsectional basis functions are in use for applications such as wire antennas or scatterers [14, 15]. A common expansion function used in electromagnetics is the sinusoidal triangle function defined as

$$S(x) = \begin{cases} \dfrac{\sin(kx - kx_1)}{\sin(kx_2 - kx_1)} & x_1 < x < x_2 \\ \dfrac{\sin(kx_3 - kx)}{\sin(kx_3 - kx_2)} & x_2 < x < x_3 \end{cases} \tag{5.39}$$

In common with the piecewise-linear triangle function defined in Equation (5.27), this function overlaps two adjacent cells. In fact, if the range of kx is small, $S(x)$ is almost identical to the triangle function in shape. If the parameter k in the function is taken equal to the electromagnetic wavenumber in the surrounding medium, the Helmholtz operator $(\nabla\nabla \cdot + k^2)$ applied to the basis function produces the simple result

$$\left(\frac{\partial^2}{\partial x^2} + k^2\right) S(x) = \delta(x - x_1)\frac{k}{\sin(kx_2 - kx_1)} + \delta(x - x_3)\frac{k}{\sin(kx_3 - kx_2)} \tag{5.40}$$
$$-\delta(x - x_2)k[\cot(kx_3 - kx_2) + \cot(kx_2 - kx_1)]$$

Because of the delta functions, the expression for the electric field produced by an electric current density $S(x)$ is easily obtained in closed form using Equation (1.52). The closed-form evaluation of the field may be an advantageous property in practice.

An alternate expansion function can be constructed from the three-term sinusoid

$$I(x) = A + B\sin(kx - kx_0) + C\cos(kx - kx_0) \tag{5.41}$$

defined locally over a cell. Two of the three coefficients in Equation (5.41) are evaluated in order to ensure the proper continuity or discontinuity of the function at the two cell boundaries, leaving one unknown coefficient that represents the amplitude of the basis function. For instance, a sinusoidal type of spline function can be obtained from this general form by employing this expansion on three adjacent cells in order to obtain a continuous function that vanishes at the outer endpoints and has first derivatives that vanish at the outer endpoints. The three-term sinusoid is well suited for representing the current on thin wires and is used in the Numerical Electromagnetics Code (NEC) [16], where it provides the flexibility to build a derivative discontinuity into the current at junctions between wires of different radii.

A variety of other basis functions are presented in the literature [10, 11, 14, 15]. In recent years, the success of *wavelets* for signal-processing applications has motivated their use as basis functions for representing currents within integral equation formulations [17–19]. Because of their oscillatory nature, wavelets produce matrices that are effectively sparse and may offer computational advantages.

The extension of the scalar Lagrangian functions to the multidimensional case is discussed in Chapter 9. Chapter 9 also considers several *vector* expansion functions, which are often useful for multidimensional electromagnetic applications. These expansion functions share the characteristics that (1) they do not usually form orthogonal sets and (2) their "completeness" property is obtained by shrinking the domain of support rather than adding functions of greater polynomial order.

5.4 INTERPOLATION ERROR

If a linear combination of the basis functions used in a discretization can exactly represent the solution, the method-of-moments process should adjust the coefficients to produce the exact solution. If the basis functions cannot exactly represent the solution, even the best possible choice of coefficients leaves some residual error, known as *interpolation error*. The interpolation error associated with a piecewise-polynomial basis function is relatively easy to characterize.

Consider the function

$$f(x) = a + bx + cx^2 \tag{5.42}$$

Suppose $f(x)$ is approximated on the interval $(-\frac{1}{2}\Delta < x < \frac{1}{2}\Delta)$ using two of the subsectional triangle functions defined in (5.27), that is,

$$f(x) \cong f_{\text{ap}}(x) = f(-\tfrac{1}{2}\Delta)B_1(x) + f(\tfrac{1}{2}\Delta)B_2(x) \tag{5.43}$$

where B_1 and B_2 are defined throughout the interval by

$$B_1(x) = \frac{1}{2} - \frac{x}{\Delta} \tag{5.44}$$

$$B_2(x) = \frac{1}{2} + \frac{x}{\Delta} \tag{5.45}$$

By direct substitution, the function f_{ap} can be expressed in the form

$$f_{\text{ap}}(x) = a + bx + c(\tfrac{1}{2}\Delta)^2 \tag{5.46}$$

The approximation captures the correct constant and linear dependence but obviously cannot represent the quadratic term. In this case, the error between f and f_{ap} reaches a maximum at $x = 0$ and has the peak value

$$|\text{Error}| = \tfrac{1}{4}c\Delta^2 \tag{5.47}$$

As the interval size shrinks to zero, the peak error decreases as $O(\Delta^2)$. In other words, a 50% decrease in Δ causes a 75% decrease in the error. The same result holds if $f(x)$ is an arbitrary polynomial and a piecewise-linear representation is employed.

This result is easily generalized to other polynomial orders (Prob. P5.6), with the result that a representation of polynomial degree p results in an interpolation error of order Δ^{p+1} as $\Delta \to 0$. (Strictly speaking, this estimate requires continuity of the derivative of the function being approximated. With Lagrangian functions, however, the estimate holds even in that case if a cell boundary coincides with the point of derivative discontinuity.) In a situation where variable-sized cells are employed, the error estimate is valid provided that Δ corresponds to the largest cell in the discretization.

Using some of the sample results presented in Chapters 2 and 3, we can investigate the solution accuracy in actual method-of-moments calculations. For example, Table 2.2 shows the error in the surface current density as the number of basis functions is increased for an integral equation formulation. As expected for piecewise-constant basis functions, this error closely follows an $O(\Delta)$ behavior as $\Delta \to 0$. From Table 3.2 and the result of Prob. P3.9, we observe that the error in the field produced by a finite-element discretization of the Helmholtz equation behaves as $O(\Delta^2)$ when piecewise-linear basis functions are employed. This also agrees with the theoretical prediction $O(\Delta^{p+1})$. Although we generally expect additional error in method-of-moments results due to the fact that the coefficients are not optimal, the primary error in these examples appears to be interpolation error.

Occasionally, however, the coefficient values at specific interpolation points are more accurate than $O(\Delta^{p+1})$ as $\Delta \to 0$. This phenomenon, known as superconvergence, occurs when errors combine in such a way that their leading order terms happen to cancel. A similar phenomenon sometimes occurs with secondary quantities derived from the numerical results of integral equation formulations and will be considered in Section 5.12.

5.5 DISPERSION ANALYSIS [20, 21]

An alternate way of investigating the error associated with a basis representation is to consider the distortion that a plane-wave solution undergoes as it propagates across the computational domain. Consider the one-dimensional scalar Helmholtz equation

$$\frac{d^2 E_z}{dx^2} + k^2 E_z(x) = 0 \tag{5.48}$$

and the use of piecewise-linear basis and testing functions. In addition, we assume that the mesh is uniform and large in extent and ignore boundaries. Under these conditions, the mth finite-element equation can be written (Prob. P5.5)

$$2E_m - E_{m-1} - E_{m+1} - k^2\Delta^2\left(\tfrac{2}{3}E_m + \tfrac{1}{6}E_{m-1} + \tfrac{1}{6}E_{m+1}\right) = 0 \tag{5.49}$$

where Δ is the cell size and E_{m-1}, E_m, and E_{m+1} are the basis function coefficients for $E_z(x)$ at $x_m - \Delta$, x_m, and $x_m + \Delta$, respectively.

Equation (5.48) has a traveling-wave solution $E_z(x) = E_0 e^{\pm jkx}$ and motivates an investigation to determine whether a similar solution exists for (5.49). In fact, the discrete solution

$$E_z(x_m) = E_0 e^{\pm j\beta x_m} \tag{5.50}$$

satisfies (5.49) exactly provided that

$$\beta = \frac{1}{\Delta}\cos^{-1}\left(\frac{1 - (k\Delta)^2/3}{1 + (k\Delta)^2/6}\right) \tag{5.51}$$

Consequently, Equations (5.50) and (5.51) constitute the numerical solutions that would be obtained for waves on an infinite, uniform mesh. In a lossless medium, β is real valued for

$$k\Delta \leq \sqrt{12} \tag{5.52}$$

and complex valued for larger $k\Delta$. Therefore, for cell sizes smaller than $\Delta \cong 0.55\lambda$, *the error in the numerical result is entirely phase error.* For larger cell sizes, a complex-valued β indicates error in both the magnitude and phase of the wave.

The phase error across a single cell is given by

$$k\Delta - \beta\Delta \tag{5.53}$$

and is easily tabulated as a function of Δ from Equation (5.51). For example, the error across a cell of width $\Delta = 0.1\lambda$ is about 0.57°. This phase error builds progressively across a mesh, so cells of width 0.1λ spanning a 10λ region would produce 57° of total error.

For this cell size, an error of 180° would be reached in about 32 wavelengths. Table 5.1 summarizes the predicted phase error per wavelength as a function of Δ.

TABLE 5.1 Predicted Phase Error per Wavelength and the Percent Error in the Field as a Function of Cell Size Δ

Δ	Phase error per λ (deg)	Percent error per λ
0.2λ	20.103	34.9
0.1λ	5.670	9.9
0.05λ	1.464	2.6
0.025λ	0.369	0.64
0.0125λ	0.0925	0.16
0.00625λ	0.0231	0.04

Note: From Equations (5.51) and (5.53).

These theoretical estimates correlate with observed numerical error. Figure 5.6 depicts the error in a finite-element solution of the Helmholtz equation for a plane-wave incident on a dielectric slab of thickness 1.0λ and relative permittivity $\varepsilon_r = 2$ [22]. A region of 20λ is included in the finite-element mesh on either side of the slab. Results are shown as a function of the uniform cell size Δ used throughout the free-space region of the computational domain. The fact that the error is primarily phase error explains the oscillatory behavior of the $\Delta = 0.2\lambda$ curve (the error increasing to a maximum, subsequently decreasing, and increasing again throughout the region). For $\Delta = 0.1\lambda$, a phase error of 180° is reached at a distance of about 32λ from the side of the region where the excitation is coupled into the equation, as predicted. Despite the fact that the region is not entirely homogeneous and is terminated by radiation boundary conditions, the observed discretization error exhibits excellent agreement with the theoretical predictions from (5.53).

From an examination of Table 5.1, it is easily discerned that the error decreases as $O(\Delta^2)$ as $\Delta \to 0$, in agreement with the theoretical interpolation error associated with piecewise-linear basis functions. In fact, although we have obtained the preceding results for the one-dimensional case, the phase error estimates arising from this dispersion analysis remain essentially unchanged for waves on triangular-cell models in two dimensions. Irregular cell sizes are typical for the two-dimensional case, and a worst-case approach involving a general triangular-cell model would assign the length of the largest cell edge to Δ, although this may overestimate the error.

Given the size of the computational domain and the phase constant of the medium, the cell density required to produce an acceptable error can be estimated in advance. For example, a cell density of 30 cells/λ would limit the maximum phase error across a free-space region spanning 10λ to about 10°. In order to model a one-dimensional region spanning 100λ with a 10° maximum phase error, however, the minimum required cell density must be increased to about 80 cells/λ. Because the phase error builds on itself, the peak error increases with the size of the region being modeled. *This suggests that to limit the growth in phase error, the cell density must increase as the size of the computational domain increases.* These findings directly impact finite-element discretizations of the Helmholtz equation, where the entire region of interest must often be contained within the computational domain.

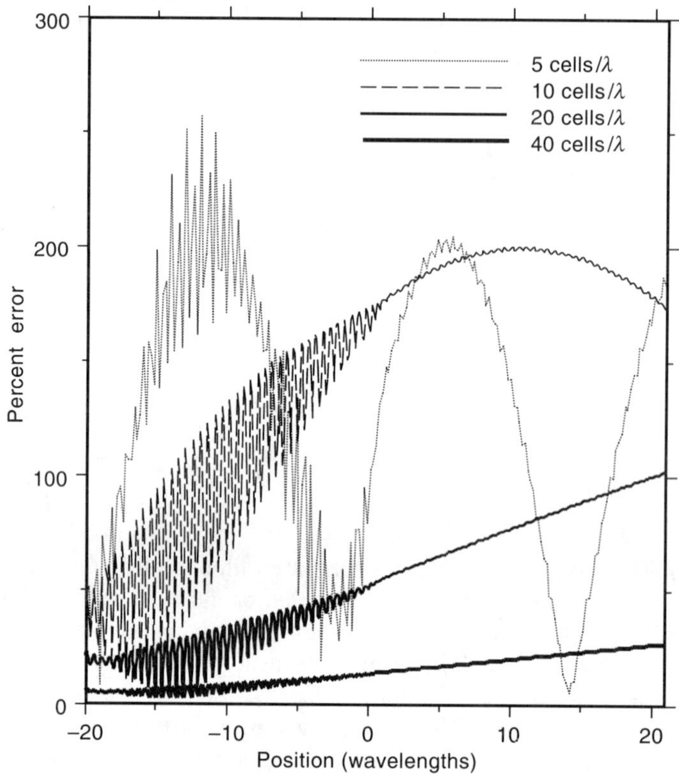

Figure 5.6 Percent error in the finite-element result for the one-dimensional scalar Helmholtz equation. The curves suggest that the error is primarily associated with the phase of the result. After [22]. ©1991 IEEE.

Integral equation formulations using an exact Green's function to span a large region will not incur this cumulative error.

The growth of error with domain size poses an obvious difficulty when attempting to model electrically large structures. However, the error can be reduced by the use of higher order polynomial interpolation functions [21], such as those introduced in Section 5.3. A dispersion analysis for quadratic basis functions is suggested in Prob. P5.10 and indicates a substantial reduction in error.

5.6 DIFFERENTIABILITY CONSTRAINTS ON BASIS AND TESTING FUNCTIONS

We now turn our attention to the role of the testing functions in the numerical solution process. Previous sections have considered the intrinsic error introduced by the basis functions, under the assumption that functions of any polynomial order can be employed. In fact, the specific linear operator to be discretized dictates a minimum polynomial degree for the basis and testing functions. In practice, this degree must increase in proportion to the number of derivatives operating on the unknown function. We will explore this issue in the context of several typical equations.

Consider an integral equation of the form

$$E^i(x) = \frac{k\eta}{4} \int_a^b J(x') H_0^{(2)}(k|x - x'|)\, dx' \qquad a < x < b \tag{5.54}$$

where $J(x)$ is an unknown to be determined and $E^i(x)$ is a known excitation. For instance, $J(x)$ could represent the TM current density induced on a conducting strip. The integrand in this case involves the Hankel function $H_0^{(2)}$, which is a weakly singular function behaving as

$$H_0^{(2)}(kx) \approx 1 - j\frac{2}{\pi} \ln\left(\frac{\gamma k x}{2}\right) \tag{5.55}$$

as $x \to 0$, where γ is defined in Equation (2.13). We wish to discretize Equation (5.54) into matrix form using the method of moments with subsectional basis and testing functions. We first define an inner product

$$\langle a, b \rangle = \int_a^b a^\dagger(x) b(x)\, dx \tag{5.56}$$

Using this inner product, a generic combination of basis and testing functions produces a matrix equation $\mathbf{L}\alpha = \beta$ having entries of the form

$$l_{mn} = \frac{k\eta}{4} \int_a^b T_m^\dagger(x) \int_a^b B_n(x') H_0^{(2)}(k|x - x'|)\, dx'\, dx \tag{5.57}$$

We are now in a position to consider what constraints, if any, must be imposed on the basis and testing functions to ensure a meaningful numerical solution.

Throughout this text, we will adhere to the heuristic assumption that to ensure a meaningful numerical solution *it is necessary to employ a combination of basis and testing functions that keep the coefficients in Equation (5.57) finite and well defined for any location of the basis or testing functions throughout the domain.* From an examination of the integral operator in Equation (5.54), we observe that if piecewise-constant or pulse basis functions are used to represent $J(x)$, the integral produces a function that is continuous and bounded. Thus, the use of Dirac delta functions as testing functions would produce a finite result for each entry of the matrix (regardless of where the testing functions were placed throughout the range of the operator). We also observe that if Dirac delta functions were used as both basis and testing functions, the entries would be infinite if the testing location coincided with a basis function location. Therefore, we conclude that pulse basis functions and Dirac delta testing functions provide the minimum degree of smoothness necessary to discretize an integral operator having a weakly singular kernel.

We will now demonstrate that, for many basis and testing functions, the entries in Equation (5.57) remain the same if the basis and testing functions are exchanged. To show this, we employ a Fourier transformation in conjunction with the convolution theorem. The Fourier transform can be defined as

$$F\{A(x)\} = \tilde{A}(k_x) = \int_{-\infty}^{\infty} A(x) e^{-jk_x x}\, dx \tag{5.58}$$

The transformation converts functions of x to functions of the variable k_x. The inverse

Fourier transform, which converts functions of k_x to functions of x, is defined as

$$F^{-1}\{\tilde{A}(k_x)\} = A(x) = \frac{1}{2\pi}\int_{-\infty}^{\infty} \tilde{A}(k_x)e^{jk_x x}\,dk_x \tag{5.59}$$

The convolution of two functions is given by

$$A(x) * B(x) = \int_{-\infty}^{\infty} A(x')B(x-x')\,dx' \tag{5.60}$$

The convolution theorem associated with the Fourier transform states that

$$A(x) * B(x) = F^{-1}\{\tilde{A}(k_x)\tilde{B}(k_x)\} \tag{5.61}$$

The integral operator of (5.54) is obviously a convolution, but it is not as obvious that the double integral appearing in (5.57) can be thought of as a double convolution. However, since subsectional functions B_n and T_m are usually shifted to the location of cells n and m, respectively, we can redefine the functions as

$$B_n(x) = B(x - x_n) \tag{5.62}$$

and

$$T_m(x) = T(x - x_m) \tag{5.63}$$

It follows that the entries of the method-of-moments matrix $\mathbf{L}$ can be written as

$$l_{mn} = \frac{k\eta}{4} T^R(x) * \left[B(x) * H_0^{(2)}(k|x|)\right]\Bigg|_{x=x_m-x_n} \tag{5.64}$$

where T^R is the space reversal of the complex conjugate of the testing function, that is,

$$T^R(x) = T^\dagger(-x) \tag{5.65}$$

For many of the real-valued subsectional functions used in practice (e.g., splines), $T^R = T$. The convolution theorem in (5.61) shows that the convolutions can be performed in any order. It follows that interchanging the basis and testing functions does not change the entries of $\mathbf{L}$.

Since the entries of $\mathbf{L}$ are the same if the basis and testing functions are interchanged, the new arrangement also satisfies our criteria for minimum differentiability. This means that it is appropriate to employ Dirac delta functions as basis functions to discretize Equation (5.54), provided that pulse functions (or functions smoother than pulse functions) are used as testing functions. In principle, we are modifying the definitions of the domain and range of the linear operator when we consider the use of delta functions as a basis for this problem. This is possible because, as a consequence of the testing function introduced during discretization, the original operator is replaced by a "weak" operator that imposes less stringent mathematical properties.

Suppose we now consider the integro-differential equation

$$E^i(x) = \frac{\eta}{4k}\left(\frac{\partial^2}{\partial x^2} + k^2\right)\int_a^b J(x')H_0^{(2)}(k|x-x'|)\,dx' \qquad a < x < b \tag{5.66}$$

where E^i and J play the same role as they did in the preceding example. This equation could represent TE scattering from conducting strips. The presence of two additional derivatives in Equation (5.66) indicates that the minimum differentiability requirements have increased by two polynomial orders. Employing the convolution notation, the matrix entries have the form

$$l_{mn} = \left[\frac{\eta}{4k} \left(\frac{\partial^2}{\partial x^2} + k^2 \right) T^R(x) * \left[B(x) * H_0^{(2)}(k|x|) \right] \right]_{x=x_m-x_n} \tag{5.67}$$

It follows from the properties associated with the Fourier transform of derivatives and Equation (5.61) that the convolution and differentiation operations in (5.67) commute. Thus, the differentiation may be transferred to either of the basis or testing functions in whole or part. For example, to satisfy our heuristic assumption of keeping the matrix elements finite for any location of basis or testing function, we might employ (a) quadratic spline basis functions with Dirac delta testing functions, (b) triangle basis functions with pulse testing functions, (c) pulse basis functions with triangle testing functions, or (d) Dirac delta basis functions with quadratic spline testing functions. Of course, smoother combinations would also satisfy the minimum criterion.

The differential operators considered in Chapter 3 require similar constraints on basis and testing functions. For example, a one-dimensional weak equation similar in form to (3.5) would require entries of the **L** matrix

$$l_{mn} = \int_a^b \left(\frac{1}{\mu_r} \frac{dT_m}{dx} \frac{dB_n}{dx} - \varepsilon_r k^2 T_m(x) B_n(x) \right) dx \tag{5.68}$$

The minimum differentiability condition suggests the use of piecewise-linear basis functions (triangles) and piecewise-constant testing functions (pulses). An equivalent combination such as quadratic spline basis functions and Dirac delta testing functions also meet the minimum smoothness condition.

Whether the operator is of the integral or differential type, the minimum smoothness condition should be satisfied if reasonable accuracy is expected in the numerical solution. Our heuristic condition places a lower limit on the net differentiability of the basis and testing functions, but not a specific constraint on either function. Since derivatives appearing in the expression for l_{mn} can be moved to the basis or testing function using the convolution idea or straightforward integration by parts, other factors can dictate the specific choice of basis and testing functions. Violation of the minimum smoothness condition will likely degrade the accuracy of the result and prevent numerical solutions from converging to the exact as the number of expansion functions is increased.

Although the exchange of basis and testing functions does not alter the entries of **L** for convolutional operator equations, it does change the entries of the excitation vector β. The need to maintain a well-defined β will also constrain the process of selecting basis and testing functions. In addition, if the representation of the current density is to be used within secondary calculations, that fact may motivate the use of smoother basis functions than the minimum identified above. (For example, it may be necessary to evaluate fields at specific points near the scatterer, for which a delta function, pulse function, or even triangle function representation of the current is likely to produce erratic near fields.) A third consideration is the combination of different geometrical structures generally requiring different types of basis and testing functions. As an example, the NEC [16] employs Dirac delta testing functions for discretizing both the EFIE for conducting wires and the MFIE for conducting surfaces, which simplifies the calculation of mutual interaction terms (the same subroutines can be used regardless of whether the testing location is on a wire or on a surface).

Other factors enter into the choice of basis and testing functions. In practice, it may not be desired to represent the unknown quantity by continuous functions. For instance, the true fields at a dielectric boundary may exhibit a discontinuity or a derivative discontinuity. Often, approximations employed to simplify the scatterer geometry may overwhelm improvements in the basis functions. In several of the examples discussed in Chapter 2, the smooth surface of a scatterer was approximated by a flat-strip model. Once that type of approximation is employed, it makes little sense to try to use basis functions having continuous derivatives at the strip edge. In the past, the numerical integration effort associated with the matrix entries of integral equation formulations sometimes dictated the use of simple integrands. Recent trends seem to favor the use of sophisticated quadrature libraries [23] to efficiently evaluate the necessary integrals to a rather arbitrary degree of accuracy, making the use of more complicated basis functions a fairly convenient task.

We have not yet considered the impact of testing function choice on the relative accuracy of the numerical solution. To illustrate the accuracy for a specific example, consider the EFIE for TM scattering from perfectly conducting cylinders (Section 2.1)

$$E_z^{\text{inc}}(t) = jk\eta \int J_z(t') \frac{1}{4j} H_0^{(2)}(kR)\, dt' \tag{5.69}$$

where

$$R = \sqrt{[x(t) - x(t')]^2 + [y(t) - y(t')]^2} \tag{5.70}$$

and t is a parametric variable around the contour of the cylinder. We wish to test the accuracy of the method-of-moments discretization of this equation using the spline basis and testing functions defined in (5.25)–(5.28).

For circular cylinders excited by a uniform plane wave, exact and numerical solutions can be systematically compared using a normalized error

$$\text{Percent error} = \frac{\left\| J_z^{\text{exact}} - J_z^{\text{numerical}} \right\|}{\left\| J_z^{\text{exact}} \right\|} \times 100 \tag{5.71}$$

based on the norm

$$\left\| J_z^{\text{exact}} - J_z^{\text{numerical}} \right\| = \sqrt{\int \left| J_z^{\text{exact}}(t) - J_z^{\text{numerical}}(t) \right|^2 dt} \tag{5.72}$$

The integration required in (5.72) will be performed over the actual basis function set used in each case to represent the current density, with no additional smoothing or interpolation after the coefficients are determined.

Table 5.2 summarizes the results for a cylinder of size $ka = 6$ as a function of the order of the splines employed as basis and testing functions [24, 25]. Results are presented for 20 basis and testing functions, corresponding to a density of only 3.3 basis functions per wavelength, and 60 basis and testing functions, corresponding to a density of 10 basis functions per wavelength. From Table 5.2, we see that the solution accuracy is primarily determined by the order of the basis functions, or equivalently by the polynomial interpolation error associated with the approximation of J_z. Although the testing functions are expected to play a role in the accuracy of the basis function coefficients, for this example the coefficients appear to be almost independent of the testing function order.

TABLE 5.2 Error in Surface Current Density

Order of Testing	Percent Error	
	20 Unknowns	60 Unknowns
	Order of Basis 1	
0	34.9	11.3
1	35.0	11.3
2	35.6	11.3
3	35.8	11.3
4	36.0	11.3
5	36.2	11.3
	Order of Basis 2	
0	11.7	0.89
1	11.7	0.89
2	11.8	0.89
3	11.9	0.89
4	11.9	0.89
	Order of Basis 3	
0	6.56	0.10
1	6.51	0.10
2	6.53	0.10
3	6.55	0.10

Note: As measured by Equation (5.71) for a TM circular cylinder with circumference of 6λ as a function of the order of the splines employed as basis and testing functions [24, 25]. Basis and testing functions of the indicated order were employed with equal-size cells to construct the matrix equation. A spline of order n has polynomial order $n - 1$ with order zero denoting a Dirac delta function.

5.7 EIGENVALUE PROJECTION THEORY

If the domain and range of the continuous operator L coincide, there may be solutions to the eigenvalue equation

$$Le = \lambda e \tag{5.73}$$

where λ is an eigenvalue and e an eigenfunction of L. There are several reasons for considering Equation (5.73). Applications involving cavities or waveguides lead naturally to eigenvalue equations, and it may be necessary to construct numerical solutions in that context. On the other hand, the behavior of the continuous equation $Lf = g$ will also depend on the eigenfunctions and eigenvalues of L, and these properties are projected in some sense onto the method-of-moments matrix **L**. Since the numerical stability of matrix solution algorithms (Chapter 4) depends on the condition number and eigenvalues of the system matrix, it is natural to inquire into the relationship between the eigenvalues of the continuous operator and those of the matrix. (Although a general linear operator L may not have eigenvalues, the matrix $\mathbf{L}$ always has N eigenvalues.)

To study the relationship, consider the discretization of the eigenvalue equation (5.73) using basis functions $\{B_n\}$ and testing functions $\{T_m\}$. In other words, we employ the expansion

$$e \cong \sum_{n=1}^{N} e_n B_n \tag{5.74}$$

and weigh the residual equations to zero with testing functions $\{T_m\}$ in order to construct the discrete equation

$$\sum_{n=1}^{N} \langle T_m, LB_n \rangle e_n = \lambda \sum_{n=1}^{N} \langle T_m, B_n \rangle e_n \qquad m = 1, 2, \ldots, N \tag{5.75}$$

This is a generalized matrix eigenvalue equation of the form $\mathbf{Le} = \lambda \mathbf{Se}$, where the entries of $\mathbf{L}$ are

$$l_{mn} = \langle T_m, LB_n \rangle \tag{5.76}$$

and the entries of $\mathbf{S}$ are

$$s_{mn} = \langle T_m, B_n \rangle \tag{5.77}$$

As long as the basis and testing functions are each linearly independent sets, $\mathbf{S}$ is nonsingular and Equation (5.75) can be written as

$$\mathbf{S}^{-1}\mathbf{Le} = \lambda \mathbf{e} \tag{5.78}$$

This is an ordinary eigenvalue equation and can be thought of as a discretization of Equation (5.73). It follows that the eigenvalues of the product matrix $\mathbf{S}^{-1}\mathbf{L}$ should approximate those of the original operator L. Furthermore, the corresponding eigenvectors of $\mathbf{S}^{-1}\mathbf{L}$ provide coefficients for Equation (5.74) to approximate the eigenfunctions of L. The accuracy of a particular matrix eigenvalue should depend on the ability of the basis functions to represent the associated eigenfunction.

Conceptually, the eigenvalues of L are projected from the continuous operator onto the method-of-moments matrix $\mathbf{L}$ by the discretization. However, the matrix $\mathbf{S}$ provides a scaling that alters the direct projection and complicates the relationship between the original and matrix eigenvalues. In the special case where the basis functions and testing functions are orthonormal, so that $\mathbf{S}$ is an identity matrix, the eigenvalues of $\mathbf{L}$ are a direct approximation to those of the continuous operator. We will study the numerical accuracy of the eigenvalue projection process in the following section using several canonical examples.

To summarize, if it is necessary to discretize the continuous eigenvalue equation, Equation (5.78) provides the matrix analog. In addition, the relationship between the eigenvalue spectrum of the continuous operator L and the matrix operator $\mathbf{L}$ can be explored in order to learn more about the deterministic equation $Lf = g$ and its numerical treatment. The knowledge gained from this relationship will be the focus for the following sections of this chapter and parts of Chapter 6.

5.8 CLASSIFICATION OF OPERATORS FOR SEVERAL CANONICAL EQUATIONS

We have seen that electromagnetic scattering problems can be posed in terms of equations involving integral or differential operators. To characterize the typical behavior of the specific linear operators of interest, we first classify them into one of several types. These classifications follow the conventions of functional analysis [1, 2, 7, 26–29].

Section 2.1 presented the EFIE for TM scattering from perfectly conducting strips or cylinders. The TM EFIE operator involves an integration over the weakly singular Hankel function $H_0^{(2)}(kR)$. Under mild assumptions concerning the smoothness of the scatterer surface, this integral is an example of a *compact operator*.

Consider the special case of a circular cylinder of radius a. The TM EFIE operator is given by

$$L_{\text{EFIE}}^{\text{TM}}(J_z) = \frac{k\eta}{4} \int_{\phi'=0}^{2\pi} J_z(\phi') H_0^{(2)}(kR) a \, d\phi' \tag{5.79}$$

where

$$R = 2a \left| \sin\left[\tfrac{1}{2}(\phi - \phi')\right] \right| \tag{5.80}$$

For this simple geometry, solutions to the eigenvalue equation $Le_n = \lambda_n e_n$ are easily found in closed form (Prob. P5.11). The eigenfunctions of this operator are the exponential functions

$$e_n(\phi) = e^{jn\phi} \tag{5.81}$$

and the corresponding eigenvalues are

$$\lambda_n^{\text{TM, EFIE}} = \tfrac{1}{2}(\eta\pi ka) J_n(ka) H_n^{(2)}(ka) \tag{5.82}$$

where J_n and H_n are the nth order Bessel and Hankel functions, respectively.

The eigenvalues in Equation (5.82) are complex valued and lie in the right-half complex plane. A plot of λ_0, λ_1, and λ_2 as a function of ka is shown in Figure 5.7. There are a number of noteworthy characteristics that can be gleaned from the behavior of the eigenvalues. For electrically small cylinders, the eigenvalues can be simplified using asymptotic formulas for the Bessel and Hankel functions as $ka \to 0$, yielding

$$\lambda_0^{\text{TM}} \approx \frac{\eta\pi ka}{2} \left[1 - j\frac{2}{\pi} \ln\left(\frac{\gamma ka}{2} \right) \right] \to 0 \tag{5.83}$$

$$\lambda_n^{\text{TM}} \approx j\frac{\eta ka}{2|n|} \to 0 \qquad n \neq 0 \tag{5.84}$$

The asymptotic behavior for large n is also given by Equation (5.84) and indicates that the eigenvalues cluster at the origin as $n \to \infty$. This is a typical characteristic of a compact operator and leads directly to the conclusion that the inverse of a compact operator is unbounded (since the reciprocal eigenvalues "cluster" at infinity).

As $ka \to \infty$, straightforward analysis yields asymptotic formulas such as

$$\lambda_0^{\text{TM}} \approx \eta \cos^2\left(ka - \tfrac{1}{4}\pi\right) - j\eta \cos\left(ka - \tfrac{1}{4}\pi\right) \sin\left(ka - \tfrac{1}{4}\pi\right) \tag{5.85}$$

which is the equation of a circle in the complex plane. For large ka, the eigenvalues tend to follow a common circular trajectory. It is interesting that they pass through the

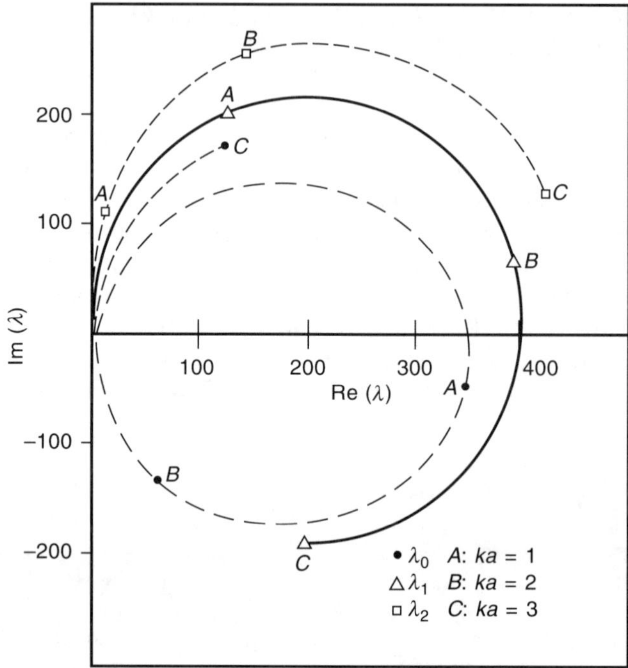

Figure 5.7 Plot of the three dominant eigenvalues of the TM EFIE as a function of ka. After [30]. ©1990 Hemisphere Publishing Corporation.

origin whenever $J_n(ka) = 0$. At these values of ka, the EFIE has homogeneous solutions associated with interior resonant cavity modes. Surface integral equations applied to closed geometries often exhibit homogeneous solutions. (A discussion of the interior solutions is deferred until Chapter 6.)

The behavior of the EFIE eigenvalues is directly indicative of the behavior of the eigenvalues associated with the method-of-moments matrix. The approach of Section 2.1 employed pulse basis functions and Dirac delta testing functions. Under these conditions, the scaling matrix **S** from Equation (5.78) is an identity matrix, and the eigenvalues of the method-of-moments matrix **L** should be a direct approximation to those of the original operator. To test this theory, Table 5.3 compares eigenvalues of **L** with the analytical eigenvalues from Equation (5.82) for a circular cylinder with $ka = 1$. Good agreement is observed between the numerical data and the exact eigenvalues.

The TE EFIE operator for perfectly conducting strips or cylinders was presented in Section 2.4. For a circular cylinder of radius a, the TE EFIE operator has the form

$$L_{\text{EFIE}}^{\text{TE}}(J_\phi) = \frac{\eta}{4k}\hat{\phi} \cdot (\nabla\nabla \cdot + k^2) \int_{\phi'=0}^{2\pi} \hat{\phi}(\phi') J_\phi(\phi') H_0^{(2)}(kR) a \, d\phi' \tag{5.86}$$

where R is defined in Equation (5.80). The eigenfunctions of this operator are also given by (5.81), with associated eigenvalues

$$\lambda_n^{\text{TE, EFIE}} = \tfrac{1}{2}(\eta\pi ka) J_n'(ka) H_n^{(2)\prime}(ka) \tag{5.87}$$

There are many similarities in form between the TE and TM eigenvalues. The TE eigen-

TABLE 5.3 First 10 Distinct Eigenvalues of TM EFIE Compared to Those of Moment-Method Matrix for Circular p.e.c. Cylinder with Circumference 1λ

TM EFIE (5.82)	30 × 30 Matrix
346.50 − j 39.96	346.43 − j 41.95
114.59 + j 203.43	114.55 + j 201.56
7.81 + j 112.24	7.82 + j 110.35
0.23 + j 67.40	0.23 + j 65.62
0.00 + j 48.77	0.00 + j 47.08
0.00 + j 38.49	0.00 + j 36.96
0.00 + j 31.85	0.00 + j 30.51
0.00 + j 27.19	0.00 + j 26.13
0.00 + j 23.74	0.00 + j 23.02
0.00 + j 21.06	0.00 + j 20.82

values also lie in the right-half complex plane, as illustrated by a plot of λ_0, λ_1, and λ_2 as a function of ka provided in Figure 5.8. As $ka \to \infty$, they follow a circular trajectory similar to that of the TM EFIE eigenvalues. The TE eigenvalues pass through the origin whenever $J_n'(ka) = 0$, indicating homogeneous solutions of the EFIE at those values of ka (see Chapter 6). However, their behavior differs from the TM EFIE eigenvalues in the limiting cases of small ka and large n. As $ka \to 0$,

$$\lambda_0^{\mathrm{TE}} \approx j\frac{\eta ka}{2} \to 0 \tag{5.88}$$

$$\lambda_n^{\mathrm{TE}} \approx -j\frac{\eta|n|}{2ka} \to -j\infty \qquad n \neq 0 \tag{5.89}$$

The large spread between the eigenvalues in the low-frequency case suggests that the TE EFIE is unstable for electrically small scatterers, an issue that will be considered in Section 10.6. The asymptotic form of the eigenvalues as $n \to \infty$ is also given by (5.89) and indicates that the TE eigenvalues cluster at infinity. This is a consequence of the derivatives appearing in Equation (5.86), and in common with differential operators this EFIE is known as an *unbounded operator*.

The method-of-moments procedure described in Section 2.4 for discretizing the TE EFIE involved the use of subsectional triangle basis functions and pulse testing functions. Because of the overlap between adjacent basis and testing functions, the scaling matrix $\mathbf{S}$ from Equation (5.78) is not an identity matrix. Table 5.4 compares the eigenvalues of $\mathbf{L}$ and $\mathbf{S}^{-1}\mathbf{L}$ with the analytical eigenvalues from Equation (5.87) for a circular cylinder with $ka = 1$. The eigenvalues of $\mathbf{S}^{-1}\mathbf{L}$ exhibit reasonable agreement with the eigenvalues of the continuous operator.

Section 2.2 presented an alternate approach for TE scattering from closed conducting cylinders using the MFIE. The MFIE operator is characteristic of a third type that we will call an "*identity-plus-compact*" *operator*. This type of operator is generally considered to possess the nicest mathematical properties of the three categories and is considered in detail in several texts [27–29].

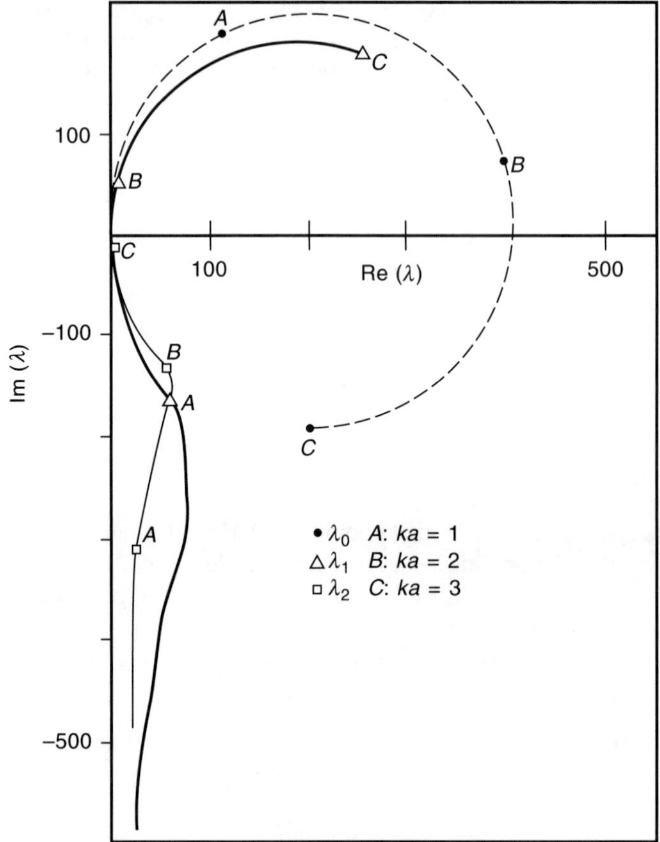

Figure 5.8 Plot of the three dominant eigenvalues of the TE EFIE as a function of ka. After [30]. ©1990 Hemisphere Publishing Corporation.

TABLE 5.4 First 10 Distinct Eigenvalues of TE EFIE Compared to Those of Moment-Method **L**-Matrix and Product Matrix $\mathbf{S}^{-1}\mathbf{L}$ for Circular p.e.c. Cylinder with Circumference 1λ

TE EFIE (5.87)	**L**-Matrix	$\mathbf{S}^{-1}\mathbf{L}$-Matrix
$114.6 + j\,203.4$	$111.6 + j\,200.1$	$112.2 + j\,201.1$
$62.6 - j\,167.3$	$61.7 - j\,166.6$	$62.4 - j\,168.4$
$26.2 - j\,313.5$	$25.6 - j\,306.4$	$26.3 - j\,314.9$
$1.9 - j\,526.1$	$1.8 - j\,500.5$	$1.9 - j\,528.5$
$0.1 - j\,727.4$	$0.1 - j\,668.9$	$0.1 - j\,733.2$
$0.0 - j\,921.9$	$0.0 - j\,813.0$	$0.0 - j\,934.3$
$0.0 - j\,1114$	$0.0 - j\,935.6$	$0.0 - j\,1137$
$0.0 - j\,1305$	$0.0 - j\,1037$	$0.0 - j\,1344$
$0.0 - j\,1495$	$0.0 - j\,1120$	$0.0 - j\,1555$
$0.0 - j\,1685$	$0.0 - j\,1185$	$0.0 - j\,1770$

Note: The order of **S** and **L** is 30.

If applied to a circular cylinder of radius a, the TE MFIE operator has the form

$$L_{\text{MFIE}}^{\text{TE}}(J_\phi) = J_\phi + \frac{1}{4j}\hat{\phi} \cdot \nabla \times \int_{\phi'=0}^{2\pi} \hat{\phi}(\phi') J_\phi(\phi') H_0^{(2)}(kR) a \, d\phi' \tag{5.90}$$

where R is defined in Equation (5.80). For the circular geometry, the eigenfunctions are the set $\{e^{jn\phi}\}$ and the corresponding eigenvalues are given by

$$\lambda_n^{\text{TE, MFIE}} = \tfrac{1}{2}(j\pi ka) J_n(ka) H_n^{(2)\prime}(ka) \tag{5.91}$$

These eigenvalues also lie in the right-half complex plane, as illustrated by a plot of λ_0, λ_1, and λ_2 as a function of ka provided in Figure 5.9. However, the limiting cases differ from the EFIE eigenvalues. For instance, as $ka \to 0$,

$$\lambda_0^{\text{TE, MFIE}} \approx 1 \tag{5.92}$$

$$\lambda_n^{\text{TE, MFIE}} \approx \tfrac{1}{2} \qquad n \neq 0 \tag{5.93}$$

As $n \to \infty$, the MFIE eigenvalues cluster at $0.5 + j0.0$ in the complex plane, a finite location bounded away from the origin.

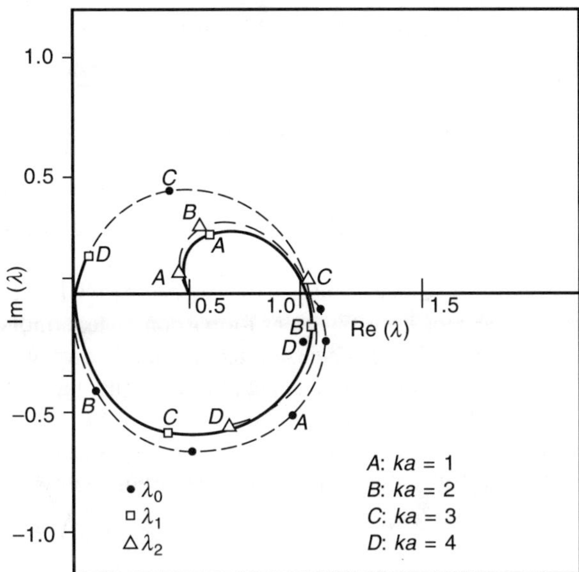

Figure 5.9 Plot of the three dominant eigenvalues of the TE MFIE as a function of ka. After [30]. ©1990 Hemisphere Publishing Corporation.

We have identified the three typical types of operators arising in electromagnetic scattering problems: the compact, unbounded, and "identity-plus-compact" operators. The distinction between these operators is easily illustrated by the eigenvalue behavior identified for the special case of circular cylinders, although the classification holds for similar operators even if no eigenvalue interpretation is possible. The characterization of the operator is important when discussing the convergence of numerical solutions, since convergence proofs assume some particular form. We now turn our attention to some of the arguments that have been proposed to prove convergence.

5.9 CONVERGENCE ARGUMENTS BASED ON GALERKIN'S METHOD [24, 31]

If the identical set of functions is used for basis and testing within a method-of-moments discretization, the procedure is denoted *Galerkin's method.* This approach is advantageous in the special case when (1) the operator L is self-adjoint with respect to the inner product, so that

$$\langle La, b\rangle = \langle a, Lb\rangle \tag{5.94}$$

and (2) the operator is positive definite, that is,

$$\langle a, La\rangle > 0 \text{ for all nonzero } a \tag{5.95}$$

In this situation, a second inner product space can be defined in terms of the new inner product

$$\langle a, b\rangle_2 = \langle a, Lb\rangle \tag{5.96}$$

If $T_m = B_m$, the orthogonal projection of the residuals onto the testing functions described by Equations (5.13)–(5.20) is equivalent to enforcing

$$\langle B_m, L(f^N - f)\rangle = 0 \tag{5.97}$$

in the original inner product space, which can be expressed as

$$\langle B_m, f^N - f\rangle_2 = 0 \tag{5.98}$$

in the new inner product space. This is a statement that the projection of the error in f^N is orthogonal to the basis in the new space.

Consequently, in the special case when L is a positive-definite, self-adjoint operator and the basis functions form a complete, orthogonal set in the *new* inner product space, the method-of-moments projection of the unknown f onto the basis functions is an orthogonal projection. It immediately follows that the numerical approximation f^N will converge to f in the limit as $N \to \infty$.

The constraint on orthogonality can be relaxed. In fact, it may be difficult to identify a practical basis satisfying the orthogonality

$$\langle B_m, LB_n\rangle = 0 \qquad m \neq n \tag{5.99}$$

However, a Gram–Schmidt procedure can be used to show the equivalence of any linearly independent set $\{B_n\}$ to an orthogonal set $\{\Psi_n\}$ satisfying (5.99). Since $\{B_1, \ldots, B_N\}$ and $\{\Psi_1, \ldots, \Psi_N\}$ span the same N-dimensional subspace, it is sufficient to force the residual error to be orthogonal to the set $\{B_n\}$. Thus, the orthogonality of the basis set is not necessary to establish the convergence of $f^N \to f$.

The preceding idea is often used to demonstrate the convergence of solutions to a finite-element discretization of Laplace's equation. Consider an inner product defined according to (5.56), for instance, with the second inner product defined by (5.96). Subsectional Lagrangian expansion functions satisfy the necessary completeness property in the new inner product space, and the operator is positive definite and self-adjoint. Unfortunately, the same argument is not directly applicable to the scalar Helmholtz equation when the operator is indefinite, since then the second inner product is not valid. For the indefinite Helmholtz equation, convergence must be established by other means [32, 33].

The integral operators arising in electromagnetics are usually not positive definite or self-adjoint except in the static limit, and consequently the above convergence proof does not apply to the EFIE or MFIE. There may, however, be persuasive reasons for choosing $T_m = B_m$ in practice. For instance, using Galerkin's method with the EFIE produces matrices with diagonal symmetry, which permits a 50% reduction in computation during matrix construction and solution (recall Prob. P2.1 and the discussion in Section 4.4.)

5.10 CONVERGENCE ARGUMENTS BASED ON DEGENERATE KERNEL ANALOGS

We now turn our attention to an integral equation of the form $f + Lf = g$, where f is the unknown to be determined and L is a compact integral operator

$$Lf = \int K(x, x') f(x')\, dx' \tag{5.100}$$

defined in terms of a weakly singular kernel K. (This equation might represent the MFIE of Section 2.2 applied to a closed, smooth, perfectly conducting surface.) We define an inner product

$$\langle a, b \rangle = \int a^\dagger(x) b(x)\, dx \tag{5.101}$$

and postulate a (complete, orthogonal) basis $\{B_n\}$ for the domain space and a second basis $\{T_m\}$ for the range space. We therefore intend to seek a solution of the form

$$f(x) \cong \sum_{n=1}^{N} \alpha_n B_n(x) \tag{5.102}$$

Instead of substituting Equation (5.102) into the integral equation, however, we proceed in a different manner. Suppose that we project the kernel of the integral operator as a function of x' onto the basis $\{B_n^\dagger\}$ constructed from the complex conjugates of $\{B_n\}$. This yields

$$K(x, x') \cong \sum_{n=1}^{N} \gamma_n(x) B_n^\dagger(x') \tag{5.103}$$

where, to obtain an orthogonal projection,

$$\gamma_n(x) = \frac{\langle B_n^\dagger, K \rangle}{\langle B_n^\dagger, B_n^\dagger \rangle} = \frac{L B_n}{\langle B_n, B_n \rangle} \tag{5.104}$$

(The integration within the inner products is carried out over the primed coordinates.) We also project the kernel as a function of x onto the set $\{T_m\}$, to obtain

$$K(x, x') \cong K_N(x, x') = \sum_{m=1}^{N} \sum_{n=1}^{N} k_{mn} T_m(x) B_n^\dagger(x') \tag{5.105}$$

Here, K_N is an example of a *degenerate* kernel. The coefficients

$$k_{mn} = \frac{\langle T_m, \gamma_n \rangle}{\langle T_m, T_m \rangle} = \frac{\langle T_m, LB_n \rangle}{\langle T_m, T_m \rangle \langle B_n, B_n \rangle} \tag{5.106}$$

provide the best representation by minimizing the norm of the error

$$\|K - K_N\| = \left\| K(x, x') - \sum_{m=1}^{N} \sum_{n=1}^{N} k_{mn} T_m(x) B_n^{\dagger}(x') \right\| \tag{5.107}$$

This error norm converges to zero as $N \to \infty$ as long as the kernel K is either a continuous or a weakly singular function [34].

We now consider the equation

$$f^N(x) + L^N(f^N) = f^N(x) + \int K_N(x, x') f^N(x')\, dx' = g(x) \tag{5.108}$$

obtained by replacing the original kernel by the degenerate kernel K_N. We want to establish the relationship between f^N and the actual solution f. First, however, note that this system can be expressed as

$$f^N(x) + \sum_{m=1}^{N} \sum_{n=1}^{N} k_{mn} T_m(x) \langle B_n, f^N \rangle = g(x) \tag{5.109}$$

which illustrates that f^N can be obtained exactly in the form

$$f^N(x) = g(x) - \sum_{m=1}^{N} \delta_m T_m(x) \tag{5.110}$$

where

$$\delta_m = \sum_{n=1}^{N} k_{mn} \langle B_n, f^N \rangle = \sum_{n=1}^{N} \frac{\langle T_m, LB_n \rangle}{\langle T_m, T_m \rangle} \frac{\langle B_n, f^N \rangle}{\langle B_n, B_n \rangle} = \frac{\langle T_m, L^N f^N \rangle}{\langle T_m, T_m \rangle} \tag{5.111}$$

If we introduce

$$c_n = \langle B_n, f^N \rangle \tag{5.112}$$

the coefficients $\{c_n\}$ and $\{\delta_n\}$ can be found by constructing an inner product of Equation (5.93) with B_j to produce

$$c_j + \sum_{n=1}^{N} c_n \left(\sum_{m=1}^{N} \frac{\langle T_m, LB_n \rangle}{\langle T_m, T_m \rangle} \frac{\langle B_j, T_m \rangle}{\langle B_n, B_n \rangle} \right) = \langle B_j, g \rangle \tag{5.113}$$

Solving this $N \times N$ linear system completes the solution. In addition to obtaining f^N in the form of Equation (5.110), we have also determined the projection of f^N onto the basis $\{B_n\}$ as

$$f^N(x) \cong \sum_{n=1}^{N} \frac{c_n}{\langle B_n, B_n \rangle} B_n(x) \tag{5.114}$$

Although this orthogonal projection is the best representation of f^N in the subspace spanned by the basis $\{B_n\}$, it is not an exact equality, as is Equation (5.110).

It is a straightforward matter to show that f^N converges to f as $N \to \infty$ [27]. Using the identity

$$f - f^N + L^N f - L^N f^N = L^N f - Lf \tag{5.115}$$

we write

$$f - f^N = (I + L^N)^{-1}(L^N - L)f \tag{5.116}$$

where I denotes the identity operator. We then construct the inequality

$$\frac{\|f - f^N\|}{\|f\|} \leq \|(I + L^N)^{-1}\| \|L^N - L\| \tag{5.117}$$

Since L and L^N are compact operators, the first norm on the right-hand side is bounded as long as -1 is not an eigenvalue of L.[1] The second norm converges to zero as $N \to \infty$ as a consequence of (5.106). Therefore (as long as -1 is not an eigenvalue of the operator L), f^N converges to f.

We now investigate the relationship between f^N and the formal method-of-moments solution of the original equation $f + Lf = g$. To establish this relationship, first substitute (5.114) into (5.109) to obtain

$$\sum_{n=1}^{N} \frac{c_n}{\langle B_n, B_n \rangle} B_n(x) + \sum_{m=1}^{N} \sum_{n=1}^{N} k_{mn} c_n T_m(x) \cong g(x) \tag{5.118}$$

where the coefficients $\{c_n\}$ are defined in Equation (5.112) and produce an orthogonal projection of f^N onto the basis functions. Because (5.114) is only exact in the subspace spanned by $\{B_1, B_2, \ldots, B_N\}$, Equation (5.118) is not an equality except in that subspace. It follows that a projection onto the testing functions

$$\sum_{n=1}^{N} \frac{c_n}{\langle B_n, B_n \rangle} \langle T_m, B_n \rangle + \sum_{n=1}^{N} k_{mn} c_n \langle T_m, T_m \rangle \cong \langle T_m, g \rangle \qquad m = 1, 2, \ldots, N \tag{5.119}$$

is also only approximate. But, by rewriting the left side of this equation to obtain

$$\sum_{n=1}^{N} \frac{c_n}{\langle B_n, B_n \rangle} \langle T_m, (I + L) B_n \rangle \cong \langle T_m, g \rangle \qquad m = 1, 2, \ldots, N \tag{5.120}$$

we arrive at the method-of-moments system representing the original equation $f + Lf = g$. This demonstrates that the coefficients produced by solving the $N \times N$ method-of-moments matrix equation

$$\sum_{n=1}^{N} \langle T_m, (I + L) B_n \rangle \alpha_n = \langle T_m, g \rangle \qquad m = 1, 2, \ldots, N \tag{5.121}$$

are *not* the best representation of f^N in the subspace spanned by $\{B_1, B_2, \ldots, B_N\}$ [except in the special case $T_m = B_m$, in which case (5.120) is an exact equality]. However, in either case the approximate equality in (5.118) becomes exact in the limit as $N \to \infty$, suggesting that the method-of-moments solution does converge to the orthogonal projection.

To summarize, assuming that the basis and testing functions are complete, orthogonal sets and that L in the equation $f + Lf = g$ is a compact operator, the method-of-moments solution converges to f^N in the limit as $N \to \infty$, and f^N in turn converges to f. Whether

[1] In practice, -1 may be an eigenvalue of L for certain geometries; refer to Chapter 6.

or not the method-of-moments solution is an orthogonal projection of f^N onto the basis functions is not particularly relevant, since there is no reason to believe that f^N is an orthogonal projection of f onto the basis functions for finite N in the first place.

We next consider the equation $Lf = g$, where L is a compact operator. (The TM EFIE representing scattering from smooth perfectly conducting cylinders is an example of such an equation.) Unfortunately, the convergence arguments outlined above cannot be applied to this equation. Replacing the original kernel by the degenerate kernel K_N leads to the equation

$$\sum_{m=1}^{N}\sum_{n=1}^{N} k_{mn} T_m(x)\langle B_n, \tilde{f}\rangle \cong g(x) \tag{5.122}$$

The approximate equality is a consequence of the fact that a finite combination of the $\{T_m\}$ may not be able to represent g exactly. We can obtain an exact equality by considering the modified equation

$$\sum_{m=1}^{N}\sum_{n=1}^{N} k_{mn} T_m(x)\langle B_n, f^N\rangle = \sum_{m=1}^{N} \frac{\langle T_m, g\rangle}{\langle T_m, T_m\rangle} T_m(x) \tag{5.123}$$

obtained by projecting the excitation $g(x)$ onto the basis $\{T_m\}$. Since this is an orthogonal projection, the representation for g converges as $N \to \infty$ as long as g is well behaved. The modified equation is just the method-of-moments system

$$\sum_{n=1}^{N} \langle T_m, LB_n\rangle \alpha_n = \langle T_m, g\rangle \qquad m = 1, 2, \ldots, N \tag{5.124}$$

In this case, we have defined f^N so that its projection into the subspace spanned by $\{B_1, B_2, \ldots, B_N\}$ is the method-of-moments solution. By definition, the $\{\alpha_n\}$ produce an orthogonal projection of f^N onto $\{B_1, B_2, \ldots, B_N\}$.

However, we are not able to show that f^N converges to f as $N \to \infty$. For an equation of the form $Lf = g$, (5.115) is modified to

$$L^N f - L^N f^N = L^N f - Lf + g - g^N \tag{5.125}$$

from which we construct

$$f - f^N = (L^N)^{-1}[(L^N - L)f + (g - g^N)] \tag{5.126}$$

The appropriate inequality obtained from this result is

$$\|f - f^N\| \le \|(L^N)^{-1}\| \|(L^N - L)f + (g - g^N)\| \tag{5.127}$$

Since L^N is compact, its inverse increases without bound as $N \to \infty$, and the inequality does not bound the error in f^N. Consequently, we cannot use this argument to prove general convergence for an equation such as the TM EFIE with the form $Lf = g$, where L is a compact operator. (In fact, the difficulty here is of a fundamental nature, and solutions to this type of equation may not exist in the absence of additional constraints on g. See Prob. P5.15.)

The convergence argument also fails to apply in the case where L represents an unbounded operator. In that situation, the kernel K is strongly singular and the projection onto the basis and testing functions in Equation (5.105) may not converge as $N \to \infty$. Therefore the degenerate kernel analog cannot be used to show convergence for unbounded integro-differential equations, such as the TE EFIE.

5.11 CONVERGENCE ARGUMENTS BASED ON PROJECTION OPERATORS

The preceding arguments require the basis and testing functions to form complete, orthogonal sets. However, the expansion functions used in practice seldom satisfy the orthogonality criterion. Furthermore, the testing functions used in practice may include Dirac delta functions, which cannot be used in the preceding convergence arguments. Fortunately, some of the assumptions needed to show convergence can be relaxed in a more general framework based on the concept of a *projection operator*. The following discussion has been adapted from Atkinson [27].

A bounded projection operator P_N can be defined to map functions in some space S to a subspace S^N in such a manner that

$$P_N f = f \quad \text{for all } f \text{ in } S^N \tag{5.128}$$

A simple example of a projection operator is that obtained from linear interpolation on a point set $\{x_1, x_2, \ldots, x_N\}$. Suppose we employ as basis functions the subsectional triangle functions defined in Equation (5.27), that is,

$$B_n = \begin{cases} \dfrac{x - x_{n-1}}{x_n - x_{n-1}} & x_{n-1} < x < x_n \\ \dfrac{x_{n+1} - x}{x_{n+1} - x_n} & x_n < x < x_{n+1} \end{cases} \tag{5.129}$$

These functions provide a linear interpolation, and the representation

$$f(x) \cong \sum_{n=1}^{N} f(x_n) B_n(x) \tag{5.130}$$

is clearly a projection onto a subspace spanned by $\{B_1, B_2, \ldots, B_N\}$. The projection also satisfies Equation (5.128). Furthermore, if f has a continuous first derivative, and if the points are spaced at equal intervals throughout the domain, the error associated with linear interpolation is known to be $O(\Delta^2)$, where $\Delta = x_n - x_{n-1}$ is the interval size. Under these conditions, (5.130) converges to $f(x)$ pointwise as $N \to \infty$, which is a stronger statement of convergence than the convergence-in-norm used in previous sections.

To place the solution of an equation $f + Lf = g$ in the context of projection operators, consider

$$f^N(x) + P_N L f^N = P_N g \tag{5.131}$$

where f^N lies in the subspace S^N. The relationship between f and f^N can be studied by combining Equation (5.131) with

$$P_N(f + Lf) = P_N f + P_N L f = P_N g \tag{5.132}$$

to obtain

$$f - f^N + P_N L f - P_N L f^N = f - P_N f \tag{5.133}$$

From Equation (5.133), we construct the inequality

$$\|f - f^N\| \le \|(I + P_N L)^{-1}\| \|f - P_N f\| \tag{5.134}$$

Therefore, convergence of f to f^N is assured only if $(I + P_N L)^{-1}$ is bounded and if $P_N f$ converges to f as $N \to \infty$.

Atkinson has shown [27, pp. 51–54] that if L is compact, the pointwise convergence of $P_N f$ to f (for all f in the space) is sufficient to establish a bound on $(I + P_N L)^{-1}$. The complete development of this result is lengthy and will be omitted. The pointwise convergence is also sufficient to show that

$$\|f - P_N f\| \to 0 \quad \text{as } N \to \infty \tag{5.135}$$

It follows that f^N converges to f.

We now demonstrate the connection between Equation (5.131) and the method-of-moments system for two examples involving compact integral operators of the form

$$Lf = \int K(x, x') f(x')\, dx' \tag{5.136}$$

In the first example, we employ subsectional triangle functions from Equation (5.27) in connection with point matching on the set of equally spaced points $\{x_1, x_2, \ldots, x_N\}$ in order to obtain the discrete system

$$f_m + \sum_{n=1}^{N} f_n \int K(x_m, x') B_n(x')\, dx' = g_m \qquad m = 1, 2, \ldots, N \tag{5.137}$$

This system is identical in form to (5.131), with $P_N f$ defined as the projection operator that produces the element of the subspace spanned by $\{B_1, B_2, \ldots, B_N\}$ that interpolates to $f(x)$ at $(x_1, x_2, \ldots, x_N)$. The convergence of f^N to f is readily established for this example from the preceding results.

Therefore, in the case of an identity-plus-compact operator, a bounded projection operator is obtained when employing the method of moments with subsectional triangle basis functions and point matching. Furthermore, the method-of-moments solution converges to f as $N \to \infty$. This shows that a convergence proof is possible even if the testing functions are Dirac delta functions.

As a second example, consider the use of a linearly independent set of functions $\{B_1, B_2, \ldots, B_N\}$, defined in some subspace of an inner product space where

$$\langle a, b\rangle = \int a^\dagger(x) b(x)\, dx \tag{5.138}$$

Specifically, we want to consider the use of subsectional spline functions of some order (Section 5.3) for $\{B_n\}$. We seek a solution in the form

$$f^N(x) = \sum_{n=1}^{N} \alpha_n B_n(x) \tag{5.139}$$

The method-of-moments system obtained by employing the same functions for basis and testing (i.e., Galerkin's method) is

$$\sum_{n=1}^{N} \langle B_m, (I + L) B_n\rangle \alpha_n = \langle B_m, g\rangle \qquad m = 1, 2, \ldots, N \tag{5.140}$$

Let $\{\Psi_n\}$ denote the orthonormal basis constructed in the subspace S^N from $\{B_n\}$ by a Gram–Schmidt procedure. The projection operator associated with this discretization can be defined as providing the orthogonal projection onto $\{\Psi_n\}$, that is,

$$P_N f = \sum_{n=1}^{N} \langle \Psi_n, f \rangle \Psi_n(x) \tag{5.141}$$

Since Equation (5.140) forces the residual

$$f^N(x) + Lf^N - g(x) \tag{5.142}$$

to be orthogonal to each entry of $\{B_1, \ldots, B_N\}$, the residual is also orthogonal to the set $\{\Psi_1, \ldots, \Psi_N\}$, and (5.140) is equivalent to

$$P_N f^N + P_N L f^N = P_N g \tag{5.143}$$

Since $P_N f^N = f^N$, this method-of-moments system is also equivalent to that obtained from the projection operator defined in (5.131).

To show convergence of f^N to f, it is again necessary to demonstrate the pointwise convergence of $P_N f$ to f for all f in the space. Atkinson outlines a proof valid for piecewise-polynomial representations, which encompass the spline functions considered here, and we refer the readers to his text for the details [27, p. 68].

The preceding discussion introduced the "projection operator" interpretation of the method-of-moments procedure for solving $f + Lf = g$, where L is a compact operator. The analysis is applicable to a variety of practical discretizations involving subsectional basis and testing functions (which, as we have seen, may not always form orthogonal sets), even if Dirac delta testing functions are involved. Consequently, this approach is somewhat more general than the degenerate kernel analysis considered in Section 5.10.

As with the degenerate kernel approach, however, the convergence of f^N to f cannot be established by the preceding arguments for equations of the form $Lf = g$, where L is either compact or unbounded. For these systems, Equation (5.134) is replaced by

$$\|f - f^N\| \le \|(P_N L)^{-1}\| \|(P_N L - L) f + (g - P_N g)\| \tag{5.144}$$

To establish convergence, $(P_N L)^{-1}$ must be bounded as $N \to \infty$, which is not the case for L compact. For unbounded L, the problem arises in showing that

$$\|P_N L - L\| \to 0 \quad \text{as } N \to \infty \tag{5.145}$$

Although this condition is certainly true in particular situations, it appears to be difficult to demonstrate for the general case.

5.12 THE STATIONARY CHARACTER OF FUNCTIONALS EVALUATED USING NUMERICAL SOLUTIONS [25]

In electromagnetic radiation and scattering problems, we are often primarily concerned with the far-zone fields needed to characterize the scattering cross section of a target or the radiation pattern of an antenna. These quantities can be expressed as quadratic functionals of the surface current density. Such functionals can sometimes be defined in a way that ensures that they have a stationary point at the true solution and consequently exhibit an error of only $O(\varepsilon^2)$ as $\varepsilon \to 0$ when the surface current used in their evaluation has an error of $O(\varepsilon)$ [35]. In this section, we show that when the surface current density is computed via the method of moments, the error in the far fields or any other continuous functional depends equally on the basis and testing functions used in the method-of-moments process. Furthermore, if the basis and testing functions have similar smoothness characteristics, any continuous functional exhibits error of the same order as a stationary expression. Therefore, there appears to be no advantage in using a strictly stationary functional.

In general, suppose we want to compute an expression

$$Q = \langle f, h \rangle \tag{5.146}$$

where f is a solution of the linear equation $Lf = g$, h is a given function, and the inner product is defined in accordance with (5.1)–(5.3). We can seek a method-of-moments approximation

$$f \cong f^N = \sum_{n=1}^{N} \alpha_n B_n \tag{5.147}$$

obtained by forcing the residual $Lf^N - g$ to be orthogonal to testing functions $\{T_m\}$. In other words, the coefficients $\{\alpha_n\}$ are obtained from the system

$$\sum_{n=1}^{N} \alpha_n^\dagger \langle LB_n, T_m \rangle = \langle g, T_m \rangle \qquad m = 1, 2, \ldots, N \tag{5.148}$$

Let us assume that the error in the result is

$$f - f^N = \varepsilon_f \tag{5.149}$$

and investigate the error in the approximation $\langle f^N, h \rangle$.

To study the error in $\langle f^N, h \rangle$, we must first consider the equation $L^A e = h$, where L^A is the adjoint of L with respect to the inner product and h is the function appearing in (5.146). The adjoint operator is defined according to the relationship $\langle La, b \rangle = \langle a, L^A b \rangle$. Suppose we construct a method-of-moments approximation for the adjoint equation using basis functions $\{T_n\}$ to represent e and testing functions $\{B_m\}$ (the opposite of what we used above). In other words, e is replaced by

$$e \cong e^N = \sum_{n=1}^{N} \beta_n T_n \tag{5.150}$$

and the associated system of equations is

$$\sum_{n=1}^{N} \beta_n \langle B_m, L^A T_n \rangle = \langle B_m, h \rangle \qquad m = 1, 2, \ldots, N \tag{5.151}$$

Let us assume that the error in e^N will be

$$e - e^N = \varepsilon_e \tag{5.152}$$

The role of the adjoint equation becomes apparent if we rewrite Q as

$$Q = \langle f, h \rangle = \langle f, L^A e \rangle = \langle Lf, e \rangle \tag{5.153}$$

and consider the approximation

$$Q \cong Q^N = \langle Lf^N, e^N \rangle \tag{5.154}$$

Because of the way that f^N and e^N were constructed, we can show that

$$Q^N = \langle Lf^N, e^N \rangle = \langle Lf^N, e \rangle = \langle Lf, e^N \rangle \tag{5.155}$$

without approximation. This follows from Equations (5.148) and (5.151), for instance,

$$\begin{aligned}\langle Lf^N, e^N\rangle &= \sum_{m=1}^{N} \beta_m \left\langle L\left(\sum_{n=1}^{N} \alpha_n B_n\right), T_m\right\rangle \\ &= \sum_{m=1}^{N} \beta_m \langle g, T_m\rangle \\ &= \sum_{m=1}^{N} \beta_m \langle Lf, T_m\rangle \\ &= \langle Lf, e^N\rangle\end{aligned} \tag{5.156}$$

where the second equality is established by (5.148). Using (5.155), the error in Q can be expressed as

$$\begin{aligned}Q - Q^N &= \langle Lf, e\rangle - \langle Lf^N, e^N\rangle \\ &= \langle Lf, e\rangle - \langle Lf^N, e\rangle \\ &= \langle L(f - f^N), e\rangle \\ &= \langle L(f - f^N), e^N\rangle + \langle L\varepsilon_f, \varepsilon_e\rangle\end{aligned} \tag{5.157}$$

The first inner product is identically zero, by (5.156). Therefore, we obtain [36]

$$Q - Q^N = \langle L\varepsilon_f, \varepsilon_e\rangle \tag{5.158}$$

Although the expression $\langle Lf^N, e^N\rangle$ was used to obtain this error estimate, (5.155) shows that the identical error estimate applies to $\langle f^N, h\rangle$ or $\langle g, e^N\rangle$. Thus, there is no need to solve two method-of-moments problems to obtain $O(\varepsilon_f \varepsilon_e)$ error.

To summarize, if $f^N \cong f$ is obtained from a method-of-moments solution of $Lf = g$ using basis functions $\{B_n\}$ and testing functions $\{T_n\}$, the approximation $\langle f^N, h\rangle$ contains error of $O(\varepsilon_f \varepsilon_e)$, where ε_f and ε_e are defined in (5.149) and (5.152). We realize an error proportional to ε_e even though we do not actually solve the adjoint equation $L^A e = h$. From the previous discussion of solution error in Sections 5.4–5.6, it is reasonable to assume that ε_f is primarily determined by the ability of the set $\{B_n\}$ to represent f, while ε_e is primarily determined by the ability of the set $\{T_n\}$ to represent e. These observations suggest that the accuracy of $\langle f, h\rangle$ is enhanced if the set $\{B_n\}$ is a good basis for the domain space of L and the set $\{T_n\}$ is a good basis for the range space of L.

In the special case when L is self-adjoint, f satisfies $Lf = g$, and the functional to be computed is $\langle f, g\rangle$, we have

$$Q = \langle f, g\rangle = \langle Lf, f\rangle \cong \langle Lf^N, f^N\rangle \tag{5.159}$$

If the same functions are used for $\{B_n\}$ and $\{T_n\}$ (Galerkin's method), the approximation in (5.159) exhibits an error of $O(\varepsilon_f^2)$. This estimate follows directly from equation (5.158), since under these assumptions $Lf = g$ and its adjoint equation are identical, and e^N is replaced by f^N throughout. The $O(\varepsilon_f^2)$ error is often reported in the literature [37, 38] and sometimes used as an argument in favor of Galerkin's method. Although Equation (5.159) satisfies the strict definition of a stationary functional, it is important to note that the error in (5.159) is not necessarily any better than the error in the more general case described by (5.158). As long as ε_f and ε_e are comparable, the general functional $\langle f^N, h\rangle$ is also a stationary quantity. Thus, there is no definitive advantage to this special case and no reason to use testing functions that are identical to the basis functions, in preference to an alternative choice that provides similar error levels.

To illustrate these calculations in practice, consider the scattered electric field produced by a point source illuminating a perfectly conducting target. A specific functional can be obtained from a reciprocity relationship. Consider two sets of sources, $(\bar{J}_1, \bar{K}_1)$ and $(\bar{J}_2, \bar{K}_2)$, and the fields produced by these in a common region, $(\bar{E}_1, \bar{H}_1)$ and $(\bar{E}_2, \bar{H}_2)$. In a manner similar to that developed in Section 1.6, Maxwell's curl equations can be written for each source–field pair and combined to produce

$$\begin{aligned} E_2 \cdot J_1 - H_2 \cdot K_1 - E_1 \cdot J_2 + H_1 \cdot K_2 & \\ = \bar{H}_2 \cdot \nabla \times \bar{E}_1 - \bar{E}_1 \cdot \nabla \times \bar{H}_2 + \bar{E}_2 \cdot \nabla \times \bar{H}_1 - \bar{H}_1 \cdot \nabla \times \bar{E}_2 & \qquad (5.160) \\ = \nabla \cdot (\bar{E}_1 \times \bar{H}_2 - \bar{E}_2 \times \bar{H}_1) & \end{aligned}$$

After integrating throughout the region and applying the divergence theorem, we obtain

$$\begin{aligned} \iiint_V \bar{E}_2 \cdot \bar{J}_1 - \bar{H}_2 \cdot \bar{K}_1 - \bar{E}_1 \cdot \bar{J}_2 + \bar{H}_1 \cdot \bar{K}_2 & \\ = \iiint_V \nabla \cdot (\bar{E}_1 \times \bar{H}_2 - \bar{E}_2 \times \bar{H}_1) & \\ = \iint_S (\bar{E}_1 \times \bar{H}_2 - \bar{E}_2 \times \bar{H}_1) \cdot \hat{n}\, dS & \qquad (5.161) \\ = 0 & \end{aligned}$$

as the closed surface S containing the region recedes to infinity, provided that all the sources are contained within S. Therefore, we obtain the reciprocity relationship

$$\iiint_V \bar{E}_2 \cdot \bar{J}_1 - \bar{H}_2 \cdot \bar{K}_1 = \iiint_V \bar{E}_1 \cdot \bar{J}_2 - \bar{H}_1 \cdot \bar{K}_2 \qquad (5.162)$$

where the volume V contains all sources. Equation (5.162) can be specialized to the case where the sources are entirely electric currents, as might describe a p.e.c. scatterer illuminated by an electric point source

$$\bar{J}_0 = \hat{u} J_0 \delta(\bar{r} - \bar{r}_0) \qquad (5.163)$$

where $\bar{r}_0$ represents a location outside the scatterer. This source produces an incident electric field $\bar{E}^{\text{inc}}$, and in turn an induced surface current density $\bar{J}_s$ on the scatterer. The $\hat{u}$-component of the scattered electric field at the point $\bar{r}_0$ is given by

$$\hat{u} \cdot \bar{E}^s|_{\bar{r}=\bar{r}_0} = \frac{1}{J_0} \iiint_V \bar{E}^s \cdot \bar{J}_0 \qquad (5.164)$$

However, using (5.162), this expression can be rewritten as

$$\hat{u} \cdot \bar{E}^s|_{\bar{r}=\bar{r}_0} = \frac{1}{J_0} \iiint_V \bar{E}^{\text{inc}} \cdot \bar{J}_s \qquad (5.165)$$

Equation (5.165) is a functional for the scattered field at the source point.

We next introduce a method-of-moments formulation for $\bar{J}_s$, in terms of the EFIE. Since the EFIE operator is not self-adjoint with respect to an inner product defined according to (5.1)–(5.3), instead we introduce a symmetric product defined as

$$(\bar{A}, \bar{B}) = \iint_S \bar{A}_{\tan} \cdot \bar{B}_{\tan}\, dS \tag{5.166}$$

These symmetric and inner products exhibit similar properties, except that the symmetric product does not produce a norm. Since $\bar{J}_s$ is confined to the surface of the scatterer, the functional in (5.165) can be written as

$$\hat{u} \cdot \bar{E}^s|_{\bar{r}=\bar{r}_0} = \frac{1}{J_0}(\bar{E}^{\text{inc}}, \bar{J}_s) = \frac{1}{J_0}(L\bar{J}_s, \bar{J}_s) \tag{5.167}$$

where L denotes the EFIE operator

$$L\bar{J}_s = \frac{\nabla\nabla \cdot + k^2}{j\omega\varepsilon} \iint \bar{J}_s(\bar{r}') \frac{e^{-jk|\bar{r}-\bar{r}'|}}{4\pi|\bar{r}-\bar{r}'|}\, dS \tag{5.168}$$

and $\bar{J}_s$ is the solution to $L\bar{J}_s = \bar{E}^{\text{inc}}$. A method-of-moments approximation $\bar{J}_s^N$ obtained using the identical functions for basis and testing (Galerkin's method) satisfies the conditions necessary for Equation (5.167) to be a stationary functional of the form of (5.159). Thus, if $\bar{\varepsilon}_J = \bar{J}_s - \bar{J}_s^N$, it follows that

$$\frac{1}{J_0}(\bar{E}^{\text{inc}}, \bar{J}_s) - \frac{1}{J_0}(\bar{E}^{\text{inc}}, \bar{J}_s^N) = \frac{1}{J_0}(L\bar{\varepsilon}_J, \bar{\varepsilon}_J) \tag{5.169}$$

The error in this approximation for $\bar{E}^s$ at $\bar{r}_0$ is $O(|\bar{\varepsilon}_J|^2)$. If the problem was discretized using a non-Galerkin selection of basis and testing functions, the error estimate would revert to $O(|\bar{\varepsilon}_J||\bar{\varepsilon}_E|)$, where $\bar{\varepsilon}_E$ denotes the error in the solution of the adjoint equation using the opposite functions for basis and testing.

To illustrate the nature of these estimates, Table 5.5 shows the error in the monostatic scattering cross section for the circular cylinder example used to generate Table 5.2, for various order splines employed as basis and testing functions. These data support the notion that the error is actually a function of the combined order of the basis and testing functions ($P + Q$ in the table). In other words, ε_E is comparable to ε_J for a given degree spline function, and the Galerkin solution ($P = Q$) is no more accurate than a non-Galerkin solution obtained with the same total $P + Q$.

TABLE 5.5 Percentage Error in Backscattered Far Field for TM Circular Cylinder with Circumference of 6λ as Function of Order of Splines Employed as Basis (P) and Testing (Q) Functions

	Percent Error				
Q	$P = 1$	$P = 2$	$P = 3$	$P = 4$	$P = 5$
1	4.1×10^{-2}	9.2×10^{-5}	2.1×10^{-5}	2.4×10^{-6}	4.5×10^{-7}
2	9.2×10^{-5}	2.1×10^{-5}	2.4×10^{-6}	4.5×10^{-7}	
3	2.1×10^{-5}	2.4×10^{-6}	4.5×10^{-7}		
4	2.4×10^{-6}	4.5×10^{-7}			
5	4.5×10^{-7}				

Note: A total of 60 basis and testing functions of the indicated order were employed with equal-size cells to construct the matrix equation. A spline of order P has polynomial order $P - 1$. After [25].

5.13 SUMMARY

This chapter has introduced several types of one-dimensional subsectional expansion sets, including polynomial spline functions and interpolative Lagrangian functions. The extension of these to the multidimensional and vector case will be considered in Chapter 9. In addition, the role of the basis and testing functions was discussed, and their impact on the accuracy of numerical solutions was explored. While the accuracy of a numerical result f^N appears to depend primarily on the ability of the basis functions to represent f, the testing functions play an equal role in the error associated with a secondary quantity $\langle f, h \rangle$.

The question of whether a particular discretization produces a numerical solution f^N that converges to the true solution f as $N \to \infty$ is fundamental. In order to introduce the reader to mathematical convergence proofs, a number of concepts from functional analysis have been reviewed. Sections 5.9–5.11 consider several approaches for establishing the convergence of f^N to f for integral equations with operators of the identity-plus-compact type. The specific arguments presented in these sections cannot easily be extended to more general operators. Our previous experience with integral equation formulations (Chapter 2) supports the notion that, if constructed with sufficient care, numerical solutions appear to converge under much more general conditions. Despite this observation, the authors are not aware of more general convergence proofs applicable to the specific integral operators arising in electromagnetic scattering.

A principal assumption required in order to relate the operators arising in electromagnetics with the "compact" or identity-plus-compact operators used within the convergence proofs is the boundedness of the inverse operators. Unfortunately, when applied to closed scatterers, surface integral equations such as the EFIE and MFIE may not always have bounded inverses. In fact, there are certain discrete frequencies where these equations do not produce unique solutions. Chapter 6 investigates this topic in detail and presents several alternate formulations that circumvent the difficulty.

REFERENCES

[1] D. H. Griffel, *Applied Functional Analysis*, New York: Wiley, 1981.

[2] I. Stakgold, *Green's Functions and Boundary Value Problems*, New York: Wiley, 1979.

[3] R. F. Harrington, *Field Computation by Moment Methods*, Malabar, FL: Krieger, 1982 Reprint.

[4] M. Becker, *The Principles and Applications of Variational Methods*, Cambridge, MA: MIT Press, 1964.

[5] B. A. Findlaysen, *The Method of Weighted Residuals and Variational Principles*, New York: Academic, 1972.

[6] L. V. Kantorovich and V. F. Krylov, *Approximate Methods of Higher Analysis*, New York: Wiley, 1964.

[7] L. V. Kantorovich and G. P. Akilov, *Functional Analysis*, Oxford: Pergamon, 1982.

[8] R. F. Harrington, "Origin and development of the method of moments for field computation," in *Applications of the Method of Moments to Electromagnetic Fields*, ed. B. Strait, St. Cloud, FL: SCEEE Press, 1981.

[9] S. G. Mikhlin, *Variational Methods in Mathematical Physics*, New York: Macmillan, 1964.

[10] O. C. Zienckiewicz and R. L. Taylor, *The Finite Element Method*, New York: McGraw-Hill, 1989.

[11] P. P. Silvester and R. L. Ferrari, *Finite Elements for Electrical Engineers*, Cambridge: Cambridge University Press, 1990.

[12] T. K. Sarkar, A. R. Djordjevic, and E. Arvas, "On the choice of expansion and weighting functions in the numerical solution of operator equations," *IEEE Trans. Antennas Propagat.*, vol. AP-33, pp. 988–996, Sept. 1985.

[13] D. Kahaner, C. Moler, and S. Nash, *Numerical Methods and Software*, Englewood Cliffs, NJ: Prentice-Hall, 1989.

[14] G. A. Thiele, "Wire antennas," in *Computer Techniques for Electromagnetics*, ed. R. Mittra, New York: Pergamon, 1973.

[15] E. K. Miller and F. J. Deadrick, "Some computational aspects of thin-wire modeling," in *Numerical and Asymptotic Techniques in Electromagnetics*, ed. R. Mittra, New York: Springer-Verlag, 1975.

[16] G. J. Burke and A. J. Poggio, *Numerical Electromagnetics Code (NEC)—Method of Moments*, Technical Document 116, Naval Ocean System Center, San Diego, Jan. 1981.

[17] B. Z. Steinberg and Y. Leviatan, "On the use of wavelet expansions in the method of moments," *IEEE Trans. Antennas Propagat.*, vol. 41, pp. 610–619, May 1993.

[18] G. Wang, "On the utilization of periodic wavelet expansions in the moment methods," *IEEE Trans. Microwave Theory Tech.*, vol. 43, pp. 2495–2498, Oct. 1995.

[19] J. C. Goswami, A. K. Chan, and C. K. Chui, "On solving first-kind integral equations using wavelets on a bounded interval," *IEEE Trans. Antennas Propagat.*, vol. 43, pp. 614–622, June 1995.

[20] R. Lee and A. C. Cangellaris, "A study of discretization error in the finite element approximation of wave solutions," *IEEE Trans. Antennas Propagat.*, vol. 40, pp. 542–549, May 1992.

[21] W. R. Scott, Jr., "Errors due to spatial discretization and numerical precision in the finite element method," *IEEE Trans. Antennas Propagat.*, vol. 42, pp. 1565–1570, Nov. 1994.

[22] A. F. Peterson and R. J. Baca, "Error in the finite element discretization of the scalar Helmholtz equation over electrically large regions," *IEEE Microwave Guided Wave Lett.*, vol. 1, pp. 219–222, Aug. 1991.

[23] R. Piessens, E. deDoncker-Kapenga, C. W. Uberhuber, and D. K. Kahaner, *QUADPACK: A Subroutine Package for Automatic Integration*, Berlin: Springer-Verlag, 1983.

[24] A. F. Peterson and R. E. Jorgenson, "Is Galerkin's method really better?" *Proceedings of the Sixth Annual Review of Progress in Applied Computational Electromagnetics*, Monterey, CA, The Applied Computational Electromagnetics Society, Monterey, CA, pp. 380–386, Mar. 1990.

[25] A. F. Peterson, D. R. Wilton, and R. E. Jorgenson, "Variational nature of Galerkin and non-Galerkin moment method solutions," *IEEE Trans. Antennas Propagat.*, vol. 44, pp. 500–503, Apr. 1996.

[26] D. Colton and R. Kress, *Integral Equation Methods in Scattering Theory*, New York: Wiley, 1983.

[27] K. E. Atkinson, *A Survey of Numerical Methods for the Solution of Fredholm Integral Equations of the Second Kind*, Philadelphia: SIAM, 1976.

[28] C. T. H. Baker, *The Numerical Treatment of Integral Equations*, Oxford: Clarendon, 1977.

[29] P. M. Anselone, *Collectively Compact Operator Approximation Theory*, Englewood Cliffs, NJ: Prentice-Hall, 1971.

[30] A. F. Peterson, "The 'interior resonance' problem associated with surface integral equations of electromagnetics: Numerical consequences and a survey of remedies," *Electromagnetics*, vol. 10, pp. 293–312, 1990.

[31] C. W. Steele, *Numerical Computation of Electric and Magnetic Fields*, New York: Van Nostrand Reinhold, 1987.

[32] G. Strang and G. J. Fix, *An Analysis of the Finite Element Method*, Englewood Cliffs, NJ: Prentice-Hall, 1973.

[33] M. H. Schultz, "L^2 error bounds for the Rayleigh-Ritz-Galerkin method," *SIAM J. Num. Anal.*, vol. 8, pp. 737–748, 1971.

[34] G. F. Roach, *Green's Functions*, Cambridge: Cambridge University Press, 1982.

[35] R. F. Harrington, *Time-Harmonic Electromagnetic Fields*, New York: McGraw-Hill, 1961, Chapter 7.

[36] J. R. Mautz, "Non-variational nature of the functional obtained by testing with a Dirac delta function," *Digest of the 1994 IEEE Antennas and Propagation International Symposium*, Seattle, WA, IEEE, NY, pp. 1169–1172, June 1994.

[37] J. H. Richmond, "On the variational aspects of the moment method," *IEEE Trans. Antennas Propagat.*, vol. 39, pp. 473–479, Apr. 1991.

[38] S. Wandzura, "Optimality of Galerkin method for scattering computations," *Microwave Opt. Technol. Lett.*, vol. 4, pp. 199–200, Apr. 1991.

PROBLEMS

P5.1 For real-valued functions defined on $0 \leq x \leq 1$, define inner products

$$\langle a, b\rangle_1 = \int_0^1 a(x)b(x)\,dx$$

$$\langle a, b\rangle_2 = \int_0^1 \sin(\pi x)a(x)b(x)\,dx$$

$$\langle a, b\rangle_3 = a\left(\tfrac{1}{2}\right) b\left(\tfrac{1}{2}\right)$$

and consider the three functions

$$f(x) = \sin(\pi x)$$

$$g(x) = \begin{cases} 2x & 0 \leq x \leq \frac{1}{2} \\ 2(1-x) & \frac{1}{2} \leq x \leq 1 \end{cases}$$

$$h(x) = 4x(1-x)$$

For each of the three inner products, evaluate the norm of each function, the metric $d(f, g)$ and the metric $d(f, h)$.

P5.2 (a) Given a set of functions $\{f_1, f_2, \ldots, f_N\}$, and an inner product $\langle f, g\rangle$, construct a set of orthogonal functions $\{g_1, g_2, \ldots, g_N\}$ from the set $\{f_n\}$. *Hint:* Choose $g_1 = f_1$, and let $g_2 = f_2 - \gamma_{21} g_1$, where γ_{21} is selected so that $\langle g_1, g_2\rangle = 0$. Continue this procedure and find a general expression for the coefficient γ_{ij}. This process is known as *Gram–Schmidt orthogonalization*.

(b) Modify the procedure in order to produce a set of orthonormal functions.

(c) Finally, use the procedure to produce a set of orthonormal functions from the set $\{1, x, x^2, x^3, x^4\}$, employing the inner product

$$\langle a, b\rangle = \int_0^1 a(x)b(x)\,dx$$

Sketch the resulting orthonormal functions.

P5.3 Consider the differential equation $Lf = g$, where

$$Lf = \frac{d^2 f}{dx^2} + \frac{df}{dx}$$

and

$$g = (1 - x)\cos x - (2 + x)\sin x$$

The equation is defined on the interval $(0 \le x \le \frac{1}{2}\pi)$ and is subject to the boundary conditions $f(0) = 0$, $f(\frac{1}{2}\pi) = 0$. The solution is $f(x) = x\cos x$.
Define an inner product

$$\langle a, b\rangle = \int_0^{\pi/2} a(x)b(x)\,dx$$

and consider the complete, orthonormal basis on $(0, \frac{1}{2}\pi)$ given by

$$B_n(x) = \sqrt{\frac{4}{\pi}}\sin(2nx) \qquad n = 1, 2, \ldots$$

(a) Calculate numerical values for the three coefficients α_1, α_2, and α_3 that give the "best" representation according to (5.9), that is,

$$\alpha_n = \int_0^{\pi/2} \sqrt{\frac{4}{\pi}}\sin(2nx)x\cos x\,dx$$

(b) Construct a method-of-moments solution using the identical functions B_1, B_2, and B_3 as basis and testing functions. In other words, let

$$f^N = \sum_{n=1}^{3} \beta_n \sqrt{\frac{4}{\pi}}\sin(2nx)$$

and construct the 3×3 system $\mathbf{L}\beta = \mathbf{b}$, where

$$L_{mn} = \langle B_m, LB_n\rangle$$

and

$$b_m = \langle B_m, g\rangle$$

Compare the numerical values obtained for β_1, β_2, and β_3 with those obtained in part (a) for α_1, α_2, and α_3. Are they the same? Discuss the implications.

P5.4 Apply the method-of-moments procedure to obtain approximate solutions for the integral equation (D. R. Wilton and S. Govind, "Incorporation of edge condition in moment method solutions," *IEEE Trans. Antennas Propagat.*, vol. AP-25, pp. 845–850, Nov. 1977).

$$\int_{-1}^{1} f(x') \ln|x - x'|\, dx' = -\pi \ln 2$$

Use pulse basis functions and Dirac delta testing functions as defined in Section 5.3. Compare your numerical results with the solution

$$f(x') = \frac{1}{\sqrt{1 - (x')^2}}$$

P5.5 Consider the equation

$$\frac{d^2 f}{dx^2} + k^2 f = g$$

on some domain $0 < x < M\Delta$, subject to boundary conditions $f(0) = f_L$ and $f(M\Delta) = f_R$.

(a) Construct a second-order finite-difference discretization using cells of dimension Δ and the central-difference formula

$$\frac{d^2 f}{dx^2} \cong \frac{f_{n+1} - 2f_n + f_{n-1}}{\Delta^2}$$

to replace the second derivative. Identify the entries of the tridiagonal finite-difference matrix.

(b) Construct a method-of-moments discretization using the subsectional triangles in (5.27) to represent $f(x)$ and pulse testing functions to enforce the equation. The basis functions are to straddle two cells of dimension Δ, while the testing functions are defined between the center of adjacent cells. How does the (mn)th entry of the resulting matrix compare with the entries obtained in part (a)? Discuss the implications.

P5.6 Use the three quadratic Lagrangian functions

$$\phi_1(x) = \frac{1}{2}\frac{x}{\Delta}\left(\frac{x}{\Delta} - 1\right)$$

$$\phi_2(x) = \left(1 + \frac{x}{\Delta}\right)\left(1 - \frac{x}{\Delta}\right)$$

$$\phi_3(x) = \frac{1}{2}\frac{x}{\Delta}\left(\frac{x}{\Delta} + 1\right)$$

to approximate the cubic polynomial

$$f(x) = a + bx + cx^2 + dx^3$$

on the interval $-\Delta \le x \le \Delta$. In other words, construct

$$f_{\text{ap}}(x) = f(-\Delta)\phi_1(x) + f(0)\phi_2(x) + f(\Delta)\phi_3(x)$$

(a) Show that the error in this approximation is

$$f - f_{\text{ap}} = d(x^3 - \Delta^2 x)$$

(b) Plot the error function over the interval $-\Delta \le x \le \Delta$, and determine the peak error. At what value of x does the peak error occur?

(c) Identify the integer exponent p that best characterizes the interpolation error as $O(\Delta^p)$ as $\Delta \to 0$.

P5.7 Repeat Prob. P5.6 using the cubic Hermitian functions defined in (5.35)–(5.38) to represent

$$f(x) = a + bx + cx^2 + dx^3 + ex^4$$

on the interval $-\Delta \le x \le \Delta$.

P5.8 (a) Show that the central finite-difference formula (Prob. P5.5), if applied to the one-dimensional scalar Helmholtz equation in (5.48), produces

$$2E_m - E_{m-1} - E_{m+1} - k^2\Delta^2 E_m = 0$$

(b) Carry out a dispersion analysis similar to that of Section 5.5 for this discrete equation. Show that a solution of the form of (5.50) satisfies the equation as long as

$$\beta = \frac{1}{\Delta}\cos^{-1}\left(1 - \frac{(k\Delta)^2}{2}\right)$$

(c) Tabulate the resulting error as a function of Δ, comparing your results with those in Table 5.1. Is the finite-difference approach more accurate than the first-order finite-element discretization?

P5.9 Quadratic interpolation functions can be defined in one dimension so that three functions

$$\Phi_1(x) = \frac{x(x-\Delta)}{2\Delta^2}$$

$$\Phi_2(x) = 1 - \left(\frac{x}{\Delta}\right)^2$$

$$\Phi_3(x) = \frac{x(x+\Delta)}{2\Delta^2}$$

overlap a cell spanning the interval $-\Delta < x < \Delta$, with Φ_2 interpolative at the center of the cell. Show that element matrices associated with a discretization of the one-dimensional scalar Helmholtz equation in (5.48) have the form

$$\left[\int \frac{d\Phi_i}{dx}\frac{d\Phi_j}{dx}\,dx\right] = \begin{bmatrix} \frac{7}{6\Delta} & \frac{-8}{6\Delta} & \frac{1}{6\Delta} \\ \frac{-8}{6\Delta} & \frac{16}{6\Delta} & \frac{-8}{6\Delta} \\ \frac{1}{6\Delta} & \frac{-8}{6\Delta} & \frac{7}{6\Delta} \end{bmatrix}$$

and

$$\left[\int \Phi_i \Phi_j\,dx\right] = \begin{bmatrix} \frac{4\Delta}{15} & \frac{2\Delta}{15} & \frac{-\Delta}{15} \\ \frac{2\Delta}{15} & \frac{16\Delta}{15} & \frac{2\Delta}{15} \\ \frac{-\Delta}{15} & \frac{2\Delta}{15} & \frac{4\Delta}{15} \end{bmatrix}$$

P5.10 (a) Extend the dispersion analysis of Section 5.5 to the quadratic Lagrangian interpolation functions in Prob. P5.9 in order to obtain a numerical phase constant

$$\beta = \frac{1}{2\Delta} \cos^{-1}\left(\frac{15 - 26(k\Delta)^2 + 3(k\Delta)^4}{15 + 4(k\Delta)^2 + (k\Delta)^4} \right)$$

(b) Identify the region where β is real valued.

(c) Tabulate the error per wavelength as a function of Δ, comparing your result with Table 5.1. Identify the integer exponent q that best fits the error, assuming that the error decreases as $O(\Delta^q)$ as $\Delta \to 0$. Is this a superconvergent result?

P5.11 Review Prob. P3.21 in order to derive the analytical eigenvalues presented in Equations (5.82), (5.87), and (5.91).

P5.12 Obtain an analytical expression for the eigenvalues of the TM MFIE (Prob. P2.9) applied to a circular conducting cylinder of radius a.

P5.13 For a circular conducting cylinder of fixed radius, show that as the number of basis functions N is increased, the ratio of the largest to smallest eigenvalue in the method-of-moments matrix behaves as $O(1)$ for the TE MFIE, $O(1/\Delta)$ for the TE EFIE, and $O(1/\Delta)$ for the TM EFIE. Assume that the cell size is given by $\Delta = ka/N$ and that no eigenvalue is identically zero for this value of ka. Use the analytical expressions in (5.82), (5.87), and (5.91) to approximate the first N matrix eigenvalues. Finally, assume that N is large enough to justify use of the asymptotic approximations

$$J_n(z) \approx \sqrt{\frac{1}{2\pi n}} \left(\frac{ez}{2n}\right)^n \qquad \text{as } n \to \infty$$

$$H_n^{(2)}(z) \approx j\sqrt{\frac{2}{\pi n}} \left(\frac{2n}{ez}\right)^n \qquad \text{as } n \to \infty$$

for the Bessel functions.

P5.14 Consider the two-dimensional scalar Helmholtz equation

$$L\Psi = -\nabla^2\Psi - k^2\Psi = g$$

on the rectangular domain $0 < x < a, 0 < y < b$, subject to a homogeneous Dirichlet boundary condition.

(a) Show that the eigenvalues of L are given by

$$\lambda_{mn} = \left(\frac{m\pi}{a}\right)^2 + \left(\frac{n\pi}{b}\right)^2 - k^2$$

(b) Assuming that this Helmholtz equation is discretized using a uniform finite-difference or finite-element grid, with cells of dimension $\Delta \times \Delta$, and that the matrix eigenvalues are well approximated by the λ_{mn} in part (a), show that the ratio of the largest to smallest matrix eigenvalue behaves as $O(\Delta^{-2})$ as $\Delta \to 0$.

P5.15 Consider the equation $Lf = g$, where L has eigenvalues $\{\lambda_n\}$ and eigenfunctions $\{e_n\}$ that form a complete, orthonormal set. The excitation g can be expressed as

$$g = \sum_n \langle g, e_n \rangle e_n$$

The solution can be sought in the form

$$f = \sum_n \alpha_n e_n$$

and determined by substitution, since

$$Lf = \sum_n \alpha_n L e_n = \sum_n \alpha_n \lambda_n e_n$$

Therefore, the solution can be written formally as

$$f = \sum_n \frac{\langle g, e_n \rangle}{\lambda_n} e_n$$

However, if L is a compact operator such as the TM EFIE in Equation (5.79), $\lambda_n \to 0$ as $n \to \infty$. Consequently, the summation for f may be divergent. For the TM EFIE applied to a circular cylinder, determine the asymptotic order of the nth term in this summation when $g(\phi)$ is (a) a delta function, (b) a subsectional triangle function with support limited to a 90° portion of the circle, and (c) a constant function over the entire circle. Use the eigenfunctions and eigenvalues presented in (5.81)–(5.84). What constraint on g is necessary to ensure convergence?

6

Alternative Surface Integral Equation Formulations

The discussion of surface integral equations in Chapters 1 and 2 avoided one potential problem with these formulations. When used to describe exterior electromagnetic scattering problems, certain surface integral equations may not produce unique solutions if applied to closed geometries that also represent resonant cavities. This chapter explores the numerical consequences of the uniqueness problem and considers several remedies.

6.1 UNIQUENESS OF SOLUTIONS TO THE EXTERIOR SURFACE EFIE AND MFIE [1]

An EFIE formulation for the scattering of a TM plane wave from an infinite, perfectly conducting cylinder was discussed in Section 2.1. The EFIE is given by

$$E_z^{\text{inc}}(t) = jk\eta \int J_z(t') \frac{1}{4j} H_0^{(2)}(kR)\, dt' \tag{6.1}$$

where

$$R = \sqrt{[x(t) - x(t')]^2 - [y(t) - y(t')]^2} \tag{6.2}$$

and t is a parametric variable along the cylinder surface. Although this surface integral equation can be applied to a cylinder of arbitrary cross-sectional shape, consider the special case of a circular cylinder. Comparisons presented in Section 2.1 demonstrated excellent agreement between numerical and exact solutions for circular cylinders. However, consider Figure 6.1, which presents the numerical solution for the magnitude of the surface current density induced on a cylinder with $ka = 2.405$. The numerical result obtained using the procedure of Section 2.1 with 40 basis and testing functions differs appreciably from the exact solution. Interestingly, if we increase the order of the discretization by adding additional basis and testing functions, the results do not improve. The behavior of the

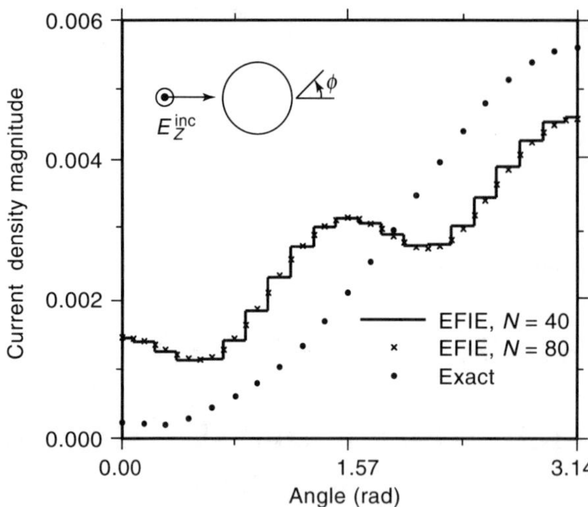

Figure 6.1 Magnitude of the surface current density produced by two discretizations of the TM EFIE for a circular cylinder with $ka = 2.405$ illuminated by a uniform plane wave. The exact solution is shown for comparison.

numerical solution is symptomatic of a fundamental difficulty with certain surface integral formulations. For reasons that will soon be apparent, the difficulty has come to be known as the *interior resonance* problem.

To understand the source of the difficulty in this example, recall the discussion in Chapter 5 concerning eigenfunctions and eigenvalues of integral operators. It was shown that the eigenfunctions of the TM EFIE operator for circular cylinders are the exponential functions $\{e^{jn\phi}\}$. Analytical expressions for the eigenvalues of the EFIE as a function of the cylinder radius a are given as

$$\lambda_n^{\text{EFIE,TM}} = \frac{1}{2}(\eta\pi ka) J_n(ka) H_n^{(2)}(ka) \tag{6.3}$$

where J_n and H_n denote the Bessel and Hankel functions of order n. A plot of the eigenvalues of orders 0, 1, and 2 as a function of ka is presented in Figure 5.7. For small values of ka (low-frequency excitation), the eigenvalues cluster at the origin. As ka increases, each eigenvalue moves in a circularlike path that passes through the origin at various nonzero values of ka. From Equation (6.3), it is clear that the eigenvalues vanish at the zeros of the Bessel function $J_n(ka)$.

What is the effect of a zero eigenvalue? In short, it means that any multiple of the corresponding eigenfunction can be added to the solution and still satisfy the equation. This eigenfunction is a source-free or homogeneous solution. Consequently, whenever $J_n(ka)$ vanishes in (6.3), the EFIE does not have a unique solution. Although the EFIE was constructed in order to represent the exterior scattering problem, the equation also admits interior cavity fields as homogeneous solutions. The root of the problem is that the surface integral equation only involves data on the mathematical surface of the scatterer and cannot distinguish between "inside" and "outside" in order to produce the desired exterior solution. The interior geometry represented by this EFIE is the circular cavity having perfect electric walls. Since nonzero cavity fields can occur only when $J_n(ka) = 0$, the EFIE has unique solutions at other frequencies.

For geometries that cannot be treated analytically (basically anything other than circular cylinders), we employ the method of moments to convert a surface integral equation into a finite-dimensional matrix equation. As a consequence of the manner in which eigenvalues are projected from a continuous operator onto a matrix operator during discretization

(Section 5.7), at values of ka where an eigenvalue of the integral operator vanishes, an eigenvalue of the matrix operator will also become very small. If the matrix eigenvalue is zero to machine precision, any multiple of the corresponding eigenvector will satisfy the matrix equation. When this occurs, the matrix is singular and cannot be inverted or factorized into the LU form. If the eigenvalue is nonzero but very small, the matrix is nearly singular. It is no surprise that numerical solutions become less stable under these conditions. In fact, since Figure 6.1 illustrates the behavior of the numerical solution when $ka = 2.405$, near the first zero of J_0 in Equation (6.3), error in the solution might be expected.

Similar behavior can be observed with an MFIE formulation. Section 2.2 discussed the MFIE representing the scattering of a TE wave from a perfectly conducting cylinder. The equation has the form

$$-H_z^{\text{inc}}(t) = J_t(t) + \hat{z} \cdot \nabla \times \int \hat{t}(t') J_t(t') \frac{1}{4j} H_0^{(2)}(kR) dt' \tag{6.4}$$

For the special case of a circular cylinder, the eigenfunctions of the MFIE operator are also the exponential functions $\{e^{jn\phi}\}$. Analytical expressions for the eigenvalues corresponding to these eigenfunctions are

$$\lambda_n^{\text{MFIE,TE}} = \tfrac{1}{2}(j\pi ka) J_n(ka) H_n^{(2)\prime}(ka) \tag{6.5}$$

where the prime denotes differentiation with respect to the argument of the Hankel function. A plot of the eigenvalues of order 0, 1, and 2 is presented in Figure 5.9. Note that the character of the MFIE eigenvalues differs from that of the EFIE eigenvalues (Figure 5.7). For small ka, the MFIE eigenvalues cluster at 0.5 or 1.0 in the right-half complex plane. However, as ka increases, these eigenvalues move in a circularlike path, in common with the EFIE eigenvalues, and occasionally pass through the origin. As indicated by Equation (6.5), the MFIE eigenvalues also vanish at the zeros of the Bessel function $J_n(ka)$. These values of ka represent the frequencies where source-free solutions exist for the interior problem, which in this case is a circular cavity having walls that are perfect magnetic conductors.

Figure 6.2 shows the magnitude of the surface current density obtained from a numerical solution of the MFIE near $ka = 2.405$ using the approach of Section 2.2. When compared to the exact solution, we observe that the numerical solution for the surface current density contains considerable error. The bistatic scattering cross section computed from these currents is shown in Table 6.1 and is grossly inaccurate.

For the example illustrated in Figure 6.2, the numerical results produced by our specific implementation are incorrect within a frequency range spanning approximately $2.401 < ka < 2.409$. Since the original integral operator has a zero eigenvalue at the resonance frequency $ka \cong 2.405$, one might expect to find that the matrix is singular at that frequency. However, this is not the case. Although an eigenvalue of the matrix approaches the origin near this frequency, the matrix condition number (see Chapter 4) computed for our implementation remains below 100. Because of discretization error, the matrix remains nonsingular and the equation apparently has a unique solution over the entire frequency range. In fact, numerical experimentation demonstrates that the incorrect numerical results displayed in Figures 6.1 and 6.2 are actually very stable. As the order of the discretization is improved (more basis and testing functions), the results appear to converge to some solution. This solution, however, is clearly not the desired solution to the exterior scattering problem. In other words, the numerical solution is incorrect, but not because of ill-conditioning or numerical round-off errors.

To understand the nature of the incorrect numerical solution, consider the magnitude of the surface current density produced by the TM EFIE at $ka = 2.405$, as displayed in

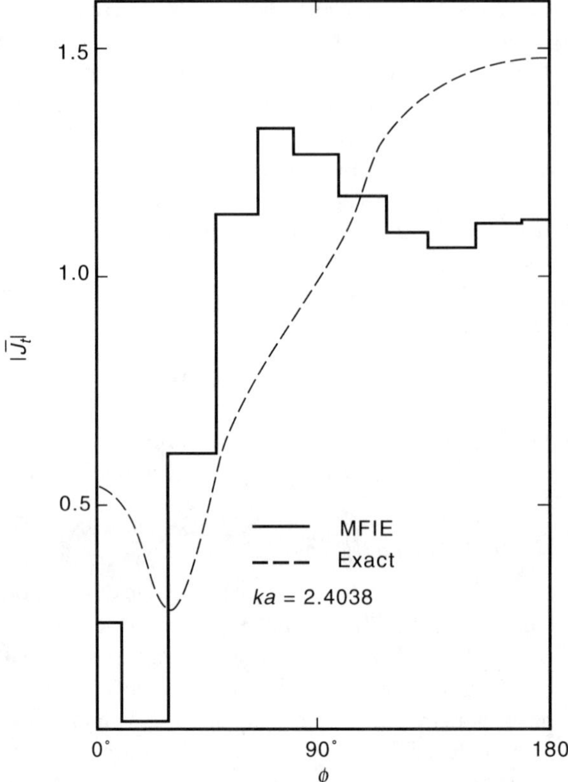

Figure 6.2 Comparison of the MFIE and exact solutions for the TE current density on a circular cylinder of radius $0.3826\lambda_0$. After [1]. ©1990 Hemisphere Publishing Corporation.

TABLE 6.1 Bistatic Scattering Cross Section for Circular Cylinder of Circumference 2.4038

Bistatic Angle (deg)	Scattering Cross Section (dB λ) MFIE	Exact
0	–2.49	3.46
30	–2.70	0.90
60	+2.51	0.98
90	–3.17	–3.68
120	–13.4	–0.72
150	–6.34	0.68
180	–6.42	0.17

Note: The MFIE result obtained using 20 pulse basis functions is corrupted by an internal resonance. The scattering cross section computed from the incorrect MFIE current density is compared with the exact solution.

Figure 6.1. The exact solution for the surface current density produced by a uniform plane wave can be written as [2]

$$J_z(\phi) = \frac{-2}{\eta\pi ka} \sum_{n=-\infty}^{\infty} j^{-n} \frac{1}{H_n^{(2)}(ka)} e^{jn\phi} \tag{6.6}$$

Numerical experimentation suggests that as the discretization is refined, the difference between the exact and numerical result at $ka = 2.405$ converges to

$$J_{\text{exact}}(\phi) - J_{\text{numerical}}(\phi) = \frac{-2}{\eta\pi ka} \frac{1}{H_0^{(2)}(ka)} \tag{6.7}$$

In other words, the numerical result is correct except for the magnitude of the resonant eigenfunction (in this case, $e^{j0\phi}$). This is as expected, since, in theory, the original integral equation would admit any multiple of this eigenfunction [i.e., $J_z(\phi) = K_0 e^{j0\phi}$ for any K_0] as a solution at resonance. What is initially surprising is that the numerical solution apparently suppressed this eigenfunction completely. Instead of producing solutions with unstable and wildly varying coefficients of the resonant eigenfunction, as might be expected, the matrix equation consistently produced a stable solution having zero for the coefficient K_0.

Two observations help explain the stability of the incorrect solution for this example. First, the discretization error associated with the matrix operator prevents the eigenvalue corresponding to the resonant eigenfunction from being as small as that of the original integral operator. This is apparent from the matrix condition number, which does not exceed several hundred in our implementation. The second observation involves the excitation employed in the problem. For the TM EFIE, a plane-wave incident electric field on the surface of the circular cylinder can be written as

$$E_z^{\text{inc}}(\rho = a, \phi) = \sum_{n=-\infty}^{\infty} j^{-n} J_n(ka) e^{jn\phi} \tag{6.8}$$

Because of the factor $J_n(ka)$, the component of the zeroth resonant eigenfunction in the excitation vanishes when ka is approximately 2.405. The eigenvalue appearing in Equation (6.3) also contains the factor $J_0(ka)$, which in an analytical solution exactly cancels the $J_0(ka)$ of the excitation. However, discretization error perturbs the matrix eigenvalue from the exact analytical eigenvalue in the vicinity of the first zero of $J_0(ka)$. For our implementation, the matrix eigenvalue is larger than the exact eigenvalue in magnitude, and the matrix solution tends to suppress the resonant eigencurrent. The numerical solution for surface current density is incorrect because of the improper balance between the tangential component of the excitation and the nearly resonant eigencurrent. In fact, the solution will only be correct if these two factors precisely cancel.

Although the difficulty in this example appears to be a consequence of numerical limitations, it is actually of a more fundamental nature. In fact, *there is not enough information contained in the tangential incident electric field on the scatterer surface at resonance to determine the true current density*. This observation is apparent from Equation (6.8). Although the zero harmonic of the tangential incident electric field vanishes at the cylinder surface, it does not vanish within the cylinder and requires some component of surface current to properly cancel the interior incident field. Additional information not contained in the tangential incident electric field on the surface is necessary to determine the proper weighting of the zero harmonic when $J_n(ka) = 0$.

The numerical behavior depends somewhat on the specific approximations employed. In addition to the discretization error introduced with the basis and testing functions, the matrix entries computed for the preceding examples contained some error due to approximations made in evaluating the integrals. These errors affect the precise location of the matrix eigenvalues. A more accurate evaluation of these integrals should improve the balance between the excitation and the matrix eigenvalue and reduce the frequency range over which the incorrect numerical behavior is observed. At the same time, improved accuracy should increase the condition number of the matrix near resonance and introduce the possibility that the matrix will appear singular to machine precision at resonance. On the other hand, if the entries of the excitation column vector are not accurately computed, it is possible that the matrix eigenvalue might be smaller in magnitude than the corresponding coefficient in the excitation. Under these circumstances, the coefficient of the resonant eigencurrent can be large and the resonant eigencurrent will not be suppressed in the numerical solution. Since the excitation entries do not require an integration over a singular Green's function, however, they are likely to be computed more accurately than the matrix eigenvalue in practice.

As observed in Figure 6.1, suppressing the resonant eigencurrent introduces significant error into the surface current density. However, because the resonant eigencurrent produced by the EFIE does not radiate exterior to the scatterer, the far-zone fields should not be affected by the incorrect weighting of that eigencurrent. In other words, accurate far fields can sometimes be computed from the incorrect EFIE surface current density.

The TE MFIE exhibits a similar behavior in the vicinity of an interior resonance. A uniform plane-wave incident magnetic field used with Equation (6.4) can be expressed as

$$H_z^{\text{inc}}(\rho = a, \phi) = \sum_{n=-\infty}^{\infty} j^{-n} J_n(ka) e^{jn\phi} \tag{6.9}$$

This field has the same form as Equation (6.8), and the coefficient of the eigenfunction $e^{jn\phi}$ vanishes whenever ka is a zero of $J_n(ka)$. Since the MFIE of (6.4) only involves the z-component of the incident field, no information is available to properly specify the resonant eigencurrent when $J_n(ka)$ vanishes. In common with the EFIE formulation, a nonzero resonant eigencurrent must be present in order to properly cancel the incident field within the cylinder. The consequence of the incorrect eigencurrent is severe in the case of the MFIE because the resonant eigencurrent does radiate and will significantly alter the far fields (Table 6.1 and Prob. P6.1). Thus, it is usually impossible to compute accurate far fields from a corrupted solution of the MFIE.

We have considered circular cylinder examples because of the ease in identifying analytical expressions for the eigenvalues of the EFIE and MFIE operators. For general scatterer shapes, incorrect numerical solutions occur whenever the scatterer surface is closed and coincides with a resonant cavity. (Solutions of the EFIE are incorrect near the resonant frequencies of cavities having perfect electric walls; solutions of the MFIE degrade near the resonant frequencies of cavities having perfect magnetic walls.) In practice, it is not trivial to predict in advance whether an arbitrary closed surface happens to coincide with a resonant cavity. However, because a matrix eigenvalue becomes relatively small near an internal resonance, abrupt changes in the matrix condition number can be used to identify probable resonant frequencies. For instance, Figure 6.3 shows the matrix condition number returned by the LINPACK LU factorization algorithm CGECO for the TE MFIE circular cylinder example. The sharp peaks in the graph agree with theoretical resonance frequencies. As a general rule, the condition number may vary considerably from one problem to another, and a stable value on the order of 1000 does not necessarily indicate a problem. However, any rapid change in the condition number is characteristic of a problem, and changes of the type depicted in Figure 6.3 are diagnostic of an interior resonance situation.

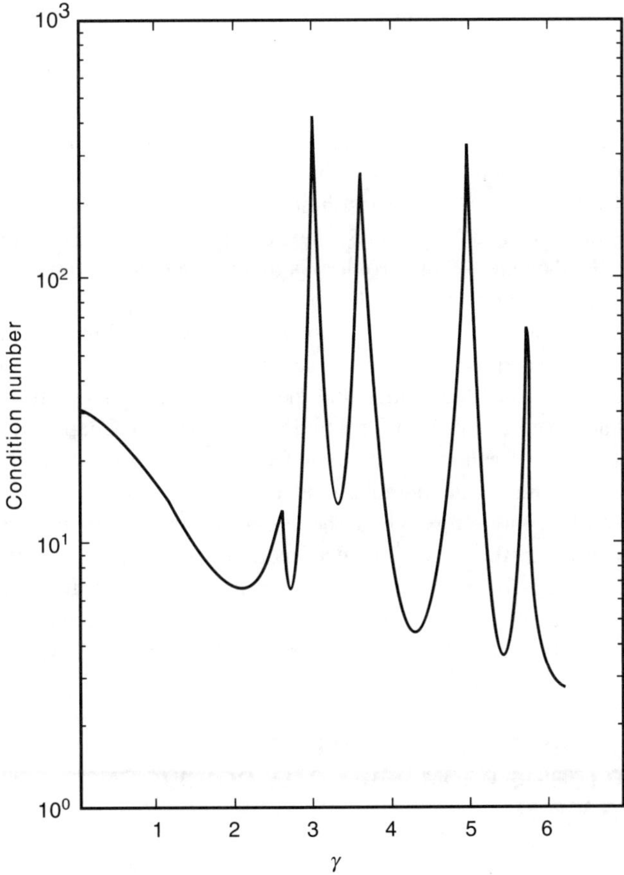

Figure 6.3 Plot of the matrix condition number of the system representing the TE MFIE for a circular cylinder of radius $a = 1$ as a function of the frequency parameter γ, where the circumference ka in wavelengths is given by $[(2\pi)^2 - \gamma^2]^{0.5}$. After [1]. ©1990 Hemisphere Publishing Corporation.

The relation between the resonant eigencurrent and the tangential incident field can be studied under general conditions using the reciprocity theorem [2] presented in Section 5.12, which can be written as

$$\iiint \bar{E}^a \cdot \bar{J}^b - \bar{H}^a \cdot \bar{K}^b = \iiint \bar{E}^b \cdot \bar{J}^a - \bar{H}^b \cdot \bar{K}^a \tag{6.10}$$

where the sources labeled a are located exterior to the scatterer and produce the incident fields associated with the exterior scattering problem (also labeled a). At a frequency where the scatterer surface coincides with the interior surface of a resonant cavity, the sources labeled b can be taken to represent any cavity eigencurrents. If the surface coincides with a cavity constructed of perfect electric walls (so that the tangential electric field vanishes on the surface), the magnetic current density $\bar{K}^b$ vanishes. Since $\bar{J}^b$ does not radiate outside the cavity, it follows that

$$\iiint \bar{E}^{\text{inc}} \cdot \bar{J}^{\text{res}} = 0 \tag{6.11}$$

where $\bar{J}^{\text{res}} = \bar{J}^b$ denotes the resonant eigencurrent. If the scatterer surface coincides with a cavity constructed of perfect magnetic walls, a similar result is obtained involving the incident magnetic field and the resonant eigencurrent $\bar{K}^{\text{res}}$. These results demonstrate that the tangential component of the incident electromagnetic field will not excite the eigencurrent associated with the interior resonant cavity. This observation is in accordance with a mathematical principle known as the Fredholm alternative [3].

To summarize, numerical solutions of the surface EFIE or MFIE for closed scatterers may be incorrect near frequencies where the scatterer surface happens to represent a resonant cavity. Fundamentally, the problem is that the tangential components of a single incident field on the scatterer surface do not contain enough information to uniquely determine the desired surface currents at an interior resonance. Incorrect numerical solutions over a range of frequencies near resonance are caused by the improper balance between the small eigenvalue of the matrix and the corresponding term in the excitation vector. All the surface integral equation formulations discussed in Chapter 2 suffer from this problem, including those for homogeneous dielectric and impedance bodies (see Probs. P6.3 and P6.4). The volume integral equations do not suffer from this type of difficulty, since they sample the excitation throughout the interior region. Alternative surface integral equations are discussed in the following sections and can be used to avoid the internal resonance problem. We will see that each of these alternative formulations incorporates additional information about the incident fields present in the scattering problem. Unfortunately, they all require more computational effort than the EFIE or MFIE to produce the desired solution.

6.2 THE COMBINED-FIELD INTEGRAL EQUATION FOR SCATTERING FROM PERFECTLY CONDUCTING CYLINDERS [4, 5]

The preceding section described the basic interior resonance problem associated with the EFIE and MFIE. Not all surface integral equations suffer from this difficulty. An alternative formulation known as the combined-field integral equation (CFIE) provides unique, stable solutions for all closed scatterers [4, 5]. The CFIE is obtained from a linear combination of the EFIE and the MFIE. To scale terms to a similar order of magnitude and employ common units, the scale factors can be taken as α and $(1-\alpha)\eta$, where α is a real value between 0 and 1 and η is the intrinsic impedance of the exterior medium. The CFIE representing a perfectly conducting cylinder illuminated by a TM wave has the form

$$\begin{aligned}\alpha E_z^{\text{inc}}(t) + (1-\alpha)\eta H_t^{\text{inc}}(t) = (1-\alpha)\eta J_z(t) + \alpha jk\eta \int J_z(t')\frac{1}{4j}H_0^{(2)}(kR)\,dt' \\ -(1-\alpha)\eta\hat{t}\cdot\nabla\times\int \hat{z}J_z(t')\frac{1}{4j}H_0^{(2)}(kR)\,dt'\end{aligned} \tag{6.12}$$

where R is defined in Equation (6.2). For TE-wave scattering from perfectly conducting cylinders, the CFIE is given by

$$\alpha E_t^{\text{inc}}(t) - (1-\alpha)\eta H_z^{\text{inc}}(t) = (1-\alpha)\eta J_t(t) + (1-\alpha)\eta \hat{z}\cdot\nabla\times\int \hat{t}(t')J_t(t')\frac{1}{4j}H_0^{(2)}(kR)\,dt' - \alpha\eta\hat{t}\cdot\frac{\nabla\nabla\cdot+k^2}{jk}\int \hat{t}(t')J_t(t')\frac{1}{4j}H_0^{(2)}(kR)\,dt' \tag{6.13}$$

The integrals in these equations are to be evaluated with the observer an infinitesimal distance outside the scatterer surface.

To illustrate the difference between these surface integral equations and the EFIE or MFIE, we again consider the special case of a circular cylinder with radius a. For the TE polarization, the eigenfunctions of the CFIE operator are $\{e^{jn\phi}\}$ and the corresponding eigenvalues are

$$\lambda_n^{\text{CFIE,TE}} = \tfrac{1}{2}(\eta\pi ka)H_n^{(2)\prime}(ka)[\alpha J_n'(ka) + j(1-\alpha)J_n(ka)] \tag{6.14}$$

A plot of the three dominant eigenvalues as a function of ka is presented in Figure 6.4 for $\alpha = 0.2$ and demonstrate that the eigenvalues never vanish. Consequently, the CFIE has unique solutions at all values of ka. The TM case exhibits similar behavior (Prob. P6.5).

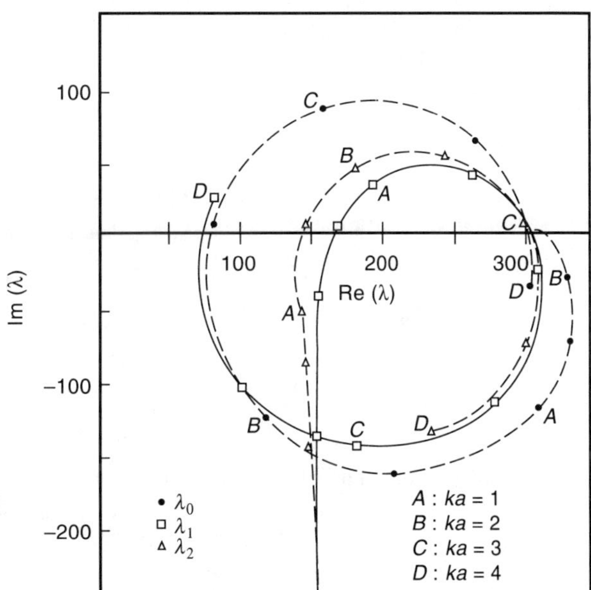

Figure 6.4 Plot of the three dominant eigenvalues of the TE CFIE as a function of ka. After [1]. ©1990 Hemisphere Publishing Corporation.

A more general uniqueness proof, adapted from [5], is as follows: Imposing Equation (6.12) in the absence of any incident field is equivalent to enforcing

$$\alpha E_z^s + (1-\alpha)\eta H_t^s = 0 \tag{6.15}$$

an infinitesimal distance inside the scatterer surface. By multiplying (6.15) with its complex conjugate and integrating the result over the scatterer surface, we obtain

$$\int \mathrm{Re}[\bar{E}_z^s \times \bar{H}_t^{s\dagger} \cdot (-\hat{n})]\, dt = \frac{-1}{\alpha(1-\alpha)\eta} \int [\alpha^2 |E_z^s|^2 + (1-\alpha)^2 \eta^2 |H_t^s|^2]\, dt \tag{6.16}$$

The first expression in (6.16) is an integral over the time-averaged Poynting vector representing the flow of power into the scatterer interior, which must vanish. Consequently, if $\alpha \neq 0$ and $\alpha \neq 1$, then

$$E_z^s = 0 \tag{6.17}$$

and

$$H_z^s = 0 \tag{6.18}$$

an infinitesimal distance inside the scatterer. This demonstrates that the CFIE has no "interior-resonance" solutions unless $\alpha = 0$ or $\alpha = 1$ (in which case we are solving the MFIE or EFIE, respectively).

Furthermore, the excitation in Equations (6.12) and (6.13) involves incident tangential components of electric *and* magnetic fields. The combined excitation does not decouple from any of the eigencurrents as did the single type of excitation in Equations (6.8) and (6.9). As a result, the CFIE formulation involves sufficient information to uniquely determine the true surface currents for any nonzero value of ka.

We now consider a numerical implementation of the CFIE for the TM polarization. Suppose that the scatterer contour is represented by flat strips, as depicted in Figure 6.5. Normal and tangential vectors can be defined at the mth cell according to the angle Ω depicted in Figure 6.6, producing

$$\hat{n}_m = \hat{x} \cos \Omega_m + \hat{y} \sin \Omega_m \tag{6.19}$$

$$\hat{t}_m = -\hat{x} \sin \Omega_m + \hat{y} \cos \Omega_m \tag{6.20}$$

For pulse basis functions and Dirac delta testing functions, the method-of-moments matrix equation is of the form

$$\begin{bmatrix} C_{11} & C_{12} & \dots & C_{1N} \\ C_{21} & C_{22} & \dots & C_{2N} \\ \cdot & & & \cdot \\ \cdot & \ddots & \ddots & \cdot \\ \cdot & & & \cdot \\ C_{N1} & \dots & \dots & C_{NN} \end{bmatrix} \begin{bmatrix} j_1 \\ j_2 \\ \cdot \\ \cdot \\ \cdot \\ j_N \end{bmatrix} = \begin{bmatrix} e_1 \\ e_2 \\ \cdot \\ \cdot \\ \cdot \\ e_N \end{bmatrix} \tag{6.21}$$

The diagonal and off-diagonal entries are given by

$$C_{mm} = \frac{(1-\alpha)\eta}{2} + \frac{\alpha k \eta}{4} \int_{\text{cell } m} H_0^{(2)}(kR_m)\, dt' \tag{6.22}$$

and

$$\begin{aligned} C_{mn} = {} & \frac{\alpha k \eta}{4} \int_{\text{cell } n} H_0^{(2)}(kR_m)\, dt' \\ & + \frac{(1-\alpha)\eta jk}{4} \int_{\text{cell } n} \left[\sin \Omega_m \left(\frac{y_m - y'}{R_m} \right) + \cos \Omega_m \left(\frac{x_m - x'}{R_m} \right) \right] H_1^{(2)}(kR_m)\, dt' \end{aligned} \tag{6.23}$$

respectively, where

$$R_m = \sqrt{[x_m - x(t')]^2 + [y_m - y(t')]^2} \tag{6.24}$$

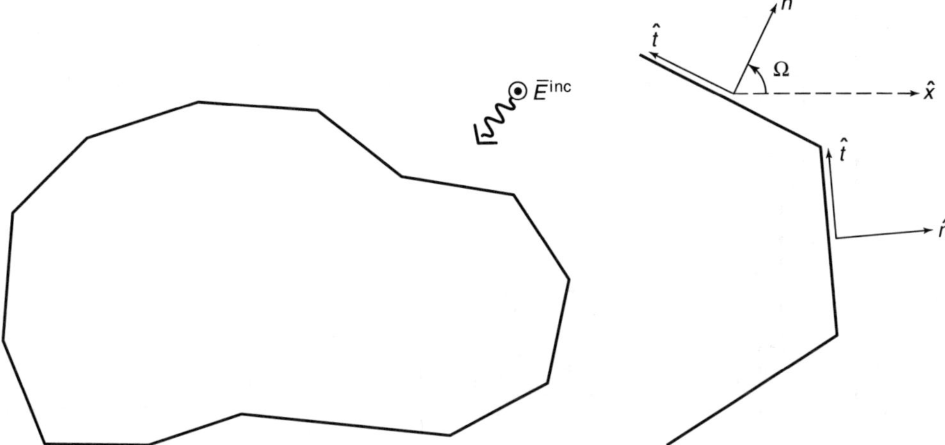

Figure 6.5 Flat-strip model of a cylinder cross section.

Figure 6.6 Tangent vector, normal vector, and polar angle Ω used to describe the orientation of the nth cell in the model.

If the incident field is a uniform plane wave propagating in the direction indicated by a conventional polar angle θ in the x–y plane, the entries of the excitation column vector are

$$e_m = [\alpha - (1-\alpha)\cos(\theta - \Omega_m)]e^{-jk(x_m \cos\theta + y_m \sin\theta)} \tag{6.25}$$

These expressions involve integrals that are identical in form to those of the EFIE and MFIE discussed in Chapter 2, and their numerical evaluation is similar. From Equation (6.23), we see that two separate integrals are required for each of the off-diagonal entries of the matrix. Thus, although the CFIE guarantees unique solutions, the computational effort required to construct a matrix equation representing the CFIE is roughly double that required for the EFIE or MFIE alone.

Figures 6.7–6.12 illustrate the performance of the CFIE. Figure 6.7 shows the EFIE result for the current density induced by a plane wave on a circular cylinder with $ka = 5.15$ (near the theoretical TM_{21} cavity resonance). Clearly, the EFIE solution is completely incorrect (eigencurrents associated with the $e^{j2\phi}$ and $e^{-j2\phi}$ eigenfunctions are suppressed in the numerical result). For comparison, Figure 6.8 shows the CFIE result for the current density obtained with $\alpha = 0.7$ for the same cylinder. The CFIE formulation has eliminated the gross errors observed in the EFIE result. Figure 6.9 shows the scattering cross section for the circular cylinder example computed from the EFIE currents. Although in theory the far-zone fields should not be affected by the suppressed resonant current, some error is observed in the scattering cross section plot. (In this case, some discretization error may be present due to the relatively large cell sizes.)

As a second example and to illustrate the use of more general scatterer shapes, Figures 6.10 and 6.11 show the EFIE result for the current density induced by a TM wave on pie-shaped cylinders of similar electrical size. The currents of Figure 6.11 deviate significantly from the previous case and the physical optics solution, suggesting a possible interior resonance problem. Figure 6.12 shows the CFIE result obtained using $\alpha = 0.7$ for comparison to Figure 6.11. The CFIE solution exhibits good agreement with the physical optics solution and the solution of Figure 6.10, supporting our conjecture that the solution

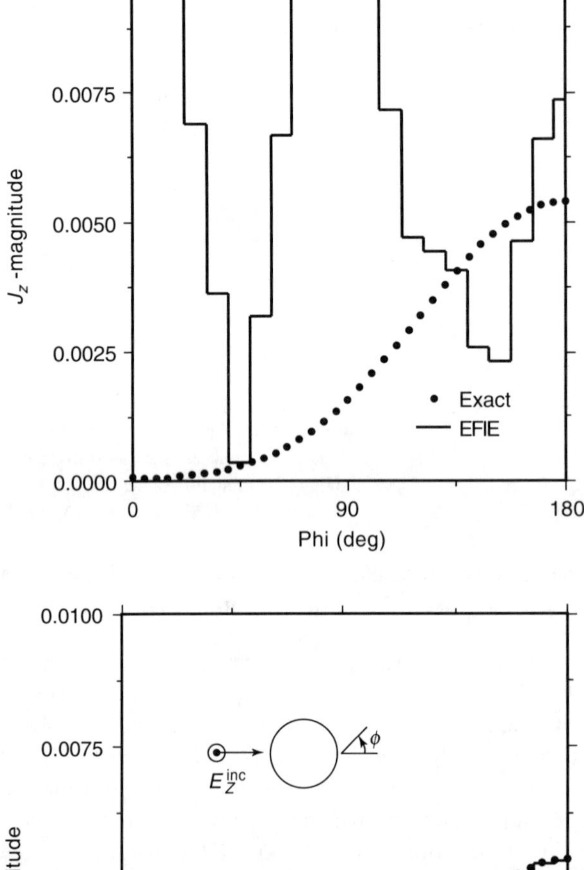

Figure 6.7 Comparison of the EFIE and exact solutions for the TM current density on a circular cylinder of radius $0.82\lambda_0$. The numerical result was obtained using 40 equal-sized cells.

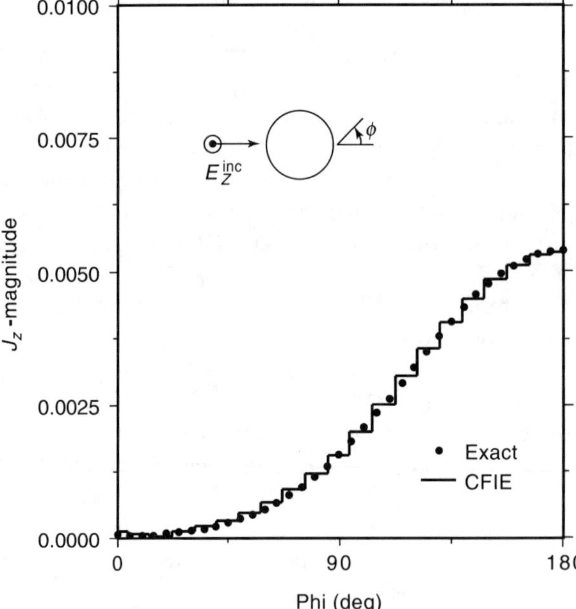

Figure 6.8 The CFIE result compared to the exact solution for the example of Figure 6.7. The CFIE parameter α equals 0.7.

of Figure 6.11 is corrupted by an interior resonance. For nonresonant pie-shaped cylinders, the CFIE result agrees with the EFIE solution.

In summary, the CFIE formulation can be used with perfectly conducting scatterers to produce the desired surface current density at all nonzero frequencies of interest. In ad-

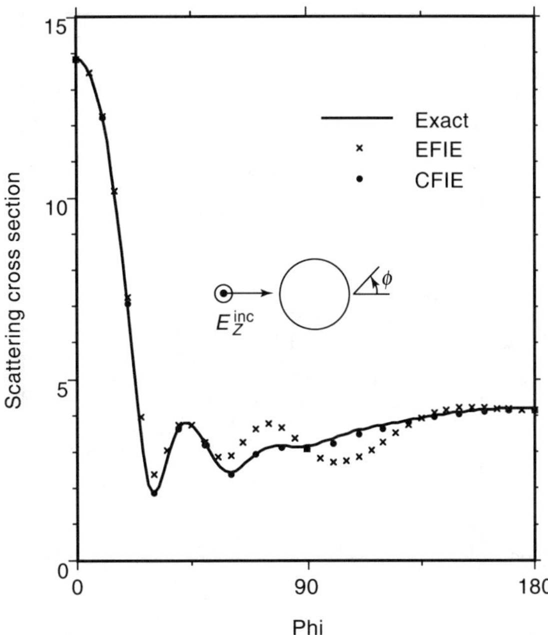

Figure 6.9 Two-dimensional scattering cross section produced by the EFIE result of Figure 6.7 compared with the exact solution and the CFIE result.

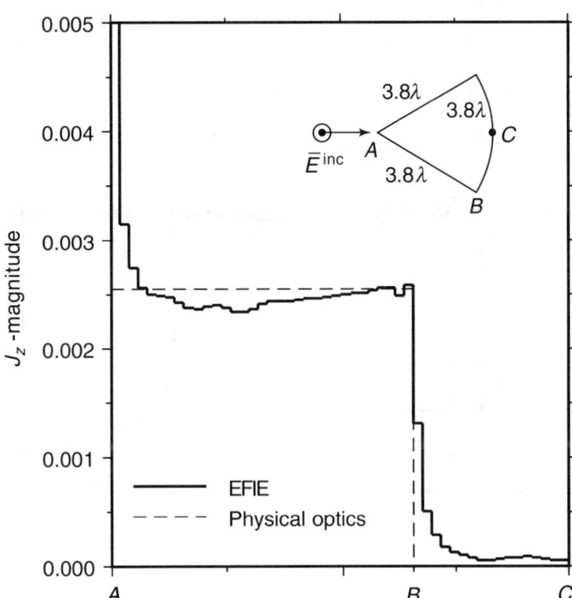

Figure 6.10 The EFIE result for the current density induced by a TM plane wave on a pie-shaped cylinder compared with the physical optics solution.

dition, the CFIE formulation is directly applicable to homogeneous dielectric scatterers and geometries incorporating an impedance boundary condition (Probs. P6.8 and P6.9). Since the matrix elements arising within the CFIE formulation are more complicated than those of the EFIE or MFIE, additional computational effort is required to implement the CFIE.

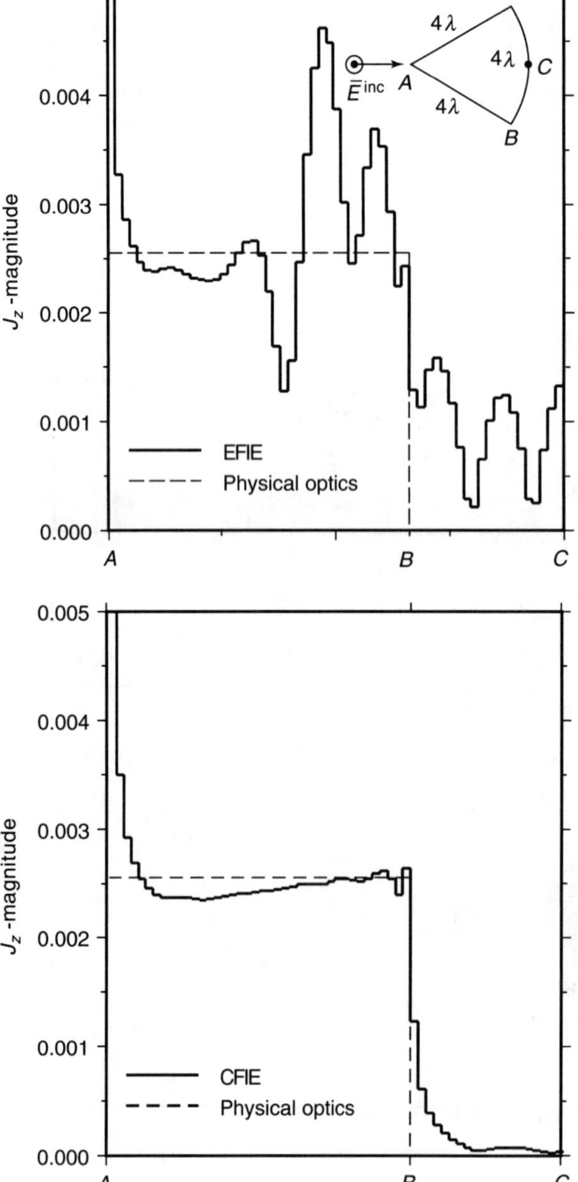

Figure 6.11 The EFIE result for the current density induced by a TM plane wave on a pie-shaped cylinder compared with the physical optics solution.

Figure 6.12 The CFIE result for the current density induced by a TM plane wave on the pie-shaped cylinder in Figure 6.11 compared with the physical optics solution.

6.3 THE COMBINED-SOURCE INTEGRAL EQUATION FOR SCATTERING FROM PERFECTLY CONDUCTING CYLINDERS [6, 7]

Integral equations are constructed by combining three factors: an equivalent mathematical source (electric or magnetic current density), a source–field relationship (expression for the field at some point in space in terms of an integral over electric or magnetic sources), and a boundary condition on the surface of the scatterer under consideration (such as forcing the tangential $\bar{E}$-field at a perfect conductor to vanish). The EFIE, MFIE, and CFIE for perfectly conducting cylinders are formulated in terms of equivalent surface current densities

$$\bar{J} = \hat{n} \times \bar{H} \tag{6.26}$$

$$\bar{K} = \bar{E} \times \hat{n} = 0 \tag{6.27}$$

where $\bar{E}$ and $\bar{H}$ are the total fields external to the scatterer, $\bar{J}$ and $\bar{K}$ are the equivalent mathematical electric and magnetic surface current densities, and $\hat{n}$ is the outward normal vector along the scatterer surface. These equivalent sources are such that the correct fields are produced external to the scatterer and null fields are produced within the scatterer.

To model a perfectly conducting cylinder, it is natural to choose equivalent sources that, if used in connection with the complete set of boundary conditions, force the interior fields to vanish. However, the fields external to a scatterer can be correctly represented by many other equivalent sources, provided that the interior fields do not vanish. [Equations (6.26) and (6.27) are the only sources that produce zero interior fields and the correct exterior fields.] Since only the exterior fields are of interest, the interior fields can be arbitrarily chosen. This arbitrary choice allows a degree of freedom in the selection of equivalent sources, which may be taken to satisfy a constraint such as

$$\bar{K} = \frac{1-\beta}{\beta}\eta(\hat{n} \times \bar{J}) \tag{6.28}$$

for some real value of β between zero and one. It has been shown that this choice of equivalent sources may be used with perfectly conducting scatterers in order to remove the uniqueness problem from the original EFIE [6, 7]. This is possible because no eigencurrents associated with the interior cavity modes satisfy the additional constraint imposed by Equation (6.28).

The combined-source integral equation (CSIE) is constructed using the boundary condition that $\hat{n} \times \bar{E} = 0$ on the surface of the perfect electric conductor. For the TE polarization, the CSIE can be written as

$$\begin{aligned} \beta E_t^{\text{inc}}(t) = -\hat{t} \cdot \frac{\nabla\nabla \cdot + k^2}{jk}\eta \int \hat{t}(t')\beta J_t(t')\frac{1}{4j}H_0^{(2)}(kR)\, dt' \\ + \hat{t} \cdot \nabla \times \int \hat{z}(1-\beta)\eta J_t(t')\frac{1}{4j}H_0^{(2)}(kR)\, dt' \end{aligned} \tag{6.29}$$

This equation is readily discretized into matrix form using the procedures discussed in Chapter 2. If specialized to a circular cylinder of radius a, the eigenvalues of this operator (associated with eigenfunctions $\{e^{jn\phi}\}$) can be found analytically to be

$$\lambda_n^{\text{CSIE,TE}} = \tfrac{1}{2}(\eta\pi ka)H_n^{(2)\prime}(ka)[\beta J_n{}'(ka) - j(1-\beta)J_n(ka)] \tag{6.30}$$

By analogy with Equation (6.14), it can be seen that these eigenvalues never vanish for real β between zero and one. In fact, for an appropriate inner product, the combined-source and combined-field operators are the adjoint of one another [7].

Although the combined-source formulation produces a unique solution, a drawback to the procedure is that the electric current density $\bar{J}$ in (6.29) is not the true electric current density induced upon the scatterer surface. In fact, since Equation (6.29) employs only the tangential incident electric field on the surface, it cannot produce the correct resonant eigencurrent associated with the exterior problem. Once the mathematical sources $\bar{J}$ and $\bar{K}$ are identified by solving (6.29), the true electric current density can be found from a secondary calculation involving the incident magnetic field (Prob. P6.12). In common with the CFIE formulation, the tangential components of both the incident electric and magnetic fields must be employed in order to determine the true current density within a CSIE formulation. The matrix equation resulting from a method-of-moments discretization of the combined-source equation is of the same order and complexity as the matrix representing the combined-field integral equation.

6.4 THE AUGMENTED-FIELD FORMULATION [8]

The EFIE, MFIE, CFIE, and CSIE formulations involve boundary conditions imposed on the tangential components of the fields on the surface of a scatterer. As discussed, the tangential electric field or magnetic field alone is insufficient to determine the correct surface current density at resonant frequencies. Alternate surface integral equations have been proposed involving both the tangential and normal components of a single type of field [8]. These equations enforce the complete set of boundary conditions associated with the particular field and involve enough information about the incident field to uniquely determine the surface currents, except in a few exceptional situations. These "augmented-field integral equations" have the advantage over the CFIE formulation that the matrix elements are only as complicated as the original EFIE or MFIE formulations. However, when reduced to a matrix form using the method of moments, the system of equations is overdetermined by factors of 2 in two dimensions and $\frac{3}{2}$ in three dimensions. A special solution algorithm and additional computational time and storage beyond that of the original EFIE and MFIE will be required to solve the overdetermined system.

The augmented-field formulation will be illustrated in Chapter 8, when considering the problem of scattering from infinite cylinders illuminated from oblique angles.

6.5 OVERSPECIFICATION OF THE ORIGINAL EFIE OR MFIE AT INTERIOR POINTS

Since the basic interior resonance problem is the presence of fields interior to the scatterer geometry at discrete eigenfrequencies, a conceptually straightforward solution to the problem is the direct enforcement of a boundary condition throughout the interior region. This approach would necessarily involve the incident field within the scatterer and circumvent the difficulty associated with the tangential incident field on the scatterer surface decoupling from the resonant eigencurrent. A variety of proposed remedies have been based on variations of this idea [9–14], and we will not consider all of these in detail. A relatively simple approach has been proposed by Mittra and Klein, who suggest enforcing the condition that the E-field or H-field vanish at a variety of points within the scatterer geometry [10]. Although their original procedure requires the overspecification of the boundary conditions and the consequential solution of an overdetermined matrix equation, it is straightforward to add unknowns to balance the additional match points [12]. The latter idea has the potential advantages that the equation is no longer overdetermined and feedback concerning the interior current density (which should vanish if a "good" solution is obtained) is readily available. In addition, only the scatterer model is changed, so that existing computer codes based on the EFIE or MFIE may not need to be modified.

Consider Equation (6.1), the EFIE for TM scattering from perfectly conducting cylinders. Theoretically, this equation suffers from uniqueness problems associated with the TM modes of a perfectly conducting cavity, since the internal cavity fields satisfy the boundary condition imposed on the tangential electric field at the scatterer surface. In order to suppress these cavity fields, consider augmenting the scatterer model (originally consisting of strips located on the desired scatterer surface) with additional perfectly conducting strips distributed throughout the interior. The idea is to employ interior strips to "short out" any interior fields that might otherwise be present. The incident field sampled at the interior match points provides additional information in order to properly determine the resonant eigencurrent.

To illustrate the performance of this approach, consider a wave normally incident on a circular cylinder having a circumference of 5.15λ (near the theoretical TM_{21} cavity resonance). The EFIE solution using pulse basis functions and Dirac delta testing functions is displayed in Figure 6.7 and exhibits considerable error due to the cavity resonance. (As discussed previously, the $e^{j2\phi}$ and $e^{-j2\phi}$ eigenfunctions are suppressed in the numerical result.) Figure 6.13 displays the result after the cylinder model was augmented with three additional interior strips whose locations were chosen using random numbers. The additional strips have completely eliminated the interior fields from the cylinder, without a large increase in computational effort. For this particular example, the procedure is much more efficient than the CFIE formulation discussed in Section 6.2.

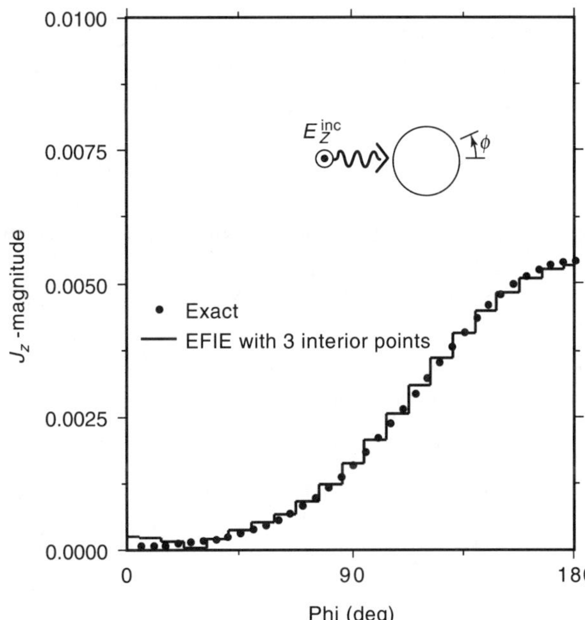

Figure 6.13 Comparison of the EFIE and exact solutions for the TM current density on a circular cylinder of radius $0.82\lambda_0$ after three interior strips were added to the model used to produce Figure 6.7.

The principal drawback to this type of remedy is the degree of user expertise required to determine the number of interior points and their precise location. Furthermore, the addition of a few interior strips may shift a resonance to a different frequency instead of completely eliminating it. Studies suggest that a uniform density of approximately 10 interior strips per square wavelength is necessary to completely suppress resonances in two-dimensional formulations [13]. Since the volume of the interior region increases faster than the surface area, this approach appears to require the introduction of a prohibitive number of additional unknowns as the size of the scatterer increases relative to the wavelength. Although this simple procedure can be useful to eliminate resonances if encountered at single frequencies, it will prove to be less efficient for large scatterers than some of the other alternatives outlined in the preceding sections.

6.6 DUAL-SURFACE INTEGRAL EQUATIONS [14, 15]

The preceding section demonstrated that it is possible to enforce additional conditions inside the scatterer to eliminate the interior resonance problem. Unfortunately, the specific procedure of Section 6.5 generally requires user expertise and a large number of additional unknowns. Recently, a more systematic approach has been proposed that is convenient to apply and requires no additional unknowns [14, 15].

Consider Figure 6.14, which depicts a closed scatterer of perfect electric conducting material whose surface is denoted S_1. A second mathematical surface S_2 resides inside the scatterer. If equivalent electric surface current $\bar{J}$ is introduced on S_1 to replace the perfectly conducting material, a tangential-field MFIE may be expressed on the surface S_1 in the form

$$\hat{n} \times \bar{H}^{\text{inc}}(\bar{r}) = \bar{J}(\bar{r}) - \hat{n} \times \nabla \times \bar{A}\big|_{\bar{r}} \tag{6.31}$$

where $\bar{A}$ is the conventional magnetic vector potential due to $\bar{J}$ and $\bar{r}$ denotes a point approaching S_1 from the exterior. Equation (6.31) suffers from the interior resonance problem if S_1 happens to coincide with the interior of a resonant cavity having walls that are perfect magnetic conductors.

Since the total fields vanish everywhere inside S_1, an alternate tangential-field MFIE can be expressed on the surface S_2 in the form

$$\hat{n} \times \bar{H}^{\text{inc}}(\bar{r} - \bar{\delta}) = -\hat{n} \times \nabla \times \bar{A}\big|_{\bar{r}-\bar{\delta}} \tag{6.32}$$

where $\bar{r} - \bar{\delta}$ denotes an observation point on S_2. Note that the current density $\bar{J}$ used in (6.32) still resides on S_1. Equation (6.32) can produce correct solutions for $\bar{J}$ as long as S_2 does not coincide with the surface of a resonant cavity. However, consider the *dual-surface equation* obtained by combining Equations (6.31) and (6.32) to produce

$$\hat{n} \times \bar{H}^{\text{inc}}(\bar{r}) + \gamma \hat{n} \times \bar{H}^{\text{inc}}(\bar{r} - \bar{\delta}) = \bar{J}(\bar{r}) - \hat{n} \times \nabla \times \bar{A}\big|_{\bar{r}} - \gamma \hat{n} \times \nabla \times \bar{A}\big|_{\bar{r}-\bar{\delta}} \tag{6.33}$$

As long as the constant γ has a nonzero imaginary part, Equation (6.33) has unique solutions at all frequencies for which the surfaces S_1 and S_2 are separated by a nonzero distance of no more than approximately $\frac{1}{2}\lambda$ [14]. Furthermore, the unique solution is the desired exterior current density. Similar dual-surface equations can be derived based on an EFIE.

To illustrate that Equation (6.33) can eliminate the interior resonance problem, we again appeal to an eigenvalue interpretation possible with circular cylinders. Consider a circular cylinder illuminated by a TE wave. Suppose that the surface S_1 is circular with radius a and the interior surface S_2 is circular with radius $a - \delta$. If the dual-surface MFIE operator is defined to map functions on S_1 to functions on S_1, its eigenfunctions are of the form $e^{jn\phi}$ and the associated eigenvalues can be expressed for the TE case as (Prob. P6.13)

$$\lambda_n = \tfrac{1}{2}(j\pi ka)H_n^{(2)\prime}(ka)[J_n(ka) - \gamma J_n(ka - k\delta)] \tag{6.34}$$

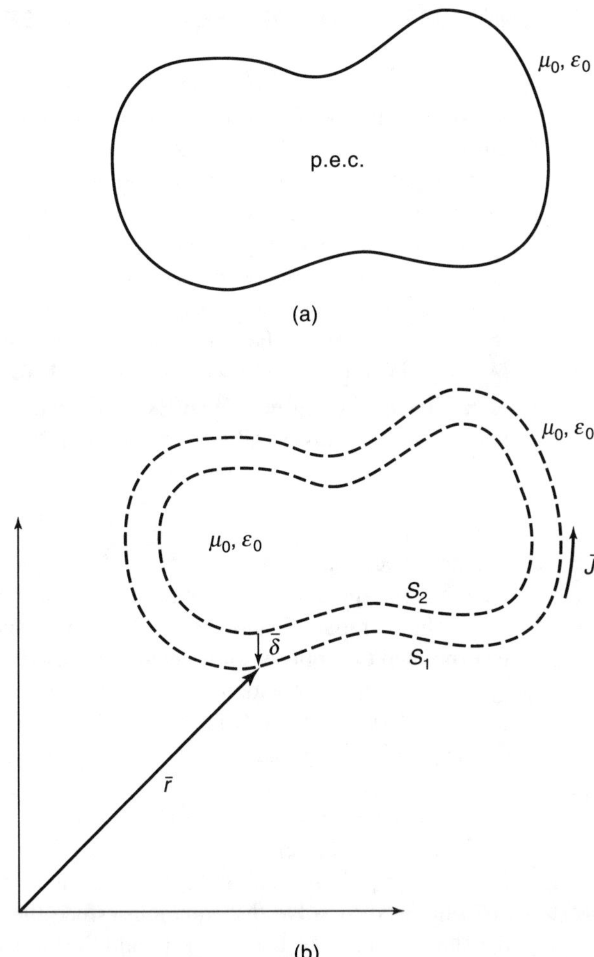

Figure 6.14 Geometry of a scatterer showing the dual surfaces S_1 and S_2.

As long as γ has a nonzero imaginary part, Equation (6.34) will not vanish unless $J_n(ka)$ and $J_n(ka - k\delta)$ vanish simultaneously. Since the zeros of $J_n(ka)$ occur at approximately half-wavelength intervals, it suffices to restrict the separation distance δ to less than that distance. (In practice, it would seem reasonable to separate S_1 and S_2 by $\frac{1}{4}\lambda$ whenever possible.)

Because the dual-surface equations sample the incident field on both S_1 and S_2, the incident field is not permitted to decouple from an individual eigenfunction (by restricting the separation distance between S_1 and S_2, we preclude the possibility that a single eigenfunction can vanish on both surfaces). In contrast to the CFIE or CSIE formulations, the dual-surface equations only require one type of integral operator (i.e., EFIE type or MFIE type), which may simplify the programming and the overall accuracy. The computational requirements to create the method-of-moments matrix are approximately twice that of the original surface integral equations. Since there are no additional unknowns, no extra computation is required for matrix solution beyond that of the conventional MFIE. The dual-surface equations are easily applied and appear to require no particular user expertise. If used over a range of frequencies, the location of S_2 must be adjusted to maintain the desired separation as a function of λ.

6.7 COMPLEXIFICATION OF THE WAVENUMBER [16, 17]

In theory, interior resonance solutions can only occur in a lossless environment. This raises the possibility of eradicating them by introducing a small amount of media loss. Unfortunately, loss sufficient to suppress resonances will also alter the desired solution to some extent. A procedure introduced in [16] avoids this problem by an extrapolation process and is easy to implement.

The idea is to solve the original EFIE or MFIE twice with enough medium loss in each case to suppress spurious interior resonances and use the two solutions to extrapolate to the lossless result. For instance, the real-valued wavenumber k can be replaced with $k_2 = k - j\delta$ and again with $k_1 = k - j\frac{1}{2}\delta$, where δ is typically on the order of a few percent of k. If the surface current densities obtained using k_1 and k_2 are denoted $\bar{J}_1$ and $\bar{J}_2$, a linear interpolation process similar to that used in Prob. P2.7 suggests that the surface current in the lossless case is approximately

$$\bar{J} \cong 2\bar{J}_1 - \bar{J}_2 \tag{6.35}$$

For better accuracy, three values of $\bar{J}_i$ can be obtained using three wavenumbers, and parabolic interpolation can be employed [17].

The "complexification" procedure can be successful as long as enough medium loss is incorporated to suppress spurious interior resonances. From the perspective of the circular cylinder examples considered in Section 6.1, the lossy medium changes the problematic factor $J_n(ka)$ to a Bessel function with complex argument, which no longer has zeros. Consequently, the incident field does not decouple from any of the matrix eigenvalues in the lossy case.

Compared with the original solution of the EFIE or MFIE, the procedure requires additional computation because of the need to solve two (or more) systems of equations. (As noted in [17], this computation can be alleviated to some extent if iterative techniques are employed to solve the equations, since the solution for one system can be used as the starting point for the next.) In addition, for the two-dimensional case, the procedure requires Hankel functions of complex argument, which are slightly more costly to compute than Hankel functions of real argument.

6.8 DETERMINATION OF THE CUTOFF FREQUENCIES AND PROPAGATING MODES OF WAVEGUIDES OF ARBITRARY SHAPE USING SURFACE INTEGRAL EQUATIONS [18–20]

In addition to treating scattering problems, the two-dimensional EFIE and MFIE can also be used to determine the cutoff frequencies of arbitrarily shaped waveguides. We have seen that the solutions to the surface integral equations are not unique at frequencies where the scatterer surface happens to coincide with a resonant cavity. Because the cutoff frequencies of the propagating modes are also the frequencies where the guides are two-dimensional resonant cavities, it is possible to use these equations to determine the cutoff frequencies and waveguide modes.

Consider a homogeneous waveguide with perfectly conducting walls. The TM cavity resonances of the corresponding infinite cylinder can be found using the EFIE of Equation (6.1), which we rewrite in the homogeneous form

$$\int J_z(t')H_0^{(2)}(k_c R)\, dt' = 0 \tag{6.36}$$

where R is defined in (6.2) and the integration is carried out over the interior surface of the cavity. In Equation (6.36), the current density $J_z(t)$ and the cutoff wavenumber k_c are unknowns to be determined. If the EFIE operator is converted to a method-of-moments matrix, the condition number or determinant of the matrix can be used to identify the wavenumber where the geometry supports a cavity resonance. This is a result of the fact that an eigenvalue of the matrix becomes relatively small near a cavity resonance. The method-of-moments matrix is a nonlinear function of the wavenumber, and generally an iterative or trial-and-error procedure must be employed to determine the wavenumber that minimizes the magnitude of the complex determinant or maximizes the condition number. (Because of discretization error, the determinant will usually not vanish.)

For illustration, Figures 6.15 and 6.16 show plots of the magnitude of the complex-valued determinant of the method-of-moments matrix as a function of the wavenumber k for triangular cavities modeled with 21 cells. The sharp dips in the determinant indicate resonant wavenumbers of the cavity. Figure 6.15 is based on an EFIE discretized according to the procedure described in Section 2.1, while Figure 6.16 is obtained from an MFIE formulation similar to that presented in Section 2.2. In order to use the MFIE to model a cavity with perfect electric walls, it is necessary to construct the method-of-moments matrix with the "outward" normal vector to the surface pointing *into* the interior of the cavity. (In the scattering problem discussed in Section 2.2 the normal vector points *out of* the closed scatterer. If the cavity surface was modeled with the normal vector pointing out,

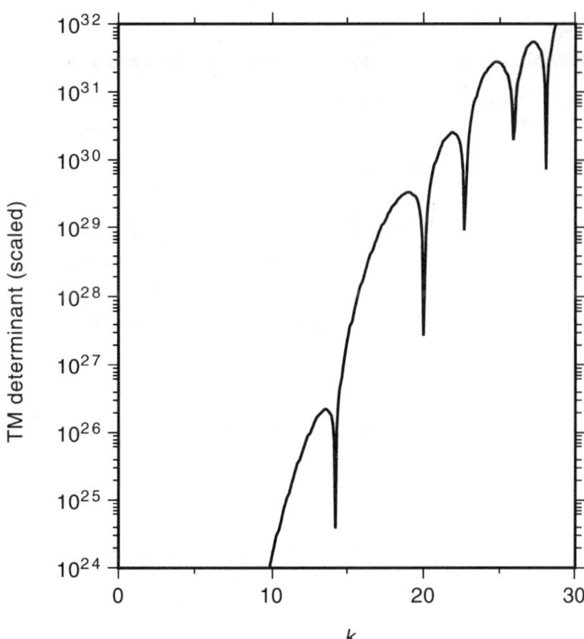

Figure 6.15 The EFIE result for the TM cutoff wavenumbers of a triangular-shaped cylinder. The triangle is isosceles with side dimension 0.5. Sharp minima correspond to solutions of Equation (6.36).

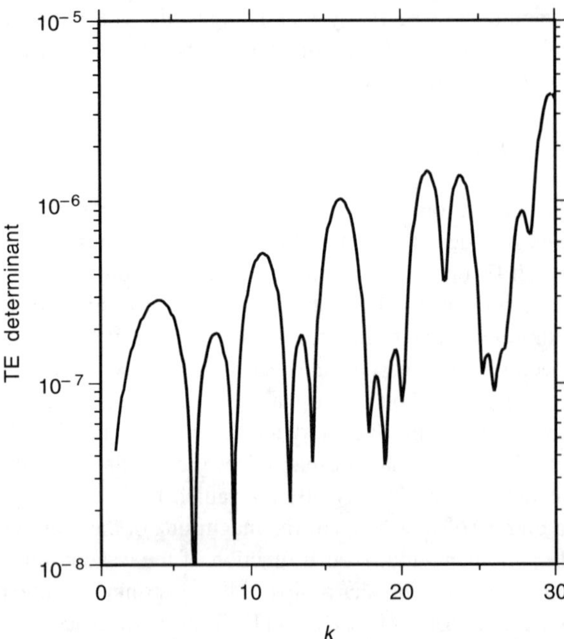

Figure 6.16 The MFIE result for the TE cutoff wavenumbers of a triangular-shaped cylinder.

the resonant frequencies obtained from the MFIE would be those for a cavity with perfect magnetic walls rather than perfect electric walls.)

Once the cutoff wavenumber is determined for a given mode, the current density can be found from the eigenvector of the method-of-moments matrix associated with the near-zero eigenvalue. The modal field distribution within the waveguide can be subsequently determined by an integration over the current density.

By combining surface integral equations for homogeneous dielectric regions with those for conductors, the idea described above can be extended to determine propagation constants in dielectric-loaded waveguides [19, 20]. A number of variations on the approach are possible, depending on whether the determinant, eigenvalues, or singular values of the matrix are to be computed and the degree of sophistication brought to bear on the problem of identifying the minimum values as a function of frequency. Because of the nonlinear behavior as a function of frequency and the need to construct the complete method-of-moments system at each frequency, the approach generally involves substantially more computational effort than a typical scattering problem. The complete characterization of a given waveguide may be time consuming.

6.9 UNIQUENESS DIFFICULTIES ASSOCIATED WITH DIFFERENTIAL EQUATION FORMULATIONS

In a lossless region, the scalar Helmholtz equation

$$\nabla \cdot \left(\frac{1}{\mu_r} \nabla E_z \right) + k^2 \varepsilon_r E_z = 0 \tag{6.37}$$

also admits homogeneous solutions when constrained by Dirichlet or Neumann boundary conditions and is often used to find the resonant frequencies of closed cavities or the cutoff frequencies of waveguide modes. Consequently, if used with these boundary conditions within a scattering formulation, the Helmholtz operator may become highly ill-conditioned near frequencies where the interior region represents a resonant cavity.

Practical difficulties arise in connection with the so-called *inward-looking formulations* described in Section 3.12, since these employ Dirichlet or Neumann boundary conditions and attempt to solve (6.37) as the initial step of the procedure [21]. For illustration, Figure 6.17 shows a plot of the matrix condition number obtained from a discretization of (6.37) with Neumann boundary conditions for a circular region of free space having radius a. The scalar Helmholtz equation is discretized with piecewise-linear interpolation functions (Chapter 3) defined on a triangular-cell mesh with 85 nodes and 131 cells. Sharp spikes in the figure indicate resonant frequencies, which for this example correspond to the TM modes of an air-filled cavity with perfect magnetic walls.

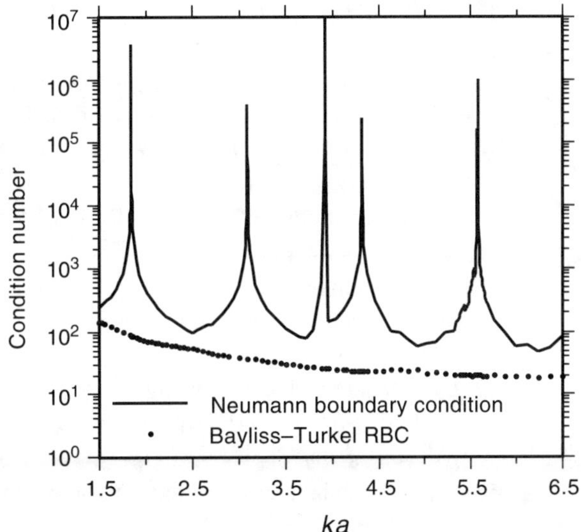

Figure 6.17 Matrix condition number for the system obtained from a discretization of the scalar Helmholtz equation for Neumann and Bayliss–Turkel boundary conditions. After [21]. ©1992 IEEE.

For comparison, Figure 6.17 also shows the matrix condition number when a local radiation boundary condition, the second-order Bayliss–Turkel RBC presented in Section 3.8, is used to constrain the scalar Helmholtz equation. Clearly, the radiation loss provided by the RBC is sufficient to eliminate any cavity resonances. Since the *outward-looking formulations* presented in Chapter 3 explicitly incorporate radiation loss, they are generally free of uniqueness difficulties arising from the scalar Helmholtz equation.

Outward-looking formulations that use either the EFIE or MFIE as an RBC (Section 3.10) remain vulnerable to internal resonance difficulties arising from the surface integral equations, however. As demonstrated previously, these formulations can be made robust by employing a CFIE as an alternative radiation boundary condition.

6.10 SUMMARY

The focus of this chapter has been the interior resonance problem and its remediation. Because of this difficulty, the conventional surface EFIE and MFIE are not recommended for analyzing closed-body scatterers large enough to support cavity resonant fields. Although the internal resonance problem may not be encountered for every geometry of interest, there are certain situations where it is difficult or impossible to avoid. One such example is the case of a finite source illuminating an infinite cylinder, which can be treated by superimposing solutions obtained over a wide range of spatial frequencies. This procedure will be considered in Chapter 8 using some of the alternative formulations discussed in this chapter.

REFERENCES

[1] A. F. Peterson, "The interior resonance problem for surface integral equations of electromagnetics: Numerical consequences and a survey of remedies," *Electromagnetics*, vol. 10, pp. 293–312, 1990.

[2] R. F. Harrington, *Time-Harmonic Electromagnetic Fields*, New York: McGraw-Hill, 1961.

[3] D. H. Griffel, *Applied Functional Analysis*, New York: Wiley, 1981.

[4] K. M. Mitzner, "Numerical solution of the exterior scattering problem at eigenfrequencies of the interior problem," *Digest of the 1968 URSI Radio Science Meeting*, Boston, MA, International Union of Radio Scientists (URSI), p. 75, Sept. 1968. This conference digest not formally available from URSI.

[5] J. R. Mautz and R. F. Harrington, "H-field, E-field, and combined-field solutions for conducting bodies of revolution," Archiv für Elektronik und Übertragungstechnik (A. E. U.), vol. 32, pp. 157–163, 1978.

[6] J. C. Bolomey and W. Tabbara, "Numerical aspects on coupling between complementary boundary value problems," *IEEE Trans. Antennas Propagat.*, vol. AP-21, pp. 356–363, May 1973.

[7] J. R. Mautz and R. F. Harrington, "A combined-source formulation for radiation and scattering from a perfectly conducting body," *IEEE Trans. Antennas Propagat.*, vol. AP-27, pp. 445–454, July 1979.

[8] A. D. Yaghjian, "Augmented electric- and magnetic-field equations," *Radio Science*, vol. 16, pp. 987–1001, Nov. 1981.

[9] H. A. Shenk, "Improved integral equation formulation for acoustic radiation problem," *J. Acoust. Soc. Am.*, vol. 44, pp. 41–58, 1968.

[10] R. Mittra and C. A. Klein, "Stability and convergence of moment method solutions," in *Numerical and Asymptotic Techniques in Electromagnetics*, ed. R. Mittra, New York: Springer-Verlag, 1975.

[11] P. C. Waterman, "Numerical solution of electromagnetic scattering problems," in *Computer Techniques for Electromagnetics*, ed. R. Mittra, New York: Hemisphere, 1987 Reprint.

[12] A. F. Peterson and R. Mittra, "On the method of conjugate gradients for scattering by PEC cylinders," Electromagnetics Lab. Tech. Rep. 84-3, UILU-ENG-84-2540, University of Illinois, Urbana, IL, Jan. 1984.

[13] A. F. Peterson and R. Mittra, "Mutual admittance between slots in cylinders of arbitrary shape," Coordinated Science Lab. Tech. Rep., UILU-ENG-87-2247, University of Illinois, Urbana, IL, Aug. 1987.

[14] M. B. Woodworth and A. D. Yaghjian, "Derivation, application, and conjugate gradient solution of dual-surface integral equations for three-dimensional, multi-wavelength

perfect conductors," in *Application of Conjugate Gradient Method to Electromagnetics and Signal Analysis*, ed. T. K. Sarkar, New York: Elsevier, 1991.

[15] M. B. Woodworth and A. D. Yaghjian, "Multiwavelength three-dimensional scattering with dual-surface integral equations," *J. Opt. Soc. Am. A*, vol. 11, pp. 1399–1413, Apr. 1994.

[16] W. D. Murphy, V. Rokhlin, and M. S. Vassiliou, "Solving electromagnetic scattering problems at resonance frequencies," *J. Appl. Phys.*, vol. 67, pp. 6061–6065, May 1990.

[17] N. Engheta, W. D. Murphy, V. Rokhlin, and M. S. Vassiliou, "The fast multipole method (FMM) for electromagnetic scattering problems," *IEEE Trans. Antennas Propagat.*, vol. 40, pp. 634–641, June 1992.

[18] B. E. Spielman and R. F. Harrington, "Waveguides of arbitrary cross section by solution of a nonlinear integral eigenvalue equation," *IEEE Trans. Microwave Theory Tech.*, vol. MTT-20, pp. 578–585, Sept. 1972.

[19] M. Swaminathan, T. K. Sarkar, and A. T. Adams, "Computation of TM and TE modes in waveguides based on a surface integral formulation," *IEEE Trans. Microwave Theory Tech.*, vol. 40., pp. 285–297, Feb. 1992.

[20] S. Shu, P. M. Goggans, and A. A. Kishk, "Computation of cutoff wavenumbers for partially-filled waveguides of arbitrary cross-section using surface integral formulations and the method of moments," *IEEE Trans. Microwave Theory Tech.*, vol. 41, pp. 1111–1118, June/July 1993.

[21] L. W. Pearson, A. F. Peterson, L. J. Bahrmasel, and R. A. Whitaker, "Inward-looking and outward-looking formulations for scattering from penetrable objects," *IEEE Trans. Antennas Propagat.*, vol. 40, pp. 714–720, June 1992.

PROBLEMS

P6.1 Consider a perfectly conducting, circular cylinder with circumference ka illuminated by a TM or TE wave.

(a) A uniform TM plane wave can be expanded in a Fourier series

$$E_z^{\text{inc}}(\rho, \phi) = e^{-jkx} = \sum_{n=-\infty}^{\infty} j^{-n} J_n(k\rho) e^{jn\phi}$$

Assuming that the scattered field can be expressed as

$$E_z^s(\rho, \phi) = \sum_{n=-\infty}^{\infty} \alpha_n H_n^{(2)}(k\rho) e^{jn\phi}$$

find the exact solution to the scattering problem by determining the coefficients $\{\alpha_n\}$ in order to satisfy the boundary condition $E_z^{\text{tot}} = 0$ at $\rho = a$.

(b) Use the result of part (a) to show that the TM current density associated with the $n = 1$ harmonic $e^{j\phi}$ does not contribute to E_z^s when $J_1(ka) = 0$. What does this imply about the far-zone fields computed from the incorrect surface current density produced by the TM EFIE at an interior resonance?

(c) Using the TE plane wave

$$H_z^{\text{inc}}(\rho, \phi) = e^{-jkx} = \sum_{n=-\infty}^{\infty} j^{-n} J_n(k\rho) e^{jn\phi}$$

and the assumed scattered field

$$H_z^s(\rho, \phi) = \sum_{n=-\infty}^{\infty} \beta_n H_n^{(2)}(k\rho) e^{jn\phi}$$

find the exact solution by determining the coefficients $\{\beta_n\}$ that satisfy the appropriate boundary condition on H_z^{tot} at $\rho = a$.

(d) Use the result of part (c) to investigate the contribution to H_z^s from the TE current density associated with the $n = 1$ harmonic $e^{j\phi}$. Does this contribution vanish when $J_1(ka) = 0$ (at the interior resonance of the TE MFIE)? Discuss the implications.

(e) The eigenvalues presented in Equation (5.87) indicate that the interior resonance frequencies of the TE EFIE applied to a perfectly conducting circular cylinder occur at the zeros of $J_n'(ka)$. Would an incorrect $n = 1$ harmonic $e^{j\phi}$ produced by the TE EFIE contribute to the far-zone fields when $J_1'(ka) = 0$?

P6.2 Derive the reciprocity theorem used in Equation (6.10) by filling in the steps in the brief development presented in Section 5.12. Assume that sources $\bar{J}^a$ and $\bar{K}^a$ produce fields $\bar{E}^a$ and $\bar{H}^a$ and that sources $\bar{J}^b$ and $\bar{K}^b$ produce fields $\bar{E}^b$ and $\bar{H}^b$, when all the sources and fields exist in unbounded empty space. Clearly state necessary assumptions and restrictions. You may wish to review the similar development used in Equations (1.60)–(1.71) from Section 1.6.

P6.3 Consider a circular homogeneous dielectric cylinder of circumference ka illuminated by a uniform TM plane wave traveling in the x direction. The problem can be described by a coupled EFIE formulation, with equivalent currents $J_z(\phi)$ and $K_\phi(\phi)$ as primary unknowns (Section 2.8). In order to investigate interior resonance difficulties associated with this EFIE formulation, develop a method-of-moments discretization using the eigenfunctions $e^{jn\phi}$ as basis and testing functions and using the inner product

$$\langle a, b \rangle = \frac{1}{2\pi} \int_0^{2\pi} a^\dagger(\phi) b(\phi) \, d\phi$$

where $a^\dagger$ is the complex conjugate of a. Since the eigenfunctions of the integral operators are used as basis functions, the matrices will be diagonal. Show that the coefficients (j_n, k_n) associated with the nth cylindrical harmonic are described by the coupled equations

$$\alpha_n j_n + \beta_n k_n = e_n \qquad \gamma_n j_n + \delta_n k_n = 0$$

where

$$\alpha_n = \tfrac{1}{2}(\eta \pi ka) J_n(ka) H_n^{(2)}(ka) \qquad \beta_n = \tfrac{1}{2}(j\pi ka) J_n(ka) H_n^{(2)\prime}(ka)$$

$$\gamma_n = \tfrac{1}{2}(\eta_d \pi k_d a) J_n(k_d a) H_n^{(2)}(k_d a) \qquad \delta_n = \tfrac{1}{2}(j\pi k_d a) J_n'(k_d a) H_n^{(2)}(k_d a)$$

and

$$e_n = j^{-n} J_n(ka)$$

Discuss the implications of these equations with regard to possible internal resonance difficulties. What do these results suggest about the solution of similar EFIE formulations for noncircular cylinders?

P6.4 Consider a circular cylinder illuminated by a TM wave. Suppose that the cylinder is constructed from a material that can be characterized by a surface impedance boundary condition (IBC)

$$E_z(\phi) = \eta_s H_\phi(\phi) \quad \text{or} \quad K_\phi(\phi) = \eta_s J_z(\phi)$$

where η_s is the constant surface impedance. This problem can be described using the EFIE formulation discussed in Section 2.9. To investigate the interior resonance difficulties associated with this surface integral, repeat the procedure from Prob. P6.3 in order to identify the entries of the matrix and the excitation vector associated with cylindrical harmonics. In general, will an interior resonance problem arise with this formulation? What does this observation suggest about the solution of similar IBC formulations for noncircular cylinders?

P6.5 By combining the results of Probs. P5.11 and P5.12, derive eigenvalues for the TM CFIE appearing in Equation (6.12) when applied to a circular conducting cylinder of radius a.

P6.6 Under what conditions will the TM CFIE matrix in Equation (6.21) exhibit symmetry across the main diagonal? How do the symmetry properties compare with those of the EFIE matrix developed in Section 2.1?

P6.7 Develop a method-of-moments discretization of the TE CFIE in Equation (6.13) using subsectional triangle basis functions and pulse testing functions (see the EFIE approach described in Section 2.4). Provide complete expressions for the matrix entries.

P6.8 Develop two different CFIE formulations for homogeneous dielectric cylinders illuminated by a TM wave. Use an appropriate combination of the surface integral equations appearing in (1.111), (1.112), (1.117), and (1.118) specialized to the TM case. (It is not necessary to provide a method-of-moments discretization of the resulting equations.) For the first CFIE formulation, use a linear combination of the two exterior equations to obtain a combined-field exterior equation similar to (6.12) and (6.13) and a linear combination of the two interior equations to obtain a combined-field interior equation. (Is the combined-field interior equation necessary for unique solutions?) For the second CFIE formulation, obtain one equation by enforcing continuity of the interior and exterior electric field and a second equation by enforcing continuity of the interior and exterior magnetic field. Can you think of any reasons why one of these approaches might be preferable to the other?

P6.9 Develop a CFIE formulation for TM scattering from cylinders described by a surface IBC. Identify the entries of the method-of-moments matrix if pulse basis functions and Dirac delta testing functions are used in the discretization in a manner similar to the Chapter 2 examples.

P6.10 Formulate a CSIE for TM-wave scattering from perfectly conducting cylinders of arbitrary shape. Develop a method-of-moments discretization using pulse basis functions and Dirac delta testing functions, and identify the entries of the method-of-moments matrix.

P6.11 Derive an expression for the eigenvalues of the TM CSIE when applied to a circular perfectly conducting cylinder of radius a.

P6.12 The equivalent current density $J_t(t)$ determined by a solution of the TE CSIE in Equation (6.29) is not the actual electric current induced upon the cylinder surface. Develop a procedure by which the true current can be determined once (6.29) is solved to produce $J_t(t)$.

P6.13 Verify Equation (6.34) using Equation (6.5) for the eigenvalues of the TE MFIE.

P6.14 Develop an MFIE formulation describing the TE cutoff frequencies for waveguides as discussed in Section 6.8. Provide entries for the method-of-moments matrix obtained using pulse basis functions and delta testing functions.

7

Strip Gratings and Other Two-Dimensional Structures with One-Dimensional Periodicity

Periodic structures find applications as electromagnetic filters, polarizers, artificial dielectrics, and radomes. These scatterers are electrically large and would be intractable if not for the fact that the computational domain can be restricted to one period of the geometry. Integral equation formulations from the preceding chapters can be systematically extended to periodic scatterers through the use of periodic Green's functions. The development of these functions is most easily understood through Fourier transforms, and we begin by reviewing elementary properties of Fourier integrals. In their original form, periodic Green's functions give rise to slowly converging summations that must be evaluated during method-of-moments analysis. Several approaches for accelerating the slowly converging summations will be considered. After illustrating several specific integral equation formulations, this chapter concludes with the development of an outward-looking differential equation formulation for inhomogeneous gratings.

7.1 FOURIER ANALYSIS OF PERIODIC FUNCTIONS [1]

The Fourier transform can be defined in several ways. For the purpose of this chapter, the transform and its inverse will be defined as

$$F\{A(x)\} = \tilde{A}(f) = \int_{-\infty}^{\infty} A(x) e^{-j2\pi f x} \, dx \tag{7.1}$$

and

$$F^{-1}\{\tilde{A}(f)\} = A(x) = \int_{-\infty}^{\infty} \tilde{A}(f)e^{j2\pi fx}\,df \tag{7.2}$$

The convolution of two functions is the operation defined by

$$A(x) * B(x) = \int_{-\infty}^{\infty} A(x')B(x-x')\,dx' \tag{7.3}$$

We have seen that convolutions arise naturally in integral equation formulations (Chapter 5). The convolution theorem associated with the Fourier transform states that

$$A(x) * B(x) = F^{-1}\{\tilde{A}(f)\tilde{B}(f)\} \tag{7.4}$$

An alternative form of the convolution theorem states that

$$A(x)B(x) = F^{-1}\{\tilde{A}(f) * \tilde{B}(f)\} \tag{7.5}$$

Thus, Fourier transforms offer an alternative way of evaluating a convolution integral. This property is especially useful when the functions involved are discrete or periodic.

A periodic function $A_p(x)$ may be constructed from the convolution operation

$$A_p(x) = A(x) * P(x) \tag{7.6}$$

where $A(x)$ is an arbitrary function and $P(x)$ is the "comb" function, defined as

$$P(x) = \sum_{i=-\infty}^{\infty} \delta(x - i\,\Delta X) \tag{7.7}$$

Performing the convolution operation yields the function

$$A_p(x) = \sum_{i=-\infty}^{\infty} A(x - i\,\Delta X) \tag{7.8}$$

which is periodic with period ΔX (Figure 7.1).

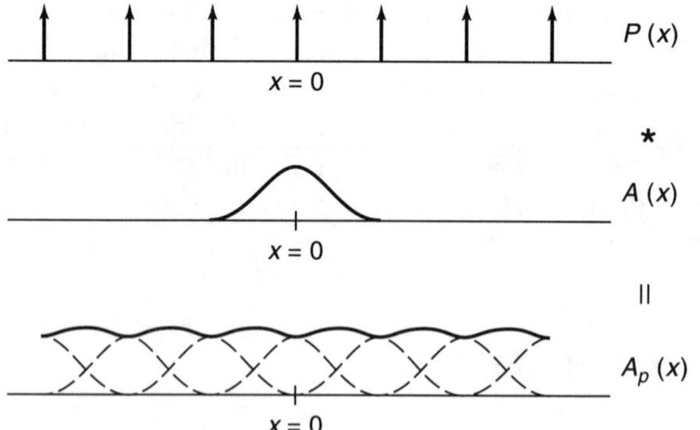

Figure 7.1 Convolution with a comb function.

The Fourier transform of the comb function $P(x)$ is another comb function

$$\tilde{P}(f) = \Delta f \sum_{i=-\infty}^{\infty} \delta(f - i\,\Delta f) \tag{7.9}$$

where

$$\Delta f = \frac{1}{\Delta X} \tag{7.10}$$

Using the convolution theorem, the transform of $A_p(x)$ can be obtained as

$$\tilde{A}_p(f) = \tilde{A}(f)\tilde{P}(f) = \Delta f \sum_{i=-\infty}^{\infty} \tilde{A}(i\Delta f)\delta(f - i\Delta f) \tag{7.11}$$

Multiplication with the comb function $\tilde{P}(f)$ is equivalent to a sampling operation, and the transform $\tilde{A}_p(f)$ is a discrete function. The Fourier transform of any periodic function is discrete, with the period and sampling interval related by Equation (7.10).

A discrete function $A_d(x)$ can be constructed by sampling a function $A(x)$ at regular intervals of Δx (Figure 7.2). This process, which can be expressed as multiplication with the comb function

$$S(x) = \sum_{i=-\infty}^{\infty} \delta(x - i\,\Delta x) \tag{7.12}$$

produces

$$A_d(x) = A(x)S(x) = \sum_{i=-\infty}^{\infty} A(i\,\Delta x)\delta(x - i\,\Delta x) \tag{7.13}$$

The Fourier transform of $S(x)$ is the comb function

$$\tilde{S}(f) = \Delta F \sum_{i=-\infty}^{\infty} \delta(f - i\,\Delta F) \tag{7.14}$$

where ΔF is related to the sampling interval Δx by

$$\Delta F = \frac{1}{\Delta x} \tag{7.15}$$

The convolution theorem of Equation (7.5) can be employed to produce the transform

$$\tilde{A}_d(f) = \tilde{A}(f) * \tilde{S}(f) = \Delta F \sum_{i=-\infty}^{\infty} \tilde{A}(f - i\,\Delta F) \tag{7.16}$$

The function $\tilde{A}(f)$ is periodic with period ΔF. We see that the Fourier transform of a discrete function is periodic, with period and sampling interval related by Equation (7.15).

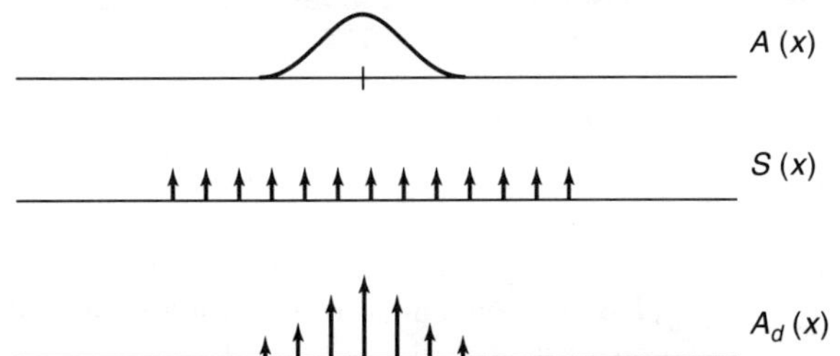

Figure 7.2 Sampling with a comb function.

Ultimately, numerical evaluation may require the discretization of both the original and transform domains. A representation that is both discrete and periodic can be obtained by the expression

$$A_{dp}(x) = [A(x) * P(x)]\, S(x) \tag{7.17}$$

which has the Fourier transform

$$\tilde{A}_{dp}(f) = \left[\tilde{A}(f) * \tilde{S}(f)\right] \tilde{P}(f) \tag{7.18}$$

Thus, the transform of a discrete, periodic function is itself discrete and periodic. For these functions, Fourier transforms reduce to discrete Fourier transforms and their exact implementation can be obtained using the fast Fourier transform (FFT) algorithm.

7.2 FLOQUET HARMONICS [2]

The cross-sectional view of a strip grating is illustrated in Figure 7.3. This two-dimensional geometry is an infinite periodic extension of a "unit cell" containing a single strip, also illustrated in the figure. The precise location of the unit cell is arbitrary, and it may straddle two strips or even contain more than one complete strip, provided that its periodic repetition reproduces the original structure. We wish to determine the electromagnetic response of the strip grating due to a plane-wave excitation of the form

$$E_z^{\text{inc}}(x, y) = E_0 e^{-jk(x\cos\theta + y\sin\theta)} \tag{7.19}$$

This uniform plane wave is a special case of a general excitation satisfying the condition

$$E_z^{\text{inc}}(x+a, y) = E_z^{\text{inc}}(x, y)e^{-jk_x a} \tag{7.20}$$

where a is the dimension of the unit cell of Figure 7.3 and $k_x a$ represents a linear progressive phase shift from one unit cell to the next.

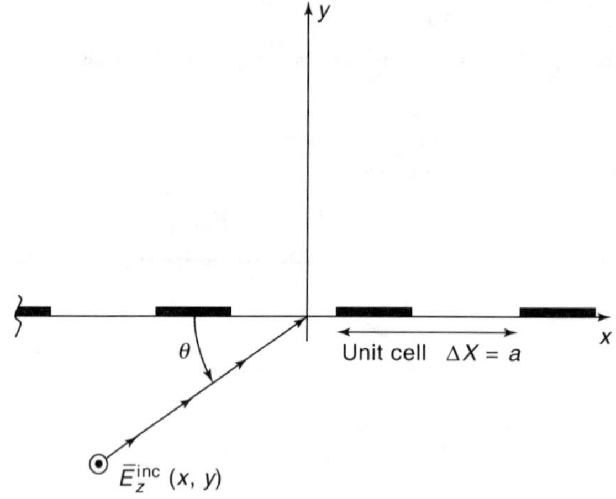

Figure 7.3 Periodic strip grating.

The electromagnetic response to such an excitation can be expressed as a superposition of Floquet harmonics,

$$E_z^s(x, y) = \sum_{n=-\infty}^{\infty} e_n \Psi_n(x, y) \tag{7.21}$$

where

$$\Psi_n(x, y) = e^{-jk_{xn}x} e^{\pm j\sqrt{k^2-k_{xn}^2}\,y} \tag{7.22}$$

and

$$k_{xn} = k_x - \frac{2\pi}{a}n \tag{7.23}$$

When $k^2 > k_{xn}^2$, Floquet harmonics are waves that propagate away from the grating [the plus or minus sign in (7.22) is used when $y < 0$ or $y > 0$, respectively]. When $k_{xn}^2 > k^2$, the branch of the square root is taken to ensure that the harmonics decay in a direction away from the grating.

The Floquet harmonics form a complete orthogonal set over one period of the geometry in x and preserve the desired progressive phase shift imposed by the excitation for all x. Some of the terms in the expansion represent waves that exhibit true propagation (i.e., the visible region $k_{xn}^2 < k^2$) while others represent evanescent waves (the invisible region $k_{xn}^2 > k^2$). Of the infinite number of Floquet harmonics, generally only a few are associated with true propagation. The set of Floquet harmonics is entirely determined by the geometric period and the assumed phase progression along x and thus does not depend on the specific dimensions of the strips in the grating. The coefficients $\{e_n\}$ in Equation (7.21), however, depend on the specific strip geometry.

As an illustration, Figure 7.4 shows the propagating Floquet harmonics for an example involving a wave incident 30° from normal upon a strip grating having a period equal to two wavelengths.

n	k_{xn}	k_{yn}
−1	k	0
0	$k/2$	$k\sqrt{3}/2$
1	0	k
2	$-k/2$	$k\sqrt{3}/2$
3	$-k$	0

$$k_{xn} = k\,(\cos\theta^{\text{inc}} - \tfrac{n}{2})$$

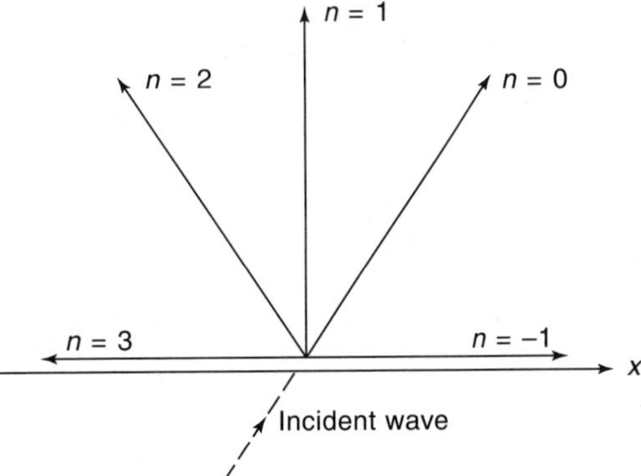

Figure 7.4 Illustration of the propagating Floquet harmonics for a grating with $\theta^{\text{inc}} = 60°$ and $a = 2\lambda_0$.

Because of the phase shift imposed by the incident field, the fields and currents of interest are not strictly periodic functions of x. Instead, they are modulated periodic functions of the general form

$$A_{mp}(x) = \sum_{i=-\infty}^{\infty} A(x - i\,\Delta X)e^{-jk_x i\,\Delta X} \tag{7.24}$$

where $\Delta X = a$. Properties of the Fourier transform dictate that

$$\tilde{A}_{mp}(f) = \tilde{A}(f)\tilde{P}(f + f_0) = \Delta f \sum_{i=-\infty}^{\infty} \tilde{A}(i\,\Delta f - f_0)\delta(f - i\,\Delta f + f_0) \tag{7.25}$$

where Δf is defined in Equation (7.10) and

$$f_0 = \frac{k_x}{2\pi} \tag{7.26}$$

As a consequence of the modulation, the transform of (7.24) is sampled at values of the transform variable f that are shifted from those of Equation (7.11).

Every periodic geometry can be associated with a *spatial lattice* of points in the original domain and a *reciprocal lattice* in the Fourier transform domain [2–4]. The spatial lattice–reciprocal lattice concept is often useful in multidimensional applications, because of the ease with which the propagating harmonics can be identified. In this trivial one-dimensional example, the strip grating imposes a lattice in x defined by the values

$$x = i\,\Delta X \tag{7.27}$$

The reciprocal lattice in the transform domain identifies the discrete spectral frequencies associated with the Floquet harmonics. These frequencies are given by the values

$$f = i\,\Delta f = \frac{i}{\Delta X} \tag{7.28}$$

For the example depicted in Figure 7.4, the spatial lattice is defined by values of x in the set $\{\ldots, -2, 0, 2, 4, 6, \ldots\}$, and the reciprocal lattice is defined by values of f in the collection $\{\ldots, -1, -\frac{1}{2}, 0, \frac{1}{2}, 1, \ldots\}$. The range $(k_x/2\pi - 1) < f < (k_x/2\pi + 1)$ defines the "visible region" of the spectrum containing the propagating Floquet harmonics (the five propagating Floquet harmonics presented in Figure 7.4 span this range).

7.3 TM SCATTERING FROM A CONDUCTING STRIP GRATING: EFIE DISCRETIZED WITH PULSE BASIS FUNCTIONS AND DELTA TESTING FUNCTIONS

Consider a TM plane wave having the form of Equation (7.19) incident on the infinite, periodic strip grating illustrated in Figure 7.3. The surface equivalence principle (Chapter 1) can be used to replace the perfect conducting strips by an equivalent electric current density $J_z(x)$. Due to the phase progression imposed by the incident field, the equivalent currents must satisfy the Floquet condition

$$J_z(x + a) = J_z(x)e^{-jk_x a} \tag{7.29}$$

A conventional EFIE formulation requires the superposition of the fields of each of the currents. If $J_z(x)$ is considered nonzero only when x is located on one of the conducting strips, the EFIE can be written as

$$E_z^{\text{inc}}(x,0) = jk\eta \int_{-\infty}^{\infty} J_z(x') \frac{1}{4j} H_0^{(2)}(k|x-x'|)\,dx' \tag{7.30}$$

where the equality holds only for values of x on the conducting strips. Since the currents from one strip to the next are related by (7.29), the domain of the integral operator can be transformed to a single unit cell to produce the equivalent equation

$$E_z^{\text{inc}}(x,0) = jk\eta \int_{\text{single strip}} J_z(x') G_p(x-x')\,dx' \tag{7.31}$$

where

$$G_p(x) = \frac{1}{4j} \sum_{i=-\infty}^{\infty} H_0^{(2)}(k|x-ia|) e^{-jik_x a} \tag{7.32}$$

The conventional free-space Green's function employed in Equation (7.30) has been replaced by its periodic counterpart. The periodic Green's function G_p differs from the Green's functions employed in previous chapters in one fundamental way: It depends on the progressive phase shift imposed by the incident field. In other words, changes in the direction of the incident plane wave will alter both sides of Equation (7.31).

By restricting the domain of the equation to a single unit cell, the electrical size of the problem has been reduced to manageable proportions. The method-of-moments discretization of the EFIE follows in the usual manner. Suppose that the conducting strip contained in the unit cell is divided into N intervals over which pulse basis functions reside. Dirac delta testing functions can be located in the center of each interval. The resulting matrix has the general form

$$\begin{bmatrix} Z_{11} & Z_{12} & \cdots & Z_{1N} \\ Z_{21} & Z_{22} & \cdots & Z_{2N} \\ \cdot & & & \cdot \\ \cdot & & \ddots & \cdot \\ \cdot & & & \cdot \\ Z_{N1} & Z_{N2} & \cdots & Z_{NN} \end{bmatrix} \begin{bmatrix} j_1 \\ j_2 \\ \cdot \\ \cdot \\ \cdot \\ j_N \end{bmatrix} = \begin{bmatrix} e_1^i \\ e_2^i \\ \cdot \\ \cdot \\ \cdot \\ e_N^i \end{bmatrix} \tag{7.33}$$

where

$$Z_{mn} = jk\eta \int_{\text{cell } n} G_p(x_m - x')\,dx' \tag{7.34}$$

and

$$e_m^i = E_0 e^{-jk_x x_m} \tag{7.35}$$

The task of computing the entries is complicated by the slowly converging nature of the summation for G_p, and the following sections discuss this calculation. Once the matrix is obtained, Equation (7.33) can be solved to produce the coefficients of the pulse expansion for $J_z(x)$.

After the coefficients of the current density on the strips have been determined, the scattered field can be expressed in terms of the Floquet harmonics. To determine the coefficients $\{e_n\}$ in Equation (7.21), observe that since the incident field is continuous across the grating,

$$J_z(x) = H_x^s(x,0^-) - H_x^s(x,0^+) \tag{7.36}$$

By combining (7.21), (7.22), (7.36), and

$$H_x(x, y) = \frac{-1}{j\omega\mu}\frac{\partial E_z}{\partial y} \tag{7.37}$$

the current density can be expressed in terms of the Floquet harmonics as

$$J_z(x) = \frac{-2}{k\eta}\sum_{n=-\infty}^{\infty} e_n\sqrt{k^2 - k_{xn}^2}\, e^{-jk_{xn}x} \tag{7.38}$$

Using the orthogonality of the Floquet harmonics, the coefficients are

$$e_n = \frac{-k\eta}{2a\sqrt{k^2 - k_{xn}^2}}\int_0^a J_z(x)e^{jk_{xn}x}\, dx \tag{7.39}$$

Reflection and transmission coefficients can be defined by the ratio of the total electric field carried by the zero-order Floquet harmonic to the incident electric field (see Section 7.9). Assuming a common reference plane located at $y = 0$, these have the form

$$R_0 = \frac{e_0}{E_0} \tag{7.40}$$

and

$$T_0 = 1 + \frac{e_0}{E_0} \tag{7.41}$$

Figure 7.5 shows a plot of the reflection coefficient for a strip grating having conducting material occupying exactly half the unit cell as a function of the unit cell size. An exact solution is available for this geometry [5, 6], and the numerical results obtained from Equation (7.33) exhibit good agreement with the exact solution.

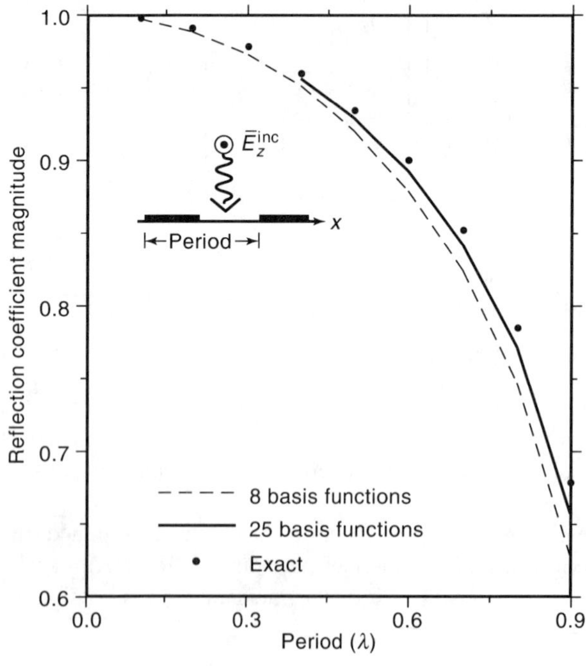

Figure 7.5 Plot of the TM reflection coefficient magnitude produced by the EFIE compared with the exact result.

7.4 SIMPLE ACCELERATION PROCEDURES FOR THE GREEN'S FUNCTION [7, 8]

The magnitude of the ith term in the periodic Green's function summation of Equation (7.32) can be estimated from the asymptotic form of the Hankel function for large arguments,

$$H_0^{(2)}(kx) \approx \sqrt{\frac{2j}{\pi kx}} e^{-jkx} \tag{7.42}$$

The ith term behaves as

$$O\left(\frac{1}{\sqrt{i}}\right) \qquad \text{as } i \to \infty \tag{7.43}$$

Obviously, the summation would diverge if not for the oscillatory behavior of the exponential function. In practice, the direct summation required in Equation (7.32) is prohibitively slow (Table 7.1 illustrates the number of terms required to produce a specified accuracy).

TABLE 7.1 Computed Decimal Places of Accuracy in $G_p(x)$

N	Using Equation (7.32)	Using Equation (7.47)	Using Equations (7.49), (7.50) with $\alpha = 0.1a$
10	1	2	3
30	1	2	4
100	1	2	4
300	1	3	4
1,000	1	3	5
3,000	2	4	5
10,000	2	4	5

Note: As a function of the upper limit N of the summation. The parameters are $a = 0.8\lambda$, $x = 0.5\lambda$, and $\theta^{\text{inc}} = 60$ deg. The correct answer is $G_p(x) \cong 0.05779617 + j\,0.13912689$.

To improve the efficiency of the calculation, we consider an acceleration technique based on the Fourier transform pair from Equations (7.1) and (7.2). This procedure is sometimes known as the *Poisson sum transformation* [7]. Using

$$F\{H_0^{(2)}(k\,|x|)\} = \frac{2}{\beta_y} \tag{7.44}$$

where

$$\beta_y = \begin{cases} \sqrt{k^2 - (2\pi f)^2} & k^2 > (2\pi f)^2 \\ -j\sqrt{(2\pi f)^2 - k^2} & k^2 < (2\pi f)^2 \end{cases} \tag{7.45}$$

the Fourier transform of $G_p(x)$ can be found from (7.25) to be

$$\tilde{G}_p(f) = \frac{1}{2ja} \sum_{i=-\infty}^{\infty} \delta\left(f - \frac{i}{a} + \frac{k_x}{2\pi}\right) \frac{1}{\beta_y} \tag{7.46}$$

Applying the inverse Fourier transform to (7.46) yields the result

$$G_p(x) = \frac{1}{2ja} \sum_{i=-\infty}^{\infty} \left[\frac{e^{j2\pi f x}}{\beta_y} \right]_{f=i/a-k_x/2\pi} \tag{7.47}$$

This summation is an alternative to Equation (7.32) for the direct calculation of the periodic Green's function. Observe that the magnitude of the ith term behaves as

$$0\left(\frac{1}{i}\right) \qquad \text{as } i \to \infty \tag{7.48}$$

Thus, as illustrated in Table 7.1, Equation (7.47) would appear to converge faster than (7.32).

Although preferable to the original summation, Equation (7.47) is not rapidly converging. A more general acceleration procedure can be obtained after introducing a new summation $G_a(x)$ having terms that are asymptotically equal to those of $G_p(x)$ for large i. Then, by combining the Poisson transformation with a decomposition known as *Kummer's transformation* [7], the Green's function can be computed according to

$$G_p(x) = \left[G_p(x) - G_a(x)\right] + F^{-1}\{\tilde{G}_a(f)\} \tag{7.49}$$

where the terms in G_p and G_a cancel for large i. For this procedure to be effective, the Fourier transform of G_a should be a rapidly converging summation.

The transform of $G_p(x)$ is a slowly decaying function as $f \to \infty$ because of the singularity in the Hankel function at $x = 0$. A more rapidly converging function can be obtained by eliminating the singularity. A possible choice for $G_a(x)$ is the function

$$G_a(x) = \frac{1}{4j} \sum_{i=-\infty}^{\infty} H_0^{(2)}\left(k\sqrt{(x-ia)^2 + \alpha^2}\right) e^{-jik_x a} \tag{7.50}$$

where the positive parameter α has been introduced in order to eliminate the singularity. (Physically, the addition of this parameter is equivalent to moving the observation point away from the $y = 0$ plane.) The effect of α on the ith term becomes negligible as $i \to \infty$, and G_a and G_p are identical in the asymptotic limit. The Fourier transform of G_a is given by

$$\tilde{G}_a(f) = \frac{1}{2ja} \sum_{i=-\infty}^{\infty} \delta\left(f - \frac{i}{a} + \frac{k_x}{2\pi}\right) \frac{e^{-j\alpha\beta_y}}{\beta_y} \tag{7.51}$$

The inverse Fourier transform produces

$$G_a(x) = \frac{1}{2ja} \sum_{i=-\infty}^{\infty} \left[\frac{e^{j2\pi f x} e^{-j\alpha\beta_y}}{\beta_y} \right]_{f=i/a-k_x/2\pi} \tag{7.52}$$

Since $-j\beta_y$ is a negative real-valued quantity for $(2\pi f)^2 > k^2$, a nonzero α introduces exponential decay into the summation (7.52), making it rapidly convergent. It can also be shown (Prob. P7.7) that for sufficiently large i the ith term of $G_p(x) - G_a(x)$ is proportional to α^2 and is of asymptotic order $O(i^{-3/2})$. Equation (7.49) can be an effective way to compute $G_p(x)$, provided that the parameter α is selected in order to balance the workload required to compute the summation $G_p - G_a$ and the inverse Fourier transform of $\tilde{G}_a$ for a desired accuracy level.

In summary, three ways of computing the periodic Green's function have been considered. The original summation of Equation (7.32) contains terms of $O(i^{-1/2})$ as $i \to \infty$ and thus is prohibitively slow to converge. The summation obtained from the inverse Fourier transform in (7.47) contains terms of $O(i^{-1})$, while Equations (7.49)–(7.52) involve a summation with terms of $O(i^{-3/2})$. The latter approach offers a substantial improvement on the original summation, as illustrated by a representative example in Table 7.1.

Our ultimate objective is not to evaluate the periodic Green's function, but instead to calculate the matrix entries defined in (7.34). One possibility is to employ numerical quadrature using one of the preceding techniques to obtain G_p at needed values of the integration variable. However, a more efficient approach can be developed that does not require (even as an intermediate step) the computation of the periodic Green's function. Observe that the matrix entries can be written as a convolution,

$$Z_{mn} = jk\eta B(x) * G_p(x)\big|_{x=x_m-x_n} \tag{7.53}$$

where $B(x)$ denotes a pulse basis function defined as

$$B(x) = \begin{cases} 1 & -\frac{1}{2}b < x < \frac{1}{2}b \\ 0 & \text{otherwise} \end{cases} \tag{7.54}$$

Using the convolution theorem

$$B(x) * G_p(x) = F^{-1}\{\tilde{B}(f)\tilde{G}_p(f)\} \tag{7.55}$$

and the Fourier transform

$$\tilde{B}(f) = \frac{\sin(\pi f b)}{\pi f} \tag{7.56}$$

Z_{mn} can be directly obtained in the form of an inverse Fourier transformation,

$$Z_{mn} = \frac{k\eta}{2a} \sum_{i=-\infty}^{\infty} \left[\frac{\sin(\pi f b)}{\pi f} \frac{e^{j2\pi f(x_m-x_n)}}{\beta_y} \right]_{f=i/a-k_x/2\pi} \tag{7.57}$$

The convolution integral in (7.53) is transformed into a multiplication in the Fourier transform domain, and the inverse transform produces a summation for Z_{mn}. Furthermore, the terms appearing in (7.57) behave asymptotically for large i as

$$O\left(\frac{1}{i^2}\right) \tag{7.58}$$

Therefore, by exploiting the convolutional nature of the matrix entries, we eliminate the need to compute G_p directly and at the same time obtain Z_{mn} in the form of a summation that converges faster than the inverse transform summation for the periodic Green's function. Consequently, Equation (7.57) will be a far more efficient alternative than numerical quadrature using the previous expressions for G_p. If other basis and testing functions are employed, with an inner product such as (5.56), analogous formulas may be developed for the general matrix elements

$$Z_{mn} = jk\eta T^{\dagger}(-x) * B(x) * G_p(x)\big|_{x=x_m-x_n} \tag{7.59}$$

The use of smoother testing functions can also help accelerate the convergence of the summation, by contributing additional factors of $1/f$ in the transform domain (Prob. P7.11).

7.5 ALTERNATE ACCELERATION PROCEDURES

The preceding section identified several methods for obtaining the periodic Green's function G_p or the matrix entries Z_{mn}. The most efficient approach of those considered is the summation for Z_{mn} in Equation (7.57), which involves terms of $O(i^{-2})$ as $i \to \infty$.

By combining the convolution of (7.55) with the Poisson and Kummer transformations defined in (7.49), the summation for Z_{mn} may be cast into the slightly more general form [8]

$$\begin{aligned} Z_{mn} = jk\eta \int_{x_n-b/2}^{x_n+b/2} \left[G_p(x_m - x') - G_a(x_m - x')\right] dx' \\ + jk\eta F^{-1} \left\{ \frac{\sin(\pi f b)}{\pi f} \tilde{G}_a(f) \right\}_{x=x_m-x_n} \end{aligned} \tag{7.60}$$

where Equation (7.50) can be used for G_a, with the parameter α selected to optimize the computational effort required to evaluate the two terms in (7.60). When $\alpha = 0$, (7.60) reduces to (7.57). When $m = n$, the singularity in the Green's function G_p can be treated by the singularity extraction procedure discussed in Chapter 2 or the generalized Gaussian quadrature algorithm of Appendix A.

In general, (7.60) divides the computational effort between the spatial domain and the Fourier transform (spectral) domain. It is noteworthy that the transform domain summation (the second term) exhibits exponential convergence for $\alpha > 0$. However, the convergence in the spatial domain term $G_p(x) - G_a(x)$ is at best algebraic (Prob. P7.7). Since the spatial domain integral must be evaluated by numerical quadrature, the division of computational effort in (7.60) places a heavy burden on the spatial domain term and a light burden on the spectral domain summation. In fact, the choice $\alpha = 0$ appears optimal [8].

A more balanced division of effort can be obtained by the alternative expression [9]

$$\begin{aligned} Z_{mn} = jk\eta \int_{x_n-b/2}^{x_n+b/2} G_b(x_m - x')\, dx' \\ + jk\eta F^{-1} \left\{ \frac{\sin(\pi f b)}{\pi f} \left[\tilde{G}_p(f) - \tilde{G}_b(f)\right] \right\}_{x=x_m-x_n} \end{aligned} \tag{7.61}$$

where

$$G_b(x) = \frac{1}{4j} \sum_{i=-\infty}^{\infty} H_0^{(2)}(-j\gamma\,|x - ia|) e^{-jik_x a} = \frac{1}{2\pi} \sum_{i=-\infty}^{\infty} K_0(\gamma\,|x - ia|) e^{-jik_x a} \tag{7.62}$$

and

$$\tilde{G}_b(f) = \frac{1}{2a} \sum_{i=-\infty}^{\infty} \delta\left(f - \frac{i}{a} + \frac{k_x}{2\pi}\right) \frac{1}{\sqrt{(2\pi f)^2 + \gamma^2}} \tag{7.63}$$

Observe that G_b is obtained by replacing the real-valued wavenumber k in $G_p(x)$ with the imaginary quantity $-j\gamma$. This substitution eliminates the singularity occurring at $(2\pi f)^2 = k^2$ in the Fourier transform $\tilde{G}_p(f)$. Since

$$K_0(u) \approx \sqrt{\frac{\pi}{2u}} e^{-u} \quad \text{as } u \to \infty \tag{7.64}$$

the summation for G_b exhibits exponential convergence. The difference summation in the second term of (7.61) involves terms that decrease as $O(i^{-4})$ as $i \to \infty$ (Prob. P7.7). Since the integrand necessary for numerical quadrature in (7.61) is exponentially convergent, it should be easier to evaluate than the summation for $G_p(x)-G_a(x)$. Consequently, Equation (7.61) should provide better numerical efficiency than (7.60).

In progressing from the original form of the periodic Green's function in (7.32) to the expression for Z_{mn} in (7.61), the convergence rates of the summations have improved from $O(i^{-1/2})$ to $O(i^{-4})$. However, the convergence of all the preceding expressions is ultimately algebraic. There are alternate approaches that provide exponential convergence, and we turn our attention to two such procedures.

The first approach involves the transformation of G_p into an integral with an exponentially decaying integrand, as suggested by Veysoglu [10]. The development can be based on the geometric series relation

$$\sum_{i=1}^{\infty} e^{-jit}e^{-iu} = \frac{e^{-u}e^{-jt}}{1-e^{-u}e^{-jt}} \tag{7.65}$$

By multiplying both sides of (7.65) with a function $f(u)$ and integrating, we obtain

$$\sum_{i=1}^{\infty} e^{-jit} \int_0^{\infty} f(u)e^{-iu}\,du = e^{-jt}\int_0^{\infty} \frac{f(u)e^{-u}}{1-e^{-u}e^{-jt}}\,du \tag{7.66}$$

The integral on the left-hand side of (7.66) can be recognized as the Laplace transform

$$L\{f\} = \tilde{f}(i) = \int_0^{\infty} f(u)e^{-iu}\,du \tag{7.67}$$

which enables us to write the general relation as

$$\sum_{i=1}^{\infty} e^{-jit} L\{f\} = e^{-jt}\int_0^{\infty} \frac{f(u)e^{-u}}{1-e^{-u}e^{-jt}}\,du \tag{7.68}$$

A Laplace transform pair suitable for use in the present application is [11, 3.364]

$$L\left\{ j\frac{2}{\pi}\frac{1}{\sqrt{u^2+j2u}} \right\} = e^{ji}H_0^{(2)}(|i|) \tag{7.69}$$

which can be generalized using the scale and shift properties of the Laplace transform to obtain

$$L\left\{ j\frac{2}{\pi}e^{jkx}\frac{e^{-xu/a}}{\sqrt{u^2+j2kau}} \right\} = e^{jika}H_0^{(2)}(k|x-ia|) \tag{7.70}$$

Substituting (7.70) into (7.68) and assigning the value

$$t = (k+k_x)a \tag{7.71}$$

produce the relation

$$\begin{aligned} &\frac{1}{4j}\sum_{i=1}^{\infty} H_0^{(2)}(k|x-ia|)e^{-jik_x a} \\ &= \frac{K_1 e^{jkx}}{2\pi}\int_0^{\infty} \frac{e^{-(1+x/a)u}}{\sqrt{u^2+j2kau}(1-K_1e^{-u})}\,du \end{aligned} \tag{7.72}$$

where

$$K_1 = e^{-j(k+k_x)a} \tag{7.73}$$

The integrand in (7.72) is singular at $u = 0$, and a change of variables,

$$u = kaw^2 \tag{7.74}$$

can be used to recast (7.72) into the form [10]

$$\begin{aligned} \frac{1}{4j}\sum_{i=1}^{\infty} H_0^{(2)}(k|x - ia|)e^{-jik_x a} \\ = \frac{K_1 e^{jkx}}{\pi}\int_0^{\infty} \frac{e^{-k(a+x)w^2}}{\sqrt{w^2 + j2(1 - K_1 e^{-kaw^2})}}\,dw \end{aligned} \tag{7.75}$$

This summation index in (7.75) runs from 1 to ∞ and therefore is only part of the complete expression for G_p. However, a similar expression can be obtained for negative values of the index and combined with (7.75) to produce

$$\begin{aligned} \frac{1}{4j}\sum_{i=-\infty}^{\infty} H_0^{(2)}(k|x - ia|)e^{-jik_x a} = {} & \frac{1}{4j}H_0^{(2)}(k|x|) \\ & + \frac{K_1 e^{jkx}}{\pi}\int_0^{\infty} \frac{e^{-k(a+x)w^2}}{\sqrt{w^2 + j2(1 - K_1 e^{-kaw^2})}}\,dw \\ & + \frac{K_2 e^{-jkx}}{\pi}\int_0^{\infty} \frac{e^{-k(a-x)w^2}}{\sqrt{w^2 + j2(1 - K_2 e^{-kaw^2})}}\,dw \end{aligned} \tag{7.76}$$

where

$$K_2 = e^{-j(k-k_x)a} \tag{7.77}$$

In order to obtain the matrix entries Z_{mn} defined in (7.34) for pulse basis functions and delta testing functions, the convolution integral in x can be carried out explicitly to produce

$$\begin{aligned} Z_{mn} = {} & \frac{k\eta}{4}\int_{x_n-\Delta/2}^{x_n+\Delta/2} H_0^{(2)}(k|x_m - x'|)\,dx' \\ & + \frac{j2\eta K_1 e^{jk(x_m-x_n)}}{\pi}\int_0^{\infty} \frac{e^{-k(a+x_m-x_n)w^2}\sinh[(j - w^2)k\Delta/2]}{\sqrt{w^2 + j2}(j - w^2)(1 - K_1 e^{-kaw^2})}\,dw \\ & + \frac{j2\eta K_2 e^{-jk(x_m-x_n)}}{\pi}\int_0^{\infty} \frac{e^{-k(a-x_m+x_n)w^2}\sinh[(j - w^2)k\Delta/2]}{\sqrt{w^2 + j2}(j - w^2)(1 - K_2 e^{-kaw^2})}\,dw \end{aligned} \tag{7.78}$$

Although the domain of integration in (7.76) and (7.78) extends to infinity, the exponential decay provided by the integrand facilitates an efficient computation by numerical quadrature.

A second approach that provides exponential convergence in the periodic Green's function was developed for the three-dimensional case by Jordan, Richter, and Sheng [12], based on earlier work by Ewald [13]. A two-dimensional adaptation of this idea [14] can be developed using the error function

$$\operatorname{erf}(u) = \frac{2}{\sqrt{\pi}}\int_0^{u} e^{-t^2}\,dt \tag{7.79}$$

and the complementary error function

$$\operatorname{erfc}(u) = 1 - \operatorname{erf}(u) = \frac{2}{\sqrt{\pi}} \int_u^\infty e^{-t^2}\, dt \tag{7.80}$$

The periodic Green's function can be written as

$$G_p(x) = G_{p1}(x) + G_{p2}(x) \tag{7.81}$$

where

$$G_{p1}(x) = \frac{1}{2a} \sum_{i=-\infty}^{\infty} \left[\operatorname{erfc}\left(\frac{j\beta_y}{E}\right) \frac{e^{j2\pi f x}}{j\beta_y} \right]_{f=i/a-k_x/2\pi} \tag{7.82}$$

$$G_{p2}(x) = \frac{1}{2a} \sum_{i=-\infty}^{\infty} \left[\operatorname{erf}\left(\frac{j\beta_y}{E}\right) \frac{e^{j2\pi f x}}{j\beta_y} \right]_{f=i/a-k_x/2\pi} \tag{7.83}$$

and where β_y is defined in (7.45). The parameter E is arbitrary, to be determined below. Since the complementary error function has the asymptotic behavior

$$\operatorname{erfc}(u) \approx O\left(\frac{e^{-u^2}}{\pi u}\right) \qquad \text{as } u \to \infty \tag{7.84}$$

the summation for G_{p1} exhibits exponential convergence. The summation for G_{p2} in (7.83) contains terms that only decrease as $O(i^{-1})$ as $i \to \infty$ and requires additional manipulation to accelerate the convergence. Using (7.79) and a change of variables, we obtain

$$\begin{aligned} G_{p2}(x) &= \frac{1}{a\sqrt{\pi}} \sum_{i=-\infty}^{\infty} \left[e^{j2\pi f x} \int_0^{1/E} e^{\beta_y^2 s^2}\, ds \right]_{f=i/a-k_x/2\pi} \\ &= \frac{1}{\sqrt{\pi}} \int_0^{1/E} \left(\frac{1}{a} \sum_{i=-\infty}^{\infty} \left[e^{j2\pi f x} e^{\beta_y^2 s^2} \right]_{f=i/a-k_x/2\pi} \right) ds \end{aligned} \tag{7.85}$$

The summation in (7.85) can be recognized as the inverse Fourier transform of a modulated periodic function $\tilde{A}_{mp}(f)$ of the form defined in (7.25), where the generating function is

$$\tilde{A}(f) = e^{\beta_y^2 s^2} = e^{-(2\pi s)^2 f^2} e^{k^2 s^2} \tag{7.86}$$

Consequently, using (7.24) and the inverse Fourier transform

$$A(x) = \frac{1}{2\sqrt{\pi}} \frac{e^{-x^2/(4s^2)} e^{k^2 s^2}}{s} \tag{7.87}$$

the expression in (7.85) can be rewritten as

$$\begin{aligned} G_{p2}(x) &= \frac{1}{2\pi} \int_0^{1/E} \left(\sum_{i=-\infty}^{\infty} \frac{e^{-(x-ia)^2/(4s^2)} e^{k^2 s^2}}{s} e^{-jik_x a} \right) ds \\ &= \frac{1}{2\pi} \sum_{i=-\infty}^{\infty} e^{-jik_x a} \int_0^{1/E} \frac{e^{-(x-ia)^2/(4s^2)} e^{k^2 s^2}}{s}\, ds \end{aligned} \tag{7.88}$$

Finally, after a change of variables

$$u = \frac{1}{4s^2} \tag{7.89}$$

we obtain the spatial domain summation

$$G_{p2}(x) = \frac{1}{4\pi} \sum_{i=-\infty}^{\infty} e^{-jik_x a} \int_{E^2/4}^{\infty} \frac{e^{-(x-ia)^2 u} e^{k^2/(4u)}}{u} du \tag{7.90}$$

The summation in (7.90) exhibits exponential convergence as $i \to \infty$ but is not in a convenient form for computation. The exponential function can be replaced by a Taylor series, converting the integrand in (7.90) according to

$$\frac{e^{-(x-ia)^2 u} e^{k^2/(4u)}}{u} = \sum_{q=0}^{\infty} \left(\frac{k}{2}\right)^{2q} \frac{1}{q!} \frac{e^{-(x-ia)^2 u}}{u^{q+1}} \tag{7.91}$$

The result can be integrated term by term to produce the expression

$$G_{p2}(x) = \frac{1}{4\pi} \sum_{i=-\infty}^{\infty} e^{-jik_x a} \sum_{q=0}^{\infty} \left(\frac{k}{E}\right)^{2q} \frac{1}{q!} E_{q+1} \left\{\tfrac{1}{4}(x-ia)^2 E^2\right\} \tag{7.92}$$

where E_{q+1} is the exponential integral

$$E_{q+1}\{z\} = \int_1^{\infty} \frac{e^{-zt}}{t^{q+1}} dt \tag{7.93}$$

The exponential integral exhibits the asymptotic behavior

$$E_{q+1}\{z\} \approx \frac{e^{-z}}{z} \qquad \text{as } z \to \infty \tag{7.94}$$

and can be evaluated using efficient polynomial approximations [14].

By combining (7.81), (7.82), and (7.92), we obtain an expression for the periodic Green's function in terms of two exponentially convergent summations. In common with the preceding acceleration techniques, the burden has been divided between the spectral domain summation in (7.82) and the spatial domain summation in (7.92). The parameter E determines the balance between these summations, with a large E placing the burden on (7.82) and a small E placing the burden on (7.92). In fact, it can be shown that the choice

$$E > \frac{3\sqrt{d}}{|x|} \tag{7.95}$$

reduces the contribution of $G_{p2}(x)$ to less than 10^{-d} for the range of parameters likely to be of interest [14]. In other words, the entire burden can be shifted to G_{p1} in most cases. Since this is a spectral domain summation, the convolution with the basis function required to compute Z_{mn} can also be incorporated, to produce

$$Z_{mn} \cong \frac{k\eta}{2a} \sum_{i=-\infty}^{\infty} \left[\operatorname{erfc}\left(\frac{j\beta_y}{E}\right) \frac{\sin(\pi f b)}{\pi f} \frac{e^{j2\pi f(x_m - x_n)}}{\beta_y} \right]_{f=i/a-k_x/2\pi} \tag{7.96}$$

for E large enough to satisfy (7.95). The treatment of the case when $x \to 0$ (the diagonal entries Z_{mm}) requires a combination of the spatial and spectral summations [14].

Figure 7.6 illustrates the performance of the exponentially convergent methods suggested in (7.78) and (7.96), compared with the algebraic summation from (7.57), for a representative example. Clearly, for high accuracy the exponentially convergent summations prove far superior to the algebraic methods considered in Section 7.4.

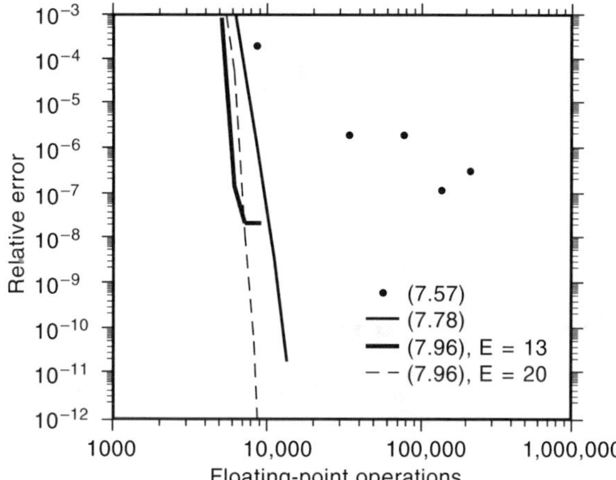

Figure 7.6 Comparison of the required computation (floating-point operations) needed to produce a given accuracy in the matrix entries Z_{mn} for a geometry with $\theta^{\text{inc}} = 60°$, $a = 0.8\lambda_0$, $b = 0.05\lambda_0$, and $x_m - x_n = 0.5\lambda_0$.

7.6 BLIND ANGLES

There are values of the period and imposed phase progression for which the periodic Green's function $G_p(x)$ diverges. This can occur if insufficient oscillation is present between successive terms of the summation (7.32); it is more readily observed in the alternative forms of $G_p(x)$ based on inverse Fourier transform summations. [The denominators of Equations (7.47), (7.52), (7.57), and (7.96) vanish if $k^2 = (2\pi f)^2$, while the expression in (7.76) has a singularity if $K_1 = 1$ or $K_2 = 1$.] For a given strip grating, propagation directions of the incident field that result in a divergent $G_p(x)$ are known as the *blind angles* of the strip grating, named after a similar effect in infinite phased-array antennas. Blind angles occur when $(k \pm k_x)a = 2n\pi$ for integer n.

Physically, blind angles arise whenever a Floquet harmonic makes a transition from the invisible to the visible region of the spectrum (from evanescent to propagating character). For example, Figure 7.4 illustrates two harmonics in such a state. Although the periodic Green's function is divergent, the current density remains finite and well behaved at the blind angles (the coefficient of a Floquet harmonic in transition equals zero). Because of a redistribution of the power carried by the various Floquet harmonics at such a transition, irregularities in the reflection and transmission coefficients (*Wood's anomalies*) can be observed as a function of frequency.

For numerical calculations, blind angles can usually be anticipated and avoided. Simple interpolation can be employed in order to determine the behavior of the current density at a blind angle.

7.7 TE SCATTERING FROM A CONDUCTING STRIP GRATING BACKED BY A DIELECTRIC SLAB: EFIE FORMULATION

Generally, strip gratings require a physical support structure that often takes the form of a dielectric slab (Figure 7.7). For analysis purposes, the effect of the slab can be taken

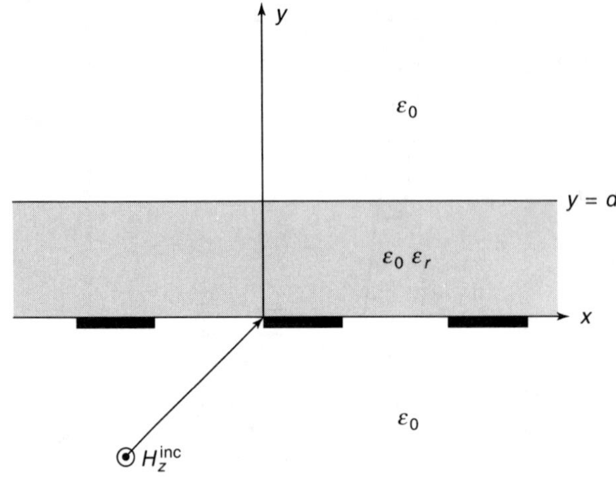

Figure 7.7 Strip grating supported by a dielectric slab illuminated by a TE field.

into account by an appropriate modification of the free-space Green's function. Consider an incident TE plane wave, which only excites the x-component of the current density on the conducting strips. Assuming that the current density explicitly vanishes off of the conductors and that the equation holds only for points on the conducting strips, the EFIE can be written as

$$E_x^{\text{inc}}(x, 0) = \frac{-1}{j\omega\varepsilon_0}\left(\frac{\partial^2}{\partial x^2} + k^2\right)\int_{-\infty}^{\infty} J_x(x')G_d(x - x')\,dx' \tag{7.97}$$

where G_d represents the Green's function for an individual line source of x-directed electric current radiating on the surface of a dielectric slab (Figure 7.8). In order to derive the Green's function, we observe that the line source can be expressed as an inverse Fourier transform

$$J_x(x, y) = \delta(x)\delta(y) = \delta(y)\int_{-\infty}^{\infty} e^{j2\pi f x}\,df \tag{7.98}$$

which can be thought of as the superposition of current sheets of the form

$$J_x(x) = e^{j2\pi f x} \tag{7.99}$$

The response of each current sheet can be obtained by the solution of a one-dimensional wave equation or, equivalently, through the use of a transmission line analogy as illustrated in Figure 7.9. The parameters employed in the transmission line analogy are the TE-wave impedances

$$Z_0 = -\frac{E_x}{H_z} = \frac{\eta k_y}{k} \tag{7.100}$$

$$Z_d = \frac{\eta_d k_{dy}}{k_d} \tag{7.101}$$

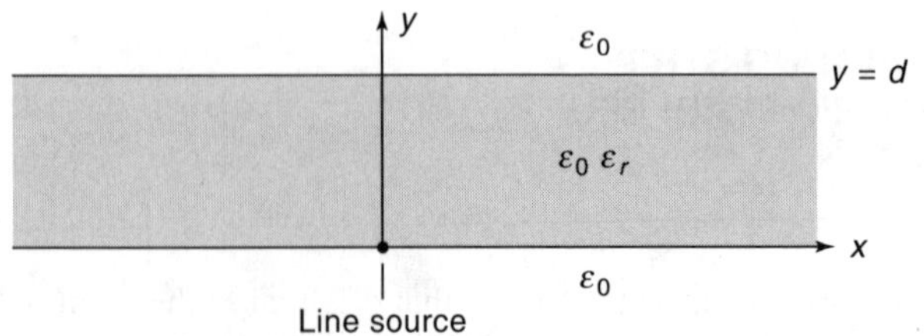

Figure 7.8 Line source radiating on the surface of a dielectric slab.

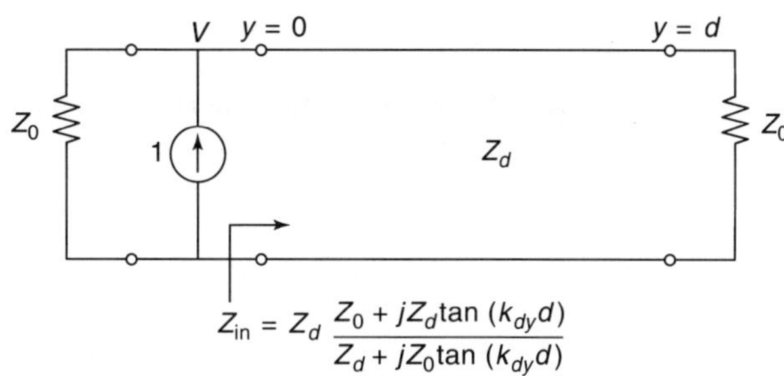

$$V = \frac{Z_0 Z_{in}}{Z_0 + Z_{in}} = \frac{Z_d\left(1 + j\frac{Z_d}{Z_0}\tan\,[k_{dy}d]\right)}{2\frac{Z_d}{Z_0} + j\left(1 + \left[\frac{Z_d}{Z_0}\right]^2\right)\tan\,[k_{dy}d]}$$

Figure 7.9 Equivalent transmission line model for a line source radiating on the surface of a dielectric slab.

where η and η_d represent the intrinsic impedance of the background medium and the dielectric slab, respectively, and k and k_d denote the wavenumber of the background medium and the dielectric slab, respectively,

$$k_y = \sqrt{k^2 - (2\pi f)^2} \tag{7.102}$$

and

$$k_{dy} = \sqrt{k_d^2 - (2\pi f)^2} \tag{7.103}$$

After solving the transmission line circuit of Figure 7.9, we obtain the electric field due to a single current sheet, which has the form

$$E_x^{\text{sheet}}(x, 0) = \frac{-k_{dy}}{2\omega\varepsilon_d} \frac{1 + j(k_{dy}/\varepsilon_r k_y)\tan(k_{dy}d)}{k_{dy}/\varepsilon_r k_y + j(1/2)\left[1 + \left(k_{dy}/\varepsilon_r k_y\right)^2\right]\tan(k_{dy}d)} e^{j2\pi f x} \tag{7.104}$$

The electric field due to a unit line source is the superposition of the fields produced by the current sheets, which can be expressed as

$$E_x(x, 0) = \int_{-\infty}^{\infty} E_x^{\text{sheet}}(x, 0)\, df \tag{7.105}$$

Since the electric field due to the line source can also be found from

$$E_x(x, 0) = \frac{1}{j\omega\varepsilon_0}\left(\frac{\partial^2}{\partial x^2} + k^2\right) G_d(x) \tag{7.106}$$

we obtain the Green's function G_d as

$$G_d(x) = \int_{-\infty}^{\infty} \frac{1}{2j\varepsilon_r k_{dy}} \frac{1 + j(k_{dy}/\varepsilon_r k_y)\tan(k_{dy}d)}{k_{dy}/\varepsilon_r k_y + j(1/2)\left[1 + \left(k_{dy}/\varepsilon_r k_y\right)^2\right]\tan(k_{dy}d)} e^{j2\pi f x}\, df \tag{7.107}$$

Because of the integral, this Green's function is cumbersome for direct evaluation. However, the Fourier transform is the algebraic expression

$$\tilde{G}_d(f) = \frac{1}{2j\varepsilon_r k_{dy}} \frac{1 + j(k_{dy}/\varepsilon_r k_y)\tan(k_{dy}d)}{k_{dy}/\varepsilon_r k_y + j(1/2)\left[1 + \left(k_{dy}/\varepsilon_r k_y\right)^2\right]\tan(k_{dy}d)} \tag{7.108}$$

Consequently, we seek a formulation where the Green's function need only be evaluated in the Fourier transform domain.

For a strip grating illuminated by a TE incident plane wave, the current density satisfies the Floquet condition

$$J_x(x+a) = J_x(x)e^{-jk_x a} \tag{7.109}$$

Therefore, the EFIE of (7.97) can be expressed as an integral over a single period of the structure, namely,

$$E_x^{\text{inc}}(x,0) = -\frac{1}{j\omega\varepsilon_0}\left(\frac{\partial^2}{\partial x^2}+k^2\right)\int_{\text{single strip}} J_x(x')G_{dp}(x-x')\,dx' \tag{7.110}$$

where we now employ the periodic Green's function

$$G_{dp}(x) = \sum_{i=-\infty}^{\infty} G_d(x-ia)e^{-jik_x a} \tag{7.111}$$

To effect a solution, we divide the strip into equal-sized cells and introduce basis and testing functions in order to approximate the current density by

$$J_x(x) \cong \sum_{n=1}^{N} j_n B(x-x_n) \tag{7.112}$$

and discretize the EFIE into the matrix equation

$$\int T^{\dagger}(x-x_m)E_x^{\text{inc}}(x,0)\,dx = \sum_{n=1}^{N} j_n Z_{mn} \qquad m=1,2,\ldots,N \tag{7.113}$$

where

$$Z_{mn} = -\frac{1}{j\omega\varepsilon_0}\left(\frac{\partial^2}{\partial x^2}+k^2\right)\int T^{\dagger}(x-x_m)\int B(x'-x_n)G_{dp}(x-x')\,dx'\,dx \tag{7.114}$$

and appropriate limits of integration are implied. Following the approach outlined in Section 7.4, the matrix entries can be expressed in general form as the inverse Fourier transformation

$$\begin{aligned} Z_{mn} &= -\frac{1}{j\omega\varepsilon_0}\left(\frac{\partial^2}{\partial x^2}+k^2\right)T^R(x)*B(x)*G_{dp}(x)|_{x=x_m-x_n} \\ &= F^{-1}\left\{\frac{(2\pi f)^2-k^2}{j\omega\varepsilon_0}\tilde{T}^R(f)\tilde{B}(f)\tilde{G}_{dp}(f)\right\}_{x=x_m-x_n} \\ &= \sum_{i=-\infty}^{\infty}\left[\frac{(2\pi f)^2-k^2}{j\omega\varepsilon_0 a}e^{j2\pi f(x_m-x_n)}\tilde{T}^R(f)\tilde{B}(f)\tilde{G}_d(f)\right]_{f=i/a-k_x/2\pi} \end{aligned} \tag{7.115}$$

where $T^R(x) = T^{\dagger}(-x)$.

In Chapter 2, subsectional triangle basis functions and pulse testing functions were employed to discretize the EFIE for TE scattering from a single strip or cylinder. It is reasonable to suppose that these same functions will provide a robust discretization for this TE strip grating example. The Fourier transforms of a triangle and pulse function (Figure 2.8) are

$$F\{t(x;-b,0,b)\} = \tilde{B}(f) = \frac{\sin^2(\pi bf)}{b(\pi f)^2} \tag{7.116}$$

and

$$F\left\{p\left(x; -\frac{1}{2}b, \frac{1}{2}b\right)\right\} = \tilde{T}^R(f) = \frac{\sin(\pi b f)}{\pi f} \tag{7.117}$$

We observe from these equations and (7.108) that for large values of the index i, the magnitude of the ith term in (7.115) behaves as

$$O\left(\frac{1}{i^2}\right) \tag{7.118}$$

Therefore, the use of these basis and testing functions will provide enough decay as $i \to \infty$ to ensure convergence of the summation. If desired, a faster scheme for the computation of the matrix elements may be developed based on the ideas discussed in Section 7.5.

7.8 APERTURE FORMULATION FOR TM SCATTERING FROM A CONDUCTING STRIP GRATING

Previous sections have considered formulations where the primary unknown is the equivalent electric surface current located on perfectly conducting material. For periodic geometries with large conducting regions and small apertures between them, it may be more efficient to treat the tangential electric field in the aperture (or the equivalent magnetic surface current in the aperture) as the primary unknown. Consider the case of TM scattering from a conducting grating such as illustrated in Figure 7.3. Based on the aperture formulation discussed in Section 1.10, we can write an integral equation for the strip grating problem as

$$H_x^{\text{inc}}(x, 0) = -\frac{2}{j\omega\mu_0}\left(\frac{\partial^2}{\partial x^2} + k^2\right)\int_{\text{single aperture}} K_x(x')G_p(x - x')\,dx' \tag{7.119}$$

where G_p is the periodic Green's function defined in (7.32) and K_x is the equivalent magnetic current in the aperture, defined so that $K_x(x) = E_z(x, 0)$. The incident magnetic field is that field produced by the original source in the absence of the grating.

The method-of-moments discretization of (7.119) follows in the usual manner and produces a matrix equation $\mathbf{Zk} = \mathbf{h}$, where the general form for the entries is given by

$$\begin{aligned} Z_{mn} &= -\frac{2}{j\omega\mu_0}\left(\frac{\partial^2}{\partial x^2} + k^2\right) T^R(x) * B(x) * G_p(x)\big|_{x=x_m-x_n} \\ &= -2F^{-1}\left\{\frac{k^2 - (2\pi f)^2}{j\omega\mu_0}\tilde{T}^R(f)\tilde{B}(f)\tilde{G}_p(f)\right\}_{x=x_m-x_n} \end{aligned} \tag{7.120}$$

and

$$h_m = \int_{\text{aperture}} T(x_m - x)H_x^{\text{inc}}(x)\,dx \tag{7.121}$$

where $T^R(x) = T^\dagger(-x)$. As in the preceding section, it is reasonable to suppose that subsectional triangle basis functions and pulse testing functions are sufficient to discretize the integral operator. The acceleration procedures discussed in Sections 7.4 and 7.5 can be incorporated as needed.

After the coefficients of the magnetic current density in the apertures have been determined, the scattered field can be expressed in terms of the Floquet harmonics. The approach is the electromagnetic dual to that considered in Prob. P7.6 for the TE polarization.

7.9 SCATTERING MATRIX ANALYSIS OF CASCADED PERIODIC SURFACES [4, 15, 16]

Because strip gratings allow complete transmission at certain frequencies and complete reflection at other frequencies, they are a simple type of *frequency-selective surface*. A wide variety of filter characteristics can be obtained by cascading multiple layers of frequency-selective surfaces. To efficiently analyze multilayered structures, each periodic surface can be modeled as a multiport network with each port representing one Floquet harmonic of a given polarization. The coefficients of each propagating Floquet harmonic determine the "far-field" reflection and transmission characteristics of periodic surfaces. A complete near-zone model will generally require a combination of propagating and evanescent harmonics. Generalized scattering and transmission matrices representing each individual surface can subsequently be employed as a building block in the electromagnetic model of a multilayered structure.

The definition of a scattering matrix is presented in the context of a two-port network in Figure 7.10. For TM excitation, the scattering matrix arising from the strip grating problem is a 2-by-2 block matrix of the form

$$\mathbf{S} = \begin{bmatrix} S_{11} & S_{12} \\ S_{21} & S_{22} \end{bmatrix} \tag{7.122}$$

where each of the blocks comprising (7.122) are infinite-dimensional matrices relating the coefficients of the scattered Floquet harmonics to those of the incident harmonics. These blocks take the form

$$S_{11} = \begin{bmatrix} s_{11}^{11} & s_{12}^{11} & \cdots \\ s_{21}^{11} & s_{22}^{11} & \\ s_{31}^{11} & s_{32}^{11} & \\ \vdots & & \end{bmatrix} \tag{7.123}$$

In practice, these matrices are truncated to finite dimension. The entries of the generalized

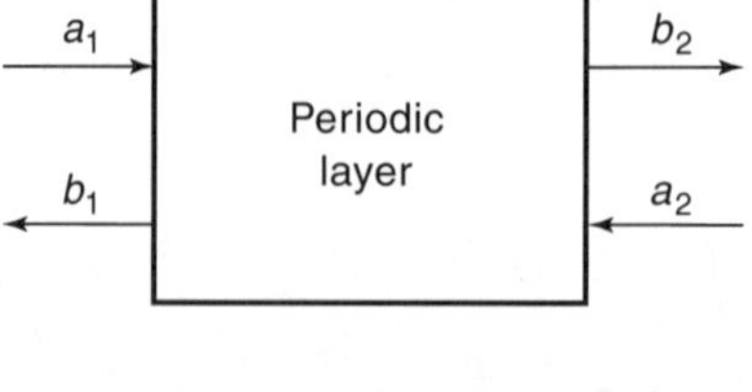

$$\begin{bmatrix} b_1 \\ b_2 \end{bmatrix} = \begin{bmatrix} S_{11} & S_{12} \\ S_{21} & S_{22} \end{bmatrix} \begin{bmatrix} a_1 \\ a_2 \end{bmatrix}$$

Figure 7.10 Two-port scattering parameters.

Figure 7.11 Location of reference planes for the definition of S-parameters.

scattering matrices depend on the location of reference planes as indicated in Figure 7.11.

There are a variety of ways to define the scattering parameters, and we use the definitions employed in microwave circuit analysis [17]. The (m, n) entry in the scattering matrix of Equation (7.122) is proportional to the square root of the ratio of the power carried by the mth reflected harmonic to the power carried by the nth incident harmonic. In general, this is a complex-valued quantity having magnitude

$$\mathrm{mag}(s_{mn}^{11}) = \sqrt{\frac{\int_0^a \left|\bar{E}_m^s \times \bar{H}_m^{s\dagger} \cdot \hat{y}\right|_{y=-d} dx}{\int_0^a \left|\bar{E}_n^{\mathrm{inc}} \times \bar{H}_n^{\mathrm{inc}\dagger} \cdot \hat{y}\right|_{y=-d} dx}} \tag{7.124}$$

and phase equal to the difference of the phase of the corresponding electric field harmonic at the output and input reference planes, that is,

$$\mathrm{phase}(s_{mn}^{11}) = \mathrm{phase}\{E_{zm}^s(0, -d)\} - \mathrm{phase}\{E_{zn}^{\mathrm{inc}}(0, -d)\} \tag{7.125}$$

Using the form of the Floquet harmonics specified in Equation (7.22), the magnitude of the (m, n) entry can be simplified to

$$\mathrm{mag}(s_{mn}^{11}) = \frac{|e_m|}{|e_n^{\mathrm{inc}}|}\sqrt{\left|\frac{\sqrt{k^2 - k_{xm}^2}}{\sqrt{k^2 - k_{xn}^2}}\right|} \tag{7.126}$$

The entries of the S_{21} matrix are defined in a similar manner, only in terms of the total transmitted fields instead of the scattered reflected fields. In other words,

$$\mathrm{mag}(s_{mn}^{21}) = \sqrt{\frac{\int_0^a \left|\bar{E}_m^{\mathrm{tot}} \times \bar{H}_m^{\mathrm{tot}\dagger} \cdot \hat{y}\right|_{y=c} dx}{\int_0^a \left|\bar{E}_n^{\mathrm{inc}} \times \bar{H}_n^{\mathrm{inc}\dagger} \cdot \hat{y}\right|_{y=-d} dx}} \tag{7.127}$$

and

$$\mathrm{phase}(s_{mn}^{21}) = \mathrm{phase}\{E_{zm}^{\mathrm{tot}}(0, c)\} - \mathrm{phase}\{E_{zn}^{\mathrm{inc}}(0, -d)\} \tag{7.128}$$

The S_{12} and S_{22} matrices are defined in an analogous manner.

For a single surface illuminated by a plane wave, generalized reflection and transmission coefficients are sometimes employed as an alternative to the scattering matrix description. The magnitudes of these expressions simplify to

$$|R_n| = \mathrm{mag}(s_{n0}^{11}) = \frac{|e_n|}{|E_0|}\sqrt{\left|\frac{\sqrt{k^2 - k_{xn}^2}}{\sqrt{k^2 - k_x^2}}\right|} \tag{7.129}$$

and

$$|T_n| = \text{mag}(s_{n0}^{21}) = \left| \delta_0^n + \frac{e_n}{E_0} \sqrt{\left| \frac{\sqrt{k^2 - k_{xn}^2}}{\sqrt{k^2 - k_x^2}} \right|} \right| \tag{7.130}$$

and the phases are defined in accordance with Equations (7.125) and (7.128).

For cascading several layers, the transmission matrix representation

$$\begin{bmatrix} b_2 \\ a_2 \end{bmatrix} = \begin{bmatrix} T_{11} & T_{12} \\ T_{21} & T_{22} \end{bmatrix} \begin{bmatrix} a_1 \\ b_1 \end{bmatrix} \tag{7.131}$$

(see Figure 7.10 for a definition of the a's and b's) is more convenient than the scattering matrix description, and can be found from the elements of the scattering matrix according to

$$T_{11} = S_{21} - S_{22} S_{12}^{-1} S_{11} \tag{7.132}$$

$$T_{12} = S_{22} S_{12}^{-1} \tag{7.133}$$

$$T_{21} = -S_{12}^{-1} S_{11} \tag{7.134}$$

$$T_{22} = S_{12}^{-1} \tag{7.135}$$

As an example, suppose we cascade three gratings represented by transmission matrices T_1, T_2, and T_3, where the excitation passes through grating 1 before gratings 2 and 3. The composite transmission matrix is given by

$$T = T_3 T_2 T_1 \tag{7.136}$$

Of course, the cascading process requires that a common set of Floquet harmonics be employed within the scattering or transmission matrix description of each grating.

The process of cascading may become numerically unstable if higher order (evanescent) harmonics are cascaded over large distances. As a consequence of the exponential decay these terms experience, the transmission matrices may become extremely ill-conditioned as the terminal planes are separated [4]. Under these conditions, it may be advantageous to construct composite scattering matrices by an alternative process discussed in the literature [4, 18, 19].

7.10 TM SCATTERING FROM A HALF-SPACE HAVING A GENERAL PERIODIC SURFACE: EFIE DISCRETIZED WITH PULSE BASIS FUNCTIONS AND DELTA TESTING FUNCTIONS [20]

Figure 7.12 depicts two regions separated by an interface whose location in y varies periodically with x. We desire to determine the reflected and transmitted fields in response to a TM plane wave

$$E_z^{\text{inc}}(x, y) = E_0 e^{-jk_1(x \cos\theta + y \sin\theta)} \tag{7.137}$$

incident from the lower region. Huygens' equivalence principle can be used to develop an equivalent model of each region in terms of electric and magnetic sources $J_z(t)$ and $K_t(t)$ on

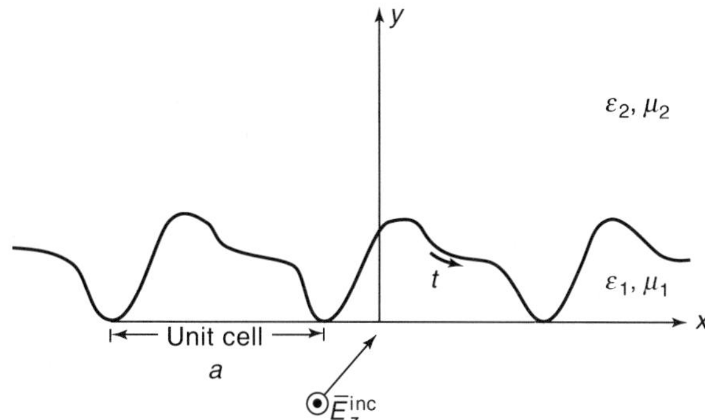

Figure 7.12 Periodic interface between two semi-infinite regions. A TM wave is incident from region 1.

the surface, where t is a parametric variable defined in Figure 7.12. Following the general procedure discussed in Section 1.9, coupled EFIEs can be constructed of the form

$$E_z^{\text{inc}}(t) = K_t(t) + jk_1\eta_1 A_z^1(t) + \hat{z} \cdot \nabla \times \bar{F}_t^1|_{S^+} \tag{7.138}$$

$$0 = -K_t(t) + jk_2\eta_2 A_z^2(t) + \hat{z} \cdot \nabla \times \bar{F}_t^2|_{S^-} \tag{7.139}$$

where

$$A_z^i(t) = \int_{-\infty}^{\infty} J_z(t') \frac{1}{4j} H_0^{(2)}(k_i R)\, dt' \tag{7.140}$$

$$\bar{F}_t^i(t) = \int_{-\infty}^{\infty} \hat{t}(t') K_t(t') \frac{1}{4j} H_0^{(2)}(k_i R)\, dt' \tag{7.141}$$

and

$$R = \sqrt{[x(t) - x(t')]^2 + [y(t) - y(t')]^2} \tag{7.142}$$

The superscript on A_z and F_t denotes the medium. For generality, region 2 may be lossy. Equation (7.138) is enforced an infinitesimal distance outside (i.e., the region 1 side) of the interface, while (7.139) is enforced an infinitesimal distance inside the interface.

The plane-wave excitation imposes a progressive phase shift with phase constant

$$k_x = k_1 \cos\theta \tag{7.143}$$

and the equivalent currents must satisfy the Floquet conditions

$$J_z(x + a, y) = J_z(x, y)e^{-jk_x a} \tag{7.144}$$

$$K_t(x + a, y) = K_t(x, y)e^{-jk_x a} \tag{7.145}$$

As a consequence of these relationships, the domain of the integral operators can be transformed to a single period so that

$$A_z^i(t) = \int_{\text{one period}} J_z(t') G_p^i(x(t) - x(t'), y(t) - y(t'))\, dt' \tag{7.146}$$

$$\bar{F}_t^i(t) = \int_{\text{one period}} \hat{t}(t') K_t(t') G_p^i(x(t) - x(t'), y(t) - y(t'))\, dt' \tag{7.147}$$

where the periodic Green's function is given by

$$G_p^i(x, y) = \frac{1}{4j} \sum_{q=-\infty}^{\infty} H_0^{(2)} \left[k_i \sqrt{(x - qa)^2 + y^2} \right] e^{-jqk_x a} \tag{7.148}$$

Numerical solutions can be obtained in the context of the method of moments. Suppose that one period of the surface is divided into N cells, as depicted in Figure 7.13. The equivalent currents can be represented by the expansions

$$J_z(t) \cong \sum_{n=1}^{N} j_n B_n(t) \tag{7.149}$$

$$K_t(t) \cong \sum_{n=1}^{N} h_n B_n(t) \tag{7.150}$$

where B_n denotes a pulse basis function having constant support over cell n. The coupled equations can be enforced at the centers of each interval (Dirac delta testing functions) to produce a matrix equation

$$\begin{bmatrix} \mathbf{A} & \mathbf{B} \\ \mathbf{C} & \mathbf{D} \end{bmatrix} \begin{bmatrix} j_1 \\ j_2 \\ \cdot \\ \cdot \\ \cdot \\ j_N \\ h_1 \\ h_2 \\ \cdot \\ \cdot \\ \cdot \\ h_N \end{bmatrix} = \begin{bmatrix} e_1 \\ e_2 \\ \cdot \\ \cdot \\ \cdot \\ e_N \\ 0 \\ 0 \\ \cdot \\ \cdot \\ \cdot \\ 0 \end{bmatrix} \tag{7.151}$$

where

$$A_{mn} = jk_1\eta_1 \int_{\text{cell } n} G_p^1(x(t_m) - x(t'), y(t_m) - y(t'))\, dt' \tag{7.152}$$

$$B_{mn} = -\int_{\text{cell } n} \hat{z} \cdot \hat{t}_n \times \nabla G_p^1|_{x(t_m)-x(t'),y(t_m)-y(t')}\, dt' \qquad m \neq n \tag{7.153}$$

$$B_{mm} \cong \frac{1}{2} - \int_{\text{cell } m} \hat{z} \cdot \hat{t}_n \times \nabla G_p^{1m}\, dt' \tag{7.154}$$

$$C_{mn} = jk_2\eta_2 \int_{\text{cell } n} G_p^2\, dt' \tag{7.155}$$

$$D_{mn} = -\int_{\text{cell } n} \hat{z} \cdot \hat{t}_n \times \nabla G_p^2\, dt' \qquad m \neq n \tag{7.156}$$

$$D_{mm} \cong -\frac{1}{2} - \int_{\text{cell } m} \hat{z} \cdot \hat{t}_n \times \nabla G_p^{2m}\, dt' \tag{7.157}$$

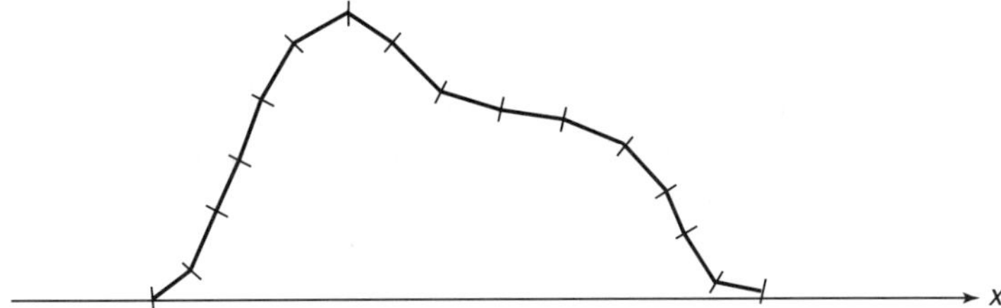

Figure 7.13 Flat-cell model of one period of the surface in Figure 7.12.

and

$$e_m = E_0 e^{-jk_1[x(t_m)\cos\theta + y(t_m)\sin\theta]} \tag{7.158}$$

In the preceding equations, t_n represents the center of cell n, and G_p^{im} denotes G_p^i with the $q = 0$ term omitted from the summation of Equation (7.148). The arguments of each integrand are the same as those in (7.152) and (7.153). Equations (7.154) and (7.157) represent the situation in which the source and observation cells coincide; these expressions were obtained by a limiting procedure that is exact if the cells are flat and a good approximation if the cells have a large radius of curvature (see Chapter 2).

Because the periodic Green's function of (7.148) is a slowly converging summation, the efficient evaluation of the matrix entries requires some sort of acceleration procedure. Consider the evaluation of A_{mn}. In Section 7.4, a similar summation was accelerated using the convolution theorem. Although the matrix entry in (7.152) is also a convolution between the pulse basis function and the Green's function, the domain of the pulse function is not necessarily located parallel to the x-axis in this case. For a general orientation, the basis function will be a Dirac delta function in x. Since the Fourier transform of a delta function is constant, the convolution will not provide a means of accelerating the summation.

The Fourier transform of the periodic Green's function has the form

$$\tilde{G}_p^i(f, y) = \frac{1}{2ja} \sum_{q=-\infty}^{\infty} \delta\left(f - \frac{q}{a} + \frac{k_x}{2\pi}\right) \frac{e^{-j\beta_{yi}|y|}}{\beta_{yi}} \tag{7.159}$$

where, for the lossless region 1,

$$\beta_{y1} = \begin{cases} \sqrt{k_1^2 - (2\pi f)^2} & k_1^2 > (2\pi f)^2 \\ -j\sqrt{(2\pi f)^2 - k_1^2} & k_1^2 < (2\pi f)^2 \end{cases} \tag{7.160}$$

Note that the function in (7.159) exhibits exponential decay as $f \to \infty$ for nonzero y, suggesting in that case that the periodic Green's function can be directly evaluated in terms of the inverse Fourier transform summation. However, a more general acceleration procedure is required if the separation in y vanishes or becomes very small, since the inverse Fourier transform summation might then be slow to converge. Any of the acceleration procedures from Sections 7.4 and 7.5 can be adapted to (7.152)–(7.157). One approach has been developed using algebraic acceleration, and the reader is referred to [20] for detailed expressions for the matrix entries. It is likely that a more computationally efficient scheme could be contrived using one of the exponential acceleration techniques (Section 7.5).

The TM-to-z formulation can be adapted to the TE polarization using the ideas of electromagnetic duality and can be obtained from the preceding expressions by exchanging the electric and magnetic fields, currents, and medium parameters according to the duality relationships presented in Section 1.5.

We will illustrate the accuracy of the procedure for the TE case. Figure 7.14 depicts a corrugated surface separating a region of free space from a region with relative permittivity $\varepsilon_r = 2.5$ and relative permeability $\mu_r = 1$. The triangular corrugation has a period equal to 1.0 λ, where λ denotes the free-space wavelength. A uniform plane wave is incident upon the interface at an angle 30° from normal. Figure 7.15 compares the magnitudes of the Floquet harmonics as a function of the corrugation depth, as produced by this approach (as implemented in reference [20]) and a "coupled-wave" formulation described in reference [21]. Good agreement is observed between the two formulations, neither of which is exact. The accuracy of the integral equation approach is primarily limited by the finite number of basis functions; the approach of reference [21] is primarily limited by the finite number of Floquet harmonics used within a field expansion throughout the corrugated region. For the data shown in Figure 7.15, a density of approximately 20 cells/λ was used within the integral equation approach and a total of nine Floquet harmonics were used with the approach of reference [21].

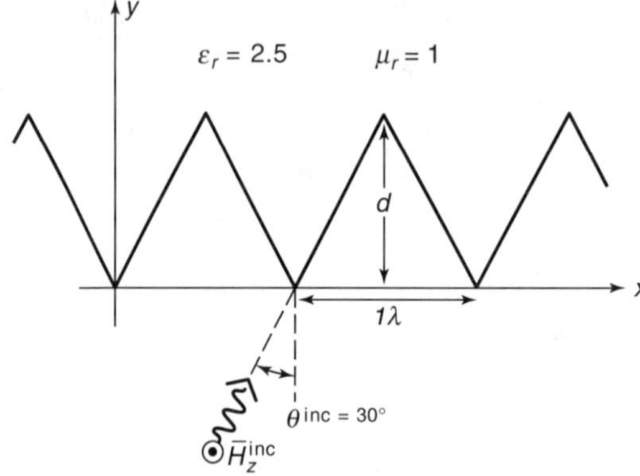

Figure 7.14 Triangular corrugated interface between two regions. Region 1 is free space, region 2 has $\epsilon_r = 2.5$. A TE wave is incident from region 1.

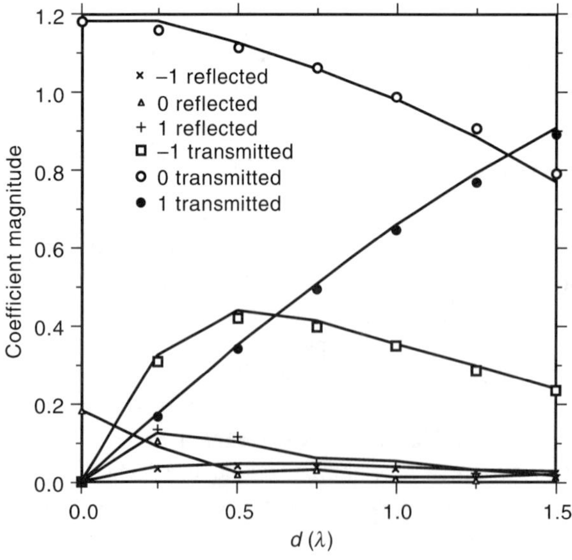

Figure 7.15 Comparison of Floquet coefficients produced by the integral equation formulation (solid line) and the coupled-mode formulation of reference [21] (markers) applied to the geometry depicted in Figure 7.14. After [20]. ©1992 S. Hirzel Verlag GmbH & Co.

Although the preceding discussion was confined to a continuous periodic surface, Equations (7.152)–(7.154) also provide the exterior matrix entries necessary to treat an array of individual two-dimensional homogeneous dielectric scatterers exhibiting one-dimensional periodicity. The above approach is easily adapted to the discrete situation by incorporating analogous integral expressions for the interior interactions, which do not require a periodic Green's function (Prob. P7.17).

7.11 TM SCATTERING FROM AN INHOMOGENEOUS GRATING: OUTWARD-LOOKING FORMULATION WITH AN INTEGRAL EQUATION RBC [22]

To treat highly heterogeneous geometries, such as the periodic structure illustrated in Figure 7.16, the integral formulations described in this chapter could be extended to volume equations. However, every matrix entry arising within a surface or volume integral formulation for periodic geometries involves a time-consuming summation over the periodic Green's function. As an alternative approach for heterogeneous structures, we turn our attention to an outward-looking differential equation formulation similar to those introduced in Chapter 3. In a differential equation formulation, the periodicity is incorporated through boundary conditions. Consequently, the matrix entries associated with interior interactions are identical to those of a general nonperiodic structure (Chapter 3). Furthermore, because the shape of the unit cell is arbitrary, it is convenient to employ planar radiation boundaries along the direction of periodicity. Planar boundaries simplify the calculations associated with an RBC, since the convolution idea described in Section 7.4 can be used to accelerate the periodic Green's function summation within the RBC.

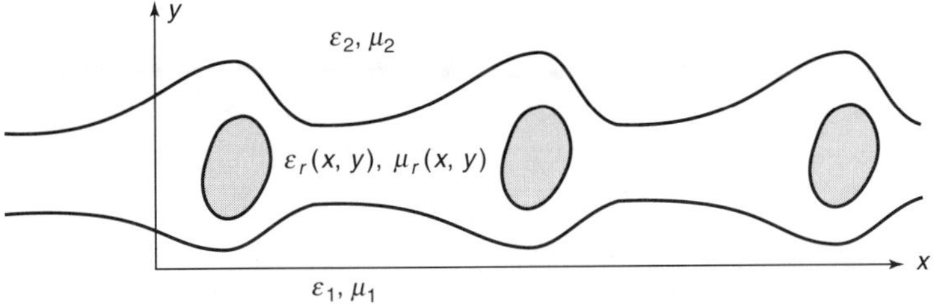

Figure 7.16 Inhomogeneous structure with periodicity along the x direction.

Suppose that the periodic structure of Figure 7.16 is illuminated by a TM-to-z wave incident from the lower ($y < 0$) region. The computational domain may be truncated at the unit cell depicted in Figure 7.17. The interior region Γ associated with a single unit cell is bounded by flat surfaces $\partial\Gamma_1$ and $\partial\Gamma_2$ parallel to the x-axis and arbitrarily shaped surfaces $\partial\Gamma_3$ and $\partial\Gamma_4$. These boundaries are translations of a common surface along the direction of periodicity, so that a TM incident wave

$$E_z^{\text{inc}}(x, y) = E_0 e^{-jk_1(x\cos\theta + y\sin\theta)} \tag{7.161}$$

imposes a Floquet condition

$$E_z(x, y)\,|_{\partial\Gamma_4} = E_z(x, y)\,|_{\partial\Gamma_3}\, e^{-jk_x a} \tag{7.162}$$

where $k_x = k_1 \cos\theta$, constraining the fields on $\partial\Gamma_3$ and $\partial\Gamma_4$.

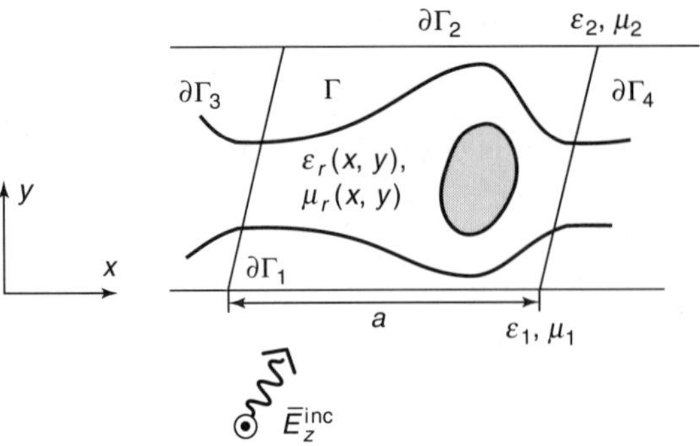

Figure 7.17 Unit cell defined by boundaries $\partial\Gamma_1$ and $\partial\Gamma_2$ parallel to the x-axis and boundaries $\partial\Gamma_3$ and $\partial\Gamma_4$ on which a Floquet condition must hold.

The fields throughout the interior region Γ must satisfy the scalar Helmholtz equation, which can be written in weak form as

$$\iint_{\Gamma} \frac{1}{\mu_r} \nabla T \cdot \nabla E_z - k^2 \varepsilon_r T E_z = \int_{\partial\Gamma_1+\partial\Gamma_2+\partial\Gamma_3+\partial\Gamma_4+\partial\Gamma_c} \frac{1}{\mu_r} T \frac{\partial E_z}{\partial n} \tag{7.163}$$

where $\partial\Gamma_c$ denotes the boundaries of any imbedded perfectly conducting regions, if present. The integral over $\partial\Gamma_c$ vanishes, as previously explained in Chapter 3. The integrals over boundaries $\partial\Gamma_3$ and $\partial\Gamma_4$ also vanish, provided that the periodicity is explicitly incorporated into the global finite-element system during the matrix construction phase of the analysis. We illustrate this process below. The boundary integrals over $\partial\Gamma_1$ and $\partial\Gamma_2$ provide a means for coupling the incident field to the Helmholtz equation and also for imposing an RBC. Specifically, we consider an RBC obtained from the EFIE.

Assuming the existence of a normal vector $\hat{n}$ pointing out of the region Γ and a tangent vector $\hat{t}$ defined so that $\hat{n} \times \hat{t} = \hat{z}$, equivalent sources can be introduced on $\partial\Gamma_1$ and $\partial\Gamma_2$ according to

$$J_z = \hat{z} \cdot \hat{n} \times \bar{H} = H_t = \frac{1}{jk\eta} \frac{\partial E_z}{\partial n} \tag{7.164}$$

$$K_t = \hat{t} \cdot \bar{E} \times \hat{n} = E_z \tag{7.165}$$

where k and η are the wavenumber and intrinsic impedance, respectively, of the medium along each boundary. Therefore, the boundary integrals over $\partial\Gamma_1$ and $\partial\Gamma_2$ can be replaced by

$$j\omega\mu_0 \int_{\partial\Gamma_1} T(t) J_z(t)\, dt \quad \text{and} \quad j\omega\mu_0 \int_{\partial\Gamma_2} T(t) J_z(t)\, dt \tag{7.166}$$

Suppose that the computational domain Γ is divided into triangular cells, over which the field is represented by a piecewise-linear basis (the "pyramid" functions of Chapter 3)

$$E_z(x, y) \cong \sum_{n=1}^{N} e_n B_n(x, y) \tag{7.167}$$

In addition, let the electric current density $J_z(t)$ along boundaries $\partial\Gamma_1$ and $\partial\Gamma_2$ be replaced by the piecewise-linear (subsectional triangle) expansion

$$J_z(t) \cong \sum_{n=1}^{M} j_n B_n(t) \tag{7.168}$$

For convenience, we choose testing functions $T_m(x, y) = B_m(x, y)$ and discretize Equation (7.163) into matrix form to obtain

$$\begin{bmatrix} \mathbf{I} & \mathbf{I}_1^T & \mathbf{I}_2^T & \mathbf{0} & \mathbf{0} \\ \mathbf{I}_1 & \mathbf{E}_1 & \mathbf{0} & \mathbf{J}_1 & \mathbf{0} \\ \mathbf{I}_2 & \mathbf{0} & \mathbf{E}_2 & \mathbf{0} & \mathbf{J}_2 \end{bmatrix} \begin{bmatrix} \mathbf{e}^{\text{int}} \\ \mathbf{e}_1 \\ \mathbf{e}_2 \\ \mathbf{e}_3 \\ \mathbf{j}_1 \\ \mathbf{j}_2 \end{bmatrix} = \begin{bmatrix} \mathbf{0} \\ \mathbf{0} \\ \mathbf{0} \end{bmatrix} \tag{7.169}$$

where the submatrix $\mathbf{I}$ contains all of the interactions between basis and testing functions centered at interior nodes of the mesh, submatrices $\mathbf{I}_1$ and $\mathbf{I}_2$ contain all of the interactions between interior basis functions and testing functions that are located on one of the radiation boundaries, and $\mathbf{E}_1$ and $\mathbf{E}_2$ contain interactions between basis and testing functions that are both located on radiation boundaries. Each of these matrices has entries of the general form

$$I_{mn} = \iint_\Gamma \frac{1}{\mu_r} \nabla T_m \cdot B_n - k^2 \varepsilon_r T_m B_n \tag{7.170}$$

Submatrices $\mathbf{J}_1$ and $\mathbf{J}_2$ have entries

$$J_{1,mn} = -j\omega\mu_0 \int_{\partial\Gamma_1} T_m(t) B_n(t)\, dt \tag{7.171}$$

and

$$J_{2,mn} = -j\omega\mu_0 \int_{\partial\Gamma_2} T_m(t) B_n(t)\, dt \tag{7.172}$$

Since we have not yet incorporated the RBC or the incident field, (7.169) is underdetermined and has no driving function.

For the TM polarization, it is convenient to obtain an RBC from the EFIE

$$E_z^{\text{inc}}(t) = K_t(t) + jk\eta A_z(t) + \hat{z} \cdot \nabla \times \bar{F}_t|_{\partial\Gamma^+} \tag{7.173}$$

which can be used to relate the incident field and equivalent sources J_z and K_t along either $\partial\Gamma_1$ or $\partial\Gamma_2$. Repeating the procedure discussed in Section 7.10, we incorporate the Floquet condition in order to adapt (7.173) to the periodic geometry illustrated in Figures 7.16 and 7.17. Consequently, on either boundary,

$$A_z(t) = \int_{\text{one period}} J_z(t') G_p(x(t) - x(t'), 0)\, dt' \tag{7.174}$$

and

$$\bar{F}_t(t) = \int_{\text{one period}} \hat{t}(t') K_t(t') G_p(x(t) - x(t'), 0)\, dt' \tag{7.175}$$

where $\hat{t} = \hat{x}$ for points along $\partial\Gamma_1$, $\hat{t} = -\hat{x}$ for points along $\partial\Gamma_2$, and

$$G_p(x, y) = \frac{1}{4j} \sum_{q=-\infty}^{\infty} H_0^{(2)} \left[k\sqrt{(x - qa)^2 + y^2} \right] e^{-jqk_x a} \tag{7.176}$$

Using the expansion (7.168), the representation

$$K_t(t) \cong \sum_{n=1}^{M} e_n B_n(t) \tag{7.177}$$

(where B_n again denotes a subsectional triangle basis function), and point matching at the nodes of the mesh located along either radiation boundary, the EFIE in (7.173) can be discretized into the form

$$\mathbf{Aj} + \mathbf{Be} = \mathbf{e}^{\text{inc}} \tag{7.178}$$

In (7.178), the entries of $\mathbf{A}$ are given by

$$A_{mn} = jk\eta \int_{\partial\Gamma} B_n(t') G_p(x(t_m) - x(t'), 0)\, dt' \tag{7.179}$$

where k and η assume the values k_1 and η_1 for points along $\partial\Gamma_1$ and k_2 and η_2 for points along $\partial\Gamma_2$. For these basis functions, the entries of $\mathbf{B}$ are

$$B_{mn} = \begin{cases} \frac{1}{2} & m = n \\ 0 & \text{otherwise} \end{cases} \tag{7.180}$$

which follows from the usual limiting procedure (Chapter 2). Assuming that the excitation is incident from the lower ($y < 0$) region in Figure 7.16,

$$e_m^{\text{inc}} = \begin{cases} E_0 e^{-jk_1(x(t_m)\cos\theta + y(t_m)\sin\theta)} & \text{for points along } \partial\Gamma_1 \\ 0 & \text{for points along } \partial\Gamma_2 \end{cases} \tag{7.181}$$

Consequently, for the boundary $\partial\Gamma_1$ a numerical RBC can be expressed as

$$\mathbf{j}_1 = \mathbf{A}_1^{-1}\mathbf{e}_1^{\text{inc}} - \mathbf{A}_1^{-1}\mathbf{B}_1\mathbf{e}_1 \tag{7.182}$$

Along $\partial\Gamma_2$, we obtain the RBC

$$\mathbf{j}_2 = -\mathbf{A}_2^{-1}\mathbf{B}_2\mathbf{e}_2 \tag{7.183}$$

These constraints can be substituted into (7.169) to produce

$$\begin{bmatrix} \mathbf{I} & \mathbf{I}_1^T & \mathbf{I}_2^T \\ \mathbf{I}_1 & \mathbf{E}_1 - \mathbf{J}_1\mathbf{A}_1^{-1}\mathbf{B}_1 & \mathbf{0} \\ \mathbf{I}_2 & \mathbf{0} & \mathbf{E}_2 - \mathbf{J}_2\mathbf{A}_2^{-1}\mathbf{B}_2 \end{bmatrix} \begin{bmatrix} \mathbf{e}^{\text{int}} \\ \mathbf{e}_1 \\ \mathbf{e}_2 \end{bmatrix} = \begin{bmatrix} \mathbf{0} \\ -\mathbf{J}_1\mathbf{A}_1^{-1}\mathbf{e}_1^{\text{inc}} \\ \mathbf{0} \end{bmatrix} \tag{7.184}$$

Observe that blocks $\mathbf{I}$, $\mathbf{I}_1$, $\mathbf{I}_2$, $\mathbf{I}_1^T$, and $\mathbf{I}_2^T$ are sparse, while the remaining two blocks along the main diagonal are fully populated. Equation (7.184) is a properly determined system that can be solved for the coefficients $\{e_n\}$ of the interior basis functions. This system of equations constitutes an outward-looking formulation, as defined in Chapter 3, since the primary unknowns represent the interior fields.

We have described the formulation in general terms and now consider some of the details associated with evaluating the entries in (7.184). Consider the boundaries $\partial\Gamma_3$ and $\partial\Gamma_4$ as shown in Figure 7.17. The nodes located along $\partial\Gamma_4$ represent "dummy" values, since the fields on $\partial\Gamma_4$ are prescribed by the Floquet condition (7.162). Thus, there will be no unknowns (and no matrix rows or columns) associated with these nodes. Furthermore, in order to enforce the desired periodicity, each node on $\partial\Gamma_4$ must be the periodic translation of a corresponding node along $\partial\Gamma_3$. The matrix entries arising from the scalar Helmholtz equation can be constructed in the usual manner using the "element matrix" concept developed in Chapter 3. As a consequence of the Floquet condition, however, any element matrix entry involving a testing function located on $\partial\Gamma_4$ must be scaled by $e^{jk_x a}$ and added to the matrix row associated with the corresponding node along $\partial\Gamma_3$. Similarly, any element matrix entry involving a basis function on $\partial\Gamma_4$ must be scaled by $e^{-jk_x a}$ and added to the matrix column associated with the corresponding node along $\partial\Gamma_3$. These operations force the fields to "wrap around" the unit cell in the desired periodic manner.

The entries of $\mathbf{J}_1$ and $\mathbf{J}_2$ also "wrap around" the unit cell. Consider the special case of N uniformly spaces nodes along $\partial\Gamma_1$, where $N+1$ denotes the index of the dummy node at the $\partial\Gamma_4$ corner. For subsectional triangle basis functions for J_z and piecewise-linear testing functions $\{T_m(t)\}$, the entries of $\mathbf{J}_1$ are given by

$$J_{1,mn} = -j\omega\mu_0 b \begin{cases} \frac{2}{3} & m = n \\ \frac{1}{6} & |m-n| = 1 \\ \frac{1}{6}e^{-jk_x a} & m = 1 \text{ and } n = N \\ \frac{1}{6}e^{jk_x a} & m = N \text{ and } n = 1 \end{cases} \tag{7.185}$$

The entries of $\mathbf{J}_2$ have a similar form.

Finally, again under the assumption of uniformly spaced nodes, the entries of $\mathbf{A}_1$ can be obtained along $\partial\Gamma_1$ as the inverse Fourier transform summation

$$A_{mn} = \frac{k_1\eta_1}{2a} \sum_{i=-\infty}^{\infty} \frac{\sin^2(\pi b f)}{b(\pi f)^2} \frac{e^{j2\pi f(x_m - x_n)}}{\beta_y} \Bigg]_{f=i/a-k_x/2\pi} \tag{7.186}$$

where a denotes the period, b denotes the uniform interval size, and β_y is defined in (7.45). A similar expression may be obtained for boundary $\partial\Gamma_2$. Observe that the ith term in (7.186) behaves as $O(i^{-3})$, ensuring convergence of the summation. We leave the generalization of this expression for nonuniformly spaced nodes as an exercise.

The numerical approach requires a scatterer model containing a list of node coordinates (x_i, y_i), a connectivity array linking cell indices to those of the surrounding nodes, and pointers to identify nodes on imbedded conducting boundaries as well as those located along $\partial\Gamma_1$, $\partial\Gamma_2$, and $\partial\Gamma_4$. An additional pointer is required to link the dummy nodes along $\partial\Gamma_4$ with their counterparts along $\partial\Gamma_3$. Finally, the usual arrays providing the material parameters of each cell are also necessary.

After Equation (7.184) is solved to produce the fields throughout the computational domain Γ, the fields along $\partial\Gamma_1$ and $\partial\Gamma_2$ are readily decomposed into a Floquet harmonic representation such as (7.21), where

$$e_n = \frac{1}{a}\int_0^a E_z(x, y_0)e^{jk_{xn}x}\,dx \tag{7.187}$$

and where k_{xn} is defined in Equation (7.23). The Floquet harmonic description is often useful for constructing scattering or transmission matrices, as discussed in Section 7.9.

The TE polarization can be obtained from the preceding expressions using the duality concept. To illustrate the performance of the procedure, we initially apply it to the corrugated surface depicted in Figure 7.14. Table 7.2 presents numerical values for the magnitudes of the Floquet harmonics in response to a TE-to-z illumination. Results from the outward-looking formulation of this section are compared with results from the coupled-wave formulation of reference [21], which were previously used to plot Figure 7.15. The numerical results were obtained using triangular-cell models with a maximum cell size of approximately $\lambda_0/20$. Excellent agreement is observed between the two procedures.

Figure 7.18 shows one period of an inhomogeneous structure consisting of a corrugated conducting region surrounded by a dielectric cladding. The response of the structure to normally incident TM and TE plane waves, obtained using the outward-looking formulation of this section, is displayed in Figures 7.19 and 7.20.

TABLE 7.2 Magnitude of nth Floquet Harmonic as a Function of Corrugation Depth d for Geometry Illustrated in Figure 7.14

Harmonic, n	Depth, d	Outward-Looking Approach	Coupled-Wave Approach
0 refl.	0.25	0.102	0.104
	0.5	0.017	0.017
	0.75	0.029	0.028
	1.0	0.009	0.008
1 refl.	0.25	0.135	0.134
	0.5	0.114	0.114
	0.75	0.052	0.048
	1.0	0.040	0.042
−1 trans.	0.25	0.306	0.308
	0.5	0.422	0.421
	0.75	0.403	0.396
	1.0	0.350	0.350
0 trans.	0.25	1.157	1.158
	0.5	1.113	1.115
	0.75	1.056	1.064
	1.0	0.983	0.987
1 trans.	0.25	0.167	0.168
	0.5	0.338	0.340
	0.75	0.503	0.495
	1.0	0.647	0.645

Note: Results obtained from the outward-looking formulation of this section (reconfigured for the TE-to-z polarization) are compared with results obtained from the "coupled-wave" approach of reference [21]. The depth is in λ.

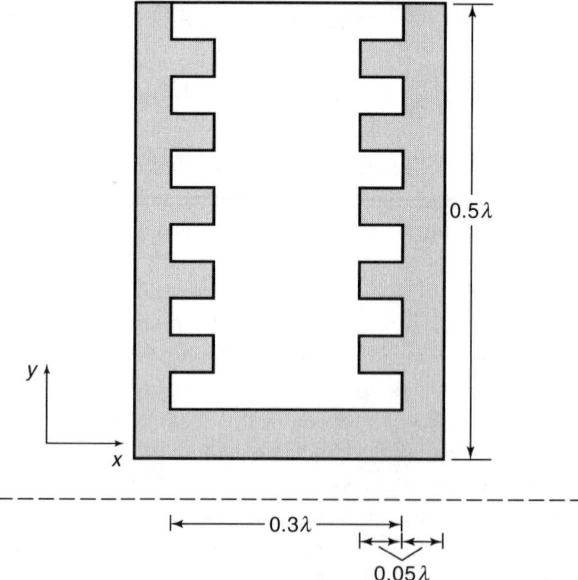

Figure 7.18 Corrugated p.e.c. cylinder partially coated with a lossless dielectric ($\epsilon_r = 3$). This structure comprises one element of an infinite periodic array along the x-axis. After [22]. ©1994 Taylor & Francis.

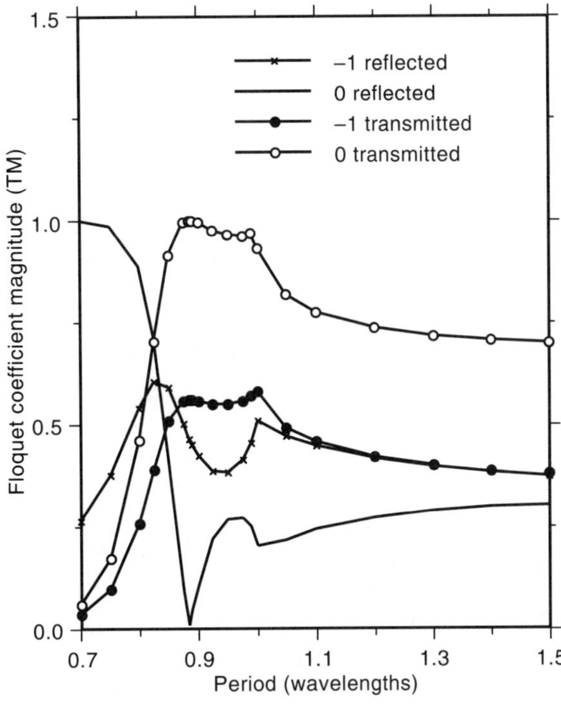

Figure 7.19 Response of the periodic array when illuminated by a TM-to-z wave propagating into the y direction. After [22]. ©1994 Taylor & Francis.

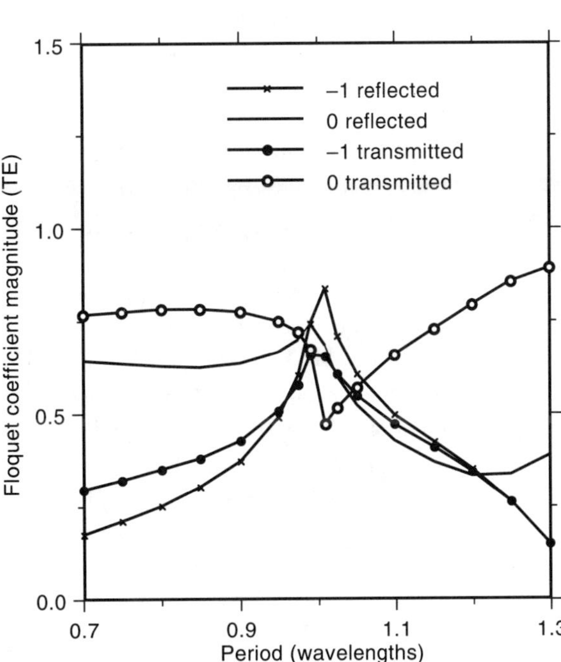

Figure 7.20 Response of the periodic array when illuminated by a TE-to-z wave propagating in the y direction. After [22]. ©1994 Taylor & Francis.

In this section, we have presented an outward-looking formulation for general two-dimensional structures having one-dimensional periodicity. We have also illustrated the use of an integral equation RBC. There are a variety of other ways in which integral and differential equations can be combined into hybrid formulations for periodic structures, as attested to by the recent literature [23].

7.12 SUMMARY

This chapter has extended and applied the integral and differential equation formulations discussed in Chapters 2 and 3 to problems exhibiting one-dimensional periodicity. Because the required computational domain can be restricted to a single unit cell, these geometries are tractable despite their electrically large nature. However, the trade-off involves a Green's function that generally appears in the form of a slowly converging summation. Several acceleration procedures have been introduced to facilitate the required computations.

REFERENCES

[1] E. O. Brigham, *The Fast Fourier Transform and Its Applications*, Englewood Cliffs, NJ: Prentice-Hall, 1988.

[2] N. Amitay, V. Galindo, and C. P. Wu, *Theory and Analysis of Phased Array Antennas*, New York: Wiley, 1972.

[3] Y. T. Lo and S. W. Lee, "Affine transformation and its application to antenna arrays," *IEEE Trans. Antennas Propagat.*, vol. AP-13, pp. 890–896, Nov. 1965.

[4] R. E. Jorgenson, "Electromagnetic scattering from a structured slab comprised of periodically placed resistive cards," Ph.D. dissertation, University of Illinois, Urbana, IL, 1989.

[5] L. A. Weinstein, *The Theory of Diffraction and the Factorization Method*, Boulder, CO: Golem, 1969, pp. 267–290.

[6] T. A. Cwik, "Scattering from general periodic surfaces," Ph.D. dissertation, University of Illinois, Urbana, IL, 1986.

[7] R. Lampe, P. Klock, and P. Mayes, "Integral transforms useful for the accelerated summation of periodic, free-space Green's functions," *IEEE Trans. Microwave Theory Tech.*, vol. MTT-33, pp. 734–736, Aug. 1985.

[8] R. E. Jorgenson and R. Mittra, "Efficient calculation of the free-space periodic Green's function," *IEEE Trans. Antennas Propagat.*, vol. AP-38, pp. 633–642, May 1990.

[9] S. Singh, W. F. Richards, J. R. Zinecker, and D. R. Wilton, "Accelerating the convergence of series representing the free space periodic Green's function," *IEEE Trans. Antennas Propagat.*, vol. AP-38, pp. 1958–1962, Dec. 1990.

[10] M. E. Veysoglu, "Polarimetric passive remote sensing of periodic surfaces and anisotropic media," M.S. Thesis, Massachusetts Institute of Technology, Cambridge, MA, May 1991.

[11] I. S. Gradshteyn and I. M. Ryzhik, *Table of Integrals, Series, and Products*, New York: Academic, 1980.

[12] K. E. Jordan, G. R. Richter, and P. Sheng, "An efficient numerical evaluation of the Green's function for the Helmholtz operator on periodic surfaces," *J. Computat. Phys.*, vol. 63, pp. 222–235, 1986.

[13] P. P. Ewald, "Die berechnung optischer und elektrostatischen giterpotentiale," *Ann. der Phys.*, vol. 64, p. 253, 1921.

[14] A. W. Mathis and A. F. Peterson, "A comparison of acceleration procedures for the two-dimensional periodic Green's function," *IEEE Trans. Antennas Propagat.*, vol. 44, pp. 567–571, Apr. 1996.

[15] T. Cwik and R. Mittra, "The cascade connection of planar periodic surfaces and lossy dielectric layers to form an arbitrary periodic screen," *IEEE Trans. Antennas Propagat.*, vol. AP-35, pp. 1397–1405, Dec. 1987.

[16] R. C. Hall, R. Mittra, and K. M. Mitzner, "Analysis of multilayered periodic structures using generalized scattering matrix theory," *IEEE Trans. Antennas Propagat.*, vol. AP-36, pp. 511–517, Apr. 1988.

[17] R. E. Collin, *Foundations for Microwave Engineering*, New York: McGraw-Hill, 1966.

[18] R. Mittra and S. W. Lee, *Analytical Techniques in the Theory of Guided Waves*, New York: Macmillan, 1971, pp. 207–211.

[19] T. Itoh, "Generalized scattering matrix technique," in *Numerical Techniques for Microwave and Millimeter-Wave Passive Structures*, ed. T. Itoh, New York: Wiley, 1989.

[20] A. F. Peterson, "Integral equation formulation for scattering from continuous or discrete dielectric targets exhibiting one-dimensional periodicity," *Arch. Elektron. Ubertragungst.*, vol. 46, pp. 336–342, Sept. 1992.

[21] M. G. Moharam and T. K. Gaylord, "Diffraction analysis of dielectric surface-relief gratings," *J. Opt. Soc. Am.*, vol. 72, pp. 1385–1392, 1982.

[22] A. F. Peterson, "An outward-looking differential equation formulation for scattering from one-dimensional periodic diffraction gratings," *Electromagnetics*, vol. 14, pp. 227–238, 1994.

[23] S. D. Gedney, J. F. Lee, and R. Mittra, "A combined FEM/MoM approach to analyze the plane wave diffraction by arbitrary gratings," *IEEE Trans. Antennas Propagat.*, vol. 40, pp. 363–370, Feb. 1992.

PROBLEMS

P7.1 (a) Using the Fourier transform of the comb function $P(x)$ defined in (7.9), derive the relationship

$$\sum_{n=-\infty}^{\infty} e^{-j2\pi nf/\Delta f} = \Delta f \sum_{n=-\infty}^{\infty} \delta(f - n\Delta f)$$

(b) By considering the Fourier transform of the product $U(x)P(x)$, where U is arbitrary and P is the comb function used in part (a), derive the *Poisson summation formula*

$$\sum_{n=-\infty}^{\infty} U(n\Delta X) = \frac{1}{\Delta X} \sum_{n=-\infty}^{\infty} \tilde{U}\left(\frac{n}{\Delta X}\right)$$

P7.2 If $\tilde{U}(f)$ is the Fourier transform of $U(x)$ according to the definition in (7.1), show that the Fourier transform of

$$U(x)e^{j2\pi f_0 x}$$

is given by $\tilde{U}(f - f_0)$.

P7.3 Identify the propagating Floquet harmonics for a strip grating with period equal to 3λ and a wave incident from an angle 45° normal to the grating. Provide a sketch similar to Figure 7.4 showing the direction of each propagating harmonic.

P7.4 A two-dimensional periodic surface is defined by the spatial lattice $x = m\Delta X$ and $y = n\Delta Y$. The excitation imposes a phase shift of $k_x \Delta X$ from one cell to another in the x direction and $k_y \Delta Y$ from one cell to another along y.

(a) Identify the values of f and g that comprise the reciprocal lattice (the discrete spectral frequencies associated with the Floquet harmonics).
(b) If $\Delta X = 0.1\lambda$, $\Delta Y = 0.5\lambda$, and the excitation is a plane wave incident at an angle 45° from normal along x and 45° from normal along y, identify the propagating Floquet harmonics.

P7.5 For the TM-to-z representation given in (7.21), show that the time-averaged power per period carried by the nth propagating Floquet harmonic in the y direction is

$$\frac{a|e_n|^2}{2k\eta}\sqrt{k^2 - k_{xn}^2}$$

assuming that the phasor representation for E_z is based on peak value rather than root-mean-square (rms) value.

P7.6 Consider the strip grating of Figure 7.3 illuminated by a TE-to-z plane wave that excites a current density J_x on the conducting strips. The scattered field can be expressed as

$$H_z^s(x, y) = \begin{cases} \sum_{n=-\infty}^{\infty} h_n^+ e^{-jk_{xn}x} e^{-j\sqrt{k^2-k_{xn}^2}\,y} & y > 0 \\ \sum_{n=-\infty}^{\infty} h_n^- e^{-jk_{xn}x} e^{j\sqrt{k^2-k_{xn}^2}\,y} & y < 0 \end{cases}$$

where k_{xn} is defined in (7.23). Find an expression for the coefficients h_n^+ and h_n^- in terms of the current density $J_x(x)$.

P7.7 (a) Using $G_p(x)$ and $G_a(x)$ defined in (7.32) and (7.50), respectively, and the asymptotic form of the Hankel function from (7.42), show that the magnitude of the nth term in the difference summation $G_p(x) - G_a(x)$ tends to

$$\left(\frac{\alpha}{a}\right)^2 n^{-3/2}\sqrt{\frac{ka}{32\pi}} \qquad \text{as } n \to \infty.$$

Hint: Use the approximation

$$\sqrt{(na-x)^2 + \alpha^2} \cong (na - x)\sqrt{1 + \frac{\alpha^2}{(na)^2}} \cong (na - x)\left(1 + \frac{\alpha^2}{2(na)^2}\right)$$

(b) An alternative expression for the periodic Green's function is

$$G_p(x) = G_b(x) + F^{-1}\{\tilde{G}_p(f) - \tilde{G}_b(f)\}$$

where $\tilde{G}_p(f)$ is defined in Equation (7.46) and $\tilde{G}_b(f)$ is defined in (7.63). Use techniques similar to those employed in part (a) to show that the nth term in the difference summation $F^{-1}\{\tilde{G}_p(f) - \tilde{G}_b(f)\}$ is $O(n^{-3})$ as $n \to \infty$.

P7.8 Using (1.46) and (7.44), evaluate the two-dimensional Fourier transform of the three-dimensional free-space Green's function to produce

$$\int_{-\infty}^{\infty}\int_{-\infty}^{\infty} \frac{e^{-jk\sqrt{x^2+y^2}}}{4\pi\sqrt{x^2+y^2}} e^{-j2\pi fx} e^{-j2\pi gy}\, dx\, dy = \frac{1}{j2\beta_z}$$

where

$$\beta_z = \begin{cases} \sqrt{k^2 - (2\pi f)^2 - (2\pi g)^2} & k^2 > (2\pi f)^2 + (2\pi g)^2 \\ -j\sqrt{(2\pi f)^2 + (2\pi g)^2 - k^2} & \text{otherwise} \end{cases}$$

P7.9 The three-dimensional Green's function describing the potential on a planar surface having a regular two-dimensional periodicity can be expressed in the spatial domain as

$$G_p(x - x', y - y') = \sum_{m=-\infty}^{\infty} \sum_{n=-\infty}^{\infty} \frac{e^{-jkR_{mn}}}{4\pi R_{mn}} e^{-jmk_x a} e^{-jnk_y b}$$

where

$$R_{mn} = \sqrt{(x - x' - ma)^2 + (y - y' - nb)^2}$$

(a) Identify the asymptotic order of the (mn)th term in the summation.

(b) Using the result of Prob. P7.8 and properties of the Fourier transform, show that an equivalent summation is given by

$$G_p(x - x', y - y') = \left. \frac{1}{ab} \sum_{m=-\infty}^{\infty} \sum_{n=-\infty}^{\infty} \frac{1}{j2\beta_z} e^{j2\pi f(x-x')} e^{j2\pi g(y-y')} \right]_{f=m/a-k_x/2\pi, g=n/b-k_y/2\pi}$$

where β_z is defined in Prob. P7.8.

(c) What is the asymptotic order of the (mn)th term in the new summation?

P7.10 (a) Using the idea expressed in Equation (7.49), develop an alternative procedure for computing the three-dimensional periodic Green's function defined in Prob. P7.9. Determine the asymptotic order of the (mn)th term in each of the new summations.

(b) Using the idea expressed in Prob. P7.7(b), develop a different procedure for computing the three-dimensional periodic Green's function, and determine the asymptotic order of the (mn)th term in each of the new summations.

P7.11 Suppose that the TM strip grating formulation in Section 7.3 is posed using pulse testing functions instead of Dirac delta testing functions. Develop an expression for Z_{mn} analogous to (7.57) reflecting the change. What is the asymptotic behavior of the ith term in the new summation?

P7.12 Create a computer program to determine the surface current density and the reflection and transmission coefficients for a planar strip grating of perfect conductors illuminated by a TE-to-z plane wave. Develop a formulation along the lines of Section 7.7 (without including the dielectric slab support). Use subsectional triangle basis functions and subsectional pulse testing functions within a method-of-moments discretization. Implement a series acceleration technique similar to that Equation (7.115) to compute the matrix entries, placing all of the weight on the inverse Fourier transform summation.

P7.13 Develop a formulation for TM-to-z scattering from a strip grating backed by a dielectric slab. Using the discussion in Section 7.7 as a guide, find the appropriate TM Green's function for a line source radiating on the surface of a slab, construct an EFIE expressed as an integral over a single strip, and identify matrix entries associated with a pulse basis function and Dirac delta testing function discretization. Write the matrix entries in the form of an inverse Fourier transform summation, similar to (7.115).

P7.14 Using the Green's function in Equation (7.107) for a TE line source radiating on the surface of a dielectric slab, develop an EFIE formulation for TE scattering from a single conducting strip on the surface of a dielectric slab. Provide expressions for the matrix entries associated with subsectional triangle basis and pulse testing functions. Discuss ways in which you might evaluate the matrix entries. Equation (7.107) is one form of a Sommerfeld integral, and a number of special procedures have been developed for its efficient evaluation.

P7.15 The planar strip grating geometries discussed in Sections 7.3, 7.7, and 7.8 are easily adapted to iterative solution procedures such as the CG–FFT approach from Chapter 4. For the TM grating formulation in Section 7.3, describe the specialization necessary to produce (7.33) in a form suitable for CG–FFT solution.

P7.16 For the aperture formulation of Section 7.8, find an expression for the coefficients of the Floquet harmonics in terms of the aperture field $E_z(x)$.

P7.17 Develop an integral equation formulation to treat TM-to-z scattering from a one-dimensional periodic array of homogeneous dielectric scatterers, where the unit cell consists of a single scatterer of general shape. Use a coupled EFIE formulation, with the exterior equation (Section 7.10) incorporating periodicity in the x direction. The interior equation does not involve a periodic Green's function and may be found in Section 2.8. Provide the equations, and develop (nonaccelerated) expressions for the matrix entries associated with pulse basis functions and Dirac delta testing functions.

P7.18 Using the TM formulation in Section 7.11 as a guide, develop an outward-looking TE-to-z formulation for scattering from an inhomogeneous grating exhibiting periodicity in the x direction. Develop RBCs from a discretization of the MFIE, and define all entries of the global finite-element matrix, assuming the use of linear interpolation functions on triangular cells.

P7.19 The Floquet expansion discussed in Section 7.2 can be used to develop an alternate RBC that can replace the integral equation RBC employed in Section 7.11. By expanding $E_z(x, y)$ in a Floquet expansion, use a development similar to that of Section 3.3 in order to obtain the RBC

$$\frac{\partial E_z^s}{\partial y} = \frac{1}{a} \int_0^a E_z^s(x', y) G(x - x', y)\, dx'$$

applicable along boundary $\partial\Gamma_2$ in Figure 7.17, where

$$G(x, y) = -j \sum_{n=-\infty}^{\infty} \sqrt{k^2 - k_{xn}^2}\, e^{-jk_{xn}x} e^{-j\sqrt{k^2 - k_{xn}^2}\, y}$$

and k denotes the wavenumber of region 2. Find a similar expression for an RBC applicable to points along boundary $\partial\Gamma_1$ in Figure 7.17. Is there a connection between this RBC and the integral equation RBC used in Section 7.11?

8

Three-Dimensional Problems with Translational or Rotational Symmetry

The electromagnetic scattering formulations considered in preceding chapters have been limited to two-dimensional, infinite cylinder geometries. These two-dimensional formulations require both the scatterer and the excitation to be invariant along a preferred axis, which may not be a realistic model of the desired physical problem. A wider class of problems may be modeled under the assumption that, while the scatterer is translationally invariant along z, the source may vary with z or may be finite in extent. The fields of an arbitrary source may be decomposed into a spectrum of plane waves incident on the infinite cylinder from oblique angles with respect to z. The response to each plane wave can be obtained by the solution of a two-dimensional problem, after which the individual responses are superimposed to give the solution to the original three-dimensional problem. Because of their translational invariance, these three-dimensional problems impose computational requirements similar to two-dimensional problems. The "building block" in this analysis is the scattering of an obliquely incident wave from an infinite cylinder, and several formulations for oblique scattering will be described in this chapter.

Axisymmetric scatterers, sometimes known as *bodies of revolution*, permit a similar reduction in computational effort. As a function of the circumferential variable, the fields and currents can be represented by cylindrical harmonics that decouple the pertinent equations and reduce the effective dimensionality of the problem. Simple examples of the axisymmetric scattering problem are illustrated in Sections 8.7 and 8.8. Section 8.7 describes a surface integral equation formulation, while Section 8.8 presents a scalar differential equation formulation.

8.1 SCATTERING FROM INFINITE CYLINDERS ILLUMINATED BY FINITE SOURCES [1]

Consider a finite source radiating in the presence of an infinite, perfectly conducting cylinder, as depicted in Figure 8.1. Using the equivalence principle from Section 1.6, the scattered fields can be expressed in terms of the equivalent electric current density $\bar{J}$ located on the cylinder surface. Because the source is finite, the electromagnetic fields vary with all three spatial variables.

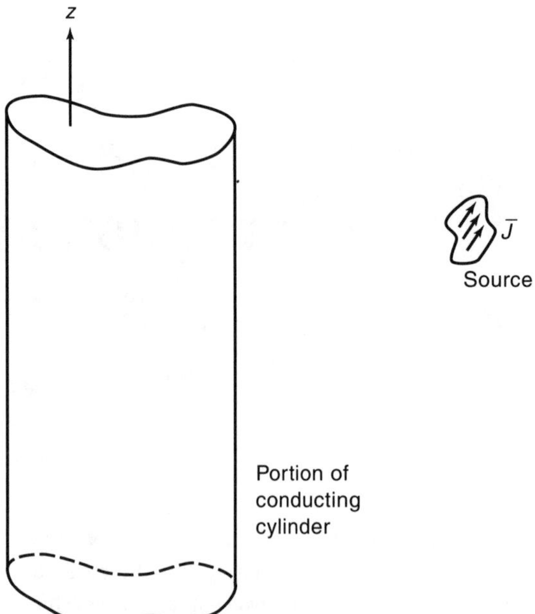

Figure 8.1 Conducting cylinder illuminated by a finite source.

Suppose the cylinder axis lies along $\hat{z}$, as illustrated in Figure 8.1. The *Fourier transform* with respect to z can be defined as

$$F_z\{A(z)\} = \tilde{A}(k_z) = \int_{-\infty}^{\infty} A(z)e^{-jk_z z}\, dz \tag{8.1}$$

This transformation converts functions of the spatial variable z to functions of the "spectral" variable k_z. If the Fourier transform is applied to all of the field quantities and equations describing the scattering problem, the solution for unknown surface current density can be obtained in the Fourier transform domain (sometimes called the spectral domain or *k-space*) as a function of k_z. Once the solution for $\bar{J}$ is obtained as a function of k_z, the *inverse Fourier transform*

$$F_z^{-1}\{\tilde{A}(k_z)\} = A(z) = \frac{1}{2\pi}\int_{-\infty}^{\infty} \tilde{A}(k_z)e^{jk_z z}\, dk_z \tag{8.2}$$

can be employed to return to the spatial domain.

The task of transforming the quantities into the Fourier transform domain would be of little use were it not for the following: As a consequence of the invariance of the cylinder geometry, the z-dependence of the wave equation can be separated in accordance with the classical separation-of-variables procedure. Because the kernel of the Fourier transform is a characteristic solution of the wave equation, the transformed equations are uncoupled as a function of k_z. In other words, the solution for the transform of $\bar{J}$ can be obtained at one value of k_z by the solution of a two-dimensional equation without regard for the solution at other values of k_z.

For example, suppose that an infinite conducting cylinder is analyzed using the three-dimensional MFIE. The $\hat{z}$-component of the MFIE only involves the transverse current density and can be expressed in the spatial domain as

$$H_z^{\text{inc}}(t, z) = -J_t(t, z) - \hat{z} \cdot \nabla \times \iint \hat{t}(t') J_t(t', z') \frac{e^{-jkR}}{4\pi R} \, dt' \, dz' \tag{8.3}$$

where

$$R = \sqrt{[x(t) - x(t')]^2 + [y(t) - y(t')]^2 + (z - z')^2} \tag{8.4}$$

and the integration is assumed to be restricted to the surface of a closed mathematical cylinder whose cross-sectional shape is defined parametrically by the variable t. Equation (8.3) involves a convolution in the variable z, and an application of the Fourier transform in conjunction with the convolution theorem

$$F_z\{A(z) * B(z)\} = \tilde{A}(k_z)\tilde{B}(k_z) \tag{8.5}$$

produces the transformed equation

$$\tilde{H}_z^{\text{inc}}(t, k_z) = -\tilde{J}_t(t, k_z) - \hat{z} \cdot \nabla \times \int \hat{t}(t') \tilde{J}_t(t', k_z) \tilde{G}(\rho; k, k_z) \, dt' \tag{8.6}$$

where

$$\tilde{G}(\rho; k, k_z) = F_z \left\{ \frac{e^{-jk\sqrt{\rho^2+z^2}}}{4\pi\sqrt{\rho^2 + z^2}} \right\} = \begin{cases} \dfrac{1}{4j} H_0^{(2)}\left(\rho\sqrt{k^2 - k_z^2}\right) & k^2 > k_z^2 \\[2ex] \dfrac{1}{2\pi} K_0\left(\rho\sqrt{k_z^2 - k^2}\right) & k_z^2 > k^2 \end{cases} \tag{8.7}$$

$$\rho = \sqrt{[x(t) - x(t')]^2 + [y(t) - y(t')]^2} \tag{8.8}$$

and $H_0^{(2)}$ and K_0 represent the Hankel and modified Bessel functions of the second kind, respectively. The Fourier transform in Equation (8.7) has effectively produced the two-dimensional free-space Green's function from Equation (1.47) with the wavenumber in the argument replaced by the transverse wavenumber

$$k_t = \begin{cases} \sqrt{k^2 - k_z^2} & k^2 > k_z^2 \\ -j\sqrt{k_z^2 - k^2} & k_z^2 > k^2 \end{cases} \tag{8.9}$$

This result can also be obtained from a direct application of the Fourier transform to the Helmholtz equation (Prob. P8.1). In the Fourier transform domain, the equation to be solved at each value of k_z is essentially the two-dimensional MFIE discussed in Section 2.2, with the wavenumber in the argument replaced by the transverse wavenumber of (8.9). An electromagnetic wave propagating at an oblique angle with respect to the cylinder axis involves the identical transverse wavenumber and suggests the interpretation that the Fourier transform decomposes the field into a spectrum of waves incident from oblique spatial angles.

The original three-dimensional problem of a finite source radiating in the presence of an infinite cylinder involves three coupled spatial variables x, y, and z. If cast into the Fourier transform domain, the solution as a function of k_z can be obtained by solving uncoupled two-dimensional equations involving the variables x and y and the parameter k_z. Consequently, this approach is much more amenable to numerical solution than the original three-dimensional problem. However, we are still faced with the task of solving the two-dimensional equations over the infinite continuum spanned by the variable k_z in order to construct the inverse transformation according to (8.2).

A numerical implementation of the inverse Fourier transform requires that the continuum spanned by k_z be truncated and discretized. In other words, we construct the solution at a finite number of values $\{k_{zi}\}$, where i may range from $-N$ to N. Assuming evenly spaced samples as a function of k_z, the inverse Fourier transform is approximated by

$$A(z) \cong \frac{\Delta k_z}{2\pi} \sum_{n=-N}^{N} \tilde{A}(n\ \Delta k_z) e^{jnz\,\Delta k_z} \tag{8.10}$$

If desired, this summation could be implemented with the aid of a fast Fourier transform (FFT) algorithm. As is well known from signal analysis, the combination of an evenly spaced sampling procedure with a Fourier transform is equivalent to working directly with the Fourier transform of a periodic extension of the original function of z. Thus, the approximations inherent in Equation (8.10) are equivalent to replacing the original finite source with an infinite, periodic array of sources distributed along the axis of the cylinder. The period of this fictitious array is related to the sampling interval Δk_z by

$$P_z = \frac{2\pi}{\Delta k_z} \tag{8.11}$$

To successfully employ the procedure, the equivalent spatial period P_z must be sufficiently large so that the contribution of the fictitious sources to the currents on the cylinder is negligible in the region of interest compared with the contribution of the desired source. In addition, the truncation of the spectrum as a function of k_z must not introduce inaccuracy. We observe that it will generally be necessary to consider both the *visible region* of the spectrum ($k_z < k$) and the *invisible region* ($k_z > k$). (Wave propagation in the transverse direction is evanescent in the invisible region.) For most problems, it will be necessary to analyze a large number of two-dimensional problems before inverse transformation. The specific number will depend on the actual source distribution.

Note that scanning the transverse wavenumber over a range of values is equivalent to varying the electrical size of the scatterer or the frequency under consideration. Therefore, it is likely that "interior resonance" frequencies discussed in Chapter 6 will be encountered throughout the range necessary for inverse transformation. It is therefore essential to employ the alternative formulations described in Chapter 6 to ensure a robust formulation, at least throughout the visible region of the spectrum.

The following sections present several different approaches for treating the scattering of an oblique wave from conducting or penetrable cylinders. While these formulations provide a complete model for a single incident wave, they also constitute the basic building block required for the analysis of a finite source radiating in the presence of an infinite cylinder.

8.2 OBLIQUE TM-WAVE SCATTERING FROM INFINITE CONDUCTING CYLINDERS: CFIE DISCRETIZED WITH PULSE BASIS FUNCTIONS AND DELTA TESTING FUNCTIONS

Consider an infinite, perfectly conducting cylinder illuminated by a uniform plane wave incident from an oblique angle with respect to the cylinder axis (Figure 8.1). For a TM uniform plane wave, the z-component of the incident electric field has the form

$$E_z^{\text{inc}}(x, y, z) = e^{-jk_t(x\cos\theta + y\sin\theta)} e^{jk_z z} \tag{8.12}$$

The surface equivalence principle can be employed to replace the conducting material by equivalent electric currents J_z. In order to ensure unique solutions at all frequencies, we consider a numerical solution of the CFIE. Equation (6.12) can be generalized for an oblique excitation by replacing the wavenumber k by the transverse wavenumber k_t defined in (8.9), to produce

$$\begin{aligned}\alpha E_z^{\text{inc}}(t) + (1-\alpha)\eta H_t^{\text{inc}}(t) = (1-\alpha)\eta J_z(t) + j\alpha\eta\frac{k_t^2}{k}\int J_z(t')\tilde{G}(\rho; k, k_z)\, dt' \\ -(1-\alpha)\eta\hat{t}\cdot\nabla\times\int \hat{z}J_z(t')\tilde{G}(\rho; k, k_z)\, dt'\end{aligned} \tag{8.13}$$

where $\tilde{G}$ is defined in Equation (8.7). Although this CFIE remains valid if k_t is purely imaginary, which would occur if $k_z > k$, there are no "interior resonance" frequencies with $k_z > k$, and consequently the conventional EFIE or MFIE alone would be a more efficient choice throughout the invisible region of the spectrum.

Following the approach of Section 6.2, the CFIE can be discretized using pulse basis functions and Dirac delta testing functions to produce a matrix equation

$$\begin{bmatrix} C_{11} & C_{12} & \cdots & C_{1N} \\ C_{21} & C_{22} & \ddots & \vdots \\ \vdots & & & \\ C_{N1} & \cdots & \cdots & C_{NN} \end{bmatrix} \begin{bmatrix} j_1 \\ j_2 \\ \vdots \\ j_N \end{bmatrix} = \begin{bmatrix} e_1 \\ e_2 \\ \vdots \\ e_N \end{bmatrix} \tag{8.14}$$

Using the conventions for normal and tangent vectors from Figure 6.6 and Equations (6.19) and (6.20), the diagonal and off-diagonal entries are

$$C_{mm} = \frac{(1-\alpha)\eta}{2} + \frac{\alpha\eta k_t^2}{4k}\int_{\text{cell } m} H_0^{(2)}(k_t R_m)\, dt' \tag{8.15}$$

and

$$C_{mn} = \frac{\alpha \eta k_t^2}{4k} \int_{\text{cell } n} H_0^{(2)}(k_t R_m)\, dt' + \frac{(1-\alpha)\eta j k_t}{4} \int_{\text{cell } n} \left[\sin \Omega_m \left(\frac{y_m - y'}{R_m} \right) + \cos \Omega_m \left(\frac{x_m - x'}{R_m} \right) \right] H_1^{(2)}(k_t R_m)\, dt' \tag{8.16}$$

respectively, where

$$R_m = \sqrt{[x_m - x(t')]^2 + [y_m - y(t')]^2} \tag{8.17}$$

If the incident field is given by Equation (8.12), the entries of the excitation column vector are

$$e_m = \left[\alpha - (1-\alpha) \frac{k^2}{k_t^2} \cos(\theta - \Omega_m) \right] e^{-jk_t(x_m \cos\theta + y_m \sin\theta)} \tag{8.18}$$

These expressions involve integrals that are identical in form to those of the EFIE and MFIE discussed in Chapter 2, and their numerical evaluation is similar. As discussed in Chapter 6, the computational effort required to construct the CFIE matrix equation is roughly double that required for the EFIE or MFIE alone.

For illustration, Figure 8.2 shows the current density induced on an ogival cylinder by a plane wave incident at an angle 45° from normal using the CFIE formulation. Table 8.1 provides the bistatic scattering cross section for several discretizations of the same geometry, defined for the oblique TM case as

$$\sigma_{\text{TM}}(\phi) = \lim_{\rho \to \infty} 2\pi\rho \frac{|E_z^s(\rho, \phi)|^2}{|E_z^{\text{inc}}(0, 0)|^2} \tag{8.19}$$

These results were obtained using a weighting parameter $\alpha = 0.2$ within the CFIE (the selection of α has been discussed in Chapter 6).

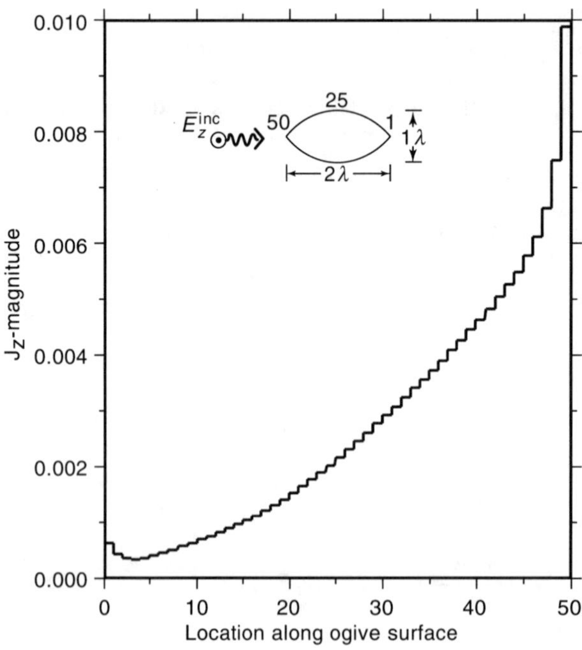

Figure 8.2 The CFIE result for the current density induced by an obliquely incident wave on an ogival p.e.c. cylinder. The incident field has unit magnitude E_z and impinges on the cylinder at an angle 45° from the x–y plane. The result was obtained using 100 pulse basis functions to represent J_z.

TABLE 8.1 Bistatic Two-Dimensional Scattering Cross Section (dB λ) for an Ogival Cylinder of Overall Dimension 2.0λ by 1.0λ Obtained Using the CFIE Formulation

ϕ (deg)	50-Cell Model	100-Cell Model	200-Cell Model
0	9.98	10.18	10.30
30	6.61	6.80	6.91
60	3.25	3.43	3.53
90	2.14	2.29	2.37
120	0.07	0.34	0.50
150	−1.28	−1.26	−1.24
180	−1.78	−1.86	−1.88

Note: The incident plane wave is TM to z and is incident at an angle that is 45° from the transverse plane and edge-on in the transverse plane. The 200-cell model represents a density of 43 cells per wavelength.

The CFIE formulation can easily be adapted to treat the TE polarization, and we leave that exercise to the reader (Prob P8.4). We note that the CFIE approach is just one of several integral equation formulations that eliminate the interior resonance difficulties. In the following section, we describe another procedure for oblique scattering from conducting cylinders: the augmented-field formulation. The augmented-field equations also eliminate spurious interior resonances.

8.3 OBLIQUE TE-WAVE SCATTERING FROM INFINITE CONDUCTING CYLINDERS: AUGMENTED MFIE DISCRETIZED WITH PULSE BASIS FUNCTIONS AND DELTA TESTING FUNCTIONS

A closed, perfectly conducting cylinder illuminated by an obliquely incident wave can also be analyzed using the augmented MFIE introduced in Section 6.4, and we turn our attention to that approach. The augmented-field formulation imposes boundary conditions on both the normal and tangential fields at the scatterer surface and, like the CFIE, guarantees unique solutions at all frequencies [2]. If the infinite conducting cylinder of Figure 8.1 is illuminated by an oblique TE wave having $\hat{z}$-component

$$H_z^{\text{inc}}(x, y, z) = e^{-jk_t(x\cos\theta + y\sin\theta)}e^{jk_z z} \tag{8.20}$$

both $\hat{z}$- and $\hat{t}$-components of $\bar{J}$ are excited. Although the $\hat{z}$-component of the MFIE only couples to the transverse current density, even in the case of oblique excitation, both the CFIE and the augmented MFIE formulations couple to both $\hat{z}$- and $\hat{t}$-components of the current density and consequently require twice the number of unknowns in this situation to ensure unique solutions at all frequencies. The augmented MFIE formulation offers the advantage that the matrix entries are simpler in form than those of the CFIE formulation but at the same time imposes the disadvantage of working with all three components of the $\bar{H}$-field equation. As a result, the augmented-field formulation produces an overdetermined system of equations and requires additional storage and a more sophisticated matrix solution algorithm.

The augmented-field equations, constructed using the boundary conditions $\hat{n} \times \bar{H} = \bar{J}$ and $\hat{n} \cdot \bar{H} = 0$ at the surface of the conducting cylinder, have the form

$$H_z^{\text{inc}}(t) = -J_t(t) - \hat{z} \cdot \nabla \times \int \hat{t}(t') J_t(t') \frac{1}{4j} H_0^{(2)}(k_t R)\, dt' \tag{8.21}$$

$$H_t^{\text{inc}}(t) = J_z(t) - \hat{t} \cdot \nabla \times \int [\hat{t}(t') J_t(t') + \hat{z} J_z(t')] \frac{1}{4j} H_0^{(2)}(k_t R)\, dt' \tag{8.22}$$

$$H_n^{\text{inc}}(t) = -\hat{n} \cdot \nabla \times \int [\hat{t}(t') J_t(t') + \hat{z} J_z(t')] \frac{1}{4j} H_0^{(2)}(k_t R)\, dt' \tag{8.23}$$

where R and k_t have been defined in the preceding section. (For the special case of a normally incident TE wave, $H_t = 0$ and $H_n = 0$. In that case, this particular formulation reduces to the tangential-field MFIE and will not necessarily guarantee unique solutions.)

Assuming that the cylinder contour is represented in terms of flat strips as depicted in Figure 6.5, pulse basis functions can be used to represent both components of the current density. It is sufficient to enforce the resulting equations by point matching in the center of each cell in the model (following an approach similar to that used in Section 2.2 for normally incident TE waves). The resulting system of equations is overdetermined and can be written as

$$\begin{bmatrix} \mathbf{A} & \mathbf{0} \\ \mathbf{B} & \mathbf{C} \\ \mathbf{D} & \mathbf{E} \end{bmatrix} \begin{bmatrix} j_t \\ j_z \end{bmatrix} = \begin{bmatrix} H_z^{\text{inc}} \\ H_t^{\text{inc}} \\ H_n^{\text{inc}} \end{bmatrix} \tag{8.24}$$

The entries of (8.24) are given by

$$A_{mm} = -\tfrac{1}{2} \tag{8.25}$$

$$A_{mn} = -\int_{\text{cell } n} \left(\cos \Omega_n \frac{\partial \tilde{G}}{\partial x} + \sin \Omega_n \frac{\partial \tilde{G}}{\partial y} \right) dt' \qquad (m \neq n) \tag{8.26}$$

$$B_{mn} = -jk_z \sin(\Omega_m - \Omega_n) \int_{\text{cell } n} \tilde{G}(R_m; k, k_z)\, dt' \tag{8.27}$$

$$C_{mm} = \tfrac{1}{2} \tag{8.28}$$

$$C_{mn} = \int_{\text{cell } n} \left(\cos \Omega_m \frac{\partial \tilde{G}}{\partial x} + \sin \Omega_m \frac{\partial \tilde{G}}{\partial y} \right) dt' \qquad (m \neq n) \tag{8.29}$$

$$D_{mn} = jk_z \cos(\Omega_m - \Omega_n) \int_{\text{cell } n} \tilde{G}(R_m; k, k_z)\, dt' \tag{8.30}$$

and

$$E_{mn} = \int_{\text{cell } n} \left(\sin \Omega_m \frac{\partial \tilde{G}}{\partial x} - \cos \Omega_m \frac{\partial \tilde{G}}{\partial y} \right) dt' \tag{8.31}$$

In these expressions, $\tilde{G}(R_m; k, k_z)$ is defined in Equation (8.7) and R_m is defined in (8.17). Although these matrix elements are simpler in form than those produced by a CFIE formulation, the augmented MFIE approach involves additional overhead associated with the solution of the overdetermined $3N \times 2N$ system in Equation (8.24).

To illustrate the augmented MFIE formulation, consider a TE wave incident on a circular cylinder with three wavelength circumference at an oblique angle 36.75° from the normal. The tangential-field MFIE result for the transverse current density is shown in Figure 8.3. This particular oblique angle happens to correspond to an interior resonance, and as a result the MFIE result contains significant error. The augmented-field MFIE result is shown in Figure 8.4 and exhibits good agreement with the exact eigenfunction solution.

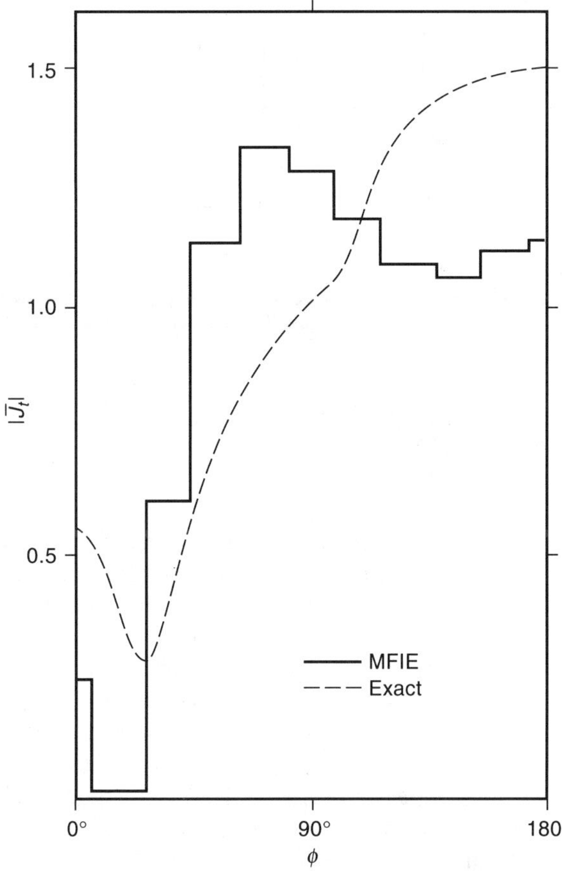

Figure 8.3 Tangential-field MFIE result for a circular cylinder with $ka = 3$ at an oblique angle $\theta^{\text{inc}} = 36.75°$.

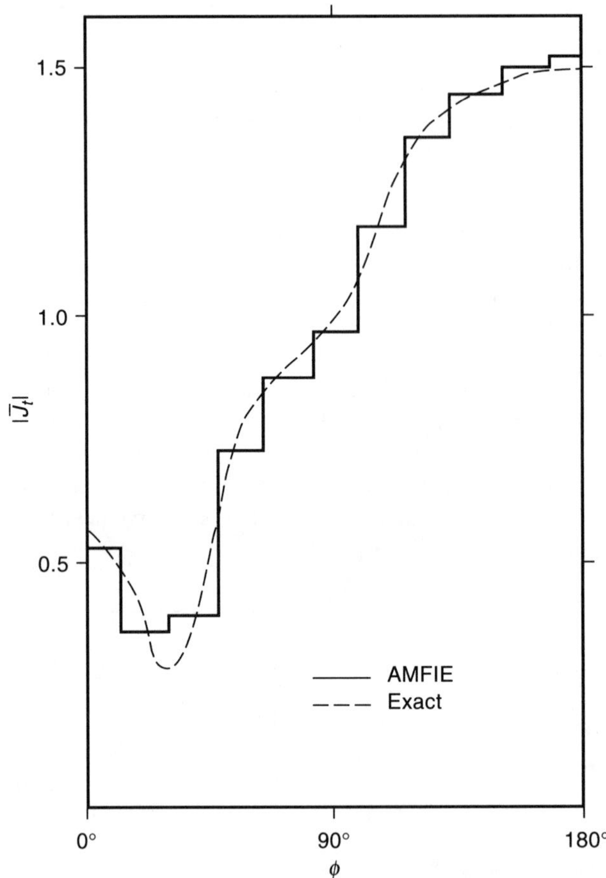

Figure 8.4 Augmented-field MFIE result for the problem in Figure 8.3.

8.4 APPLICATION: MUTUAL ADMITTANCE BETWEEN SLOT ANTENNAS [1]

To illustrate an application involving finite sources illuminating an infinite cylinder, consider the coupling between two slot antennas located on a conducting cylinder. The mutual admittance between two slots can be defined in accordance with Richmond's reaction theorem [3] and requires knowledge of the electric currents produced on a closed cylinder (at the location of aperture 2) by the source slot (aperture 1). Assuming that the tangential electric field in aperture 1 is known, an equivalent magnetic current density can be defined exterior to aperture 1 and used to generate the "incident" electric and magnetic fields in the vicinity of the cylinder as a function of the spectral variable k_z. In other words, using the Fourier transform from (8.1), the field produced by slot 1 is decomposed into a spectrum of waves incident upon the cylinder from oblique angles. Surface integral equation formulations similar to those described in Sections 8.2 and 8.3 can be used to produce the current distribution induced by each incident wave. The currents can be superimposed according to the inverse Fourier transform in Equation (8.2) to yield the total current density needed for the mutual admittance calculation.

Reference [1] describes the details of the calculation and presents a variety of numerical results for coupling between two axial slots and two circumferential slots. (For axial slots the formulation requires the treatment of the TE polarization; for circumferential slots a combination of the TM and TE polarizations is required. Therefore, the actual approach differs slightly from the purely TM and TE situations discussed in Sections 8.2 and 8.3.) For illustration, Table 8.2 presents admittance data for axial slots in a circular cylinder as a function of the displacement along the cylinder axis. For this circular cylinder example, the numerical solutions exhibit good agreement with analytical values obtained using a cylindrical eigenfunction expansion [4]. Of course, the formulation is equally applicable to arbitrarily shaped cylinders.

TABLE 8.2 Mutual Admittance between Axial Slots in Circular Cylinder of Radius 1.0λ

Separation, Δz	Numerical Result		Eigenfunction Solution [4]	
	Magnitude (dB)	Angle (deg)	Magnitude (dB)	Angle (deg)
1λ	−86.22	−168.6	−87.1	−171
2λ	−99.13	−174.3	−100.0	−174
4λ	−111.66	−174.9	−112.4	−175
8λ	−124.32	−151.0	−124.3	−174

Note: The slots have height of 0.214λ and width of 0.5λ. Sampling imposes a fictitious spatial period P_z equal to 40.5λ; the spectrum included in the inverse Fourier transformation is truncated at $k_z = 20$. The cylinder is modeled with 88 equal-sized cells for moment-method analysis based on a CFIE formulation with pulse basis functions and Dirac delta testing functions [1].

The slot coupling example provides insight into the range of some of the parameters involved in the Fourier transformation process. In the actual implementation used in reference [1], the Fourier transform of the mutual admittance function $\tilde{Y}(z)$ was computed directly at each value of k_z as $\tilde{Y}(k_z)$, then transformed back to the spatial domain. Figures 8.5 and 8.6 show $\tilde{Y}(k_z)$ for circumferential and axial slots, respectively, and demonstrate that the support of the transform quantity is bandlimited to approximately 4 or 5 times the size of the visible region ($k_z \leq 2\pi$). Results from reference [1] also suggest that the fictitious period defined in Equation (8.11) must be at least 5 times the slot spacing to ensure reasonable accuracy for closely spaced slots.

Note that the overall computational requirements associated with this procedure can be considerable, since in practice the matrix equation (8.14) must be solved at several hundred different values of k_z in order to produce the required data prior to inverse Fourier transformation!

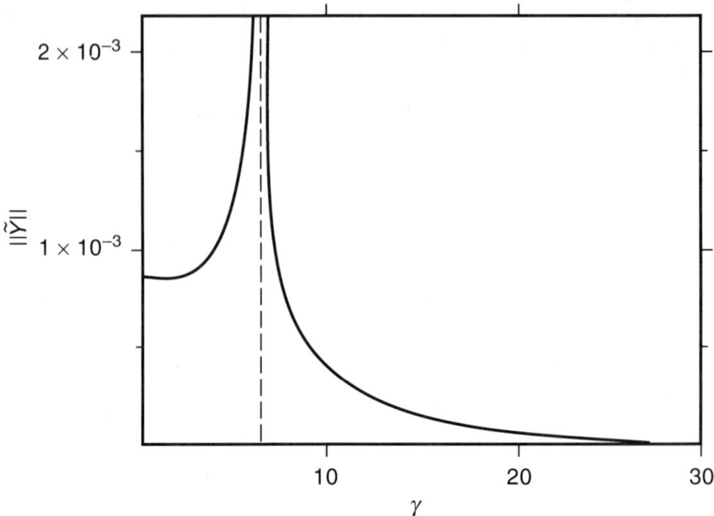

Figure 8.5 Norm of the spectral admittance function for a problem involving circumferential slots of height $0.2\lambda_0$ in a cylinder with $1.0\lambda_0$ radius. After [1]. ©1989 IEEE.

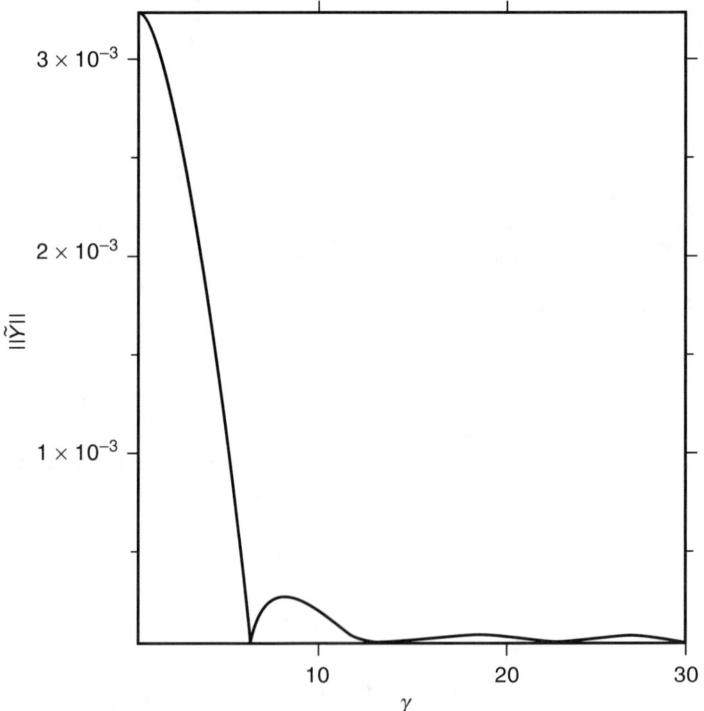

Figure 8.6 Norm of the spectral admittance function for a problem involving axial slots of width $0.686\lambda_0$ in a cylinder with $0.758\lambda_0$ radius.

8.5 OBLIQUE SCATTERING FROM INHOMOGENEOUS CYLINDERS: VOLUME INTEGRAL EQUATION FORMULATION [5]

The previous sections considered the scattering of oblique waves from perfectly conducting cylinders. In this section, we consider a volume integral equation in order to model heterogeneous material. If the cylinder is perfectly conducting, there is no coupling between the TM and TE cases. However, if the cylinder is inhomogeneous, an oblique TE or TM excitation will be depolarized into a combination of both types. Consequently, we combine the TM and TE polarizations in the following approach.

Consider an infinite, z-invariant cylinder characterized by complex-valued relative permittivity and permeability functions $\varepsilon_r(x, y)$ and $\mu_r(x, y)$. The cylinder is illuminated by a source that, in the absence of the cylinder, produces fields with z-components:

$$\hat{z} \cdot \bar{E}^{\text{inc}}(x, y, z) = E_z^{\text{inc}}(x, y)e^{jk_z z} \tag{8.32}$$

$$\hat{z} \cdot \bar{H}^{\text{inc}}(x, y, z) = H_z^{\text{inc}}(x, y)e^{jk_z z} \tag{8.33}$$

Suppose that the cylinder cross section is modeled by triangular cells of constant permittivity and permeability. The dielectric and magnetic material may be replaced by equivalent volumetric polarization currents defined in each cell according to

$$J_z = j\omega\varepsilon_0(\varepsilon_r - 1)E_z \tag{8.34}$$

$$K_z = j\omega\mu_0(\mu_r - 1)H_z \tag{8.35}$$

$$\bar{J}_t = \frac{\varepsilon_r - 1}{k^2\varepsilon_r\mu_r - k_z^2}(-k_z\omega\varepsilon_0\nabla_t E_z - k^2\mu_r\hat{z} \times \nabla_t H_z) \tag{8.36}$$

$$\bar{K}_t = \frac{\mu_r - 1}{k^2\varepsilon_r\mu_r - k_z^2}(k^2\varepsilon_r\hat{z} \times \nabla_t E_z - k_z\omega\mu_0\nabla_t H_z) \tag{8.37}$$

where ∇_t is the transverse gradient $\hat{x}\,\partial/\partial x + \hat{y}\,\partial/\partial y$ and $E_z(x, y)$ and $H_z(x, y)$ are unknown functions that provide the transverse dependence of the total field. Note that all components of the equivalent polarization current density are obtained in terms of just the z-components of the unknown fields, and therefore these z-components are the primary unknowns to be determined.

The equivalent currents of Equations (8.34)–(8.37) radiate in free space but are unknown quantities since they depend on the total field components E_z and H_z. These unknown quantities must satisfy the coupled scalar integral equations

$$E_z^{\text{inc}}(x, y) = E_z(x, y) - \frac{k^2 - k_z^2}{j\omega\varepsilon_0}(J_z * G) - \frac{k_z}{\omega\varepsilon_0}(\nabla_t \cdot \bar{J}_t * G) + (\hat{z} \cdot \nabla_t \times \bar{K}_t * G) \tag{8.38}$$

$$H_z^{\text{inc}}(x, y) = H_z(x, y) - \frac{k^2 - k_z^2}{j\omega\mu_0}(K_z * G) - \frac{k_z}{\omega\mu_0}(\nabla_t \cdot \bar{K}_t * G) - (\hat{z} \cdot \nabla_t \times \bar{J}_t * G) \tag{8.39}$$

where the asterisk denotes two-dimensional convolution, that is,

$$a * b = \int_{-\infty}^{\infty}\int_{-\infty}^{\infty} a(x', y')b(x - x', y - y')\,dx'\,dy' \tag{8.40}$$

and the Green's function is given by

$$G(x, y) = \frac{1}{4j} H_0^{(2)} \left(\sqrt{k^2 - k_z^2} \sqrt{x^2 + y^2} \right) \tag{8.41}$$

where $H_0^{(2)}$ is the zero-order Hankel function of the second kind. Equations (8.38) and (8.39) can be obtained from (1.99) and (1.100) by exchanging all z-derivatives for multiplications with jk_z.

For an arbitrary scatterer, (8.38) and (8.39) can be discretized in the context of the method of moments. We consider an approach similar to that presented in Section 2.7 and employ a piecewise-linear representation of E_z and H_z throughout triangular cells. From a global perspective, linear interpolation between the corners of a triangular cell is equivalent to the expansion of E_z and H_z in the linear pyramid basis functions depicted in Figure 2.20. If $B_n(x, y)$ denotes a pyramid basis function centered at the nth corner or node within the triangular-cell model of the cylinder cross section, the total fields can be expressed as

$$E_z(x, y) \cong \sum_{n=1}^{N} e_n B_n(x, y) \tag{8.42}$$

$$H_z(x, y) \cong \sum_{n=1}^{N} h_n B_n(x, y) \tag{8.43}$$

Since the z-components of the scattered E_z- and H_z-fields will be continuous across cell boundaries, the integral equations can be enforced by point matching at the corners of each cell. This procedure produces a matrix equation of the form

$$\begin{bmatrix} \mathbf{EE} & \mathbf{EH} \\ \mathbf{HE} & \mathbf{HH} \end{bmatrix} \begin{bmatrix} e_1 \\ \vdots \\ e_N \\ h_1 \\ \vdots \\ h_N \end{bmatrix} = \begin{bmatrix} E_z^{\text{inc}}(x_1, y_1) \\ \vdots \\ E_z^{\text{inc}}(x_N, y_N) \\ H_z^{\text{inc}}(x_1, y_1) \\ \vdots \\ H_z^{\text{inc}}(x_N, y_N) \end{bmatrix} \tag{8.44}$$

having entries

$$\begin{aligned} EE_{mn} = {} & \delta_n^m - (k^2 - k_z^2)[(\varepsilon_r - 1) B_n * G]_{x=x_m, y=y_m} \\ & + k_z^2 \left[\nabla_t \cdot \left(\frac{\varepsilon_r - 1}{k^2 \varepsilon_r \mu_r - k_z^2} \nabla_t B_n \right) * G \right]_{x=x_m, y=y_m} \\ & + k^2 \left[\hat{z} \cdot \nabla_t \times \left(\frac{\varepsilon_r (\mu_r - 1)}{k^2 \varepsilon_r \mu_r - k_z^2} \hat{z} \times \nabla_t B_n \right) * G \right]_{x=x_m, y=y_m} \end{aligned} \tag{8.45}$$

$$\begin{aligned} EH_{mn} = {} & \frac{k^2 k_z}{\omega \varepsilon_0} \left[\nabla_t \cdot \left(\frac{(\varepsilon_r - 1)\mu_r}{k^2 \varepsilon_r \mu_r - k_z^2} \hat{z} \times \nabla_t B_n \right) * G \right]_{x=x_m, y=y_m} \\ & - k_z \omega \mu_0 \left[\hat{z} \cdot \nabla_t \times \left(\frac{\mu_r - 1}{k^2 \varepsilon_r \mu_r - k_z^2} \nabla_t B_n \right) * G \right]_{x=x_m, y=y_m} \end{aligned} \tag{8.46}$$

$$HE_{mn} = -\frac{k^2 k_z}{\omega\mu_0}\left[\nabla_t \cdot \left(\frac{(\mu_r - 1)\varepsilon_r}{k^2\varepsilon_r\mu_r - k_z^2}\hat{z} \times \nabla_t B_n\right) * G\right]_{x=x_m, y=y_m}$$
$$+ k_z\omega\varepsilon_0\left[\hat{z} \cdot \nabla_t \times \left(\frac{\varepsilon_r - 1}{k^2\varepsilon_r\mu_r - k_z^2}\nabla_t B_n\right) * G\right]_{x=x_m, y=y_m} \quad (8.47)$$

and

$$HH_{mn} = \delta_n^m - (k^2 - k_z^2)[(\mu_r - 1)B_n * G]_{x=x_m, y=y_m}$$
$$+ k_z^2\left[\nabla_t \cdot \left(\frac{\mu_r - 1}{k^2\varepsilon_r\mu_r - k_z^2}\nabla_t B_n\right) * G\right]_{x=x_m, y=y_m} \quad (8.48)$$
$$+ k^2\left[\hat{z} \cdot \nabla_t \times \left(\frac{\mu_r(\varepsilon_r - 1)}{k^2\varepsilon_r\mu_r - k_z^2}\hat{z} \times \nabla_t B_n\right) * G\right]_{x=x_m, y=y_m}$$

where δ_n^m is the Kronecker delta function. Because the z-components of the fields are linear functions within this representation, transverse fields and transverse current densities are constant within a cell. Furthermore, derivatives of the piecewise-constant functions must be interpreted as generalized functions. As a result, the differential operators in (8.45)–(8.48) produce Dirac delta functions having support limited to the edges between cells. Both convolutions appearing in Equations (8.46) and (8.47) and the last two convolutions in Equations (8.45) and (8.48) involve functions with support only over cell edges. Thus, these are actually one-dimensional convolution integrals. The first convolution in Equation (8.45) and the first convolution in (8.48) are two-dimensional integrals. In general, all of the convolution integrals must be evaluated by numerical quadrature.

Each entry in (8.45)–(8.48) represents a contribution from a source that is distributed over several adjacent cells. The direct evaluation of these expressions, although straightforward, involves redundant computations since the same integrals arise again and again. As previously illustrated in Chapters 2 and 3, it is usually more efficient to adopt an indirect approach in which the necessary integrals are evaluated only once over each cell face or cell edge and then added to the appropriate locations throughout the matrix. The indirect approach is conceptually more difficult to follow, but it requires the minimum data structure to link cells, edges, and corner nodes within a general triangular-cell model.

The coupled integral equation formulation is directly applicable to arbitrary inhomogeneous cylinders, and a variety of results illustrating the accuracy of the formulation have been presented in the literature [5]. The bistatic scattering cross section, defined as

$$\sigma(\phi) = \lim_{\rho\to\infty} 2\pi\rho\frac{|E_z^s(\rho,\phi)|^2 + \eta^2|H_z^s(\rho,\phi)|^2}{|E_z^{\text{inc}}(0,0)|^2 + \eta^2|H_z^{\text{inc}}(0,0)|^2} \quad (8.49)$$

can be computed as a secondary calculation. As an example, consider the oblique scattering from a finite rectangular slab of dimension 2.0λ by 0.1λ, a geometry for which there is no known exact solution. Results for the bistatic scattering cross section are presented in Figure 8.7 for TM incidence and Figure 8.8 for TE incidence for two permittivity profiles, the homogeneous case with $\varepsilon_r = 3 - j0.3$ and $\mu_r = 2 - j0.1$ and an inhomogeneous slab with permittivity and permeability profiles given by

$$\varepsilon_r = 3 - j0.3 + 2\cos(0.5\pi y) \quad (8.50)$$

$$\mu_r = 1 + (1 - j0.05)|y| \quad (8.51)$$

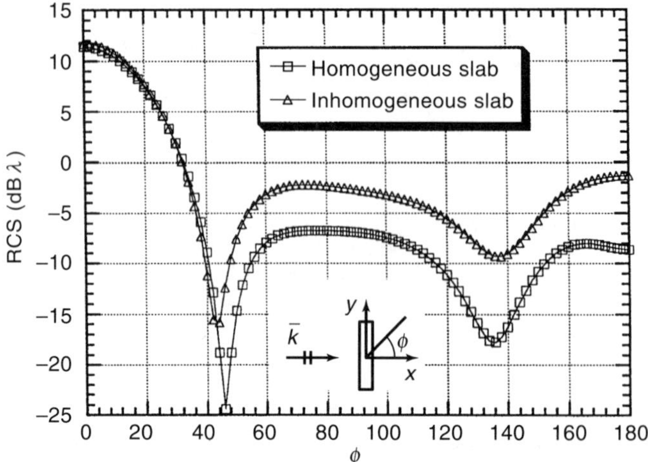

Figure 8.7 Bistatic scattering cross section for an inhomogeneous dielectric cylinder and a homogeneous cylinder when illuminated by a TM wave incident at an oblique angle of 45°. The cylinders are of cross-sectional dimension $2\lambda_0$ by $0.1\lambda_0$. After [5]. ©1991 IEEE.

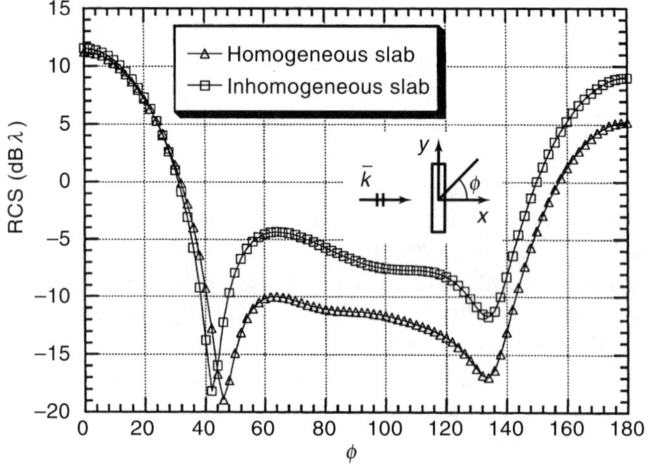

Figure 8.8 Bistatic scattering cross section for the cylinders of Figure 8.7 when illuminated by a TE wave incident at an oblique angle of 45°. After [5]. ©1991 IEEE.

where y is specified in units of wavelength (λ). The y-axis is located along the long dimension of the slab cross section with the origin in the center of the slab. The incident field propagates in the x–z plane (broadside incidence). A triangular-cell model having 50 nodes along the long dimension and 3 nodes across the short dimension is employed, to yield 300 unknowns and a density (for the homogeneous-slab problem) of approximately 250 unknowns per square dielectric wavelength. These curves are in agreement with data from Figure 6 of reference [6].

The preceding formulation may be extended to include closed, perfectly conducting regions imbedded within the cylinder geometry. Perfectly conducting material may be modeled by an equivalent electric surface current density radiating in the presence of the equivalent polarization currents defined in (8.34)–(8.37). For oblique incidence, both $\hat{z}$- and $\hat{t}$-components of current are excited. Equations (8.38) and (8.39) can be modified to include appropriate surface integrals over the additional sources, and we leave this modification as an exercise for the reader.

If a perfectly conducting region is in direct contact with the dielectric or magnetic material, the H_z-field component at the surface directly defines the transverse electric current density. The linear pyramid function representation for H_z translates into a piecewise-linear representation along the surface, and it is not necessary to introduce additional unknowns to represent transverse currents on the conductor. Although the z-component of the electric current density could in principle be defined by the normal derivative of E_z at the surface of the conductor, approximations inherent in this type of numerical derivative have been found to introduce appreciable error. Thus, an independent expansion can be introduced for the z-component of the surface current in terms of pulse (piecewise-constant) or triangle (piecewise-linear) basis functions defined along each edge of the triangular-cell model bordering the conductor. The coefficients of these basis functions are additional unknowns to be determined. However, E_z should vanish at the surface of a perfect electric conductor, meaning that the coefficients of the pyramid basis functions for E_z along the conductor surface may be set equal to zero. Consequently, the additional unknowns for J_z are balanced by an equal reduction in the number of unknown coefficients for E_z. In summary, the total number of unknowns is given by twice the number of nodes in the triangular-cell model, despite any imbedded conductors.

8.6 OBLIQUE SCATTERING FROM INHOMOGENEOUS CYLINDERS: SCALAR DIFFERENTIAL EQUATION FORMULATION [7]

Volume integral equation formulations, such as the approach of the preceding section, involve fully populated matrix equations and require a relatively large computational effort to treat scatterers of moderate electrical size. Our previous investigation in Chapter 3 suggests that the computational effort associated with differential equation formulations can be significantly less than that required for volume integral formulations, because of the sparse nature of the matrices, and motivates an extension of the scalar differential equation methods to treat oblique scattering.

Figure 8.9 depicts the cross section of an infinite, z-invariant cylinder characterized by complex-valued relative permittivity and permeability functions $\varepsilon_r(x, y)$ and $\mu_r(x, y)$. A conducting region having boundary $\partial\Gamma_c$ is also imbedded within the cylinder. The cylinder is illuminated by an incident electromagnetic field with z-components described by the general form in Equations (8.32) and (8.33). In common with the volume integral formulation of Section 8.5, we suppose that the cylinder cross section is modeled by triangular cells of constant permittivity and permeability. Furthermore, we wish to employ the total fields E_z and H_z as the primary unknowns to be determined and obtain the transverse field components within a homogeneous cell via the relations

$$\bar{E}_t = \frac{1}{k^2\varepsilon_r\mu_r - k_z^2}(j\omega\mu_0\mu_r\hat{z} \times \nabla_t H_z + jk_z\nabla_t E_z) \tag{8.52}$$

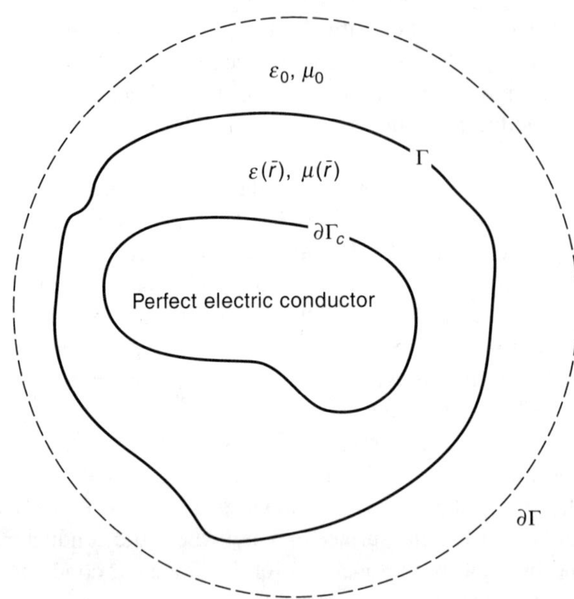

Figure 8.9 Cross section of a general inhomogeneous cylinder.

and

$$\bar{H}_t = \frac{1}{k^2 \varepsilon_r \mu_r - k_z^2}(jk_z \nabla_t H_z - j\omega\varepsilon_0\varepsilon_r \hat{z} \times \nabla_t E_z) \tag{8.53}$$

where ∇_t is the transverse gradient.

To derive weak differential equations for E_z and H_z, observe that the fields in the vicinity of the scatterer must satisfy Maxwell's curl equations:

$$\nabla \times \bar{H} = j\omega\varepsilon_0\varepsilon_r \bar{E} \tag{8.54}$$

$$\nabla \times \bar{E} = -j\omega\mu_0\mu_r \bar{H} \tag{8.55}$$

The scalar product of (8.54) and (8.55) with the testing function

$$\bar{T}(x, y, z) = \hat{z}T(x, y) \tag{8.56}$$

can be integrated over the domain Γ to produce

$$\iint_\Gamma \bar{T} \cdot \nabla \times \bar{H} - j\omega\varepsilon_0\varepsilon_r \bar{T} \cdot \bar{E} = 0 \tag{8.57}$$

$$\iint_\Gamma \bar{T} \cdot \nabla \times \bar{E} + j\omega\mu_0\mu_r \bar{T} \cdot \bar{H} = 0 \tag{8.58}$$

The two-dimensional divergence theorem

$$\iint_\Gamma \nabla \cdot (\bar{T} \times \bar{H}) = \int_{\partial\Gamma + \partial\Gamma_c} (\bar{T} \times \bar{H}) \cdot \hat{n} \tag{8.59}$$

and the vector identities

$$\bar{T} \cdot \nabla \times \bar{H} = \nabla \times \bar{T} \cdot \bar{H} - \nabla \cdot (\bar{T} \times \bar{H}) \tag{8.60}$$

and

$$(\bar{T} \times \bar{H}) \cdot \hat{n} = -\bar{T} \cdot (\hat{n} \times \bar{H}) \tag{8.61}$$

can be employed to convert Equations (8.57) and (8.58) into

$$\iint_\Gamma \nabla \times \bar{T} \cdot \bar{H}_t - j\omega\varepsilon_0\varepsilon_r T E_z = -\int_{\partial\Gamma+\partial\Gamma_c} \bar{T} \cdot (\hat{n} \times \bar{H}) \tag{8.62}$$

$$\iint_\Gamma \nabla \times \bar{T} \cdot \bar{E}_t + j\omega\mu_0\mu_r T H_z = -\int_{\partial\Gamma+\partial\Gamma_c} \bar{T} \cdot (\hat{n} \times \bar{E}) \tag{8.63}$$

where $\partial\Gamma + \partial\Gamma_c$ denotes the entire boundary of the region Γ, with normal $\hat{n}$ pointing out of the region (in accordance with Figure 8.9, $\partial\Gamma_c$ denotes the surface of the imbedded conductor; $\partial\Gamma$ denotes the exterior boundary on which the computational domain is to be truncated). It is assumed that all inhomogeneities are completely enclosed by the exterior boundary.

Following an approach similar to that introduced in Section 3.2, boundary conditions on the imbedded perfect conductor will be incorporated via the surface integrals over $\partial\Gamma_c$. Since the tangential electric field must vanish on the perfectly conducting surface, E_z is a known function on $\partial\Gamma_c$ and satisfies a homogeneous Dirichlet boundary condition. Consequently, all the testing functions used to discretize (8.62) will vanish on $\partial\Gamma_c$, and the boundary integral will contribute nothing to the matrix equation. In the case of the similar boundary integral appearing in Equation (8.63), it is also desired that the tangential electric field vanish on $\partial\Gamma_c$. This is equivalent to a homogeneous Neumann boundary condition and is usually enforced by ignoring the integral over $\partial\Gamma_c$. As a result, the integrals over $\partial\Gamma_c$ in (8.62) and (8.63) will be ignored throughout the rest of the development. Note that an impedance boundary condition can be incorporated on $\partial\Gamma_c$ in an obvious way, in which case the integrals would not vanish. Along a perfectly conducting boundary $\partial\Gamma_c$, the basis function representation for E_z must explicitly satisfy the zero boundary condition. In contrast, no special conditions are to be imposed on the expansion for H_z along $\partial\Gamma_c$. While E_z is known along the conducting boundary, H_z is an unknown function on the surface.

The expressions for transverse fields in Equations (8.52) and (8.53) may be substituted into (8.62) and (8.63) and combined with the vector identity

$$\nabla \times \bar{T} = -\hat{z} \times \nabla_t T \tag{8.64}$$

to cast the resulting equations into the form

$$\iint_\Gamma \frac{\varepsilon_r}{\varepsilon_r\mu_r - (k_z^2/k^2)} \nabla_t T \cdot \nabla_t E_z - k^2\varepsilon_r T E_z + \iint_\Gamma \frac{-k_z}{\omega\varepsilon_0} \frac{1}{\varepsilon_r\mu_r - (k_z^2/k^2)} \hat{z} \cdot \nabla_t T \times \nabla_t H_z = -\int_{\partial\Gamma} \bar{T} \cdot \hat{n} \times \nabla \times \bar{E} \tag{8.65}$$

and

$$\iint_\Gamma \frac{\mu_r}{\varepsilon_r\mu_r - (k_z^2/k^2)} \nabla_t T \cdot \nabla_t H_z - k^2\mu_r T H_z + \iint_\Gamma \frac{k_z}{\omega\mu_0} \frac{1}{\varepsilon_r\mu_r - (k_z^2/k^2)} \hat{z} \cdot \nabla_t T \times \nabla_t E_z = -\int_{\partial\Gamma} \bar{T} \cdot \hat{n} \times \nabla \times \bar{H} \tag{8.66}$$

Equations (8.65) and (8.66) constitute coupled weak differential equations describing the electromagnetic fields throughout the region Γ in terms of the total fields on the boundary $\partial\Gamma$. For a normally incident excitation, these equations reduce to (3.5) for the TM polarization and (3.6) for the TE polarization. In order to represent the complete scattering problem, it is necessary to augment these equations with additional information about the fields on $\partial\Gamma$. For instance, an inward-looking formulation can be obtained by combining these equations with an appropriate surface integral equation or eigenfunction expansion representing the exterior region, as described in Section 3.12. On the other hand, an outward-looking formulation can be obtained by incorporating some type of RBC directly into the boundary integrals.

We limit our investigation to one specific outward-looking approach and for simplicity assume that the boundary $\partial\Gamma$ is circular with radius a. An exact RBC can be derived from an exterior eigenfunction expansion following the procedure of Section 3.3, and we leave the details of the development to the reader (Prob. P8.9). For the case of oblique excitation, the pertinent RBCs can be written as

$$\hat{z} \cdot \hat{\rho} \times (\nabla \times \bar{E})|_{(a,\phi,0)} = \frac{1}{2\pi} \int_0^{2\pi} E_z(a, \phi') K_1(\phi - \phi')\, d\phi' + \eta \frac{1}{2\pi} \int_0^{2\pi} H_z(a, \phi') K_2(\phi - \phi')\, d\phi' + \frac{1}{2\pi} \int_0^{2\pi} E_z^{\text{inc}}(a, \phi') K_3(\phi - \phi')\, d\phi' \tag{8.67}$$

and

$$\hat{z} \cdot \hat{\rho} \times (\nabla \times \bar{H})|_{(a,\phi,0)} = \frac{1}{2\pi} \int_0^{2\pi} H_z(a, \phi') K_1(\phi - \phi')\, d\phi' - \frac{1}{\eta} \frac{1}{2\pi} \int_0^{2\pi} E_z(a, \phi') K_2(\phi - \phi')\, d\phi' + \frac{1}{2\pi} \int_0^{2\pi} H_z^{\text{inc}}(a, \phi') K_3(\phi - \phi')\, d\phi' \tag{8.68}$$

where the kernels are defined as

$$K_1(\phi) = -\frac{k^2}{k_t} \sum_{n=-\infty}^{\infty} \frac{H_n^{(2)\prime}(k_t a)}{H_n^{(2)}(k_t a)} e^{jn\phi} \tag{8.69}$$

$$K_2(\phi) = \frac{jkk_z}{k_t^2 a} \sum_{n=-\infty}^{\infty} n e^{jn\phi} \tag{8.70}$$

$$K_3(\phi) = -\frac{j2k^2}{\pi k_t^2 a} \sum_{n=-\infty}^{\infty} \frac{1}{J_n(k_t a) H_n^{(2)}(k_t a)} e^{jn\phi} \tag{8.71}$$

and where

$$k_t = \sqrt{k^2 - k_z^2} \tag{8.72}$$

Although the summations appearing in Equations (8.69) and (8.70) are divergent, the required calculations in Equations (8.67) and (8.68) involve only integrals over these kernels. Because the fields are well behaved on $\partial\Gamma$, the cylindrical harmonic content of the functions $E_z(\phi)$ and $H_z(\phi)$ ensures that the integrals are convergent and relatively easy to compute.

Equations (8.67) and (8.68) may be substituted into the boundary integrals in (8.65) and (8.66) to complete the formulation. The resulting equations can be discretized following the finite-element procedure using a piecewise-linear representation for E_z and H_z and piecewise-linear testing functions, to produce the matrix equation

$$\begin{bmatrix} \mathbf{A} & \mathbf{B} \\ \mathbf{C} & \mathbf{D} \end{bmatrix} \begin{bmatrix} e \\ h \end{bmatrix} = \begin{bmatrix} e^{\text{inc}} \\ h^{\text{inc}} \end{bmatrix} \tag{8.73}$$

where

$$\begin{aligned} A_{mn} = &\iint_\Gamma \frac{\varepsilon_r}{\varepsilon_r \mu_r - (k_z^2/k^2)} \nabla_t B_m \cdot \nabla_t B_n - k^2 \varepsilon_r B_m B_n \\ &- \frac{k^2 a}{2\pi k_t} \sum_{q=-\infty}^{\infty} \frac{H_q^{(2)\prime}(k_t a)}{H_q^{(2)}(k_t a)} I_m(q) I_n(-q) \end{aligned} \tag{8.74}$$

$$\begin{aligned} B_{mn} = &\iint_\Gamma \frac{-k_z}{\omega \varepsilon_0} \frac{1}{\varepsilon_r \mu_r - (k_z^2/k^2)} \hat{z} \cdot \nabla_t B_m \times \nabla_t B_n \\ &+ \frac{jkk_z \eta}{2\pi k_t^2} \sum_{q=-\infty}^{\infty} q I_m(q) I_n(-q) \end{aligned} \tag{8.75}$$

$$\begin{aligned} C_{mn} = &\iint_\Gamma \frac{k_z}{\omega \mu_0} \frac{1}{\varepsilon_r \mu_r - (k_z^2/k^2)} \hat{z} \cdot \nabla_t B_m \times \nabla_t B_n \\ &- \frac{jkk_z}{2\pi k_t^2 \eta} \sum_{q=-\infty}^{\infty} q I_m(q) I_n(-q) \end{aligned} \tag{8.76}$$

and

$$\begin{aligned} D_{mn} = &\iint_\Gamma \frac{\mu_r}{\varepsilon_r \mu_r - (k_z^2/k^2)} \nabla_t B_m \cdot \nabla_t B_n - k^2 \mu_r B_m B_n \\ &- \frac{k^2 a}{2\pi k_t} \sum_{q=-\infty}^{\infty} \frac{H_q^{(2)\prime}(k_t a)}{H_q^{(2)}(k_t a)} I_m(q) I_n(-q) \end{aligned} \tag{8.77}$$

where

$$I_m(q) = \int_{\partial\Gamma} B_m(\phi) e^{jq\phi} \, d\phi \tag{8.78}$$

only contributes to the preceding expressions if node m is located on the boundary $\partial\Gamma$.

Assuming the excitation is a uniform plane wave of the form

$$\begin{aligned} \begin{Bmatrix} E_z^{\text{inc}} \\ H_z^{\text{inc}} \end{Bmatrix} &= \begin{Bmatrix} e_0 \\ h_0 \end{Bmatrix} e^{-jk_t(x\cos\theta + y\sin\theta)} e^{jk_z z} \\ &= \begin{Bmatrix} e_0 \\ h_0 \end{Bmatrix} \sum_{q=-\infty}^{\infty} j^{-q} J_q(k_t a) e^{jq\phi} e^{-jq\theta} e^{jk_z z} \end{aligned} \tag{8.79}$$

the entries of the excitation vector are given by

$$e_m^{\text{inc}} = \frac{j2k^2 e_0}{\pi k_t^2} \sum_{q=-\infty}^{\infty} \frac{j^{-q} e^{-jq\theta}}{H_q^{(2)}(k_t a)} I_m(q) \tag{8.80}$$

and

$$h_m^{\text{inc}} = \frac{j2k^2 h_0}{\pi k_t^2} \sum_{q=-\infty}^{\infty} \frac{j^{-q} e^{-jq\theta}}{H_q^{(2)}(k_t a)} I_m(q) \tag{8.81}$$

The element matrix entries required for the finite-element system have been previously discussed in Chapter 3, with the sole exception of the integral

$$E_{mn} = \iint_{\Gamma} \hat{z} \cdot \nabla_t T_m \times \nabla_t B_n \, dx \, dy \tag{8.82}$$

Equation (8.82) can be evaluated over a triangular cell using the simplex coordinates (L_1, L_2, L_3) defined in Equations (3.48)–(3.51) of Section 3.7. Using Equation (3.57), we immediately obtain

$$E_{mn} = \frac{b_m c_n - c_m b_n}{4A} \tag{8.83}$$

where A denotes the area of the cell and $\{b_n\}$ and $\{c_n\}$ are functions of the cell shape defined in (3.50) and (3.51).

We will illustrate the performance of this outward-looking formulation by comparing some numerical results with those obtained using the volume integral approach from Section 8.5. Table 8.3 shows numerical values of the E_z-field within a circular, homogeneous dielectric cylinder having $k_t a = 1$ and relative permittivity $\varepsilon_r = 4 - j1$. Results obtained from the volume integral formulation of Section 8.5 and the volume differential formulation of this section using an identical model for the cylinder cross section are compared with the exact solution. The 121-node, 204-cell model contains approximately 369 nodes/λ_d^2. The excitation is purely TM and is incident at an oblique angle of 45° with respect to the cylinder axis. The two numerical results exhibit excellent agreement with each other and the exact solution and demonstrate once again that the accuracy of differential and integral equation formulations is comparable for a similar basis function set.

TABLE 8.3 E_z-Field Produced by Incident TM Plane Wave Along $y = 0$ Cut Through Center of Circular, Homogeneous Dielectric Cylinder Having $k_t a = 1$ and Relative Permittivity $\varepsilon_r = 4 - j1$.

x	Magnitude of E_z			Phase of E_z		
	Integral	Differential	Exact	Integral	Differential	Exact
–0.159	0.934	0.937	0.931	30.74	30.49	30.61
–0.128	0.925	0.927	0.923	11.83	11.61	11.72
–0.096	0.903	0.905	0.902	–8.09	–8.25	–8.22
–0.064	0.891	0.892	0.889	–29.49	–29.59	–29.63
–0.032	0.911	0.912	0.908	–51.59	–51.63	–51.75
0.0	0.974	0.974	0.969	–72.38	–72.41	–72.64
0.032	1.066	1.066	1.060	–90.24	–90.27	–90.61
0.064	1.159	1.158	1.152	–104.80	–104.83	–105.27
0.096	1.222	1.222	1.215	–116.58	–116.61	–117.12
0.128	1.234	1.234	1.227	–126.32	–126.35	–126.91
0.159	1.179	1.182	1.174	–134.62	–134.72	–135.23

Note: Results obtained from the volume integral formulation of Section 8.5 and the volume differential formulation of this section using the identical 161-node model are compared with the exact solution. The excitation is incident at an oblique angle of 45° with respect to the cylinder axis.

As a second example, consider a high-contrast circular cylinder with $k_t a = 0.25$ and relative permittivity $\varepsilon_r = 50 - j20$ (a possible model for muscle tissue at microwave frequencies). Numerical results for this geometry also exhibit excellent agreement with the integral equation formulation of Section 8.5 and with the exact eigenfunction solution. For illustration, Figure 8.10 shows a plot of the E_z-field produced along the outer surface of the cylinder in response to a TE wave incident at an oblique angle of 30°. These data were obtained using a cylinder model containing 138 triangular cells and a density of 317 nodes/λ_d^2. Even for the large-magnitude relative permittivity employed in this example, both numerical results exhibit excellent agreement with the exact eigenfunction series solution. In addition, the nonzero E_z-component clearly shows the depolarization occurring for oblique scattering, since the excitation in this case is purely TE.

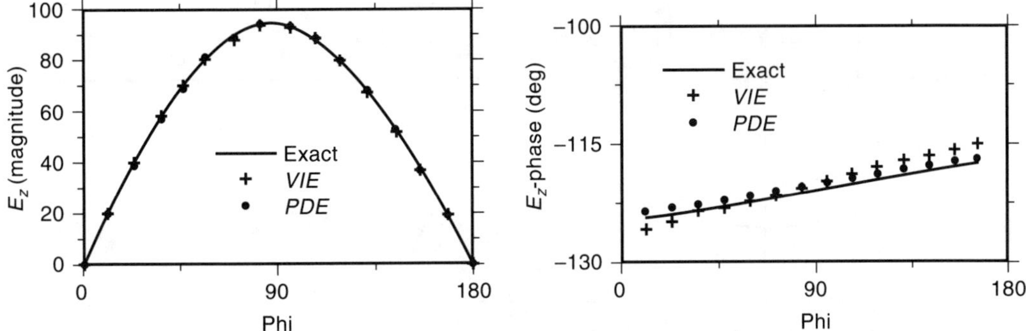

Figure 8.10 The E_z-field produced on the surface of a circular dielectric cylinder with $ka = 0.25$ and $\varepsilon_r = 50 - j20$ by a TE wave incident at an oblique angle of 30°. After [8]. ©1994 IEEE.

8.7 SCATTERING FROM A FINITE-LENGTH, HOLLOW CONDUCTING RIGHT-CIRCULAR CYLINDER: THE BODY-OF-REVOLUTION EFIE FORMULATION [9]

Preceding examples have considered infinite cylinders illuminated by a source that is a function of z. Although the resulting problem is three dimensional in nature, a Fourier decomposition along the cylinder axis can be used to reduce the problem to a superposition of two-dimensional problems more amenable to numerical solution. Axisymmetric scatterers illuminated by a general incident field can also be treated by the superposition of two-dimensional problems. Since the surface of an axisymmetric geometry can be obtained by rotating a "generating arc" around an axis, these structures are often described as *bodies of revolution.*[1] Assuming that the axis of revolution coincides with the z-axis in a cylindrical system, a Fourier series in ϕ can be employed to recast the problem into one involving uncoupled equations for each harmonic.

The body-of-revolution formulation will be illustrated by one simple geometry: a finite-length, hollow, perfectly conducting circular cylinder. The perfectly conducting material may be modeled by equivalent electric currents radiating in free space. Using the coordinate system defined in Figure 8.11, the current density contains both $\hat{z}$- and $\hat{\phi}$-components. Figure 8.11 also depicts the discretization of the cylinder surface as a function

[1]The term "body of revolution" is not intended to imply physical motion of the scatterer.

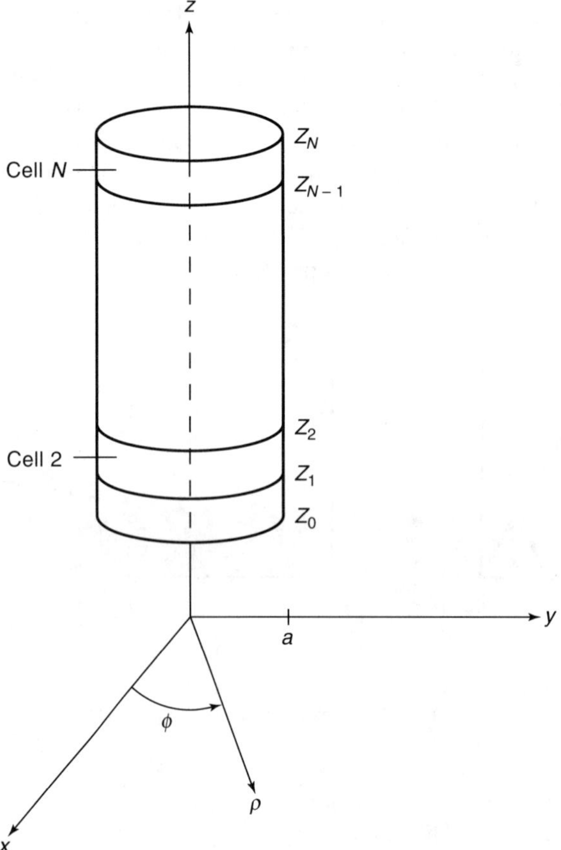

Figure 8.11 Geometry of the hollow cylinder under consideration.

of z. A suitable form of the EFIE is given by

$$\hat{n} \times \bar{E}^{\text{inc}} = -\hat{n} \times (-j\omega\mu_0 \bar{A} - \nabla\Phi_e) \tag{8.84}$$

where $\hat{n}$ is the outward normal vector at the cylinder surface. The magnetic vector and electric scalar potentials are defined as

$$\bar{A}(a, \phi, z) = \int_{z'=z_0}^{z_N} \int_{\phi'=0}^{2\pi} \bar{J}(\phi', z') \frac{e^{-jkR}}{4\pi R} a \, d\phi' \, dz' \tag{8.85}$$

and

$$\Phi_e(a, \phi, z) = \frac{-1}{j\omega\varepsilon_0} \int_{z'=z_0}^{z_N} \int_{\phi'=0}^{2\pi} (\nabla' \cdot \bar{J}) \frac{e^{-jkR}}{4\pi R} a \, d\phi' \, dz' \tag{8.86}$$

where

$$R = \sqrt{(z - z')^2 + 4a^2 \sin^2[\tfrac{1}{2}(\phi - \phi')]} \tag{8.87}$$

Note that Equation (8.84) is valid only on the surface of the original cylinder.

Because the unknown current densities, the incident fields, and the Green's function are periodic in ϕ, these quantities can each be expressed as a Fourier series according to

$$J_\phi(\phi', z') = \sum_{m=-\infty}^{\infty} J_{\phi m}(z') e^{jm\phi'} \tag{8.88}$$

$$J_z(\phi', z') = \sum_{m=-\infty}^{\infty} J_{zm}(z') e^{jm\phi'} \tag{8.89}$$

$$E_\phi^{\text{inc}}(\phi, z) = \sum_{m=-\infty}^{\infty} E_{\phi m}^{\text{inc}}(z) e^{jm\phi} \tag{8.90}$$

$$E_z^{\text{inc}}(\phi, z) = \sum_{m=-\infty}^{\infty} E_{zm}^{\text{inc}}(z) e^{jm\phi} \tag{8.91}$$

and

$$\frac{e^{-jkR}}{4\pi R} = \sum_{m=-\infty}^{\infty} G_m(z - z') e^{jm(\phi-\phi')} \tag{8.92}$$

where $J_{\phi m}(z')$ and $J_{zm}(z')$ are unknowns to be determined,

$$E_{\phi m}^{\text{inc}}(z) = \frac{1}{2\pi} \int_{\alpha=-\pi}^{\pi} E_\phi^{\text{inc}}(a, \alpha, z) e^{-jm\alpha} \, d\alpha \tag{8.93}$$

$$E_{zm}^{\text{inc}}(z) = \frac{1}{2\pi} \int_{\alpha=-\pi}^{\pi} E_z^{\text{inc}}(a, \alpha, z) e^{-jm\alpha} \, d\alpha \tag{8.94}$$

and

$$G_m(z - z') = \frac{1}{2\pi} \int_{\alpha=-\pi}^{\pi} \frac{e^{-jk\tilde{R}}}{4\pi \tilde{R}} e^{-jm\alpha} \, d\alpha \tag{8.95}$$

Equation (8.95) can be simplified to

$$G_m(z - z') = \frac{1}{4\pi^2} \int_{\alpha=0}^{\pi} \frac{e^{-jk\tilde{R}}}{\tilde{R}} \cos(m\alpha) \, d\alpha \tag{8.96}$$

where

$$\tilde{R} = \sqrt{(z - z')^2 + 4a^2 \sin^2(\tfrac{1}{2}\alpha)} \tag{8.97}$$

By substituting the above expansions into the integral equation and taking an inner product of both sides with an individual Fourier harmonic such as

$$\frac{1}{2\pi} e^{-jp\phi} \tag{8.98}$$

the original equation is separated into independent equations for each of the Fourier harmonics $J_{\phi m}$ and J_{zm}. The coupled system for the mth harmonic can be written as

$$-E_{\phi m}^{\text{inc}}(z) = \frac{2\pi a}{j\omega\varepsilon_0} \int_{z'=z_0}^{z_N} \left\{ k^2 J_{\phi m}(z') \frac{G_{m-1}(z-z') + G_{m+1}(z-z')}{2} + \left[-\frac{m^2}{a^2} J_{\phi m}(z') + j\frac{m}{a}\frac{\partial J_{zm}}{\partial z'} \right] G_m(z-z') \right\} dz' \tag{8.99}$$

$$-E_{zm}^{\text{inc}}(z) = \frac{2\pi a}{j\omega\varepsilon_0} \int_{z'=z_0}^{z_N} k^2 J_{zm}(z') G_m(z-z')\, dz' + \frac{2\pi a}{j\omega\varepsilon_0}\frac{\partial}{\partial z} \int_{z'=z_0}^{z_N} \left[j\frac{m}{a} J_{\phi m}(z') + \frac{\partial J_{zm}}{\partial z'} \right] G_m(z-z')\, dz' \tag{8.100}$$

We will consider a general plane-wave incident field of the form

$$\bar{E}^{\text{inc}} = [u_y \hat{y} + u_\theta(\hat{x}\cos\theta_i + \hat{z}\sin\theta_i)] e^{-jk(z\cos\theta_i - x\sin\theta_i)} \tag{8.101}$$

where θ is the usual spherical angle measured from the z-axis. After substituting (8.101) into Equations (8.93) and (8.94), we obtain

$$E_{zm}^{\text{inc}}(z) = u_\theta \sin\theta_i\, j^m J_m(ka\sin\theta_i) e^{-jkz\cos\theta_i} \tag{8.102}$$

and

$$E_{\phi m}^{\text{inc}}(z) = \tfrac{1}{2}\{u_\theta \cos\theta_i\, j^m [J_{m-1}(ka\sin\theta_i) + J_{m+1}(ka\sin\theta_i)] + u_y j^{m-1}[J_{m-1}(ka\sin\theta_i) - J_{m+1}(ka\sin\theta_i)]\} e^{-jkz\cos\theta_i} \tag{8.103}$$

as explicit expressions for the mth harmonics. These expression were obtained using

$$\int_0^\pi e^{jka\cos\phi\sin\theta_i} \cos(m\phi)\, d\phi = j^m \pi J_m(ka\sin\theta_i) \tag{8.104}$$

where J_m is the ordinary Bessel function.

Equations (8.99) and (8.100) must be solved for each of the significant harmonics excited by the incident field, after which the solutions for $J_{\phi m}$ and J_{zm} may be superimposed to produce the total current density.

If the cylinder is divided into cells along the z direction as illustrated in Figure 8.11, the unknown current densities may be expanded in basis functions

$$J_{\phi m}(z) \cong \sum_{n=1}^{N} j_{\phi n}\, p(z; z_{n-1}, z_n) \tag{8.105}$$

$$J_{zm}(z) \cong \sum_{n=1}^{N-1} j_{zn}\, t(z; z_{n-1}, z_n, z_{n+1}) \tag{8.106}$$

where $p(z; z_1, z_2)$ and $t(z; z_1, z_2, z_3)$ are subsectional pulse and triangle functions defined in Figure 5.1. If these expansions are substituted into (8.99) and (8.100), the resulting equations may be enforced approximately via an inner product with the testing functions

$$T_\phi^p(z) = (\Delta z)_p \delta[z - \tfrac{1}{2}(z_{p-1} + z_p)] \tag{8.107}$$

$$T_z^p(z) = p(z; \tfrac{1}{2}(z_{p-1} + z_p), \tfrac{1}{2}(z_p + z_{p+1})) \tag{8.108}$$

to produce the $(2N-1)\times(2N-1)$ matrix equation

$$\begin{bmatrix} \mathbf{A} & \mathbf{B} \\ \mathbf{C} & \mathbf{D} \end{bmatrix}\begin{bmatrix} \mathbf{j}_\phi \\ \mathbf{j}_z \end{bmatrix} = \begin{bmatrix} \mathbf{E}_\phi^{\text{inc}} \\ \mathbf{E}_z^{\text{inc}} \end{bmatrix} \tag{8.109}$$

where

$$E_{\phi p}^{\text{inc}} = -(\Delta z)_p E_{\phi m}^{\text{inc}}[\tfrac{1}{2}(z_{p-1}+z_p)] \qquad p = 1, 2, \ldots, N \tag{8.110}$$

$$E_{zp}^{\text{inc}} = -\int_{(z_{p-1}+z_p)/2}^{(z_p+z_{p+1})/2} E_{zm}^{\text{inc}}(z)\, dz \qquad p = 1, 2, \ldots, N-1 \tag{8.111}$$

$$\begin{aligned} A_{pn} = -ja\eta(\Delta z)_p \Bigg\{ &\frac{m^2}{a^2}\int_{z_{n-1}}^{z_n} G_m\left(\frac{z_{p-1}+z_p}{2} - z'\right) dz' \\ &- k^2 \int_{z_{n-1}}^{z_n} \frac{1}{2}\left[G_{m-1}\left(\frac{z_{p-1}+z_p}{2} - z'\right)\right. \\ &\left. + G_{m+1}\left(\frac{z_{p-1}+z_p}{2} - z'\right)\right] dz' \Bigg\} \\ & n = 1, 2, \ldots, N \qquad p = 1, 2, \ldots, N \end{aligned} \tag{8.112}$$

$$\begin{aligned} B_{pn} = (\Delta z)_p m\eta \Bigg\{ &\frac{1}{(\Delta z)_n}\int_{z_{n-1}}^{z_n} G_m\left(\frac{z_{p-1}+z_p}{2} - z'\right) dz' \\ &- \frac{1}{(\Delta z)_{n+1}}\int_{z_n}^{z_{n+1}} G_m\left(\frac{z_{p-1}+z_p}{2} - z'\right) dz' \Bigg\} \\ & n = 1, 2, \ldots, N-1 \qquad p = 1, 2, \ldots, N \end{aligned} \tag{8.113}$$

$$\begin{aligned} C_{pn} = m\eta \left\{ \int_{z_{n-1}}^{z_n} G_m\left(\frac{z_p+z_{p+1}}{2} - z'\right) dz' - \int_{z_{n-1}}^{z_n} G_m\left(\frac{z_{p-1}+z_p}{2} - z'\right) dz' \right\} \\ n = 1, 2, \ldots, N \qquad p = 1, 2, \ldots, N-1 \end{aligned} \tag{8.114}$$

and

$$\begin{aligned} D_{pn} = -ja\eta \Bigg\{ &k^2 \int_{(z_{p-1}+z_p)/2}^{(z_p+z_{p+1})/2}\int_{z_{n-1}}^{z_{n+1}} t(z'; z_{n-1}, z_n, z_{n+1}) G_m(z-z')\, dz'\, dz \\ &+ \frac{1}{(\Delta z)_n}\int_{z_{n-1}}^{z_n} G_m\left(\frac{z_p+z_{p+1}}{2} - z'\right) - G_m\left(\frac{z_{p-1}+z_p}{2} - z'\right) dz' \\ &+ \frac{1}{(\Delta z)_{n+1}}\int_{z_n}^{z_{n+1}} G_m\left(\frac{z_{p-1}+z_p}{2} - z'\right) - G_m\left(\frac{z_p+z_{p+1}}{2} - z'\right) dz' \Bigg\} \\ & p = 1, 2, \ldots, N-1 \qquad n = 1, 2, \ldots, N-1 \end{aligned} \tag{8.115}$$

where G_m is defined in Equation (8.96).

Because of the regular nature of the scatterer, it is possible to restrict the cells in the model to identical lengths Δz without loss of generality. Under this condition, the above expressions can be simplified to produce

$$A_{pn} = -ja\eta\left(k^2(\Delta z)K^m_{p-n} - \frac{m^2}{a^2}(\Delta z)I^m_{p-n}\right) \tag{8.116}$$

$$B_{pn} = m\eta(I^m_{p-n} - I^m_{p-n-1}) \tag{8.117}$$

$$C_{pn} = m\eta(I^m_{p-n+1} - I^m_{p-n}) \tag{8.118}$$

$$D_{pn} \cong -ja\eta\left(k^2(\Delta z)I^m_{p-n} + \frac{1}{(\Delta z)}(I^m_{p-n+1} - 2I^m_{p-n} + I^m_{p-n-1})\right) \tag{8.119}$$

where

$$I^m_q = \int_{-\Delta z/2}^{\Delta z/2} G_m(q\,\Delta z - z')\,dz' \tag{8.120}$$

and

$$K^m_q = \tfrac{1}{2}(I^{m-1}_q + I^{m+1}_q) \tag{8.121}$$

Equation (8.119) involves an approximation similar to that used with the EFIE formulation described in Section 2.4. The triple integral appearing in (8.115) has been replaced by a double integral for convenience. Although this approximation is not essential, it permits all the matrix entries to be obtained from a common subroutine implementing Equation (8.120). The error associated with this approximation should decrease as the cell size Δ_z shrinks.

The integrand in (8.120) is singular for the $q = 0$ case and requires special treatment. For $q = 0$, the integral can be expressed as the sum of three parts according to

$$\begin{aligned} I^m_0 &= \frac{1}{2\pi^2}\int_0^{\Delta z/2}\int_0^{\pi} \frac{\cos(m\alpha)e^{-jk\tilde{R}} - 1}{\tilde{R}}\,d\alpha\,dz \\ &+ \frac{1}{2\pi^2}\int_0^{\Delta z/2}\left[\int_0^{\pi}\frac{1}{\tilde{R}}\,d\alpha + \frac{1}{a}\ln\left(\frac{z}{8a}\right)\right]dz \\ &- \frac{1}{2\pi^2}\int_0^{\Delta z/2}\frac{1}{a}\ln\left(\frac{z}{8a}\right)dz \end{aligned} \tag{8.122}$$

The first integral can be evaluated by numerical quadrature, since for small values of $\tilde{R}$ the integrand becomes

$$-\frac{\tilde{R}}{2}\left(k^2 + \frac{m^2\alpha^2}{z^2 + a^2\alpha^2}\right) - jk\cos(m\alpha) \to -jk\cos(m\alpha) \tag{8.123}$$

which is finite. The second integral can be simplified to

$$\frac{1}{2\pi^2}\int_0^{\Delta z/2}\left[\frac{\sqrt{w}}{a}E(w) + \frac{1}{a}\ln\left(\frac{z}{8a}\right)\right]dz \tag{8.124}$$

where

$$E(w) = \int_0^{\pi/2}\frac{d\beta}{\sqrt{1 - w\sin^2\beta}} \tag{8.125}$$

is the complete elliptic integral of the first kind, and

$$w = \frac{4a^2}{z^2 + 4a^2} \tag{8.126}$$

Here, $E(w)$ can be calculated using a polynomial approximation [10]. It can be shown that the two terms in (8.124) cancel as $z \to 0$, leaving a finite integrand that can be evaluated by quadrature. The final integral in (8.122) is easily evaluated analytically, to yield

$$-\frac{1}{2\pi^2}\int_0^{\Delta z/2}\frac{1}{a}\ln\left(\frac{z}{8a}\right)dz = \frac{\Delta z}{4\pi^2 a}\left[1-\ln\left(\frac{\Delta z}{16a}\right)\right] \tag{8.127}$$

For the case where the integrand is nonsingular, (8.120) may be evaluated using two-dimensional quadrature.

If the cell size Δz is identical throughout the model, the submatrices of the system in (8.109) are highly structured due to the manner in which the integral equation was discretized. In fact, **A** and **D** are symmetric Toeplitz matrices, that is,

$$\mathbf{A} = \begin{bmatrix} a_0 & a_1 & a_2 & \cdots & a_{N-1} \\ a_1 & a_0 & a_1 & \ddots & \cdot \\ a_2 & a_1 & a_0 & & \cdot \\ \vdots & & & & \vdots \\ a_{N-1} & \cdots & \cdots & \cdots & a_0 \end{bmatrix} \tag{8.128}$$

The **B**-matrix can be expressed as

$$\mathbf{B} = \begin{bmatrix} -b_1 & -b_2 & -b_3 & \cdots & -b_{N-1} \\ b_1 & -b_1 & -b_2 & \ddots & \cdot \\ b_2 & b_1 & -b_1 & & \cdot \\ \cdot & & & & \cdot \\ \cdot & & & & \cdot \\ \cdot & & & & -b_1 \\ b_{N-1} & \cdots & \cdots & \cdots & b_1 \end{bmatrix} \tag{8.129}$$

and **C** and **B** are related by

$$\mathbf{C} = -\mathbf{B}^T \tag{8.130}$$

Because of these symmetries, all of the elements of the above matrices can be generated from the first rows of the **A**, **B**, and **D** systems. This amounts to a substantial degree of redundancy that can be exploited to reduce the computation associated with generating the matrix entries. In addition, iterative algorithms such as the CG–FFT procedure (Section 4.12) are easily incorporated into the formulation in order to minimize memory requirements [9].

Equation (8.109) must be solved for all of the significant Fourier harmonics excited by the incident field, including both positive and negative values of m. From an inspection of (8.120), we observe that

$$I_q^m = I_q^{-m} = I_{-q}^m = I_{-q}^{-m} \tag{8.131}$$

and

$$K_q^m = K_q^{-m} = K_{-q}^m = K_{-q}^{-m} \tag{8.132}$$

Thus, the **A** and **D** submatrices are independent of the sign of m, while the **B** and **C** submatrices are proportional to the sign of m. In addition, Bessel functions have the property

$$J_{-m}(x) = (-1)^m J_m(x) \tag{8.133}$$

which means that

$$E_{z(-m)}^{\text{inc}} = E_{z(m)}^{\text{inc}} \tag{8.134}$$

$$E_{\phi m}^{\text{inc}} = E_\theta^m u_\theta + E_y^m u_y \tag{8.135}$$

and

$$E_{\phi(-m)}^{\text{inc}} = -E_\theta^m u_\theta + E_y^m u_y \tag{8.136}$$

These relations can be used to minimize the required computations.

After the harmonics of the current density are found for all necessary values of m, other parameters such as the scattering cross section can be computed. If the bistatic scattering cross section is to be computed at many observation angles, it is probably desirable to first find the total current density by summing over the harmonics and then compute the scattering cross section according to the formulas of Chapter 1 [Equations (1.157)–(1.159)]. The current must be sampled at a sufficient density of points around the cylinder circumference. As an alternative, the Fourier harmonics of the far-zone fields can be computed directly from the solutions $J_{\phi m}$ and J_{zm} and superimposed to compute the scattering cross section at the observation angles of interest.

As an application of this approach, Figure 8.12 shows the current density induced by an axially incident plane wave on conducting cylinders of radius 0.1λ and 0.6λ. Since the geometry is an infinitesimally thin conducting shell, the EFIE results show the superposition of the current density on the inside and outside of the cylinder. In the first case, the cylinder radius is small enough to suppress any interior circular waveguide modes, and the current exhibits good agreement with the physical optics approximation

$$\bar{J} \cong 2\hat{n} \times \bar{H}^{\text{inc}} \tag{8.137}$$

In the second case, the cylinder radius is larger and an interior mode is excited by the incident field. The waveguide mode propagates with a different effective wavenumber than the exterior field, and the interference between the interior and exterior currents is easily observed in Figure 8.12.

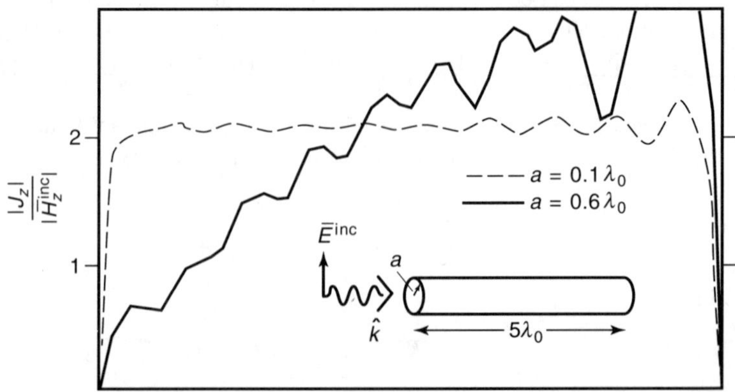

Figure 8.12 Current density (EFIE result) on hollow p.e.c. cylinders.

This example illustrates a surface integral body-of-revolution formulation for a very simple geometry. We will leave the extension to arbitrary axisymmetric scatterers as an exercise for the reader, after noting that general approaches are well established and may be found in the literature [11–14].

8.8 DIFFERENTIAL EQUATION FORMULATION FOR AXISYMMETRIC SCATTERERS [15]

To complement the integral equation formulation of the preceding section, we also present a thumbnail sketch of a differential equation approach for axisymmetric scatterers. For simplicity, we make the additional assumption that the incident field is also axisymmetric, which permits us to employ a pseudoscalar formulation in terms of the ϕ-components of the fields even in the presence of inhomogeneous materials. The approach is general enough to treat problems such as a monopole antenna radiating into radially inhomogeneous media [15], as illustrated in Figure 8.13. (The treatment of a nonsymmetric excitation requires a vector formulation and will be deferred until Chapter 11.)

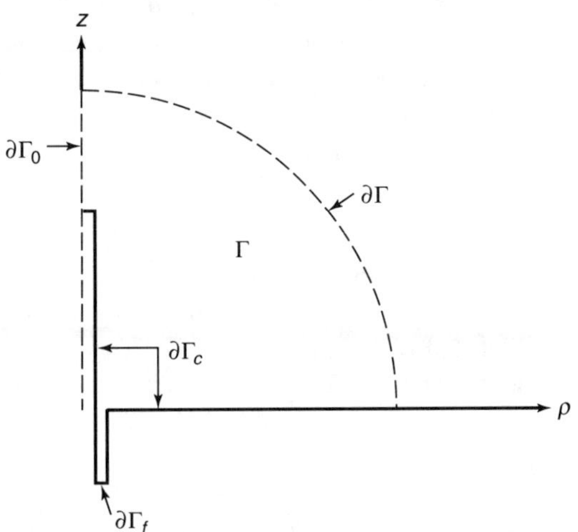

Figure 8.13 Generating sector associated with an axisymmetric monopole antenna geometry.

We will assume that the only nonzero component of the magnetic field is H_ϕ. The ϕ-component of the curl–curl equation

$$\nabla \times \left(\frac{1}{\varepsilon_r} \nabla \times \bar{H} \right) - k^2 \mu_r \bar{H} = 0 \tag{8.138}$$

can be expressed in the form

$$\tilde{\nabla} \cdot \left[\frac{1}{\rho \varepsilon_r} \nabla(\rho H_\phi) \right] + k^2 \frac{\mu_r}{\rho} (\rho H_\phi) = 0 \tag{8.139}$$

where $\tilde{\nabla}$ is the pseudodivergence operator

$$\tilde{\nabla} \cdot \bar{A} = \frac{\partial A_\rho}{\partial \rho} + \frac{\partial A_z}{\partial z} \tag{8.140}$$

A testing function $T(\rho, z)$ can be employed to convert (8.139) into the weak equation

$$\iint_\Gamma \frac{1}{\rho\varepsilon_r}\nabla T \cdot \nabla(\rho H_\phi) - k^2\frac{\mu_r}{\rho}T(\rho H_\phi) = \int_{\partial\Gamma+\partial\Gamma_c+\partial\Gamma_0+\partial\Gamma_f} \frac{1}{\varepsilon_r}T\frac{1}{\rho}\frac{\partial(\rho H_\phi)}{\partial n} \tag{8.141}$$

where $\partial\Gamma$ denotes the radiation boundary, $\partial\Gamma_c$ denotes the surface of any perfectly conducting material that may be present (Figure 8.13), $\partial\Gamma_0$ denotes the $\rho = 0$ axis, and $\partial\Gamma_f$ denotes a feed region (as might be part of a radiating antenna). The form of (8.141) suggests the use of the product ρH_ϕ as the primary unknown rather than H_ϕ alone, and we adopt that approach.

Along perfectly conducting material, the Neumann boundary condition

$$\frac{1}{\rho}\frac{\partial(\rho H_\phi)}{\partial n} = 0 \tag{8.142}$$

must be imposed. In addition, along the $\rho = 0$ axis, the field must satisfy the Dirichlet condition

$$\rho H_\phi = 0 \tag{8.143}$$

Equations (8.142) and (8.143) eliminate the boundary integrals over $\partial\Gamma_c$ and $\partial\Gamma_0$, respectively. We will assume that the radiation boundary $\partial\Gamma$ is circular with radius r and centered at the origin in the ρ–z plane. In addition, we define a local orthogonal coordinate system (n, t) along the boundary so that t is a tangential variable along $\partial\Gamma$. We will impose a second-order local RBC adapted from reference [16], which is derived in Chapter 11. This RBC can be written as

$$\frac{1}{\rho}\frac{\partial(\rho H_\phi^s)}{\partial n} \cong -\frac{jk}{\rho}(\rho H_\phi^s) + \frac{1}{2jk + 2/r}\frac{\partial}{\partial t}\left(\frac{1}{\rho}\frac{\partial}{\partial t}(\rho H_\phi^s)\right) \tag{8.144}$$

If the incident field is exterior to $\partial\Gamma$, Equation (8.144) can be combined with similar expressions involving the incident field and substituted into the weak differential equation to complete the formulation. If the excitation is incorporated via the boundary $\partial\Gamma_f$ depicted in Figure 8.13, it appears as either a nonhomogeneous Dirichlet or Neumann boundary condition and can be incorporated by prescribing field values or by adding the appropriate integral to the right side of (8.141), respectively.

The primary unknown may be expanded in basis functions according to

$$\rho H_\phi(\rho, z) \cong \sum_{n=1}^{N} h_n B_n(\rho, z) \tag{8.145}$$

If identical functions are used for testing, the resulting matrix equation can be written as $\mathbf{Ah} = \mathbf{b}$, where

$$\begin{aligned} A_{mn} = &\iint_\Gamma \frac{1}{\rho\varepsilon_r}\nabla B_m \cdot \nabla B_n - k^2\frac{\mu_r}{\rho}B_m B_n \\ &+ \int_{\partial\Gamma} \frac{jk}{\rho}B_m B_n + \frac{1}{2jk + 2/r}\frac{1}{\rho}\frac{\partial B_m}{\partial t}\frac{\partial B_n}{\partial t} \end{aligned} \tag{8.146}$$

Since the variable ρ appears in the denominator of these expressions, they do not yield to simple closed-form evaluation in terms of the local-area coordinates defined in Chapter 3. However, their evaluation by numerical quadrature is straightforward.

An approach similar to the above formulation has been used to analyze monopole antennas coated with dielectric layers, and the numerical results appear to exhibit excellent agreement with available data [15].

8.9 SUMMARY

This chapter has considered electromagnetic scattering from translationally invariant or rotationally invariant three-dimensional geometries. Usually, these problems can be posed in such a manner as to reduce their computational requirements to those of purely two-dimensional structures. Examples include finite sources radiating in the presence of infinite cylinders and axisymmetric bodies. As might be expected, the price to be paid for the increased computational efficiency is a slight increase in the complexity of the formulation.

REFERENCES

[1] A. F. Peterson and R. Mittra, "Mutual admittance between slots in cylinders of arbitrary shape," *IEEE Trans. Antennas Propagat.*, vol. 37, pp. 858–864, July 1989.

[2] A. D. Yaghjian, "Augmented electric- and magnetic-field equations," *Radio Science*, vol. 16, pp. 987–1001, Nov. 1981.

[3] J. H. Richmond, "A reaction theorem and its application to antenna impedance calculations," *IEEE Trans. Antennas Propagat.*, vol. AP-9, pp. 515–520, Nov. 1961.

[4] S. W. Lee, S. Safavi-Naini, and R. Mittra, "Mutual admittance between slots on a cylinder," Electromagnetics Laboratory Report 77-8, UILU-ENG-77-2549, Dept. of Electrical and Computer Engineering, University of Illinois, Urbana, IL, Mar. 1977.

[5] E. Michielssen, A. F. Peterson, and R. Mittra, "Oblique scattering from inhomogeneous cylinders using a coupled integral equation formulation with triangular cells," *IEEE Trans. Antennas Propagat.*, vol. 39, pp. 485–490, Apr. 1991.

[6] R. G. Rojas, "Scattering by an inhomogeneous dielectric/ferrite cylinder of arbitrary cross-section shape—oblique incidence case," *IEEE Trans. Antennas Propagat.*, vol. AP-36, pp. 238–246, Feb. 1988.

[7] R. B. Wu and C. H. Chen, "Variational reaction formulation of scattering problem for anisotropic dielectric cylinders," *IEEE Trans. Antennas Propagat.*, vol. AP-34, pp. 640–645, May 1986.

[8] A. F. Peterson, "Application of volume discretization methods to oblique scattering from high-contrast penetrable cylinders," *IEEE Trans. Microwave Theory Tech.*, vol. 42, pp. 686–689, Apr. 1994.

[9] A. F. Peterson, S. L. Ray, C. H. Chan, and R. Mittra, "Numerical implementations of the conjugate gradient method and the CG-FFT for electromagnetic scattering," in *Application of Conjugate Gradient Method to Electromagnetics and Signal Analysis (PIER 5)*, ed. T. K. Sarkar, New York: Elsevier, 1991.

[10] M. Abramowitz and I. A. Stegun, *Handbook of Mathematical Functions*, New York: Dover, 1965, p. 591.

[11] M. G. Andreasen, "Scattering from bodies of revolution," *IEEE Trans. Antennas Propagat.*, vol. AP-13, pp. 303–310, Mar. 1965.

[12] J. R. Mautz and R. F. Harrington, "Radiation and scattering from bodies of revolution," *Appl. Sci. Res.*, vol. 20, pp. 405–435, 1969.

[13] A. W. Glisson and D. R. Wilton, "Simple and efficient numerical techniques for treating bodies of revolution," RADC-TR-79-22, Rome Air Development Center, Griffis AFB, NY, March 1979.

[14] S. Govind, D. R. Wilton, and A. W. Glisson, "Scattering from inhomogeneous penetrable bodies of revolution," *IEEE Trans. Antennas Propagat.*, vol. AP-32, pp. 1163–1173, Nov. 1984.

[15] E. Sumbar, F. E. Vermeulen, and F. S. Chute, "Implementation of radiation boundary conditions in the finite element analysis of electromagnetic wave propagation," *IEEE Trans. Microwave Theory Tech.*, vol. 39, pp. 267–273, Feb. 1991.

[16] A. F. Peterson, "Absorbing boundary conditions for the vector wave equation," *Microwave Opt. Technol. Lett.*, vol. 1, pp. 62–64, Apr. 1988.

PROBLEMS

P8.1 Suppose fields $\bar{E}(x, y, z)$ and $\bar{H}(x, y, z)$ are produced by an arbitrary source in free space radiating in the presence of a cylindrical scatterer. A Fourier transformation along z, following (8.1) and (8.2), can be used to replace these fields by $\tilde{\bar{E}}(x, y)e^{jk_z z}$ and $\tilde{\bar{H}}(x, y)e^{jk_z z}$. By combining the $\hat{z}$-component of the appropriate curl–curl equations (1.13) and (1.14) with the associated divergence condition, demonstrate the following:

(a) Assuming that the scatterer is perfectly conducting, show that the equations simplify to

$$\nabla_t^2 \tilde{E}_z + (k^2 - k_z^2)\tilde{E}_z = 0$$

$$\nabla_t^2 \tilde{H}_z + (k^2 - k_z^2)\tilde{H}_z = 0$$

away from the source region, where ∇_t is the transverse gradient defined in Section 8.5. Therefore, in the Fourier transform domain, the three-dimensional problem can be described by an equivalent two-dimensional equation that is a function of the parameter k_z. The TM and TE polarizations are only coupled by the source.

(b) Assuming that the scatterer is a penetrable cylinder described by suitably differentiable functions $\varepsilon_r(x, y)$ and $\mu_r(x, y)$, show that the equations reduce to

$$\nabla_t \cdot \left(\frac{1}{\mu_r}\nabla_t \tilde{E}_z\right) + \frac{k^2\mu_r\varepsilon_r - k_z^2}{\mu_r}\tilde{E}_z = jk_z \nabla_t \left(\frac{1}{\varepsilon_r \mu_r}\right) \cdot \varepsilon_r \tilde{\bar{E}}_t$$

$$\nabla_t \cdot \left(\frac{1}{\varepsilon_r}\nabla_t \tilde{H}_z\right) + \frac{k^2\mu_r\varepsilon_r - k_z^2}{\varepsilon_r}\tilde{H}_z = jk_z \nabla_t \left(\frac{1}{\varepsilon_r \mu_r}\right) \cdot \mu_r \tilde{\bar{H}}_t$$

outside the source region. In fact, these equations also hold if the differentiability of ε_r and μ_r is relaxed to include the case of jump discontinuities in the medium, provided that $\nabla_t((\varepsilon_r\mu_r)^{-1})$ is treated as a generalized function. Therefore, whenever ε_r or μ_r vary with position and $k_z \neq 0$, the transverse and longitudinal fields are coupled and the TM and TE polarizations cannot be treated independently. However, the three-dimensional problem can still be described in the Fourier transform domain by coupled two-dimensional equations that are a function of k_z.

P8.2 Equation (8.10) can be expressed in terms of an inverse discrete Fourier transform, defined in (4.90), and evaluated using an inverse FFT algorithm. Describe the steps necessary to incorporate the FFT and discuss advantages and disadvantages of using the FFT instead of the direct summation of (8.10).

P8.3 Consider a perfectly conducting, circular cylinder with radius a illuminated by an oblique TE wave beyond the visible range of the transform ($k_z^2 > k^2$). Use the eigenfunctions and eigenvalues of the MFIE from Equation (6.5) and the relationship between the Hankel function and the modified Bessel function of the second kind,

$$H_n^{(2)}\left(-j\sqrt{k_z^2-k^2}\rho\right)=\frac{2}{\pi}j^{n+1}K_n\left(\sqrt{k_z^2-k^2}\rho\right)$$

to derive an explicit expression for the eigenvalues of the MFIE when $k_z^2 > k^2$. Show that there are no zeros in the eigenvalue spectrum and thus there is no interior resonance problem when $k_z^2 > k^2$.

P8.4 Derive the TE CFIE describing currents induced on a perfectly conducting cylinder by an obliquely incident plane wave for a range of oblique angles within the visible region of the transform variable ($k_z^2 < k^2$). Develop a method-of-moments discretization using subsectional triangle basis functions and pulse testing functions (see the EFIE approach described in Section 2.4). Provide complete expressions for the matrix entries.

P8.5 Develop an explicit integral expression for the TM bistatic scattering cross section defined in (8.19).

P8.6 Derive component integral equations comprising an augmented MFIE for currents induced on a perfectly conducting cylinder by an obliquely incident TM plane wave within the visible region of the transform variable ($k_z^2 < k^2$). Develop a method-of-moments discretization, providing complete expressions for the matrix entries associated with subsectional pulse basis functions and Dirac delta testing functions.

P8.7 Beginning with Maxwell's equations and assuming an $e^{jk_z z}$-dependence in all field components, derive Equations (8.36) and (8.37).

P8.8 Suppose it is desired to modify the volume integral formulation of Section 8.5 in order to add imbedded perfectly conducting material to the inhomogeneous dielectric cylinder. Derive the generalization of Equations (8.38) and (8.39) that includes the contribution from equivalent electric current on the surface of the imbedded conductor. Leave the result in terms of currents J_z and J_t.

P8.9 Generalize the derivation of the exact eigenfunction RBC from Chapter 3 for the case of oblique incidence and carry out the development to produce Equations (8.67) and (8.68).

P8.10 For obliquely incident waves within the visible region of the transform variable ($k_z^2 < k^2$), a local RBC can be derived by a procedure analogous to that used in Chapter 3 to obtain the Bayliss–Turkel RBC family. Develop first- and second-order RBCs that could be used with the formulation of Section 8.6, assuming a circular boundary of radius a. Leave the result in the form of Equations (8.67) and (8.68). *Hint:* Show that the desired expression for (8.67) can be written in terms of E_z and H_z as follows:

$$\hat{z}\cdot\hat{\rho}\times(\nabla\times\bar{E})|_{(a,\phi,0)}=-\frac{k^2}{k_t^2}\frac{\partial E_z}{\partial\rho}+\frac{\eta k k_z}{k_t^2 a}\frac{\partial H_z}{\partial\phi}$$

after which E_z can be replaced by a Bayliss–Turkel-like condition, to produce (for instance) the second-order RBC

$$\begin{aligned}\hat{z}\cdot\hat{\rho}\times(\nabla\times\bar{E})|_{(a,\phi,0)}=&-\frac{k^2}{k_t^2}\left(\frac{\partial E_z^{\text{inc}}}{\partial\rho}-\alpha(a)E_z^{\text{inc}}-\beta(a)\frac{\partial^2E_z^{\text{inc}}}{\partial\phi^2}\right)\\&-\frac{k^2}{k_t^2}\left(\alpha(a)E_z+\beta(a)\frac{\partial^2E_z}{\partial\phi^2}\right)+\frac{\eta k k_z}{k_t^2 a}\frac{\partial H}{\partial\phi}\end{aligned}$$

where α and β are defined in Chapter 3. Is it also necessary to replace H_z?

P8.11 Discuss the modifications necessary to extend the differential equation formulation of Section 8.6 to the invisible region of the transform variable ($k_z^2 > k^2$).

P8.12 A hollow linear dipole antenna excited by a ϕ-symmetric feed can be modeled using a specialization of the procedure developed in Section 8.7.

(a) For the ϕ-symmetric case, identify the current components present on the hollow cylinder and specialize the EFIE presented in Equations (8.99) and (8.100) to this case.

(b) Assuming that the linear dipole is discretized into equal-size cells along the axis of the cylinder, simplify the matrix entries given in Equations (8.116)–(8.121) for the ϕ-symmetric case.

(c) A magnetic frill model is often used for the dipole antenna feed. Referring to the frill equations developed in Prob. P1.17, generate an explicit integral expression for the "tested" incident electric field originally defined in (8.111). Discuss the evaluation of any complicating singularities in the integrand.

P8.13 Suppose the hollow cylinder analyzed in Section 8.7 is illuminated with a uniform plane wave incident along the z-axis. How many ϕ-harmonics are necessary to represent this excitation? Develop a procedure for solving the problem that only requires a single matrix solution in this special case.

P8.14 A general three-dimensional local RBC developed for a spherical boundary shape has the form ([16] and Chapter 11)

$$\hat{r} \times (\nabla \times \bar{H}) = jk\bar{H}^{\tan} + \frac{1}{2jk + 2/r} \nabla \times [\hat{r}(\hat{r} \cdot \nabla \times \bar{H})]$$
$$+ \frac{jk}{2jk + 2/r} \nabla^{\tan}(\hat{r} \cdot \bar{H})$$

where $\bar{H}$ refers to the scattered part of the field and $\nabla^{\tan}$ is the tangential part of the gradient, that is, the θ- and ϕ-components. Derive Equation (8.144) by specializing this RBC to the axisymmetric situation described in Section 8.8.

9

Subsectional Basis Functions for Multidimensional and Vector Problems

The numerical techniques discussed in the preceding chapters are deliberately limited to the use of the simplest, lowest order basis and testing functions that are believed to provide converging results. The accuracy of a particular solution usually improves with the polynomial degree of the expansion functions, and it is therefore worthwhile to consider functions smoother than a piecewise-constant or piecewise-linear polynomial. Several types of higher order basis functions were introduced for one-dimensional expansions in Chapter 5. This chapter develops higher order subsectional functions for multidimensional applications, illustrates their implementation in several two-dimensional formulations, and considers some of the computational trade-offs associated with their use.

Despite the fact that electromagnetic fields are vector quantities, previous chapters have only employed a scalar representation. While it is possible to use a scalar basis in the general three-dimensional case, by employing the obvious remedy of separately expanding the x, y, and z components, it is not always desirable. At medium interfaces, for instance, tangential-field components are subject to different continuity conditions than normal components, and a boundary condition can be difficult to impose on an expansion that does not naturally separate into normal and tangential components along the boundary. In addition, scalar representations tend to be continuous, and with certain vector quantities the imposition of complete continuity can produce erroneous results (Section 9.7). Consequently, several types of subsectional *vector* basis functions are introduced for two- and three-dimensional electromagnetic applications. For illustration, one class of functions is used to discretize the curl–curl form of the two-dimensional vector Helmholtz equation for cavity, waveguide, and scattering applications. These functions are extended to the three-dimensional case in this chapter and used for three-dimensional applications in Chapter 11. Other vector expansion functions introduced in this chapter appear appropriate for surface patch discretizations of three-dimensional integral equations and will be employed in Chapter 10.

As the polynomial order of expansion functions is increased, providing greater accuracy with fewer cells, the modeling error associated with simple cell shapes may exceed the discretization error associated with the expansion functions. A remedy is obtained by the use of parametric basis functions defined on cells having curved edges or surfaces. Parametric mappings are discussed for scalar and vector functions.

9.1 HIGHER ORDER LAGRANGIAN BASIS FUNCTIONS ON TRIANGLES [1, 2]

We initially restrict our attention to expansion functions defined on triangular cells. As illustrated in Chapter 3, triangle-based analysis is facilitated through the use of *simplex coordinates* $\{L_1, L_2, L_3\}$ related to the Cartesian coordinates by

$$x = L_1x_1 + L_2x_2 + L_3x_3 \tag{9.1}$$

$$y = L_1y_1 + L_2y_2 + L_3y_3 \tag{9.2}$$

where (x_i, y_i) are the coordinates of the ith vertex of the triangle. These coordinates specify the position of a point within a triangle by giving the relative perpendicular distance measured from each side to the point, with the distance expressed as a fraction of the triangle altitude (Figure 3.9). Since the resulting interior triangles divide the original cell into three parts, each having a fraction of the total area in proportional to their respective coordinate, the additional condition

$$L_1 + L_2 + L_3 = 1 \tag{9.3}$$

must always hold. Equations (9.1)–(9.3) can be inverted to produce

$$L_i = \frac{1}{2A}(a_i + b_ix + c_iy) \tag{9.4}$$

where

$$a_i = x_{i+1}y_{i+2} - x_{i+2}y_{i+1} \tag{9.5}$$

$$b_i = y_{i+1} - y_{i-1} \tag{9.6}$$

$$c_i = x_{i-1} - x_{i+1} \tag{9.7}$$

and A denotes the area of the triangle. The index i in Equations (9.4)–(9.7) assumes values 1, 2, and 3 cyclically, so that if $i = 3$, then $i + 1 = 1$.

Chapters 2 and 3 included examples in which a two-dimensional scalar function was represented by the superposition of linear pyramid basis functions defined on triangular cells (Figure 3.4). The pyramid functions straddle several cells and provide a linear interpolation between values at the cell corners. The representation is automatically continuous across cell boundaries and is therefore especially convenient for representing continuous quantities such as the z-components of the two-dimensional electromagnetic fields.

Since it straddles several cells, the pyramid function depicted in Figure 3.4 is conceptually a global entity. If we restrict our attention to a single cell, it is more convenient to employ a local description in simplex coordinates. Within a particular triangle, a pyramid function centered at node i of the triangle is a linear function having unity value at corner i and vanishing at the other corners. It has the obvious representation

$$B_i(L_1, L_2, L_3) = L_i \tag{9.8}$$

A local picture of the three overlapping functions is provided in Figure 9.1.

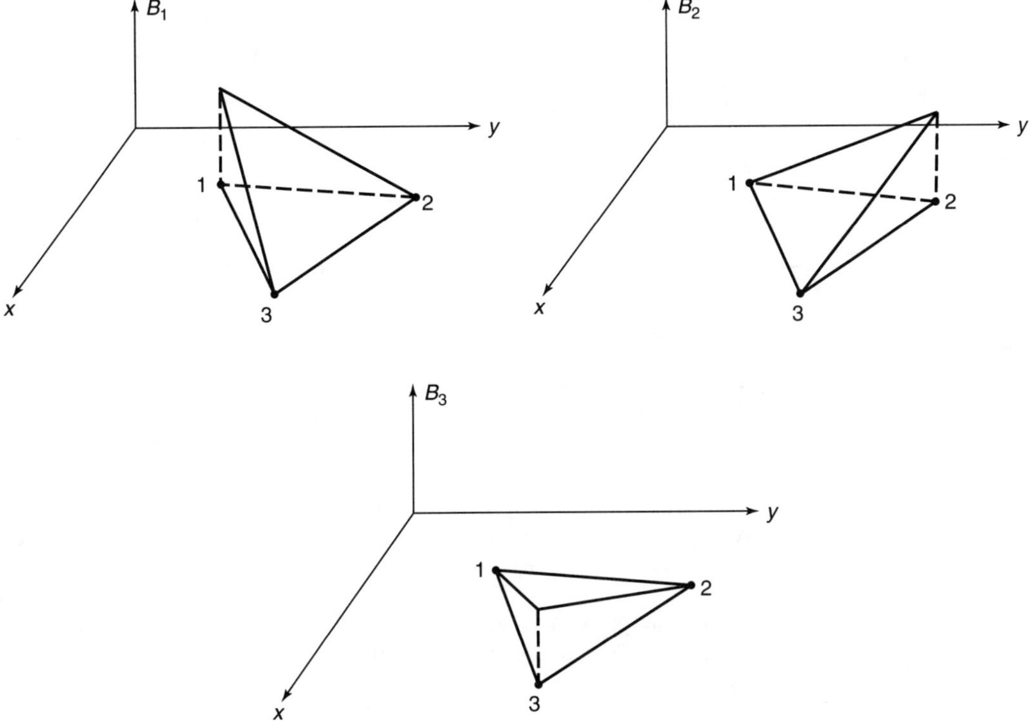

Figure 9.1 Linear Lagrangian functions within a triangular cell.

The linear pyramid functions are actually the simplest members of the family of Lagrangian functions defined on triangular cells. From the introductory discussion in Section 5.3, we recall that Lagrangian functions interpolate field values between some number of nodes throughout an interval of interest. As in the one-dimensional case, higher order basis functions on triangles are defined to be interpolatory at a number of points or nodes within each triangular cell. The linear pyramid functions interpolate between the three corner nodes. Quadratic, cubic, and other higher order expansion functions require the introduction of additional nodes within each cell. In general, an Mth-order polynomial can be defined over a triangular cell containing m nodes, where m and M are related through

$$m = \tfrac{1}{2}(M+1)(M+2) \tag{9.9}$$

For example, a quadratic basis ($M = 2$) involves 6 nodes. A cubic basis requires 10 nodes. These nodes can be located at a regularly spaced set of points defined by the simplex coordinate values

$$\left\{\frac{i}{M}, \frac{j}{M}, \frac{k}{M}\right\} \qquad i, j, k = 0, 1, \ldots, M \tag{9.10}$$

where the indices are restricted to integers satisfying $i + j + k = M$. For ease of notation when using higher order interpolation functions on triangles, the triple index (i, j, k) is often employed to denote a particular node. Figure 9.2 illustrates the arrangement of nodes and the triple-index scheme for $M = 2$ and $M = 3$.

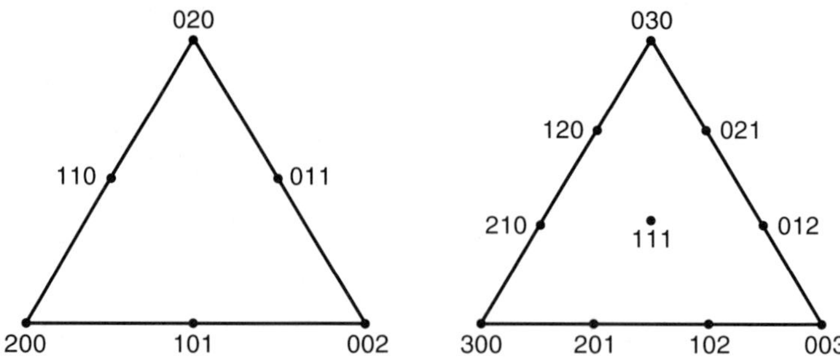

Figure 9.2 Triple indices for $M = 2$ and $M = 3$.

In order to define an Mth-order polynomial basis function B_{ijk} directly in terms of the coordinates $\{L_1, L_2, L_3\}$, Silvester and Ferrari introduce the auxiliary polynomials [2]

$$R_s(M, L) = \frac{1}{s!}\prod_{k=0}^{s-1}(ML - k) \qquad (s > 0) \tag{9.11}$$

$$R_0(M, L) = 1 \tag{9.12}$$

The polynomial R_s has exactly s equally spaced zeros at $L = 0, 1/M, \ldots, (s-1)/M$ and equals one at $L = s/M$. For illustration, Figure 9.3 depicts these polynomials for $M = 2$ and $M = 3$. The basis functions can be defined in terms of R_s as

$$B_{ijk}(L_1, L_2, L_3) = R_i(M, L_1)R_j(M, L_2)R_k(M, L_3) \tag{9.13}$$

Basis function B_{ijk} has unity value at node ijk and vanishes at every other node in the triangle. Since the basis functions are interpolatory, a function $E(x, y)$ of polynomial order M can be exactly represented throughout the triangle by the superposition

$$E(L_1, L_2, L_3) = \sum_{i=0}^{M}\sum_{j=0}^{M-i} e_{ijk}B_{ijk}(L_1, L_2, L_3) \tag{9.14}$$

where the coefficient e_{ijk} is the value of E at node (i, j, k).

As an example, consider the set of quadratic basis functions ($M = 2$). The six functions provided by Equation (9.13) are

$$B_{200}(L_1, L_2, L_3) = (2L_1 - 1)L_1 \tag{9.15}$$

$$B_{020}(L_1, L_2, L_3) = (2L_2 - 1)L_2 \tag{9.16}$$

$$B_{002}(L_1, L_2, L_3) = (2L_3 - 1)L_3 \tag{9.17}$$

$$B_{110}(L_1, L_2, L_3) = 4L_1L_2 \tag{9.18}$$

$$B_{101}(L_1, L_2, L_3) = 4L_1L_3 \tag{9.19}$$

$$B_{011}(L_1, L_2, L_3) = 4L_2L_3 \tag{9.20}$$

These functions are illustrated in Figure 9.4. Observe that the first three functions have unity value at nodes 200, 020, and 002, respectively, while the latter three functions have unity value in the center of one of the three sides. Each of the basis functions vanishes at five of the six nodes. The superposition of all six functions, appropriately weighted, provides a quadratic representation with coefficients associated with the field values at each

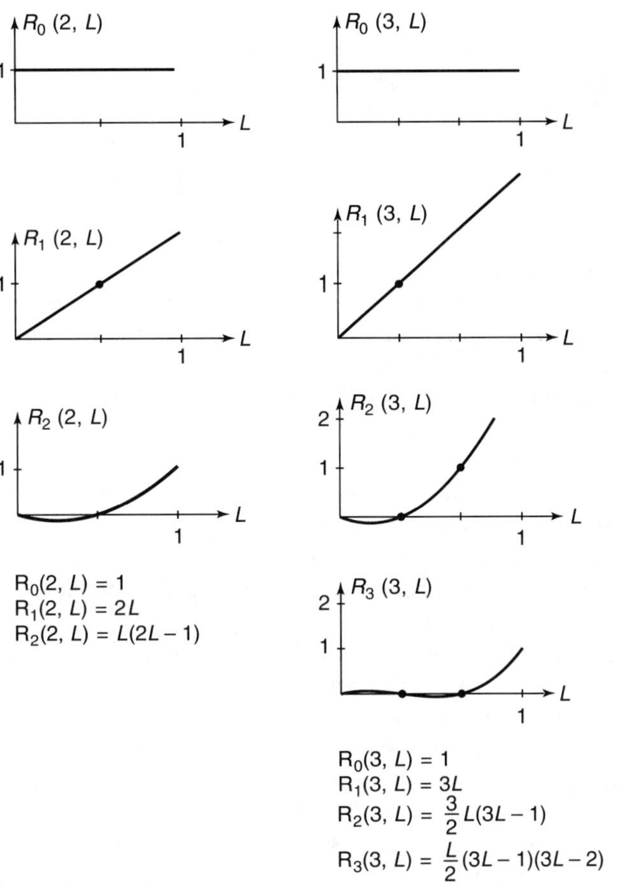

Figure 9.3 Polynomials used to build Lagrangian functions.

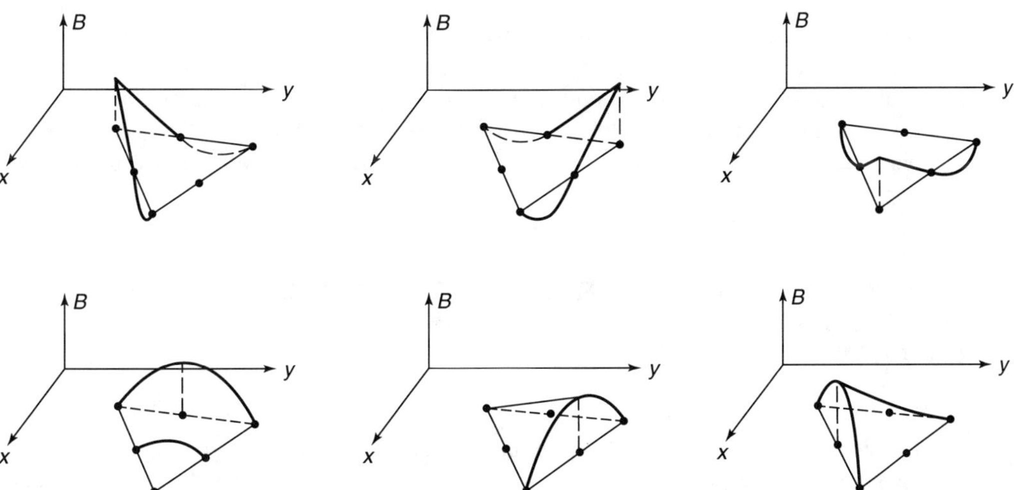

Figure 9.4 Quadratic Lagrangian functions within a triangular cell.

of the six nodes. A description of these functions in Cartesian coordinates is not particularly illuminating but can be obtained using Equation (9.4) if desired.

The higher order Lagrangian polynomials permit a smoother representation of the field within a cell than possible using the linear pyramid functions employed in previous chapters. As discussed in Chapter 5, the interpolation error associated with these functions behaves as $O(\Delta^{p+1})$ as $\Delta \to 0$, where Δ is the longest edge in the triangle and p is the polynomial degree [3]. Although the actual solution error in a particular situation may be dominated by other sources (such as the use of approximate radiation boundary conditions), observations support the notion that the overall error is primarily interpolation error [4]. In any case, for fixed computational resources, a far greater reduction in interpolation error for small Δ can be obtained by increasing the polynomial order p than by decreasing the cell size.

Although the higher order Lagrangian polynomials permit a smoother representation of the field within a cell, the smoother representation is not imposed across cell boundaries. Continuity of the global expansion is ensured by the common nodes along each boundary, but the normal derivative of the field expansion at a cell boundary is generally not continuous. The discontinuity of the derivatives is usually a second order effect producing an error that disappears in the limit as the cell sizes shrink to zero. In fact, whether or not the derivative discontinuity is undesirable depends on the specific problem under consideration. As an example, the z-components of the true electric and magnetic fields in a two-dimensional problem exhibit discontinuous normal derivatives at material interfaces, and consequently it might be desirable that the expansion functions permit this behavior.

If derivative continuity is desired, the Hermitian representation introduced in Section 5.3 can be extended to the multidimensional case [3]. In the two-dimensional situation, it is necessary to adopt an expansion of polynomial order 5 in order to ensure the continuity of a scalar function and its tangential and normal derivatives across the cell boundaries. If used with triangular cell shapes, at least two different representations are possible, one employing 18 functions [5] and the other employing 21 functions. The 18-function expansion is only complete to polynomial order 4 and will produce an interpolation error of $O(\Delta^5)$ as $\Delta \to 0$, while the 21-function expansion is complete to order 5 and produces an interpolation error of $O(\Delta^6)$ [3]. Since all the Hermitian coefficients are shared by multiple cells, the Hermitian type of expansion offers the possibility of a high-order expansion with far fewer unknowns than a Lagrangian representation of the same order. However, the reduction in unknowns is offset by an increase in the number of nonzero entries per row of the matrix.

To illustrate the implementation of higher order Lagrangian functions, we turn our attention to an example employing the scalar Helmholtz equation for two-dimensional scattering.

9.2 EXAMPLE: USE OF HIGHER ORDER BASIS FUNCTIONS WITH THE TWO-DIMENSIONAL SCALAR HELMHOLTZ EQUATION

Consider the two-dimensional scattering problem posed at the beginning of Chapter 3, involving a TM wave incident on a cylinder containing dielectric and magnetic inhomogeneities. A weak equation incorporating the Bayliss–Turkel RBC was provided in Equation

(3.87), which we repeat for convenience:

$$\iint_{\Gamma}\left(\frac{1}{\mu_r}\nabla T\cdot\nabla E_z^{\text{tot}}-k^2\varepsilon_r T E_z^{\text{tot}}\right)dx\,dy-\int_{\partial\Gamma}\left(\alpha T E_z^{\text{tot}}-\beta\frac{\partial T}{\partial\phi}\frac{\partial E_z^{\text{tot}}}{\partial\phi}\right)\rho\,d\phi$$
$$=\int_{\partial\Gamma}T\left(\frac{\partial E_z^{\text{inc}}}{\partial\rho}-\alpha E_z^{\text{inc}}-\beta\frac{\partial^2 E_z^{\text{inc}}}{\partial\phi^2}\right)\rho\,d\phi \tag{9.21}$$

The variables α and β are defined in Equations (3.81) and (3.82). The finite-element discretization of this equation produces a matrix equation

$$\mathbf{Ae}=\mathbf{b} \tag{9.22}$$

with matrix entries given by

$$A_{mn}=\iint_{\Gamma}\left(\frac{1}{\mu_r}\nabla B_m\cdot\nabla B_n-k^2\varepsilon_r B_m B_n\right)dx\,dy$$
$$-\int_{\partial\Gamma}\left(\alpha B_m B_n-\beta\frac{\partial B_m}{\partial\phi}\frac{\partial B_n}{\partial\phi}\right)\rho\,d\phi \tag{9.23}$$

and

$$b_m=\int_{\partial\Gamma}B_m\left(\frac{\partial E_z^{\text{inc}}}{\partial\rho}-\alpha E_z^{\text{inc}}-\beta\frac{\partial^2 E_z^{\text{inc}}}{\partial\phi^2}\right)\rho\,d\phi \tag{9.24}$$

where $\{B_n\}$ denote the basis functions. Chapter 3 presented a detailed description of the computational implementation of this procedure for linear basis functions.

We wish to evaluate these expressions using the higher order Lagrangian functions developed in the preceding section. Following the procedure employed in Chapter 3, we initially construct *element matrices* on a cell-by-cell basis containing the interaction terms. The element matrices will be of order m, where m is a specific function of the polynomial order M as described by Equation (9.9). Linear, quadratic, and cubic expansion functions give rise to element matrices of order 3, 6, and 10, respectively. For any polynomial order, the two-dimensional integrals to be evaluated are

$$A_{pq}^{(1)}=\iint_{\Gamma}\nabla B_p\cdot\nabla B_q\,dx\,dy \tag{9.25}$$

$$A_{pq}^{(2)}=\iint_{\Gamma}B_p B_q\,dx\,dy \tag{9.26}$$

The required boundary integrals are

$$A_{pq}^{(3)}=\int_{\partial\Gamma}B_p B_q\,dt \tag{9.27}$$

and

$$\int_{\partial\Gamma}\frac{\partial B_p}{\partial\phi}\frac{\partial B_q}{\partial\phi}\rho\,d\phi \tag{9.28}$$

Using the change of variable

$$\frac{\partial}{\partial\phi}=\rho\frac{\partial}{\partial t} \tag{9.29}$$

Equation (9.28) can be written in terms of the integral

$$A_{pq}^{(4)} = \int_{\partial\Gamma} \frac{\partial B_p}{\partial t} \frac{\partial B_q}{\partial t} \, dt \tag{9.30}$$

For a triangular-cell model, these integrals can be evaluated in closed form for any polynomial order. Since the evaluation becomes tedious for higher order functions [neglecting symmetry across the main diagonal, the $M = 4$ element matrices for (9.25) and (9.26) each require 225 entries!], it is often accomplished by numerical quadrature. Purists who prefer closed-form evaluation will probably find the approach of P. P. Silvester [2, 6] to be the most convenient. We summarize this procedure below.

Consider the evaluation of Equation (9.25) over one triangular cell denoted Γ_t. The integrand can be written explicitly as

$$\frac{\partial B_p}{\partial x} \frac{\partial B_q}{\partial x} + \frac{\partial B_p}{\partial y} \frac{\partial B_q}{\partial y} \tag{9.31}$$

Using the chain rule of differentiation and Equation (9.4), we find

$$\frac{\partial B_q}{\partial x} = \sum_{i=1}^{3} \frac{\partial B_q}{\partial L_i} \frac{\partial L_i}{\partial x} = \sum_{i=1}^{3} \frac{\partial B_q}{\partial L_i} \frac{b_i}{2A} \tag{9.32a}$$

$$\frac{\partial B_q}{\partial y} = \sum_{i=1}^{3} \frac{\partial B_q}{\partial L_i} \frac{\partial L_i}{\partial y} = \sum_{i=1}^{3} \frac{\partial B_q}{\partial L_i} \frac{c_i}{2A} \tag{9.32b}$$

The integral can now be written as

$$A_{pq}^{(1)} = \frac{1}{4A^2} \iint_{\Gamma_t} \sum_{i=1}^{3} \sum_{j=1}^{3} (b_i b_j + c_i c_j) \frac{\partial B_p}{\partial L_i} \frac{\partial B_q}{\partial L_j} \, dx \, dy \tag{9.33}$$

Geometric properties of a triangle dictate that [2]

$$b_i b_j + c_i c_j = -2A \cot\theta_k \qquad (i \neq j) \tag{9.34}$$

$$b_i^2 + c_i^2 = 2A(\cot\theta_j + \cot\theta_k) \tag{9.35}$$

where θ_i is the interior angle at node i and the indices i, j, and k assume the values 1, 2, and 3 cyclically. It follows that the integral of Equation (9.25) reduces to

$$A_{pq}^{(1)} = \frac{1}{2A} \iint_{\Gamma_t} \sum_{i=1}^{3} \left(\frac{\partial B_p}{\partial L_{i+1}} - \frac{\partial B_p}{\partial L_{i-1}} \right) \left(\frac{\partial B_q}{\partial L_{i+1}} - \frac{\partial B_q}{\partial L_{i-1}} \right) \cot\theta_i \, dx \, dy \tag{9.36}$$

It remains to transform the integral from Cartesian coordinates to the local-area coordinates $\{L_1, L_2, L_3 = 1 - L_1 - L_2\}$ using the transformation

$$dx \, dy = dL_1 \, dL_2 \frac{\partial(x, y)}{\partial(L_1, L_2)} = 2A \, dL_1 \, dL_2 \tag{9.37}$$

Equation (9.25) can be written in the local coordinates as

$$A_{pq}^{(1)} = \sum_{i=1}^{3} Q_{pq}^{(i)} \cot\theta_i \tag{9.38}$$

where

$$Q_{pq}^{(i)} = \iint_{\Gamma_t} \left(\frac{\partial B_p}{\partial L_{i+1}} - \frac{\partial B_p}{\partial L_{i-1}} \right) \left(\frac{\partial B_q}{\partial L_{i+1}} - \frac{\partial B_q}{\partial L_{i-1}} \right) dL_1 \, dL_2 \tag{9.39}$$

Note that $Q_{pq}^{(i)}$ is independent of the specific shape of the triangle Γ_t. In fact, Equation (9.39) consists entirely of integrals having the form [2]

$$I = \iint_{\Gamma_t} L_1^a L_2^b L_3^c \, dL_1 dL_2 = \frac{a!b!c!}{(a+b+c+2)!} \tag{9.40}$$

where a, b, and c represent integer powers.

A similar transformation can be applied to Equation (9.26) to produce

$$A_{pq}^{(2)} = 2A \iint_{\Gamma_t} B_p B_q \, dL_1 \, dL_2 \tag{9.41}$$

which can also be evaluated using Equation (9.40).

In the simplex coordinate system, the integrals arising in the matrix elements depend only on the order of the interpolation functions and not on the specific shape or size of the triangular cells. The evaluation of these integrals may be done once and for all prior to the matrix fill and tabulated in a systematic way for easy access. The remaining calculations required for matrix element computation are limited to the area A of each cell and the three cotangents necessary for Equation (9.38). The three Q-matrices required for Equation (9.38) are simple permutations of each other, and Silvester's approach employs rotation matrices to construct the entries of each Q-matrix from a minimal-sized table [2, 6]. FORTRAN subroutines producing the complete element matrices for (9.25) and (9.26) are available for polynomial orders $M = 1$ through $M = 4$ [2].

The entries of the boundary integral terms in Equations (9.27) and (9.30) can be found in a similar manner. Their evaluation will be left as an exercise for the reader.

The complete finite-element system may be constructed by scanning through the mesh, obtaining element matrices on a cell-by-cell basis, and subsequently adding these entries to the global system in Equation (9.22). To facilitate the required bookkeeping, we employ a *connectivity matrix* mapping the local node numbers to the global node numbers. The primary connectivity array has dimension N_c by m, where N_c is the number of cells in the model and m is the number of nodes per cell. A second connectivity array is required to link nodes on the outer boundary with the adjacent cells, in order to facilitate the evaluation of the boundary integral terms. These connectivity arrays are part of the scatterer model generated prior to finite-element analysis. (As we have seen in previous chapters, the scatterer model also includes a list of node coordinates and the constitutive parameters of each cell.)

The use of higher order expansion functions increases the required storage relative to linear expansion functions as a consequence of larger element matrices and a larger connectivity matrix. In addition, the sparsity of the resulting global system will decrease as the order of the expansion functions increases, since more basis functions interact over each cell. For a scatterer model having a fixed number of cells, the number of nodes and the number of nonzero entries can grow quickly with increasing polynomial order. For the idealized situation of N equilateral triangles arranged in an asymptotically large hexagonal mesh, the number of nodes grows as $N/2$, $4N/2$, and $9N/2$ for linear, quadratic, and cubic Lagrangian functions. Data shown in Figure 9.5 from a variety of triangular-cell models closely tracks this 1:4:9 ratio. From a similar analysis, the number of nonzero entries in the **A**-matrix can be estimated as $3.5N$, $23N$, and $76.5N$ for linear, quadratic, and cubic Lagrangian functions on an idealized mesh containing N cells. Figure 9.6 illustrates actual data for linear and quadratic expansions that are in reasonable agreement with these estimates. The overall computational requirements, however, may not always grow as

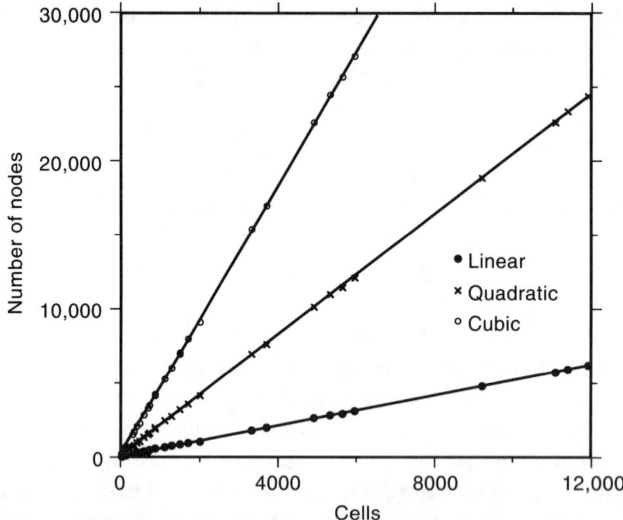

Figure 9.5 Growth of nodes with element order versus number of cells in a triangular-cell mesh.

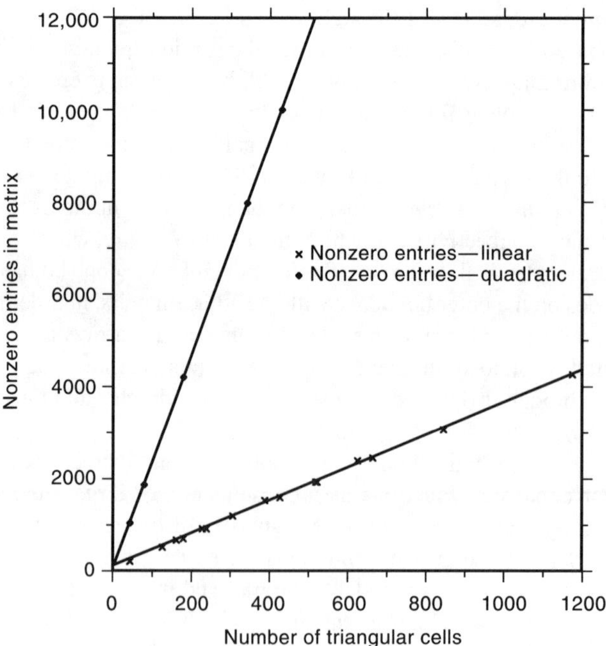

Figure 9.6 Growth of matrix entries with element order versus number of cells in a triangular-cell mesh.

quickly as these figures seem to suggest. Figure 9.7 compares the number of nonzero entries in the **A**-matrix after LU factorization via the sparse matrix program Y12M [7] for linear and quadratic expansions. These data suggest that there is often little additional storage in the use of quadratic expansion functions compared to linear for an equal number

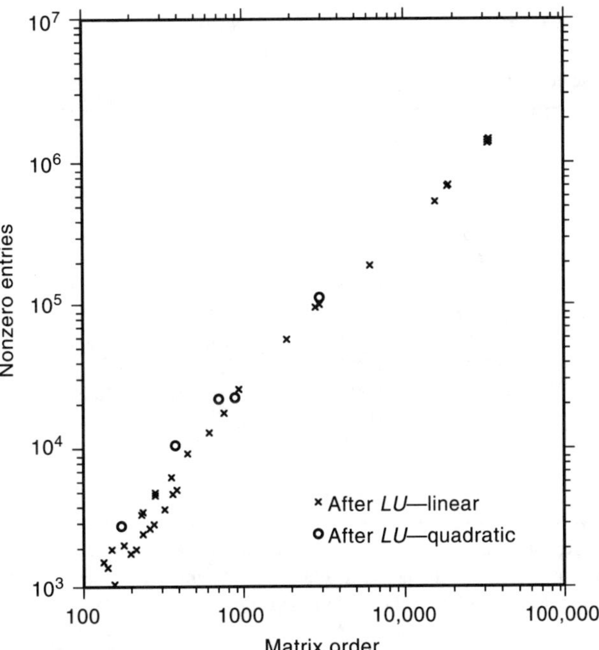

Figure 9.7 Nonzero entries after LU factorization versus matrix order: Y12M performance for scalar bases.

of nodes. Furthermore, we expect a substantial improvement in accuracy when using higher order expansion functions.

As an illustration, consider a lossy dielectric cylinder of size $ka = 2$ and relative permittivity $\varepsilon_r = 2.56 - j2.56$. Table 9.1 shows the interior E_z-field produced by an incident TM plane wave along a cut through the center of the cylinder. Results obtained using linear basis and testing functions are compared with results from a quadratic Lagrangian representation for a mesh containing 1480 triangular cells. The linear expansion involves one unknown for each of the 781 corner nodes in the mesh, yielding a density within the cylinder of 105 nodes/λ_d^2. For the quadratic expansion functions, the additional midside nodes bring the total to 3041 unknowns. The second-order Bayliss–Turkel RBC is imposed on a circular boundary having radius $\rho = 0.98\lambda_0$, approximately three times the cylinder radius. The accuracy obtained with the quadratic basis and testing functions is quite good, especially in light of the fact that the RBC is approximate.

As a second example, Figure 9.8 shows the electric field magnitude within a cylinder having $ka = 3.5449$ and $\varepsilon_r = 6$. Linear and quadratic expansions were employed using different models adjusted to produce a similar number of unknowns (2905 for the linear functions, 3041 for the quadratic). The second-order Bayliss–Turkel RBC was imposed at $\rho = 1.36\lambda_0$, a location approximately 2.4 times the radius of the cylinder. The linear result employed a mesh with 5656 triangular cells, 2905 nodes, and a maximum cell dimension of $0.068\lambda_0$. The quadratic result employed a mesh with 1480 triangular cells, 3041 nodes (781 corner nodes), and a maximum cell dimension of $0.135\lambda_0$. Although the linear result contains gross errors, the quadratic result appears to exhibit excellent agreement with the exact solution. (The quadratic basis coefficients used to plot Figure 9.8 actually have a total error that varies between 0.4 and 12% along this cut through the cylinder.)

TABLE 9.1 E_z-Field Produced by Incident TM Plane Wave Along $y = 0$ Cut Through Center of Circular, Homogeneous Dielectric Cylinder Having $ka = 2$ and Relative Permittivity $\varepsilon_r = 2.56 - j2.56$

	Magnitude of E_z			Phase of E_z (deg)		
x	Linear	Quadratic	Exact	Linear	Quadratic	Exact
−0.319	0.7163	0.7046	0.7039	123.75	123.28	123.32
−0.255	0.5683	0.5568	0.5566	83.39	82.32	82.35
−0.191	0.4508	0.4388	0.4386	43.13	41.62	41.64
−0.128	0.3568	0.3445	0.3443	1.80	0.05	0.07
−0.064	0.2902	0.2768	0.2767	−40.65	−42.66	−42.63
0.0	0.2414	0.2276	0.2274	−82.36	−84.81	−84.78
0.064	0.1962	0.1833	0.1831	−126.16	−129.01	−128.98
0.128	0.1837	0.1698	0.1696	178.41	178.09	178.12
0.191	0.2412	0.2237	0.2235	139.68	135.48	135.50
0.255	0.3169	0.2964	0.2961	120.23	116.14	116.15
0.319	0.3511	0.3303	0.3299	115.98	112.54	112.57

Note: Results obtained using linear and quadratic expansion functions with the same 1480-cell model are compared with the exact solution. The numerical solutions were obtained with the second-order Bayliss–Turkel RBC imposed at $\rho = 0.98\lambda_0$.

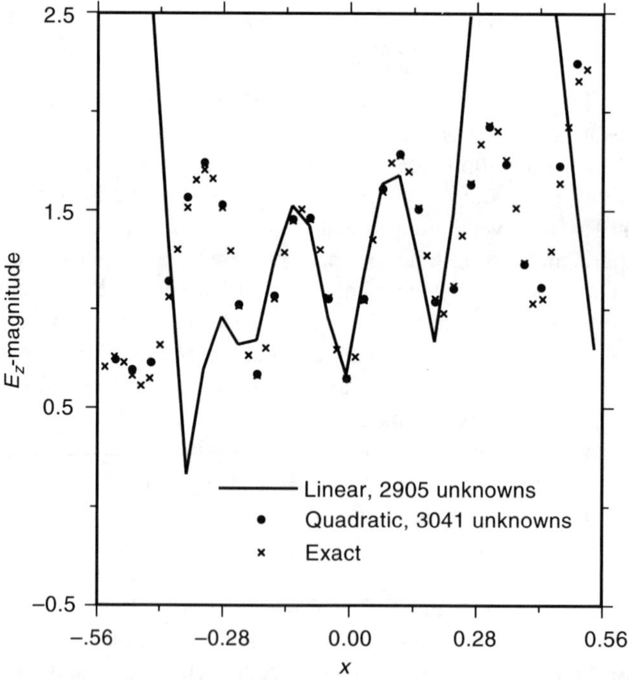

Figure 9.8 Electric field within a circular dielectric cylinder obtained using linear (solid line) and quadratic (dots) expansion functions. The exact solution is also shown (crosses). Both results required about the same number of unknowns. The cylinder has $ka = 3.5449$, $\varepsilon_r = 6$, $\mu_r = 1$, and the radiation boundary is located at $\rho = 1.36\lambda_0$.

Numerical results for the electric field in the center of a circular dielectric cylinder with $\varepsilon_r = 3$ and $ka = 1$ were presented in Table 3.2 based on linear Lagrangian basis functions. For purposes of comparison, Table 9.2 shows the error in the electric field for quadratic basis functions defined on triangular-cell models similar to those used to create Table 3.2. In this formulation, the approximate Bayliss–Turkel RBC is used to truncate the mesh at a radius of 0.3λ (less than twice the cylinder radius). Although the approximate RBC places a limit on the accuracy possible with a large density of cells, the error due to the RBC appears to be less than 0.5% in this case.

TABLE 9.2 Numerical Values of E_z-Field Produced at Center of Circular Dielectric Cylinder with $\varepsilon_r = 3$ and $ka = 1$

Model	Longest Edge in Mesh	Magnitude of E_z	Phase of E_z (deg)	Error (%)
97-Node	$0.166\lambda_0$	0.909	−57.36	1.88
273-Node	$0.098\lambda_0$	0.897	−57.39	0.59
565-Node	$0.071\lambda_0$	0.895	−57.37	0.37
901-Node	$0.053\lambda_0$	0.895	−57.36	0.30
Exact	—	0.892	−57.36	—

Note: Results from four triangular-cell models are compared with the exact solution. The Bayliss–Turkel RBC is imposed at a 0.3λ radius and places a limit on the accuracy in this case. Results are obtained with quadratic basis functions.

9.3 LAGRANGIAN BASIS FUNCTIONS FOR RECTANGULAR AND QUADRILATERAL CELLS

While triangular cells offer modeling flexibility, a rectangular-cell mesh is sometimes preferred because of ease of generation and visualization. Often, an irregular region can be represented by a combination of rectangular and triangular cells, provided that the continuity of the expansion is maintained across cell boundaries. This section considers the development of Lagrangian basis functions for rectangular domains. In addition, basis functions defined on quadrilateral cells can be obtained by a straightforward transformation of the rectangular-cell functions. This transformation is described.

The simplest representation on rectangles is bilinear interpolation between the four cell corners. A suitable expansion consists of four overlapping functions, each having the general form

$$B_i(x, y) = \alpha_i + \beta_i x + \gamma_i y + \delta_i xy \tag{9.42}$$

The four degrees of freedom in each function allow interpolation to one of the four corners of the cell, that is, $B_i(x, y) = 1$ at node i and $B_i(x, y) = 0$ at the other three nodes. Because the representation along every edge of the rectangular cell is linear, this expansion automatically provides continuity to a linear expansion on an adjacent cell, whether it be rectangular or triangular in shape. These bilinear functions are the lowest order member of a Lagrangian family for rectangular cells.

In order to develop a general formula for these expansion functions, we repeat the expression originally introduced in Equation (5.33) for the jth Lagrangian function of order

$N-1$ on an interval $[x_1, x_N]$:

$$\phi_j(x) = \frac{(x-x_1)(x-x_2)\cdots(x-x_{j-1})(x-x_{j+1})\cdots(x-x_N)}{(x_j-x_1)(x_j-x_2)\cdots(x_j-x_{j-1})(x_j-x_{j+1})\cdots(x_j-x_N)} \tag{9.43}$$

This expression defines N independent polynomials $\{\phi_1, \phi_2, \ldots, \phi_N\}$ spanning the interval. It is convenient to introduce a local coordinate system (η, ξ), where

$$\eta = \frac{2x - x_1 - x_2}{x_2 - x_1} \tag{9.44}$$

$$\xi = \frac{2y - y_1 - y_2}{y_2 - y_1} \tag{9.45}$$

These coordinates are illustrated in Figure 9.9. We can construct a general formula for the Lagrangian expansion functions using

$$B_{ij}(\eta, \xi) = \phi_i(\eta)\phi_j(\xi) \tag{9.46}$$

where $\phi_i(\eta)$ denotes a polynomial of order M spanning the interval $[-1, 1]$, and i and j may assume values $1, 2, \ldots, M+1$.

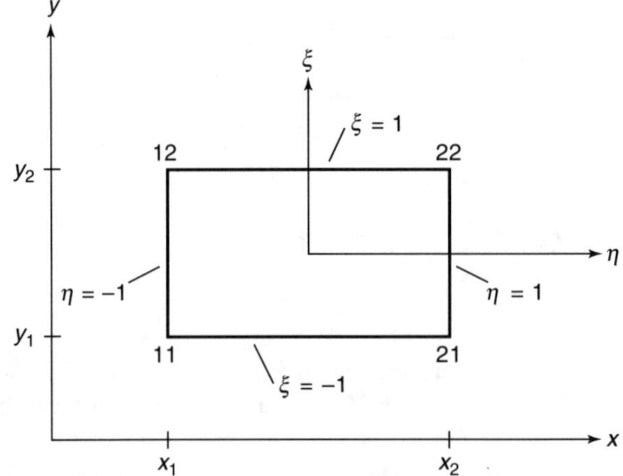

Figure 9.9 Local (η, ξ) coordinate system.

If expressed as explicit functions of η and ξ over the domain $(-1 < \eta < 1, -1 < \xi < 1)$, the bilinear expansion functions have the form

$$B_{11}(\eta, \xi) = \frac{1-\eta}{2}\frac{1-\xi}{2} \tag{9.47}$$

$$B_{12}(\eta, \xi) = \frac{1-\eta}{2}\frac{\xi+1}{2} \tag{9.48}$$

$$B_{21}(\eta, \xi) = \frac{\eta+1}{2}\frac{1-\xi}{2} \tag{9.49}$$

$$B_{22}(\eta, \xi) = \frac{\eta+1}{2}\frac{\xi+1}{2} \tag{9.50}$$

Element matrices for integrals such as (9.25) and (9.26) can be evaluated directly in the

(η, ξ)-coordinate system. The mapping from (η, ξ) to (x, y) involves a transformation

$$\begin{bmatrix} \dfrac{\partial}{\partial \eta} \\ \dfrac{\partial}{\partial \xi} \end{bmatrix} = \mathbf{J} \begin{bmatrix} \dfrac{\partial}{\partial x} \\ \dfrac{\partial}{\partial y} \end{bmatrix} \tag{9.51}$$

where $\mathbf{J}$ is the Jacobian matrix

$$\mathbf{J} = \begin{bmatrix} \dfrac{\partial x}{\partial \eta} & \dfrac{\partial y}{\partial \eta} \\ \dfrac{\partial x}{\partial \xi} & \dfrac{\partial y}{\partial \xi} \end{bmatrix} = \begin{bmatrix} \dfrac{x_2 - x_1}{2} & 0 \\ 0 & \dfrac{y_2 - y_1}{2} \end{bmatrix} \tag{9.52}$$

Therefore, since

$$dx\, dy = (\det \mathbf{J})\, d\eta\, d\xi = \tfrac{1}{4} A\, d\eta\, d\xi \tag{9.53}$$

where A is the area of the rectangular cell, (9.26) can be written as

$$A_{pq}^{(2)} = \int_x \int_y B_p B_q\, dx\, dy = \frac{A}{4} \int_{\eta=-1}^{1} \int_{\xi=-1}^{1} B_p B_q\, d\eta\, d\xi \tag{9.54}$$

In a similar manner, (9.51) can be used to transform (9.25) into the expression

$$\begin{aligned} A_{pq}^{(1)} &= \int_x \int_y \begin{bmatrix} \dfrac{\partial B_p}{\partial x} & \dfrac{\partial B_p}{\partial y} \end{bmatrix} \begin{bmatrix} \dfrac{\partial B_q}{\partial x} \\ \dfrac{\partial B_q}{\partial y} \end{bmatrix} dx\, dy \\ &= \frac{A}{4} \int_{\eta=-1}^{1} \int_{\xi=-1}^{1} \begin{bmatrix} \dfrac{\partial B_p}{\partial \eta} & \dfrac{\partial B_p}{\partial \xi} \end{bmatrix} \mathbf{J}^{-T} \mathbf{J}^{-1} \begin{bmatrix} \dfrac{\partial B_q}{\partial \eta} \\ \dfrac{\partial B_q}{\partial \xi} \end{bmatrix} d\eta\, d\xi \\ &= \frac{y_2 - y_1}{x_2 - x_1} \int_{\eta=-1}^{1} \int_{\xi=-1}^{1} \frac{\partial B_p}{\partial \eta} \frac{\partial B_q}{\partial \eta} + \frac{x_2 - x_1}{y_2 - y_1} \int_{\eta=-1}^{1} \int_{\xi=-1}^{1} \frac{\partial B_p}{\partial \xi} \frac{\partial B_q}{\partial \xi} \end{aligned} \tag{9.55}$$

The generation of the actual element matrix entries for the bilinear basis functions follows directly from (9.54) and (9.55) and will be left as an exercise.

Biquadratic functions may be defined to interpolate between nine nodes on a rectangular cell. The appropriate general form is obtained from Equation (9.46), with quadratic polynomials used for each ϕ. Specific expressions in local coordinates are

$$B_{11}(\eta, \xi) = \frac{\eta(\eta - 1)}{2} \frac{\xi(\xi - 1)}{2} \tag{9.56}$$

$$B_{12}(\eta, \xi) = \frac{\eta(\eta - 1)}{2} (1 + \xi)(1 - \xi) \tag{9.57}$$

$$B_{13}(\eta, \xi) = \frac{\eta(\eta - 1)}{2} \frac{\xi(\xi + 1)}{2} \tag{9.58}$$

$$B_{21}(\eta, \xi) = (1 + \eta)(1 - \eta) \frac{\xi(\xi - 1)}{2} \tag{9.59}$$

$$B_{22}(\eta, \xi) = (1+\eta)(1-\eta)(1+\xi)(1-\xi) \tag{9.60}$$

$$B_{23}(\eta, \xi) = (1+\eta)(1-\eta)\frac{\xi(\xi+1)}{2} \tag{9.61}$$

$$B_{31}(\eta, \xi) = \frac{\eta(\eta+1)}{2}\frac{\xi(\xi-1)}{2} \tag{9.62}$$

$$B_{32}(\eta, \xi) = \frac{\eta(\eta+1)}{2}(1+\xi)(1-\xi) \tag{9.63}$$

$$B_{33}(\eta, \xi) = \frac{\eta(\eta+1)}{2}\frac{\xi(\xi+1)}{2} \tag{9.64}$$

Each of these functions has unity value at one of the nine nodes and vanishes at the others. Element matrix entries for these functions can also be obtained from Equations (9.54) and (9.55). Extensions to higher order polynomials follow in an obvious manner. The general form employed in Equation (9.46) allows flexibility in deriving other expansion functions; in fact, a large variety may be constructed that are of mixed polynomial order (one degree along η and a different degree along ξ) [1].

Lagrangian functions on rectangular cells lack one important property associated with Lagrangian functions on triangles. On triangular cells, the expansion functions are complete to an exact polynomial order. The polynomial terms required for completeness to a particular degree are best illustrated through the Pascal triangle (Figure 9.10). Unfortunately, the Lagrangian functions used on rectangular cells are not complete to the highest order appearing in the representation. As an example, the biquadratic functions contain terms that vary as x^2y^2 but do not contain terms of y^4 or even y^3! Because the accuracy of the numerical solution is constrained by the first "missing" term in the expansion, the associated interpolation error may only decrease as $O(\Delta^3)$ rather than as $O(\Delta^5)$ as $\Delta \to 0$. Consequently, the rectangular Lagrangian functions might be construed to contain wasted degrees of freedom.

						Interpolation error as $\Delta \to 0$
		1			Order 0	$O(\Delta)$
	x		y		Order 1	$O(\Delta^2)$
	x^2	xy	y^2		Order 2	$O(\Delta^3)$
x^3	x^2y		xy^2	y^3	Order 3	$O(\Delta^4)$
x^4	x^3y	x^2y^2	xy^3	y^4	Order 4	$O(\Delta^5)$

Figure 9.10 Pascal's triangle to illustrate completeness properties of polynomial expansions.

Obviously, a mesh restricted to rectangular cells is only useful for simple geometries. General quadrilateral cells offer much more flexibility while providing a conceptual similarity to rectangular cells. A bilinear transformation can be used to map basis functions from rectangular cells to quadrilateral cells.

Consider a quadrilateral cell in the (x, y) plane with vertices at (x_i, y_i), $i = 1, 2, 3, 4$ (Figure 9.11). The region $(-1 < \eta < 1, -1 < \xi < 1)$ can be mapped into the quadrilateral using the transformation

$$x = x_1 B_{22} + x_2 B_{12} + x_3 B_{11} + x_4 B_{21} \tag{9.65}$$

$$y = y_1 B_{22} + y_2 B_{12} + y_3 B_{11} + y_4 B_{21} \tag{9.66}$$

where B_{11}, B_{12}, B_{21}, and B_{22} are the bilinear basis functions defined in (9.47)–(9.50). The transformation can be written in expanded form as

$$\begin{aligned} x = \tfrac{1}{4}[&\eta\xi(x_1 - x_2 + x_3 - x_4) + \eta(x_1 - x_2 - x_3 + x_4) \\ &+\xi(x_1 + x_2 - x_3 - x_4) + (x_1 + x_2 + x_3 + x_4)] \end{aligned} \tag{9.67}$$

$$\begin{aligned} y = \tfrac{1}{4}[&\eta\xi(y_1 - y_2 + y_3 - y_4) + \eta(y_1 - y_2 - y_3 + y_4) \\ &+\xi(y_1 + y_2 - y_3 - y_4) + (y_1 + y_2 + y_3 + y_4)] \end{aligned} \tag{9.68}$$

An application of the chain rule produces

$$\begin{bmatrix} \dfrac{\partial}{\partial \eta} \\ \dfrac{\partial}{\partial \xi} \end{bmatrix} = \begin{bmatrix} \dfrac{\partial x}{\partial \eta} & \dfrac{\partial y}{\partial \eta} \\ \dfrac{\partial x}{\partial \xi} & \dfrac{\partial y}{\partial \xi} \end{bmatrix} \begin{bmatrix} \dfrac{\partial}{\partial x} \\ \dfrac{\partial}{\partial y} \end{bmatrix} = \mathbf{J} \begin{bmatrix} \dfrac{\partial}{\partial x} \\ \dfrac{\partial}{\partial y} \end{bmatrix} \tag{9.69}$$

where $\mathbf{J}$ is the Jacobian of the transformation from the (η, ξ) system to the (x, y) system. For the mapping in (9.67) and (9.68), the matrix $\mathbf{J}$ has the form

$$\mathbf{J} = \frac{1}{4}\begin{bmatrix} \xi(x_1 - x_2 + x_3 - x_4) + (x_1 - x_2 - x_3 + x_4) & \xi(y_1 - y_2 + y_3 - y_4) + (y_1 - y_2 - y_3 + y_4) \\ \eta(x_1 - x_2 + x_3 - x_4) + (x_1 + x_2 - x_3 - x_4) & \eta(y_1 - y_2 + y_3 - y_4) + (y_1 + y_2 - y_3 - y_4) \end{bmatrix} \tag{9.70}$$

An explicit expression for the bilinear basis functions in the (x, y)-coordinate system may be unnecessary, since knowledge of the linear behavior along cell edges is often sufficient for visualization purposes. However, a finite-element implementation requires a convenient way of computing the 4×4 element matrices, and again we seek a local evaluation directly in the (η, ξ) system.

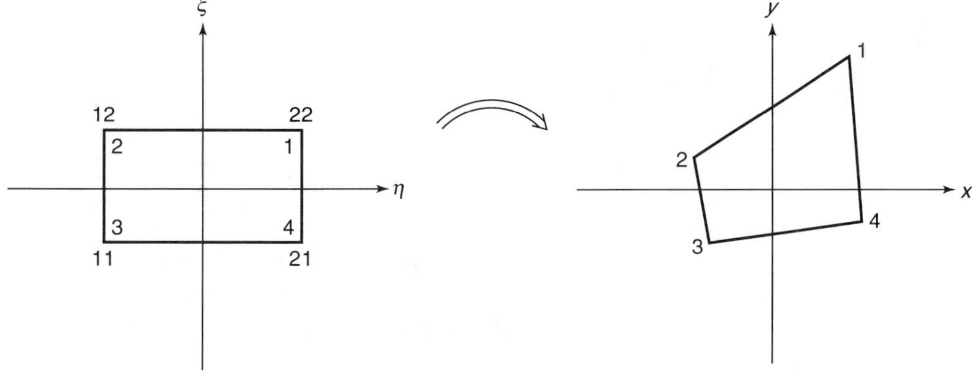

Figure 9.11 Mapping from a reference cell to a quadrilateral cell.

Using

$$dx\,dy = (\det \mathbf{J})\,d\eta\,d\xi \tag{9.71}$$

the entries of (9.26) are given by

$$A_{pq}^{(2)} = \int_x \int_y B_p B_q \,dx\,dy = \int_{\eta=-1}^{1} \int_{\xi=-1}^{1} B_p B_q (\det \mathbf{J})\,d\eta\,d\xi \tag{9.72}$$

where det **J** is generally a function of (η, ξ) and cannot be extracted from the integrand. The evaluation of

$$A_{pq}^{(1)} = \int_x \int_y \begin{bmatrix} \dfrac{\partial B_p}{\partial x} & \dfrac{\partial B_p}{\partial y} \end{bmatrix} \begin{bmatrix} \dfrac{\partial B_q}{\partial x} \\ \dfrac{\partial B_q}{\partial y} \end{bmatrix} dx\,dy \tag{9.73}$$

requires the Jacobian of the inverse mapping, that is,

$$\begin{bmatrix} \dfrac{\partial}{\partial x} \\ \dfrac{\partial}{\partial y} \end{bmatrix} = \begin{bmatrix} \dfrac{\partial \eta}{\partial x} & \dfrac{\partial \xi}{\partial x} \\ \dfrac{\partial \eta}{\partial y} & \dfrac{\partial \xi}{\partial y} \end{bmatrix} \begin{bmatrix} \dfrac{\partial}{\partial \eta} \\ \dfrac{\partial}{\partial \xi} \end{bmatrix} = \mathbf{J}^{-1} \begin{bmatrix} \dfrac{\partial}{\partial \eta} \\ \dfrac{\partial}{\partial \xi} \end{bmatrix} \tag{9.74}$$

Since the mapping defines x and y in terms of η and ξ, it is usually easier to compute $\mathbf{J}^{-1}$ from **J** at the necessary quadrature points by matrix inversion than it is to construct $\mathbf{J}^{-1}$ from the explicit inverse transformation. It follows that

$$A_{pq}^{(1)} = \int_{\eta=-1}^{1} \int_{\xi=-1}^{1} \begin{bmatrix} \dfrac{\partial B_p}{\partial \eta} & \dfrac{\partial B_p}{\partial \xi} \end{bmatrix} \mathbf{J}^{-T} \mathbf{J}^{-1} \begin{bmatrix} \dfrac{\partial B_q}{\partial \eta} \\ \dfrac{\partial B_q}{\partial \xi} \end{bmatrix} (\det \mathbf{J})\,d\eta\,d\xi \tag{9.75}$$

Once the integrals are recast into the (η, ξ) system, (9.72) and (9.75) may be evaluated by numerical quadrature.

We have considered a bilinear mapping in order to create basis functions on quadrilateral cells with straight sides. For numerical stability, the transformation must map points in (η, ξ) uniquely to points in (x, y) and vice versa and must not allow interior angles of the quadrilateral to approach or exceed 180°. A similar procedure will be used in the following section in order to produce basis functions on cells with curved sides.

9.4 SCALAR BASIS FUNCTIONS FOR TWO-DIMENSIONAL CELLS WITH CURVED SIDES

Higher order polynomial basis functions generally produce better accuracy than simple functions at a modest additional cost. However, improvements in interpolation accuracy can be offset by errors in modeling the region of interest, especially if the geometry has curved boundaries. These modeling errors can be reduced by the use of basis functions defined on cells with curved sides. Parametric basis functions can be obtained by a mapping procedure similar to that introduced in Section 9.3 to generate basis functions for quadrilateral cells. The bilinear transformation used in Section 9.3 maps straight lines into straight lines. In this section, we consider a quadratic transformation capable of mapping straight lines into parabolic curves.

Figure 9.12 depicts a distorted triangular cell. The figure also shows six nodes, denoted by the triple index (i, j, k) introduced in Section 9.1. The curved geometry can be described by a transformation from simplex coordinates (L_1, L_2, L_3) to the (x, y) system

$$x = \sum_{i=0}^{2}\sum_{j=0}^{2} x_{ijk} B_{ijk}(L_1, L_2, L_3) \tag{9.76}$$

$$y = \sum_{i=0}^{2}\sum_{j=0}^{2} y_{ijk} B_{ijk}(L_1, L_2, L_3) \tag{9.77}$$

where (x_{ijk}, y_{ijk}) are the coordinates of the appropriate nodes, $\{B_{ijk}\}$ are the six quadratic functions defined in (9.15)–(9.20), and $k = 2 - i - j$. The cell shape defined by this transformation is entirely determined by the six-node coordinates, and the sides have at most a parabolic curvature.

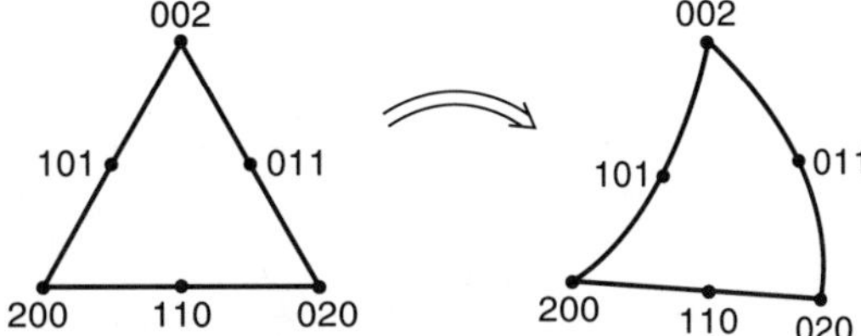

Figure 9.12 Mapping from a reference cell to a curved triangular cell.

Quadratic basis functions of the Lagrangian type may be defined on the curved cell using the same transformation, that is, (9.76) and (9.77). The use of quadratic basis functions on a cell whose shape is defined by a quadratic mapping is known as an *isoparametric* expansion. (In contrast, our previous use of quadratic Lagrangian functions on a triangular cell is a *subparametric* expansion.) Since the cell shapes are defined by the required nodes, the isoparametric expansion on a curved-cell mesh imposes no additional bookkeeping to describe the model beyond that required for the same order expansion on a triangular-cell mesh.

Using $L_1 + L_2 + L_3 = 1$, the transformation can be written entirely in terms of the variables L_1 and L_2 to produce

$$\begin{aligned} x = x_{002} &+ L_1(4x_{101} - 3x_{002} - x_{200}) + L_2(4x_{011} - 3x_{002} - x_{020}) \\ &+ L_1^2(2x_{200} + 2x_{002} - 4x_{101}) + L_1L_2(4x_{002} + 4x_{110} - 4x_{101} - 4x_{011}) \\ &+ L_2^2(2x_{020} + 2x_{002} - 4x_{011}) \end{aligned} \tag{9.78}$$

$$\begin{aligned} y = y_{002} &+ L_1(4y_{101} - 3y_{002} - y_{200}) + L_2(4y_{011} - 3y_{002} - y_{020}) \\ &+ L_1^2(2y_{200} + 2y_{002} - 4y_{101}) + L_1L_2(4y_{002} + 4y_{110} - 4y_{101} - 4y_{011}) \\ &+ L_2^2(2y_{020} + 2y_{002} - 4y_{011}) \end{aligned} \tag{9.79}$$

The chain rule dictates that

$$\begin{bmatrix} \dfrac{\partial}{\partial L_1} \\ \dfrac{\partial}{\partial L_2} \end{bmatrix} = \mathbf{J} \begin{bmatrix} \dfrac{\partial}{\partial x} \\ \dfrac{\partial}{\partial y} \end{bmatrix} \tag{9.80}$$

where

$$\mathbf{J} = \begin{bmatrix} \dfrac{\partial x}{\partial L_1} & \dfrac{\partial y}{\partial L_1} \\ \dfrac{\partial x}{\partial L_2} & \dfrac{\partial y}{\partial L_2} \end{bmatrix} \tag{9.81}$$

is the Jacobian matrix. The entries of $\mathbf{J}$ are given by

$$\begin{aligned} \frac{\partial x}{\partial L_1} &= 4x_{101} - 3x_{002} - x_{200} + L_1(4x_{200} + 4x_{002} - 8x_{101}) \\ &\quad + L_2(4x_{002} + 4x_{110} - 4x_{101} - 4x_{011}) \end{aligned} \tag{9.82}$$

$$\begin{aligned} \frac{\partial y}{\partial L_1} &= (4y_{101} - 3y_{002} - y_{200}) + L_1(4y_{200} + 4y_{002} - 8y_{101}) \\ &\quad + L_2(4y_{002} + 4y_{110} - 4y_{101} - 4y_{011}) \end{aligned} \tag{9.83}$$

$$\begin{aligned} \frac{\partial x}{\partial L_2} &= (4x_{011} - 3x_{002} - x_{020}) + L_1(4x_{002} + 4x_{110} - 4x_{101} - 4x_{011}) \\ &\quad + L_2(4x_{020} + 4x_{002} - 8x_{011}) \end{aligned} \tag{9.84}$$

$$\begin{aligned} \frac{\partial y}{\partial L_2} &= (4y_{011} - 3y_{002} - y_{020}) + L_1(4y_{002} + 4y_{110} - 4y_{101} - 4y_{011}) \\ &\quad + L_2(4y_{020} + 4y_{002} - 8y_{011}) \end{aligned} \tag{9.85}$$

Required integrals for element matrices associated with the scalar Helmholtz operations can be obtained directly in the (L_1, L_2) system using

$$\begin{aligned} A_{pq}^{(1)} &= \int_x \int_y \begin{bmatrix} \dfrac{\partial B_p}{\partial x} & \dfrac{\partial B_p}{\partial y} \end{bmatrix} \begin{bmatrix} \dfrac{\partial B_q}{\partial x} \\ \dfrac{\partial B_q}{\partial y} \end{bmatrix} dx\, dy \\ &= \int_{L_1} \int_{L_2} \begin{bmatrix} \dfrac{\partial B_p}{\partial L_1} & \dfrac{\partial B_p}{\partial L_2} \end{bmatrix} \mathbf{J}^{-T} \mathbf{J}^{-1} \begin{bmatrix} \dfrac{\partial B_q}{\partial L_1} \\ \dfrac{\partial B_q}{\partial L_2} \end{bmatrix} (\det \mathbf{J})\, dL_1\, dL_2 \end{aligned} \tag{9.86}$$

$$\begin{aligned} A_{pq}^{(2)} &= \int_x \int_y B_p B_q \, dx\, dy \\ &= \int_{L_1} \int_{L_2} B_p B_q (\det \mathbf{J})\, dL_1\, dL_2 \end{aligned} \tag{9.87}$$

where $\mathbf{J}^{-1}$ is the inverse of $\mathbf{J}$. The integrals in (9.86) and (9.87) are generally computed by numerical quadrature. Of course, for numerical stability the mapping from the (L_1, L_2, L_3) system to the (x, y) system must be one to one, and the interior angles of the distorted triangle must not approach or exceed 180° [8]. In addition, the location of the midside nodes must not deviate much from the halfway point along the edges of the curved cell; see Prob. P9.10 and reference [8].

A similar transformation can be used to define distorted nine-node quadrilateral cells (Figure 9.13) and map quadratic Lagrangian functions to those domains. The mapping has the form

$$x = \sum_{i=1}^{3} \sum_{j=1}^{3} x_{ij} B_{ij}(\eta, \xi) \tag{9.88}$$

$$y = \sum_{i=1}^{3} \sum_{j=1}^{3} y_{ij} B_{ij}(\eta, \xi) \tag{9.89}$$

where (x_{ij}, y_{ij}) are the coordinates of the appropriate nodes and $\{B_{ij}\}$ are the nine quadratic functions defined throughout the standard cell $(-1 < \eta < 1, -1 < \xi < 1)$ in Equations (9.56)–(9.64). The associated Jacobian relationship and element matrix entries can be obtained by calculations similar to those in (9.80)–(9.87). As an aid to implementation, a FORTRAN subroutine providing the element matrices associated with an eight-node quadrilateral expansion is available in reference [2]. (The eight-node expansion is slightly different from the nine-node Lagrangian basis used here.)

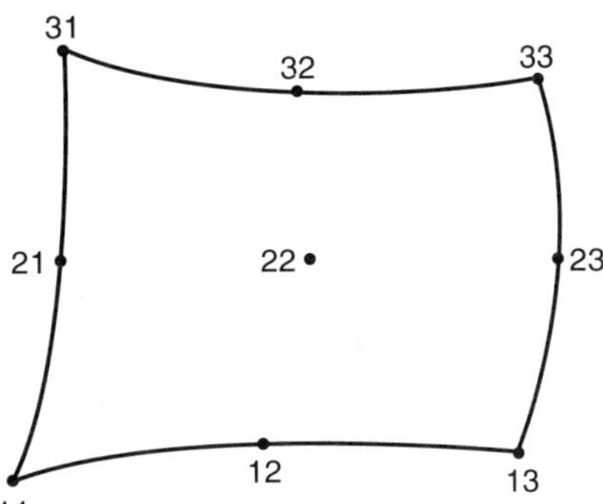

Figure 9.13 A nine-node curved quadrilateral cell.

Because of the straightforward nature of the transformation, curved cells are widely used when discretizing second-order differential equations [1, 2]. Scalar isoparametric expansions on curved cells have also been investigated for use with volume integral equation formulations [9, 10]. These formulations often require the integration of singular functions, and systematic procedures for treating the most common singularities have been developed [9]. The procedure can also be applied to create curved cells for surface integral equation discretizations, as illustrated in the following section.

9.5 DISCRETIZATION OF TWO-DIMENSIONAL SURFACE INTEGRAL EQUATIONS USING AN ISOPARAMETRIC QUADRATIC REPRESENTATION [11]

Chapters 2 and 6 presented a number of two-dimensional discretization schemes employing flat-cell models of the cylinder contours and piecewise-constant or piecewise-linear basis functions. The scatterer models can be improved by using cells with parabolic curvature. Suppose t is a parametric variable with the interval $-1 \leq t \leq 1$ used to describe a single cell. The cell can be defined by the three points (x_1, y_1), (x_2, y_2), and (x_3, y_3), and the mapping

$$x(t) = x_1 B_1(t) + x_2 B_2(t) + x_3 B_3(t) \tag{9.90}$$

$$y(t) = y_1 B_1(t) + y_2 B_2(t) + y_3 B_3(t) \tag{9.91}$$

where

$$B_1(t) = \tfrac{1}{2}t(t-1) \tag{9.92}$$

$$B_2(t) = 1 - t^2 \tag{9.93}$$

$$B_3(t) = \tfrac{1}{2}t(t+1) \tag{9.94}$$

are the quadratic Lagrangian functions defined in Section 5.3. It is convenient to also use quadratic Lagrangian interpolation polynomials to represent the surface current density (an isoparametric expansion). For the TM polarization, the current density within a cell can be replaced by

$$J_z(t) \cong \sum_{n=1}^{3} j_n B_n(t) \qquad -1 \le t \le 1 \tag{9.95}$$

Consider the TM EFIE and the use of Dirac delta testing functions to complete the discretization. The entries of the system matrix involve integrals of the form

$$I_{mn}^{\text{cell } p} = \frac{k\eta}{4} \int_{\text{cell } p} B_n(t') H_0^{(2)}(kR_m) J(t')\, dt' \tag{9.96}$$

where m and n now denote global indices,

$$R_m = \sqrt{[x_m - x(t')]^2 + [y_m - y(t')]^2} \tag{9.97}$$

and J is the Jacobian

$$J(t) = \sqrt{\left(\frac{dx}{dt}\right)^2 + \left(\frac{dy}{dt}\right)^2} \tag{9.98}$$

The Jacobian can be evaluated using the mapping in (9.90) and (9.91), which yields

$$\frac{dx}{dt} = \left(\frac{x_3 - x_1}{2}\right) + (x_1 - 2x_2 + x_3)t \tag{9.99}$$

$$\frac{dy}{dt} = \left(\frac{y_3 - y_1}{2}\right) + (y_1 - 2y_2 + y_3)t \tag{9.100}$$

within a particular cell. In general, the integrals defined by (9.96) must be evaluated by numerical quadrature. In the case where R_m vanishes within the interval of integration, the integral can be evaluated by the procedures described in Chapter 2 or Appendix A.

Because the Lagrangian functions interpolatory at the intercell nodes (x_1, y_1) or (x_3, y_3) actually span two cells, two integrals of the general form (9.96) are required for each matrix entry. Functions interpolatory at the interior point (x_2, y_2) are confined to a single cell, and the corresponding matrix entries are limited to a single integral of the form (9.96).

For the TE MFIE, a similar discretization can be developed. The off-diagonal matrix entries involve integrals of the form

$$I_{mn}^{\text{cell } p} = \frac{k}{4j} \int_{\text{cell } p} B_n(t') \left(\sin\Omega(t') \frac{x_m - x(t')}{R_m} - \cos\Omega(t') \frac{y_m - y(t')}{R_m} \right) H_1^{(2)}(kR_m) J(t')\, dt' \tag{9.101}$$

where J is defined in (9.98), R_m is defined in (9.97), and Ω denotes the angle defined in Section 2.2 (except that Ω is now a continuous function since the cells are curved). If node m lies at an intercell node, the diagonal matrix entries have the form

$$Z_{mm} = -\frac{\Gamma_m}{2\pi} + I_{mm}^{\text{cell } p} + I_{mm}^{\text{cell } q} \tag{9.102}$$

where Γ_m denotes the total interior wedge angle formed by the conductor at node m and $I_{mm}^{\text{cell } p}$ and $I_{mm}^{\text{cell } q}$ have the form of (9.101), except that a small region in the vicinity of node m is excluded from the integral. If node m lies in the cell interior, the associated diagonal entries are

$$Z_{mm} = -\tfrac{1}{2} + I_{mm}^{\text{cell } p} \tag{9.103}$$

where again a small region around node m is excluded from the integral.

To illustrate the accuracy of the isoparametric Lagrangian approach, Table 9.3 shows the TE surface current density induced on a circular cylinder by a uniform plane wave. Results are repeated from Table 2.8 for comparison. For 40 unknowns/λ, there is a consistent improvement in accuracy as the basis function order is increased.

TABLE 9.3 Comparison of Current Density Induced on Circular Cylinder with Circumference $1\lambda_0$ by TE Plane Wave Propagating in $\phi = 0$ Direction

	Magnitude				Phase (deg)			
ϕ (deg)	MFIE Pulse Basis, Flat Cells	MFIE Linear Basis, Flat Cells	MFIE Quadratic Basis, Parabolic Cells	Exact	MFIE Pulse Basis, Flat Cells	MFIE Linear Basis, Flat Cells	MFIE Quadratic Basis, Parabolic Cells	Exact
0	0.8907	0.8891	0.8883	0.8882	66.29	66.66	66.56	66.56
45	0.6733	0.6729	0.6722	0.6722	113.41	113.57	113.56	113.56
90	1.1751	1.1708	1.1714	1.1713	–164.88	–164.80	–164.82	–164.82
135	1.6232	1.6201	1.6199	1.6199	–125.83	–125.82	–125.84	–125.84
180	1.7094	1.7076	1.7073	1.7071	–110.77	–110.82	–110.83	–110.83

Note: MFIE results obtained with pulse, linear, and quadratic basis functions are compared with the exact solution for a 40-unknown discretization. The quadratic case employs parabolic cells defined by (9.90) and (9.91); the other results were obtained using flat cells. After [11].

9.6 SCALAR LAGRANGIAN FUNCTIONS IN THREE DIMENSIONS [1, 2]

The rectangular/quadrilateral-cell scalar Lagrangian functions introduced in Section 9.3 can be easily extended to the three-dimensional situation in order to represent fields using hexahedral cell shapes [1]. We will leave the full development of these functions as an exercise for the reader. The representation on tetrahedrons is perhaps somewhat more complicated, because of the use of simplex coordinates, and we briefly summarize it below.

Consider a point (x, y, z) within a tetrahedron. The point divides the cell into four smaller tetrahedra, each with a volume that is some fraction of the total. Simplex coordinates (L_1, L_2, L_3, L_4) may be defined by the respective ratio of the volume of each of the smaller cells to that of the original. Equivalently,

$$x = L_1x_1 + L_2x_2 + L_3x_3 + L_4x_4 \tag{9.104}$$

$$y = L_1y_1 + L_2y_2 + L_3y_3 + L_4y_4 \tag{9.105}$$

$$z = L_1z_1 + L_2z_2 + L_3z_3 + L_4z_4 \tag{9.106}$$

where (x_i, y_i, z_i) is the coordinate of vertex i. Since the sum of the volumes of the smaller cells must equal that of the original tetrahedron,

$$1 = L_1 + L_2 + L_3 + L_4 \tag{9.107}$$

A Cartesian representation may be obtained from

$$\begin{bmatrix} L_1 \\ L_2 \\ L_3 \\ L_4 \end{bmatrix} = \begin{bmatrix} x_1 & x_2 & x_3 & x_4 \\ y_1 & y_2 & y_3 & y_4 \\ z_1 & z_2 & z_3 & z_4 \\ 1 & 1 & 1 & 1 \end{bmatrix}^{-1} \begin{bmatrix} x \\ y \\ z \\ 1 \end{bmatrix} \tag{9.108}$$

and written in the form

$$L_i = \frac{1}{6V}(a_i + b_i x + c_i y + d_i z) \tag{9.109}$$

We leave the task of constructing the explicit Cartesian representation to the reader.

Lagrangian expansion functions having polynomial order M can be defined on tetrahedral cells containing m regularly spaced nodes, where

$$m = \tfrac{1}{6}(M+1)(M+2)(M+3) \tag{9.110}$$

In terms of a quadruple index (i, j, k, l), subject to $i + j + k + l = M$, these nodes are located at

$$(L_1, L_2, L_3, L_4) = \left(\frac{i}{M}, \frac{j}{M}, \frac{k}{M}, \frac{l}{M}\right) \tag{9.111}$$

Basis functions can be defined in general according to

$$B_{ijkl}(L_1, L_2, L_3, L_4) = R_i(M, L_1)R_j(M, L_2)R_k(M, L_3)R_l(M, L_4) \tag{9.112}$$

where $R_s(M, L)$ is the auxiliary polynomial defined in Equations (9.11) and (9.12). The function B_{ijkl} has unity value at node $ijkl$ and vanishes at every other node in the cell.

There are 4 linear functions ($M = 1$) spanning a cell. These have the simple representation

$$B_i = L_i \tag{9.113}$$

A quadratic representation requires the superposition of 10 functions. The 4 quadratic functions that interpolate to the corner nodes can be expressed as

$$B_i = (2L_i - 1)L_i \tag{9.114}$$

The 6 remaining quadratic functions interpolate to nodes located in the middle of each edge and are given by

$$B_{1100} = 4L_1L_2 \tag{9.115}$$

$$B_{1010} = 4L_1L_3 \tag{9.116}$$

$$B_{1001} = 4L_1L_4 \tag{9.117}$$

$$B_{0101} = 4L_2L_4 \tag{9.118}$$

$$B_{0110} = 4L_2L_3 \tag{9.119}$$

$$B_{0011} = 4L_3L_4 \tag{9.120}$$

Basis functions of cubic and greater order may be constructed using Equation (9.112).

The scalar Lagrangian functions for hexahedral or tetrahedral cells are widely used in electrostatic and acoustic applications, where the primary unknowns are scalar in nature. Since the primary unknowns in three-dimensional electromagnetic applications are usually vector functions, a scalar representation can only be employed if each component is represented independently. In addition, if the electric or magnetic field is the primary unknown, it is generally necessary to permit jump discontinuities at medium interfaces. Since the Lagrangian functions impose continuity between cells, they have limited applicability in three-dimensional problems. However, they can be used if the primary unknowns are auxiliary vector potential functions instead of the fields [12]. When used to model electric or magnetic fields, even if the primary unknown field is a continuous function throughout the region of interest, there are difficulties with a Lagrangian representation. The following section illustrates some of the difficulties and provides the motivation for special vector basis functions developed in Sections 9.8 and 9.9.

9.7 SCALAR LAGRANGIAN DISCRETIZATION OF THE VECTOR HELMHOLTZ EQUATION FOR CAVITIES: SPURIOUS EIGENVALUES AND OTHER DIFFICULTIES

To illustrate several drawbacks associated with the scalar representation of vector quantities, we consider a discretization of the curl–curl form of the vector Helmholtz equation

$$\nabla \times \left(\frac{1}{\varepsilon_r} \nabla \times \bar{H} \right) = k^2 \mu_r \bar{H} \tag{9.121}$$

within a two-dimensional cavity containing material with relative permittivity $\varepsilon_r(x, y)$ and constant relative permeability μ_r. The cavity interior Γ is bounded by perfect electric walls. Here, $\bar{H}$ denotes the transverse magnetic field

$$\bar{H} = \hat{x} H_x + \hat{y} H_y \tag{9.122}$$

throughout cavity interior. By multiplying (9.121) with a transverse testing function $\bar{T}$ and employing the vector identities

$$\bar{T} \cdot \nabla \times \bar{H} = \nabla \times \bar{T} \cdot \bar{H} - \nabla \cdot (\bar{T} \times \bar{H}) \tag{9.123}$$

$$(\bar{T} \times \bar{H}) \cdot \hat{n} = -\bar{T} \cdot (\hat{n} \times \bar{H}) \tag{9.124}$$

and the divergence theorem

$$\iint_{\Gamma} \nabla \cdot (\bar{T} \times \bar{H}) \, dx \, dy = \int_{\partial\Gamma} (\bar{T} \times \bar{H}) \cdot \hat{n} \, dt \tag{9.125}$$

the vector Helmholtz equation can be recast as the weak equation

$$\iint_\Gamma \frac{1}{\varepsilon_r} \nabla \times \bar{T} \cdot \nabla \times \bar{H}\, dx\, dy = k^2 \iint_\Gamma \mu_r \bar{T} \cdot \bar{H}\, dx\, dy - \int_{\partial\Gamma} \frac{1}{\varepsilon_r} \bar{T} \cdot \hat{n} \times (\nabla \times \bar{H})\, dt \tag{9.126}$$

where $\partial\Gamma$ denotes the cavity boundary. Along perfect electric walls, the magnetic field must satisfy the natural boundary condition

$$\hat{n} \times \nabla \times \bar{H} = 0 \tag{9.127}$$

which eliminates the boundary integral on the right-hand side of (9.126). The remainder of (9.126) constitutes an eigenvalue equation for the resonant wavenumber k and the eigenfunction $\bar{H}$ associated with the TM cavity modes. Note that the TM cavity modes are also described by the scalar Helmholtz equation for E_z; here we deliberately investigate the vector formulation to gain an appreciation for issues that do not arise in the scalar case.

Assuming that the cavity cross section is modeled with triangular cells, each with constant μ_r and ε_r, consider an independent expansion for the x- and y-components of the magnetic field according to

$$H_x(x, y) \cong \sum_{n=1}^{N} h_{xn} B_n(x, y) \tag{9.128}$$

$$H_y(x, y) \cong \sum_{n=1}^{N} h_{yn} B_n(x, y) \tag{9.129}$$

where B_n denotes a Lagrangian expansion function (Section 9.1) interpolatory at node n. The Lagrangian functions provide a continuous expansion and would not be suitable for representing the transverse magnetic field in regions where μ_r exhibits jump discontinuities. The restriction to continuous μ_r is one limitation of this specific scalar discretization.

Following common practice, we choose testing functions identical to the basis functions and therefore test the equation with independent functions $\hat{x}T_{xn} = \hat{x}B_{xn}$ and $\hat{y}T_{yn} = \hat{y}B_{yn}$. This procedure produces the generalized matrix eigenvalue equation

$$\begin{bmatrix} A^{11} & A^{12} \\ A^{21} & A^{22} \end{bmatrix} \begin{bmatrix} H_x \\ H_y \end{bmatrix} = k^2 \begin{bmatrix} B^{11} & 0 \\ 0 & B^{22} \end{bmatrix} \begin{bmatrix} H_x \\ H_y \end{bmatrix} \tag{9.130}$$

with entries

$$A_{mn}^{11} = \iint \nabla \times \bar{T}_{xm} \cdot \nabla \times \bar{B}_{xn}\, dx\, dy \tag{9.131}$$

$$A_{mn}^{12} = \iint \nabla \times \bar{T}_{xm} \cdot \nabla \times \bar{B}_{yn}\, dx\, dy \tag{9.132}$$

$$A_{mn}^{21} = \iint \nabla \times \bar{T}_{ym} \cdot \nabla \times \bar{B}_{xn}\, dx\, dy \tag{9.133}$$

$$A_{mn}^{22} = \iint \nabla \times \bar{T}_{ym} \cdot \nabla \times \bar{B}_{yn}\, dx\, dy \tag{9.134}$$

$$B_{mn}^{11} = \iint \bar{T}_{xm} \cdot \bar{B}_{xn}\, dx\, dy \tag{9.135}$$

$$B_{mn}^{22} = \iint \bar{T}_{ym} \cdot \bar{B}_{yn}\, dx\, dy \tag{9.136}$$

where all integrals are over the region Γ.

Suppose we employ linear-order Lagrangian functions within each cell, that is,

$$T_{xm} = B_{xm} = T_{ym} = B_{ym} = L_m \qquad m = 1, 2, 3 \tag{9.137}$$

where $\{L_i\}$ are defined in Section 9.1. Using Equation (9.4), we observe that

$$\nabla \times \bar{T}_{xm} = \nabla \times \bar{B}_{xm} = -\hat{z}\frac{\partial L_m}{\partial y} = -\hat{z}\frac{c_m}{2A} \tag{9.138}$$

$$\nabla \times \bar{T}_{ym} = \nabla \times \bar{B}_{ym} = \hat{z}\frac{\partial L_m}{\partial x} = \hat{z}\frac{b_m}{2A} \tag{9.139}$$

and with (9.37) and (9.40) immediately obtain the element matrix entries for a single cell as

$$A_{mn}^{11} = \frac{1}{4A}c_m c_n \tag{9.140}$$

$$A_{mn}^{12} = \frac{-1}{4A}c_m b_n \tag{9.141}$$

$$A_{mn}^{21} = \frac{-1}{4A}b_m c_n \tag{9.142}$$

$$A_{mn}^{22} = \frac{1}{4A}b_m b_n \tag{9.143}$$

and

$$B_{mn}^{11} = B_{mn}^{22} = \iint L_m L_n \, dx \, dy = \begin{cases} \frac{1}{6}A & m = n \\ \frac{1}{12}A & \text{otherwise} \end{cases} \tag{9.144}$$

where $\{b_i\}$ and $\{c_i\}$ are defined in (9.6) and (9.7) and A denotes the cell area.

Since there are two unknown coefficients (h_{xn} and h_{yn}) associated with each node of the triangular-cell mesh, the boundary condition in (9.127) is not sufficient to uniquely determine the solution. We must also enforce the Dirichlet condition

$$\hat{n} \cdot \bar{H} = 0 \tag{9.145}$$

on the boundary $\partial\Gamma$. The process of enforcing Equation (9.145) would be trivial if the basis set happened to separate into normal and tangential components along $\partial\Gamma$, since the appropriate coefficient could simply be set to zero. Instead, the coefficients in (9.128) and (9.129) are associated with the x- and y-components, and the formulation is complicated by the fact that for a general boundary orientation (9.145) couples several coefficients together.

The entries of the finite-element system in (9.140)–(9.144) are each purely local in character. In contrast, the essential boundary condition in (9.145) acts as a constraint on the global system. Suppose that the normal vector to the boundary is given by

$$\hat{n} = \hat{x}\cos\theta + \hat{y}\sin\theta \tag{9.146}$$

where $\theta(t)$ is an ordinary polar angle defined along $\partial\Gamma$. At any point along the boundary, Equation (9.145) can be written as

$$H_x \cos\theta + H_y \sin\theta = 0 \tag{9.147}$$

which is a constraint between the x- and y-components of the field. If enforced at a node on the boundary, (9.147) involves two coefficients; if imposed along a cell edge, it involves four coefficients. In order to implement this type of constraint in a manner that systematically reduces the rank of the matrix eigensystem, we rewrite Equation (9.130) as

$$\mathbf{A}_{2N\times 2N}\mathbf{h}_{2N\times 1} = k^2\mathbf{B}_{2N\times 2N}\mathbf{h}_{2N\times 1} \tag{9.148}$$

Suppose that there are $2N - M$ constraints from (9.147). These can be collected together in matrix form as

$$\mathbf{h}_{2N\times 1} = \mathbf{C}_{2N\times M}\tilde{\mathbf{h}}_{M\times 1} \tag{9.149}$$

where $\tilde{\mathbf{h}}$ contains the M coefficients to be retained in the eigensystem. Equation (9.149) can be used to reduce the original $2N \times 2N$ system to the $M \times M$ eigensystem

$$\mathbf{C}^T_{M\times 2N}\mathbf{A}_{2N\times 2N}\mathbf{C}_{2N\times M}\tilde{\mathbf{h}}_{M\times 1} = k^2\mathbf{C}^T_{M\times 2N}\mathbf{B}_{2N\times 2N}\mathbf{C}_{2N\times M}\tilde{\mathbf{h}}_{M\times 1} \tag{9.150}$$

The process of implementing Equation (9.145) does not require the explicit matrix operations indicated above; instead rows and columns can be combined in order to impose one constraint at a time. However, this type of global constraint is always much more cumbersome to implement than the local constraints encountered in our previous study of scalar finite-element methods (e.g., Chapter 3, Sections 8.6 and 9.2) and substantially complicates a sparse storage scheme. The need to impose global conditions is a consequence of the fact that the primary unknowns represent the x- and y-field components, instead of tangential and normal components of the field at cell edges, and is a second limitation of this specific discretization.

After the eigensystem in Equation (9.130) is constructed and reduced to (9.150), the resulting generalized eigenvalue equation can be solved by standard algorithms [13]. However, the results are far from satisfactory. Table 9.4 shows the smallest resonant wavenumbers generated by this procedure for a homogeneous, circular cavity of unit radius. The triangular-cell model contained 31 nodes, producing an original eigensystem of rank 62 that was reduced to a system of rank 44 after imposing the essential boundary condition in (9.145). From an inspection of Table 9.4, it is clear that most of the numerical eigenvalues are incorrect. Interestingly, the solutions do not improve with smaller cells. In this case, the numerical data are not just inaccurate but are completely corrupted by the presence of spurious eigenvalues and eigenfunctions. The few values that appear to represent true wavenumbers are interspersed with spurious eigenvalues and are impossible to identify without extensive postprocessing.

In summary, we have identified three difficulties with the approach outlined in Equations (9.128)–(9.150): the inability of this expansion to model discontinuous magnetic material, the cumbersome procedure needed to implement the essential boundary condition, and (by far the most serious) the appearance of spurious eigenfunctions and eigenvalues in the spectrum of the discrete operator.

The source of the "spurious" eigenvalues in Table 9.4 is the curl–curl operator itself, which admits eigenfunctions that do not satisfy the complete set of Maxwell's equations. In fact, the eigenfunctions of this equation can generally be separated into two families, one of which is a valid electromagnetic field and the other of which has the form $\{\bar{H} = \nabla\Phi\}$. Both eigenfamilies satisfy the boundary conditions as well as the Helmholtz equation. The gradient $\nabla\Phi$ is a valid mathematical solution to (9.121) for $k = 0$, since $\nabla \times \nabla\Phi = 0$, but is not required to satisfy Gauss' law

$$\nabla \cdot (\mu_0 \mu_r \bar{H}) = 0 \tag{9.151}$$

Therefore, $\nabla\Phi$ cannot represent a time-varying magnetic field in a source-free region. Since their associated eigenvalues are zero, eigenfunctions of this form comprise the *nullspace* of

TABLE 9.4 Smallest TM Resonant Wavenumbers for Homogeneous Circular (Two-Dimensional) Cavity of Unit Radius Produced by Equation (9.150)

Numerical	Exact
0.00 (5)	2.405 (TM_{01})
0.49 (2)	3.832 (TM_{11})
0.73 (2)	5.136 (TM_{21})
1.13 (2)	5.520 (TM_{02})
1.47 (1)	6.380 (TM_{31})
2.40 (2)	7.016 (TM_{12})
2.47 (1)	7.588 (TM_{41})
2.59 (2)	8.417 (TM_{22})
3.25 (1)	⋮
4.12 (2)	
4.51 (1)	
5.56 (2)	
5.67 (2)	
6.37 (1)	
7.06 (1)	
7.20 (1)	
7.75 (2)	
8.00 (2)	
⋮	

Note: Numerical solutions are obtained using piecewise-linear Lagrangian interpolation functions. The cavity model consists of a relatively coarse triangular-cell mesh with 31 nodes and 42 cells and produced an eigensystem of order 44. The resonant wavenumbers are listed in order of increasing values with their degree of multiplicity in parentheses. The boldface results are believed to represent the first four genuine solutions; the rest appear to be spurious.

the curl–curl operator. Problem P9.13 illustrates an example of nullspace eigenfunctions for a rectangular cavity.

The solution family $\{\nabla\Phi\}$ is of interest, even when k is not zero, because a general discretization of the Helmholtz operator will capture eigenfunctions from both families (Section 5.7). In other words, unless the basis functions are orthogonal to all functions in the nullspace, a matrix representing the curl–curl operator will have some eigenvectors that approximate those functions. The difficulty is that the spurious eigenvalues in Table 9.4 are not zero, apparently a consequence of the fact that the low-order Lagrangian basis cannot adequately represent the nullspace eigenfunctions. It has been reported that fifth-order polynomial expansions do adequately model the nullspace eigenfunctions and produce associated eigenvalues of zero [14]. (A fifth-order representation has enough degrees of freedom so that the function, its tangential derivative, and its normal derivative are independent at cell boundaries; refer to the discussion at the end of Section 9.1.) It has also been observed that representations not imposing normal-field continuity between cells produce the correct nullspace eigenvalues of zero [14, 15]. It appears that relaxing the continuity conditions (by only imposing tangential continuity) also alleviates one of the other difficulties encountered above, namely that the resulting expansion can represent jump discontinuities in the normal-field components at material interfaces.

It is of interest to consider properties of the nullspace eigensolutions in more detail. First, note that both eigensolution families must maintain tangential continuity across any mathematical boundary, and in the absence of medium discontinuities the true electromagnetic fields are expected to exhibit normal continuity. However, the nullspace functions $\{\nabla\Phi\}$ may exhibit jump discontinuities in their normal component, even in homogeneous media, while still maintaining the property that $\nabla \times \nabla\Phi = 0$. As an illustration, the function

$$\bar{H}^{\text{spurious}} = \begin{cases} (1-y)\hat{x} + (1-x)\hat{y} & \text{(quadrant 1)} \\ (y-1)\hat{x} + (1+x)\hat{y} & \text{(quadrant 2)} \\ -(1+y)\hat{x} - (1+x)\hat{y} & \text{(quadrant 3)} \\ (1+y)\hat{x} + (x-1)\hat{y} & \text{(quadrant 4)} \end{cases} \tag{9.152}$$

defined in the usual quadrants of the x–y plane is the gradient of a scalar function and has identically zero curl. This function can be terminated discontinuously at $x = \pm 1$ and $y = \pm 1$, as though it resided on a square-cell mesh surrounding the origin and had its domain of support truncated to these cells. The resulting function (Figure 9.14) is a nullspace eigenfunction of the curl–curl operator for electric wall boundary conditions applied at some location removed from these cells or magnetic wall boundary conditions applied at $x = \pm 1$ and $y = \pm 1$. Note that Equation (9.152) has a strong discontinuity at the origin in addition to the discontinuous normal component at the cell edges. Although (9.152) has zero curl, its projection onto a continuous basis set does not (Prob. P9.14). If the field expansion permits discontinuous normal components, functions similar to (9.152) are observed in the set of numerical eigenfunctions.

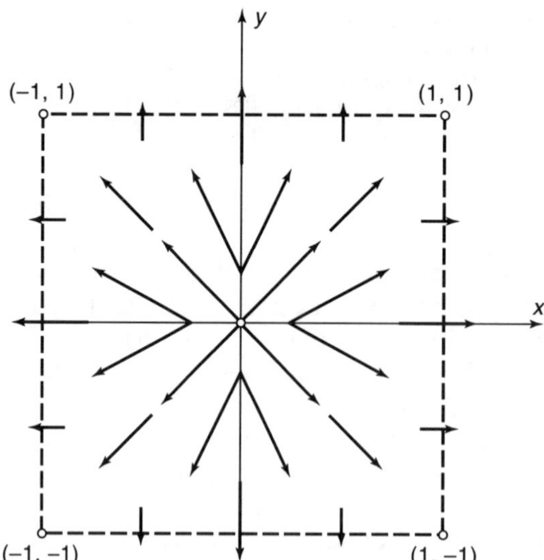

Figure 9.14 Discontinuous vector function defined in (9.152).

In summary, we conclude that a discretization of the vector Helmholtz equation (9.121) should employ a basis set that imposes tangential continuity but not normal continuity between cells. On a fine enough mesh, such an expansion should be able to accurately represent the lower order electromagnetic eigenfunctions in continuous or discontinuous media as well as the nullspace eigenfunctions. The following sections will investigate the performance of several vector basis sets of this type. In the finite-element literature, basis functions that maintain continuity between cells are known as *conforming* functions. Similarly, vector basis functions that only impose tangential continuity between cells have been described as *curl conforming* [16].

9.8 POLYNOMIAL-COMPLETE VECTOR BASIS FUNCTIONS THAT IMPOSE TANGENTIAL CONTINUITY BUT NOT NORMAL CONTINUITY BETWEEN TRIANGULAR CELLS [17]

The previous section identified several difficulties associated with the use of Lagrangian-type interpolation polynomials to independently represent the Cartesian components of the vector function. The major difficulty is that associated with the continuity of the representation: The expansion functions should impose tangential continuity but not normal continuity. A second difficulty identified in Section 9.7 is the cumbersome imposition of boundary conditions at general interfaces when the primary unknowns are the Cartesian components of the vector field. The second difficulty can be eliminated by developing basis functions with coefficients that represent the normal or tangential components along cell edges. In this section, a polynomial-complete family of vector basis functions are described with these properties.

Consider a triangular cell in the x–y plane and an associated vector basis function expressed as a linear polynomial

$$\bar{B}(x, y) = \hat{x}(A + Bx + Cy) + \hat{y}(D + Ex + Fy) \tag{9.153}$$

The six coefficients provide six degrees of freedom in the expansion, and we are free to impose six conditions on the representation in order to generate specific basis functions. For triangular cells, this might translate into two constraints per edge or two constraints per corner node. By appropriate constraints, several different basis sets can be obtained from Equation (9.153).

For instance, suppose we constrain the tangential component of the basis function $\bar{B}$ to equal unity at one node of a cell edge while simultaneously vanishing at the opposite node. In addition, the tangential component along the other two edges can be constrained to vanish at both nodes of each edge (and therefore vanish entirely along both edges). Since the basis function is at most linear, these six constraints completely specify its shape within the triangle. Five other basis functions can be defined in a similar fashion to produce a set of six that each interpolate to the tangential component along one end of an edge (Figure 9.15). These basis functions can be forced to maintain tangential continuity between cells by the simple expedient of sharing coefficients with the analogous functions in the adjacent cell. From a global perspective, each basis function straddles two triangular cells and interpolates to the tangential component along one end of the common edge. Thus, there are a total of two basis functions per edge throughout the model. The normal component of each global basis function is discontinuous at the common edge, and therefore these functions do not impose normal continuity between cells. The simplex-coordinate representation of these basis functions within a cell is given by

$$w_i L_j \nabla L_k \qquad i \neq j \neq k \tag{9.154}$$

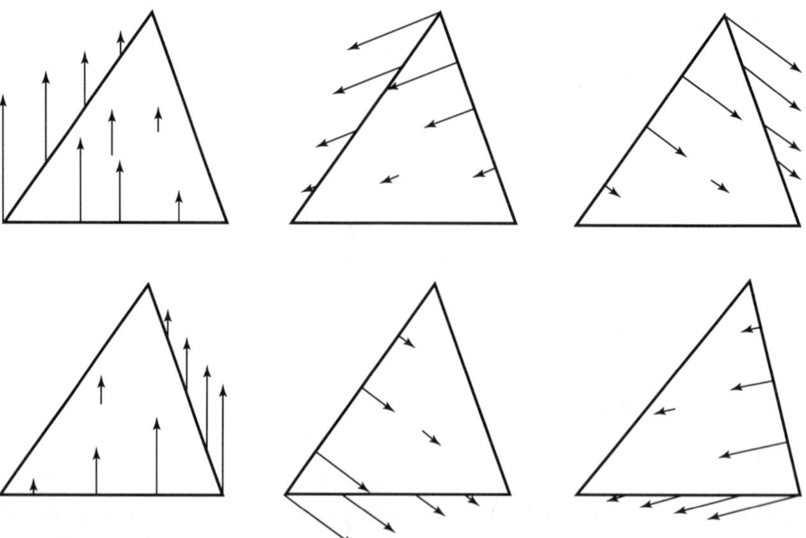

Figure 9.15 Set of six polynomial-complete curl-conforming linear vector basis functions.

where w_i denotes the length of the edge opposite node i, and i, j, and k assume values between 1 and 3. The normalization in (9.154) is investigated in Prob. P9.16.

Since the coefficients of the vector basis functions in (9.154) represent the tangential component along cell edges, boundary conditions such as $\hat{n} \times \bar{E} = 0$ can be implemented in a purely local fashion by zeroing the appropriate coefficient. Therefore, these basis functions make it easier to incorporate essential boundary conditions into the solution process. Although these functions are complete to linear polynomial order, they do not produce the same representation as a linear Lagrangian expansion of the Cartesian vector components since they are not completely continuous between cells. Consequently, there are more degrees of freedom (and more unknowns) associated with the vector functions than with the usual Lagrangian expansion on the same mesh. However, as demonstrated below, the vector basis functions in (9.154) eliminate the spurious nonzero eigenvalues encountered in Section 9.7.

Analogous basis functions can be created that provide a complete quadratic representation. An expansion of the form

$$\begin{aligned}\bar{B}(x, y) = \hat{x}(A + Bx + Cy + Dx^2 + Exy + Fy^2) \\ +\hat{y}(G + Hx + Iy + Jx^2 + Kxy + Ly^2)\end{aligned} \tag{9.155}$$

contains 12 degrees of freedom, and basis functions can be defined that interpolate to the tangential components at three locations along each edge of a triangular cell and the normal component at the middle of each edge. For instance, in simplex coordinates, we can define six functions of the form

$$w_i L_j(2L_j - 1)\,\nabla L_k \qquad i \neq j \neq k \tag{9.156}$$

that interpolate to a unity tangential component at the outside nodes of each edge, three functions of the form

$$2w_i L_j L_k(\nabla L_j - \nabla L_k) \qquad i \neq j \neq k \tag{9.157}$$

that interpolate to a unity tangential component at the midside nodes, and three functions of the form

$$L_i L_j \nabla L_k \qquad i \neq j \neq k \tag{9.158}$$

that provide a linearly independent normal component along each edge. The three functions in (9.158) are not interpolatory, since in general the functions of (9.157) also contribute a normal component at the midside nodes. The possible normalization of (9.158) is left as an exercise, as is the task of sketching the set of 12 functions (Prob. P9.17).

To maintain tangential continuity between cells, the nine quadratic basis functions in (9.156) and (9.157) share coefficients with the analogous functions in the adjacent cells. The three basis functions per cell in (9.158) are entirely local and thus have coefficients that are independent from those of the neighboring cells. Consequently, the normal component can be discontinuous between cells. The global representation produced by the quadratic functions requires three unknowns per edge and three unknowns per cell.

Table 9.5 shows numerical eigenvalues produced when the basis functions from (9.154) and (9.156)–(9.158) are used to discretize the vector Helmholtz equation

$$\nabla \times \left(\frac{1}{\mu_r} \nabla \times \bar{E} \right) - k^2 \varepsilon_r \bar{E} = 0 \tag{9.159}$$

for a circular, homogeneous cavity bounded by a p.e.c. wall. The solution procedure can be organized along the lines of the previous finite-element formulations; that is, the field can be represented by an expansion

$$\bar{E}(x, y) \cong \sum_{n=1}^{N} e_n \bar{B}_n(x, y) \tag{9.160}$$

where $\bar{B}_n$ denotes a vector expansion function and is substituted into the weak vector

TABLE 9.5 Smallest Eigenvalues Produced by Discretization of Circular Cavity with $\varepsilon_r = 1$, $\mu_r = 1$, and Unit Radius Using Polynomial-Complete Linear[a] and Polynomial-Complete Quadratic[b] Basis Functions for TE polarization

LT/LN	QT/QN	Exact
0.0 (67)	0.0 (163)	
1.87 (2)	1.84 (2)	1.841 (2)
3.20 (2)	3.06 (2)	3.054 (2)
4.13 (1)	3.84 (1)	3.832 (1)
4.6 (2)	4.21 (2)	4.201 (2)
6.0 (2)	5.35 (2)	5.318 (2)
6.1 (2)	5.37 (2)	5.331 (2)

[a] LT/LN; from Equation (9.154).

[b] QT/QN; from Equations (9.156)–(9.158).

Note: The model consisted of 42 triangular cells and resulted in a matrix of order 108 for the LT/LN functions and 288 for the QT/QN functions.

equation

$$\iint_\Gamma \left(\frac{1}{\mu_r} \nabla \times \bar{T} \cdot \nabla \times \bar{E} - k^2 \varepsilon_r \bar{T} \cdot \bar{E} \right) dx\, dy = - \int_{\partial\Gamma} \frac{1}{\mu_r} \bar{T} \cdot \hat{n} \times (\nabla \times \bar{E})\, dt \tag{9.161}$$

obtained in a manner analogous to (9.126). For the TE polarization, the boundary condition $\hat{n} \times \bar{E} = 0$ on the p.e.c. wall is enforced by omitting any basis function with a nonzero tangential value on the boundary from the system of equations. Since there are no basis functions remaining with nonzero tangential components along the boundary $\partial\Gamma$, there will be no need for testing functions with tangential components along $\partial\Gamma$ and the boundary integral in (9.161) will not contribute to the system. The discretization process yields a matrix eigenvalue equation $\mathbf{Ae} = k^2\mathbf{Be}$ with entries

$$A_{mn} = \iint \nabla \times \bar{T}_m \cdot \nabla \times \bar{B}_n\, dx\, dy \tag{9.162}$$

and

$$B_{mn} = \iint \bar{T}_m \cdot \bar{B}_n\, dx\, dy \tag{9.163}$$

We will employ the same vector functions for expansion and testing.

In common with our previous finite-element implementations, the region of interest is divided into triangular cells, with the mesh terminating on the cavity boundary $\partial\Gamma$. The matrices **A** and **B** are constructed on a cell-by-cell basis using a connectivity array that identifies the rows and columns of the matrix associated with a particular cell. The ordinary connectivity array employed with "node-based" scalar finite-element analysis identifies the three nodes associated with each cell. However, the coefficients of the vector basis functions represent the tangential or normal fields along cell edges. In this "edge-based" formulation, it is much more convenient to employ a connectivity array that identifies the three edges associated with each cell. (A second pointer array is also necessary to identify the two nodes associated with each edge.) The matrices may be constructed by scanning through the global model, computing 6×6 element matrices from the integrals in (9.162)–(9.163) for the linear basis functions and 12×12 element matrices for the quadratic basis functions. The connectivity arrays identify the appropriate locations of the element matrix entries within the global system. To ensure that the basis functions (those that overlap two cells) have a common orientation on either side of a given edge, we adopt the convention that the vector basis functions tangential to an edge always point from the smaller node index to the larger node index, according to the global numbering. The details of the element matrix calculations for the functions of (9.154) are suggested in Prob. P9.18.

The nonzero eigenvalues in Table 9.5 appear to have a one-to-one correlation with analytical results for the electromagnetic cavity modes. The results also contain a large number of zero eigenvalues that represent the nullspace. Overall, these results support the hypothesis that the spurious eigenvalues observed previously can be eliminated by an expansion that only imposes tangential continuity between cells. In addition to remediating the spurious eigenvalue problem, the vector basis functions simplify the task of imposing boundary conditions since the coefficients directly represent the tangential field at the boundary. The polynomial-complete vector basis functions developed in (9.154) and (9.156)–(9.158) appear to be equivalent to those used in various electromagnetic applications previously described in the literature [17–19].

Unfortunately, Table 9.5 also shows that more than half of the available degrees of freedom in the vector expansions are used to capture nullspace eigensolutions. Since these are nonphysical results, a substantial amount of computational effort is wasted. It would be desirable to eliminate some or all of these wasted degrees of freedom. The following section considers alternative vector basis functions that eliminate some of the nullspace eigenfunctions.

9.9 MIXED-ORDER VECTOR BASIS FUNCTIONS THAT IMPOSE TANGENTIAL BUT NOT NORMAL CONTINUITY FOR TRIANGULAR AND RECTANGULAR CELLS [16, 20]

As illustrated in Table 9.5, a polynomial-complete discretization of the curl–curl operator captures many nullspace eigensolutions. Some of the wasted degrees of freedom can be eliminated by the use of special mixed-order polynomial basis functions proposed in 1980 by Nedelec [16]. In common with the polynomial-complete functions used in Section 9.8, the mixed-order functions are curl conforming: They impose tangential continuity between cells but do not impose normal continuity. Thus, if used to discretize the curl–curl operator, they should not produce spurious nonzero eigenvalues.

Consider a vector representation for triangular cells. Within a cell, a complete linear representation has the form

$$\hat{x}(A + Bx + Cy) + \hat{y}(D + Ex + Fy) \tag{9.164}$$

Equation (9.164) can be decomposed into two parts, one with three degrees of freedom

$$\bar{B}(x, y) = \hat{x}\left(A + \frac{1}{2}[C - E]y\right) + \hat{y}\left(D + \frac{1}{2}[E - C]x\right) \tag{9.165}$$

and a complementary representation, also containing three degrees of freedom,

$$\bar{B}_{\text{grad}}(x, y) = \hat{x}\left(Bx + \frac{1}{2}[C + E]y\right) + \hat{y}\left(\frac{1}{2}[E + C]x + Fy\right) \tag{9.166}$$

The functions in (9.166) are actually the gradient of the quadratic-order function

$$\Phi(x, y) = \left(\frac{1}{2}Bx^2 + \frac{1}{2}[E + C]xy + \frac{1}{2}Fy^2\right) \tag{9.167}$$

and therefore have identically zero curl within the cell. If functions of the form (9.166) are constrained to have tangential continuity from cell to cell, their curl will be identically zero over the entire problem domain. Therefore, if used to discretize the curl–curl operator, the degrees of freedom in (9.166) can only represent nullspace eigenfunctions. Furthermore, the linear degrees of freedom in (9.166) are not needed to balance terms in the discretized vector Helmholtz equation, since the curl of (9.165) only produces a constant. By restricting the basis set to the functions in (9.165) and discarding the degrees of freedom in (9.166), we obtain a vector representation with only half the unknowns as the polynomial-complete linear basis functions introduced in Section 9.7. Below, we demonstrate that the use of (9.165) reduces the number of nullspace eigensolutions by exactly the number of discarded degrees of freedom without a detrimental effect on accuracy.

Nedelec presented constraints that can be imposed on a polynomial-complete expansion of any order to produce the desired mixed-order representation. For the two-dimensional linear case, the three constraints are [16]

$$\frac{\partial B_x}{\partial x} = 0 \qquad \frac{\partial B_y}{\partial y} = 0 \qquad \frac{\partial B_x}{\partial y} + \frac{\partial B_y}{\partial x} = 0 \tag{9.168}$$

which reduce (9.164) to (9.165). It is interesting to observe that the identical representation can be obtained from (9.164) and the constraint that the tangential component of the basis function $\bar{B}$ equal unity at both endpoints of a common edge while simultaneously vanishing along the other two edges. In fact, within a triangular domain (9.165) is a vector function with a constant-tangential component along one edge and zero tangential component along the other two edges (Figure 9.16). There is also a nonzero normal component along each of the edges that varies linearly. Consequently, these basis functions provide a constant-tangential, linear-normal (CT/LN) representation of the field. To obtain tangential continuity between cells, the coefficients of the three basis functions can be shared by similar functions in the adjacent cells. In the global model, the support of this type of vector basis function would be the two triangles associated with each edge (Figure 9.16). These basis functions offer a simple approach for representing tangentially continuous vector quantities and have been widely used under the name "edge elements." Although first introduced for finite-element formulations by Nedelec [16], they appear to have been developed independently by other researchers for magnetics applications [21, 22].

Within a cell, the CT/LN basis functions can be expressed in terms of the simplex coordinates as

$$\bar{B}_i = w_i(L_{i+1}\nabla L_{i+2} - L_{i+2}\nabla L_{i+1}) \tag{9.169}$$

where the index i denotes the edge opposite vertex i in the triangle and w_i denotes the length of edge i. These functions are oriented so that they point from node $i+1$ to node $i+2$ along edge i. Using the relation

$$\nabla L_i = \frac{1}{2A}(\hat{x}b_i + \hat{y}c_i) \tag{9.170}$$

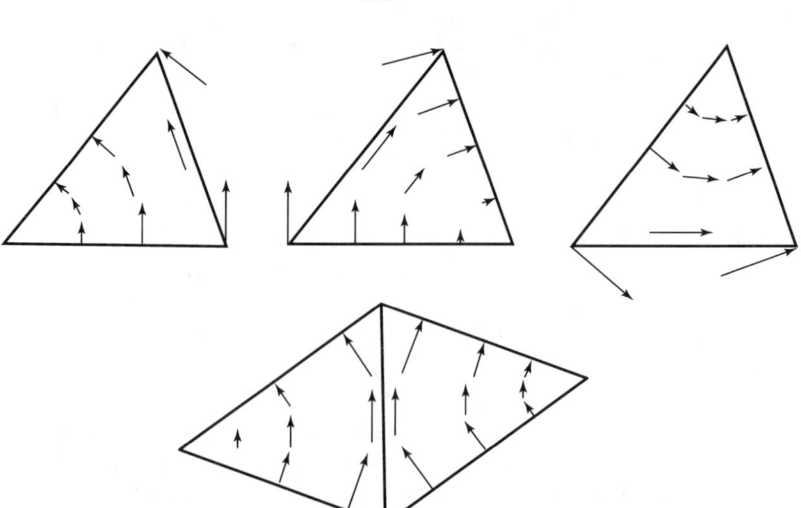

Figure 9.16 Constant tangential/linear normal (CT/LN) curl-conforming vector basis functions.

where b_i and c_i are defined in Section 9.1 and A is the cell area, we obtain the Cartesian representation

$$\begin{aligned}\bar{B}_i = \frac{w_i}{4A^2} &\{\hat{x}[a_j b_k - a_k b_j + (c_j b_k - c_k b_j)y] \\ &+\hat{y}[a_j c_k - a_k c_j + (b_j c_k - b_k c_j)x]\}\end{aligned} \tag{9.171}$$

confirming that (9.169) is equivalent to (9.165). We also observe that

$$\nabla \times \bar{B}_i = \hat{z}\frac{w_i}{2A^2}(b_j c_k - b_k c_j) = \pm\hat{z}\frac{w_i}{A} \tag{9.172}$$

and

$$\nabla \cdot \bar{B}_i = 0 \tag{9.173}$$

Thus, the CT/LN functions have constant curl and zero divergence within a cell. It is important to note, however, that because the normal component is discontinuous across edges, their divergence is not zero (or even finite) at cell boundaries. As an aid to visualization, consider a triangle having one node located at the origin of a polar coordinate system (ρ, ϕ). The basis function oriented along the edge opposite the node at the origin can be written as

$$\bar{B}(\rho, \phi) = K\rho\hat{\phi} \tag{9.174}$$

where K is a normalization constant.

A similar procedure can be used to develop vector basis functions of higher polynomial orders. In the quadratic case, the polynomial-complete representation in (9.155) can be decomposed into an expansion

$$\begin{aligned}\bar{B}(x, y) = \hat{x}&\left(A + Bx + Cy + \frac{1}{3}[E - 2J]xy + \frac{1}{3}[2F - K]y^2\right) \\ &+\hat{y}\left(G + Hx + Iy + \frac{1}{3}[2J - E]x^2 + \frac{1}{3}[K - 2F]xy\right)\end{aligned} \tag{9.175}$$

containing eight degrees of freedom and a complementary representation

$$\begin{aligned}\bar{B}_{\text{grad}}(x, y) = \hat{x}&\left(Dx^2 + \frac{2}{3}[E + J]xy + \frac{1}{3}[F + K]y^2\right) \\ &+\hat{y}\left(\frac{1}{3}[E + J]x^2 + \frac{2}{3}[K + F]xy + Ly^2\right)\end{aligned} \tag{9.176}$$

containing four degrees of freedom. Equation (9.176) is the gradient of a cubic-order function and consequently has identically zero curl. Therefore, those degrees of freedom can only represent functions in the nullspace of the curl–curl operator. Since they are not needed to balance terms in the discretized vector Helmholtz equation, they can be discarded to improve computational efficiency. Equation (9.175) can also be obtained from (9.155) by imposing the four conditions [16]

$$\frac{\partial^2 B_x}{\partial x^2} = 0 \tag{9.177}$$

$$\frac{\partial^2 B_y}{\partial y^2} = 0 \tag{9.178}$$

$$\frac{\partial^2 B_x}{\partial y^2} + 2\frac{\partial^2 B_y}{\partial x\,\partial y} = 0 \tag{9.179}$$

$$\frac{\partial^2 B_y}{\partial x^2} + 2\frac{\partial^2 B_x}{\partial x\,\partial y} = 0 \tag{9.180}$$

which eliminate the degrees of freedom associated with (9.176).

Equation (9.175) provides a linear-tangential, quadratic-normal (LT/QN) representation along any cut through a triangular cell. Several equivalent sets of LT/QN basis functions can be constructed. It is important to note that these functions should be distributed around the triangle in a symmetric way, and a symmetric distribution of eight basis functions may not be immediately obvious. One possibility is to define six basis functions that are entirely linear, having the simplex-coordinate description

$$w_i L_j \nabla L_k \qquad i \neq j \neq k \tag{9.181}$$

where w_i denotes the length of the edge opposite node i. These are the polynomial-complete basis functions introduced in Section 9.8, and they provide the simplest way of interpolating to a linear-tangential component on the edges. The remaining two functions must provide a quadratic-normal component along three edges, and a suitable choice for these functions is

$$L_2 L_3 \nabla L_1 - L_1 L_2 \nabla L_3 \tag{9.182}$$

$$L_1 L_3 \nabla L_2 - L_1 L_2 \nabla L_3 \tag{9.183}$$

It can be verified that the functions in (9.182) and (9.183) satisfy the conditions in (9.177)–(9.180). These eight basis functions are depicted in Figure 9.17. The six functions in (9.181) interpolate to a unity-tangential component at the cell edges and will share a coefficient with similar functions in adjacent cells to maintain tangential continuity. The functions in (9.182) and (9.183) each contribute a quadratic-normal component to two of the three edges and have no tangential component along any of the edges. Their coefficients are independent of those in neighboring cells, allowing them to reproduce a discontinuity in the normal component of the vector field at cell junctions. Since the functions in (9.181) also contribute a nonzero normal component along the cell edges, (9.182) and (9.183) are not interpolatory.

To verify that the mixed-order basis functions in (9.169) and (9.181)–(9.183) provide a robust discretization of the curl–curl operator, we again consider a numerical solution of the vector Helmholtz equation (9.159) for a circular, homogeneous cavity bounded by a p.e.c. wall. The formulation parallels that of Equations (9.159)–(9.163) and produces a matrix eigenvalue equation $\mathbf{A}\mathbf{e} = k^2\mathbf{B}\mathbf{e}$. The boundary condition $\hat{n} \times \hat{E} = 0$ is imposed along the p.e.c. wall by omitting from the system of equations those basis functions that represent the tangential electric field along the conductor. For the CT/LN basis and testing functions, the matrix entries in (9.162) and (9.163) can be evaluated on a cell-by-cell basis using the expressions

$$\iint \nabla \times \bar{B}_m \cdot \nabla \times \bar{B}_n \, dx\, dy = \frac{w_m w_n}{4A^3}(b_{m+1}c_{m+2} - b_{m+2}c_{m+1})(b_{n+1}c_{n+2} - b_{n+2}c_{n+1}) \tag{9.184}$$

$$\iint \bar{B}_m \cdot \bar{B}_n \, dx\, dy = \frac{w_m w_n}{2A} \sum_{i=1}^{2} \sum_{j=1}^{2} \alpha_{ij} \beta_{mn}^{ij} (b_{m+3-i} b_{n+3-j} + c_{m+3-i} c_{n+3-j}) \tag{9.185}$$

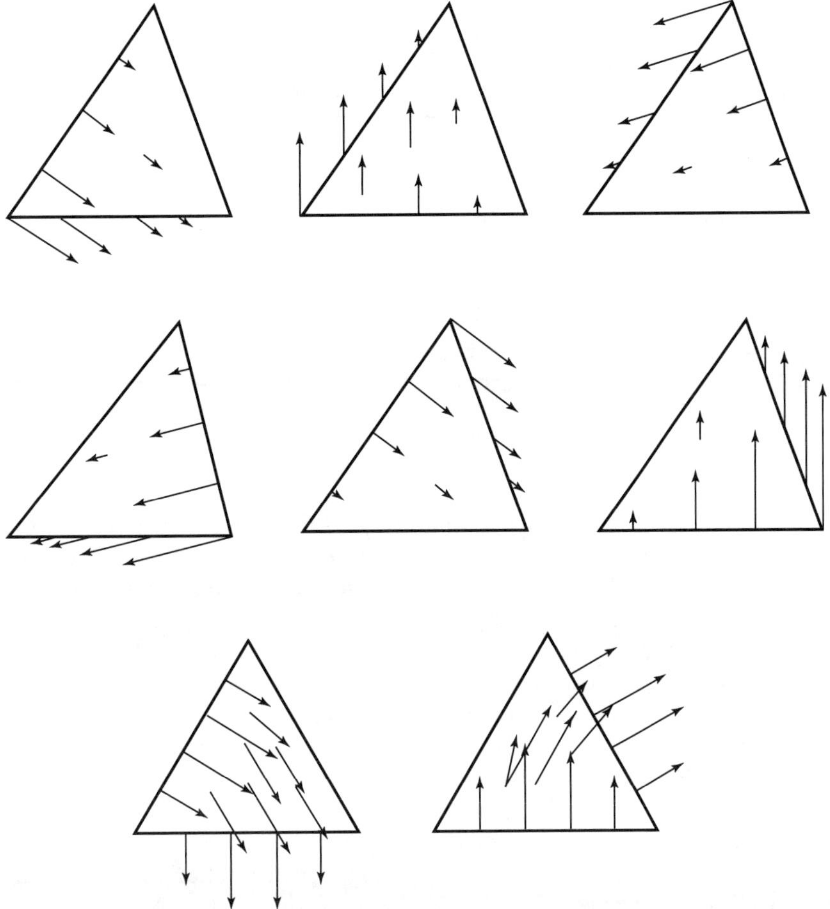

Figure 9.17 Set of eight linear tangential/quadratic normal (LT/QN) curl-conforming vector basis functions.

for the integrals over a single triangular cell, where w_i is the width of edge i, A is the cell area, $\{b_i\}$ and $\{c_i\}$ are defined in (9.6) and (9.7), and

$$\alpha_{ij} = \begin{cases} +1 & \text{if } i = j \\ -1 & \text{otherwise} \end{cases} \tag{9.186}$$

$$\beta_{mn}^{ij} = \begin{cases} \frac{1}{12} & \text{if } m + i = n + j \\ \frac{1}{24} & \text{otherwise} \end{cases} \tag{9.187}$$

The evaluation of element matrix entries for the LT/QN functions is left as an exercise. The generalized matrix eigenvalue equation is solved using standard techniques [13]. Table 9.6 presents numerical eigenvalues obtained using the CT/LN discretization for the identical cavity model used to generate Table 9.5. Results are also shown for the LT/QN functions from (9.181)–(9.183). The mixed-order vector basis functions clearly separate the true electromagnetic eigenvalues from the zero eigenvalues associated with the nullspace. Furthermore, the CT/LN results in Table 9.6 and the polynomial-complete linear results in Table 9.5 contain the same number of nonzero eigenvalues, but the CT/LN functions produce far

TABLE 9.6 Lowest Eigenvalues Produced by Discretization of Circular Cavity with $\varepsilon_r = 1$, $\mu_r = 1$, and Unit Radius Using Mixed-Order Basis Functions for TE Polarization

CT/LN	LT/QN	Exact
0.0 (13)	0.0 (67)	
1.86 (2)	1.84 (2)	1.841 (2)
3.10 (2)	3.05 (2)	3.054 (2)
3.82 (1)	3.84 (1)	3.832 (1)
4.28 (2)	4.20 (2)	4.201 (2)
5.27 (2)	5.32 (2)	5.318 (2)
5.39 (2)	5.35 (2)	5.331 (2)

Note: The 42-cell model produced a matrix with order 54 for the CT/LN functions and 192 for the LT/QN functions.

fewer zero eigenvalues. The number of zero eigenvalues is reduced by exactly the number of basis functions excluded by imposing (9.168), which is half the original matrix order. The LT/QN results also contain the same number of nonzero eigenvalues but far fewer zero eigenvalues than produced by the polynomial-complete quadratic expansion in Table 9.5. For the LT/QN functions, the number of zero eigenvalues is reduced by exactly the number of basis functions excluded by imposing (9.177)–(9.180).

It is noteworthy that the six polynomial-complete linear basis functions in (9.154) form a subset of the LT/QN functions. By comparing Tables 9.5 and 9.6, we observe that the LT/QN expansion produces exactly the same number of zero eigenvalues as the complete linear expansion. This suggests that the additional basis functions used to build up the set of 8 LT/QN functions do not contribute to nullspace eigensolutions. However, their addition to the set of six polynomial-complete linear functions clearly improves the accuracy of the nonzero eigenvalues, as can be seen by comparing the LT/LN results in Table 9.5 with the LT/QN results in Table 9.6.

To illustrate the implementation of the CT/LN and LT/QN basis functions for the TM polarization, we repeat the circular cavity analysis while treating the transverse magnetic field as the unknown to be determined. The magnetic field can be represented as

$$\bar{H}(x, y) \cong \sum_{n=1}^{N} h_n \bar{B}_n(x, y) \tag{9.188}$$

and substituted into the weak equation defined in (9.126). For these expansion functions, the boundary condition $\hat{n} \times (\nabla \times \bar{H}) = 0$ along $\partial\Gamma$ eliminates the boundary integral in (9.126) and is sufficient. The resulting eigensystem has the matrix form $\mathbf{Ah} = k^2\mathbf{Bh}$. Table 9.7 shows results obtained using CT/LN basis and testing functions with the element matrix entries presented in (9.184)–(9.187) for two different triangular-cell meshes. Eigenvalues obtained using LT/QN basis and testing functions are presented in Table 9.8. These results show that the different type of boundary condition imposed in the TM case does not affect the basic character of the numerical eigenvalues. For a given mesh, the LT/QN basis functions produce numerical eigenvalues that are more accurate than those of the CT/LN functions.

TABLE 9.7 Smallest TM Resonant Wavenumbers for Homogeneous Circular (Two-Dimensional) Cavity of Unit Radius Produced by Discretization of Equation (9.121) Using CT/LN Basis Functions

31-Node, 72-Edge Model	55-Node, 138-Edge Model	Exact
0.0 (30)	0.0 (54)	
2.45	2.42	2.405 (TM_{01})
3.87 (2)	3.85 (2)	3.832 (TM_{11})
5.08 (2)	5.14 (2)	5.136 (TM_{21})
5.41	5.50	5.520 (TM_{02})
6.14 (2)	6.36 (2)	6.380 (TM_{31})
6.32 (2)	6.91 (2)	7.016 (TM_{12})
7.13 (2)	7.49 (2)	7.588 (TM_{41})

Note: The resonant wavenumbers are listed in order of increasing values with their degree of multiplicity in parentheses.

TABLE 9.8 Smallest TM Resonant Wavenumbers for Homogeneous Circular (Two-Dimensional) Cavity of Unit Radius Produced by Discretization of Equation (9.121) Using LT/QN Basis Functions

31-Node, 72-Edge Model	55-Node, 138-Edge Model	Exact
0.0 (102)	0.0 (192)	
2.405	2.404	2.405 (TM_{01})
3.837 (2)	3.833 (2)	3.832 (TM_{11})
5.16 (2)	5.141 (2)	5.136 (TM_{21})
5.55	5.526	5.520 (TM_{02})
6.46 (2)	6.394 (2)	6.380 (TM_{31})
7.19 (2)	7.036 (2)	7.016 (TM_{12})
7.66 (2)	7.616 (2)	7.588 (TM_{41})

Note: The resonant wavenumbers are listed in order of increasing values with their degree of multiplicity in parentheses.

In summary, the triangular-cell CT/LN and LT/QN basis functions developed in this section enable a reduction in the degrees of freedom used to represent nullspace eigensolutions without reducing the number of true electromagnetic eigenfunctions. Furthermore, they appear to produce eigenvalues that are no less accurate than those obtained using a polynomial-complete expansion of comparable order.

Vector basis functions of greater polynomial order can be developed following the procedure described above and the analogous Nedelec constraints [16]. In general, a two-dimensional vector expansion of polynomial degree k will have $k(k+2)$ degrees of freedom remaining after eliminating the gradient of a higher polynomial order. For example, a general cubic-polynomial representation of a vector function in two-dimensions contains 20 degrees of freedom. There are 5 degrees of freedom associated with the gradient of a fourth-order polynomial, which can be excluded to reduce the Cartesian form of the basis functions to

$$\begin{aligned}\bar{B}(x, y) = \hat{x}(A + Bx + Cy + Dx^2 + Exy + Fy^2 + Gx^2y + Hxy^2 + 3Iy^3) \\ +\hat{y}(J + Kx + Ly + Mx^2 + Nxy + Oy^2 - 3Gx^3 - Hx^2y - Ixy^2)\end{aligned} \tag{9.189}$$

Within a cell, one possible form of the 15 basis functions is given in simplex coordinates in Table 9.9. Nine of these functions interpolate to the tangential vector component along cell edges, while six functions build up the normal component. Together, these 15 basis functions provide a representation with quadratic-tangential and cubic-normal components (QT/CuN).

TABLE 9.9 Summary of Simplex-Coordinate Representation of First Three Types of Mixed-Order Basis Functions within Triangular Cell

CT/LN	LT/QN	QT/CuN
$L_1\nabla L_2 - L_2\nabla L_1$	$L_1\nabla L_2$	$L_2(2L_2 - 1)\nabla L_1$
$L_1\nabla L_3 - L_3\nabla L_1$	$L_2\nabla L_1$	$L_3(2L_3 - 1)\nabla L_1$
$L_2\nabla L_3 - L_3\nabla L_2$	$L_1\nabla L_3$	$L_1(2L_1 - 1)\nabla L_2$
	$L_3\nabla L_1$	$L_3(2L_3 - 1)\nabla L_2$
	$L_2\nabla L_3$	$L_1(2L_1 - 1)\nabla L_3$
	$L_3\nabla L_2$	$L_2(2L_2 - 1)\nabla L_3$
	$L_2L_3\nabla L_1 - L_1L_2\nabla L_3$	$L_2L_3(\nabla L_2 - \nabla L_3)$
	$L_1L_3\nabla L_2 - L_1L_2\nabla L_3$	$L_1L_3(\nabla L_3 - \nabla L_1)$
		$L_1L_2(\nabla L_1 - \nabla L_2)$
		$\nabla(L_1L_2L_3)$
		$L_2(2L_2 - 1)(L_3\nabla L_1 - L_1\nabla L_3)$
		$L_3(2L_3 - 1)(L_1\nabla L_2 - L_2\nabla L_1)$
		$L_1^2(L_2\nabla L_3 - L_3\nabla L_2)$
		$L_2^2(L_3\nabla L_1 - L_1\nabla L_3)$
		$L_3^2(L_1\nabla L_2 - L_2\nabla L_1)$

Note: The last two LT/QN and the last six QT/CuN functions are entirely local. The functions are not normalized.

The CT/LN, LT/QN, and QT/CuN basis functions for triangular cells are summarized in Table 9.9. Note that there are equivalent basis functions that differ from those presented here but still provide a representation consistent with Nedelec's spaces (Prob. P9.21 and [23]). On the other hand, there have been alternative mixed-order functions proposed in the literature [14, 24] that are not equivalent to those presented here and do not eliminate the same degrees of freedom associated with the nullspace of the curl operator.

Mixed-order basis functions for rectangular and quadrilateral cells are also possible and in fact were also originally proposed by Nedelec [16]. These functions have a different number of degrees of freedom and a different mathematical expression in Cartesian coordinates than those used on triangles. However, they are based on the common idea of discarding degrees of freedom associated with nullspace eigensolutions.

The mixed-order functions for triangular cells discard all the degrees of freedom associated with the gradient of a higher order polynomial function. In the case of rectangular cells, some of these degrees of freedom must be kept to provide a symmetric distribution of basis functions. For instance, consider the quadratic case. As an alternative to the decomposition defined by (9.175) and (9.176), a quadratic expansion can be separated into

$$\bar{B}(x, y) = \hat{x}(A + Bx + Cy + Dxy + Ey^2) + \hat{y}(F + Gx + Hy + Ix^2 + Jxy) \tag{9.190}$$

and

$$\bar{B}_{\text{grad}}(x, y) = \hat{x}Kx^2 + \hat{y}Ly^2 \tag{9.191}$$

This decomposition eliminates 2 degrees of freedom, leaving 10 in (9.190). However, 12 degrees of freedom are required to build up an LT/QN component along the sides of a rectangular cell. The polynomial components in (9.191) do not contribute to an LT/QN behavior, and instead it is convenient to add two cubic-order degrees,

$$\hat{x}Kxy^2 + \hat{y}Lx^2y \tag{9.192}$$

to (9.190). The resulting function provides an LT/QN representation along the cell edges, although, unlike (9.175), this expansion is not purely LT/QN along any cut within a cell.

To develop specific vector basis functions, consider a standard rectangular cell ($-1 < \eta < 1$, $-1 < \xi < 1$). The mathematical form of the LT/QN functions obtained by combining (9.190) and (9.192) is equivalent to the 12 functions [16, 25]

$$\bar{B}^{\eta}_{ij} = \hat{\eta}\phi^{(1)}_i(\eta)\phi^{(2)}_j(\xi) \qquad i = 1, 2 \qquad j = 1, 2, 3 \tag{9.193}$$

$$\bar{B}^{\xi}_{ij} = \hat{\xi}\phi^{(2)}_i(\eta)\phi^{(1)}_j(\xi) \qquad i = 1, 2, 3 \qquad j = 1, 2, \tag{9.194}$$

where $\phi^{(N)}_j$ is an Nth order Lagrangian polynomial, defined on the interval $[x_1, x_{N+1}]$ by

$$\phi^{(N)}_j = \frac{(x - x_1)(x - x_2)\cdots(x - x_{j-1})(x - x_{j+1})\cdots(x - x_N)(x - x_{N+1})}{(x_j - x_1)\cdots(x_j - x_{j-1})(x_j - x_{j+1})\cdots(x_j - x_{N+1})} \tag{9.195}$$

Figure 9.18 illustrates the arrangement of these functions. The six η-component functions in (9.193) can be written explicitly as

$$\bar{B}^{\eta}_{11} = \hat{\eta}\frac{1-\eta}{2}\frac{(\xi - 1)\xi}{2} \tag{9.196}$$

$$\bar{B}^{\eta}_{12} = \hat{\eta}\frac{1-\eta}{2}(1 - \xi)(1 + \xi) \tag{9.197}$$

$$\bar{B}^{\eta}_{13} = \hat{\eta}\frac{1-\eta}{2}\frac{\xi(1 + \xi)}{2} \tag{9.198}$$

$$\bar{B}^{\eta}_{21} = \hat{\eta}\frac{1+\eta}{2}\frac{(\xi - 1)\xi}{2} \tag{9.199}$$

$$\bar{B}^{\eta}_{22} = \hat{\eta}\frac{1+\eta}{2}(1 - \xi)(1 + \xi) \tag{9.200}$$

$$\bar{B}^{\eta}_{23} = \hat{\eta}\frac{1+\eta}{2}\frac{\xi(1 + \xi)}{2} \tag{9.201}$$

These basis functions produce an LT/QN component along each edge, similar to the LT/QN functions for triangular cells. The absence of the highest polynomial order in the principal vector direction is consistent with the discarded degrees of freedom in (9.191).

The four basis functions in (9.196), (9.198), (9.199), and (9.201) interpolate to tangential fields along the cell edges. To maintain tangential continuity, the coefficients of these four functions will be shared with similar functions tangential to the same edge in

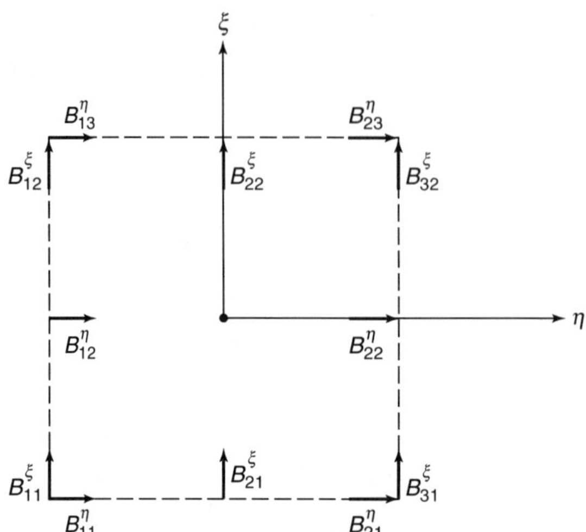

Figure 9.18 Locations of 12 LT/QN curl-conforming vector basis functions on a standard cell.

the adjacent cell. The functions $\bar{B}_{12}$ and $\bar{B}_{22}$ are associated with normal fields at the middle of the cell side, however, and will not share coefficients with neighboring cells. At a node where four cells come together in the global model, there will be four independent coefficients, one associated with the tangential field along each incoming or outgoing edge. Therefore, the rectangular-cell LT/QN expansion involves a total of two basis functions per edge and four additional basis functions per cell.

Similar basis functions can be developed that provide a CT/LN or QT/CuN representation [16, 25]. Table 9.10 summarizes the three lowest order functions for rectangular cells. Equivalent basis sets can be constructed that distribute the degrees of freedom in a different manner (Prob. P9.23 and [23]). As described in Section 9.14, a covariant mapping can be used to generalize these basis functions to quadrilateral or curvilinear cell shapes while preserving the property of tangential continuity between cells.

As indicated in Section 5.4, a subsectional basis function that is complete to polynomial order p provides a representation with interpolation error of $O(\Delta^{p+1})$ as $\Delta \rightarrow 0$, where Δ is the maximum cell dimension. These estimates also apply in general to the multidimensional vector case. Consequently, the polynomial-complete vector basis functions

TABLE 9.10 Summary of Local-Coordinate Representation of First Three Types of Mixed-Order Basis Functions within Rectangular Cell ($-1 < \eta < 1, -1 < \xi < 1$)

CT/LN	LT/QN	QT/CuN
$\hat{\eta}(1 \pm \xi)/2$	$\hat{\eta}(1 \pm \eta)\xi(\xi \pm 1)/4$	$\hat{\eta}\ \eta(\eta \pm 1)(\xi \pm 1)(\xi + 1/3)(\xi - 1/3)\ 9/32$
$\hat{\xi}(1 \pm \eta)/2$	$\hat{\xi}(1 \pm \xi)\eta(\eta \pm 1)/4$	$\hat{\xi}\ \xi(\xi \pm 1)(\eta \pm 1)(\eta + 1/3)(\eta - 1/3)\ 9/32$
	$\hat{\eta}(1 \pm \eta)(1 - \xi)(\xi + 1)/2$	$\hat{\eta}\ (\eta + 1)(\eta - 1)(\xi \pm 1)(\xi + 1/3)(\xi - 1/3)\ 9/16$
	$\hat{\xi}(1 \pm \xi)(1 - \eta)(\eta + 1)/2$	$\hat{\xi}\ (\xi + 1)(\xi - 1)(\eta \pm 1)(\eta + 1/3)(\eta - 1/3)\ 9/16$
		$\hat{\eta}\ \eta(\eta \pm 1)(\xi + 1)(\xi \pm 1/3)(\xi - 1)\ 27/32$
		$\hat{\xi}\ \xi(\xi \pm 1)(\eta + 1)(\eta \pm 1/3)(\eta - 1)\ 27/32$
		$\hat{\eta}\ (\eta + 1)(\eta - 1)(\xi + 1)(\xi \pm 1/3)(\xi - 1)\ 27/16$
		$\hat{\xi}\ (\xi + 1)(\xi - 1)(\eta + 1)(\eta \pm 1/3)(\eta - 1)\ 27/16$

Note: The last 4 LT/QN and last 12 QT/CuN functions are entirely local.

introduced in Section 9.8 are expected to provide $O(\Delta^{p+1})$ interpolation error. However, since the mixed-order vector basis functions of the present section are not complete to the highest polynomial order, the error analysis suggests that they provide one less order of interpolation accuracy. For instance, the LT/QN functions would be expected to provide $O(\Delta^2)$ accuracy as $\Delta \to 0$, not $O(\Delta^3)$.

However, data presented in the previous sections suggest that, when used within a discretization of the vector Helmholtz equation, the LT/QN functions actually provide slightly more accurate eigenvalues than the polynomial-complete quadratic basis functions (Tables 9.5 and 9.6). These results suggest that the interpolation error analysis might not take into account the nature of the discarded degrees of freedom in the mixed-order expansion. Consequently, it is possible that the analysis produces overly pessimistic conclusions when applied to the mixed-order functions.

In fact, a very different conclusion can be obtained from a dispersion analysis similar to that described in Section 5.5. The procedure considers the error associated with a uniform plane wave propagating in an arbitrary direction on an infinite regular mesh when the wave is represented by some type of basis function. Despite the complexity of dispersion analysis in the multidimensional case, some progress has been made in extending it to the two- and three-dimensional vector basis functions considered in this chapter [26, 27]. The results of this analysis show that mixed-order vector basis functions (containing orders up to $p-1$ and p on triangles or tetrahedra and orders $p-1$, p, and $p+1$ on quadrilateral and hexahedral cells) appear to produce a discretization error of $O(\Delta^{p+1})$ when used to represent a plane wave. In common with scalar basis functions, the error appears entirely as phase error for electrically small cells in a homogeneous region, and the phase constant converges at the superconvergent rate of $O(\Delta^{2p})$. These findings confirm that the mixed-order vector basis functions are well suited for representing electromagnetic fields and suggest that the pessimistic result of the interpolation error analysis should be discounted, at least in source-free regions.

A dispersion analysis permits a direct comparison of the efficiency of several different expansions. Warren compared the CT/LN and LT/QN representations for triangular and quadrilateral cells (Tables 9.9 and 9.10) with the polynomial-complete representation in (9.154) [27]. Table 9.11 shows the unknown density needed to limit the phase error to 0.1 deg/λ in the worst-case direction for an infinite square-cell mesh. For the triangular-cell expansions, the square cells are divided diagonally, with all the diagonals running in the same direction. Despite the fact that such a triangular-cell mesh produces relatively large errors in the worst-case direction, the triangular-cell expansions are about as efficient as the quadrilateral-cell expansions. For high accuracies, the quadratic-order basis functions are far more efficient than the linear-order basis functions. It is interesting that the polynomial-complete expansion from (9.154) requires four times as many unknowns to produce dispersion error as low as that of the CT/LN representation for triangles.

The curl-conforming vector basis functions introduced in this section eliminate some of the degrees of freedom associated with the nullspace of the curl–curl operator. One might wonder if the entire nullspace could be eliminated. The answer is yes, but the process of separating the nullspace involves global matrix operations that may not be compatible with sparse matrix storage and will generally prevent a purely local approach to constructing the finite-element system. For example, Manges and Cendes propose a partitioning based on the tree and cotree of the finite-element mesh [28]. A similar partitioning can be constructed by combining the standard mixed-order basis functions into *loop* and *star* functions, as suggested by Wilton [29]. Each loop function encircles a cell of the mesh, while star functions

TABLE 9.11 Approximate Cell Size Δ and Unknown Density Required to Limit Phase Error to 0.1 deg/λ for Plane-Wave Dispersion Analysis

Basis Function Type	Δ	Unknowns/λ^2
LT/LN, triangular, (9.154)	0.013λ	35570
CT/LN, quadrilateral	0.013λ	11860
CT/LN, triangular	0.018λ	8880
LT/QN, quadrilateral	0.127λ	246
LT/QN, triangular	0.140λ	307

Note: Several two-dimensional vector representations are compared on an infinite mesh of $\Delta \times \Delta$ square cells divided diagonally into triangular cells. The unknown density calculation assumes that the interior LT/QN unknowns are eliminated. Adapted from Warren's data [27].

are directed outward from each node. In the simplest case, the loop and star functions are a linear combination of the standard CT/LN functions and provide an equivalent representation. The star functions have zero curl and exhibit a one-to-one correspondence with nullspace eigenfunctions. However, either approach to eliminating the nullspace requires matrix transformations similar in form to (9.148)–(9.150) to collapse the system order to the non–nullspace eigenrank.

The use of the mixed-order vector basis functions described in this section resolves the three difficulties identified in Section 9.7. Not only do these functions eliminate the spurious nonzero eigenvalues, but boundary conditions are easy to incorporate since the coefficients are primarily associated with tangential fields at cell edges. Because they permit a jump discontinuity in the normal field, they place no restriction on the medium continuity from cell to cell. Furthermore, the mixed-order nature of these functions suppresses some of the degrees of freedom associated with the nullspace of the curl operator, enhancing their efficiency for modeling fields in source-free regions. In the following sections, we demonstrate that these functions are also easily adapted to scattering and waveguide formulations.

9.10 TE SCATTERING USING THE VECTOR HELMHOLTZ EQUATION WITH CT/LN AND LT/QN VECTOR BASIS FUNCTIONS DEFINED ON TRIANGULAR CELLS [30]

Prior to attacking a three-dimensional formulation (Chapter 11), we will consider the use of curl-conforming vector basis functions for discretizing the vector Helmholtz equation for TE scattering from infinite, inhomogeneous cylinders. Although a scalar formulation is likely to be more efficient for the two-dimensional situation, we turn to the vector formulation in order to study issues associated with the three-dimensional problem without the computational requirements of three-dimensional analysis. Since the use of CT/LN and LT/QN functions has already been demonstrated for cavity analysis, the principal focus of this section is the incorporation of an RBC into the formulation. We would also like to explore the extent to which these vector basis functions can represent jump discontinuities in the normal-field components at medium interfaces.

The curl–curl form of the vector Helmholtz equation for $\bar{E}$ can be written as

$$\nabla \times \left(\frac{1}{\mu_r} \nabla \times \bar{E} \right) - k^2 \varepsilon_r \bar{E} = 0 \tag{9.202}$$

We will develop an outward-looking formulation incorporating a local RBC. Following a procedure similar to that used to develop the scalar Bayliss–Turkel conditions in Section 3.8, we seek an RBC that forces the scattered field to have the form of the asymptotic expansion

$$\bar{E}^s(\rho, \phi) \cong \frac{e^{-jk\rho}}{\sqrt{\rho}} \sum_{n=0}^{\infty} \frac{\bar{E}_n(\phi)}{\rho^n} \tag{9.203}$$

It is easily verified that the condition

$$\hat{\rho} \times \nabla \times \bar{E}^s = \hat{\phi} \left(jk - \frac{1}{2\rho} \right) E^s_\phi \tag{9.204}$$

forces the scattered field to agree with the first two terms in (9.203). A second-order RBC can be obtained as

$$\left(\hat{\rho} \times (\nabla \times \ ) - jk - \frac{3}{2\rho} \right) \left(\hat{\rho} \times (\nabla \times \bar{E}^s) - jk\bar{E}^s_\phi + \frac{1}{2\rho} \bar{E}^s_\phi \right) = 0 \tag{9.205}$$

and, using (9.202), can be rewritten as

$$\hat{\rho} \times \nabla \times \bar{E}^s = \alpha(\rho) \bar{E}^s_\phi + \frac{jk - 1/2\rho}{2jk + 2/\rho} \nabla^{\tan} (\hat{\rho} \cdot \bar{E}^s) \tag{9.206}$$

where

$$\alpha(\rho) = \frac{-2k^2 + jk/\rho - 1/4\rho^2}{2jk + 2/\rho} \tag{9.207}$$

Equation (9.206) forces the scattered field to agree with the first four terms in the preceding expansion and reduces the error to $O(\rho^{-9/2})$. However, the single derivative in this RBC gives rise to a nonsymmetric matrix operator when incorporated into the weak form of the vector Helmholtz equation. Consequently, after observing

$$E_\rho = \frac{1}{k^2} \hat{\rho} \cdot \nabla \times \nabla \times \bar{E} = -\frac{1}{k^2} \nabla \cdot (\hat{\rho} \times \nabla \times \bar{E}) \tag{9.208}$$

we are motivated to substitute the RBC recursively into itself to recast the first-order derivative into a second-order derivative [31]. Neglecting higher order terms, the result can be written as

$$\hat{n} \times \nabla \times \bar{E}^s = \alpha(\rho) \bar{E}^s_\phi + \beta(\rho) \frac{\partial^2 \bar{E}^s_\phi}{\partial \phi^2} \tag{9.209}$$

where $\alpha(\rho)$ is defined in (9.207) and

$$\beta(\rho) = -\frac{\alpha(\rho)}{k^2 \rho^2} \frac{jk - 1/2\rho}{2jk + 2/\rho} \tag{9.210}$$

The RBC in (9.209) is to be imposed on a circular boundary of radius ρ, as illustrated in Figure 9.19. After combining this RBC with the incident field and substituting the result

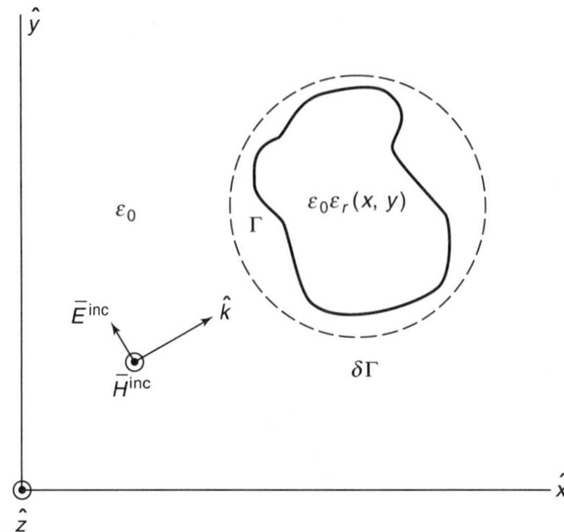

Figure 9.19 Cross section of an inhomogeneous cylinder enclosed within a mathematical radiation boundary $\partial\Gamma$. After [30]. ©1994 IEEE.

into the weak equation associated with (9.202), we obtain

$$\iint_{\Gamma}\left(\frac{1}{\mu_r}\nabla\times\bar{T}\cdot\nabla\times\bar{E}-k^2\varepsilon_r\bar{T}\cdot\bar{E}\right)dx\,dy+\int_{\partial\Gamma}\left(\alpha\bar{T}\cdot\bar{E}_{\phi}-\beta\rho^2\frac{\partial T_{\phi}}{\partial t}\frac{\partial E_{\phi}}{\partial t}\right)dt$$
$$=-\int_{\partial\Gamma}\bar{T}\cdot\left(\hat{n}\times\nabla\times\bar{E}^{\text{inc}}-\alpha\bar{E}_{\phi}^{\text{inc}}-\beta\frac{\partial^2\bar{E}_{\phi}^{\text{inc}}}{\partial\phi^2}\right)dt \tag{9.211}$$

as the equation describing the scattering problem.

Within the region Γ, the unknown electric field can be expressed in terms of the vector expansion

$$\bar{E}(x,y)\cong\sum_{n=1}^{N}e_n\bar{B}_n(x,y) \tag{9.212}$$

and substituted into (9.211). Testing functions identical to the basis functions can be employed to construct the $N\times N$ matrix equation $\mathbf{Ae}=\mathbf{b}$, where

$$A_{mn}=\iint_{\Gamma}\left(\frac{1}{\mu_r}\nabla\times\bar{B}_m\cdot\nabla\times\bar{B}_n-k^2\varepsilon_r\bar{B}_m\cdot\bar{B}_n\right)dx\,dy$$
$$+\int_{\partial\Gamma}\left(\alpha\bar{B}_m\cdot\bar{B}_n-\beta\rho^2\frac{\partial\bar{B}_m}{\partial t}\cdot\frac{\partial\bar{B}_n}{\partial t}\right)dt \tag{9.213}$$

and

$$b_m=-\int_{\partial\Gamma}\bar{B}_m\cdot\left(\hat{n}\times\nabla\times\bar{E}^{\text{inc}}-\alpha\bar{E}_{\phi}^{\text{inc}}-\beta\frac{\partial^2\bar{E}_{\phi}^{\text{inc}}}{\partial\phi^2}\right)dt \tag{9.214}$$

Two different types of basis and testing functions will be considered, the CT/LN and the LT/QN functions introduced in Section 9.9.

In common with our previous finite-element implementations, the region of interest is divided into triangular cells, with the mesh terminating on a circular outer boundary. The matrix $\mathbf{A}$ is constructed on a cell-by-cell basis using a connectivity array that identifies the rows and columns of the matrix associated with a particular cell. As suggested in

Section 9.8, we find it convenient to employ a connectivity array that identifies the three edges associated with each cell. A second pointer array is used to identify the two nodes associated with each edge. For the CT/LN functions, 3×3 element matrices are required for the volumetric integrals in (9.213), while 3×3 element matrices also arise from the boundary integrals as explained below. To ensure a common orientation on either side of a given edge, we adopt the convention that the tangential vector basis functions always point from a smaller node index to a larger node index, according to the global numbering.

Equations (9.184)–(9.187) provide element matrix entries for the volumetric integrals in (9.213). The boundary integrals appearing in (9.213) require a tangential differentiation of the field along $\partial\Gamma$. The CT/LN basis functions are discontinuous along $\partial\Gamma$, and consequently the tangential differentiation produces Dirac delta functions between cells. Since basis and testing function discontinuities coincide, the boundary integral involves the product of two delta functions! To alleviate this difficulty, we approximate the problematic Dirac delta functions by piecewise-constant functions straddling two cell edges, as illustrated in Figure 9.20. This approximation is justified since the fictitious discontinuities are introduced by the basis representation, while the true field along $\partial\Gamma$ is continuous. Using this approximation and assuming that the basis functions $\bar{B}_m$ and $\bar{B}_n$ have a common vector orientation along $\partial\Gamma$, element matrix entries providing the boundary integrals over a single cell are obtained as

$$\int_{\text{edge } m} B_m B_n \, dt = \begin{cases} w_m & m = n \\ 0 & \text{otherwise} \end{cases} \tag{9.215}$$

and

$$\int_{\text{edge } m} \frac{\partial B_m}{\partial t} \frac{\partial B_n}{\partial t} \, dt \cong \begin{cases} \dfrac{-2}{w_{m-1} + w_m} & n = m - 1 \\[2ex] \dfrac{2}{w_{m-1} + w_m} + \dfrac{2}{w_m + w_{m+1}} & n = m \\[2ex] \dfrac{-2}{w_m + w_{m+1}} & n = m + 1 \\[2ex] 0 & \text{otherwise} \end{cases} \tag{9.216}$$

where w_m is the width of edge m and the ordered triple $(m - 1, m, m + 1)$ describes three adjacent edges along $\partial\Gamma$ with the index increasing in the ϕ direction.

The treatment of imbedded perfectly conducting material requires the homogeneous Dirichlet condition $\hat{n} \times \bar{E} = 0$, which can be imposed by equating coefficients of tangential basis functions along the conducting boundary to zero. Usually, these coefficients are not included in the global system, and contributions from the element matrices associated with these edges are simply ignored when constructing the global matrix. In any case, a list of edges located on conducting boundaries is a necessary part of the scatterer model.

As a general rule, there are between two and three times as many edges as nodes in a typical triangular-cell model. This result suggests that the system arising from the CT/LN basis functions will always be of larger order than the matrix equation arising from a linear Lagrangian discretization of the scalar Helmholtz equation using the same mesh. (This might be expected, since the transverse fields exhibit more degrees of freedom than E_z or H_z simply because they are vector rather than scalar entities.) The increase in unknowns for a particular model is offset by the fact that the triangular-cell CT/LN functions produce exactly five nonzero entries per row of matrix, except for cells adjacent to conductors, as

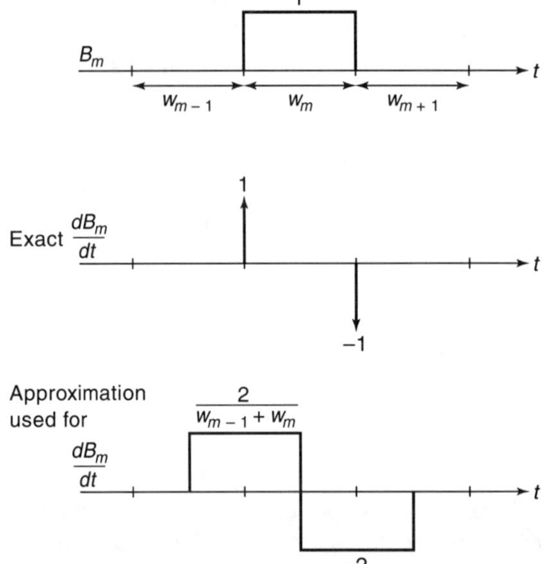

Figure 9.20 Approximation employed to replace the Dirac delta functions with pulse functions. After [30]. ©1994 IEEE.

compared to some larger (and highly variable) number for node-based expansions. Thus, the matrix arising from the edge-based expansion is generally sparser than that produced by a node-based expansion and is more amenable to packed storage because of its predictable sparsity pattern.

To illustrate the accuracy of the procedure, Figure 9.21 shows the dominant component of the (transverse) electric field induced along a cut through a cylinder having core material of relative permittivity $\varepsilon_r = 7 - j7$ and radius $0.15\lambda_0$ surrounded by a cladding of $\varepsilon_r = 3 - j0$ and radius $0.25\lambda_0$, where λ_0 denotes the free-space wavelength. Figure 9.21 is a composite that depicts two numerical results. The left half of the figure shows data obtained using a triangular-cell model with 84 cells in the core region, 112 cells in the cladding, and 320 cells outside the cylinder. The cells are roughly uniform in size, and the longest cell edge in the mesh is $0.075\lambda_0$. The cell densities in the model vary from 112 cells/λ_d^2 in the core region to 727 cells/λ_d^2 outside the cylinder, where λ_d is the dielectric wavelength and the number of edges (matrix order) is 798. To demonstrate numerical convergence, the right half of the figure shows data obtained with cells that are reduced in linear dimension to approximately one-fourth of that used in the 798-edge model, to produce a mesh with 6956 cells and 10,520 edges. The triangular-cell model is aligned so that the field component shown in Figure 9.21 is entirely tangential to cell edges and is therefore a piecewise-constant function. Plots of the phase of the electric field show similar agreement with exact solutions. From the jump discontinuities present at cell edges, it is apparent that the CT/LN representation is poor at resolving field discontinuities at medium interfaces unless electrically small cells are employed. The RBC used in this approach was imposed at a radius of 0.41λ and for these examples appears to produce a residual error that is much lower than the discretization error associated with the basis function expansion.

We next consider a discretization obtained from the LT/QN basis functions depicted in Figure 9.17. The procedure parallels that of the CT/LN functions and, if naively imple-

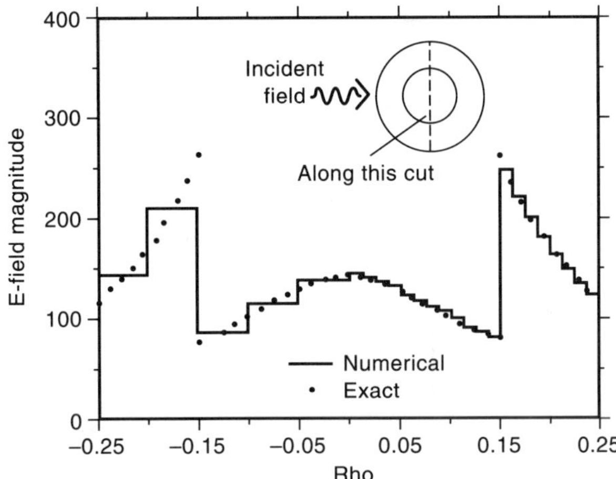

Figure 9.21 Composite result showing the electric field within a layered dielectric cylinder. The left half of the figure shows data obtained with a 798-edge mesh, while the right half shows data obtained with a 10,520-edge mesh. Both results were obtained using a CT/LN expansion on triangular cells. After [30]. ©1994 IEEE.

mented, produces a matrix with 14, 10, or 8 nonzero entries per row, depending on whether the basis functions are located tangential to an interior edge, tangential to an edge on the radiation boundary, or normal to an edge, respectively. (Fewer entries will be produced in the vicinity of perfectly conducting boundaries.) Both the normal and tangential components of the LT/QN basis functions are discontinuous at cell corners, and the representation along $\partial\Gamma$ is therefore discontinuous from cell to cell. Consequently, an approximation similar to that illustrated in Figure 9.20 is necessary to implement the boundary integral calculations required in Equation (9.184). We leave the development of the element matrix entries as an exercise for the reader (Probs. P9.28 and P9.29).

Some reduction in the total number of LT/QN unknowns is possible, however, since the two interior basis functions per cell only interact with other functions within that cell. These functions can be eliminated from the system of equations at the element matrix level, permitting a reduction in the matrix order without an increase in the number of nonzero entries per row or column. (In traditional finite-element literature, the analogous procedure is named *condensation*.) Consequently, the number of unknowns can be reduced to two per edge throughout the mesh, and the resulting sparse system will have at most 10 nonzero entries per row and column.

Figure 9.22 shows the transverse electric field produced by the LT/QN expansion within a layered dielectric cylinder having core material $\varepsilon_r = 7 - j7$ and radius 0.15λ surrounded by a cladding with $\varepsilon_r = 3 - j0$ and radius 0.25λ. The RBC is imposed at a radius of 0.45λ. The left half of Figure 9.22 shows data obtained with a 1556-order system based on a 304-cell mesh, with longest cell edge $0.1\lambda_0$, while the right half shows results from an order-4904 system based on a 968-cell mesh with longest edge $0.055\lambda_0$. The field component displayed in Figure 9.22 is entirely tangential to cell edges and is therefore a piecewise-linear function that is not constrained to be continuous from one cell to the next. Despite the freedom in the representation, the discontinuities between cells are very small, except at the dielectric interface where they track the exact solution very closely.

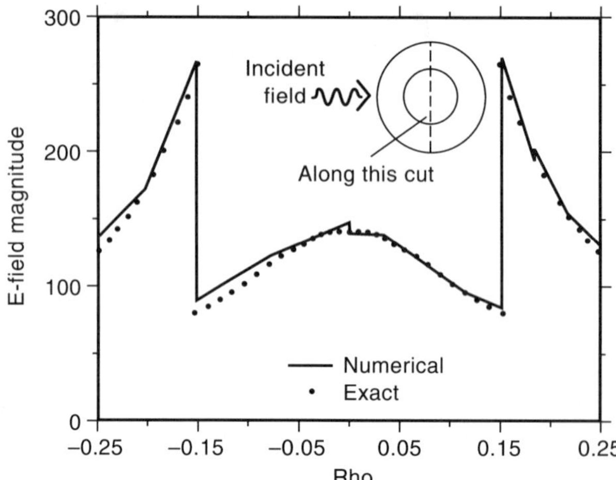

Figure 9.22 Composite result showing the electric field within a layered dielectric cylinder. The left half of the figure shows data obtained with a 304-cell mesh, while the right half shows data obtained with a 968-cell mesh. Both results were obtained using a LT/QN expansion on triangular cells. After [30]. ©1994 IEEE.

This section has illustrated the implementation of CT/LN and LT/QN basis functions within a two-dimensional scattering formulation. The formulation is straightforward, with one exception. Because a local RBC is incorporated into the formulation, the evaluation of boundary integrals is complicated by cell-to-cell discontinuities in the expansion. A simple approximation (replacing the delta functions by pulse functions) is introduced to circumvent this difficulty. Overall, numerical results for the interior fields exhibit excellent accuracy. Of particular interest is the fact that observed discontinuities in the LT/QN representation are small in homogeneous regions and closely track the desired discontinuous behavior at material interfaces.

9.11 ANALYSIS OF DIELECTRIC-LOADED WAVEGUIDES USING CURL-CONFORMING VECTOR BASIS FUNCTIONS

Consider a waveguide structure aligned with the z-axis and a waveguide mode propagating with z-dependence $e^{j\beta z}$. The task of waveguide analysis is to determine the field distribution and the phase constant β as a function of frequency associated with a waveguide mode of interest. Homogeneously filled waveguiding structures can be formulated in terms of the scalar Helmholtz equation for E_z or H_z and solved by a purely two-dimensional analysis, since the propagation constant β can be found directly from the cutoff wavenumber k_c of a mode according to

$$\beta = \sqrt{k^2 - k_c^2} \tag{9.217}$$

The cutoff wavenumbers are the resonant wavenumbers of the analogous two-dimensional cavity. For an inhomogeneously filled waveguide, Equation (9.217) no longer applies. Nor does the TM/TE characterization that permits an independent solution for E_z and H_z. Consequently, in the general case we are unable to pose the analysis in terms of a scalar Helmholtz equation.

A number of alternative waveguide formulations are described in the literature, and we focus our attention on one approach based in part on reference [24]. For illustration, we consider an inhomogeneous waveguide bounded by p.e.c. walls and base the analysis on the vector Helmholtz equation

$$\nabla \times \left(\frac{1}{\varepsilon_r} \nabla \times \bar{H} \right) = k^2 \mu_r \bar{H} \tag{9.218}$$

where ε_r and μ_r are functions of transverse position throughout the waveguide cross section. The magnetic field can be written in the form

$$\bar{H}(x, y, z) = [\bar{H}_t(x, y) + \hat{z} H_z(x, y)] e^{-j\beta z} \tag{9.219}$$

After expanding the curl according to

$$\nabla \times \bar{H} = \nabla_t \times \bar{H}_t - \hat{z} \times (\nabla_t H_z + j\beta \bar{H}_t) \tag{9.220}$$

and

$$\nabla \times \left(\frac{1}{\varepsilon_r} \nabla \times \bar{H} \right) = \nabla \times \left(\frac{1}{\varepsilon_r} \nabla_t \times \bar{H}_t \right) - \nabla \times \left(\frac{1}{\varepsilon_r} \hat{z} \times (\nabla_t H_z + j\beta \bar{H}_t) \right) \tag{9.221}$$

where ∇_t is the transverse part of the ∇ operator, the vector Helmholtz equation can be separated into its transverse components

$$\nabla_t \times \left(\frac{1}{\varepsilon_r} \nabla_t \times \bar{H}_t \right) - \frac{1}{\varepsilon_r} (j\beta \nabla_t H_z - \beta^2 \bar{H}_t) = k^2 \mu_r \bar{H}_t \tag{9.222}$$

and its z-component

$$-\nabla_t \cdot \left(\frac{1}{\varepsilon_r} (\nabla_t H_z + j\beta \bar{H}_t) \right) = k^2 \mu_r H_z \tag{9.223}$$

In order to facilitate the use of either k or β as the eigenvalue, Lee, Sun, and Cendes introduced the scaling [24]

$$\bar{h}_t = \beta \bar{H}_t \tag{9.224}$$

They also proposed the substitution

$$h_z = -j H_z \tag{9.225}$$

which produces a real-valued system of equations in the lossless case. These substitutions convert (9.222) and (9.223) into

$$\nabla_t \times \left(\frac{1}{\varepsilon_r} \nabla_t \times \bar{h}_t \right) + \frac{1}{\varepsilon_r} (\beta^2 \nabla_t h_z + \beta^2 \bar{h}_t) = k^2 \mu_r \bar{h}_t \tag{9.226}$$

$$-\nabla_t \cdot \left(\frac{1}{\varepsilon_r} (\nabla_t h_z + \bar{h}_t) \right) = k^2 \mu_r h_z \tag{9.227}$$

and can be used to construct an eigenvalue formulation for the waveguide problem.

Equation (9.226) is a vector equation with the same leading-order derivatives as the two-dimensional vector Helmholtz equation. This equation can be converted into a weak form by introducing a transverse vector testing function $\bar{T}(x, y)$ and applying the operations indicated in Equations (9.123)–(9.125) to produce

$$\iint_\Gamma \frac{1}{\varepsilon_r}(\nabla_t \times \bar{T} \cdot \nabla_t \times \bar{h}_t + \beta^2 \bar{T} \cdot \bar{h}_t + \beta^2 \bar{T} \cdot \nabla_t h_z) = k^2 \iint_\Gamma \mu_r \bar{T} \cdot \bar{h}_t - \int_{\partial\Gamma} \frac{1}{\varepsilon_r} \bar{T} \cdot \hat{n} \times \nabla_t \times \bar{h}_t \tag{9.228}$$

where Γ denotes the guide interior and $\partial\Gamma$ denotes the guide walls. Equation (9.227) is a scalar equation similar in form to the two-dimensional scalar Helmholtz equation, and after introducing a scalar testing function $T(x, y)$, an analogous weak equation can be written in the form

$$\iint_\Gamma \frac{1}{\varepsilon_r}(\nabla_t T \cdot \nabla_t h_z + \nabla_t T \cdot \bar{h}_t) = k^2 \iint_\Gamma \mu_r T h_z + \int_{\partial\Gamma} \frac{1}{\varepsilon_r}\left(T \frac{\partial h_z}{\partial n} + T \hat{n} \cdot \bar{h}_t\right) \tag{9.229}$$

Since $\partial\Gamma$ consists of p.e.c. material, appropriate boundary conditions dictate that

$$\hat{n} \times \nabla_t \times \bar{h}_t = 0 \tag{9.230}$$

$$\frac{\partial h_z}{\partial n} = 0 \tag{9.231}$$

and

$$\hat{n} \cdot \bar{h}_t = 0 \tag{9.232}$$

along $\partial\Gamma$. Imposing these conditions eliminates the boundary integrals in (9.228) and (9.229).

Suppose the waveguide interior is divided into triangular cells and the function $\bar{h}_t(x, y)$ is represented by mixed-order vector basis functions of the form introduced in Section 9.9. Similarly, the function h_z can be represented by ordinary scalar Lagrangian basis functions. Given a set of linearly independent vector testing functions and a set of linearly independent scalar testing functions, Equations (9.228) and (9.229) can be written in the form of the matrix eigenvalue equation

$$\begin{bmatrix} \mathbf{A}^{tt} & \mathbf{A}^{tz} \\ \mathbf{A}^{zt} & \mathbf{A}^{zz} \end{bmatrix} \begin{bmatrix} \mathbf{h}_t \\ \mathbf{h}_z \end{bmatrix} = k^2 \begin{bmatrix} \mathbf{B}^{tt} & \mathbf{0} \\ \mathbf{0} & \mathbf{B}^{zz} \end{bmatrix} \begin{bmatrix} \mathbf{h}_t \\ \mathbf{h}_z \end{bmatrix} \tag{9.233}$$

where

$$A_{mn}^{tt} = \iint \frac{1}{\varepsilon_r}(\nabla_t \times \bar{T}_m \cdot \nabla_t \times \bar{B}_n + \beta^2 \bar{T}_m \cdot \bar{B}_n) \tag{9.234}$$

$$A_{mn}^{tz} = \beta^2 \iint \frac{1}{\varepsilon_r}(\bar{T}_m \cdot \nabla_t B_n) \tag{9.235}$$

$$A_{mn}^{zt} = \beta^2 \iint \frac{1}{\varepsilon_r}(\nabla_t T_m \cdot \bar{B}_n) \tag{9.236}$$

$$A_{mn}^{zz} = \beta^2 \iint \frac{1}{\varepsilon_r}(\nabla_t T_m \cdot \nabla_t B_n) \tag{9.237}$$

$$B_{mn}^{tt} = \iint \mu_r \bar{T}_m \cdot \bar{B}_n \tag{9.238}$$

$$B_{mn}^{zz} = \beta^2 \iint \mu_r T_m B_n \tag{9.239}$$

and where the integrals encompass the appropriate cells of the mesh. The scalar equation has been scaled by a factor β^2 to improve the symmetry in (9.233). If the vector testing functions are identical to the vector basis functions and the scalar testing functions are identical to scalar basis functions, element matrices for (9.234), (9.237), (9.238), and (9.239) are identical to those discussed in previous sections. The calculation of element matrices for (9.235) and (9.236) is left as an exercise.

To illustrate the approach, Figure 9.23 shows a portion of the k–β diagram for a circular guide concentrically loaded with a circular dielectric rod having $\varepsilon_r = 2$. This result was obtained using linear Lagrangian basis functions for h_z and CT/LN basis functions for $\bar{h}_t$. The mesh contained 124 triangular cells and resulted in a sparse eigenvalue equation of order 273. For this magnetic field formulation, there were 75 nullspace eigenvalues (with value zero) computed, one for each node of the finite-element mesh.

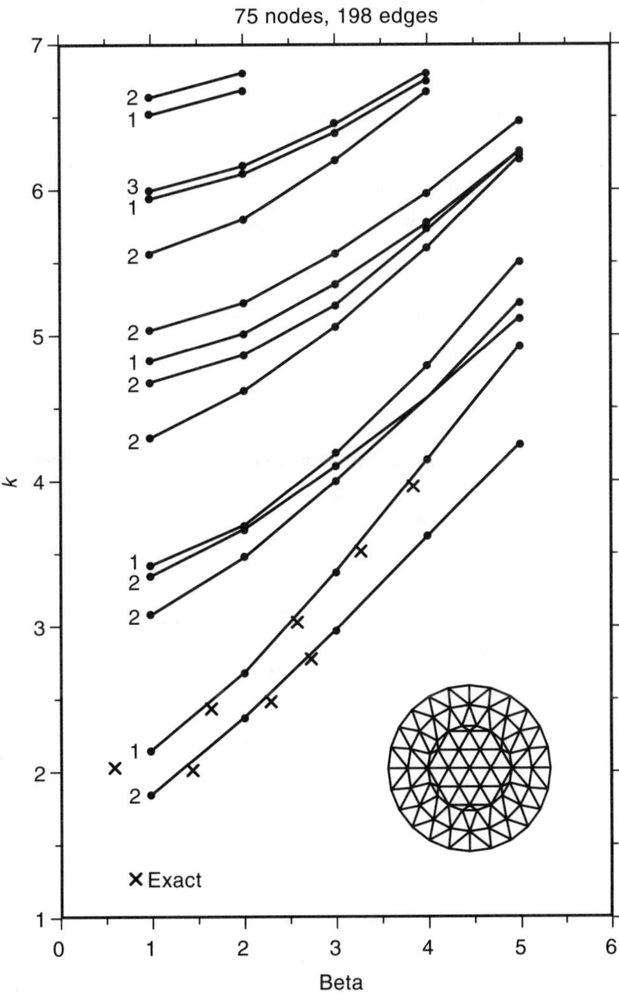

Figure 9.23 The k–β diagram for a concentrically loaded circular waveguide. (The model contained 75 nodes and 198 edges).

In summary, this waveguide formulation incorporates ordinary scalar basis functions to represent the z-component of the field and mixed-order vector basis functions to represent the transverse component, following [24]. The approach offers the possibility of treating material with simultaneous discontinuities in both ε_r and μ_r, and it eliminates spurious eigenvalues since the vector basis functions permit jump discontinuities in the normal-field component. Although Equation (9.233) is a generalized matrix eigenvalue equation for k^2 as the eigenvalue, an alternate arrangement of entries in (9.233) can be used to produce an eigenvalue equation for β^2 instead of k^2 (Prob. P9.30). The approach can also be modified to treat the electric field as the primary unknown, in which case the boundary conditions of (9.230)–(9.232) must be replaced by conditions appropriate for the electric field.

9.12 MIXED-ORDER CURL-CONFORMING VECTOR BASIS FUNCTIONS FOR TETRAHEDRAL AND HEXAHEDRAL CELLS [20]

Three-dimensional counterparts of the mixed-order curl-conforming vector basis functions (e.g., CT/LN) can be developed for tetrahedral and hexahedral cells. The general procedure follows the approach employed in Section 9.9. A polynomial-complete linear representation of a vector in three dimensions has the form

$$\hat{x}(A + Bx + Cy + Dz) + \hat{y}(E + Fx + Gy + Hz) + \hat{z}(I + Jx + Ky + Lz) \qquad (9.240)$$

containing 12 degrees of freedom. To eliminate degrees of freedom that can only represent the gradient of a quadratic function, we impose the six Nedelec constraints [16]

$$\frac{\partial B_x}{\partial x} = 0 \qquad (9.241)$$

$$\frac{\partial B_y}{\partial y} = 0 \qquad (9.242)$$

$$\frac{\partial B_z}{\partial z} = 0 \qquad (9.243)$$

$$\frac{\partial B_x}{\partial y} + \frac{\partial B_y}{\partial x} = 0 \qquad (9.244)$$

$$\frac{\partial B_x}{\partial z} + \frac{\partial B_z}{\partial x} = 0 \qquad (9.245)$$

$$\frac{\partial B_y}{\partial z} + \frac{\partial B_z}{\partial y} = 0 \qquad (9.246)$$

and reduce the linear representation to the form

$$\bar{B}(x, y, z) = \hat{x}(A + Cy + Dz) + \hat{y}(E - Cx + Hz) + \hat{z}(I - Dx - Hy) \qquad (9.247)$$

An identical Cartesian representation is obtained by constraining the tangential component of a linear basis function to be constant along one edge of a tetrahedron and vanish along the other five edges. Six such functions, overlapping a tetrahedral cell, provide a CT/LN representation. To maintain tangential continuity throughout a tetrahedral-cell model, the coefficient for the basis function defined at an edge must be shared by each cell adjacent to that edge. Note that in a general tetrahedral-cell mesh, there may be many cells sharing a single edge, and thus the sparsity pattern is not as predictable as it was in the two-dimensional examples considered in Sections 9.9 and 9.10.

A convenient representation within a tetrahedron is given in terms of the simplex coordinates (L_1, L_2, L_3, L_4) defined in Section 9.6. The basis function localized along the edge from node i to node j can be written as

$$\bar{B}_{ij} = w_{ij}(L_i \nabla L_j - L_j \nabla L_i) \tag{9.248}$$

where w_{ij} is the length of the edge between nodes i and j. By comparing Equations (9.248) and (9.169), we observe that the CT/LN type of basis function has the same simplex representation in two and three dimensions. As in the two-dimensional case, the three-dimensional functions do not impose continuity of the normal-field component and therefore have the freedom to properly model nullspace eigenfunctions of the curl–curl operator. It is a straightforward matter to show that the three-dimensional CT/LN functions have zero divergence and constant curl within a tetrahedral cell, with a delta function divergence at cell boundaries.

An LT/QN basis can be obtained from a complete quadratic expansion in three dimensions by imposing the 10 conditions [16]

$$\frac{\partial^2 B_i}{\partial i^2} = 0 \qquad (i = x, y, z) \tag{9.249}$$

$$\frac{\partial^2 B_i}{\partial j^2} + 2\frac{\partial^2 B_j}{\partial_i \partial_j} = 0 \qquad (i, j = x, y, z;\ i \neq j) \tag{9.250}$$

$$\frac{\partial^2 B_x}{\partial y \partial z} + \frac{\partial^2 B_y}{\partial x \partial z} + \frac{\partial^2 B_z}{\partial x \partial y} = 0 \tag{9.251}$$

These serve to eliminate degrees of freedom associated with the gradient of a cubic function and reduce the number of coefficients in a three-dimensional quadratic expansion function from 30 to 20. A quadratic basis set can be defined on tetrahedral cells in three dimensions so that 12 basis functions interpolate to a linear tangential component along each edge of the tetrahedron (again, 2 per edge) while 8 additional functions provide a quadratic normal vector component along each edge and face (two degrees of freedom per face). The edge-based functions straddle all the cells sharing a common edge, while the face-based functions have their domain limited to the two cells that share the appropriate face. One specific form of the LT/QN basis functions in simplex coordinates is presented in Table 9.12.

Vector basis functions with a QT/CuN behavior can be obtained by imposing the 15 constraints embodied in the general expression [16]

$$\frac{\partial^3 B_i}{\partial u_j \partial u_k \partial u_l} + \frac{\partial^3 B_j}{\partial u_i \partial u_k \partial u_l} + \frac{\partial^3 B_k}{\partial u_i \partial u_j \partial u_l} + \frac{\partial^3 B_l}{\partial u_i \partial u_j \partial u_k} = 0 \tag{9.252}$$

where indices i, j, k, and l run from 1 to 3, u_1 denotes x, u_2 denotes y, and u_3 denotes z, and B_1 denotes B_x, and so on. The resulting 45 QT/CuN basis functions are summarized in Table 9.12 and consist of edge-based, face-based, and cell-based functions. In the general case, a three-dimensional expansion of polynomial degree k has

$$N = \tfrac{1}{2}[k(k+2)(k+3)] \tag{9.253}$$

coefficients within a tetrahedral cell after eliminating the degrees of freedom that can represent the gradient of a $(k+1)$-order polynomial. Element matrix entries for the CT/LN and LT/QN functions on tetrahedral cells are developed in Chapter 11.

TABLE 9.12 Simplex-Coordinate Definition of Mixed-Order Basis Functions within Tetrahedral Cell

CT/LN (6 Functions, All Edge-Based)	LT/QN (20 Functions)	QT/CuN (45 Functions)
	Edge based	Edge based
$L_1\nabla L_2 - L_2\nabla L_1$	$L_1\nabla L_2$	12 of the form $L_i(2L_i - 1)\nabla L_j,\ i \neq j$
$L_1\nabla L_3 - L_3\nabla L_1$	$L_2\nabla L_1$	6 of the form $L_iL_j(\nabla L_i - \nabla L_j),\ i \neq j$
$L_1\nabla L_4 - L_4\nabla L_1$	$L_1\nabla L_3$	Face based
$L_2\nabla L_3 - L_3\nabla L_2$	$L_3\nabla L_1$	8 of the form $L_i(2L_i - 1)(L_j\nabla L_k - L_k\nabla L_j)$,
$L_2\nabla L_4 - L_4\nabla L_2$	$L_1\nabla L_4$	$i \neq j \neq k$, also omit combinations
$L_3\nabla L_4 - L_4\nabla L_3$	$L_4\nabla L_1$	$ijk = 123, 124, 134, 234$
	$L_2\nabla L_3$	4 of the form
	$L_3\nabla L_2$	$\nabla(L_1L_2L_3)$
	$L_2\nabla L_4$	$\nabla(L_1L_2L_4)$
	$L_4\nabla L_2$	$\nabla(L_1L_3L_4)$
	$L_3\nabla L_4$	$\nabla(L_2L_3L_4)$
	$L_4\nabla L_3$	12 of the form
	Face based	$L_i^2(L_j\nabla L_k - L_k\nabla L_j)$,
	$L_1L_2\nabla L_3 - L_1L_3\nabla L_2$	$i \neq j \neq k$
	$L_2L_3\nabla L_1 - L_1L_3\nabla L_2$	Cell based
	$L_1L_2\nabla L_4 - L_1L_4\nabla L_2$	$L_1L_2L_3\nabla L_4 - L_2L_3L_4\nabla L_1$
	$L_2L_4\nabla L_1 - L_1L_4\nabla L_2$	$L_1L_2L_4\nabla L_3 - L_2L_3L_4\nabla L_1$
	$L_2L_3\nabla L_4 - L_2L_4\nabla L_3$	$L_1L_3L_4\nabla L_2 - L_2L_3L_4\nabla L_1$
	$L_3L_4\nabla L_2 - L_2L_4\nabla L_3$	
	$L_1L_3\nabla L_4 - L_1L_4\nabla L_3$	
	$L_3L_4\nabla L_1 - L_1L_4\nabla L_3$	

Note: These functions are not normalized.

Mixed-order vector basis functions can also be defined for general hexahedral cell shapes, by transforming standard functions defined on a cube as described in Section 9.14. The standard pth-order functions can be constructed according to

$$\bar{B}_{ijk}^{\eta} = \hat{\eta}\phi_i^{(p-1)}(\eta)\phi_j^{(p)}(\xi)\phi_k^{(p)}(\nu) \qquad i = 1, \ldots, p;\ j = 1, \ldots, p+1;\ k = 1, \ldots, p+1 \tag{9.254}$$

$$\bar{B}_{ijk}^{\xi} = \hat{\xi}\phi_i^{(p)}(\eta)\phi_j^{(p-1)}(\xi)\phi_k^{(p)}(\nu) \qquad i = 1, \ldots, p+1;\ j = 1, \ldots, p;\ k = 1, \ldots, p+1 \tag{9.255}$$

$$\bar{B}_{ijk}^{\nu} = \hat{\nu}\phi_i^{(p)}(\eta)\phi_j^{(p)}(\xi)\phi_k^{(p-1)}(\nu), \qquad i = 1, \ldots, p+1;\ j = 1, \ldots, p+1;\ k = 1, \ldots, p \tag{9.256}$$

where $\phi_j^{(N)}$ is the Nth-order Lagrangian polynomial defined in Equation (9.195), and the domain of interest spans $(-1 < \eta < 1, -1 < \xi < 1, -1 < \nu < 1)$.

The lowest order ($p = 1$) basis set consists of 12 functions exhibiting a CT/LN behavior along the edges of the standard cube. Each basis function interpolates to the tangential field at an edge of the global mesh, and the coefficient for that function is shared by all the cells adjacent to that edge to ensure tangential continuity between cells. The $p = 2$ basis set consists of 54 functions that provide a LT/QN representation along the faces of the cube, with 24 functions associated with tangential components along edges, 24 functions associated with tangential components on cell faces, and 6 functions that are confined to a single cell and provide a linearly independent normal-field component at the center of each face.

Vector basis functions of this type were first proposed by Nedelec [16]. A subroutine for calculating the 54×54 element matrix entries for these functions of order $p = 2$ when used to discretize the vector Helmholtz equation for general hexahedral cell shapes was presented by Crowley [25]. Recently, fully interpolatory vector basis functions were developed for tetrahedral and hexahedral cells [23]. These functions are a linear combination of those presented above and satisfy the Nedelec conditions. The interpolatory form is particularly advantageous for functions of order $p \geq 2$.

9.13 DIVERGENCE-CONFORMING VECTOR BASIS FUNCTIONS FOR DISCRETIZATIONS OF THE EFIE

Previous sections introduced curl-conforming vector basis functions. These functions ensure the continuity of the tangential vector component from cell to cell and are well suited for discretizing the vector Helmholtz equation. Because they do not impose normal continuity between cells, curl-conforming functions properly represent eigenfunctions belonging to the nullspace of the curl operator and therefore resolve the spurious mode problem.

In EFIE formulations, the integro-differential operator involves the divergence of the unknown surface current density or volume current density. Consequently, it is necessary to employ expansion functions that ensure a finite divergence across cell boundaries, or equivalently basis functions that maintain normal continuity between cells. To properly represent eigenfunctions from the nullspace of the divergence operator with relatively low-order expansion functions, it appears necessary to allow the current density to have a discontinuous tangential component. Basis functions having finite divergence and possibly discontinuous tangential components are known as *divergence-conforming* functions.

Consider a triangular cell in the x-y plane and an associated vector basis function expressed as a linear polynomial

$$\hat{x}(A + Bx + Cy) + \hat{y}(D + Ex + Fy) \tag{9.257}$$

The six coefficients provide six degrees of freedom in the expansion, and we are free to impose six conditions on the representation in order to generate specific basis functions. Suppose we constrain the normal component of the basis function $\bar{B}$ to equal unity at both endpoints of one edge of the triangle while simultaneously vanishing along the other two edges. The result of imposing these six constraints on (9.257) is a vector function having constant normal component along one edge and zero normal component along the other two edges. There is also a linear-tangential component along each of the edges, but there are no additional degrees of freedom to constrain the tangential component. Thus, in general, the tangential component will be discontinuous between cells. A simple vector basis can be obtained by superimposing three such functions per triangle and associating the coefficients of each with the normal component at the appropriate edge. When two triangles are adjacent to a common edge, continuity of the normal component can be maintained by assigning the negative coefficient to the basis function in the adjacent triangle. In the global model, the support of this type of vector basis function would be the two triangles associated with each edge (Figure 9.24). The basis provides a CN/LT divergence-conforming representation.

The CN/LT representation can also be obtained by discarding degrees of freedom in (9.257) associated with the curl of a quadratic vector

$$\bar{V} = \hat{z}(Gx^2 + Hxy + Iy^2) \tag{9.258}$$

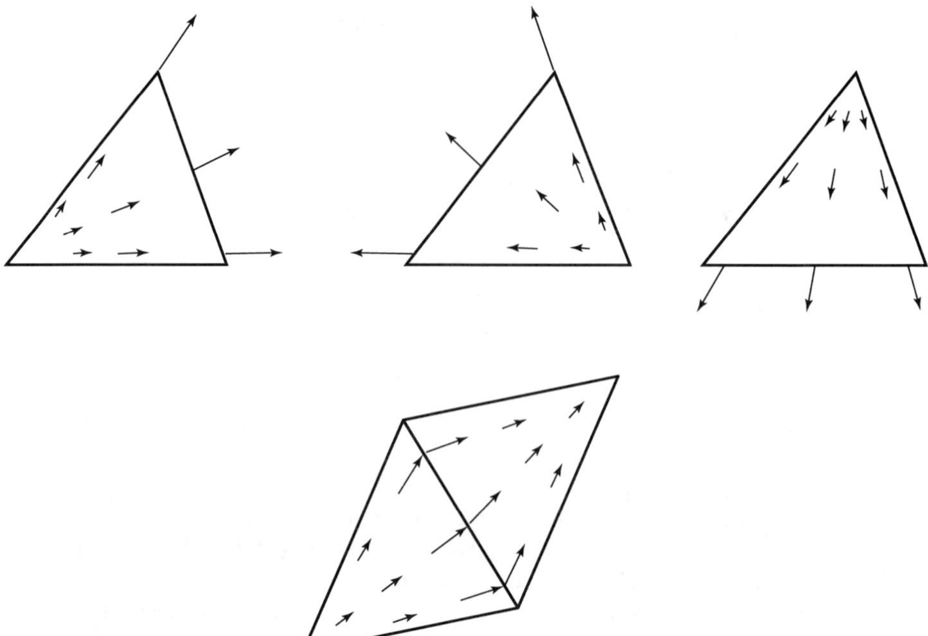

Figure 9.24 Divergence-conforming CN/LT triangular rooftop functions.

The degrees of freedom

$$\nabla \times \bar{V} = \hat{x}(Hx + 2Iy) - \hat{y}(2Gx + Hy) \tag{9.259}$$

have zero divergence and therefore can only represent nullspace eigenfunctions of the divergence operator. They also are not needed to balance terms in an equation involving the field and its divergence. Discarding these degrees of freedom reduces (9.257) to the form

$$\bar{B} = \hat{x}(A + Bx) + \hat{y}(D + By) \tag{9.260}$$

Note that the vector $\bar{B}$ defined in (9.260) is equivalent to $\hat{z} \times \bar{B}_{\text{curl}}$, where $\bar{B}_{\text{curl}}$ is the mixed-order curl-conforming basis function defined in (9.165).

Within a cell, a representation of the CN/LT functions is easily obtained using the simplex coordinates defined in Equations (9.1)–(9.7). Each basis function can be written as

$$\bar{B}_i = w_i \hat{z} \times (L_{i+2} \nabla L_{i+1} - L_{i+1} \nabla L_{i+2}) \tag{9.261}$$

where w_i is the length of edge i and edge i is located opposite from node i in the triangle. If the nodes are numbered sequentially in a counterclockwise fashion around the triangle, the basis function $\bar{B}_i$ points away from node i, as depicted in Figure 9.24.

Since

$$\nabla L_i = \frac{1}{2A}(\hat{x}b_i + \hat{y}c_i) \tag{9.262}$$

we obtain the Cartesian representation

$$\bar{B}_i = \frac{w_i}{4A^2}\{\hat{x}[a_j c_k - a_k c_j + (b_j c_k - b_k c_j)x] \\ -\hat{y}[a_j b_k - a_k b_j + (c_j b_k - c_k b_j)y]\} \tag{9.263}$$

demonstrating the equivalence of (9.260) and (9.261). Consequently,

$$\nabla \cdot \bar{B}_i = \frac{w_i}{2A^2}(b_j c_k - b_k c_j) = \pm\frac{w_i}{A} \tag{9.264}$$

and

$$\nabla \times \bar{B}_i = 0 \tag{9.265}$$

The CN/LT basis functions exhibit constant divergence and zero curl within a cell, although the curl assumes a Dirac delta behavior at the cell edges and is therefore not zero globally. As an aid in visualizing these functions, consider a cell having one node located at the origin of a polar coordinate system (ρ, ϕ). The basis function oriented away from the origin can be written

$$\bar{B}(\rho, \phi) = K\rho\hat{\rho} \tag{9.266}$$

where K is a normalization constant.

The CN/LT basis functions, sometimes known as *triangular rooftop* or *Rao–Wilton–Glisson (RWG)* functions, were proposed by Glisson in 1978 for representing the surface current density within EFIE formulations [32] and subsequently implemented for this purpose [33]. Because they are divergence conforming, these functions permit no charge accumulation at the edge between two cells. The nonphysical charge accumulation was identified as a source of difficulty in EFIE formulations (Section 2.3). Several examples illustrating the CN/LT functions appear in Chapter 10.

Nedelec presented a general family of divergence-conforming functions that include the CN/LT functions as the lowest order member [16]. In the two-dimensional case, the pth-order function can be obtained from the pth-order curl-conforming basis functions presented in Section 9.9, since within a cell

$$\bar{B}_{\text{div-conforming}} = \hat{z} \times \bar{B}_{\text{curl-conforming}} \tag{9.267}$$

As an example, LN/QT functions for triangular cells can be obtained from (9.181)–(9.183) and are depicted in Figure 9.25. The property in (9.267) can also be used to generate divergence-conforming basis functions for rectangular cells. The lowest order members of this family, sometimes known as *rooftop* functions, can be defined with an η-component of the form

$$\bar{B}(\eta, \xi) = \hat{\eta}t(\eta; \eta_3, \eta_4, \eta_5)p(\xi; \xi_1, \xi_2) \tag{9.268}$$

and a ξ-component

$$\bar{B}(\eta, \xi) = \hat{\xi}p(\eta; \eta_1, \eta_2)t(\xi; \xi_3, \xi_4, \xi_5) \tag{9.269}$$

where p and t denote the scalar pulse and subsectional triangle functions discussed in Section 5.3. The coefficients of the rooftop functions represent the vector component normal to an edge, and the expansion provides normal continuity and has a CN/LT behavior along the edges of a cell. These functions are widely used for representing surface currents on

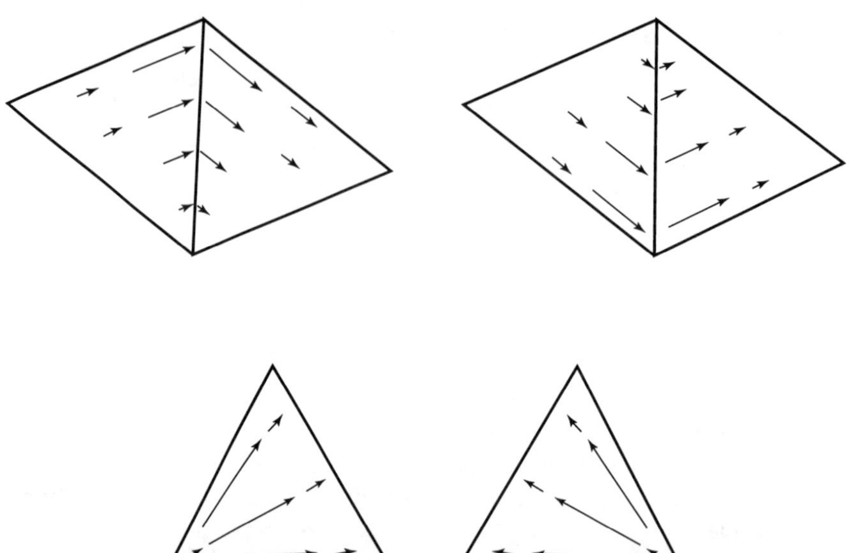

Figure 9.25 Divergence-conforming LN/QT basis functions for triangular cells. The complete set includes two per edge and two per cell.

planar scatterers and microwave devices [34, 35] and will be illustrated in Section 10.1 for expanding the surface current density on flat conducting plates.

Similar vector basis functions can be developed in three dimensions in order to represent a volume current density. Consider a tetrahedral cell and assume that within the tetrahedron the basis function is a linear polynomial

$$\bar{B}(x, y, z) = \hat{x}(A + Bx + Cy + Dz) + \hat{y}(E + Fx + Gy + Hz) + \hat{z}(I + Jx + Ky + Lz) \tag{9.270}$$

Equation (9.270) contains 12 coefficients to be determined by imposing constraints on the function. One possibility is to realize a vanishing normal component along three of the four faces and a constant normal component along the fourth face. (The normal component must be specified at the three corner nodes defining a face of the tetrahedron; thus this procedure uses all 12 degrees of freedom.) The resulting tetrahedral functions provide a CN/LT behavior and exhibit zero curl and constant divergence within a cell. The identical representation can be obtained by discarding the degrees of freedom associated with the curl of a general quadratic-polynomial vector (Prob. P9.34). By assigning one basis function per cell face throughout the mesh and sharing coefficients to ensure normal continuity across cell faces, one obtains a divergence-conforming representation. Vector functions of the divergence-conforming type are well suited for representing volume currents, since they do not permit charge accumulation at the common face between cells. The CN/LT tetrahedral rooftop functions have been used to represent the volumetric polarization current density within a three-dimensional EFIE formulation for dielectric scatterers [36]. Their simplex-coordinate description is investigated in Prob. P9.35.

Families of mixed-order divergence-conforming vector basis functions for tetrahedral and hexahedral cells were originally developed in reference [16]. The CN/LT functions are the lowest order members of these families. A systematic procedure for generating fully interpolatory basis functions of this type was recently proposed [23].

9.14 MAPPING VECTOR BASIS FUNCTIONS TO CURVILINEAR CELLS IN TWO AND THREE DIMENSIONS

The vector basis functions introduced in preceding sections can be mapped to curved cells in two-dimensional or curved surfaces or volumes in three dimensions. In the two-dimensional case, the domain can be mapped according to the scalar procedure described in Sections 9.3 and 9.4, which involves a transformation from (η, ξ)-coordinates to (x, y)-coordinates defined via Lagrangian interpolation polynomials and makes use of the Jacobian relationship in Equation (9.69). For scalar basis functions defined in (η, ξ)-coordinates, a transformation defined by this process uniquely specifies the mapped functions on the curved cells. In addition, the continuity of a scalar basis function is maintained across curved-cell boundaries, although derivative continuity is not.

When transforming vector basis functions, however, there is an additional degree of freedom embodied in the vector direction of the basis function. A local mapping that describes the curved-cell shape via Lagrangian polynomials will generally not be able to maintain the complete continuity of a vector basis function across cell boundaries. In addition, as discussed in Section 9.7, complete continuity may not be desired. Consequently, it is critical to define the vector projection in a way that provides the desired continuity properties. Suppose we have a reference cell described by (η, ξ, ν)-coordinates, and a specific mapping into (x, y, z)-space. In order to define and manipulate vector quantities within a curvilinear cell, we introduce the base vectors

$$\bar{\eta} = \frac{\partial x}{\partial \eta}\hat{x} + \frac{\partial y}{\partial \eta}\hat{y} + \frac{\partial z}{\partial \eta}\hat{z} \tag{9.271}$$

$$\bar{\xi} = \frac{\partial x}{\partial \xi}\hat{x} + \frac{\partial y}{\partial \xi}\hat{y} + \frac{\partial z}{\partial \xi}\hat{z} \tag{9.272}$$

$$\bar{\nu} = \frac{\partial x}{\partial \nu}\hat{x} + \frac{\partial y}{\partial \nu}\hat{y} + \frac{\partial z}{\partial \nu}\hat{z} \tag{9.273}$$

and the reciprocal base vectors

$$\bar{\eta}' = \frac{\partial \eta}{\partial x}\hat{x} + \frac{\partial \eta}{\partial y}\hat{y} + \frac{\partial \eta}{\partial z}\hat{z} = \nabla \eta \tag{9.274}$$

$$\bar{\xi}' = \frac{\partial \xi}{\partial x}\hat{x} + \frac{\partial \xi}{\partial y}\hat{y} + \frac{\partial \xi}{\partial z}\hat{z} = \nabla \xi \tag{9.275}$$

$$\bar{\nu}' = \frac{\partial \nu}{\partial x}\hat{x} + \frac{\partial \nu}{\partial y}\hat{y} + \frac{\partial \nu}{\partial z}\hat{z} = \nabla \nu \tag{9.276}$$

In general, neither the base vectors nor the reciprocal base vectors are mutually orthogonal within a curvilinear cell. However, they always satisfy

$$\bar{\eta} \cdot \bar{\eta}' = 1 \qquad \bar{\xi} \cdot \bar{\xi}' = 1 \qquad \bar{\nu} \cdot \bar{\nu}' = 1 \tag{9.277}$$

and

$$\bar{\eta} \cdot \bar{\xi}' = 0 \qquad \bar{\eta} \cdot \bar{\nu}' = 0 \tag{9.278}$$

and so on. Furthermore, the base and reciprocal base vectors are related by

$$\bar{\eta}' = \frac{1}{\det \mathbf{J}} \bar{\xi} \times \bar{\nu} \tag{9.279}$$

$$\bar{\xi}' = \frac{1}{\det \mathbf{J}} \bar{\nu} \times \bar{\eta} \tag{9.280}$$

$$\bar{\nu}' = \frac{1}{\det \mathbf{J}} \bar{\eta} \times \bar{\xi} \tag{9.281}$$

$$\bar{\eta} = (\det \mathbf{J}) \bar{\xi}' \times \bar{\nu}' \tag{9.282}$$

$$\bar{\xi} = (\det \mathbf{J}) \bar{\nu}' \times \bar{\eta}' \tag{9.283}$$

$$\bar{\nu} = (\det \mathbf{J}) \bar{\eta}' \times \bar{\xi}' \tag{9.284}$$

where the Jacobian matrix is defined by

$$\begin{bmatrix} \dfrac{\partial}{\partial \eta} \\ \dfrac{\partial}{\partial \xi} \\ \dfrac{\partial}{\partial \nu} \end{bmatrix} = \begin{bmatrix} \dfrac{\partial x}{\partial \eta} & \dfrac{\partial y}{\partial \eta} & \dfrac{\partial z}{\partial \eta} \\ \dfrac{\partial x}{\partial \xi} & \dfrac{\partial y}{\partial \xi} & \dfrac{\partial z}{\partial \xi} \\ \dfrac{\partial x}{\partial \nu} & \dfrac{\partial y}{\partial \nu} & \dfrac{\partial z}{\partial \nu} \end{bmatrix} \begin{bmatrix} \dfrac{\partial}{\partial x} \\ \dfrac{\partial}{\partial y} \\ \dfrac{\partial}{\partial z} \end{bmatrix} = \mathbf{J} \begin{bmatrix} \dfrac{\partial}{\partial x} \\ \dfrac{\partial}{\partial y} \\ \dfrac{\partial}{\partial z} \end{bmatrix} \tag{9.285}$$

The base vectors and reciprocal base vectors for a two-dimensional cell are depicted in Figure 9.26. As indicated, the base vectors are tangential to the cell boundaries (constant-coordinate surfaces) while the reciprocal base vectors are normal to the cell boundaries.

Returning to the two-dimensional case, consider a vector basis function defined in the reference cell ($-1 < \eta < 1, -1 < \xi < 1$), with the transformation defined so that

$$\begin{bmatrix} \dfrac{\partial}{\partial \eta} \\ \dfrac{\partial}{\partial \xi} \end{bmatrix} = \begin{bmatrix} \dfrac{\partial x}{\partial \eta} & \dfrac{\partial y}{\partial \eta} \\ \dfrac{\partial x}{\partial \xi} & \dfrac{\partial y}{\partial \xi} \end{bmatrix} \begin{bmatrix} \dfrac{\partial}{\partial x} \\ \dfrac{\partial}{\partial y} \end{bmatrix} = \mathbf{J} \begin{bmatrix} \dfrac{\partial}{\partial x} \\ \dfrac{\partial}{\partial y} \end{bmatrix} \tag{9.286}$$

This vector can be represented by its covariant components

$$\bar{B} = (\bar{B} \cdot \bar{\eta}) \bar{\eta}' + (\bar{B} \cdot \bar{\xi}) \bar{\xi}' \tag{9.287a}$$

or its contravariant components

$$\bar{B} = (\bar{B} \cdot \bar{\eta}') \bar{\eta} + (\bar{B} \cdot \bar{\xi}') \bar{\xi} \tag{9.287b}$$

When constructing a curl-conforming basis function on a curvilinear cell, the appropriate mapping is given by

$$\begin{bmatrix} B_x \\ B_y \end{bmatrix} = \mathbf{J}^{-1} \begin{bmatrix} B_\eta \\ B_\xi \end{bmatrix} \tag{9.288}$$

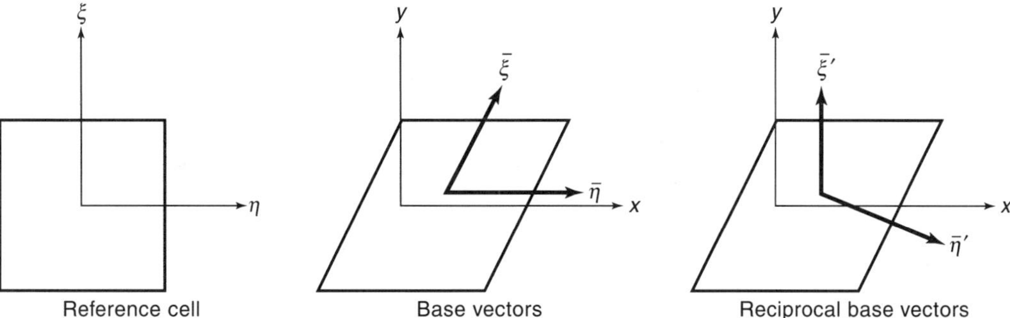

Figure 9.26 Base vectors and reciprocal base vectors for a skewed cell.

where B_η and B_ξ denote the covariant components in the reference cell. The projection defined by (9.288) is a covariant mapping [25] and maintains tangential-vector continuity between cells in the x–y space. When constructing a divergence-conforming basis function, the appropriate mapping is given by

$$\begin{bmatrix} B_x \\ B_y \end{bmatrix} = \frac{1}{\det \mathbf{J}} \mathbf{J}^T \begin{bmatrix} B_\eta \\ B_\xi \end{bmatrix} \tag{9.289}$$

where B_η and B_ξ denote the contravariant components in the reference cell. The projection defined by (9.289) is a contravariant mapping [25] and maintains normal-vector continuity across cell boundaries in the x–y space.

To demonstrate these ideas, consider an example where we map the square domain $(-1 < \eta < 1, -1 < \xi < 1)$ to the quadrilateral region defined by the points $(x = 0, y = 1)$, $(x = 2, y = 1)$, $(x = -1, y = -1)$, and $(x = 1, y = -1)$. The mapping defined by Equations (9.65) and (9.66) simplifies to

$$x = \eta + \tfrac{1}{2}\xi + \tfrac{1}{2} \tag{9.290}$$

$$y = \xi \tag{9.291}$$

and produces a Jacobian matrix

$$\mathbf{J} = \begin{bmatrix} 1 & 0 \\ \frac{1}{2} & 1 \end{bmatrix} \tag{9.292}$$

For illustration, consider the vector basis function

$$\bar{B}(\eta, \xi) = \tfrac{1}{4}(\eta + 1)(\xi + 1)\bar{\eta}' \tag{9.293}$$

defined on $(-1 < \eta < 1, -1 < \xi < 1)$. If mapped according to the covariant projection of (9.288), the Cartesian components of the resulting basis function are

$$B_x = \tfrac{1}{4}(\eta + 1)(\xi + 1) \tag{9.294}$$

$$B_y = -\tfrac{1}{8}(\eta + 1)(\xi + 1) \tag{9.295}$$

On the other hand, the same function on the reference cell can be expressed as

$$\bar{B}(\eta, \xi) = \tfrac{1}{4}(\eta + 1)(\xi + 1)\bar{\eta} \tag{9.296}$$

since the covariant and contravariant components are identical in the reference cell. Equation (9.296) can be mapped according to the contravariant projection of (9.289) to produce the Cartesian components

$$B_x = \tfrac{1}{4}(\eta + 1)(\xi + 1) \tag{9.297}$$

$$B_y = 0 \tag{9.298}$$

These two projections are sketched in Figure 9.27.

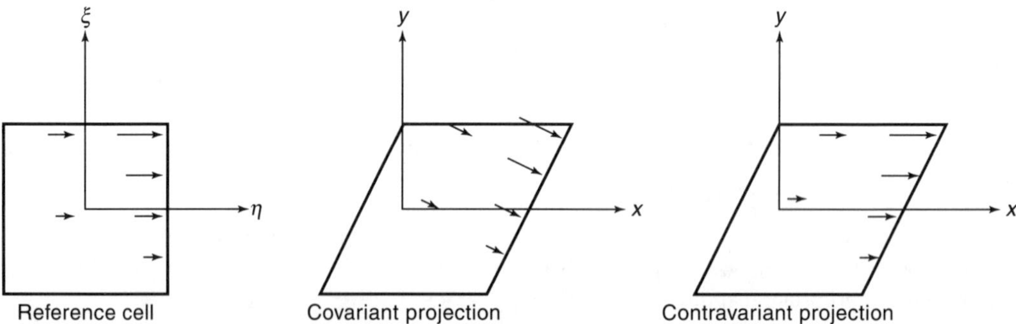

Figure 9.27 Covariant and contravariant projections of a vector function.

Consider the basis function produced by the covariant projection. Along the edge at $y = 1$ in the skewed cell, the tangential component is B_x in (9.294), which is identical to the tangential component of the original basis function along the edge $\xi = 1$. Along the edge of the skewed cell that passes through ($x = 1.5$, $y = 0$), the tangential component of the basis function is

$$\left[\frac{\hat{x} + 2\hat{y}}{\sqrt{5}} \cdot \bar{B} \right]_{\eta=1} = 0 \tag{9.299}$$

Along the other two edges, the tangential projection is also zero. Therefore, the covariant mapping does not produce a tangential component except along the cell edge where one exists in the original (η, ξ) domain. (The normal components are not preserved by the mapping, as indicated by the normal component along $y = 1$ that was not present in the reference cell.) Consequently, individual basis functions can be projected onto irregular regions using (9.288) while maintaining the tangential interpolation properties typically associated with curl-conforming functions. By considering two adjacent cells, it can be shown that the continuity of the tangential component across cell boundaries is also maintained (Prob. P9.36).

In a similar manner, the contravariant projection can be shown to preserve normal interpolation properties and normal continuity along cell boundaries. Along the edge at $y = 1$ in the skewed cell, for instance, the normal component is B_y, which is zero according to (9.298). Along the edge of the skewed cell that passes through ($x = 1.5$, $y = 0$), the normal component is given by

$$\left[\frac{2\hat{x} - \hat{y}}{\sqrt{5}} \cdot \bar{B} \right]_{\eta=1} = \frac{\xi + 1}{\sqrt{5}} = \frac{y + 1}{\sqrt{5}} \tag{9.300}$$

Along the other two edges, the normal component is zero. Therefore, the contravariant mapping has produced a vector basis function on the quadrilateral cell with no nonzero normal components except where they occurred in the reference cell. Consequently, normal-component interpolation properties are preserved by (9.289). The continuity properties of the mapping are also maintained, as demonstrated by the example in Prob. P9.37.

When working with mapped basis functions, it is convenient to perform most of the calculations directly in the (η, ξ) system. Consider the element matrix calculations necessary for the vector Helmholtz equation. Since a discretization of the vector Helmholtz equation involves curl-conforming basis functions, the mapping is performed using the covariant projection of (9.288). In the two-dimensional case, the curl operation may be evaluated according to (Prob. P9.38)

$$\hat{z} \cdot \nabla \times \bar{B} = \left(\frac{\partial B_y}{\partial x} - \frac{\partial B_x}{\partial y} \right) = \frac{1}{(\det \mathbf{J})} \left(\frac{\partial B_\xi}{\partial \eta} - \frac{\partial B_\eta}{\partial \xi} \right) \tag{9.301}$$

Therefore, element matrix entries for the two-dimensional vector Helmholtz operator can be computed directly in local coordinates via the integrals

$$\begin{aligned} A_{mn}^{(1)} &= \int_x \int_y \nabla \times \bar{B}_m \cdot \nabla \times \bar{B}_n \, dx\, dy \\ &\int_{\eta=-1}^{1} \int_{\xi=-1}^{1} \frac{1}{(\det \mathbf{J})} \left(\frac{\partial B_{m\xi}}{\partial \eta} - \frac{\partial B_{m\eta}}{\partial \xi} \right) \left(\frac{\partial B_{n\xi}}{\partial \eta} - \frac{\partial B_{n\eta}}{\partial \xi} \right) d\eta\, d\xi \end{aligned} \tag{9.302}$$

$$\begin{aligned} A_{mn}^{(2)} &= \int_x \int_y \bar{B}_m \cdot \bar{B}_n \, dx\, dy = \int_x \int_y [B_{mx} B_{my}] \begin{bmatrix} B_{nx} \\ B_{ny} \end{bmatrix} dx\, dy \\ &\int_{\eta=-1}^{1} \int_{\xi=-1}^{1} [B_{m\eta} B_{m\xi}] \mathbf{J}^{-T} \mathbf{J}^{-1} \begin{bmatrix} B_{n\eta} \\ B_{n\xi} \end{bmatrix} (\det \mathbf{J})\, d\eta\, d\xi \end{aligned} \tag{9.303}$$

where we have used the relation $dx\, dy = (\det \mathbf{J})\, d\eta\, d\xi$.

In the three-dimensional case, the curl can be expressed in the form

$$\nabla \times \bar{B} = \frac{1}{(\det \mathbf{J})} \left[\bar{\eta} \left(\frac{\partial B_\nu}{\partial \xi} - \frac{\partial B_\xi}{\partial \nu} \right) + \bar{\xi} \left(\frac{\partial B_\eta}{\partial \nu} - \frac{\partial B_\nu}{\partial \eta} \right) + \bar{\nu} \left(\frac{\partial B_\xi}{\partial \eta} - \frac{\partial B_\eta}{\partial \xi} \right) \right] \tag{9.304}$$

where the three-dimensional Jacobian matrix is defined in (9.285), the base vectors are defined in (9.271)–(9.273), and B_η, B_ξ, and B_ν denote the covariant components in the reference cell. Note that (9.304) involves the contravariant components of the curl and therefore must be transformed via the mapping

$$\begin{bmatrix} V_x \\ V_y \\ V_z \end{bmatrix} = \mathbf{J}^T \begin{bmatrix} V_\eta \\ V_\xi \\ V_\nu \end{bmatrix} \tag{9.305}$$

For notational purposes, let us introduce a column matrix $\mathbf{V}_n$ to represent the contravariant components of the curl of the nth basis function, scaled by the determinant, namely,

$$\mathbf{V}_n^T = \left[\left(\frac{\partial B_{n\nu}}{\partial \xi} - \frac{\partial B_{n\xi}}{\partial \nu} \right) \left(\frac{\partial B_{n\eta}}{\partial \nu} - \frac{\partial B_{n\nu}}{\partial \eta} \right) \left(\frac{\partial B_{n\xi}}{\partial \eta} - \frac{\partial B_{n\eta}}{\partial \xi} \right) \right] \tag{9.306}$$

The necessary element matrix entries can be computed from the integrals

$$\begin{aligned} A_{mn}^{(1)} &= \int_x \int_y \int_z \nabla \times \bar{B}_m \cdot \nabla \times \bar{B}_n \, dx\, dy\, dz \\ &= \int_{\eta=-1}^{1} \int_{\xi=-1}^{1} \int_{\nu=-1}^{1} \frac{1}{(\det \mathbf{J})} \mathbf{V}_m^T \mathbf{J} \mathbf{J}^T \mathbf{V}_n \, d\eta\, d\xi\, d\nu \end{aligned} \tag{9.307}$$

$$
\begin{aligned}
A_{mn}^{(2)} &= \int_x \int_y \int_z \bar{B}_m \cdot \bar{B}_n \, dx\, dy\, dz = \int_x \int_y \int_z [B_{mx} B_{my} B_{mz}] \begin{bmatrix} B_{nx} \\ B_{ny} \\ B_{nz} \end{bmatrix} dx\, dy\, dz \\
&= \int_{\eta=-1}^{1} \int_{\xi=-1}^{1} \int_{\nu=-1}^{1} [B_{m\eta} B_{m\xi} B_{m\nu}] \mathbf{J}^{-T} \mathbf{J}^{-1} \begin{bmatrix} B_{n\eta} \\ B_{n\xi} \\ B_{n\nu} \end{bmatrix} (\det \mathbf{J})\, d\eta\, d\xi\, d\nu
\end{aligned}
\tag{9.308}
$$

In practice, the entries of the Jacobian matrix $\mathbf{J}$ are obtained directly from the mapping, which typically specifies (x, y, z) as functions of (η, ξ, ν). The inverse Jacobian and the determinant vary with location within a cell and are usually computed numerically from $\mathbf{J}$. The required integrations can be performed by numerical quadrature.

For volume EFIE calculations, divergence-conforming basis functions are suggested. It is also convenient to perform the calculations in the reference coordinates. In the two-dimensional case, assuming that the Cartesian components of the basis function are defined by the contravariant mapping in (9.289), the divergence of the basis function is given by (Prob. P9.39)

$$
\nabla \cdot \bar{B} = \frac{\partial B_x}{\partial x} + \frac{\partial B_y}{\partial y} = \frac{1}{\det \mathbf{J}} \left(\frac{\partial B_\eta}{\partial \eta} + \frac{\partial B_\xi}{\partial \xi} \right) \tag{9.309}
$$

where B_η and B_ξ denote the contravariant components in the reference cell. In the three-dimensional case, the divergence is

$$
\nabla \cdot \bar{B} = \frac{1}{\det \mathbf{J}} \left(\frac{\partial B_\eta}{\partial \eta} + \frac{\partial B_\xi}{\partial \xi} + \frac{\partial B_\nu}{\partial \nu} \right) \tag{9.310}
$$

We may also develop expressions for the matrix entries associated with a discretization of the EFIE on a curved surface. Since divergence-conforming basis functions have the appropriate properties for representing surface currents, the mapping is performed using a contravariant projection similar to (9.289). However, in this situation the mapping to the curved surface is defined by a transformation of the form

$$
x = \sum_n x_n B_n(\eta, \xi) \tag{9.311}
$$

$$
y = \sum_n y_n B_n(\eta, \xi) \tag{9.312}
$$

$$
z = \sum_n z_n B_n(\eta, \xi) \tag{9.313}
$$

where $\{B_n\}$ represents a set of scalar Lagrangian interpolation functions. Therefore, the Jacobian relationship is given by

$$
\begin{bmatrix} \dfrac{\partial}{\partial \eta} \\ \dfrac{\partial}{\partial \xi} \end{bmatrix} = \begin{bmatrix} \dfrac{\partial x}{\partial \eta} & \dfrac{\partial y}{\partial \eta} & \dfrac{\partial z}{\partial \eta} \\ \dfrac{\partial x}{\partial \xi} & \dfrac{\partial y}{\partial \xi} & \dfrac{\partial z}{\partial \xi} \end{bmatrix} \begin{bmatrix} \dfrac{\partial}{\partial x} \\ \dfrac{\partial}{\partial y} \\ \dfrac{\partial}{\partial z} \end{bmatrix} = \mathbf{J} \begin{bmatrix} \dfrac{\partial}{\partial x} \\ \dfrac{\partial}{\partial y} \\ \dfrac{\partial}{\partial z} \end{bmatrix} \tag{9.314}
$$

The differential surface area on the curved surface can be written in (η, ξ)-coordinates using the scaling

$$
dS = Q\, d\eta\, d\xi \tag{9.315}
$$

where

$$Q = \sqrt{\left(\frac{\partial y}{\partial \eta}\frac{\partial z}{\partial \xi} - \frac{\partial z}{\partial \eta}\frac{\partial y}{\partial \xi}\right)^2 + \left(\frac{\partial z}{\partial \eta}\frac{\partial x}{\partial \xi} - \frac{\partial x}{\partial \eta}\frac{\partial z}{\partial \xi}\right)^2 + \left(\frac{\partial x}{\partial \eta}\frac{\partial y}{\partial \xi} - \frac{\partial y}{\partial \eta}\frac{\partial x}{\partial \xi}\right)^2} \tag{9.316}$$

The basis function must be tangential to the curved patch at every point and ensure normal continuity between adjacent patches. These characteristics can be obtained if the Cartesian components of the basis function are defined by the mapping

$$\begin{bmatrix} B_x \\ B_y \\ B_z \end{bmatrix} = \frac{1}{Q}\mathbf{J}^T \begin{bmatrix} B_\eta \\ B_\xi \end{bmatrix} \tag{9.317}$$

where B_η and B_ξ denote the contravariant components in the reference cell. For a basis function defined by this contravariant mapping, the surface divergence operation happens to simplify to

$$\nabla_s \cdot \bar{B} = \frac{1}{Q}\left(\frac{\partial B_\eta}{\partial \eta} + \frac{\partial B_\xi}{\partial \xi}\right) \tag{9.318}$$

Therefore, the scalar potential integral from the EFIE is given by

$$\begin{aligned} \Phi &= \frac{1}{j\omega\varepsilon}\iint (\nabla_s{}' \cdot \bar{B}_n) G \, dS' \\ &= \frac{1}{j\omega\varepsilon}\int_{\eta'=-1}^{1}\int_{\xi'=-1}^{1}\left(\frac{\partial B_\eta}{\partial \eta'} + \frac{\partial B_\xi}{\partial \xi'}\right) G \, d\eta' \, d\xi' \end{aligned} \tag{9.319}$$

where G denotes the Green's function. Since the testing functions are also defined by the transformation in (9.317), a general form for the complete matrix entry is

$$\begin{aligned} \iint \bar{T}_m \cdot \nabla\Phi \, dS &= -\iint (\nabla \cdot \bar{T}_m)\Phi \, dS \\ &= -\int_{\eta=-1}^{1}\int_{\xi=-1}^{1}\left(\frac{\partial T_{m\eta}}{\partial \eta} + \frac{\partial T_{m\xi}}{\partial \xi}\right)\Phi \, d\eta \, d\xi \end{aligned} \tag{9.320}$$

The matrix entry involving the magnetic vector potential term in the EFIE can be written in terms of the integral

$$\begin{aligned} &\iint \bar{T}_m(\bar{r}) \cdot \left(\iint \bar{B}_n(\bar{r}')G \, dS'\right) dS \\ &= \int_{\eta=-1}^{1}\int_{\xi=-1}^{1}\int_{\eta'=-1}^{1}\int_{\xi'=-1}^{1} [T_{m\eta} T_{m\xi}]\mathbf{J}\mathbf{J}^T \begin{bmatrix} B_{n\eta} \\ B_{n\xi} \end{bmatrix} G \, d\eta' \, d\xi' \, d\eta \, d\xi \end{aligned} \tag{9.321}$$

Thus, in these integrals the scale factors arising from the basis functions cancel those arising from the differential surface areas. These expressions provide a convenient way of computing the matrix entries when a piecewise-parametric representation is used to define the curved surface. All calculations can be performed in the (η, ξ)-coordinate system. An example incorporating this procedure will be considered in Chapter 10.

9.15 SUMMARY

This chapter has introduced higher order polynomial basis functions for a variety of applications. In practice, the accuracy of most numerical solutions is limited by the interpolation error associated with the expansion, and the use of higher order functions usually provides better accuracy and faster convergence.

Scalar Lagrangian basis functions for expansions on triangular and quadrilateral cells are widely used for discretizing the scalar Laplacian and scalar Helmholtz operators and have been reviewed. An isoparametric mapping of these functions to cells with curved sides permits the systematic treatment of realistic geometries without an increase in model complexity.

Three-dimensional electromagnetic applications are vector in nature, and this chapter has also introduced a number of vector basis functions for discretizing integral and differential equations. Of particular concern is the appearance of spurious nonzero eigenvalues sometimes obtained in the spectrum of the matrix for the vector Helmholtz operator. Curl-conforming vector basis functions that do not impose normal-vector continuity have been shown to eliminate spurious eigenvalues. Basis functions with this property are presented and incorporated into cavity, scattering, and waveguide formulations to demonstrate their utility. We also introduced divergence-conforming vector basis functions for integral equation formulations. Chapters 10 and 11 will illustrate the application of vector basis functions.

REFERENCES

[1] O. C. Zienkiewicz and R. L. Taylor, *The Finite Element Method*, London: McGraw-Hill, 1988.

[2] P. P. Silvester and R. L. Ferrari, *Finite Elements for Electrical Engineers*, Cambridge: Cambridge University Press, 1990.

[3] G. Strang and G. J. Fix, *An Analysis of the Finite Element Method*, Englewood Cliffs, NJ: Prentice-Hall, 1973.

[4] L. J. Bahrmasel and R. A. Whitaker, "Convergence of the finite element method as applied to electromagnetic scattering problems in the presence of inhomogeneous media," *IEEE Trans. Magnetics*, vol. 27, pp. 3845–3847, Sept. 1991.

[5] J.-C. Sabonnadiere and J.-L. Coulomb, *Finite Element Methods in CAD: Electric and Magnetic Fields*, New York: Springer-Verlag, 1987.

[6] P. Silvester, "Construction of triangular finite element universal matrices," *Int. J. Num. Methods Eng.*, vol. 12, pp. 237–244, 1978.

[7] Z. Zlatev, J. Wasniewski, and K. Schaumburg, *Y12M—Solution of Large and Sparse Systems of Linear Algebraic Equations*, Berlin: Springer-Verlag, 1981.

[8] D. S. Burnett, *Finite Element Analysis*, Reading, MA: Addison-Wesley, 1987.

[9] R. D. Graglia, "The use of parametric elements in the moment method solution of static and dynamic volume integral equations," *IEEE Trans. Antennas Propagat.*, vol. 36, pp. 636–646, May 1988.

[10] J.-M. Jin, J. L. Volakis, and V. V. Liepa, "A moment method solution of a volume-surface integral equation using isoparametric elements and point matching," *IEEE Trans. Antennas Propagat.*, vol. 37, pp. 1641–1645, Oct. 1989.

[11] A. F. Peterson and K. R. Aberegg, "Parametric mapping of vector basis functions for surface integral equation formulations," *Appl. Computat. Electromagnet. Soc. (ACES) J.*, vol. 10., pp. 107–115, Nov. 1995.

[12] W. E. Boyse, D. R. Lynch, K. D. Paulsen, and G. N. Minerbo, "Nodal based finite element modeling of Maxwell's equations," *IEEE Trans. Antennas Propagat.*, vol. 40, pp. 642–651, June 1992.

[13] B. T. Smith, J. M. Boyle, J. J. Dongarra, B. S. Garbow, Y. Ikebe, V. C. Klema, and C. B. Moler, *Matrix Eigensystem Routines—EISPACK Guide*, New York: Springer-Verlag, 1976.

[14] Z. J. Cendes, "Vector finite elements for electromagnetic field computation," *IEEE Trans. Magnetics*, vol. 27, pp. 3958–3966, Sept. 1991.

[15] D. Sun, J. Manges, X. Yuan, and Z. Cendes, "Spurious modes in finite element methods," *IEEE Antennas Propagat. Mag.*, vol. 37, no. 5, pp. 12–24, Oct. 1995.

[16] J. C. Nedelec, "Mixed finite elements in R3," *Num. Math.*, vol. 35, pp. 315–341, 1980.

[17] J. C. Nedelec, "A new family of mixed finite elements in R3," *Num. Math.*, vol. 50, pp. 57–81, 1986.

[18] G. Mur and A. T. de Hoop, "A finite element method for computing three-dimensional electromagnetic fields in inhomogeneous media," *IEEE Trans. Magnetics*, vol. MAG-21, pp. 2188–2191, Nov. 1985.

[19] G. Mur, "Edge elements, their advantages and their disadvantages," *IEEE Trans. Magnetics*, vol. 30, pp. 3552–3557, Sept. 1994.

[20] A. F. Peterson and D. R. Wilton, "Curl-conforming mixed-order edge elements for discretizing the 2D and 3D vector Helmholtz equation," in *Finite Element Software for Microwave Engineering*, eds. T. Itoh, G. Pelosi, and P. P. Silvester, New York: Wiley, 1996.

[21] A. Bossavit and J. C. Verite, "A mixed FEM-BIEM method to solve 3D eddy current problems," *IEEE Trans. Magnetics*, vol. MAG-18, pp. 431–435, Mar. 1982.

[22] M. L. Barton and Z. J. Cendes, "New vector finite elements for three-dimensional magnetic field computation," *J. Appl. Phys.*, vol. 61, pp. 3919–3921, Apr. 1987.

[23] R. D. Graglia, D. R. Wilton, and A. F. Peterson, "Higher order interpolatory vector bases for computational electromagnetics," *IEEE Trans. Antennas Propagat.*, vol. 45, pp. 329–342, Mar. 1997.

[24] J.-F. Lee, D.-K. Sun, and Z. J. Cendes, "Full-wave analysis of dielectric waveguides using tangential vector finite elements," *IEEE Trans. Microwave Theory Tech.*, vol. 39, pp. 1262–1271, Aug. 1991.

[25] C. W. Crowley, "Mixed-order covariant projection finite elements for vector fields," Ph.D. dissertation, McGill University, Montreal, Quebec, Feb. 1988.

[26] G. S. Warren and W. R. Scott, "An investigation of numerical dispersion in the vector finite element method," *IEEE Trans. Antennas Propagat.*, vol. 42, pp. 1502–1508, Nov. 1994.

[27] G. S. Warren, "The analysis of numerical dispersion in the finite element method using nodal and vector elements," Ph.D. dissertation, Georgia Institute of Technology, Atlanta, GA, 1995.

[28] J. B. Manges and Z. J. Cendes, "A generalized tree-cotree gauge for magnetic field computations," *IEEE Trans. Magnetics*, vol. 31, pp. 1342–1347, May 1995.

[29] D. R. Wilton, "Topological considerations in surface patch and volume cell modeling of electromagnetic scatterers," *Proceedings of the URSI Symposium on Electromagnetic Theory*, Santiago de Compostela (Spain), International Union of Radio Scientists (URSI), pp. 65–68, August 23–26, 1983.

[30] A. F. Peterson, "Vector finite element formulation for scattering from two-dimensional heterogeneous bodies," *IEEE Trans. Antennas Propagat.*, vol. 43, pp. 357–365, Mar. 1994.

[31] J. Parker, R. D. Ferraro, and P. C. Liewer, "A comparison of isoparametric edge elements and DnEt elements for 3D electromagnetic scattering problems," *Abstracts of the URSI Radio Science Meeting*, Dallas, TX, International Union of Radio Scientists (USRI), p. 377, May 1990.

[32] A. W. Glisson, "On the development of numerical techniques for treating arbitrarily shaped surfaces," Ph.D. dissertation, University of Mississippi, 1978.

[33] S. M. Rao, D. R. Wilton, and A. W. Glisson, "Electromagnetic scattering by surfaces of arbitrary shape," *IEEE Trans. Antennas Propagat.*, vol. AP-30, pp. 409–418, May 1982.

[34] A. W. Glisson and D. R. Wilton, "Simple and efficient numerical methods for problems of electromagnetic radiation and scattering from surfaces," *IEEE Trans. Antennas Propagat.*, vol. AP-28, pp. 593–603, Sept. 1980.

[35] J. C. Rautio and R. F. Harrington, "An electromagnetic time-harmonic analysis of shielded microstrip circuits," *IEEE Trans. Microwave Theory Tech.*, vol. MTT-35, pp. 726–730, Aug. 1987.

[36] D. H. Schaubert, D. R. Wilton, and A. W. Glisson, "A tetrahedral modeling method for electromagnetic scattering by arbitrarily shaped inhomogeneous dielectric bodies," *IEEE Trans. Antennas Propagat.*, vol. AP-32, pp. 77–85, Jan. 1984.

PROBLEMS

P9.1 Using the procedure described in Equations (9.9)–(9.14), find an expression for the 10 cubic-order Lagrangian basis functions defined within triangular cells. Provide a sketch of these functions similar to Figure 9.4.

P9.2 Using Maxwell's equations, obtain an equation that describes the proper derivative discontinuity in E_z and H_z at an interface between two two-dimensional regions with material parameters ε_1, μ_1 and ε_2, μ_2.

P9.3 Verify the formulas in (9.34) and (9.35).

P9.4 (a) Find general expressions for the element matrix entries associated with (9.27) and (9.30) for one-dimensional Lagrangian basis and testing functions of any polynomial order. In other words, outline a procedure similar to the development in Equations (9.31)–(9.41), but for the one-dimensional case.

(b) Find the explicit entries of the element matrices in part (a) for quadratic-order functions.

P9.5 Sketch the nine biquadratic Lagrangian functions in (9.56)–(9.64).

P9.6 Estimate the number of nodes and the number of nonzero entries as a function of the number of cells for the bilinear and biquadratic Lagrangian interpolation functions defined in Equations (9.47)–(9.50) and (9.56)–(9.64). Assume an idealized (infinite) mesh of rectangular cells.

P9.7 By evaluating the integrals in (9.54) and (9.55), obtain the 4×4 element matrix entries for the bilinear Lagrangian functions defined on a rectangular cell ($x_1 < x < x_2$, $y_1 < y < y_2$).

P9.8 Consider the transformation from the standard cell ($-1 < \eta < 1, -1 < \xi < 1$) to a quadrilateral region as described by (9.65)–(9.75). Write a computer program to calculate the element matrix entries defined in (9.72) and (9.75) for the bilinear Lagrangian functions in (9.47)–(9.50). Use this program to determine the element matrix entries for a region with corners at (0, 0), (1, 0), (2, 1), and (1, 1).

P9.9 Develop a computer subroutine to evaluate the integrals in (9.86) and (9.87) for quadratic Lagrangian basis functions defined on a curved triangular cell, assuming that the cell shape is defined by an isoparametric mapping using the same quadratic functions.

P9.10 Consider the mapping defined by

$$x = \sum_{i=1}^{3} x_i B_i(t)$$

where $\{B_i(t)\}$ denote the one-dimensional quadratic Lagrangian basis functions defined over the interval $-1 < t < 1$ in Equations (9.92)–(9.94) and x_2 is located a variable distance between x_1 and x_3 in the interval $x_1 < x_2 < x_3$. Show that values of x_2 outside the range

$$x_1 + \tfrac{1}{4}\Delta < x_2 < x_3 - \tfrac{1}{4}\Delta$$

where $\Delta = x_3 - x_1$ produce an unacceptable representation of the interval. Discuss the impact of this result on the placement of nodes in a one-dimensional isoparametric representation, such as that considered in Section 9.5. Can you suggest a procedure for obtaining similar guidelines with multidimensional mappings?

P9.11 Based on the two-dimensional development of the bilinear and biquadratic Lagrangian functions on a standard cell, find expressions for the 8 linear and 27 quadratic Lagrangian basis functions defined on a three-dimensional cell ($-1 < \eta < 1, -1 < \xi < 1, -1 < \upsilon < 1$).

P9.12 Expand Equation (9.108) to obtain an explicit expression with the form

$$L_i = \frac{1}{6V}(a_i + b_i x + c_i y + d_i z)$$

for the simplex coordinates (L_1, L_2, L_3, L_4) as a function of the four corners (x_i, y_i, z_i) and volume V of a tetrahedron.

P9.13 Show that mathematical functions of the form

$$\bar{V}_{mn} = \nabla\left[\sin\left(\frac{m\pi x}{a}\right)\sin\left(\frac{n\pi y}{b}\right)\right]$$

satisfy the boundary condition $\hat{n} \times \bar{V}_{mn} = 0$ along the four walls of a two-dimensional cavity ($0 < x < a, 0 < y < b$). Therefore, they constitute nullspace eigenfunctions of the two-dimensional vector Helmholtz equation $\nabla \times \nabla \times \bar{E} = k^2\bar{E}$ for a homogeneous cavity with electric walls. Compare the nullspace eigenfunctions with the usual TE cavity modes. How do they differ? Calculate the divergence of both eigenfunction families.

P9.14 Consider Equation (9.152) as though the function was defined within the four square cells surrounding the origin (Figure 9.14). A bilinear Lagrangian basis set defined in these four cells consists of nine functions interpolating to the x-components and nine basis functions interpolating to the y-components at cell corners. Construct the projection of (9.152) onto the bilinear expansion by equating the x- and y-components at the nine nodes. To obtain a continuous representation at the origin, where (9.152) is discontinuous, use the average value (zero). Sketch the resulting function. Compare the curl of (9.152) and the curl of the Lagrangian approximation.

P9.15 Using the property $\nabla \cdot \bar{E} = 0$, the vector Helmholtz equation $\nabla \times \nabla \times \bar{E} = k^2 \bar{E}$ for a homogeneous region can be cast into the form $\nabla^2 \bar{E} + k^2 \bar{E} = 0$. Consider the two-dimensional cavity described in Prob. P9.13. Are the functions $\bar{V}_{mn}$ defined in Prob. P9.13 solutions of $\nabla^2 \bar{E} + k^2 \bar{E} = 0$? If so, find their eigenvalues. Discuss the implication of your results.

P9.16 By considering the geometry in Figure 9.28, show that the basis function $w_3 L_2 \nabla L_1$ from the set defined in (9.154) has a linear tangential projection along edge 12 with value zero at node 1 and one at node 2.

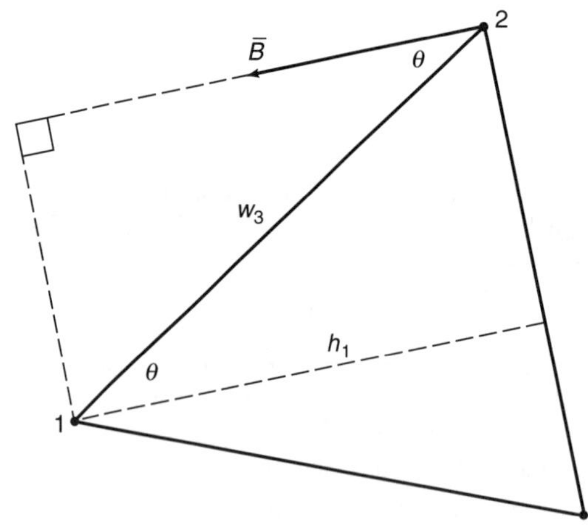

Figure 9.28 Geometry to be used for Prob. P9.16.

P9.17 (a) Sketch the 12 vector basis functions defined in Equations (9.156)–(9.158), to produce a figure similar to Figure 9.15.
(b) Show that the normalization used in (9.156) and (9.157) produces unit tangential values at the appropriate interpolation points.

P9.18 By integrating the linear vector basis functions defined in (9.154) in simplex coordinates, obtain the entries of the 6 × 6 element matrices for (9.162) and (9.163) evaluated over a single triangular cell. For instance, if

$$\bar{B}_{ij} = w_{ij} L_i \nabla L_j$$

use Equations (9.4)–(9.7) to show that

$$\iint \nabla \times \bar{B}_{ij} \cdot \nabla \times \bar{B}_{mn}\, dx\, dy = \frac{w_{ij} w_{mn}}{16A^3}(b_i c_j - c_i b_j)(b_m c_n - c_m b_n)$$

and

$$\iint \bar{B}_{ij} \cdot \bar{B}_{mn}\, dx\, dy = \frac{w_{ij} w_{mn}}{2A}(b_j b_n + c_j c_n) \begin{cases} \frac{1}{12} & i = m \\ \frac{1}{24} & \text{otherwise} \end{cases}$$

where w_{ij} is the length of the edge between nodes i and j and A is the area of the cell.

P9.19 Consider the CT/LN basis function $\bar{B} = w_3(L_2 \nabla L_1 - L_1 \nabla L_2)$ and the triangular cell depicted in Figure 9.28. Show that the tangential component of $\bar{B}$ has unit value along edge 12 and vanishes along the other two edges of the cell. How does the normal component behave along edge 12?

P9.20 Show that the basis functions in (9.182) and (9.183) satisfy Equations (9.177)–(9.180).

P9.21 A vector representation equivalent to the LT/QN functions defined in (9.181)–(9.183) can be constructed numerically for a given triangular cell in such a manner that all 8 functions are interpolatory at specific locations along the cell edges. Beginning with the complete quadratic polynomial in Equation (9.155), identify 12 conditions that must be imposed to construct each interpolatory basis function in the form of (9.175).

P9.22 The first nine QT/CuN basis functions in Table 9.9 interpolate to the tangential component at the endpoint and middle of each edge of the triangular cell. Develop a normalization of these functions that results in each having a unity-tangential component at the interpolation point.

P9.23 Construct an alternate set of LT/QN basis functions for a standard rectangular cell with the property that the eight functions that interpolate to the tangential values along the edges are entirely linear (like the LT/QN functions developed for triangular cells in Section 9.9).

P9.24 Chapter 3 summarized the development of the Bayliss–Turkel type of local RBC. By applying a similar procedure to (9.203), obtain the first- and second-order conditions presented in (9.204) and (9.205). In addition, show that (9.205) can be rewritten as (9.206).

P9.25 Using an eigenfunction expansion of the exterior fields and a development that parallels that given in Section 3.3, show that an exact RBC can be expressed in the form

$$(\hat{\rho} \times \nabla \times \bar{E}^s)(a, \phi) = \hat{\phi} \frac{1}{2\pi} \int_0^{2\pi} E_\phi^s(a, \phi') K(\phi - \phi')\, d\phi'$$

where

$$K(\phi - \phi') = k \sum_{n=-\infty}^{\infty} \frac{H_n^{(2)}(ka)}{H_n^{(2)\prime}(ka)} e^{jn(\phi - \phi')}$$

and a is the radius of the circular radiation boundary. Discuss the implementation of this RBC in connection with the TE scattering procedure of Section 9.10.

P9.26 By modifying the development in Section 3.10, obtain an integral equation RBC that can be used with the two-dimensional scattering formulation in Section 9.10 in order to treat arbitrary boundary shapes.

P9.27 (a) Consider an idealized infinite mesh of equilateral triangles arranged in a hexagonal pattern. Determine the number of unknowns per cell associated with the CT/LN and LT/QN representations introduced in Section 9.9. How do these numbers compare with the estimates obtained in Section 9.2 for the linear and quadratic Lagrangian representations for a scalar quantity?

(b) Repeat part (a) to determine the number of nonzero matrix entries per cell for the CT/LN and LT/QN basis functions. How do these estimates compare with those for the scalar basis functions shown in Figure 9.6?

(c) Combine the estimates of parts (a) and (b) to obtain an estimate of the number of nonzero matrix entries as a function of the number of unknowns for the CT/LN and LT/QN functions. Obtain a similar estimate for the linear and quadratic scalar Lagrangian functions on a triangular-cell mesh.

(d) The two interior LT/QN functions only interact with functions within their cell and in a deterministic problem can be eliminated from the global system of equations prior to matrix construction. Repeat the estimates in parts (a)–(c) without including the interior LT/QN functions.

P9.28 Obtain the entries of the 8×8 element matrices for Equations (9.162)–(9.163) and the LT/QN basis functions defined on triangular cells. (*Hint:* Part of the result is given in Prob. P9.18.)

P9.29 The scattering formulation discussed in Section 9.10 requires the calculation of boundary integrals

$$\int_{\partial\Gamma} \bar{B}_m \cdot \bar{B}_n \, dt$$

and

$$\int_{\partial\Gamma} \frac{\partial \bar{B}_m}{\partial t} \cdot \frac{\partial \bar{B}_n}{\partial t} \, dt$$

Show that the first integral can be evaluated for LT/QN functions to produce

$$\int_{\text{edge } m} \bar{B}_m^+ \cdot \bar{B}_m^+ \, dt = \int_{\text{edge } m} \bar{B}_m^- \cdot \bar{B}_m^- \, dt = \tfrac{1}{3} w_m$$

$$\int_{\text{edge } m} \bar{B}_m^+ \cdot \bar{B}_m^- \, dt = \tfrac{1}{6} w_m$$

where w_m denotes the length of edge m, $\bar{B}_m^+$ denotes a basis function whose tangential value increases in magnitude in the positive ϕ direction along edge m, and $\bar{B}_m^-$ denotes a basis function whose tangential value decreases in magnitude in the positive ϕ direction. In addition, by employing an approximation similar to that described in Figure 9.20, show that the nonzero contributions from the second integral are

$$\int_{\text{edge } m} \frac{\partial \bar{B}_m^+}{\partial t} \cdot \frac{\bar{B}_n^+}{\partial t} \, dt = \int_{\text{edge } m} \frac{\partial \bar{B}_m^-}{\partial t} \cdot \frac{\bar{B}_n^-}{\partial t} \, dt \cong \begin{cases} \dfrac{-1}{w_{m-1} + w_m} & n = m - 1 \\[2ex] \dfrac{1}{w_m} & n = m \\[2ex] \dfrac{-1}{w_m + w_{m+1}} & n = m + 1 \end{cases}$$

and

$$\int_{\text{edge } m} \frac{\partial \bar{B}_m^+}{\partial t} \cdot \frac{\bar{B}_n^+}{\partial t} \, dt \cong \frac{1}{w_m + w_{m-1}} - \frac{1}{w_m} + \frac{1}{w_m + w_{m+1}}$$

where w_{m-1}, w_m, and w_{m+1} denote the length of three adjacent edges ordered so that the index increases in the ϕ direction.

P9.30 By rearranging terms in Equation (9.233), construct a matrix eigenvalue equation with β^2 as the eigenvalue.

P9.31 Develop a formulation similar to that given in Section 9.11 for dielectric-loaded waveguides that uses the electric field as the primary unknown. Identify appropriate boundary conditions for p.e.c. walls and discuss the impact of these on the boundary integrals appearing in the E-field versions of (9.228) and (9.229).

P9.32 Develop expressions for the 6×6 element matrices associated with the CT/LN basis functions and the integrals

$$\iiint \nabla \times \bar{B}_m \cdot \nabla \times \bar{B}_n$$

$$\iiint \bar{B}_m \cdot \bar{B}_n$$

defined over a tetrahedral cell.

P9.33 Consider an idealized cubical cell mesh in three dimensions and the LT/QN basis functions described in Section 9.12. Estimate the number of unknowns per cell, the number of nonzero entries per cell, and the number of nonzero entries per unknown for this representation under the assumptions that (a) the entire set of 54 basis functions per cell is retained in the global system and (b) the 6 basis functions that are entirely local in each cell are eliminated prior to constructing the global system, leaving 48 basis functions per cell.

P9.34 (a) Show that a three-dimensional vector $\bar{Q}$ whose components are polynomials of exactly quadratic order has curl with the general form

$$\nabla \times \bar{Q} = \hat{x}(Ax + By + Cz) + \hat{y}(Dx + Ey + Fz) + \hat{z}[Gx + Hy - (A + E)z]$$

and therefore embodies exactly eight degrees of freedom.

(b) By eliminating these eight degrees of freedom from the complete linear vector in Equation (9.270), obtain the Cartesian representation

$$\bar{B}_{\text{div}} = \hat{x}(\alpha + \delta x) + \hat{y}(\beta + \delta y) + \hat{z}(\gamma + \delta z)$$

(c) Confirm that $\bar{B}_{\text{div}}$ provides a CN/LT behavior within a tetrahedral cell.

P9.35 Consider the function

$$L_i \nabla L_j \times \nabla L_k + L_j \nabla L_k \times \nabla L_i + L_k \nabla L_i \times \nabla L_j \qquad i \neq j \neq k$$

defined within a tetrahedral domain, in terms of simplex coordinates (L_1, L_2, L_3, L_4). Does this function satisfy the conditions for the CN/LT expansion described in Section 9.13?

P9.36 Verify that the mapping in (9.288) maintains tangential continuity between cells for the specific cell pair given by

Cell A: corners at $(-1, -1)$, $(1, -1)$, $(2, 1)$, $(0, 1)$

Cell B: corners at $(1, -1)$, $(3, -2)$, $(3, 1)$, $(2, 1)$

Determine the Jacobian matrix for each cell, and consider the basis function

$$\bar{B}_A(\eta, \xi) = \frac{(1+\eta)(1+\xi)}{4}\bar{\xi}'$$

in cell A and

$$\bar{B}_B(\eta, \xi) = \frac{(1-\eta)(1+\xi)}{4}\bar{\xi}'$$

in cell B. Show that the mapped functions $\bar{B}_A(x, y)$ and $\bar{B}_B(x, y)$ have the same tangential component along the common edge.

P9.37 Verify that the mapping in (9.289) maintains normal continuity between cells for the specific cell pair given in Prob P9.36 and the basis functions

$$\bar{B}_A(\eta, \xi) = \frac{(1+\eta)(1+\xi)}{4}\bar{\eta}$$

in cell A and

$$\bar{B}_B(\eta, \xi) = \frac{(1-\eta)(1+\xi)}{4}\bar{\eta}$$

in cell B. Show that the mapped functions $\bar{B}_A(x, y)$ and $\bar{B}_B(x, y)$ have the same normal component along the common edge.

P9.38 Verify (9.301) by direct calculation. *Hint:*

$$\bar{B} = \left(\frac{\partial \eta}{\partial x}B_\eta + \frac{\partial \xi}{\partial x}B_\xi\right)\hat{x} + \left(\frac{\partial \eta}{\partial y}B_\eta + \frac{\partial \xi}{\partial y}B_\xi\right)\hat{y}$$

$$\frac{\partial}{\partial x} = \frac{\partial \eta}{\partial x}\frac{\partial}{\partial \eta} + \frac{\partial \xi}{\partial x}\frac{\partial}{\partial \xi}$$

$$\frac{\partial}{\partial y} = \frac{\partial \eta}{\partial y}\frac{\partial}{\partial \eta} + \frac{\partial \xi}{\partial y}\frac{\partial}{\partial \xi}$$

Use $\mathbf{J}\mathbf{J}^{-1} = \mathbf{I}$ to obtain relations such as

$$\frac{\partial \eta}{\partial x}\frac{\partial x}{\partial \eta} + \frac{\partial \xi}{\partial y}\frac{\partial y}{\partial \xi} = 1$$

in order to simplify the expression.

P9.39 Verify (9.309) by direct calculation using

$$\bar{B} = \frac{1}{\det \mathbf{J}}\left(\frac{\partial x}{\partial \eta}B_\eta + \frac{\partial x}{\partial \xi}B_\xi\right)\hat{x} + \frac{1}{\det \mathbf{J}}\left(\frac{\partial y}{\partial \eta}B_\eta + \frac{\partial y}{\partial \xi}B_\xi\right)\hat{y}$$

and ideas similar to those suggested in Prob. P9.38.

10

Integral Equation Methods for Three-Dimensional Bodies

While formulations for the interaction of electromagnetic fields with three-dimensional bodies are generally similar to the two-dimensional approaches of previous chapters, there are several differences. Obvious differences are that the zero-order Hankel function used as a two-dimensional Green's function is replaced by the three-dimensional Green's function $e^{-jkr}/4\pi r$ and the dimensionality of the integrals increases. Other differences that impact the implementation of these methods include the proportionally larger amount of data required to describe a three-dimensional scatterer model and the rapid growth of the number of unknowns with three-dimensional scatterer size. Three-dimensional scatterer models are also more difficult to generate and visualize than their two-dimensional counterparts. Finally, and perhaps most significantly, three-dimensional problems are inherently vector in nature.

This chapter begins by considering EFIE formulations applied to flat conducting plates and objects of arbitrary shape. Approaches using rectangular-cell models and triangular-cell models are used to illustrate vector basis functions of the CN/LT and LN/QT variety (Chapter 9). The triangular-cell EFIE formulation is extended to a combined-field (CFIE) formulation for closed conducting bodies in Sections 10.4 and 10.5 and to general homogeneous dielectric objects in Section 10.7. Section 10.6 considers the treatment of electrically small scatterers, where a conventional EFIE fails. Section 10.8 summarizes an approach for treating wires and wire-grid models, while Section 10.9 considers planar periodic structures. The analysis of microstrip antennas and scatterers requires a different Green's function to account for the substrate material (Section 10.10). Finally, Section 10.11 summarizes several volume integral formulations for three-dimensional heterogeneous scatterers.

10.1 SCATTERING FROM FLAT PERFECTLY CONDUCTING PLATES: EFIE DISCRETIZED WITH CN/LT ROOFTOP BASIS FUNCTIONS DEFINED ON RECTANGULAR CELLS

Consider one or more infinitesimally thin flat conducting plates located in the $z = 0$ plane and illuminated by an electromagnetic wave. The perfectly conducting material may be replaced by equivalent electric currents radiating in free space, where the equivalent currents represent the superposition of the current densities on both sides of the plate. In anticipation of a rectangular-cell model, components J_x and J_y are selected as the primary unknowns to be determined. The EFIE can also be separated into $\hat{x}$ and $\hat{y}$ components to produce the coupled equations

$$E_x^{\text{inc}}(x, y) = -\hat{x} \cdot \frac{\nabla\nabla \cdot + k^2}{j\omega\varepsilon_0} \bar{A} \tag{10.1}$$

$$E_y^{\text{inc}}(x, y) = -\hat{y} \cdot \frac{\nabla\nabla \cdot + k^2}{j\omega\varepsilon_0} \bar{A} \tag{10.2}$$

where the magnetic vector potential is

$$\bar{A}(x, y) = \iint [\hat{x} J_x(x', y') + \hat{y} J_y(x', y')] \frac{e^{-jkR}}{4\pi R} \, dx' \, dy' \tag{10.3}$$

and

$$R = \sqrt{(x - x')^2 + (y - y')^2} \tag{10.4}$$

Equations (10.1) and (10.2) are only valid on the location of the original plate(s).

Initially, we restrict the plate size and shape to a geometry that can be represented by equal-size rectangular cells, as illustrated in Figure 10.1, and approximate the current density with a superposition of the CN/LT "rooftop" basis functions defined in Section 9.13 and illustrated in Figure 10.2. Each rooftop function spans two adjacent cells on the plate and is centered at the edge between the two cells (Figure 10.3). Rooftop basis functions for the $\hat{x}$-component of the current density vary linearly in the x direction and

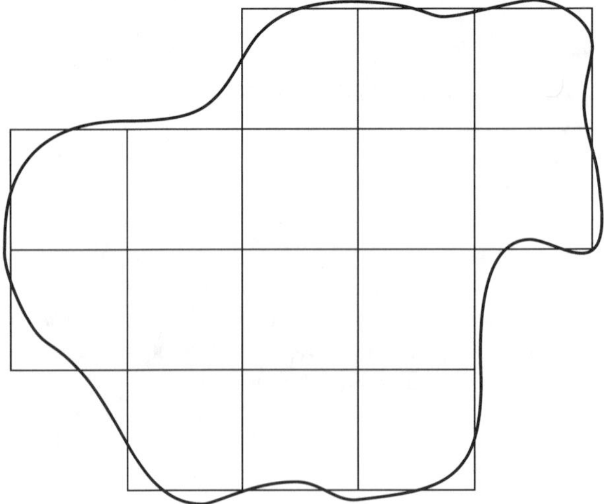

Figure 10.1 Rectangular-cell representation of an arbitrarily shaped plate.

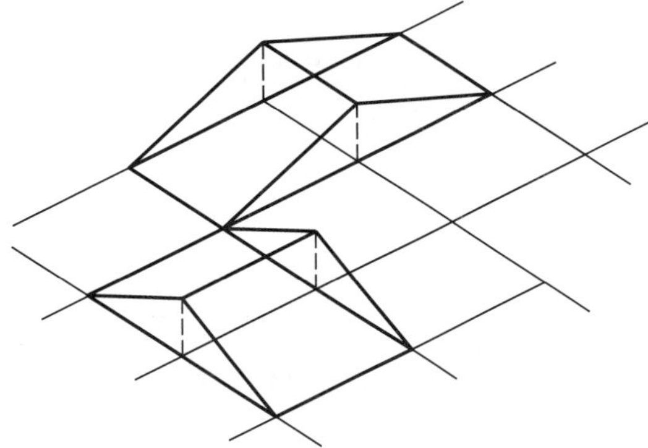

Figure 10.2 The CN/LT rooftop basis function defined on a rectangular cell.

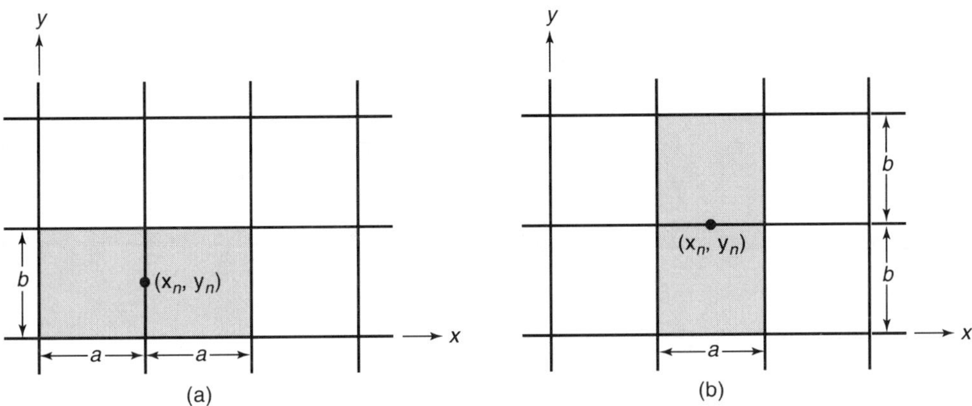

Figure 10.3 Location of rooftop basis functions within the rectangular-cell mesh.

are piecewise constant in the y direction, and vice versa. Using the fundamental triangle and pulse functions (Figure 5.1) and assuming that all cells in the model have dimensions $a \times b$, the rooftop functions can be written as

$$B_{xn}(x, y) = t(x; x_n - a, x_n, x_n + a)p(y; y_n - \tfrac{1}{2}b, y_n + \tfrac{1}{2}b) \tag{10.5}$$

$$B_{yn}(x, y) = t(y; y_n - b, y_n, y_n + b)p(x; x_n - \tfrac{1}{2}a, x_n + \tfrac{1}{2}a) \tag{10.6}$$

where (x_n, y_n) denotes the center of edge n. Using these basis functions, the current density is represented as

$$J_x(x, y) \cong \sum_{n=1}^{M} j_{xn} B_{xn}(x, y) \tag{10.7}$$

$$J_y(x, y) \cong \sum_{n=M+1}^{N} j_{yn} B_{yn}(x, y) \tag{10.8}$$

where the plate model consists of M interior cell edges associated with the $\hat{x}$-component of the current and $N - M$ interior cell edges associated with J_y. No basis functions are assigned to cell edges along the plate boundary, since the superimposed currents must vanish at the boundary.

In order to enforce the EFIE, we employ the "razor blade" testing functions depicted in Figure 10.4, which are defined as

$$T_{xm} = p(x; x_m - \tfrac{1}{2}a, x_m + \tfrac{1}{2}a)\delta(y - y_m) \tag{10.9}$$

$$T_{ym} = p(y; y_m - \tfrac{1}{2}b, y_m + \tfrac{1}{2}b)\delta(x - x_m) \tag{10.10}$$

and are spatially centered at the same location as the basis functions. The $\hat{x}$ and $\hat{y}$ components of the integral equation are tested with $T_x(x, y)$ and $T_y(x, y)$, respectively, in order to provide an additional degree of differentiability in directions where it is needed to absorb the derivatives in (10.1) and (10.2). This combination of basis and testing functions was suggested by Glisson and Wilton [1]. The testing process produces the $N \times N$ discrete system

$$\begin{bmatrix} \mathbf{A} & \mathbf{B} \\ \mathbf{C} & \mathbf{D} \end{bmatrix} \begin{bmatrix} \mathbf{j}_x \\ \mathbf{j}_y \end{bmatrix} = \begin{bmatrix} \mathbf{e}_x \\ \mathbf{e}_y \end{bmatrix} \tag{10.11}$$

Because the testing functions are even functions about $(x - x_m, y - y_m)$, the entries of the matrix can be expressed in convolution notation (Chapters 1 and 5) as

$$A_{mn} = \frac{-1}{j\omega\varepsilon_0} T_{xm} * \left(\frac{\partial^2}{\partial x^2} + k^2 \right) (B_{xn} * G) \tag{10.12}$$

$$B_{mn} = \frac{-1}{j\omega\varepsilon_0} T_{xm} * \left(\frac{\partial^2}{\partial x \partial y} \right) (B_{yn} * G) \tag{10.13}$$

$$C_{mn} = \frac{-1}{j\omega\varepsilon_0} T_{ym} * \left(\frac{\partial^2}{\partial y \partial x} \right) (B_{xn} * G) \tag{10.14}$$

$$D_{mn} = \frac{-1}{j\omega\varepsilon_0} T_{ym} * \left(\frac{\partial^2}{\partial y^2} + k^2 \right) (B_{yn} * G) \tag{10.15}$$

where the asterisk denotes two-dimensional convolution

$$U * V = \iint U(x', y') V(x - x', y - y') \, dx' \, dy' \tag{10.16}$$

The operations on the right-hand sides of (10.12)–(10.15) produce functions of x and y that are to be evaluated at $x = 0$ and $y = 0$. The Green's function appearing in these expressions is

$$G(x, y) = \frac{e^{-jk\sqrt{x^2+y^2}}}{4\pi\sqrt{x^2 + y^2}} \tag{10.17}$$

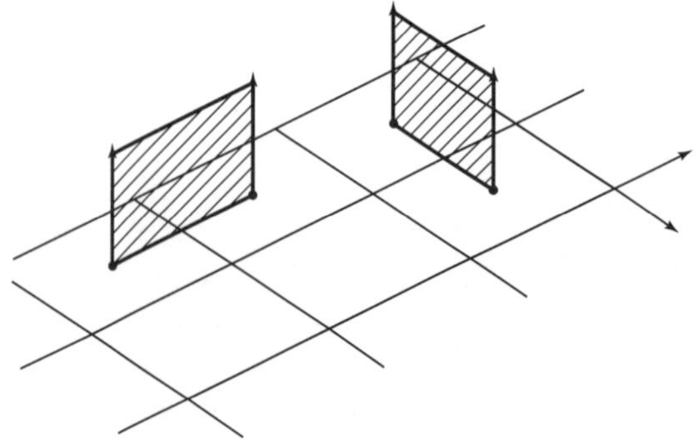

Figure 10.4 Razor-blade testing functions.

The right-hand side of the matrix equation contains entries

$$e_{xm} = \int_{x_m - a/2}^{x_m + a/2} E_x^{\text{inc}}(x, y_m, 0)\, dx \tag{10.18}$$

$$e_{ym} = \int_{y_m - b/2}^{y_m + b/2} E_y^{\text{inc}}(x_m, y, 0)\, dy \tag{10.19}$$

Because the geometry under consideration only involves basis and testing functions oriented along the x- or y-axis in the $z = 0$ plane, properties of the convolution operator can be exploited to enhance the efficiency of the impedance matrix calculations and reduce the number of required integrations. Specifically, since differentiation and convolution operators commute, Equations (10.12)–(10.15) are equivalent to

$$A_{mn} = \frac{-1}{j\omega\varepsilon_0} \left[\left(\frac{\partial^2}{\partial x^2} + k^2 \right) (T_{xm} * B_{xn}) \right] * G \tag{10.20}$$

$$B_{mn} = \frac{-1}{j\omega\varepsilon_0} \left[\frac{\partial^2}{\partial x \partial y} (T_{xm} * B_{yn}) \right] * G \tag{10.21}$$

$$C_{mn} = \frac{-1}{j\omega\varepsilon_0} \left[\frac{\partial^2}{\partial x \partial y} (T_{ym} * B_{xn}) \right] * G \tag{10.22}$$

$$D_{mn} = \frac{-1}{j\omega\varepsilon_0} \left[\left(\frac{\partial^2}{\partial y^2} + k^2 \right) (T_{ym} * B_{yn}) \right] * G \tag{10.23}$$

where the functions are evaluated at $x = 0$ and $y = 0$ after the final convolution. The basis and testing functions are simple subsectional polynomials, and both the inner convolutions in (10.20)–(10.23) and the indicated differentiations are easily evaluated. The convolutions produce

$$T_{xm} * B_{xn} = aq(x; x_m - x_n - \tfrac{3}{2}a, x_m - x_n + \tfrac{3}{2}a) p(y; y_m - y_n - \tfrac{1}{2}b, y_m - y_n + \tfrac{1}{2}b) \tag{10.24}$$

$$T_{xm} * B_{yn} = at(x; x_m - x_n - a, x_m - x_n, x_m - x_n + a) t(y; y_m - y_n - b, y_m - y_n, y_m - y_n + b) \tag{10.25}$$

$$T_{ym} * B_{xn} = bt(x; x_m - x_n - a, x_m - x_n, x_m - x_n + a) t(y; y_m - y_n - b, y_m - y_n, y_m - y_n + b) \tag{10.26}$$

$$T_{ym} * B_{yn} = bp(x; x_m - x_n - \tfrac{1}{2}a, x_m - x_n + \tfrac{1}{2}a) q(y; y_m - y_n - \tfrac{3}{2}b, y_m - y_n + \tfrac{3}{2}b) \tag{10.27}$$

where the function q denotes a quadratic spline

$$q\left(x; -\frac{3a}{2}, \frac{3a}{2}\right) = \begin{cases} 0 & x < -\dfrac{3a}{2} \\ \dfrac{9}{8} + \dfrac{3x}{2a} + \dfrac{x^2}{2a^2} & -\dfrac{3a}{2} < x < -\dfrac{a}{2} \\ \dfrac{3}{4} - \dfrac{x^2}{a^2} & -\dfrac{a}{2} < x < \dfrac{a}{2} \\ \dfrac{9}{8} - \dfrac{3x}{2a} + \dfrac{x^2}{2a^2} & \dfrac{a}{2} < x < \dfrac{3a}{2} \\ 0 & x > \dfrac{3a}{2} \end{cases} \tag{10.28}$$

After carrying out the required differentiations, the expressions simplify to

$$\left(\frac{\partial^2}{\partial x^2}+k^2\right)(T_{xm}*B_{xn})=\left\{\frac{1}{a}p\left(x;-\frac{3a}{2},-\frac{a}{2}\right)-\frac{2}{a}p\left(x;-\frac{a}{2},\frac{a}{2}\right)\right.$$
$$\left.+\frac{1}{a}p\left(x;\frac{a}{2},\frac{3a}{2}\right)+k^2aq\left(x;-\frac{3a}{2},\frac{3a}{2}\right)\right\}\bigg|_{x=x-x_m+x_n}$$
$$p\left(y;y_m-y_n-\frac{b}{2},y_m-y_n+\frac{b}{2}\right) \tag{10.29}$$

$$\frac{\partial^2}{\partial x\partial y}(T_{xm}*B_{yn})=\frac{1}{b}\{p(x;-a,0)p(y;-b,0)-p(x;-a,0)p(y;0,b)$$
$$-p(x;0,a)p(y;-b,0)+p(x;0,a)p(y;0,b)\}|_{x=x-x_m+x_n,y=y-y_m+y_n} \tag{10.30}$$

$$\frac{\partial^2}{\partial x\partial y}(T_{ym}*B_{xn})=\frac{1}{a}\{p(x;-a,0)p(y;-b,0)-p(x;-a,0)p(y;0,b)$$
$$-p(x;0,a)p(y;-b,0)+p(x;0,a)p(y;0,b)\}|_{x=x-x_m+x_n,y=y-y_m+y_n} \tag{10.31}$$

$$\left(\frac{\partial^2}{\partial y^2}+k^2\right)(T_{ym}*B_{yn})=p\left(x;x_m-x_n-\frac{a}{2},x_m-x_n+\frac{a}{2}\right)$$
$$\left\{\frac{1}{b}p\left(y;-\frac{3b}{2},-\frac{b}{2}\right)-\frac{2}{b}p\left(y;-\frac{b}{2},\frac{b}{2}\right)+\frac{1}{b}p\left(y;-\frac{b}{2},-\frac{3b}{2}\right)\right. \tag{10.32}$$
$$\left.+k^2bq\left(y;-\frac{3b}{2},\frac{3b}{2}\right)\right\}\bigg|_{y=y-y_m+y_n}$$

Finally, the entries of the impedance matrix can be obtained after convolving each of (10.29)–(10.32) with the Green's function of (10.17). To ensure high accuracy, these convolutions must be done by two-dimensional numerical quadrature. Only two types of integrals appear, those involving pulse functions and those involving the quadratic spline function. By exploiting the convolution properties, the dimension of the final integral is reduced to 2 without approximation.

The quadrature process is straightforward except in those cases where the source and observation regions overlap, resulting in a singularity in the integrand. This singularity has the form $1/R$ and can be extracted and integrated analytically, using

$$\int_{-\alpha}^{\alpha}dx'\int_{-\beta}^{\beta}dy'\frac{1}{\sqrt{(x-x')^2+(y-y')^2}}$$
$$=(x+\alpha)\ln\left(\frac{y+\beta+R_{22}}{y-\beta+R_{21}}\right)+(x-\alpha)\ln\left(\frac{y-\beta+R_{11}}{y+\beta+R_{12}}\right) \tag{10.33}$$
$$+(y+\beta)\ln\left(\frac{x+\alpha+R_{22}}{x-\alpha+R_{12}}\right)+(y-\beta)\ln\left(\frac{x-\alpha+R_{11}}{x+\alpha+R_{21}}\right)$$

where

$$R_{11}=\sqrt{(x-\alpha)^2+(y-\beta)^2} \tag{10.34}$$
$$R_{12}=\sqrt{(x-\alpha)^2+(y+\beta)^2} \tag{10.35}$$
$$R_{21}=\sqrt{(x+\alpha)^2+(y-\beta)^2} \tag{10.36}$$
$$R_{22}=\sqrt{(x+\alpha)^2+(y+\beta)^2} \tag{10.37}$$

Once the current density is determined from the solution of Equation (10.11), the bistatic scattering cross section can be calculated according to the approach outlined in Chapter 1. For the plate geometry, assuming that the incident electric field has unity amplitude and the cell dimensions are smaller than $\lambda/10$, the scattering cross section can be approximated as

$$\sigma(\theta,\phi) \cong \frac{k^2\eta^2}{4\pi}\left|\sum_{n=1}^{M} j_{xn}\cos\phi\cos\theta\Psi_n(\theta,\phi) + \sum_{n=M+1}^{N} j_{yn}\sin\phi\cos\theta\Psi_n(\theta,\phi)\right|^2 + \frac{k^2\eta^2}{4\pi}\left|\sum_{n=1}^{M} -j_{xn}\sin\phi\Psi_n(\theta,\phi) + \sum_{n=M+1}^{N} j_{yn}\cos\phi\Psi_n(\theta,\phi)\right|^2 \tag{10.38}$$

where

$$\Psi_n(\theta,\phi) = abe^{jk\sin\theta(x_n\cos\phi + y_n\sin\phi)} \tag{10.39}$$

Although the use of equal-sized rectangular cells is somewhat restrictive, this approach has considerable computational advantages when treating plates that are on the order of a wavelength or greater in size. Consider a plate geometry conforming to the parallelogram-lattice illustrated in Figure 4.12. Because the translational symmetry in the plate geometry is preserved by the use of equal-sized cells arranged along a lattice, the submatrices **A**, **B**, **C**, and **D** in Equation (10.11) are *block Toeplitz* in form, that is, **A** has the structure

$$\mathbf{A} = \begin{bmatrix} \mathbf{A}_0 & \mathbf{A}_1 & \cdots & \mathbf{A}_{P-1} \\ \mathbf{A}_1 & \mathbf{A}_0 & \cdots & \mathbf{A}_{P-2} \\ \cdot & \cdot & & \cdot \\ \cdot & \cdot & \ddots & \cdot \\ \cdot & \cdot & & \cdot \\ \mathbf{A}_{P-1} & \mathbf{A}_{P-2} & \cdots & \mathbf{A}_0 \end{bmatrix} \tag{10.40}$$

while each entry of (10.40) is itself a $Q \times Q$ Toeplitz matrix of the form

$$\mathbf{A}_i = \begin{bmatrix} a_0 & a_1 & \cdots & a_{Q-1} \\ a_1 & a_0 & \cdots & a_{Q-2} \\ \cdot & \cdot & & \cdot \\ \cdot & \cdot & \ddots & \cdot \\ \cdot & \cdot & & \cdot \\ a_{Q-1} & a_{Q-2} & \cdots & a_0 \end{bmatrix} \tag{10.41}$$

Even if the plate geometry does not fill the entire lattice, there is a considerable amount of structure built into the matrix equation by the choice of equal-sized basis and testing functions. The redundancy can be exploited in two ways. First, only one row of each of the **A**, **B**, **C**, and **D** submatrices needs to be computed, as the other entries can be found from symmetry considerations. Since the two-dimensional numerical integration used to compute each matrix element is very time consuming, the computational savings are substantial. Second, due to the Toeplitz structure, the normal computer storage requirements can be greatly reduced during the solution process. This can be accomplished either by the use of specialized block Toeplitz algorithms [2] or by the use of an iterative algorithm (i.e., the CG–FFT approach discussed in Chapter 4). For illustration, Figure 10.5 presents the surface current density induced on a square plate of side dimension 2.23λ. This result was obtained using 1984 rooftop basis functions and a CG-FFT solution of Equation (10.11) [3].

To reduce the interpolation error associated with the basis expansion, the CN/LT rooftop basis functions used to represent the current density can be replaced with higher order functions, such as functions that provide a linear normal and quadratic tangential

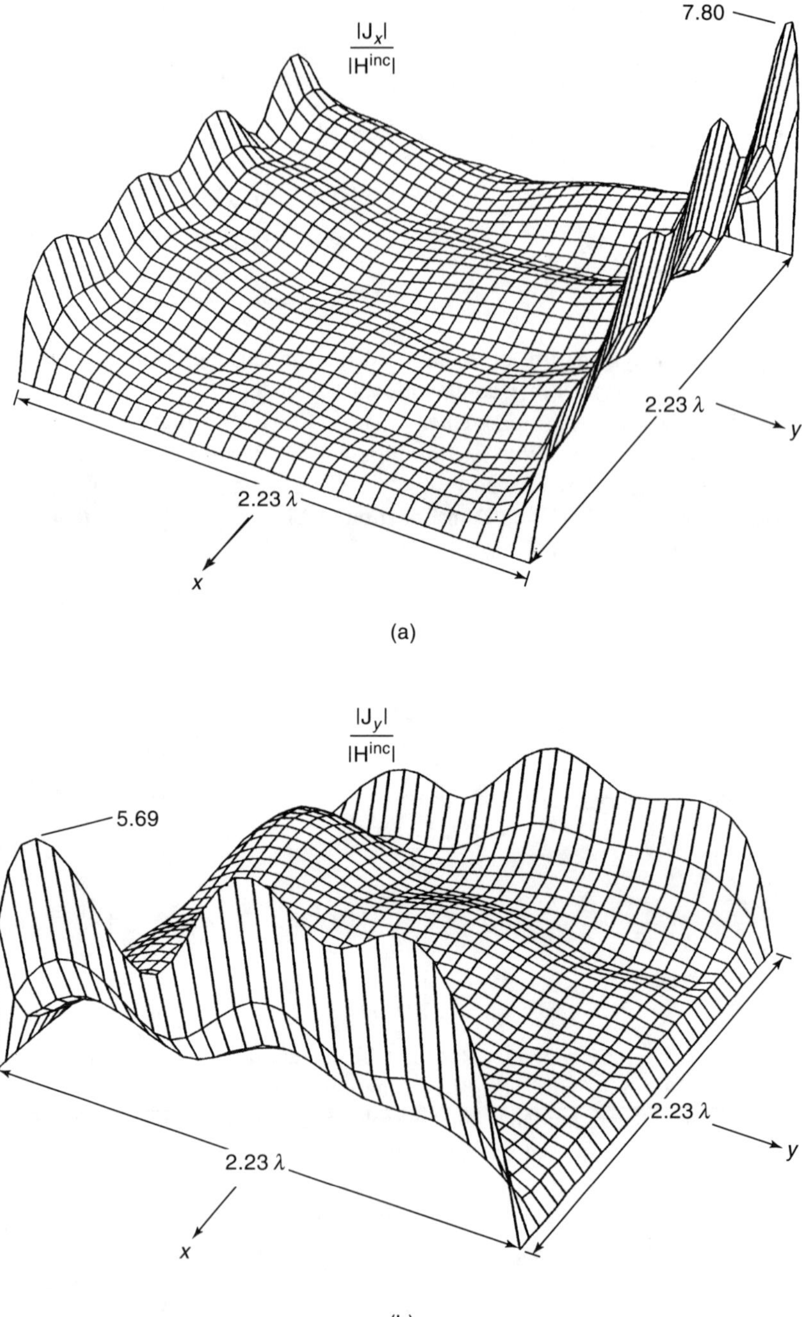

Figure 10.5 The EFIE result for the surface current density on a square plate. The top view shows the x-component; the lower view shows the y-component. A plane wave with the electric field parallel to the plate is incident from the viewing angle. A total of 1984 CN/LT rooftop basis functions are used. After [3].

(LN/QT) behavior. The LN/QT functions within a rectangular cell consist of 12 functions (Section 9.13). For convenience, they can be defined on the reference cell ($-1 < \eta < 1, -1 < \xi < 1$) and mapped to general cell shapes by the procedure of Section 9.14. On the standard cell, four of these functions interpolate to the η-directed current along the cell edges at $\xi = -1$ and $\xi = 1$ and have the form

$$\hat{\eta}\left[\tfrac{1}{2}(1 \pm \xi)\tfrac{1}{2}(1 \pm \eta)\right] \tag{10.42}$$

Similarly, there are four basis functions representing the ξ-component:

$$\hat{\xi}\left[\tfrac{1}{2}(1 \pm \xi)\tfrac{1}{2}(1 \pm \eta)\right] \tag{10.43}$$

These eight functions each straddle two cells, with the normal component constrained to be continuous across the appropriate edge to ensure the absence of fictitious charge densities and maintain a finite surface divergence. There are four additional functions with support confined to the cell, which can be expressed as

$$\hat{n}\left[\tfrac{1}{2}(1 \pm \xi)\right](1 + \eta)(1 - \eta) \tag{10.44}$$

$$\hat{\xi}(1 + \xi)(1 - \xi)\left[\tfrac{1}{2}(1 \pm \eta)\right] \tag{10.45}$$

These four functions contribute a quadratic tangential component along the cell edges and provide zero normal component. Therefore, they do not interfere with the interpolation properties of the functions defined in Equations (10.42) and (10.43). From a global perspective, there are two basis functions per non–boundary edge and two basis functions per cell throughout the plate model.

Razor-blade testing functions can be used to enforce the equations arising from the LN/QT expansion and can be defined with their domain of support as depicted in Figure 10.6. To illustrate the comparison of the CN/LT and LN/QT expansions, Figures 10.7 and 10.8 show the current density induced on a $1\lambda \times 1\lambda$ plate by a uniform plane wave [4]. Both expansions appear to produce similar results for $N = 480$. In fact, for the plate geometry, the accuracy seems to be limited primarily by the need to incorporate the proper edge singularity into the representation. Since neither CN/LT nor LN/QT basis functions provide a tangential current with the proper square root singularity [5]

$$J_{\tan} \approx O\left(\frac{1}{\sqrt{u}}\right) \quad \text{as } u \to 0 \tag{10.46}$$

near the plate's edge, it is difficult to judge relative accuracy.

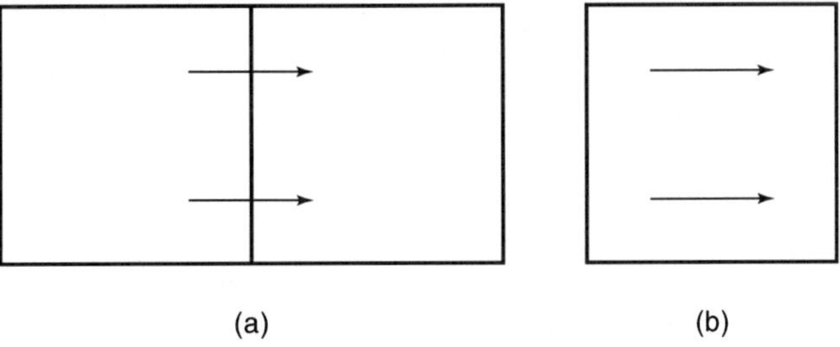

Figure 10.6 Domain of support for the razor-blade testing functions used with the LN/QT basis functions on a standard square cell: (a) paths for the two edge-based functions; (b) paths for two of the four cell-based functions.

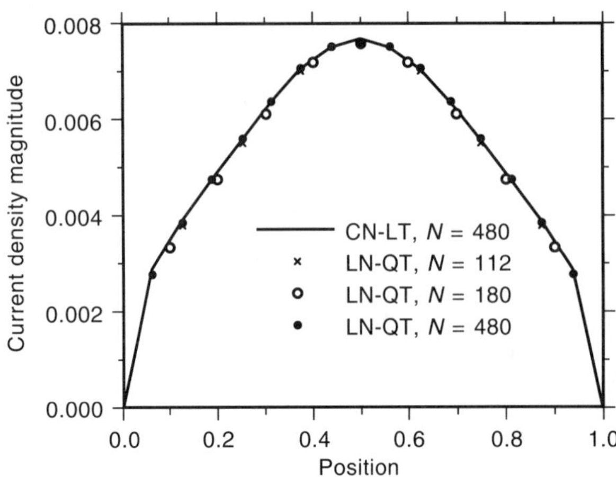

Figure 10.7 Comparison of the copolarized current density tangential to a line through the center of a $1\lambda \times 1\lambda$ plate. The EFIE result obtained using 480 CN/LT basis functions is compared with results obtained using 112, 180, and 480 LN/QT basis functions. After [4]. ©1996 American Geophysical Union.

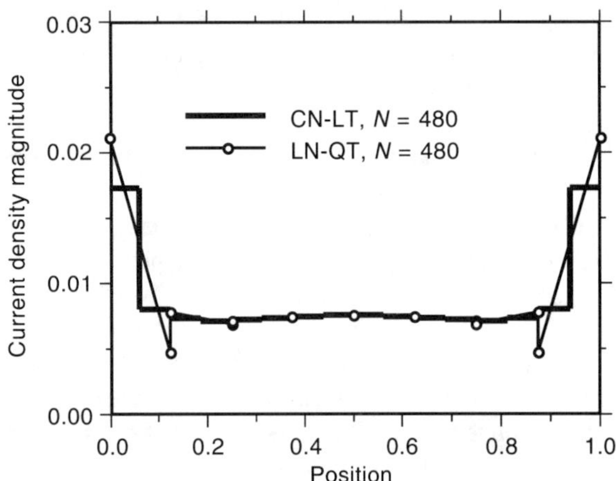

Figure 10.8 Comparison of the copolarized current density normal to a line through the center of a $1\lambda \times 1\lambda$ plate for 480 CN/LT and 480 LN/QT functions. After [4]. ©1996 American Geophysical Union.

Both the CN/LT and LN/QT functions provide a discontinuous representation of the current density, since they only impose continuity of the normal component between cells. For a regular discretization of the plate geometry, such as used to obtain the data shown in Figures 10.7 and 10.8, the tangential discontinuities tend to be small except near the plate edge, where the erratic behavior of the numerical solution is expected due to the incorrect edge singularity [5]. For irregular discretizations, the numerical solution is likely to contain noticeable discontinuities in the tangential current component at cell boundaries. Of course, these discontinuities should disappear as the cell dimensions are reduced.

The preceding EFIE formulation illustrates several special properties arising in the analysis of planar structures, including the use of convolutions to improve the efficiency of the matrix fill and discrete convolutions to improve the matrix solution efficiency. These properties have been exploited when modeling planar devices such as microstrip antennas and feed structures [6, 7] and frequency-selective surfaces [8]. However, for nonplanar objects greater flexibility is provided through the use of triangular-cell shapes. The following section considers a triangular-cell EFIE approach for scattering from plates and arbitrary conducting bodies.

10.2 SCATTERING FROM PERFECTLY CONDUCTING BODIES: EFIE DISCRETIZED WITH CN/LT TRIANGULAR-CELL ROOFTOP BASIS FUNCTIONS [9]

For scatterers of arbitrary shape, triangular cells offer the possibility of a self-consistent three-dimensional model with variable cell sizes (Figure 10.9). Consider a generalization of the previous formulation using the tangential projection of the mixed-potential EFIE

$$\bar{E}^{\text{inc}}(u, v) = jk\eta\bar{A}(u, v) + \nabla\Phi_e \tag{10.47}$$

where

$$\bar{A}(u, v) = \iint \bar{J}(u', v')\frac{e^{-jkR}}{4\pi R}\, du'\, dv' \tag{10.48}$$

$$\Phi_e(u, v) = \frac{1}{\varepsilon_0}\iint \rho_e(u', v')\frac{e^{-jkR}}{4\pi R}\, du'\, dv' \tag{10.49}$$

$$R = \sqrt{[x(u, v) - x(u', v')]^2 + [y(u, v) - y(u', v')]^2 + [z(u, v) - z(u', v')]^2} \tag{10.50}$$

and (u, v) denote parametric coordinates along the surface of the conducting scatterer. The surface charge density is related to the current by the continuity equation

$$\rho_e = \frac{1}{j\omega}\nabla_s \cdot \bar{J} \tag{10.51}$$

The mixed-potential form of the EFIE proves advantageous for general-shaped scatterers, since it facilitates the explicit transfer of one derivative to the basis functions for $\bar{J}$ and another to the testing functions.

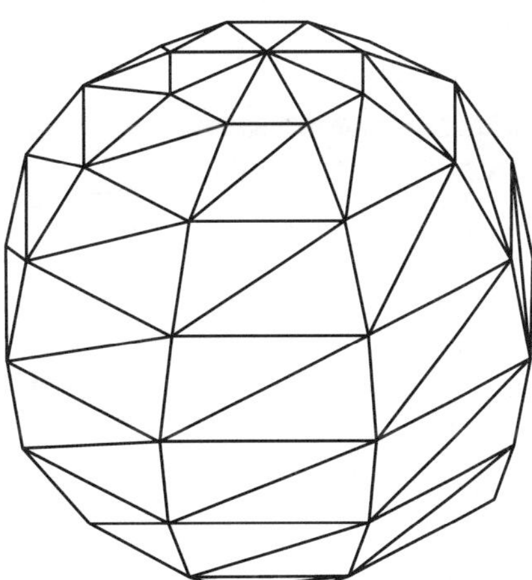

Figure 10.9 Triangular-patch model of a sphere.

A triangular-cell representation of the surface current density can be obtained using the *triangular-rooftop* CN/LT vector basis function introduced in Section 9.13 (Figure 9.24). Each rooftop function spans two adjacent triangles and interpolates to the vector component of the current density normal to the common edge. For a scatterer model with N edges, the current density can be expanded in N basis functions

$$\bar{J}(u, v) \cong \sum_{n=1}^{N} j_n \bar{B}_n(u, v) \tag{10.52}$$

where $\{j_n\}$ represents the unknown coefficients. Each coefficient can be interpreted as the current flowing across a particular edge. In addition to imposing continuity of the normal component along the common edge of the cell pair, each basis function also has a vanishing normal component along the other four edges. Thus, a triangular-rooftop function eliminates jump discontinuities in $\bar{J}$ that produce fictitious charges at cell edges [9]. In other words, because the normal component of the current density is continuous along the surface, the surface divergence of the expansion is always finite. Consider basis function $\bar{B}_n(u, v)$ associated with edge n, spanning cells i and j, with the vector direction from cell i into cell j. From Equation (9.264), we obtain

$$\nabla_s \cdot \bar{B}_n = \frac{w_n}{A_i} p_i(u, v) - \frac{w_n}{A_j} p_j(u, v) \tag{10.53}$$

where w_n is the length of edge n, A_i is the area of cell i, and $p_i(u, v)$ denotes a pulse function with support confined to cell i. Thus, the surface charge density associated with a rooftop basis function is a piecewise-constant charge doublet.

To discretize the EFIE, we seek testing functions that (1) provide the proper differentiability requirements and (2) are compatible with the triangular-cell representation. Ideally, the testing functions should also be centered at cell edges. The razor-blade functions used in the preceding section can be generalized to meet these requirements. Suppose cells i and j share edge m. The domain of a razor-blade function $\bar{T}_m(u, v)$ can be restricted to the straight-line path along the scatterer surface from the centroid of cell i to the midpoint of edge m and then along a second straight-line path to the centroid of cell j (Figure 10.10). Each testing function is a vector tangential to the path, oriented in a direction specified by the scatterer model. (In practice, the model includes a connectivity array linking edge indices to the two adjacent cell indices and simultaneously specifying the vector direction of the basis and testing function defined at each edge.)

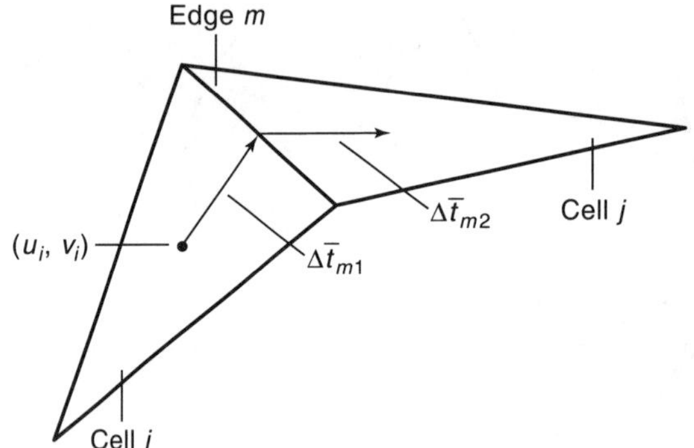

Figure 10.10 Domain of the razor-blade testing function associated with edge m in a triangular-patch model of a surface.

After substituting basis and testing functions into the EFIE, we obtain the matrix equation $\mathbf{Aj} = \mathbf{b}$, with entries

$$\begin{aligned} A_{mn} = {} & jk\eta \int_{C_m} \left(\iint \bar{B}_n(u', v') \frac{e^{-jkR}}{4\pi R} \, du' \, dv' \right) \cdot d\bar{t} \\ & + \frac{\eta}{jk} \int_{C_m} \nabla \left(\iint [\nabla_s \cdot \bar{B}_n] \frac{e^{-jkR}}{4\pi R} \, du' \, dv' \right) \cdot d\bar{t} \end{aligned} \tag{10.54}$$

$$b_m = \int_{C_m} \bar{E}^{\text{inc}} \cdot d\bar{t} \tag{10.55}$$

where C_m denotes the particular path from (u_i, v_i) to (u_j, v_j) associated with the testing function at edge m (Figure 10.10). In (10.54) and throughout this chapter, the divergence $[\nabla_s \cdot \bar{B}_n]$ is carried out in primed coordinates.

The second integral in (10.54) can be simplified using the identity

$$\int_{C_m} \nabla\Phi \cdot d\bar{t} = \Phi(u_j, v_j) - \Phi(u_i, v_i) \tag{10.56}$$

to eliminate the integral along C_m and the gradient operation. The first integral in (10.54) can also be simplified using the approximation

$$\int_{C_m} \bar{A} \cdot d\bar{t} \cong \bar{A} \cdot \Delta\bar{t}_{m1}\Big|_{u_i,v_i} + \bar{A} \cdot \Delta\bar{t}_{m2}\Big|_{u_j,v_j} \tag{10.57}$$

to eliminate the integral along C_m. In (10.57), $\Delta\bar{t}_{m1}$ denotes the vector from the centroid of cell i to the center of edge m, and $\Delta\bar{t}_{m2}$ denotes the vector from the center of edge m to the centroid of cell j, where i and j are the two cells associated with edge m (Figure 10.10). With these simplifications, the matrix entries can be written as

$$\begin{aligned} A_{mn} \cong jk\eta \Bigg(\Delta\bar{t}_{m1} \cdot \iint \bar{B}_n(u', v') \frac{e^{-jkR}}{4\pi R} \, du' \, dv' \Big|_{u_i,v_i} \\ + \Delta\bar{t}_{m2} \cdot \iint \bar{B}_n(u', v') \frac{e^{-jkR}}{4\pi R} \, du' \, dv' \Big|_{u_j,v_j} \Bigg) \\ + \frac{\eta}{jk} \Bigg(\iint [\nabla_s \cdot \bar{B}_n] \frac{e^{-jkR}}{4\pi R} \, du' \, dv' \Big|_{u_j,v_j} \\ - \iint [\nabla_s \cdot \bar{B}_n] \frac{e^{-jkR}}{4\pi R} \, du' \, dv' \Big|_{u_i,v_i} \Bigg) \end{aligned} \tag{10.58}$$

where (u_i, v_i) denotes the centroid of cell i. The integrals in (10.58) are expressed over the entire scatterer surface, but they obviously collapse to the support of the two cells associated with the basis function at edge n. In general, these integrals must be evaluated by numerical quadrature. When the source and observation regions overlap, the $1/R$ singularity may be extracted and integrated analytically, in a manner similar to that described in Section 10.1. The procedure has been detailed in recent publications [10, 11].

The scatterer model required to implement the above procedure consists of a list of the coordinates (x, y, z) of the corner of each cell (the *nodes*) and several pointer arrays specifying the connectivity between cells, edges, and nodes. Specifically, the direct evaluation of A_{mn} requires a pointer to identify the two cells adjacent to each edge where a basis function resides (and the orientation of that basis function) as well as a second pointer to identify the three nodes associated with each cell. Other needed parameters such as the centroid coordinates and the vectors $\Delta\bar{t}_{m1}$ and $\Delta\bar{t}_{m2}$ can be obtained from the given information.

The entry A_{mn} in (10.58) involves four integrals over each of the two cells adjacent to edge n, with observer locations fixed at the centroids of the two cells adjacent to edge m. By considering the possible combinations of source and observer locations, it is easily concluded that many of these individual integrals are repeated within the matrix entries for basis and testing functions at nearby edges. In fact, the independent calculation of each A_{mn} requires approximately nine times the computation that would be necessary if each integral was performed only once [9]. To minimize the time-consuming numerical quadrature, each integral can be evaluated once for each cell and observer location and added to the appropriate entry in the **A**-matrix using the connectivity information provided by the scatterer model. (This indirect way of constructing the matrix has been described in detail in connection with the finite-element procedure in Chapter 3.) This approach may require a scatterer model with slightly different connectivity arrays than those required for the naive calculation of each A_{mn}.

To illustrate the approach, Figure 10.11 shows the surface current density induced on a one-wavelength plate using the triangular-cell EFIE formulation with CN/LT basis functions. For comparison, results from a rectangular-cell CN/LT approach similar to that presented in Section 10.1 are also shown. Other results demonstrating the validity of the triangular-cell approach for a variety of scatterer shapes may be found in the literature [8, 9]. We leave the development of expressions for the scattering cross section to the readers.

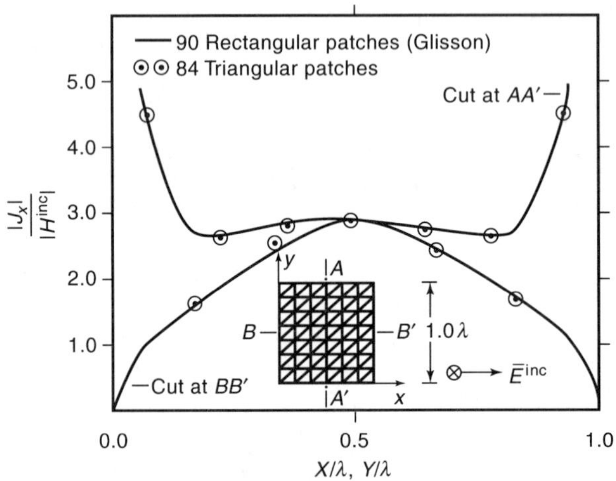

Figure 10.11 The EFIE result for the dominant current component on a 1λ × 1λ plate, obtained with triangular-cell CN/LT functions and rectangular-cell CN/LT functions. After [9]. ©1982 IEEE.

In practice, the substantial computational requirements of three-dimensional analysis motivates a reduction in the number of unknowns by any available means. Often, geometrical symmetries such as reflection symmetry or rotational symmetry can be exploited. Reference [12] provides an overview of a number of common geometrical symmetries that can be used to enhance the efficiency of solution.

10.3 SCATTERING FROM PERFECTLY CONDUCTING BODIES: MFIE DISCRETIZED WITH TRIANGULAR-CELL CN/LT BASIS FUNCTIONS

For closed p.e.c. bodies, a solution can also be obtained using the MFIE

$$\hat{n} \times \bar{H}^{\text{inc}} = \bar{J} - \hat{n} \times \nabla \times \bar{A}\big|_{S+} \tag{10.59}$$

where $\hat{n}$ denotes the outward normal vector, $\bar{A}$ is the magnetic vector potential

$$\bar{A}(u, v) = \iint \bar{J}(u', v') \frac{e^{-jkR}}{4\pi R}\, du'\, dv' \tag{10.60}$$

R is given by

$$R = \sqrt{[x(u, v) - x(u', v')]^2 + [y(u, v) - y(u', v')]^2 + [z(u, v) - z(u', v')]^2} \tag{10.61}$$

and (u, v) are parametric variables defined along the scatterer surface. The S^+ in (10.59) is a reminder that the MFIE is to be evaluated an infinitesimal distance outside the surface.

Suppose that the scatterer surface is discretized into triangular cells and the CN/LT triangular-rooftop functions employed in Section 10.2 are used to represent the current, namely,

$$\bar{J}(u, v) \cong \sum_{n=1}^{N} j_n \bar{B}_n(u, v) \tag{10.62}$$

Since the rooftop functions interpolate to the normal vector component at each edge, the coefficient j_n can be interpreted as the surface current density flowing across edge n.

A simple method-of-moments discretization can be obtained by point matching Equation (10.59) at the center of each edge in the vector direction given by the razor-blade functions used in Section 10.2. Such a testing procedure involves the components of $\bar{H}^{\text{inc}}$ and $\bar{H}^s$ parallel to each edge and the component of $\bar{J}$ perpendicular to each edge. These quantities are all continuous and well defined at the edges. This approach produces the matrix equation $\mathbf{Cj} = \mathbf{d}$, where

$$\begin{aligned} C_{mn} &= -\hat{e}_m \cdot \nabla \times \iint \bar{B}_n(u', v') \frac{e^{-jkR}}{4\pi R} \, du' \, dv' \bigg|_{u_m, v_m} \\ &= \hat{e}_m \cdot \iint \bar{B}_n(u', v') \times \nabla \left(\frac{e^{-jkR}}{4\pi R} \right) du' \, dv' \bigg|_{u_m, v_m} \qquad m \neq n \end{aligned} \tag{10.63}$$

$$\begin{aligned} C_{mm} &= 1 - \lim_{u, v \to u_m^+, v_m^+} \left(\hat{e}_m \cdot \nabla \times \iint \bar{B}_m(u', v') \frac{e^{-jkR}}{4\pi R} \, du' \, dv' \right) \\ &= \frac{2\pi - \Omega_m}{2\pi} \end{aligned} \tag{10.64}$$

and

$$d_m = \hat{e}_m \cdot \bar{H}^{\text{inc}} \Big|_{u_m, v_m} \tag{10.65}$$

where Ω_m denotes the interior angle subtended by the cell pair at edge m and $\hat{e}_m$ denotes a unit vector parallel to edge m and oriented so that $\hat{n} \times \hat{e}_m$ points in the same direction as the basis function $\bar{B}_m$.

The numerical evaluation of (10.63) is straightforward for flat-cell models since the integrand is never singular. [In the case of basis and testing functions located at different edges of the same cell, the contribution to (10.63) vanishes due to the property that there is no tangential magnetic field produced in the same plane as the source current.] The scatterer model can be described by a list of nodes and several connectivity arrays, as discussed in Section 10.2. It may be convenient to include the direction of each vector $\hat{e}_m$ within the model; otherwise some mechanism must be provided to specify the outward normal direction for every cell.

10.4 SCATTERING FROM PERFECTLY CONDUCTING BODIES: CFIE DISCRETIZED WITH TRIANGULAR-CELL CN/LT BASIS FUNCTIONS

To circumvent internal resonance difficulties (Chapter 6) that might occur with closed three-dimensional scatterers, the EFIE formulation from Section 10.2 and the MFIE approach from Section 10.3 can be brought together to produce a CFIE formulation. In common with the two-dimensional CFIE, the three-dimensional CFIE is a simple linear combination of the EFIE and MFIE. The simplest approach is to work directly with the matrix entries for the EFIE and MFIE presented in the two preceding sections. The CFIE approach produces the $N \times N$ system $\mathbf{Ej} = \mathbf{f}$, where

$$E_{mn} = \alpha A_{mn} + (1 - \alpha)\eta \, \Delta t_m C_{mn} \tag{10.66}$$

and

$$f_m = \alpha b_m + (1 - \alpha)\eta \, \Delta t_m \, d_m \tag{10.67}$$

In these equations, η denotes the intrinsic impedance of the background medium and Δt_m is the length of path C_m associated with a razor-blade testing function from Section 10.2. These factors serve to scale the numerical entries of the EFIE and MFIE to obtain similar values regardless of the cell sizes. Matrix entries A_{mn}, b_m, C_{mn}, and d_m are defined in Equations (10.58), (10.55), and (10.63)–(10.65). The parameter α is a variable in the range $0 < \alpha < 1$ that can be used to adjust the relative weighting of the EFIE and MFIE as explained in Chapter 6.

Alternative approaches for eliminating the internal resonances (such as the combined-source and the dual-surface formulations introduced in Chapter 6) are also possible for three-dimensional conducting bodies, and we leave their development to the reader.

10.5 PERFORMANCE OF THE CFIE WITH LN/QT BASIS FUNCTIONS AND CURVED PATCHES [12]

In this section, the CFIE formulation is extended to incorporate a curved-patch scatterer model. The EFIE matrix entries are computed according to the expressions in (9.319)–(9.321), while the MFIE matrix entries are computed according to

$$\iint \bar{T} \cdot \bar{H}^s \, dS = \int_\eta \int_\xi \int_{\eta'} \int_{\xi'} [T_\eta T_\xi] \mathbf{J} \begin{bmatrix} 0 & -\dfrac{\partial G}{\partial z} & \dfrac{\partial G}{\partial y} \\ \dfrac{\partial G}{\partial z} & 0 & -\dfrac{\partial G}{\partial x} \\ -\dfrac{\partial G}{\partial y} & \dfrac{\partial G}{\partial x} & 0 \end{bmatrix} \mathbf{J}^T \begin{bmatrix} B_\eta \\ B_\xi \end{bmatrix} d\eta' \, d\xi' \, d\eta \, d\xi \tag{10.68}$$

where $\mathbf{J}$ is the Jacobian matrix defined in (9.314) and B_η and B_ξ denote the contravariant components of the basis function in the reference cell. The testing and basis functions are defined using the scaled contravariant projection in (9.317). For a triangular reference cell, the integration limits in (9.319)–(9.321) and (10.68) are modified accordingly. The Green's function singularity within the EFIE can be extracted and integrated in closed form over a planar cell tangent to the curved patch, as described in previous sections.

Curved patches can be realized using a mapping based on the triangular-cell quadratic scalar Lagrangian interpolation polynomials in (9.15)–(9.20). The cell location is defined by

$$x = \sum_{i=0}^{2} \sum_{j=0}^{2} x_{ijk} B_{ijk}(L_1, L_2, L_3) \tag{10.69}$$

$$y = \sum_{i=0}^{2} \sum_{j=0}^{2} y_{ijk} B_{ijk}(L_1, L_2, L_3) \tag{10.70}$$

$$z = \sum_{i=0}^{2} \sum_{j=0}^{2} z_{ijk} B_{ijk}(L_1, L_2, L_3) \tag{10.71}$$

where (L_1, L_2, L_3) denote simplex coordinates (Section 9.1), $k = 2 - i - j$ (with $k = 0, 1, 2$), and the nodes $(x_{ijk}, y_{ijk}, z_{ijk})$ describe the corners and midside coordinates of the curved patch. This mapping produces a doubly parabolic patch shape.

For illustration, a specific discretization can be obtained using the LN/QT basis functions introduced in Section 9.13 (Figure 9.25) and razor-blade testing functions defined along paths that roughly correspond to the basis function locations (Figure 10.12). A 112-cell model of a sphere is shown in Figure 10.9; Figures 10.13 and 10.14 show the dominant currents induced on a perfectly conducting sphere of radius 0.6λ using a similar 112-cell curved-patch model. On average, a density of approximately 124 unknowns/λ^2 was provided by this model.

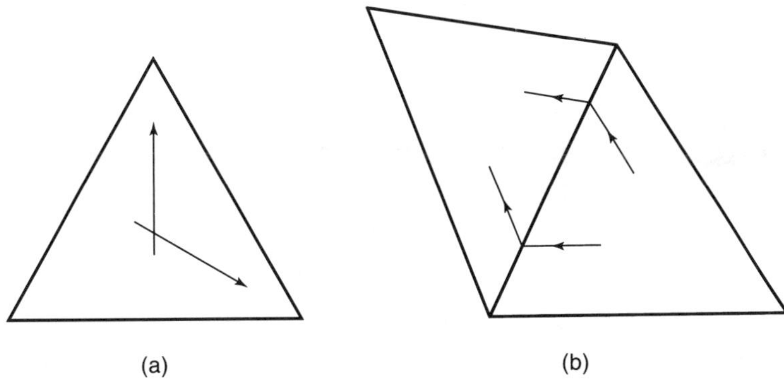

Figure 10.12 Domain of support for the razor-blade testing functions used with the LN/QT basis functions on triangular cells: (a) paths for the two cell-based functions; (b) paths for the edge-based functions.

Figures 10.13 and 10.14 show the magnitude of $\bar{J}_\theta$ and $\bar{J}_\phi$ as a function of θ, assuming that the incident field propagates in the $\theta = 180°$ direction and has an electric field polarized in the $-\hat{x}$ direction. To produce the maximum values of $\bar{J}_\theta$ and $\bar{J}_\phi$, respectively, Figure 10.13 depicts the $\phi = 0°$ cut while Figure 10.14 depicts the $\phi = 90°$ cut. The results generally exhibit good agreement with the exact solutions, although they contain slight discontinuities (more pronounced in $\bar{J}_\phi$) at cell junctions. As discussed in previous sections, the divergence-conforming basis functions ensure the continuity of the normal current density across cells but allow jump discontinuities in the tangential current density. In an irregular-grid model, a general cut through the mesh is likely to cross cell boundaries at skew angles, and thus the currents displayed in these plots contain both normal and tangential components. Consequently, some discontinuities are expected in the numerical results.

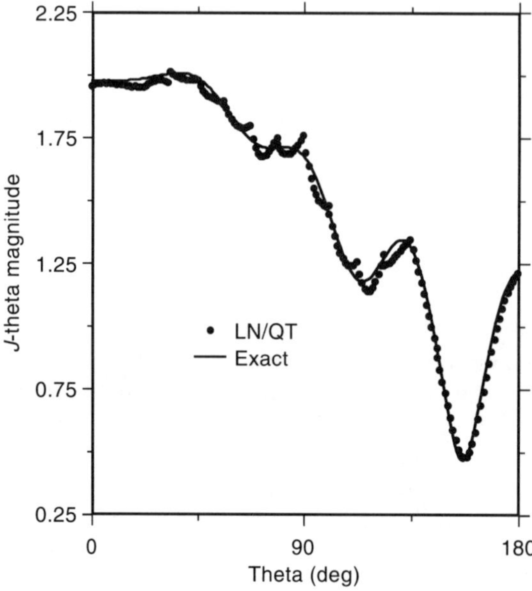

Figure 10.13 Dominant θ-component of the current density induced on a sphere of radius 0.6λ by an incident plane wave. The CFIE result is obtained using a 112-cell curved triangular-patch model involving a density of approximately 124 LN/QT unknowns per square wavelength. After [4]. ©1996 American Geophysical Union.

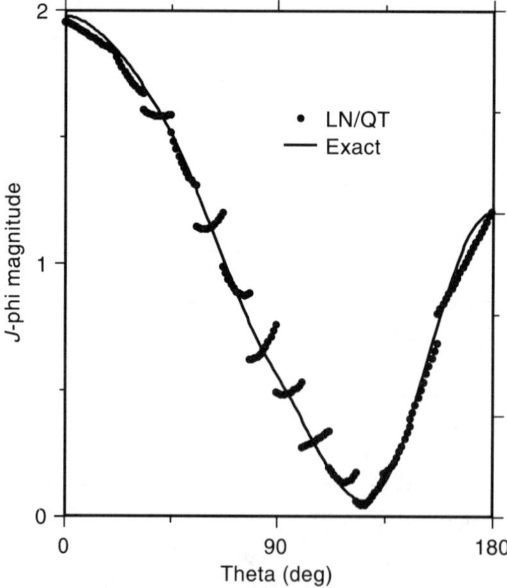

Figure 10.14 Dominant ϕ-component of the current density induced on a sphere of radius 0.6λ by an incident plane wave. The CFIE result is obtained using a 112-cell curved triangular-patch model involving a density of approximately 124 LN/QT unknowns per square wavelength. After [4]. ©1996 American Geophysical Union.

10.6 TREATMENT OF ELECTRICALLY SMALL SCATTERERS USING SURFACE INTEGRAL EQUATIONS

Chapter 6 discussed a difficulty with EFIE formulations when used to analyze a closed scatterer near an interior resonance frequency. Another difficulty with EFIE formulations arises when the equation is used near the low-frequency limit (i.e., for a scatterer geometry that is very small in terms of wavelengths). If the geometry is such that current is permitted to flow in closed loops, the EFIE will fail as the electrical size of the geometry tends to zero. Consequently, for geometries much smaller than a wavelength, a conventional EFIE discretization produces a highly ill-conditioned system of equations. In the two-dimensional situation, the difficulty only arises for the TE polarization. An eigenvalue interpretation for circular cylinders (Chapter 5) suggests that as the cylinder radius approaches zero, one eigenvalue of the TE EFIE operator approaches the origin while the rest tend to infinity! Clearly, numerical solution methods will not be able to cope with such a situation, in two or three dimensions, without special features.

The difficulty can be illustrated by considering the "mixed-potential" form of the EFIE

$$\hat{n} \times \bar{E}^{\text{inc}} = jk\eta\hat{n} \times \bar{A} + \hat{n} \times \nabla\Phi \tag{10.72}$$

As the scatterer becomes small compared to the wavelength, the numerical contribution from the $\nabla\Phi$ term of Equation (10.72) dominates the contribution from the magnetic vector potential $\bar{A}$. Since the scalar potential Φ depends only on the surface charge density, the EFIE decouples from the current density $\bar{J}$ as the contribution from $\bar{A}$ decreases. Fundamentally, this behavior is related to the decoupling of the electric field and charge density from the magnetic field and current density in the static limit. The conventional EFIE involves the incident electric field and can sometimes be used to describe the electrostatic situation; it does not incorporate the incident magnetic field and therefore cannot be used to analyze the magnetostatic situation. For electrically small scatterers whose characterization requires both electrostatic and magnetostatic contributions, the EFIE will fail.

If the scatterer geometry is such that the current density can flow in closed loops, the ordinary method-of-moments matrix operator obtained from the EFIE will become singular as the scatterer size shrinks to zero. Although the matrix is constructed using N linearly independent basis functions for $\bar{J}$, the corresponding representation of the charge density does not generally have N independent degrees of freedom. As an illustration, consider a plate discretized into rectangular cells (Figure 10.15). If rooftop basis functions are employed to discretize the current within the EFIE following the approach discussed in Section 10.1, the specific geometry of Figure 10.15 requires 37 basis functions for $\bar{J}$. The charge density associated with the rooftop functions is constant within each cell. However, there are only 24 cells in the plate (and since the total net charge is constrained to vanish because of the nature of the rooftop functions, there are only 23 degrees of freedom associated with the charge representation). Therefore, as the scatterer size tends to zero and the contribution from the magnetic vector potential is dominated by that of the scalar potential, the rank of the method-of-moments matrix will collapse from 37 to 23 for the plate example.

An alternative interpretation of the difficulty is found by examining the testing process used to discretize the EFIE. For the plate formulation of Section 10.1, razor-blade testing

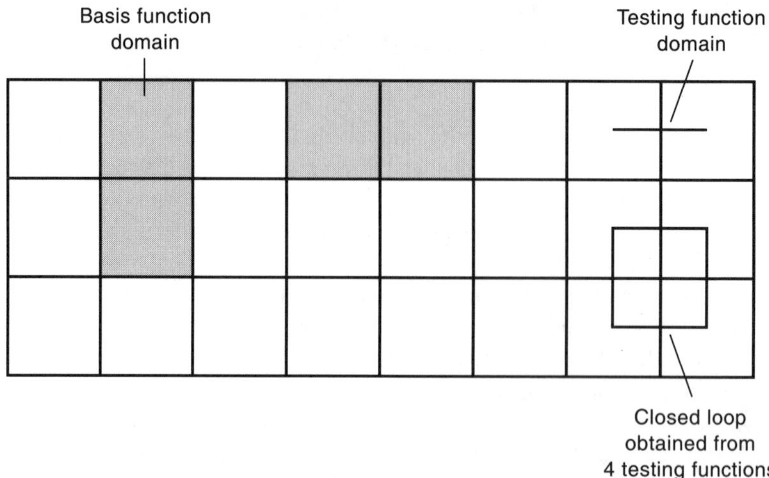

Figure 10.15 Plate discretized into 24 cells, showing the domain of a single testing function and the combined domain of four testing functions in a loop.

functions were employed with CN/LT rooftop basis functions. Each testing function straddles adjacent cells, so that there are 37 required with the plate of Figure 10.15. As the contribution from the magnetic vector potential is reduced, the net effect of "testing" the EFIE from the center of one cell to the center of the next is that the gradient $\nabla\Phi$ is integrated around closed loops. However, a vector identity states that

$$\int_{\text{any closed path}} \nabla\Phi = 0 \tag{10.73}$$

Thus, not all of the matrix rows are independent. (The number of closed loops, 14, is exactly the difference between the number of independent current basis functions and the effective number of basis functions used for the charge density.)

As a general rule, most surface EFIE formulations will fail as the scatterer size is reduced beyond some limit because of the lack of sufficient information concerning the magnetostatic part of the problem. Several remedies have been proposed [13–16].

One remedy is motivated by a modification of the testing procedure used with the EFIE for the plate example [13, 14]. In the modified procedure, three of the four testing functions used to construct each closed loop are employed as usual, but the fourth equation is constructed by integrating only the magnetic vector potential term around the entire closed loop. This approach eliminates the round-off error due to the difference in size of the scalar and vector potential contributions to the matrix equation. In addition, by integrating the EFIE around a closed loop, the incident electric field can be explicitly replaced with its circulation (a component of the incident magnetic field). In essence, some of the matrix rows of the original EFIE formulation are replaced by equations obtained from a normal-component MFIE. Consequently, this modified formulation explicitly incorporates information enabling it to treat the magnetostatic part of the problem. The procedure generalizes to arbitrarily shaped scatterers. Furthermore, this approach can be employed regardless of scatterer size and requires no increase in computation over the conventional EFIE for a given scatterer model [13, 14].

As discussed in Chapter 5, an eigenvalue interpretation of the EFIE for circular TE cylinders indicates that one eigenvalue vanishes as the cylinder radius tends to zero while the others become infinite. For the circular cylinder example, the eigenfunction associated with the zero eigenvalue is a constant function. In general, the magnetostatic part of the current density always involves functions having zero divergence. Alternative formulations for electrically small scatterers [15, 16] typically employ a basis that explicitly separates the magnetostatic part of the current (which has zero divergence) from the electrostatic part (consisting of the rest of the current distribution). In the vector case, divergence-conforming loop and star functions (Section 9.9) can be used to facilitate this partitioning. By explicitly separating the magnetostatic part of the current density, a variety of stable numerical formulations can be constructed. Additional details may be found in the literature [15, 16].

10.7 SCATTERING FROM HOMOGENEOUS DIELECTRIC BODIES: CFIE DISCRETIZED WITH TRIANGULAR-CELL CN/LT BASIS FUNCTIONS [17]

A homogeneous three-dimensional dielectric body with permittivity ε_d and permeability μ_d can be characterized by coupled surface integral equations in terms of equivalent electric and magnetic currents on the scatterer surface. To avoid difficulties associated with interior resonances, a CFIE formulation can be constructed as an alternative to the coupled electric or magnetic field equations introduced in Section 1.9. Recall that the surface integral formulation is based on two equivalent problems, an exterior problem and an interior problem. By rearranging terms in (1.111), (1.112), (1.117), and (1.118), we obtain the equations associated with the exterior equivalent problem in the form

$$\bar{K} = -\hat{n} \times \bar{E}^{\text{inc}} - \hat{n} \times \left\{ \frac{\eta}{jk}(\nabla\nabla \cdot \bar{A} + k^2 \bar{A}) - \nabla \times \bar{F} \right\}_{S^+} \tag{10.74}$$

$$\bar{J} = \hat{n} \times \bar{H}^{\text{inc}} + \hat{n} \times \left\{ \nabla \times \bar{A} + \frac{\nabla\nabla \cdot \bar{F} + k^2 \bar{F}}{jk\eta} \right\}_{S^+} \tag{10.75}$$

and the equations associated with the interior equivalent problem in the form

$$\bar{K} = \hat{n} \times \left\{ \frac{\eta_d}{jk_d}(\nabla\nabla \cdot \bar{A}_d + k_d^2 \bar{A}_d) - \nabla \times \bar{F}_d \right\}_{S^-} \tag{10.76}$$

$$\bar{J} = -\hat{n} \times \left\{ \nabla \times \bar{A}_d + \frac{\nabla\nabla \cdot \bar{F}_d + k_d^2 \bar{F}_d}{jk_d\eta_d} \right\}_{S^-} \tag{10.77}$$

where $\bar{J}$ and $\bar{K}$ are the equivalent exterior electric and magnetic surface current densities, $\hat{n}$ is the outward normal vector, $\bar{A}$ and $\bar{F}$ are the magnetic and electric vector potential functions

$$\bar{A}(u, v) = \iint \bar{J}(u', v') \frac{e^{-jkR}}{4\pi R} \, du' \, dv' \tag{10.78}$$

$$\bar{F}(u, v) = \iint \bar{K}(u', v') \frac{e^{-jkR}}{4\pi R} \, du' \, dv' \tag{10.79}$$

in the exterior medium, and $\bar{A}_d$ and $\bar{F}_d$ are the vector potential functions

$$\bar{A}_d(u, v) = \iint \bar{J}(u', v') \frac{e^{-jk_d R}}{4\pi R} \, du' \, dv' \tag{10.80}$$

$$\bar{F}_d(u, v) = \iint \bar{K}(u', v') \frac{e^{-jk_d R}}{4\pi R} \, du' \, dv' \tag{10.81}$$

in the dielectric material. Equations (10.74) and (10.75) are to be evaluated an infinitesimal distance *outside* the scatterer surface (S^+), while Equations (10.76) and (10.77) are to be evaluated an infinitesimal distance *inside* the surface (S^-).

To obtain combined-field equations, we equate (10.74) and (10.76) to obtain

$$\begin{aligned} -\hat{n} \times \bar{E}^{\text{inc}} = \hat{n} \times & \left\{ \frac{\eta}{jk} (\nabla\nabla \cdot \bar{A} + k^2 \bar{A}) - \nabla \times \bar{F} \right\}_{S^+} \\ + \hat{n} \times & \left\{ \frac{\eta_d}{jk_d} (\nabla\nabla \cdot \bar{A}_d + k_d^2 \bar{A}_d) - \nabla \times \bar{F}_d \right\}_{S^-} \end{aligned} \tag{10.82}$$

and equate (10.75) and (10.77) to obtain

$$\begin{aligned} -\hat{n} \times \bar{H}^{\text{inc}} = \hat{n} \times & \left\{ \nabla \times \bar{A} + \frac{\nabla\nabla \cdot \bar{F} + k^2 \bar{F}}{jk\eta} \right\}_{S^+} \\ + \hat{n} \times & \left\{ \nabla \times \bar{A}_d + \frac{\nabla\nabla \cdot \bar{F}_d + k_d^2 \bar{F}_d}{jk_d \eta_d} \right\}_{S^-} \end{aligned} \tag{10.83}$$

Together, Equations (10.82) and (10.83) constitute a combined-field formulation that will produce unique solutions even at frequencies where the conventional EFIE or MFIE fails because of interior resonance difficulties.

We again consider a triangular-cell model for the scatterer surface and represent the unknown electric and magnetic surface current densities by the CN/LT triangular-rooftop basis functions used in previous sections. For a model with N edges, the currents can be written

$$\bar{J}(u, v) \cong \sum_{n=1}^{N} j_n \bar{B}_n(u, v) \tag{10.84}$$

$$\bar{K}(u, v) \cong \sum_{n=1}^{N} k_n \bar{B}_n(u, v) \tag{10.85}$$

Equations (10.82) and (10.83) can be discretized using the razor-blade testing functions illustrated in Figure 10.10, which each have support along a path from the centroid of one cell to the centroid of an adjacent cell. Because the razor-blade testing functions are only capable of absorbing one derivative, the vector potential terms $\nabla \cdot \bar{A}$ and $\nabla \cdot \bar{F}$ in (10.82) and (10.83) can be replaced with equivalent scalar potential functions (Section 1.4), which is essentially the same as replacing the source–field relation embodied in these equations in order to transfer one derivative onto the surface currents. The result of the discretization is a matrix equation having the form

$$\begin{bmatrix} \mathbf{A} & \mathbf{B} \\ \mathbf{C} & \mathbf{D} \end{bmatrix} \begin{bmatrix} \mathbf{j} \\ \mathbf{k} \end{bmatrix} = \begin{bmatrix} \mathbf{e} \\ \mathbf{h} \end{bmatrix} \tag{10.86}$$

where

$$\begin{aligned}A_{mn} &= jk\eta \int_{C_m} \left\{ \iint \bar{B}_n(u', v') \frac{e^{-jkR}}{4\pi R} \, du' \, dv' \right\} \cdot d\bar{t} \\ &+ \frac{\eta}{jk} \int_{C_m} \nabla \left\{ \iint [\nabla_s \cdot \bar{B}_n] \frac{e^{-jkR}}{4\pi R} \, du' \, dv' \right\} \cdot d\bar{t} \\ &+ jk_d\eta_d \int_{C_m} \left\{ \iint \bar{B}_n(u', v') \frac{e^{-jk_dR}}{4\pi R} \, du' \, dv' \right\} \cdot d\bar{t} \\ &+ \frac{\eta_d}{jk_d} \int_{C_m} \nabla \left\{ \iint [\nabla_s \cdot \bar{B}_n] \frac{e^{-jk_dR}}{4\pi R} \, du' \, dv' \right\} \cdot d\bar{t} \end{aligned} \tag{10.87}$$

$$\begin{aligned}B_{mn} &= -\int_{C_m} \left\{ \nabla \times \iint \bar{B}_n(u', v') \frac{e^{-jkR}}{4\pi R} \, du' \, dv' \right\}_{S^+} \cdot d\bar{t} \\ &- \int_{C_m} \left\{ \nabla \times \iint \bar{B}_n(u', v') \frac{e^{-jk_dR}}{4\pi R} \, du' \, dv' \right\}_{S^-} \cdot d\bar{t} \end{aligned} \tag{10.88}$$

$$\begin{aligned}C_{mn} &= \int_{C_m} \left\{ \nabla \times \iint \bar{B}_n(u', v') \frac{e^{-jkR}}{4\pi R} \, du' \, dv' \right\}_{S^+} \cdot d\bar{t} \\ &+ \int_{C_m} \left\{ \nabla \times \iint \bar{B}_n(u', v') \frac{e^{-jk_dR}}{4\pi R} \, du' \, dv' \right\}_{S^-} \cdot d\bar{t} \end{aligned} \tag{10.89}$$

$$\begin{aligned}D_{mn} &= \frac{jk}{\eta} \int_{C_m} \left\{ \iint \bar{B}_n(u', v') \frac{e^{-jkR}}{4\pi R} \, du' \, dv' \right\} \cdot d\bar{t} \\ &+ \frac{1}{jk\eta} \int_{C_m} \nabla \left\{ \iint [\nabla_s \cdot \bar{B}_n] \frac{e^{-jkR}}{4\pi R} \, du' \, dv' \right\} \cdot d\bar{t} \\ &+ \frac{jk_d}{\eta_d} \int_{C_m} \left\{ \iint \bar{B}_n(u', v') \frac{e^{-jk_dR}}{4\pi R} \, du' \, dv' \right\} \cdot d\bar{t} \\ &+ \frac{1}{jk_d\eta_d} \int_{C_m} \nabla \left\{ \iint [\nabla_s \cdot \bar{B}_n] \frac{e^{-jk_dR}}{4\pi R} \, du' \, dv' \right\} \cdot d\bar{t} \end{aligned} \tag{10.90}$$

The entries of the right-hand side are

$$e_m = -\int_{C_m} \bar{E}^{\text{inc}} \cdot d\bar{t} \tag{10.91}$$

$$h_m = -\int_{C_m} \bar{H}^{\text{inc}} \cdot d\bar{t} \tag{10.92}$$

The double integrals in (10.86)–(10.90) denote integration over the entire scatterer surface; in actuality, these integrals collapse to the two cells in which the nth basis function is nonzero.

For efficient implementation, we consider some of the approximations introduced in Section 10.2 to simplify the matrix entries. Entries A_{mn} and D_{mn} can be simplified using

$$\int_{C_m} \nabla\Phi \cdot d\bar{t} = \Phi(u_j, v_j) - \Phi(u_i, v_i) \tag{10.93}$$

$$\int_{C_m} \bar{A} \cdot d\bar{t} \cong \bar{A} \cdot \Delta\bar{t}_{m1}\Big|_{u_i, v_i} + \bar{A} \cdot \Delta\bar{t}_{m2}\Big|_{u_j, v_j} \tag{10.94}$$

where, assuming that i and j are the two cells associated with edge m as depicted in Figure

10.10, $\Delta\bar{t}_{m1}$ denotes the vector from the centroid of cell i to the center of edge m and $\Delta\bar{t}_{m2}$ denotes the vector from the center of edge m to the centroid of cell j. Using (10.93) and (10.94), A_{mn} can be written entirely in terms of double integrals as

$$\begin{aligned}
A_{mn} \cong\ & jk\eta \left\{ \Delta\bar{t}_{m1} \cdot \iint \bar{B}_n(u', v') \frac{e^{-jkR}}{4\pi R}\, du'\, dv' \bigg|_{u_i, v_i} \right. \\
& \left. + \Delta\bar{t}_{m2} \cdot \iint \bar{B}_n(u', v') \frac{e^{-jkR}}{4\pi R}\, du'\, dv' \bigg|_{u_j, v_j} \right\} \\
& + \frac{\eta}{jk} \left\{ \iint [\nabla_s \cdot \bar{B}_n] \frac{e^{-jkR}}{4\pi R}\, du'\, dv' \bigg|_{u_j, v_j} \right. \\
& \left. - \iint [\nabla_s \cdot \bar{B}_n] \frac{e^{-jkR}}{4\pi R}\, du'\, dv' \bigg|_{u_i, v_i} \right\} \\
& + jk_d\eta_d \left\{ \nabla\bar{t}_{m1} \cdot \iint \bar{B}_n(u', v') \frac{e^{-jk_dR}}{4\pi R}\, du'\, dv' \bigg|_{u_i, v_i} \right. \\
& \left. + \Delta\bar{t}_{m2} \cdot \iint \bar{B}_n(u', v') \frac{e^{-jk_dR}}{4\pi R}\, du'\, dv' \bigg|_{u_j, v_j} \right\} \\
& + \frac{\eta_d}{jk_d} \left\{ \iint [\nabla_s \cdot \bar{B}_n] \frac{e^{-jk_dR}}{4\pi R}\, du'\, dv' \bigg|_{u_j, v_j} \right. \\
& \left. - \iint [\nabla_s \cdot \bar{B}_n] \frac{e^{-jk_dR}}{4\pi R}\, du'\, dv' \bigg|_{u_i, v_i} \right\}
\end{aligned} \tag{10.95}$$

Because of the similarity between (10.87) and (10.90), an analogous expression for D_{mn} is immediately obtained. When source and observation regions coincide, the $1/R$ singularity may be extracted, integrated analytically, and added back to the numerically computed residual.

For the situation where the path C_m does not traverse either of the two source cells (the cells adjacent to edge n), B_{mn} may be simplified to

$$\begin{aligned}
B_{mn} =\ & \int_{C_m} \left\{ \iint \bar{B}_n(u', v') \times \nabla\left(\frac{e^{-jkR}}{4\pi R}\right) du'\, dv' \right\}_{S^+} \cdot d\bar{t} \\
& + \int_{C_m} \left\{ \iint \bar{B}_n(u', v') \times \nabla\left(\frac{e^{-jk_dR}}{4\pi R}\right) du'\, dv' \right\}_{S^-} \cdot d\bar{t} \\
\cong\ & \iint \bar{B}_n(u', v') \times \nabla\left(\frac{e^{-jkR}}{4\pi R}\right) du'\, dv' \bigg|_{u_i, v_i} \cdot \Delta\bar{t}_1 \\
& + \iint \bar{B}_n(u', v') \times \nabla\left(\frac{e^{-jkR}}{4\pi R}\right) du'\, dv' \bigg|_{u_j, v_j} \cdot \Delta\bar{t}_2 \\
& + \iint \bar{B}_n(u', v') \times \nabla\left(\frac{e^{-jk_dR}}{4\pi R}\right) du'\, dv' \bigg|_{u_i, v_i} \cdot \Delta\bar{t}_1 \\
& + \iint \bar{B}_n(u', v') \times \nabla\left(\frac{e^{-jk_dR}}{4\pi R}\right) du'\, dv' \bigg|_{u_j, v_j} \cdot \Delta\bar{t}_2
\end{aligned} \tag{10.96}$$

For the case where the path C_m lies within a source cell, a limiting procedure may be used to obtain

$$\int_{C_m^i} \left\{ \nabla \times \iint_{\text{cell } i} \bar{B}_n(u', v') \frac{e^{-jkR}}{4\pi R} \, du' \, dv' \right\}_{S^+} \cdot d\bar{t} = \int_{C_m^i} \frac{\bar{B}_n(u, v) \times \hat{n}}{2} \cdot d\bar{t} \qquad (10.97)$$

where C_m^i denotes the part of path C_m within cell i. A similar limiting procedure produces

$$\int_{C_m^i} \left\{ \nabla \times \iint_{\text{cell } i} \bar{B}_n(u', v') \frac{e^{-jk_d R}}{4\pi R} \, du' \, dv' \right\}_{S^-} \cdot d\bar{t} = -\int_{C_m^i} \frac{\bar{B}_n(u, v) \times \hat{n}}{2} \cdot d\bar{t} \qquad (10.98)$$

Equations (10.97) and (10.98) can be used to evaluate B_{mn} and C_{mn} when the source and observation regions coincide.

Other aspects of the CFIE implementation for homogeneous dielectric bodies such as the specific pointer arrays required within the scatterer model, the procedures used to eliminate redundant numerical calculations while constructing the matrix, and the details of the scattering cross section calculation are similar to those described in preceding sections of this chapter and will be left to the reader. To illustrate the approach, Figure 10.16 shows the surface currents associated with a finite dielectric cylinder produced by a triangular-cell

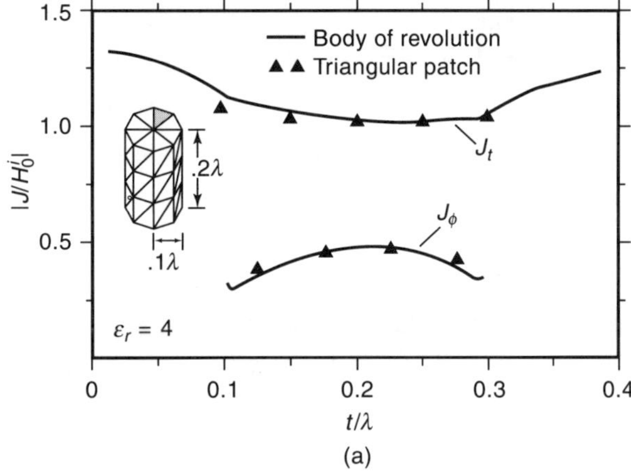

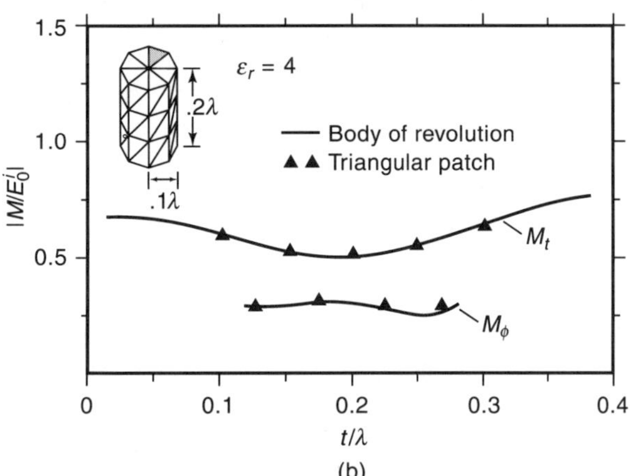

Figure 10.16 Current distribution on the surface of a finite dielectric cylinder due to an axially incident plane wave. The triangular-patch CN/LT results are compared to a body of revolution formulation. (a) Electric surface current. (b) Magnetic surface current. After [17]. ©1986 IEEE.

CFIE formulation similar to that outlined above [17]. The results exhibit good agreement with an alternative body-of-revolution approach.

10.8 RADIATION AND SCATTERING FROM THIN WIRES

The analysis of thin conducting wires is one of the most mature applications of computational electromagnetics, with hundreds of articles and a number of texts primarily devoted to this topic (see, e.g., [18, 19]). Yet wire modeling remains challenging, since often the parameters of interest such as antenna input impedance can be very sensitive to the detailed geometry of the feed region and resolution limitations govern the extent to which these regions can be modeled. In addition to antenna analysis, wire-grid models have been widely used to represent three-dimensional conducting scatterers. This section provides a brief overview of a typical thin-wire formulation based on the EFIE.

Consider a single wire having a circular cross section of radius a whose surface is described by local coordinates (s, Ψ) as illustrated in Figure 10.17. The principal assumptions of thin-wire analysis are that (1) circumferential currents around the wire are negligible and (2) the current density is not a function of Ψ. Under these assumptions, the surface current density $\bar{J}(s, \Psi)$ can be replaced by the total current

$$\hat{s}I(s) = 2\pi a \bar{J}(s, \Psi) \tag{10.99}$$

The mixed-potential form of the EFIE can be specialized to this situation to produce

$$E_s^{\text{inc}}(s) = j\omega\mu \int_s I(s')G(s, s')\, ds' - \frac{1}{j\omega\varepsilon}\frac{d}{ds}\int_s \frac{dI}{ds'}G(s, s')\, ds' \tag{10.100}$$

where

$$G(s, s') = \frac{1}{2\pi a}\int_{\Psi'=0}^{2\pi} \frac{e^{-jkR}}{4\pi R} a\, d\Psi' \tag{10.101}$$

$$R = \sqrt{[x(s, 0) - x(s', \Psi')]^2 + [y(s, 0) - y(s', \Psi')]^2 + [z(s, 0) - z(s', \Psi')]^2} \tag{10.102}$$

and where $x(s, \Psi)$, $y(s, \Psi)$, and $z(s, \Psi)$ provide the coordinates of a point along the wire surface. The integrals in (10.100) extend over the length of the wire.

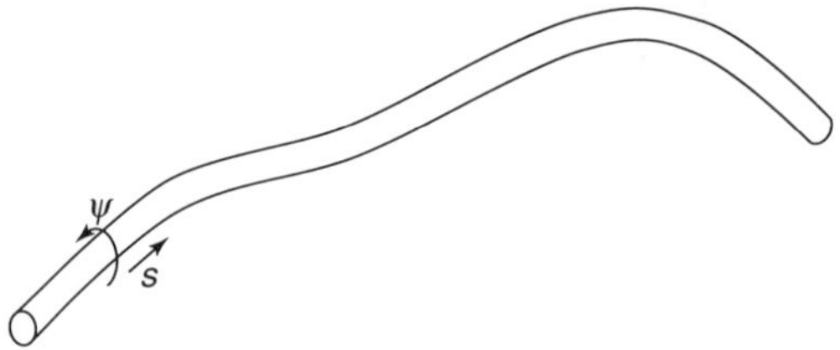

Figure 10.17 Thin-wire geometry.

Suppose that the wire is modeled by N straight, cylindrical segments (Figure 10.18) and the current $I(s)$ is represented by subsectional triangle basis functions (Figure 5.1)

$$I(s) \cong \sum_{n=1}^{N-1} I_n B_n(s) = \sum_{n=1}^{N-1} I_n t(s; s_{n-1}, s_n, s_{n+1}) \tag{10.103}$$

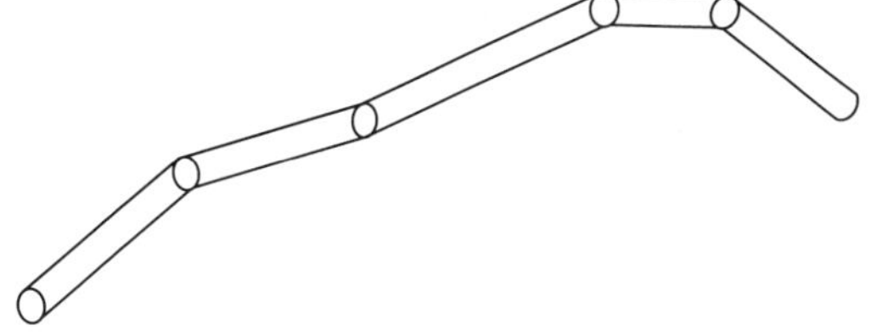

Figure 10.18 Model of thin wire consisting of straight cylindrical segments.

In other words, $I(s)$ is represented by a continuous, linear expansion. Assuming that the wire is hollow or that the endcap currents are negligible, $I(s)$ vanishes at the wire ends, and no basis functions are assigned to locations $n = 0$ and $n = N$.

The EFIE in (10.100) can be discretized using subsectional pulse testing functions

$$T_m(s) = p\{s; \tfrac{1}{2}(s_{m-1} + s_m), \tfrac{1}{2}(s_m + s_{m+1})\} \qquad m = 1, 2, \ldots, N-1 \tag{10.104}$$

to produce the matrix equation $\mathbf{ZI} = \mathbf{E}^i$. The entries of $\mathbf{I}$ are the coefficients in (10.103), while the entries of $\mathbf{Z}$ are given by

$$\begin{aligned} Z_{mn} = j\omega\mu &\int_{(s_{m-1}+s_m)/2}^{(s_m+s_{m+1})/2} \int_{s_{n-1}}^{s_{n+1}} B_n(s')G(s, s')\, ds'\, ds \\ &- \frac{1}{j\omega\varepsilon} \int_{s_{n-1}}^{s_{n+1}} \frac{dB_n}{ds'} G\left(\frac{s_{m-1} + s_m}{2}, s'\right) ds' \\ &+ \frac{1}{j\omega\varepsilon} \int_{s_{n-1}}^{s_{n+1}} \frac{dB_n}{ds'} G\left(\frac{s_m + s_{m+1}}{2}, s'\right) ds' \end{aligned} \tag{10.105}$$

where the pulse testing function in the scalar potential term is used to eliminate the derivative with respect to s. The expression

$$E_m^i = \int_{(s_{m-1}+s_m)/2}^{(s_m+s_{m+1})/2} E_s^{\text{inc}}(s)\, ds \tag{10.106}$$

provides the entries of the excitation vector $\mathbf{E}^i$.

If the incident field is a uniform plane wave, as in a scattering problem, the evaluation of (10.106) needs little elaboration. If the wire represents a radiating antenna, then one of a number of feed models can be employed as the source of E_s^{inc}. For example, a "magnetic frill" is often employed as a model for a wire fed by a coaxial transmission line through a ground plane (Chapter 1). Probably the most widely used model, however, is the so-called delta-gap feed, which consists of an idealized constant electric field in a small gap in the wire, scaled (like a Dirac delta function) so that the integral in (10.106) over the feed cell always has a constant value, regardless of how small the cell is. (The other cells receive no contribution from the feed.) The delta-gap feed can produce reasonable results if the cells are not too small; however, as the cell sizes shrink, the physical feed model changes and the resulting input impedance does not converge. A more realistic model is obtained by evaluating (by quadrature if necessary) the fields produced by the equivalent magnetic current density in the aperture of a small gap, where the finite gap size is maintained as the cells in the model are refined.

The most challenging aspect of the thin-wire formulation is the efficient and accurate evaluation of $G(s, s')$. Consider the evaluation of G when the observation point s resides on the surface of one cylindrical segment and the source point s' resides on another. The spherical angles (θ_p, ϕ_p) and (θ_q, ϕ_q) define the orientation of the two cylinders, and (x_p, y_p, z_p)

describes the location along the axis of the cylinder p at s while (x_q, y_q, z_q) denotes the location along the axis of cylinder q at s' (Figure 10.19). Without loss of generality, the observation point at $(s, 0)$ can be assigned the coordinates

$$x = x_p + a_p \cos\theta_p \cos\phi_p \tag{10.107}$$

$$y = y_p + a_p \cos\theta_p \sin\phi_p \tag{10.108}$$

$$z = z_p - a_p \sin\theta_p \tag{10.109}$$

while the source point at (s', Ψ') can be assigned

$$x' = x_q + a_q \cos\theta_q \cos\phi_q \cos\Psi' - a_q \sin\phi_q \sin\Psi' \tag{10.110}$$

$$y' = y_q + a_q \cos\theta_q \sin\phi_q \cos\Psi' + a_q \cos\phi_q \sin\Psi' \tag{10.111}$$

$$z' = z_q - a_q \sin\theta_q \cos\Psi' \tag{10.112}$$

Using these expressions within R in (10.102), $G(s, s')$ can be evaluated by quadrature to construct the off-diagonal entries of the matrix $\mathbf{Z}$.

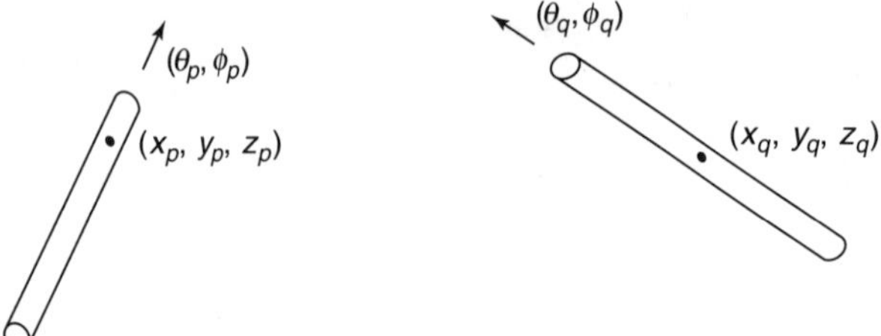

Figure 10.19 Coordinates and orientation parameters for the source and observer segments of the thin-wire model.

The diagonal matrix entries must be computed separately because of the singularity in G when $s = s'$. However, in that case Z_{mm} is identical to the matrix expressions previously developed for the hollow-cylinder example in Section 8.7 and can be found by specializing Equations (8.122)–(8.127) to the ϕ-invariant case.

Formulations similar to the above have been well documented in the literature, and representative results can be found in many publications. The triangle basis–pulse testing scheme is similar to the approach used in the commercial code MININEC [20], while other widely used discretizations include sinusoidal triangle basis and testing functions [21] and the three-term sinusoid basis function [22] (both of which are described in Chapter 5). Since the one-dimensional representation is easily realized, a wide variety of other basis functions have also been studied for thin-wire modeling [23].

Because of the complexity of $G(s, s')$, the numerical evaluation of (10.101) is usually modified to incorporate approximations that simplify the calculations. A number of accurate approximations are possible [18–24], although others that have been proposed are inaccurate. A widely used approximation is to replace (10.101) by the "thin-wire," or "reduced," kernel obtained by locating the observation point on the axis of the cylinder instead of the surface. For instance, when $p = q$, the integral is replaced by

$$\int_{\Psi'=0}^{2\pi} \frac{e^{-jkR}}{4\pi R} a \, d\Psi' \cong 2\pi a \frac{e^{-jk\sqrt{(s-s')^2+a^2}}}{4\pi\sqrt{(s-s')^2+a^2}} \tag{10.113}$$

This type of approximation eliminates the need to integrate over the variable Ψ' and removes the singularity from the kernel. Although (10.113) is a poor approximation and has been identified as the cause of irregular numerical results, the accuracy of Z_{mm} can be acceptable if the cylindrical sections of the model have a length-to-radius ratio of at least 10 [25]. Actually, the results for currents and antenna impedance based on (10.113) have a much wider range of validity than one would expect from the accuracy of the approximation, apparently because the resulting system of equations is equivalent to the discrete system obtained from the so-called extended boundary condition formulation for the wire geometry [25, 26].

It is not necessary to make unwarranted approximations. Accurate approximations are well documented and, in fact, a highly accurate evaluation of G is possible and relatively efficient [27]. There are a number of other issues associated with wire antenna modeling that are beyond the scope of the present discussion. As mentioned above, the feed region geometry may require a level of resolution not possible with a particular formulation. Ideally, the geometrical features of the desired feed model should be incorporated into the formulation. Another application limited by resolution is the treatment of wire junctions. Several approaches have been proposed and include a model for junctions between wires of different radii [18, 22]. The use of a wire grid to represent the surface of a conducting scatterer is a powerful tool, but certain constraints must be placed on the wire dimensions to ensure an equivalence. Practical guidelines are available in the literature for assigning the wire radius and spacing in order to represent solid targets [18, 28].

10.9 SCATTERING FROM PLANAR PERIODIC GEOMETRIES

Figure 10.20 shows a doubly periodic planar structure consisting of a number of conducting patches. Structures of this type, known as *frequency-selective surfaces*, are used as filters, polarizers, and artificial dielectrics [8]. The analysis of doubly periodic scatterers is similar to the singly periodic case considered in Chapter 7, and this section provides a cursory overview of the planar three-dimensional case for completeness. Suppose that the unit cell is rectangular with dimensions $a \times b$, corresponding to the x and y directions. The currents induced by a uniform plane wave satisfy the Floquet condition

$$\bar{J}(x + ma, y + nb) = \bar{J}(x, y)e^{-jk_x^{\text{inc}}ma}e^{-jk_y^{\text{inc}}nb} \tag{10.114}$$

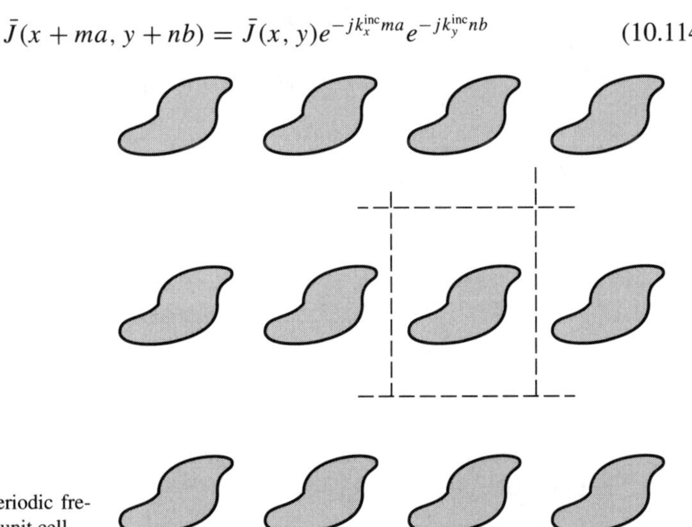

Figure 10.20 Portion of a doubly periodic frequency-selective surface showing the unit cell.

where k_x^{inc} and k_y^{inc} are the projection of the phase constants associated with the incident wave. In a situation where the conducting patches occupy less than half the area of the unit cell, it is convenient to employ an EFIE formulation using the equivalent electric current density on the conductors as the primary unknown. (If the patches occupy most of the unit cell, a formulation employing the tangential electric field in the aperture as the primary unknown would be more efficient.) The components of the EFIE for the periodic structure can be written as

$$E_x^{\text{inc}}(x, y) = -\hat{x} \cdot \frac{\nabla\nabla \cdot +k^2}{j\omega\varepsilon_0} \iint \bar{J}(x', y') G_p(x - x', y - y')\, dx'\, dy' \tag{10.115}$$

$$E_y^{\text{inc}}(x, y) = -\hat{y} \cdot \frac{\nabla\nabla \cdot +k^2}{j\omega\varepsilon_0} \iint \bar{J}(x', y') G_p(x - x', y - y')\, dx'\, dy' \tag{10.116}$$

where, using the Floquet condition, the periodic Green's function can be deduced to have the form

$$G_p(x, y) = \sum_{m=-\infty}^{\infty} \sum_{n=-\infty}^{\infty} \frac{e^{-jkR_{mn}}}{4\pi R_{mn}} e^{-jk_x^{\text{inc}} ma} e^{-jk_y^{\text{inc}} nb} \tag{10.117}$$

where

$$R_{mn} = \sqrt{(x - ma)^2 + (y - nb)^2} \tag{10.118}$$

Equations (10.115) and (10.116) are only valid for points (x, y) on the conducting patch.

Suppose that the conductor geometry is approximated by rectangular cells, as previously described for the aperiodic plate formulation (Figure 10.1). The method-of-moments analysis can be implemented in a manner almost identical to that for the individual conducting plate in Section 10.1 using rooftop basis functions for the current density

$$\bar{J}(x, y) \cong \hat{x} \sum_{p=1}^{P} j_{xp} B_{xp}(x, y) + \hat{y} \sum_{q=1}^{Q} j_{yq} B_{yq}(x, y) \tag{10.119}$$

and razor-blade testing functions to enforce the equations. The primary difference between the single plate and the periodic structure implementations is the need to accelerate the convergence of the periodic Green's function, since the terms in the summation in (10.117) only decay as $O(m^{-1})$ as $m \to \infty$ and $O(n^{-1})$ as $n \to \infty$. Problems P7.8 and P7.9 suggest the use of the Poisson sum transformation to produce the alternative summation

$$G_p(x, y) = \frac{1}{2ab} \sum_{m=-\infty}^{\infty} \sum_{n=-\infty}^{\infty} \frac{1}{j\beta z} e^{j2\pi f x} e^{j2\pi g y} \Bigg|_{f=m/a-k_x^{\text{inc}}/2\pi,\, g=n/b-k_y^{\text{inc}}/2\pi} \tag{10.120}$$

where

$$\beta_z = \begin{cases} \sqrt{k^2 - (2\pi f)^2 - (2\pi g)^2} & k^2 > (2\pi f)^2 + (2\pi g)^2 \\ -j\sqrt{(2\pi f)^2 + (2\pi g)^2 - k^2} & \text{otherwise} \end{cases} \tag{10.121}$$

Unfortunately, this summation also converges at an $O(m^{-1})$ and $O(n^{-1})$ rate.

Some acceleration can be provided by the basis and testing functions, as explained in Section 7.4, since the entries of the matrix equation have the convolutional form illustrated in Equations (10.20)–(10.23) [8]. However, a more effective approach can be based on a three-dimensional extension of the error function transformation discussed for the two-dimensional case in Section 7.5 [29, 30].

Using the definition of the error function and the complementary error function in Equations (7.79) and (7.80), respectively, the Green's function in (10.120) can be divided into two parts according to

$$G_p(x, y) = G_1(x, y) + G_2(x, y) \tag{10.122}$$

where

$$G_1(x, y) = \frac{1}{2ab} \sum_{m=-\infty}^{\infty} \sum_{n=-\infty}^{\infty} \frac{\operatorname{erfc}(j\beta_z/2E)}{j\beta z} e^{j2\pi f x} e^{j2\pi g y} \Bigg|_{f=m/a-k_x^{inc}/2\pi,\, g=n/b-k_y^{inc}/2\pi} \tag{10.123}$$

$$G_2(x, y) = \frac{1}{2ab} \sum_{m=-\infty}^{\infty} \sum_{n=-\infty}^{\infty} \frac{\operatorname{erf}(j\beta_z/2E)}{j\beta z} e^{j2\pi f x} e^{j2\pi g y} \Bigg|_{f=m/a-k_x^{inc}/2\pi,\, g=n/b-k_y^{inc}/2\pi} \tag{10.124}$$

The summation for G_1 is exponentially convergent. Using the two-dimensional Poisson sum transformation, G_2 can be converted into an equivalent spatial domain summation

$$G_2(x, y) = \frac{1}{2\pi^{3/2}} \sum_{m=-\infty}^{\infty} \sum_{n=-\infty}^{\infty} e^{-jk_x^{inc} ma} e^{-jk_y^{inc} nb} \int_0^{1/2E} \frac{\exp(R_{mn}^2/4u^2 + k^2u^2)}{u^2} du \tag{10.125}$$

where R_{mn} is defined in (10.118). After a change of variables $w = 1/4u^2$, the integration can be performed analytically to produce

$$G_2(x, y) = \frac{1}{8\pi} \sum_{m=-\infty}^{\infty} \sum_{n=-\infty}^{\infty} \frac{e^{-jk_x^{inc} ma} e^{-jk_y^{inc} nb}}{R_{mn}} \times \left[\operatorname{erfc}\left(\frac{R_{mn}}{E} + \frac{jkE}{2}\right) e^{jkR_{mn}} + \operatorname{erfc}\left(\frac{R_{mn}}{E} - \frac{jkE}{2}\right) e^{-jkR_{mn}} \right] \tag{10.126}$$

which is also exponentially convergent. Therefore, the periodic Green's function can be obtained from the combination of (10.123) and (10.126). Additional details of the error function transformation and the more general Ewald transformation can be found in the literature [29–31].

The acceleration procedure permits a highly accurate evaluation of G_p, from which the entries of the method-of-moments matrix can be determined. The solution of the system of equations yields the basis function coefficients in (10.119) and subsequently reflection and transmission coefficients (Chapter 7). The literature contains a wide variety of numerical results for periodic surfaces and illustrates their applications [8, 32].

10.10 ANALYSIS OF MICROSTRIP STRUCTURES

Conducting patches on the surface of a grounded dielectric slab ("microstrip" patches) are widely used as antennas [33]. Their radiation and scattering properties can be determined by the numerical solution of an EFIE [6, 7, 34–38]. In order to limit the unknowns to the electric surface currents on the patch, the formulation must incorporate the Green's function for a horizontal electric point source on the surface of the grounded slab (Figure 10.21).

In this section, the appropriate Green's function is obtained using Fourier transform theory and a transmission line analogy.

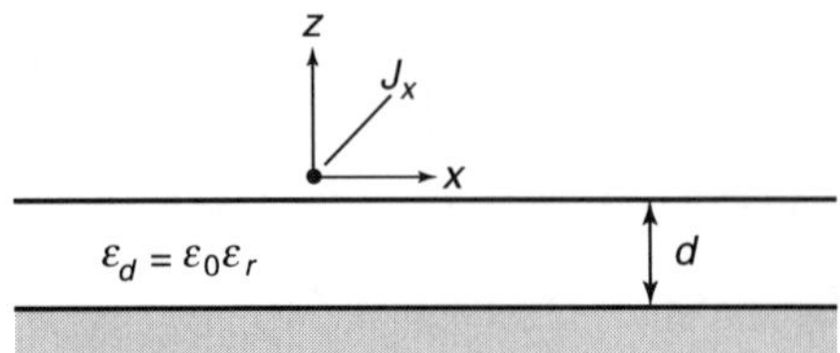

Figure 10.21 Point source on the surface of a grounded dielectric slab.

To obtain a Green's function involving a planar, layered geometry, it is convenient to work with Fourier transforms. The two-dimensional transform pair can be defined according to

$$\tilde{F}(k_x, k_y) = \int_{-\infty}^{\infty}\int_{-\infty}^{\infty} f(x, y)e^{-jk_x x}e^{-jk_y y}\, dx\, dy \tag{10.127}$$

$$f(x, y) = \frac{1}{(2\pi)^2}\int_{-\infty}^{\infty}\int_{-\infty}^{\infty} \tilde{F}(k_x, k_y)e^{jk_x x}e^{jk_y y}\, dk_x\, dk_y \tag{10.128}$$

(This is the identical transform used in Section 10.9 with $2\pi f$ replaced by k_x and $2\pi g$ replaced by k_y.) By applying the two-dimensional Fourier transform to the electric and magnetic fields, Maxwell's equations can be simplified to equations for the x- and y-field components as a function of the single variable z (and parameters k_x and k_y). Problem P10.14 reviews the development of the simplified equations. The one-dimensional equations are the same as those used to describe voltages and currents on transmission lines, and a transmission line analogy can be used to obtain the solutions [7, 39]. The Fourier transform can also be used to convert a point source at the origin (Figure 10.21) into a superposition of current sheets, according to

$$\begin{aligned}\bar{J}(x, y, z) &= \hat{x}\delta(x)\delta(y)\delta(z)\\ &= \hat{x}\delta(z)\frac{1}{(2\pi)^2}\int_{-\infty}^{\infty}\int_{-\infty}^{\infty} e^{jk_x x}e^{jk_y y}\, dk_x\, dk_y\end{aligned} \tag{10.129}$$

In the transform domain, the appropriate transmission line equations can be solved individually for each current sheet, then superimposed via the inverse transform operation to produce the fields of the original point source in the spatial domain.

The equations obtained in Prob. P10.14 involve the coupled-field components E_x, E_y, H_x, and H_y. Additional simplification is possible by changing to polar coordinates and using [7]

$$k_x = k_\rho \cos\phi \tag{10.130}$$

$$k_y = k_\rho \sin\phi \tag{10.131}$$

where

$$k_\rho = \sqrt{k_x^2 + k_y^2} \tag{10.132}$$

and

$$\phi = \arctan\left(\frac{k_y}{k_x}\right) \tag{10.133}$$

The equations obtained in Prob. P10.14 decouple into two sets, the first given by

$$\frac{d\tilde{E}_\rho}{dz} - \frac{k^2 - k_\rho^2}{j\omega\varepsilon}\tilde{H}_\phi = 0 \tag{10.134}$$

$$\frac{d\tilde{H}_\phi}{dz} + j\omega\varepsilon\tilde{E}_\rho = 0 \tag{10.135}$$

and the second given by

$$\frac{d\tilde{E}_\phi}{dz} - j\omega\mu\tilde{H}_\rho = 0 \tag{10.136}$$

$$\frac{d\tilde{H}_\rho}{dz} + \frac{k^2 - k_\rho^2}{j\omega\mu}\tilde{E}_\phi = 0 \tag{10.137}$$

Each set of equations can be converted into an equivalent transmission line problem. Equations (10.134) and (10.135) can be rewritten using $V_1 = \tilde{E}_\rho$ and $I_1 = \tilde{H}_\phi$, to produce

$$\frac{dV_1}{dz} + j\beta Z_1 I_1 = 0 \tag{10.138}$$

$$\frac{dI_1}{dz} + j\frac{\beta}{Z_1}V_1 = 0 \tag{10.139}$$

where

$$\beta = \begin{cases} \sqrt{k^2 - k_\rho^2} & k^2 > k_\rho^2 \\ -j\sqrt{k_\rho^2 - k^2} & \text{otherwise} \end{cases} \tag{10.140}$$

and

$$Z_1 = \frac{\beta}{\omega\varepsilon} \tag{10.141}$$

Equations (10.136) and (10.137) can be rewritten using $V_2 = -\tilde{E}_\phi$ and $I_2 = \tilde{H}_\rho$ to produce

$$\frac{dV_2}{dz} + j\beta Z_2 I_2 = 0 \tag{10.142}$$

$$\frac{dI_2}{dz} + j\frac{\beta}{Z_2}V_2 = 0 \tag{10.143}$$

where

$$Z_2 = \frac{\omega\mu}{\beta} \tag{10.144}$$

The transmission line systems described by these equations are uncoupled except by the source. An x-component current sheet such as (10.129) contributes a source

$$I_{s1} = -J_x \cos\phi \tag{10.145}$$

to the system in (10.138) and (10.139) and a source

$$I_{s2} = -J_x \sin\phi \tag{10.146}$$

to the system in (10.142) and (10.143). These current sources are located at $z = 0$ and are depicted in the transmission lines shown in Figure 10.22.

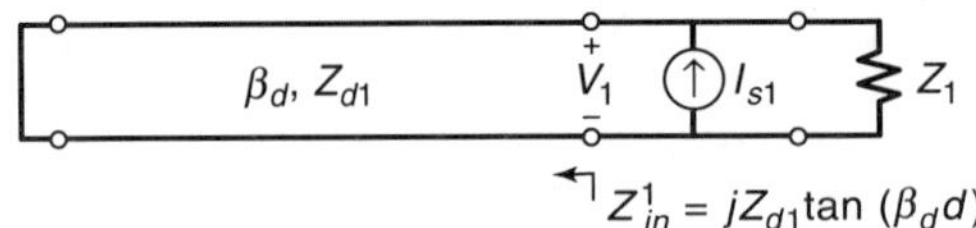

d

β_d, Z_{d2} $\quad V_2 \quad I_{s2} \quad Z_2$

$Z_{in}^2 = jZ_{d2}\tan(\beta_d d)$

$$V_1 = \frac{Z_{in}^1 Z_1}{Z_{in}^1 + Z_1} I_{s1}, \qquad V_2 = \frac{Z_{in}^2 Z_2}{Z_{in}^2 + Z_2} I_{s2}$$

Figure 10.22 Transmission line analogies.

From the transmission line analogy in Figure 10.22, we obtain

$$\tilde{E}_\rho(k_x, k_y) = V_1 = \frac{Z_1 Z_{d1}}{Z_{d1} - jZ_1 \cot(\beta_d d)}(-J_x \cos\phi) \tag{10.147}$$

$$-\tilde{E}_\phi(k_x, k_y) = V_2 = \frac{Z_2 Z_{d2}}{Z_{d2} - jZ_2 \cot(\beta_d d)}(-J_x \sin\phi) \tag{10.148}$$

where β_d is defined in (10.140) with k^2 replaced by $k^2\varepsilon_r$,

$$Z_{d1} = \frac{\beta_d}{\omega\varepsilon_d} \tag{10.149}$$

$$Z_{d2} = \frac{\omega\mu}{\beta_d} \tag{10.150}$$

and $\varepsilon_d = \varepsilon\varepsilon_r$ is the substrate permittivity. Consequently, the EFIE can be expressed in the form

$$E_x^{\text{inc}}(x, y) = -J_x * G_{xx} - J_y * G_{xy} \tag{10.151}$$

$$E_y^{\text{inc}}(x, y) = -J_x * G_{yx} - J_y * G_{yy} \tag{10.152}$$

where J_x and J_y are the components of the equivalent electric current density on the microstrip patch,

$$G_{xx}(x, y) = \frac{1}{(2\pi)^2}\frac{-1}{\omega\varepsilon}\int_{-\infty}^{\infty}\int_{-\infty}^{\infty}\frac{k^2 D_1 - k_x^2 D_3}{D_1 D_2} e^{jk_x x} e^{jk_y y}\, dk_x\, dk_y \tag{10.153}$$

$$G_{xy}(x, y) = G_{yx}(x, y) = \frac{1}{(2\pi)^2}\frac{-1}{\omega\varepsilon}\int_{-\infty}^{\infty}\int_{-\infty}^{\infty}\frac{-k_x k_y D_3}{D_1 D_2} e^{jk_x x} e^{jk_y y}\, dk_x\, dk_y \tag{10.154}$$

$$D_1 = \beta_d - j\beta\varepsilon_r \cot(\beta_d d) \tag{10.155}$$

$$D_2 = \beta - j\beta_d \cot(\beta_d d) \tag{10.156}$$

$$D_3 = \beta_d - j\beta \cot(\beta_d d) \tag{10.157}$$

and G_{yy} is the same as G_{xx} with k_x and k_y exchanged [6]. The asterisk in (10.151) and (10.152) denotes two-dimensional convolution, and it is implied that the result of the convolution is to be evaluated at (x, y). Equations (10.151) and (10.152) hold only on the location of the conducting patch. The incident electric field is evaluated in the presence of the substrate and ground plane, but in the absence of the patch.

For implementation purposes, it is sometimes convenient to convert the EFIE into the form of a mixed-potential integral equation [36, 37]

$$\bar{E}_{\tan}^{\text{inc}}(x, y) = \left(j\omega\mu\{\bar{J} * G_A\} + \nabla\left\{\frac{\rho_s}{\varepsilon} * G_\Phi\right\}\right)_{\tan} \tag{10.158}$$

where

$$G_A(x, y) = \frac{1}{(2\pi)^2}\int_{-\infty}^{\infty}\int_{-\infty}^{\infty}\frac{-j}{D_2}e^{jk_x x}e^{jk_y y}\,dk_x\,dk_y$$
$$= \frac{1}{2\pi}\int_0^{\infty}\frac{-jk_\rho J_0(k_\rho\rho)}{D_2}\,dk_\rho \tag{10.159}$$

$$G_\Phi(x, y) = \frac{1}{(2\pi)^2}\int_{-\infty}^{\infty}\int_{-\infty}^{\infty}\frac{-jD_3}{D_1 D_2}e^{jk_x x}e^{jk_y y}\,dk_x\,dk_y$$
$$= \frac{1}{2\pi}\int_0^{\infty}\frac{-jk_\rho J_0(k_\rho\rho)D_3}{D_1 D_2}\,dk_\rho \tag{10.160}$$

and

$$\rho_s = \frac{-\nabla_s \cdot \bar{J}}{j\omega} \tag{10.161}$$

The change of variables used to eliminate the double integrals is similar to that proposed in (10.130)–(10.133). The integrals in (10.159) and (10.160) are known as Sommerfeld integrals and are to be evaluated along a path of integration in the complex plane slightly above the real axis in order to avoid the singularities at zeros of the denominators. These integrals are further complicated by the growing oscillatory nature of the function $k_\rho J_0(k_\rho\rho)$ in the numerators. A special procedure known as the method of averages has been developed to combat the oscillation; other procedures can be applied to treat the integral in the vicinity of the surface wave poles [36, 37, 40, 41].

The numerical solution of the EFIE for microstrip proceeds as discussed in previous formulations and can be closely patterned after the treatment of conducting scatterers in free space (Sections 10.1 and 10.2). The essential difference is the evaluation of the Green's functions in (10.159) and (10.160).

Microstrip antenna analysis requires a model for the particular feed of interest (such as a probe feed or a microstrip line feed). In common with wire antennas, the input impedance is very sensitive to the specific feed region geometry, so the feed model must be tailored to the antenna under consideration. A current element can be used as a simple probe feed model if the substrate thickness is less than 0.02 wavelengths [35]. Microstrip line feeds that reside in the plane of the patch can be modeled as part of the integral equation analysis using additional basis functions to represent their currents. The representation can be terminated some distance from the patch where the currents become uniform [7, 40]. Other, more sophisticated methods for modeling the feeds have also been investigated [42, 43].

Arrays of microstrip patches can be analyzed by combining the ideas of this section with those of Section 10.9. The Fourier transform methodology is easily adapted to develop Green's functions for an infinite periodic array on a dielectric substrate. We refer the reader to the literature for a detailed discussion of the treatment of antenna arrays [6, 43]. Green's functions for other multilayered devices can be developed in a similar manner [38, 39].

10.11 A BRIEF SURVEY OF VOLUME INTEGRAL FORMULATIONS FOR HETEROGENEOUS DIELECTRIC BODIES

To close this chapter, we briefly consider the analysis of three-dimensional heterogeneous scatterers using volume integral equations. Volume discretizations require numerous unknowns and are quite limited in the three-dimensional case. While largely superseded by the three-dimensional differential equation formulations (Chapters 11 and 12), volume integral methods may be preferred in situations involving electrically small structures in the vicinity of other scatterers being analyzed with integral equations or in situations where a thin heterogeneous structure is not practical for analysis using differential equation approaches. They also may be useful for validating differential equation formulations.

Suppose a three-dimensional dielectric scatterer can be characterized by a relative permittivity function $\varepsilon_r(x, y, z)$. Using the volumetric equivalence principle (Chapter 1), the dielectric material may be replaced by an equivalent electric current density

$$\bar{J}(x, y, z) = j\omega\varepsilon_0[\varepsilon_r(x, y, z) - 1]\bar{E}(x, y, z) \tag{10.162}$$

radiating in free space. An electric field integral equation for the unknown current is

$$\bar{E}^{\text{inc}} = \bar{E} - \frac{\nabla\nabla\cdot + k^2}{j\omega\varepsilon_0}\bar{A} \tag{10.163}$$

where

$$\bar{A}(x, y, z) = \iiint \bar{J}(x', y', z')\frac{e^{-jkR}}{4\pi R}\, dx'\, dy'\, dz' \tag{10.164}$$

and

$$R = \sqrt{(x - x')^2 + (y - y')^2 + (z - z')^2} \tag{10.165}$$

Equation (10.163) is valid both inside and outside the original scatterer but is imposed throughout the interior to find the equivalent current density.

A number of different approaches have been proposed for discretizing Equation (10.163). The most widely used procedure employs a cubical-cell model for the scatterer, a pulse basis expansion for the current $\bar{J}$ or the electric field $\bar{E}$, and Dirac delta testing functions [44]. Unfortunately, in common with the two-dimensional case presented in Section 2.6, a cubical-cell pulse basis discretization proves unstable as the relative permittivity increases in magnitude [45–47]. However, since the approach has been widely used, we summarize it below.

Suppose that the three-dimensional scatterer is divided into N homogeneous cells that are approximately cubical and the polarization current density is represented by the superposition of $3N$ subsectional pulse basis functions

$$\bar{J}(x, y, z) \cong \sum_{n=1}^{N}(\hat{x}j_{xn} + \hat{y}j_{yn} + \hat{z}j_{zn})p_n(x, y, z) \tag{10.166}$$

where

$$p_n(x, y, z) = \begin{cases} 1 & \text{if } (x, y, z) \in \text{cell } n \\ 0 & \text{otherwise} \end{cases} \tag{10.167}$$

Equation (10.163) may be separated into components and enforced at the center of each cell to yield a $3N \times 3N$ system

$$\begin{bmatrix} \mathbf{G}^{xx} & \mathbf{G}^{xy} & \mathbf{G}^{xz} \\ \mathbf{G}^{yx} & \mathbf{G}^{yy} & \mathbf{G}^{yz} \\ \mathbf{G}^{zx} & \mathbf{G}^{zy} & \mathbf{G}^{zz} \end{bmatrix} \begin{bmatrix} \mathbf{j}_x \\ \mathbf{j}_y \\ \mathbf{j}_z \end{bmatrix} = \begin{bmatrix} \mathbf{E}_x^{\text{inc}} \\ \mathbf{E}_y^{\text{inc}} \\ \mathbf{E}_z^{\text{inc}} \end{bmatrix} \tag{10.168}$$

with off-diagonal entries

$$G_{mn}^{uv} = \frac{-1}{j\omega\varepsilon_0} \frac{\partial^2 g_n}{\partial u \, \partial v}\bigg|_{(x=x_m, y=y_m, z=z_m)} \qquad m \neq n \qquad u \neq v \tag{10.169}$$

$$G_{mn}^{uu} = \frac{-1}{j\omega\varepsilon_0} \left(\frac{\partial^2}{\partial u} + k^2 \right) g_n \bigg|_{(x=x_m, y=y_m, z=z_m)} \qquad m \neq n \tag{10.170}$$

and

$$G_{mm}^{uv} = 0 \qquad u \neq v \tag{10.171}$$

where u and v represent x, y, or z, and

$$g_n(x, y, z) = \iiint_{\text{cell } n} \frac{e^{-jkR}}{4\pi R} \, dx' \, dy' dz' \tag{10.172}$$

Assuming that each cell can be approximated by a sphere (so that cell m has radius a_m), the diagonal entries may be evaluated approximately to produce (Prob. P1.15)

$$G_{mm}^{uu} \cong \frac{1}{j\omega\varepsilon_0(\varepsilon_r - 1)} + \frac{1}{j\omega\varepsilon_0} \left[1 - \tfrac{2}{3} e^{-jka_m}(1 + jka_m) \right] \tag{10.173}$$

The remaining integrands in (10.169) and (10.170) are nonsingular and can be evaluated by numerical quadrature. We leave the details of the necessary calculations to the reader.

The above formulation is similar in many respects to the approach presented in Section 2.6 for TE-wave scattering from infinite dielectric cylinders. Section 2.6 demonstrated that the pulse basis discretization of the EFIE did not work well for cylinders having large values of the relative permittivity. Similar instabilities occur in the three-dimensional case [45–47] and limit the application of this particular formulation to relatively low values of ε_r. Because of these limitations, several alternative formulations have been proposed. One such procedure uses cubical cells to discretize the EFIE but places the match points on cell faces instead of cell centers [47]. By retaining the cubical cell models, the approach of [47] is able to exploit perturbed Toeplitz symmetries within a CG–FFT solution (Section 4.12), which is advantageous since the matrix order grows quickly with scatterer size. [The matrix has order 3N, where N is the number of cells; to maintain accuracy, the cell density should be at least 1000 per cubic dielectric wavelength.]

Volume formulations with more flexible cell shapes have also been proposed. One approach employs the EFIE with piecewise linear basis functions defined on polyhedral cells [48]. The approach of Schaubert, Wilton, and Glisson uses a divergence-conforming CN/LT function defined on tetrahedral cells to discretize the EFIE [49]. Both of these approaches attempt to use higher order basis functions to better represent the equivalent charge distribution. Parametric basis functions, defined on curved cells, may provide a better model [50]. The MFIE, which may be less sensitive to charge modeling, has also been studied for scattering from three-dimensional dielectric bodies modeled with tetrahedral cells [51]. By explicitly incorporating the surface charge density into the formulation, it is possible to convert volume integral equations into volume–surface equations, which separate the current and charge densities and may make it easier to properly model these quantities [52, 53].

Since all volume integral equation formulations suffer from a rapid growth of the associated computational complexity with increasing scatterer size, the application of these techniques in three-dimensions has been extremely limited. It appears that differential equation formulations that produce sparse matrices are more practical when three-dimensional volume discretizations are required, and Chapter 11 considers that type of formulation.

10.12 SUMMARY

The numerical solution of several types of integral equations have been considered for three-dimensional scatterers and antennas. Approaches were developed for conducting and homogeneous bodies of general shape using subsectional vector basis functions. The treatment of wire structures was also briefly reviewed. To illustrate the use of a variety of three-dimensional Green's functions, a planar frequency-selective surface formulation was presented, and the Green's functions for a microstrip patch on a grounded substrate were developed using a transmission line analogy. Finally, volume integral equations for modeling three-dimensional heterogeneous bodies were summarized.

REFERENCES

[1] A. W. Glisson and D. R. Wilton, "Simple and efficient numerical methods for problems of electromagnetic radiation and scattering from surfaces," *IEEE Trans. Antennas Propagat.*, vol. AP-28, pp. 593–603, Sept. 1980.

[2] H. Akaike, "Block-Toeplitz matrix inversion," *SIAM J. Appl. Math.*, vol. 24, pp. 234–241, Mar. 1973.

[3] M. Hurst and R. Mittra, "Scattering center analysis for radar cross section modification," Electromagnetic Communication Laboratory Technical Report 84-12, UILU-ENG-84-2551, University of Illinois, Urbana, July 1984.

[4] K. R. Aberegg, A. Taguchi, and A. F. Peterson, "Application of higher-order vector basis functions to surface integral formulations," *Radio Sci.*, vol. 31, pp. 1207–1213, 1996.

[5] D. R. Wilton and S. Govind, "Incorporation of edge condition in moment method solutions," *IEEE Trans. Antennas Propagat.*, vol. AP-25, pp. 845–850, Nov. 1977.

[6] S. M. Wright, "Efficient analysis of infinite microstrip arrays on electrically thick substrates," Ph.D. dissertation, University of Illinois, Urbana, 1984.

[7] D. R. Tanner, "Numerical methods for the electromagnetic modeling of microstrip antennas and feed systems," Ph.D. dissertation, University of Illinois, Urbana, 1988.

[8] R. Mittra, C.-H. Chan, and T. Cwik, "Techniques for analyzing frequency selective surfaces—a review," *Proc. IEEE*, vol. 76, pp. 1593–1615, Dec. 1988.

[9] S. M. Rao, D. R. Wilton, and A. W. Glisson, "Electromagnetic scattering by surfaces of arbitrary shape," *IEEE Trans. Antennas Propagat.*, vol. AP-30, pp. 409–418, May 1982.

[10] S. Caorsi, D. Moreno, and F. Sidoti, "Theoretical and numerical treatment of surface integrals involving the free-space Green's function," *IEEE Trans. Antennas Propagat.*, vol. 41, pp. 1296–1301, Sept. 1993.

[11] R. D. Graglia, "On the numerical integration of the linear shape functions times the 3-D Green's function or its gradient on a plane triangle," *IEEE Trans. Antennas Propagat.*, vol. 41, pp. 1448–1455, Oct. 1993.

[12] D. R. Wilton, "Review of current status and trends in the use of integral equations in computational electromagnetics," *Electromagnetics*, vol. 12, pp. 287–341, 1992.

[13] D. R. Wilton, "Topological considerations in surface patch modeling," *Abstracts of the 1981 URSI Radio Science Meeting*, Boulder, CO, p. 9, Jan. 1981.

[14] D. R. Wilton and A. W. Glisson, "On improving the stability of the electric field integral equation at low frequencies," *Abstracts of the 1981 URSI Radio Science Meeting*, Los Angeles, CA, p. 24, June 1981.

[15] J. R. Mautz and R. F. Harrington, "An E-field solution for a conducting surface small or comparable to the wavelength," *IEEE Trans. Antennas Propagat.*, vol. AP-32, pp. 330–339, Apr. 1984.

[16] E. Arvas, R. F. Harrington, and J. R. Mautz, "Radiation and scattering from electrically small conducting bodies of arbitrary shape," *IEEE Trans. Antennas Propagat.*, vol. AP-34, pp. 66–77, Jan. 1986.

[17] K. Umashankar, A. Taflove, and S. M. Rao, "Electromagnetic scattering by arbitrary shaped three-dimensional homogeneous lossy dielectric objects," *IEEE Trans. Antennas Propagat.*, vol. AP-34, pp. 758–766, June 1986.

[18] J. Moore and R. Pizer, eds., *Moment Methods in Electromagnetics*, Letchworth: Research Studies Press, 1984.

[19] B. D. Popovic, M. B. Dragovic, and A. R. Djordjevic, *Analysis and Synthesis of Wire Antennas*, New York: Wiley, 1982.

[20] S. T. Li, J. C. Logan, J. W. Rockway, and D. W. S. Tam, *The MININEC System: Microcomputer Analysis of Wire Antennas*, Boston: Artech House, 1988.

[21] J. H. Richmond, "Radiation and scattering by thin-wire structures in the complex frequency domain," in *Computational Electromagnetics*, eds. E. K. Miller, L. Medgyesi-Mitschang, and E. H. Newman, New York: IEEE Press, 1992.

[22] G. J. Burke and A. J. Poggio, *Numerical Electromagnetics Code (NEC)—Method of Moments*, Technical Document 116, Naval Ocean System Center, San Diego, Jan. 1981.

[23] G. A. Thiele, "Wire antennas," in *Computer Techniques for Electromagnetics*, ed. R. Mittra, New York: Pergamon, 1973.

[24] A. J. Poggio, "Numerical solutions of integral equations of dipoles and slot antennas including active and passive loading," Ph.D. dissertation, University of Illinois, Urbana, 1969.

[25] A. F. Peterson, "Difficulties encountered when attempting to validate thin-wire formulations for linear dipole antennas," *Appl. Computat. Electromagnet. Soc. (ACES) J.*, vol. 4, Special Issue on Electromagnetics Computer Code Validation, pp. 25–40, 1989.

[26] C. D. Taylor and D. R. Wilton, "The extended boundary condition solution of the dipole antenna of revolution," *IEEE Trans. Antennas Propagat.*, vol. AP-20, pp. 772–776, Nov. 1972.

[27] D. H. Werner, J. A. Huffman, and P. L. Werner, "Techniques for evaluating the uniform current vector potential at the isolated singularity of the cylindrical wire kernel," *IEEE Trans. Antennas Propagat.*, vol. 42, pp. 1549–1553, Nov. 1994.

[28] A. C. Ludwig, "Wire grid modeling of surfaces," *IEEE Trans. Antennas Propagat.*, vol. AP-35, pp. 1045–1048, Sept. 1987.

[29] K. E. Jordan, G. R. Richter, and P. Sheng, "An efficient numerical evaluation of the Green's function for the Helmholtz operator on periodic structures," *J. Comp. Phys.*, vol. 63, pp. 222–235, 1986.

[30] A. W. Mathis and A. F. Peterson, "A comparison of acceleration procedures for the two-dimensional periodic Green's function," *IEEE Trans. Antennas Propagat.*, vol. 44, pp. 567–571, Apr. 1996.

[31] E. Cohen, "An Ewald transformation of frequency-domain integral formulations," *Electromagnetics*, vol. 15, pp. 427–439, 1995.

[32] R. S. Zich, ed., "Special issue on frequency selective surfaces," *Electromagnetics*, vol. 5, 1985.

[33] D. M. Pozar and D. H. Schaubert, eds., *Microstrip Antennas*, New York: IEEE Press, 1995.

[34] M. C. Bailey and M. D. Deshpande, "Integral equation formulation of microstrip antennas," *IEEE Trans. Antennas Propagat.*, vol. AP-30, pp. 615–656, July 1982.

[35] D. M. Pozar, "Input impedance and mutual coupling of rectangular microstrip antennas," *IEEE Trans. Antennas Propagat.*, vol. AP-30, pp. 1191–1196, Nov. 1982.

[36] J. R. Mosig and F. E. Gardiol, "Analytical and numerical techniques in the Green's function treatment of microstrip antennas and scatterers," *IEE Proc., Part H*, vol. 130, pp. 175–182, 1983.

[37] J. R. Mosig, "Arbitrarily shaped microstrip structures and their analysis with a mixed potential integral equation," *IEEE Trans. Microwave Theory Tech.*, vol. 36, pp. 314–323, Feb. 1988.

[38] V. W. Hansen, *Numerical Solution of Antennas in Layered Media*, Taunton: Research Studies Press, 1989.

[39] A. K. Bhattacharyya, *Electromagnetic Fields in Multilayered Structures*, Boston: Artech House, 1994.

[40] J. R. Mosig, "Integral equation technique," in *Numerical Techniques for Microwave and Millimeter-Wave Passive Structures*, ed. T. Itoh, New York: Wiley, 1989.

[41] L. T. Hildebrand and D. A. McNamara, "A guide to implementational aspects of the spatial-domain integral equation analysis of microstrip antennas," *ACES J.*, vol. 10, pp. 40–51, Mar. 1995.

[42] M. Davidovitz, "Feed analysis for microstrip antennas," Ph.D. dissertation, University of Illinois, Urbana, 1985.

[43] J. T. Aberle and D. M. Pozar, "Analysis of infinite arrays of probe-fed rectangular microstrip patches using a rigorous feed model," *IEE Proc., Part H*, vol. 136, pp. 110–119, 1989.

[44] D. E. Livesay and K. M. Chen, "Electromagnetic fields induced inside arbitrarily shaped biological bodies," *IEEE Trans. Microwave Theory Tech.*, vol. MTT-22, pp. 1273–1280, Dec. 1974.

[45] H. Massoudi, C. H. Durney, and M. F. Iskander, "Limitations of the cubical block model of man in calculating SAR distributions," *IEEE Trans. Microwave Theory Tech.*, vol. MTT-32, pp. 746–751, Aug. 1984.

[46] M. J. Hagmann and R. L. Levin, "Criteria for accurate usage of block models," *J. Microwave Power*, vol. 22, pp. 19–27, January 1987.

[47] C.-C. Su, "The three-dimensional algorithm of solving the electric-field integral equation using face-centered node points, conjugate gradient method, and FFT," *IEEE Trans. Microwave Theory Tech.*, vol. 41, pp. 510–515, Mar. 1993.

[48] C. T. Tsai, H. Massoudi, C. H. Durney, and M. F. Iskander, "A procedure for calculating fields inside arbitrarily shaped, inhomogeneous dielectric bodies using linear basis functions with the moment method," *IEEE Trans. Microwave Theory Tech.*, vol. MTT-34, pp. 1131–1139, Nov. 1986.

[49] D. H. Schaubert, D. R. Wilton, and A. W. Glisson, "A tetrahedral modeling method for electromagnetic scattering by arbitrarily shaped inhomogeneous dielectric bodies," *IEEE Trans. Antennas Propagat.*, vol. AP-32, pp. 77–85, Jan. 1984.

[50] R. D. Graglia, P. L. E. Uslenghi, and R. S. Zich, "'Moment method with isoparametric elements for three-dimensional anisotropic scatterers," *Proc. IEEE*, vol. 77, pp. 750–760, May 1989.

[51] A. F. Peterson, "A magnetic field integral equation formulation for electromagnetic scattering from inhomogeneous 3D dielectric bodies," *Proceedings of the Fifth Annual Review of Progress in Applied Computational Electromagnetics*, Monterey, CA, pp. 387–403, Mar. 1989.

[52] J. L. Volakis, "Alternative field representations and integral equations for modeling inhomogeneous dielectrics," *IEEE Trans. Microwave Theory Tech.*, vol. 40, pp. 604–608, Mar. 1992.

[53] A. D. Yaghjian and R. W. Wang, "Electric- and magnetic-field volume-surface integral equations for 3-D inhomogeneous scatterers," *Abstracts of the 1994 URSI Radio Science Meeting*, Seattle, WA, p. 15, June 1994.

PROBLEMS

P10.1 Using the aperture formulation in Chapter 1, Section 1.10, for guidance, modify the conducting-plate approach of Section 10.1 to obtain a method-of-moments equation for scattering from an aperture in an infinite p.e.c. plane. Treat the equivalent magnetic current density in the aperture as the primary unknown and use CN/LT basis functions on rectangular cells and razor-blade testing functions to discretize the integral equation. Describe in detail how the entries of the matrix equation (10.11) must be modified to convert the plate formulation into an aperture formulation.

P10.2 Modify the flat-plate formulation in Section 10.1 in order to use rectangular-cell CN/LT functions for both expansion and testing (Galerkin's method). Use convolution properties to express all of the necessary matrix entries in terms of two-dimensional integrals.

Provide expressions for the new matrix entries in terms of pulse, triangle, quadratic, and cubic spline functions (Chapter 5).

P10.3 Under what conditions is the matrix in (10.11), obtained using rectangular-cell CN/LT basis functions and razor-blade testing functions, symmetric across its main diagonal? Is the matrix produced in Prob. P10.2 with Galerkin's method always symmetric?

P10.4 The p.e.c. plate formulation of Section 10.1 can be extended in order to model material plates that satisfy the resistive shell boundary condition

$$\bar{E} = R\bar{J}_s$$

where R is defined in Prob. P2.36. Describe the modifications to the entries of (10.11) needed to incorporate this boundary condition.

P10.5 The CN/LT basis, razor-blade testing formulation from Section 10.1 involves only two fundamentally different expressions for the entries of the method-of-moments system, Equations (10.29) and (10.30). By studying the LN/QT basis functions introduced at the end of Section 10.1, determine how many different expressions arise in an LN/QT basis/razor-blade testing scheme for flat conducting plates discretized into uniform rectangular cells.

P10.6 Equation (10.33) is an expression for the integral of 1/R over a rectangular domain. Provide a similar expression for a triangular domain, with the observation point (x, y) arbitrarily located within the triangle.

P10.7 (a) A scatterer with top-to-bottom symmetry illuminated by a source with the same symmetry enables a 50% reduction in the required number of unknowns. Describe the process of constructing the method-of-moments matrix in order to exploit the symmetry.

(b) Suppose the scatterer in part (a) is symmetric but the illumination is not. Can the order of the method-of-moments matrix still be reduced by exploiting symmetry? What additional price must be paid in this case?

P10.8 Section 10.3 presents an MFIE formulation using CN/LT basis functions. A simpler approach similar to that used in [22] to model closed conducting bodies can be obtained using pulse basis functions and Dirac delta testing functions to discretize the three-dimensional MFIE. Suppose the scatterer is represented by triangular patches, so that the outward normal vector to the nth patch is

$$\hat{n}_n = \sin\theta_n \cos\phi_n \hat{x} + \sin\theta_n \sin\phi_n \hat{y} + \cos\theta_n \hat{z}$$

where (θ_n, ϕ_n) denote conventional spherical coordinate angles. On each cell, the current density is decomposed into two orthogonal components J_u and J_v, where

$$\hat{u}_n = -\sin\phi_n \hat{x} + \cos\phi_n \hat{y}$$

$$\hat{v}_n = \hat{n}_n \times \hat{u}_n = -\cos\theta_n \cos\phi_n \hat{x} - \cos\theta_n \sin\phi_n \hat{y} + \sin\theta_n \hat{z}$$

Suppose that pulse basis functions are used to represent $\bar{J}$ according to

$$\bar{J}(u, v) \cong \sum_{n=1}^{N} (\hat{u}_n j_{un} + \hat{v}_n j_{vn}) p_n(u, v)$$

where

$$p_n(u, v) = \begin{cases} 1 & \text{if } (u, v) \in \text{cell } n \\ 0 & \text{otherwise} \end{cases}$$

and Dirac delta testing functions are used to enforce the $\hat{u}$- and $\hat{v}$-components of the

MFIE

$$\hat{n} \times \bar{H}^{\text{inc}} = \bar{J}(u, v) - \hat{n} \times (\nabla \times \bar{A})_{S^+}$$

where

$$\bar{A}(x, y, z) = \iint \bar{J}(u', v') \frac{e^{-jkR}}{4\pi R} \, du' \, dv'$$

and where R is the distance from the source point (u', v') to the observation point (x, y, z), also located on the surface. Show that the entries of the method-of-moments system

$$\begin{bmatrix} \mathbf{A}^{uu} & \mathbf{A}^{uv} \\ \mathbf{A}^{vu} & \mathbf{A}^{vv} \end{bmatrix} \begin{bmatrix} \mathbf{j}_u \\ \mathbf{j}_v \end{bmatrix} = \begin{bmatrix} \mathbf{H}_u^{\text{inc}} \\ \mathbf{H}_v^{\text{inc}} \end{bmatrix}$$

are given by

$$\begin{aligned}
A_{mn}^{uu} &= \cos\theta_m \cos(\phi_m - \phi_n) \frac{\partial g_n}{\partial z} \\
&\quad + \sin\theta_m \left(\sin\phi_n \frac{\partial g_n}{\partial y} + \cos\phi_n \frac{\partial g_n}{\partial x} \right) \qquad m \neq n \\
A_{mn}^{uv} &= \cos\theta_m \cos\theta_n \sin(\phi_m - \phi_n) \frac{\partial g_n}{\partial z} \\
&\quad - \cos\theta_m \sin\theta_n \left(\cos\phi_m \frac{\partial g_n}{\partial y} - \sin\phi_m \frac{\partial g_n}{\partial x} \right) \\
&\quad + \sin\theta_m \cos\theta_n \left(\cos\phi_n \frac{\partial g_n}{\partial y} - \sin\phi_n \frac{\partial g_n}{\partial x} \right) \qquad m \neq n \\
A_{mn}^{vu} &= -\sin(\phi_m - \phi_n) \frac{\partial g_n}{\partial z} \qquad m \neq n \\
A_{mn}^{vv} &= \sin\theta_n \left(\sin\phi_m \frac{\partial g_n}{\partial y} + \cos\phi_m \frac{\partial g_n}{\partial x} \right) \\
&\quad + \cos\theta_n \cos(\phi_m - \phi_n) \frac{\partial g_n}{\partial z} \qquad m \neq n \\
A_{mm}^{uu} &= A_{mm}^{vv} = \tfrac{1}{2}
\end{aligned}$$

and

$$A_{mm}^{uv} = A_{mm}^{vu} = 0$$

where

$$g_n(x, y, z) = \iint_{\text{patch } n} \frac{e^{-jkR}}{4\pi R} \, du' \, dv'$$

[*Hint:* First show that the scattered magnetic field at patch m can be written in the form

$$\begin{aligned}
\hat{n}_m \times \bar{H}^s &= \hat{u}_m (\cos\theta_m \cos\phi_m H_x^s + \cos\theta_m \sin\phi_m H_y^s - \sin\theta_m H_z^s) \\
&\quad + \hat{v}_m (-\sin\phi_m H_x^s + \cos\phi_m H_y^s)
\end{aligned}$$

Then, combine this result with the expressions

$$H_x^s = \sin\theta_n j_{vn}\frac{\partial g_n}{\partial y} - (\cos\phi_n j_{un} - \cos\theta_n \sin\phi_n j_{vn})\frac{\partial g_n}{\partial z}$$

$$H_y^s = -(\sin\phi_n j_{un} + \cos\theta_n \cos\phi_n j_{vn})\frac{\partial g_n}{\partial z} - \sin\theta_n j_{vn}\frac{\partial g_n}{\partial x}$$

$$H_z^s = (\cos\phi_n j_{un} - \cos\theta_n \sin\phi_n j_{vn})\frac{\partial g_n}{\partial x} + (\sin\phi_n j_{un} + \cos\theta_n \cos\phi_n j_{vn})\frac{\partial g_n}{\partial y}$$

to complete the derivation.] Finally, provide a brief discussion of the numerical implementation of this approach.

P10.9 An alternate CFIE for homogeneous dielectric scatterers can be obtained by constructing a linear combination of the exterior EFIE and MFIE and a linear combination of the interior EFIE and MFIE. How does such a formulation differ from Equations (10.82) and (10.83)?

P10.10 Verify the coordinate transformation used in (10.107)–(10.112).

P10.11 Suppose the frequency-selective surface (FSS) formulation of Section 10.9 is implemented using rectangular-cell CN/LT basis functions and razor-blade testing functions with uniform cell sizes (as in the aperiodic plate formulation of Section 10.1). Suppose further that the only acceleration procedure employed is that obtained by combining the Fourier transform of the basis and testing functions with G_p in (10.120). Provide explicit expressions for the matrix entries, and identify the rate of convergence of each summation.

P10.12 Recast the conducting patch FSS formulation in Section 10.9 in order to treat a doubly periodic array of apertures using the equivalent magnetic current density in the aperture of the unit cell as the primary unknown. Provide the new integral equation in a form similar to (10.115) and (10.116) in terms of the periodic Green's function G_p defined in (10.117).

P10.13 Using the exponentially convergent summations in (10.123) and (10.126), provide expressions for the matrix entries associated with the conducting patch FSS formulation using rectangular-cell CN/LT basis functions and razor-blade testing functions.

P10.14 After decomposing the fields into longitudinal and transverse parts,

$$\bar{E} = \hat{z}E_z + \bar{E}_t$$

$$\bar{H} = \hat{z}H_z + \bar{H}_t$$

Maxwell's curl equations for a uniform medium with sources $(\bar{J}, \bar{K})$ can be rewritten in the form

$$-\nabla_t \cdot (\hat{z} \times \bar{E}_t) = -j\omega\mu H_z - K_z$$

$$-\nabla_t \cdot (\hat{z} \times \bar{H}_t) = j\omega\varepsilon E_z + J_z$$

$$-\frac{\partial \bar{E}_t}{\partial z} + \nabla_t E_z = -j\omega\mu(\hat{z} \times \bar{H}_t) - \hat{z} \times \bar{K}_t$$

$$-\frac{\partial \bar{H}_t}{\partial z} + \nabla_t H_z = j\omega\varepsilon(\hat{z} \times \bar{E}_t) + \hat{z} \times \bar{J}_t$$

where the transverse operators are defined as

$$\nabla_t \cdot \bar{A} = \frac{\partial A_x}{\partial x} + \frac{\partial A_y}{\partial y} \qquad \nabla_t \Psi = \hat{x}\frac{\partial \Psi}{\partial x} + \hat{y}\frac{\partial \Psi}{\partial y}$$

By applying the two-dimensional Fourier transform of Equation (10.127) to the above equations and eliminating the z-components of the fields, show that the transformed equations can be expressed as

$$-\frac{\partial \tilde{E}_x}{\partial z} - \frac{j}{\omega\varepsilon}(k^2 - k_x^2)\tilde{H}_y - \frac{jk_x k_y}{\omega\varepsilon}\tilde{H}_x = \tilde{K}_y + \frac{k_x}{\omega\varepsilon}\tilde{J}_z$$

$$-\frac{\partial \tilde{E}_y}{\partial z} + \frac{j}{\omega\varepsilon}(k^2 - k_y^2)\tilde{H}_x + \frac{jk_x k_y}{\omega\varepsilon}\tilde{H}_y = -\tilde{K}_x + \frac{k_y}{\omega\varepsilon}\tilde{J}_z$$

$$-\frac{\partial \tilde{H}_x}{\partial z} + \frac{j}{\omega\mu}(k^2 - k_x^2)\tilde{E}_y + \frac{jk_x k_y}{\omega\mu}\tilde{E}_x = -\tilde{J}_y + \frac{k_x}{\omega\mu}\tilde{K}_z$$

$$-\frac{\partial \tilde{H}_y}{\partial z} - \frac{j}{\omega\mu}(k^2 - k_y^2)\tilde{E}_x - \frac{jk_x k_y}{\omega\mu}\tilde{E}_y = \tilde{J}_x + \frac{k_y}{\omega\mu}\tilde{K}_z$$

where $k^2 = \omega^2\mu\varepsilon$. These equations are functions of the variable z and the parameters k_x and k_y and are similar in form to coupled transmission line equations.

P10.15 Use the transmission line analogy developed in Section 10.10 to derive the Green's function for the electric field produced by an x-directed point source on the surface of a dielectric slab of thickness d. Then, use the Green's function to extend the conducting patch FSS formulation to the case of p.e.c. patches located on the surface of a dielectric slab. First provide the integral equation in a form similar to (10.151) and (10.152), with the appropriate Green's functions expressed in terms of inverse Fourier transform integrals. Then, introduce basis and testing functions $B(x, y)$ and $T(x, y)$ and combine the Poisson sum transformation with the convolution operation to obtain the matrix entries in the form of a double summation over the transforms $\tilde{B}$ and $\tilde{T}$.

11

Frequency-Domain Differential Equation Formulations for Open Three-Dimensional Problems

In the two-dimensional formulations considered in Chapters 2 and 3, differential equations are shown to offer computational advantages over volume integral equations because of the sparsity of the resulting matrix. Similar advantages arise in three dimensions. In this chapter, the three-dimensional vector Helmholtz equation is used as the basis for several open-region formulations. Radiation boundary conditions are developed for spherical and nonspherical boundary shapes. Discretizations are obtained using mixed-order curl-conforming basis functions, whose properties have been previously explored in Chapter 9. A formulation for general axisymmetric scatterers is developed. Finally, an alternative approach incorporating traditional node-based Lagrangian basis functions is reviewed. While there has been much progress in this field during recent years, three-dimensional differential equation implementations are far from mature, and their performance has not been evaluated as extensively as the methods described in previous chapters.

11.1 WEAK VECTOR HELMHOLTZ EQUATION AND BOUNDARY CONDITIONS

Three-dimensional time-harmonic electromagnetic fields can be expressed in terms of Maxwell's equations, the vector Helmholtz equations, or several different formulations involving scalar and vector potential functions. Although numerical approaches can be developed for any of these equations, this chapter emphasizes schemes involving the curl–curl form of the vector Helmholtz equations. These equations involve only one type of unknown field, $\bar{E}$ or $\bar{H}$, and are similar in appearance to the scalar Helmholtz equations used in previous chapters. In a source-free isotropic region, the vector Helmholtz equations are

$$\nabla \times \left(\frac{1}{\mu_r} \nabla \times \bar{E} \right) = k^2 \varepsilon_r \bar{E} \tag{11.1}$$

$$\nabla \times \left(\frac{1}{\varepsilon_r} \nabla \times \bar{H} \right) = k^2 \mu_r \bar{H} \tag{11.2}$$

These equations are obtained from Maxwell's curl equations (1.1) and (1.2) and sometimes admit solutions that fail to satisfy Gauss' laws in (1.3) and (1.4). This issue is explored in Chapter 9, where it is shown that the eigenfunctions of the curl–curl operator fall into two families, one of which has zero curl and cannot represent source-free electromagnetic fields (the "nullspace" family). The other eigenfunction family represents true solutions of Maxwell's equations. Although both eigenfunction families may be present in the discrete matrix operator arising from a numerical discretization, the nullspace eigenfunctions can be separated from the true solutions through the use of special curl-conforming basis functions developed in Chapter 9. Consequently, we use these basis functions throughout this chapter.

The vector Helmholtz equation (11.1) can be converted into a weak form by constructing a dot product with a vector testing function $\bar{T}$ and using the vector identity

$$\bar{T} \cdot \nabla \times \bar{A} = \nabla \times \bar{T} \cdot \bar{A} - \nabla \cdot (\bar{T} \times \bar{A}) \tag{11.3}$$

the divergence theorem

$$\iiint_{\Gamma} \nabla \cdot (\bar{T} \times \bar{A})\, dv = \iint_{\partial\Gamma} (\bar{T} \times \bar{A}) \cdot \hat{n}\, dS \tag{11.4}$$

and the vector identity

$$(\bar{T} \times \bar{A}) \cdot \hat{n} = -\bar{T} \cdot (\hat{n} \times \bar{A}) \tag{11.5}$$

to produce

$$\iiint_{\Gamma} \frac{1}{\mu_r} \nabla \times \bar{T} \cdot \nabla \times \bar{E} - k^2 \varepsilon_r \bar{T} \cdot \bar{E} = -\iint_{\partial\Gamma} \frac{1}{\mu_r} \bar{T} \cdot \hat{n} \times (\nabla \times \bar{E}) \tag{11.6}$$

$$\iiint_{\Gamma} \frac{1}{\varepsilon_r} \nabla \times \bar{T} \cdot \nabla \times \bar{H} - k^2 \mu_r \bar{T} \cdot \bar{H} = -\iint_{\partial\Gamma} \frac{1}{\varepsilon_r} \bar{T} \cdot \hat{n} \times (\nabla \times \bar{H}) \tag{11.7}$$

where Γ denotes the region of interest, $\partial\Gamma$ denotes the boundary of that region, and $\hat{n}$ is the outward normal vector along the boundary. As in the two-dimensional case, the weak equation relaxes the differentiability requirements imposed on the field representation. In deriving (11.6) and (11.7), we have assumed the boundedness of the curl operations $\nabla \times \bar{T}$, $\nabla \times \bar{E}$, and $\nabla \times \bar{H}$. These derivatives are bounded if the field representation maintains tangential continuity throughout the computational domain. If the basis functions ultimately used for these quantities do not ensure tangential continuity between cells, Dirac delta functions occur along the interfaces and must be included in (11.6) and (11.7). [Equivalently, additional boundary integrals similar in form to the right-hand sides of (11.6) and (11.7) must be included along the cell interfaces.]

The boundary integrals on the right-hand side of (11.6) and (11.7) provide the means for incorporating boundary conditions. If $\partial\Gamma$ is a perfect electric conductor, the boundary condition

$$\hat{n} \times \bar{E} = 0 \tag{11.8}$$

can be used with the weak equation (11.6). The imposition of (11.8) requires that the field representation used for $\bar{E}$ satisfy (11.8) along $\partial\Gamma$ in the strong sense (exactly). In essence, this is equivalent to the Dirichlet type of boundary condition used in the scalar case. It is also known as an *essential* boundary condition since it must be imposed in order to be satisfied in the limit of an infinite number of unknowns. Since the tangential component of the field representation is specified along $\partial\Gamma$, there will be no need for testing functions with a nonzero tangential component along $\partial\Gamma$. In other words, the testing functions used to construct the system of equations should also satisfy (11.8). It follows that the boundary integral in (11.6) does not contribute to the system of equations and can be ignored.

The boundary condition

$$\hat{n} \times (\nabla \times \bar{H}) = 0 \tag{11.9}$$

also describes a perfect electric conductor and is applicable to the weak equation (11.7). The imposition of (11.9) is accomplished by discarding the boundary integral on the right-hand side of (11.7). This condition is similar in form to the Neumann type of boundary condition used in the scalar case and is also known as a *natural* boundary condition since it does not have to be exactly satisfied by the field representation in order to be realized in the limiting case (an infinite number of unknowns). By carving a region out of the mesh and using the same field representation within the surrounding cells as is used throughout the interior volume (in other words, doing nothing special at the boundary), (11.9) will be automatically imposed and the procedure will treat the region as a perfect electric conductor.

Scattering problems are usually posed in the context of an RBC on $\partial\Gamma$. In practice, the RBC is of the form

$$\hat{n} \times (\nabla \times \bar{E}^s) = L(\bar{E}^s) \tag{11.10}$$

or

$$\hat{n} \times (\nabla \times \bar{H}^s) = L(\bar{H}^s) \tag{11.11}$$

where L is some linear operator. These expressions are substituted directly into the boundary integrals appearing on the right-hand side of the weak equations, after which the scattered field may be combined with the incident field to leave the undetermined quantities entirely in terms of the total field. As in the two-dimensional case discussed in Chapter 3, exact RBCs are global in nature and couple the field around the entire boundary. Thus, they create fill-in within the global matrix equation. For improved numerical efficiency, approximate local RBCs have been developed that preserve the matrix sparsity. Several different RBCs will be discussed in the following sections.

11.2 DISCRETIZATION USING CT/LN AND LT/QN FUNCTIONS FOR THREE-DIMENSIONAL CAVITIES [1]

To illustrate the discretization procedure, we initially consider the numerical solution of the weak equation in (11.6) or (11.7) for closed cavities bounded by perfect electric walls. We will also use this example to illustrate the relative accuracy of the CT/LN and LT/QN functions defined in Table 9.12 for tetrahedral cells. These basis functions belong to Nedelec's curl-conforming spaces [2], and they provide tangential continuity between cells and ensure that the nullspace eigenfunctions are separated from the true physical solutions.

Consider a closed cavity bounded by perfect electric walls. Equation (11.6) can be written in the form of an eigenvalue equation for the resonant wavenumber

$$\iiint_\Gamma \frac{1}{\mu_r} \nabla \times \bar{T} \cdot \nabla \times \bar{E} = k^2 \iiint_\Gamma \varepsilon_r \bar{T} \cdot \bar{E} \tag{11.12}$$

where the boundary integral has been dropped since it does not contribute for p.e.c. walls as long as (11.8) is imposed on the expansion functions used for $\bar{E}$. The cavity interior can be discretized into tetrahedral cells, and the electric field represented by vector basis functions

$$\bar{E}(x, y, z) \cong \sum_{n=1}^{N} e_n \bar{B}_n(x, y, z) \tag{11.13}$$

Consider the CT/LN basis functions from Table 9.12, which can be written in the form

$$\bar{B}_n = w_n (L_i \nabla L_j - L_j \nabla L_i) \tag{11.14}$$

where (L_1, L_2, L_3, L_4) are simplex coordinates (Section 9.6) and w_n denotes the length of the cell edge n, which is located between nodes i and j. Each CT/LN function interpolates to the tangential component along an edge of the tetrahedral-cell mesh. If the CT/LN functions are also used for testing, and the cavity is homogeneous, an eigenvalue equation $\mathbf{Ae} = k^2 \mathbf{Be}$ is obtained, where

$$A_{mn} = \iiint_\Gamma \nabla \times \bar{B}_m \cdot \nabla \times \bar{B}_n \, dx\, dy\, dz \tag{11.15}$$

$$B_{mn} = \iiint_\Gamma \bar{B}_m \cdot \bar{B}_n \, dx\, dy\, dz \tag{11.16}$$

As in the two-dimensional case, the global matrices can be constructed on a cell-by-cell basis using 6×6 element matrices that represent the evaluation of (11.15) and (11.16) over a single tetrahedron. Entries of the element matrices can be found using simplex coordinates, which for a given cell can be defined by the parameters (Section 9.6) associated with the transformation

$$L_i = \frac{1}{6V}(a_i + b_i x + c_i y + d_i z) \tag{11.17}$$

from Cartesian coordinates to simplex coordinates, where V denotes the cell volume. It follows that

$$\nabla L_i = \frac{1}{6V}(b_i \hat{x} + c_i \hat{y} + d_i \hat{z}) \tag{11.18}$$

For convenience, define

$$\bar{v}_{ij} = \nabla L_i \times \nabla L_j = \frac{\hat{x}(c_i d_j - d_i c_j) + \hat{y}(d_i b_j - b_i d_j) + \hat{z}(b_i c_j - c_i b_j)}{36V^2} \tag{11.19}$$

$$\Phi_{ij} = \nabla L_i \cdot \nabla L_j = \frac{b_i b_j + c_i c_j + d_i d_j}{36V^2} \tag{11.20}$$

These quantities are constant within a cell and can be determined from the coordinates of the four nodes that define the tetrahedron. Note that $\Phi_{ij} = \Phi_{ji}$ and $\bar{v}_{ij} = -\bar{v}_{ji}$.

The curl of a CT/LN basis function is given by

$$\nabla \times \bar{B}_n = w_n \nabla \times (L_i \nabla L_j - L_j \nabla L_i) = 2 w_n \bar{v}_{ij} \tag{11.21}$$

Therefore, the element matrix entries for **A** in (11.15) are

$$A_{mn} = 4V w_m w_n \bar{v}_{pq} \cdot \bar{v}_{ij} \tag{11.22}$$

where p and q denote the node indices associated with edge m and i and j are the node indices for edge n. The element matrix entries for **B** can be obtained using

$$\begin{aligned} \bar{B}_m \cdot \bar{B}_n &= w_m w_n (L_p \nabla L_q - L_q \nabla L_p) \cdot (L_i \nabla L_j - L_j \nabla L_i) \\ &= w_m w_n (L_p L_i \Phi_{qj} - L_q L_i \Phi_{pj} - L_p L_j \Phi_{qi} + L_q L_j \Phi_{pi}) \end{aligned} \tag{11.23}$$

and have the form

$$B_{mn} = w_m w_n V (\Phi_{qj} M_{pi} - \Phi_{pj} M_{qi} - \Phi_{qi} M_{pj} + \Phi_{pi} M_{qj}) \tag{11.24}$$

where

$$M_{ij} = \frac{1}{V} \iiint_{\text{cell}} L_i L_j \, dv = \frac{1}{20} \begin{cases} 2 & i = j \\ 1 & \text{otherwise} \end{cases} \tag{11.25}$$

Equation (11.25) is evaluated using the integral formula [3]

$$\iiint (L_1)^i (L_2)^j (L_3)^k (L_4)^m \, dv = V \frac{3!i!j!k!m!}{(3+i+j+k+m)!} \tag{11.26}$$

This completes the development of the element matrix entries for the CT/LN functions. As the global matrix is constructed in a cell-by-cell manner, entries that correspond to edges along the p.e.c. boundary are omitted to impose the essential boundary condition in (11.8). It is also necessary to ensure that the vector orientation of a CT/LN function is the same in all the cells bordering a common edge; a simple approach is to assign nodes i and j in (11.14) so that the vector always points from the smaller index to the larger.

Element matrix entries of order 20 can be constructed for LT/QN basis and testing functions from Table 9.12 in a similar manner. The development is complicated by the fact that there are two different types of LT/QN functions. There are two functions associated with each edge that can be expressed as

$$\bar{B}_n^{e1} = w_n L_i \nabla L_j \tag{11.27}$$

$$\bar{B}_n^{e2} = w_n L_j \nabla L_i \tag{11.28}$$

where i and j denote the node indices associated with edge n. There are also two functions associated with each face having the form

$$\bar{B}_n^{f1} = L_i L_j \nabla L_k - L_i L_k \nabla L_j \tag{11.29}$$

$$\bar{B}_n^{f2} = L_i L_j \nabla L_k - L_j L_k \nabla L_i \tag{11.30}$$

where i, j, k denote the indices of the nodes associated with face n. (The specific node-numbering scheme used in the construction to follow is depicted in Figure 11.1.)

From the numbering scheme in Figure 11.1 and the previous expressions, one can deduce that

$$\nabla \times \bar{B}_n^{e1} = w_n \nabla \times (L_i \nabla L_j) = w_n \bar{v}_{ij} \tag{11.31}$$

$$\nabla \times \bar{B}_n^{e2} = -w_n \bar{v}_{ij} \tag{11.32}$$

$$\nabla \times \bar{B}_n^{f1} = 2L_i \bar{v}_{jk} + L_j \bar{v}_{ik} - L_k \bar{v}_{ij} \tag{11.33}$$

$$\nabla \times \bar{B}_n^{f2} = L_i \bar{v}_{jk} + 2L_j \bar{v}_{ik} + L_k \bar{v}_{ij} \tag{11.34}$$

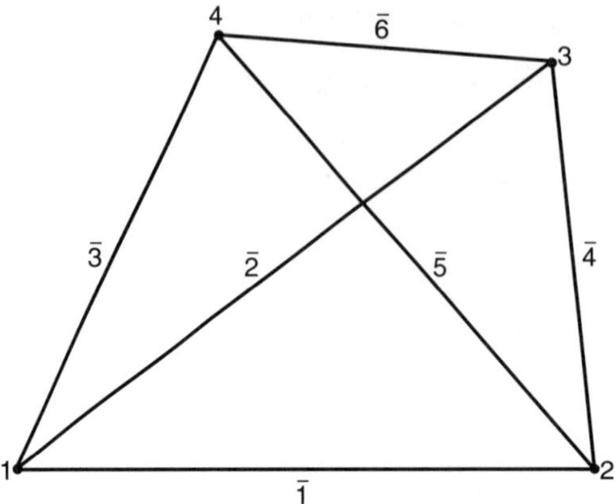

Local face, edge, and node conventions

Edge number	Node 1	Node 2	
1	1	2	
2	1	3	
3	1	4	
4	2	3	
5	2	4	
6	3	4	
Face number	Node 1	Node 2	Node 3
1	1	2	3
2	1	2	4
3	1	3	4
4	2	3	4

Figure 11.1 Node and edge indices for a tetrahedral cell. After [1]. ©1996 IEEE.

These relations are used to obtain the entries of the element matrix **A**, which are summarized in Table 11.1. Analogous expressions can be obtained for the entries of element matrix **B** using the quantities

$$N_{ijk} = \frac{1}{V} \iiint_{\text{cell}} L_i L_j L_k \, dv \tag{11.35}$$

$$P_{ijkm} = \frac{1}{V} \iiint_{\text{cell}} L_i L_j L_k L_m \, dv \tag{11.36}$$

obtained from (11.26). The entries of **B** are provided in Table 11.2.

To illustrate the performance of the tetrahedral-cell basis functions, when used with the electric field form of the vector Helmholtz equation, Table 11.3 shows the resonant frequencies of a three-dimensional cavity of dimension $1 \times 0.5 \times 0.75$ m. Results obtained using the CT/LN and the LT/QN basis functions for a tetrahedral-cell mesh with 300 cells, 120 nodes, 513 edges, and 694 faces are compared. The longest cell edge in the mesh was 0.274 m. This model produced a sparse matrix of order 231 for the CT/LN functions, containing 2835 nonzero entries, and order 1474 for the LT/QN functions, containing 49,124 nonzero entries. The results were obtained using a sparse-matrix eigenanalysis procedure similar to that proposed in [4]. From a number of similar examples, it has been observed

TABLE 11.1 Entries of Element Matrix **A** in (11.15) for LT/QN Basis and Testing Functions

$$A_{mn}^{e1e1} = A_{mn}^{e2e2} = -A_{mn}^{e1e2} = -A_{nm}^{e2e1} = Vw_m w_n \bar{v}_{pq} \cdot \bar{v}_{ij}$$

$$A_{mn}^{e1f1} = A_{nm}^{f1e1} = -A_{mn}^{e2f1} = -A_{nm}^{f1e2} = \frac{Vw_m}{4} \bar{v}_{pq} \cdot (2\bar{v}_{jk} + \bar{v}_{ik} - \bar{v}_{ij})$$

$$A_{mn}^{e1f2} = A_{nm}^{f2e1} = -A_{mn}^{e2f2} = -A_{nm}^{f2e2} = \frac{Vw_m}{4} \bar{v}_{pq} \cdot (2\bar{v}_{ik} + \bar{v}_{jk} + \bar{v}_{ij})$$

$$\begin{aligned} A_{mn}^{f1f1} = V(&4M_{pi}\bar{v}_{qr} \cdot \bar{v}_{jk} + 2M_{pj}\bar{v}_{qr} \cdot \bar{v}_{ik} - 2M_{pk}\bar{v}_{qr} \cdot \bar{v}_{ij} \\ &+ 2M_{qi}\bar{v}_{pr} \cdot \bar{v}_{jk} + M_{qj}\bar{v}_{pr} \cdot \bar{v}_{ik} - M_{qk}\bar{v}_{pr} \cdot \bar{v}_{ij} \\ &- 2M_{ri}\bar{v}_{pq} \cdot \bar{v}_{jk} - M_{rj}\bar{v}_{pq} \cdot \bar{v}_{ik} + M_{rk}\bar{v}_{pq} \cdot \bar{v}_{ij}) \end{aligned}$$

$$\begin{aligned} A_{mn}^{f1f2} = A_{nm}^{f2f1} = V(&2M_{pi}\bar{v}_{qr} \cdot \bar{v}_{jk} + 4M_{pj}\bar{v}_{qr} \cdot \bar{v}_{ik} + 2M_{pk}\bar{v}_{qr} \cdot \bar{v}_{ij} \\ &+ M_{qi}\bar{v}_{pr} \cdot \bar{v}_{jk} + 2M_{qj}\bar{v}_{pr} \cdot \bar{v}_{ik} + M_{qk}\bar{v}_{pr} \cdot \bar{v}_{ij} \\ &- M_{ri}\bar{v}_{pq} \cdot \bar{v}_{jk} - 2M_{rj}\bar{v}_{pq} \cdot \bar{v}_{ik} - M_{rk}\bar{v}_{pq} \cdot \bar{v}_{ij}) \end{aligned}$$

$$\begin{aligned} A_{mn}^{f2f2} = V(&M_{pi}\bar{v}_{qr} \cdot \bar{v}_{jk} + 2M_{pj}\bar{v}_{qr} \cdot \bar{v}_{ik} + M_{pk}\bar{v}_{qr} \cdot \bar{v}_{ij} \\ &+ 2M_{qi}\bar{v}_{pr} \cdot \bar{v}_{jk} + 4M_{qj}\bar{v}_{pr} \cdot \bar{v}_{ik} + 2M_{qk}\bar{v}_{pr} \cdot \bar{v}_{ij} \\ &+ M_{ri}\bar{v}_{pq} \cdot \bar{v}_{jk} + 2M_{rj}\bar{v}_{pq} \cdot \bar{v}_{ik} + M_{rk}\bar{v}_{pq} \cdot \bar{v}_{ij}) \end{aligned}$$

Note: **A** has the block structure

$$\mathbf{A} = \begin{bmatrix} A^{e1e1} & A^{e1e2} & A^{e1f1} & A^{e1f2} \\ A^{e2e1} & A^{e2e2} & A^{e2f1} & A^{e2f2} \\ A^{f1e1} & A^{f1e2} & A^{f1f1} & A^{f1f2} \\ A^{f2e1} & A^{f2e2} & A^{f2f1} & A^{f2f2} \end{bmatrix}$$

In the expressions in the table, *p* and *q* denote the node indices for edge *m*, while *p*, *q*, and *r* denote the node indices for face *m*. Similarly, *i* and *j* denote the node indices for edge *n*, while*i*, *j*, and *k* denote the three node indices associated with face *n*.

TABLE 11.2 Entries of Element Matrix **B** in (11.16) for LT/QN Basis and Testing Functions

$$B_{mn}^{e1e1} = Vw_m w_n \phi_{qj} M_{pi}$$

$$B_{mn}^{e1e2} = Vw_m w_n \phi_{qi} M_{pj}$$

$$B_{mn}^{e2e2} = Vw_m w_n \phi_{pi} M_{qj}$$

$$B_{mn}^{e1f1} = Vw_m (\phi_{qk} N_{pij} - \phi_{qj} N_{pik})$$

$$B_{mn}^{e2f1} = Vw_m (\phi_{pk} N_{qij} - \phi_{pj} N_{qik})$$

$$B_{mn}^{e1f2} = Vw_m (\phi_{qk} N_{pij} - \phi_{qi} N_{pjk})$$

$$B_{mn}^{e2f2} = Vw_m (\phi_{pk} N_{qij} - \phi_{pi} N_{qjk})$$

$$B_{mn}^{f1f2} = V(\phi_{rk} P_{pqij} - \phi_{rj} P_{pqik} - \phi_{qk} P_{prij} + \phi_{qj} P_{prik})$$

$$B_{mn}^{f1f2} = V(\phi_{rk} P_{pqij} - \phi_{ri} P_{pqjk} - \phi_{qk} P_{prij} + \phi_{qi} P_{prjk})$$

$$B_{mn}^{f2f2} = V(\phi_{rk} P_{pqij} - \phi_{ri} P_{pqjk} - \phi_{pk} P_{qrij} + \phi_{pi} P_{qrjk})$$

Note: **B** has the block structure of **A** in Table 11.1 and is symmetric across the main diagonal. Only entries for blocks above the main diagonal are shown.

TABLE 11.3 Lowest Nonzero Eigenvalues for Rectangular Brick-Shaped Cavity with $\varepsilon_r = 1$, $\mu_r = 1$, and dimensions 1.0 by 0.5 by 0.75 m

CT/LN	LT/QN	Exact
5.260	5.237	5.236 (TE_{101})
6.985	7.026	7.025 (TM_{110})
7.461	7.552	7.551 (TE_{011})
7.613	7.557	7.551 (TE_{201})
8.149	8.184	8.179 (TM_{111})
8.214	8.187	8.179 (TE_{111})
8.835	8.889	8.886 (TM_{210})

Note: The model consisted of 300 tetrahedral cells, with the longest cell edge 0.274 m. The discretization of the vector Helmholtz equation for the electric field required 231 unknowns with CT/LN basis functions and 1474 unknowns with LT/QN basis functions.

that the tetrahedral-cell CT/LN functions generally produce less than 15 nonzero entries per row of the **A** and **B** matrices, while the LT/QN functions typically produce about 35 nonzero entries per row (and can involve 70 or more).

To show accuracy trends, Figure 11.2 provides a comparison of the error in the resonant frequencies of the 1 × 0.5 × 0.75 m cavity. In this case, six different unstructured tetrahedral-cell meshes of various density were used. Figure 11.2 shows the average of the error in the first eight resonant frequencies versus the average edge length in each tetrahedral-cell mesh. The curve associated with the CT/LN functions has a slope near 2, while that associated with the LT/QN functions has a slope near 4. In this situation, the

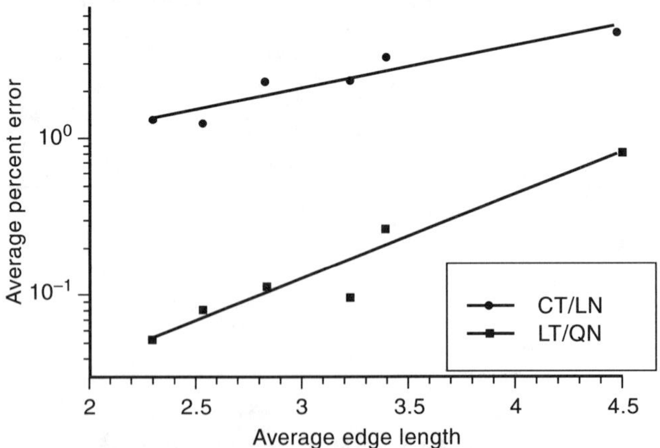

Figure 11.2 Comparison of the error produced by CT/LN and LT/QN discretizations as a function of the average edge length for the first eight nonzero resonant frequencies of a rectangular cavity. A curve fit through the actual data exhibits a slope of 1.98 for the CT/LN results and a slope of 3.86 for the LT/QN results. After [1]. ©1996 IEEE.

three-dimensional basis functions produce error behavior of $O(\Delta^2)$ for the CT/LN functions and $O(\Delta^4)$ for the LT/QN functions, similar to the two-dimensional functions investigated in Chapter 9.

11.3 EIGENFUNCTION RBC FOR SPHERICAL BOUNDARY SHAPES

An exact RBC can be obtained from an eigenfunction expansion of the scattered field. In the three-dimensional case, the scattered field at some point outside the scatterer can be expressed in the form

$$\bar{E}^s(r,\theta,\phi) = \sum_{n=1}^{\infty} \sum_{m=-n}^{n} A_{mn}\bar{M}_{mn} + B_{mn}\bar{N}_{mn} \tag{11.37}$$

where $\bar{M}_{mn}$ and $\bar{N}_{mn}$ are outward-propagating spherical vector wave functions defined as

$$\begin{aligned} \bar{M}_{mn} = & -\frac{jm}{r}\frac{P_n^m(\cos\theta)}{\sin\theta}\hat{H}_n^{(2)}(kr)e^{jm\phi}\hat{\theta} \\ & -\frac{1}{r}\sin\theta\, P_n^{m\prime}(\cos\theta)\hat{H}_n^{(2)}(kr)e^{jm\phi}\hat{\phi} \end{aligned} \tag{11.38}$$

$$\begin{aligned} \bar{N}_{mn} = & \frac{j}{r}\sin\theta\, P_n^{m\prime}(\cos\theta)\hat{H}_n^{(2)\prime}(kr)e^{jm\phi}\hat{\theta} \\ & +\frac{m}{r}\frac{P_n^m(\cos\theta)}{\sin\theta}\hat{H}_n^{(2)\prime}(kr)e^{jm\phi}\hat{\phi} \\ & +\frac{n(n+1)}{jkr^2}P_n^m(\cos\theta)\hat{H}_n^{(2)}(kr)e^{jm\phi}\hat{r} \end{aligned} \tag{11.39}$$

In these expressions, P_n^m denotes the associated Legendre polynomial and $\hat{H}_n^{(2)}$ denotes the alternative form of the spherical Hankel function of the second kind, as defined by Harrington [5]. (The vector wave functions differ slightly in form from those introduced by Stratton [6] and Ludwig [7] but provide an equivalent representation.) The $\bar{M}_{mn}$ and $\bar{N}_{mn}$ functions satisfy orthogonality relationships [7]

$$\int_{\phi=0}^{2\pi}\int_{\theta=0}^{\pi} \bar{M}_{mn} \times \bar{M}_{pq} \cdot \hat{r}\sin\theta\, d\theta\, d\phi = 0 \tag{11.40}$$

$$\int_{\phi=0}^{2\pi}\int_{\theta=0}^{\pi} \bar{N}_{mn} \times \bar{N}_{pq} \cdot \hat{r}\sin\theta\, d\theta\, d\phi = 0 \tag{11.41}$$

and

$$\int_{\phi=0}^{2\pi}\int_{\theta=0}^{\pi} \bar{M}_{mn} \times \bar{N}_{pq} \cdot \hat{r}\sin\theta\, d\theta\, d\phi = \begin{cases} 0 & q \neq n \text{ or } p \neq -m \\ \alpha_{mn}(r) & q = n \text{ and } p = -m \end{cases} \tag{11.42}$$

where

$$\alpha_{mn}(r) = -j\frac{\hat{H}_n^{(2)}(kr)\hat{H}_n^{(2)\prime}(kr)}{r^2}\frac{4\pi n(n+1)}{2n+1}\frac{(n+|m|)!}{(n-|m|)!} \tag{11.43}$$

Using (11.40)–(11.42), the coefficients in (11.37) can be determined to be

$$A_{mn} = \frac{1}{\alpha_{mn}(a)} \int_{\phi=0}^{2\pi} \int_{\theta=0}^{\pi} \bar{E}^s(a,\theta,\phi) \times \bar{N}_{(-m)n} \cdot \hat{r} \sin\theta \, d\theta \, d\phi \tag{11.44}$$

$$B_{mn} = \frac{1}{\alpha_{mn}(a)} \int_{\phi=0}^{2\pi} \int_{\theta=0}^{\pi} \bar{M}_{(-m)n} \times \bar{E}^s(a,\theta,\phi) \cdot \hat{r} \sin\theta \, d\theta \, d\phi \tag{11.45}$$

where a denotes the radius of a sphere enclosing the scatterer.

From (11.37) and with some manipulation, it is possible to show that

$$\hat{r} \times \nabla \times \bar{E}^s = jk \sum_{n=1}^{\infty} \sum_{m=-n}^{n} B_{mn}\hat{r} \times \bar{M}_{mn} - A_{mn}\hat{r} \times \bar{N}_{mn} \tag{11.46}$$

By substituting (11.44) and (11.45) into (11.46), we obtain

$$\begin{aligned} \hat{r} \times \nabla \times \bar{E}^s = jk \sum_{n=1}^{\infty} \sum_{m=-n}^{n} \frac{1}{\alpha_{mn}(a)} & \\ \Bigg(\hat{r} \times \bar{M}_{mn}(r,\theta,\phi) \int_{\phi'=0}^{2\pi} \int_{\theta'=0}^{\pi} & \bar{M}_{(-m)n}(a,\theta',\phi') \times \bar{E}^s(a,\theta',\phi') \cdot \hat{r} \sin\theta' \, d\theta' \, d\phi' \\ - \hat{r} \times \bar{N}_{mn}(r,\theta,\phi) \int_{\phi'=0}^{2\pi} \int_{\theta'=0}^{\pi} & \bar{E}^s(a,\theta',\phi') \times \bar{N}_{(-m)n}(a,\theta',\phi') \cdot \hat{r} \sin\theta' \, d\theta' \, d\phi' \Bigg) \end{aligned} \tag{11.47}$$

This expression is an exact RBC that can be used to truncate the computational domain in a three-dimensional finite-element formulation. It is similar in form to the two-dimensional RBC developed in Section 3.3 and must be imposed on a spherical boundary surrounding the scatterer. By combining this expression with the incident electric field, (11.47) can be rewritten in terms of the total field and substituted into the boundary integral in (11.6) to complete the scattering formulation. A similar RBC has been derived in [8].

Although (11.47) is an exact condition, to the authors' knowledge, it has not been implemented in practice. There are several reasons for this, including the general complexity of the vector spherical wave functions. The exact RBC has the drawback that it must be imposed on a spherical boundary, which is generally undesirable since many additional unknowns are introduced by the process of padding a computational domain out to a sphere. Thus, the implementation of (11.47) may lead to an unacceptable increase in the order of the finite-element system. Furthermore, (11.47) couples the fields around the entire boundary, creating fill-in that increases the computational effort required to create and solve the finite-element system. In other words, not only are there additional unknowns required to pad to a spherical boundary, but the resulting system contains a fully populated submatrix that couples all the boundary unknowns. In the following sections, alternative RBCs are developed that attempt to avoid these drawbacks.

11.4 SURFACE INTEGRAL EQUATION RBC FOR GENERAL BOUNDARY SHAPES

Flexibility in the shape of the outer boundary can be ensured through the use of an RBC based on a surface integral equation. The tangential fields along the boundary of the computational domain Γ can be related to equivalent surface current densities through

$$\bar{J}_s = \frac{1}{-j\omega\mu}\hat{n} \times (\nabla \times \bar{E}) \tag{11.48}$$

$$\bar{K}_s = \bar{E} \times \hat{n} \tag{11.49}$$

which are suitable if the electric field is the primary unknown and (11.6) is the equation to be solved, or

$$\bar{J}_s = \hat{n} \times \bar{H} \tag{11.50}$$

$$\bar{K}_s = \frac{1}{j\omega\varepsilon}(\nabla \times \bar{H}) \times \hat{n} \tag{11.51}$$

which are suitable for use with (11.7). A robust formulation that avoids the interior resonance problem (Chapter 6) can be obtained from the combined-field equation

$$\begin{aligned} \alpha \bar{E}^{\text{inc}}_{\tan} + (1-\alpha)\eta\hat{n} \times \bar{H}^{\text{inc}} = \alpha &\left(-\frac{\nabla\nabla \cdot \bar{A} + k^2 \bar{A}}{j\omega\varepsilon} + \hat{n} \times \bar{K}_s + \nabla \times \bar{F} \right)_{\tan} \\ &+ (1-\alpha)\eta \left(\bar{J}_s - \hat{n} \times \nabla \times \bar{A} - \hat{n} \times \frac{\nabla\nabla \cdot \bar{F} + k^2 \bar{F}}{j\omega\mu} \right) \end{aligned} \tag{11.52}$$

where $\bar{A}$ and $\bar{F}$ are the vector potential functions defined in Chapter 1, α is a parameter in the range $0 < \alpha < 1$, and the medium impedance η is used to convert to consistent units and balance the numerical values. The discretization of this type of surface integral equation for a general boundary shape is discussed in Chapter 10, in conjunction with divergence-conforming CN/LT and LN/QT basis functions for the equivalent surface current densities. Using (11.49) and (11.50), a CN/LT representation for the surface currents can be shown to be equivalent to a CT/LN representation for the tangential fields on the boundary. Similarly, an LN/QT representation for the current densities is equivalent to an LT/QN representation for the fields. The compatibility between the divergence-conforming representations introduced for surface integral equations and the curl-conforming basis functions used to discretize the vector Helmholtz equation facilitates the use of an integral equation RBC.

The discretization of (11.52) produces a matrix equation of the form

$$\mathbf{Lk} + \mathbf{Mj} = \mathbf{V}^{\text{inc}} \tag{11.53}$$

where $\mathbf{L}$ and $\mathbf{M}$ are square matrix operators, $\mathbf{k}$ and $\mathbf{j}$ are column vectors containing the coefficients of the basis functions used for the magnetic and electric current densities, respectively, and $\mathbf{V}^{\text{inc}}$ contains weighted samples of the incident fields appearing on the left-hand side of (11.52). Terms in this matrix equation can be rearranged in the form

$$\mathbf{j} = -\mathbf{M}^{-1}\mathbf{Lk} + \mathbf{M}^{-1}\mathbf{V}^{\text{inc}} \tag{11.54}$$

or

$$\mathbf{k} = -\mathbf{L}^{-1}\mathbf{Mj} + \mathbf{L}^{-1}\mathbf{V}^{\text{inc}} \tag{11.55}$$

which provide the desired numerical RBC depending on whether the electric or magnetic field is the primary unknown. The reader is referred to Chapters 9 and 10 for additional information about possible basis and testing functions and their impact on the entries of the system in (11.53). The specific choice of basis functions does not affect the general form of Equations (11.54) or (11.55).

Using (11.48), the weak Helmholtz equation in (11.6) can be rewritten as

$$\iiint_{\Gamma} \frac{1}{\mu_r} \nabla \times \bar{T} \cdot \nabla \times \bar{E} - k^2 \varepsilon_r \bar{T} \cdot \bar{E} = j\omega\mu_0 \iint_{\partial\Gamma} \bar{T} \cdot \bar{J}_s \tag{11.56}$$

Suppose that the computational domain Γ is divided into tetrahedral cells, and the electric field is represented by a superposition of vector basis functions

$$\bar{E}(x, y, z) \cong \sum_{n=1}^{N_{\text{int}}} e_n \bar{B}_n + \sum_{n=N_{\text{int}}+1}^{N_{\text{int}}+N_{\text{bound}}} e_n \bar{B}_n \tag{11.57}$$

where the basis functions contributing to a tangential field on the boundary $\partial\Gamma$ are separated from the others for convenience. We will assume that the basis functions $\{\bar{B}_n\}$ provide a tangentially continuous representation for $\bar{E}$. It is also necessary to introduce an independent representation for the equivalent surface current density

$$\bar{J}_s \cong \sum_{n=1}^{M_{\text{bound}}} j_n \bar{b}_n \tag{11.58}$$

where the vector basis functions $\{\bar{b}_n\}$ are generally different from those used in (11.57) to represent the electric field. (In most cases, the basis functions $\{\bar{B}_n\}$ are curl conforming while $\{\bar{b}_n\}$ are divergence conforming.) A discretization of the weak Helmholtz equation in (11.56) using testing functions identical to the basis functions in (11.57) throughout the region Γ produces an underdetermined matrix equation

$$\begin{bmatrix} \mathbf{I} & \mathbf{I}_b^T & \mathbf{0} \\ \mathbf{I}_b & \mathbf{B} & \mathbf{J} \end{bmatrix} \begin{bmatrix} \mathbf{e}^{\text{int}} \\ \mathbf{e}^{\text{bound}} \\ \mathbf{j} \end{bmatrix} = \begin{bmatrix} \mathbf{0} \\ \mathbf{0} \end{bmatrix} \tag{11.59}$$

where $\mathbf{I}$, $\mathbf{I}_b$, and $\mathbf{B}$ have entries of the common form

$$I_{mn} = \iiint_{\Gamma} \frac{1}{\mu_r} \nabla \times \bar{B}_m \cdot \nabla \times \bar{B}_n - k^2 \varepsilon_r \bar{B}_m \cdot \bar{B}_n \tag{11.60}$$

and the submatrix $\mathbf{J}$ has entries

$$J_{mn} = -j\omega\mu_0 \iint_{\partial\Gamma} \bar{B}_m \cdot \bar{b}_n \tag{11.61}$$

By incorporating the RBC in (11.54) into the matrix equation, we arrive at the properly determined system

$$\begin{bmatrix} \mathbf{I} & \mathbf{I}_b^T \\ \mathbf{I}_b & \mathbf{B} - \mathbf{J}\mathbf{M}^{-1}\mathbf{L} \end{bmatrix} \begin{bmatrix} \mathbf{e}^{\text{int}} \\ \mathbf{e}^{\text{bound}} \end{bmatrix} = \begin{bmatrix} \mathbf{0} \\ -\mathbf{J}\mathbf{M}^{-1}\mathbf{V}^{\text{inc}} \end{bmatrix} \tag{11.62}$$

where it is assumed that the coefficients $\mathbf{k}$ in (11.54) are the same as those in $\mathbf{e}^{\text{bound}}$ in (11.58). (As an example, this would be possible if a CN/LT representation is used for the magnetic surface current and a CT/LN representation is used for the electric field.) If desired, the procedure can be generalized to use completely unrelated basis functions for these two quantities (Prob. P11.10).

Equation (11.62) is an outward-looking formulation that can be used to describe a general scatterer, with the mesh terminated on an arbitrary boundary enclosing all inhomogeneities. The advantage of an integral equation RBC is that the boundary shape can be arbitrary and can easily conform to the surface of the scatterer under consideration. Thus, there is no need to pad the mesh to a spherical boundary. However, the submatrices **L** and **M** in (11.62) are fully populated in general, and as a consequence there is still a substantial amount of fill-in associated with this RBC.

The relative efficiency of a surface integral equation RBC imposed on an arbitrary boundary and the eigenfunction RBC of Section 11.3 imposed on a spherical boundary involves two factors: The eigenfunction RBC requires additional unknowns in order to pad the mesh to a spherical boundary shape, while the surface integral equation RBC involves additional computation associated with the LU factorization required to produce $\mathbf{M}^{-1}$ in (11.62). In an outward-looking formulation, both approaches involve a fully populated submatrix of order equal to the number of boundary unknowns. For a geometry whose boundary is more than a minor perturbation from a sphere, the larger global system required with an eigenfunction RBC is likely to offset the additional computation required to produce $\mathbf{M}^{-1}$ with an integral equation RBC. The disadvantages of these procedures can be mitigated to some extent by an RBC specialized to an axisymmetric radiation boundary (Section 11.6). Furthermore, the trade-offs between these approaches may be somewhat different if they are implemented as inward-looking formulations.

11.5 OUTWARD-LOOKING VERSUS INWARD-LOOKING FORMULATIONS

The preceding sections described outward-looking formulations, where an RBC of some type is used to constrain the vector Helmholtz equation. Since the resulting system has the form of (11.62), containing a dense submatrix, it complicates the matrix solution procedure. Perhaps as a consequence of the partially full, partially sparse matrix structure, most of the three-dimensional outward-looking formulations discussed in the literature employ iterative solution algorithms instead of direct algorithms to solve Equation (11.62).

The alternative inward-looking formulations (Chapter 3) avoid the partially full and partially sparse matrix structure. Inward-looking formulations use the solution of the vector Helmholtz equation to obtain a system constraining tangential fields (or equivalent currents) on the surface of the boundary $\partial\Gamma$, which is then combined with the integral equation in (11.53). For example, given the equivalent electric current $\bar{J}_s$ as the forcing function, (11.59) can be solved to find the equivalent magnetic current $\bar{K}_s$ (or the tangential electric field on $\partial\Gamma$). A repetitive solution for linearly independent currents (say, one basis function at a time) can be used to construct a matrix equation relating $\bar{J}_s$ and $\bar{K}_s$. This system is underdetermined since both $\bar{J}_s$ and $\bar{K}_s$ are unknowns, but it can be combined with (11.53) to produce a properly determined matrix equation. After $\bar{J}_s$ and $\bar{K}_s$ are obtained by the solution of the combined system, (11.59) can be solved again to produce the interior fields. The computational requirements of this approach are that a sparse system representing (11.59) must be solved repetitively to generate a smaller fully populated system relating $\bar{J}_s$ and $\bar{K}_s$, which is then combined with the fully populated system representing (11.53). Thus, the procedure involves a large-order sparse system and a smaller order dense system.

With an inward-looking formulation, an exterior eigenfunction expansion may be more advantageous than it is in the outward-looking case. This is because an eigenfunction series generally provides the minimum number of expansion functions needed to represent the fields in the general case (as compared with typical subsectional representations). Thus, the inward-looking eigenfunction expansion procedure involves a smaller dense matrix than an inward-looking approach based on a general integral equation representation of the exterior region. (Of course, a spherical eigenfunction expansion requires a mesh that terminates on a spherical boundary, so the sparse matrix for the interior region might be of considerably larger order than that required with an integral equation representation and a general boundary shape.) In both inward- and outward-looking formulations, the use of an eigenfunction expansion eliminates the need to obtain $\mathbf{M}^{-1}$ numerically in order to solve (11.53).

The differences between outward-looking and inward-looking approaches are summarized in Chapter 3 in the context of two-dimensional formulations and are generally that while the inward-looking formulations are easier to implement because they avoid a matrix that is partially sparse and partially full, they are susceptible to uniqueness difficulties since the part of the problem solved first is essentially a closed cavity that might be resonant. (The uniqueness issue is explored for the scalar Helmholtz equation in Section 6.9.) As an alternative to either outward-looking or inward-looking formulations, a third possibility is the combination of (11.53) and (11.59) to produce the system

$$\begin{bmatrix} \mathbf{I} & \mathbf{I}_b^T & \mathbf{0} \\ \mathbf{I}_b & \mathbf{B} & \mathbf{J} \\ \mathbf{0} & \mathbf{L} & \mathbf{M} \end{bmatrix} \begin{bmatrix} \mathbf{e}^{\text{int}} \\ \mathbf{e}^{\text{bound}} \\ \mathbf{j} \end{bmatrix} = \begin{bmatrix} \mathbf{0} \\ \mathbf{0} \\ \mathbf{V}^{\text{inc}} \end{bmatrix} \tag{11.63}$$

Equation (11.63) is neither inward looking nor outward looking and is sparse except for matrix blocks **L** and **M**. This matrix equation eliminates the inward-looking uniqueness difficulties but involves a larger system than (11.62) to solve. The possible advantage to (11.63) is that it simplifies the matrix construction if iterative solution algorithms are used to solve the system. It is particularly advantageous if an axisymmetric radiation boundary (Section 11.6) is employed.

The literature describes several finite-element implementations similar to those considered above. Yuan [9] appears to have been the first to employ the three-dimensional tetrahedral-cell CT/LN functions within an inward-looking scattering formulation, terminated with an integral equation RBC. Yuan presented results for layered spheres that exhibited good agreement with analytical solutions provided that about 10 cells per wavelength were used in the models. Although the uniqueness issue did not appear to impede accurate solutions, it would likely affect the results for electrically larger problems. Reliable inward-looking formulations might be possible, however, if the condition number of the large, sparse system can be monitored to detect fictitious resonances.

Jin and Volakis employed CT/LN functions on hexahedral and tetrahedral cells to represent scattering from a thick slot in a conducting plane, cavity-backed microstrip patch antennas, and other structures [10, 12]. Their outward-looking approach used integral equation RBCs to terminate the computational domain along planar boundaries, producing a system similar in matrix structure to (11.62). The use of planar boundaries facilitates an iterative solution of the resulting system, as described in Chapter 4, and they employed a BCG–FFT algorithm. (For general boundary shapes, they modified their approach to use a local RBC similar to those discussed in Section 11.7 instead of the surface integral equation RBC [12].)

Antilla and Alexopoulos developed a similar approach employing CT/LN functions on curved hexahedral cells for the interior fields and CN/LT functions on curved quadrilateral cells for the equivalent boundary currents [13]. (The treatment of curved cells is described in Section 9.14.) They presented a number of examples showing the accuracy of the formulation and reported that good accuracy was usually obtained with cell densities in the range of 9–12 cells per wavelength.

The preceding approaches involve general three-dimensional scatterers. Lucas and Fontana analyzed periodic structures using an integral equation that incorporated the appropriate Floquet conditions to limit the computational domain to a three-dimensional unit cell [14]. They employed a polynomial-complete quadratic vector representation to discretize the vector Helmholtz equation.

11.6 INTEGRAL EQUATION RBC FOR AXISYMMETRIC BOUNDARY SHAPES

Since the shape of the outer boundary $\partial\Gamma$ used to terminate the computational domain is arbitrary, as long as it fully encloses the scatterer, an alternate possibility is to use an axisymmetric boundary that resides as close to the scatterer as possible. This idea was first proposed by Boyse and Seidl [15] and combines the advantages of the eigenfunction RBC of Section 11.3 and the integral equation RBC of Section 11.4 (the boundary shape remains rather flexible and the numerical inversion of the matrix $\mathbf{M}$ is accomplished more efficiently). As illustrated in Chapter 8, integral equation formulations imposed on axisymmetric surfaces ("body-of-revolution" formulations) can be discretized using cylindrical harmonics to represent the equivalent surface currents. The cylindrical harmonics decouple the resulting equations, creating a block-diagonal structure that substantially reduces the computational complexity required to construct $\mathbf{M}^{-1}$ in an outward-looking formulation (or the dense exterior system arising in an inward-looking approach).

If Equation (11.53) is specialized to an axisymmetric boundary and cylindrical harmonics are used in the representation (Section 8.7), then the equation for the mth harmonic has the form

$$\mathbf{L}_m\mathbf{k} + \mathbf{M}_m\mathbf{j} = \mathbf{V}_m^{\text{inc}} \tag{11.64}$$

and is independent of the equations for the other harmonics. Consequently, the RBC can be found from the solution of the block-diagonal matrix equation

$$\begin{bmatrix} \ddots & \mathbf{0} & & & \\ \mathbf{0} & \mathbf{L}_{m-1} & \mathbf{0} & & \\ & \mathbf{0} & \mathbf{L}_m & \mathbf{0} & \\ & & \mathbf{0} & \mathbf{L}_{m+1} & \mathbf{0} \\ & & & \mathbf{0} & \ddots \end{bmatrix} \mathbf{k} + \begin{bmatrix} \ddots & \mathbf{0} & & & \\ \mathbf{0} & \mathbf{M}_{m-1} & \mathbf{0} & & \\ & \mathbf{0} & \mathbf{M}_m & \mathbf{0} & \\ & & \mathbf{0} & \mathbf{M}_{m+1} & \mathbf{0} \\ & & & \mathbf{0} & \ddots \end{bmatrix} \mathbf{j} = \mathbf{V}^{\text{inc}} \tag{11.65}$$

Clearly, there are substantial savings arising from (11.65) compared to the larger order, dense matrix equation associated with a general-purpose surface integral equation.

The use of an axisymmetric boundary to truncate the computational domain was first reported by Boyse and Seidl [15, 16] in the context of an inward-looking formulation. Their approach employed node-based Lagrangian expansion functions throughout the interior and cubic-order Hermitian interpolation polynomials to discretize the surface integral equation on the radiation boundary. Cwik et al. [17–19] investigated an inward-looking implementation as well as a formulation based on the matrix structure of (11.63) and used CT/LN functions on tetrahedral cells to discretize the curl–curl equation. Figure 11.3 illustrates the

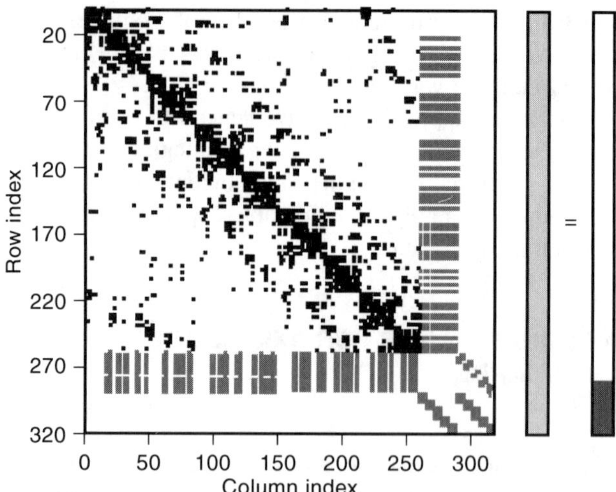

Figure 11.3 Typical matrix structure associated with the formulation illustrated in (11.63) when an axisymmetric integral equation RBC is incorporated. After [19]. ©1996 John Wiley & Sons, Inc.

typical matrix structure arising from the approach. Because of the block-diagonal structure of (11.65), the resulting global system of equations is almost as sparse as a system obtained from a local RBC. The additional unknowns required to pad the domain to an axisymmetric boundary are likely to be far less than those associated with a spherical boundary and possibly comparable to the small degree of padding required with local RBCs.

11.7 LOCAL RBCS FOR SPHERICAL BOUNDARIES

The computation and storage requirements associated with an eigenfunction RBC (Section 11.3), a general surface integral equation RBC (Section 11.4), or an axisymmetric surface integral equation RBC (Section 11.6) can be quite different and are all largely problem dependent. However, all of these truncation conditions involve global coupling of unknowns around the boundary. To some extent, the overall matrix order and/or the presence of the dense submatrix in (11.62) may limit the problem size amenable to numerical solution using global RBCs. Chapter 3 demonstrated that local RBCs minimize fill-in and can perform well in a variety of situations as long as high accuracy is not required. In this section, local RBCs are developed for three-dimensional scatterers enclosed in spherical boundaries.

Following the procedure that produced the Bayliss–Turkel RBC in the two-dimensional case, consider the outward-propagating field expansion

$$\bar{E}^s(r,\theta,\phi) = \frac{e^{-jkr}}{4\pi r}\sum_{n=0}^{\infty}\frac{\bar{E}_n(\theta,\phi)}{r^n} \tag{11.66}$$

Equation (11.66), known as the Wilcox expansion [20], can be used to represent any outward-propagating field in the region exterior to a three-dimensional scatterer or antenna. In contrast to the similar two-dimensional expansion used in Chapter 3 (which was an asymptotic approximation), Equation (11.66) can be differentiated term by term and remains uniformly convergent [20].

A family of local RBCs can be developed so that the expansion of (11.66) satisfies the RBC up to some power of $1/r$ [21]. For instance, the standard radiation condition operator applied to (11.66) leaves a residual

$$\hat{r} \times \nabla \times \bar{E}^s - jk\bar{E}^s = \frac{e^{-jkr}}{4\pi r} \sum_{n=0}^{\infty} \left(-jk \frac{\hat{r}(\hat{r} \cdot \bar{E}_n)}{r^n} + \frac{r\nabla(\hat{r} \cdot \bar{E}_n) + n\bar{E}_n^{\tan}}{r^{n+1}} \right) \tag{11.67}$$

Using the vector identities

$$\hat{r} \times (r\nabla \times \bar{E}_n) = r\nabla(\hat{r} \cdot \bar{E}_n) - \bar{E}_n^{\tan} \tag{11.68}$$

$$\hat{r} \times (\hat{r} \times \bar{E}_n) = -\bar{E}_n^{\tan} \tag{11.69}$$

where

$$\bar{E}_n^{\tan} = \bar{E}_n - \hat{r}(\hat{r} \cdot \bar{E}_n) \tag{11.70}$$

and the fact that $\bar{E}$ has zero divergence in this region, which implies that $\hat{r} \cdot \bar{E}_0 = 0$, it follows that the residual on the right-hand side of (11.67) is $O(r^{-2})$ as $r \to \infty$.

The leading-order behavior of the residual suggests modifications in the operator that can reduce the residual to a more rapidly decaying power of r and ensure a more accurate RBC for large r. For instance, the alternative operator

$$\hat{r} \times \nabla \times \bar{E}^s - jk\bar{E}^s_{\tan} = \frac{e^{-jkr}}{4\pi r} \sum_{n=0}^{\infty} \frac{r\nabla(\hat{r} \cdot \bar{E}_n) + n\bar{E}_n^{\tan}}{r^{n+1}} \tag{11.71}$$

has a residual with leading-order behavior $O(r^{-3})$. The operator

$$\begin{aligned} &[\hat{r} \times (\nabla\times\ ) - jk][\hat{r} \times \nabla \times \bar{E}^s - jk\bar{E}^s_{\tan}] \\ &= \frac{e^{-jkr}}{4\pi r} \sum_{n=1}^{\infty} (n+1) \frac{r\nabla(\hat{r} \cdot \bar{E}_n) + n\bar{E}_n^{\tan}}{r^{n+2}} \end{aligned} \tag{11.72}$$

produces a residual of $O(r^{-4})$. Furthermore, the modified operator

$$\begin{aligned} &\left[\hat{r} \times (\nabla\times\ ) - jk - \frac{2}{r}\right] [\hat{r} \times \nabla \times \bar{E}^s - jk\bar{E}^s_{\tan}] \\ &= \frac{e^{-jkr}}{4\pi r} \sum_{n=1}^{\infty} (n-1) \frac{r\nabla(\hat{r} \cdot \bar{E}_n) + n\bar{E}_n^{\tan}}{r^{n+2}} \end{aligned} \tag{11.73}$$

produces a residual of $O(r^{-5})$. This process can be continued indefinitely [21].

A "first-order" RBC is obtained by equating the left-hand side of (11.71) to zero. The second-order RBC can be obtained from the left-hand side of (11.73), which can be rewritten in the form

$$\hat{r} \times \nabla \times \bar{E}^s = jk\bar{E}^s_{\tan} + \beta(r)\nabla \times [\hat{r}(\hat{r} \cdot \nabla \times \bar{E}^s)] + jk\beta(r)\nabla^{\tan}(\hat{r} \cdot \bar{E}^s) \tag{11.74}$$

where

$$\beta(r) = \frac{1}{2jk + 2/r} \tag{11.75}$$

Because of the first-derivative term in (11.74), its direct implementation within a finite-element discretization of the weak equation in (11.6) produces a nonsymmetric matrix equation. However, the relation

$$E_r^s = \frac{1}{k^2}\hat{r} \cdot \nabla \times \nabla \times \bar{E}^s = -\frac{1}{k^2}\nabla \cdot [\hat{r} \times (\nabla \times \bar{E}^s)] \cong -\frac{j}{k}\nabla \cdot \bar{E}^s_{\tan} \tag{11.76}$$

can be incorporated into the final term in (11.74) to produce

$$\hat{r} \times \nabla \times \bar{E}^s \cong jk\bar{E}^s_{\tan} + \beta(r)\nabla \times [\hat{r}(\hat{r} \cdot \nabla \times \bar{E}^s)] + \beta(r)\nabla^{\tan}(\nabla \cdot \bar{E}^s_{\tan}) \tag{11.77}$$

Equation (11.77) preserves the symmetry of the finite-element matrix operator and is only slightly less accurate than (11.74) [22]. [However, as discussed below, the third term on the right-hand side of (11.77) is problematic for use with curl-conforming basis functions.]

The second-order RBCs in (11.74) and (11.77) can be thought of as special cases of a more general RBC suggested by Webb and Kanellopoulos [23], which has the form

$$\begin{aligned}\hat{r} \times \nabla \times \bar{E}^s \cong{}& jk\bar{E}^s_{\tan} + \beta(r)\nabla \times [\hat{r}(\hat{r} \cdot \nabla \times \bar{E}^s)]\\ &+(s-1)\beta(r)\nabla^{\tan}(\nabla \cdot \bar{E}^s_{\tan}) + (2-s)jk\beta(r)\nabla^{\tan}(\hat{r} \cdot \bar{E}^s)\end{aligned} \tag{11.78}$$

where s is an arbitrary parameter. The choice $s = 1$ produces (11.74), while the choice $s = 2$ yields (11.77).

The relative accuracy of the RBCs in (11.74), (11.77), and (11.78) can be investigated by applying them to the spherical vector wave functions in (11.38) and (11.39), which represent outward-propagating fields and should satisfy an exact RBC. The error in all the RBCs is independent of the index m in (11.38) and (11.39) but increases as the index n increases for a fixed boundary radius [22]. The performance of the second-order RBCs is generally far better than the first-order RBC obtained from (11.71). The part of the second-order condition involving the operator

$$\nabla \times [\hat{r}(\hat{r} \cdot \nabla \times \bar{E}^s)] \tag{11.79}$$

is only associated with the TE-to-r part of the field, while the operators

$$\nabla^{\tan}(\nabla \cdot \bar{E}^s_{\tan}) \tag{11.80}$$

$$\nabla^{\tan}(\hat{r} \cdot \bar{E}^s) \tag{11.81}$$

only involve the TM-to-r part of the field. Thus, these terms are of essentially equal importance in these conditions. The conditions (11.74), (11.77), and (11.78) only differ in the TM-to-r part of the field. Furthermore, the RBC in (11.78) exhibits the best accuracy for a choice of s somewhere between 0 and 1, rather than either $s = 1$ or $s = 2$ [22]. Figure 11.4 shows the error in the various RBCs for a mode with $n = 5$ as a function of the boundary location r.

The local RBCs in (11.74)–(11.78) must be imposed on a spherical boundary surrounding the scatterer. Consider (11.77). If substituted into the boundary integral in (11.6), under the assumption that $\mu_r = 1$ along the boundary, the boundary integral assumes the

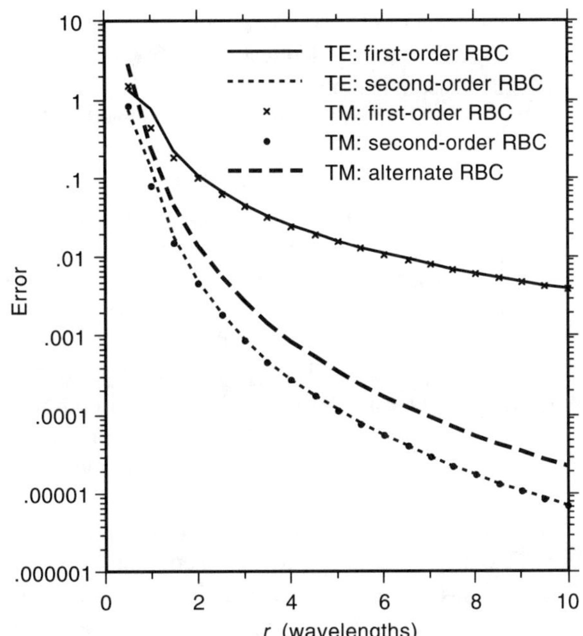

Figure 11.4 Error in the TE and TM parts of the first- and second-order local RBCs plotted as a function of the radius of the radiation boundary for the $n = 5$ spherical harmonic. After [22]. ©1992 IEEE.

form

$$\iint_{\partial\Gamma} \bar{T} \cdot \hat{n} \times \nabla \times \bar{E}^s = jk \iint_{\partial\Gamma} \bar{T} \cdot \bar{E}^s_{\text{tan}} + \beta(r) \iint_{\partial\Gamma} \bar{T} \cdot \nabla \times [\hat{r}(\hat{r} \cdot \nabla \times \bar{E}^s)] + \beta(r) \iint_{\partial\Gamma} \bar{T} \cdot \nabla^{\text{tan}}(\nabla \cdot \bar{E}^s_{\text{tan}}) \tag{11.82}$$

Using the vector identities

$$\bar{T} \cdot \nabla \times \bar{A} = \nabla \times \bar{T} \cdot \bar{A} - \nabla \cdot (\bar{T} \times \bar{A}) \tag{11.83}$$

$$\nabla \cdot (\Psi \bar{T}) = \Psi \nabla \cdot \bar{T} + \nabla \Psi \cdot \bar{T} \tag{11.84}$$

and invoking the two-dimensional divergence theorem, Equation (11.82) can be rewritten as

$$\iint_{\partial\Gamma} \bar{T} \cdot \hat{n} \times \nabla \times \bar{E}^s = jk \iint_{\partial\Gamma} \bar{T} \cdot \bar{E}^s_{\text{tan}} + \beta(r) \iint_{\partial\Gamma} (\hat{r} \cdot \nabla \times \bar{T})(\hat{r} \cdot \nabla \times \bar{E}^s) - \beta(r) \iint_{\partial\Gamma} (\nabla \cdot \bar{T}_{\text{tan}})(\nabla \cdot \bar{E}^s_{\text{tan}}) \tag{11.85}$$

The form of (11.85) spreads the derivatives among the basis and testing functions and, for testing functions identical to the basis functions, produces a symmetrical matrix operator after discretization.

In the development of (11.85), we tacitly assume that the necessary derivatives exist. However, if we subsequently expand the field in terms of curl-conforming basis functions such as the CT/LN or LT/QN functions considered in Section 11.2, this assumption is violated. The basis and testing functions do not maintain normal continuity from cell to cell along the radiation boundary, and thus the third integral on the right-hand side of (11.85) contains the product of two Dirac delta–like functions located along cell edges. This behavior is a direct consequence of the curl-conforming representation and not a fundamental limitation of the equations leading to (11.85). One way of avoiding this difficulty is to simply omit the third integral. As noted above, however, the second and third expressions are expected to be equally important in the RBC, and omitting the third integral should reduce the overall accuracy to that of the first-order RBC. A better remedy might be to employ a continuous representation for the fields along the radiation boundary, as suggested by Kanellopoulos and Webb [24]. Since a continuous representation throughout the interior might severely distort the nullspace eigenfunctions, a curl-conforming interior expansion must transition to a continuous basis confined to $\partial\Gamma$. The continuous boundary expansion permits the evaluation of all terms in (11.85) without difficulty.

A third remedy is suggested by Savage [25], who implemented a procedure for projecting the curl-conforming representation onto a divergence-conforming representation defined on the same finite-element mesh. The problematical term in (11.85) is evaluated using the divergence-conforming functions, and the result is projected back onto the curl-conforming space. Although the accuracy of the RBC implementation improves, the projection process causes the support of an individual basis function to spread into neighboring cells, resulting in additional matrix fill-in. [If the projection process is used with tetrahedral-cell CT/LN functions, there may be 27 nonzero matrix entries per row due to the third integral in (11.85), compared with 5 nonzero entries due to the first and second integrals.]

A more general remedy would be obtained through the use of dual interlocking meshes simultaneously supporting both curl-conforming and divergence-conforming expansions that share common coefficients. Except for structured quadrilateral grids, such a scheme would require vector expansion functions defined on general polygonal shapes (for instance, the dual grid associated with a triangular-cell mesh would not be triangular).

To summarize the local RBC approach, the weak equation can be combined with the RBC in (11.77), after which terms involving the incident field can be added and subtracted from the boundary integral to produce

$$\begin{aligned}
&\iiint_{\Gamma} \frac{1}{\mu_r} \nabla \times \bar{T} \cdot \nabla \times \bar{E} - k^2 \varepsilon_r \bar{T} \cdot \bar{E} + jk \iint_{\partial\Gamma} \bar{T} \cdot \bar{E}_{\tan} \\
&\quad + \beta(r) \iint_{\partial\Gamma} (\hat{r} \cdot \nabla \times \bar{T})(\hat{r} \cdot \nabla \times \bar{E}) - (\nabla \cdot \bar{T}_{\tan})(\nabla \cdot \bar{E}_{\tan}) \\
&\quad = -\iint_{\partial\Gamma} [\bar{T} \cdot \hat{r} \times \nabla \times \bar{E}^{\text{inc}} - jk \bar{T} \cdot \bar{E}^{\text{inc}}_{\tan} \\
&\qquad\qquad - \beta(r) \bar{T} \cdot \nabla \times (\hat{r}\hat{r} \cdot \nabla \times \bar{E}^{\text{inc}}) + \beta(r) \bar{T} \cdot \nabla^{\tan} (\nabla \cdot \bar{E}^{\text{inc}}_{\tan})]
\end{aligned} \tag{11.86}$$

which is entirely in terms of the total electric field as the primary unknown. The electric field can be expanded in curl-conforming basis functions, and the same functions can be used for the testing functions, provided that some means of implementing the divergence operator is developed. The remaining restriction on (11.86) is the spherical shape of the radiation boundary.

Preliminary numerical testing of local RBCs for spherical targets was reported in [26, 27], and the first extensive application of the approach appeared in [28] and is summarized in [8] and [12]. Figure 11.5 shows the bistatic scattering cross section of a perfectly conducting cube of side dimension 0.755λ, obtained using approximately 33,000 tetrahedral-cell CT/LN basis functions and the RBC in (11.77). In this case, the RBC was located only 0.1λ from the corners of the cube. Although references [8, 12, 28] claim to employ the second-order RBC in (11.77), it appears that the implementation omitted the third integral in (11.85) and thus may only be as accurate as a first-order RBC.

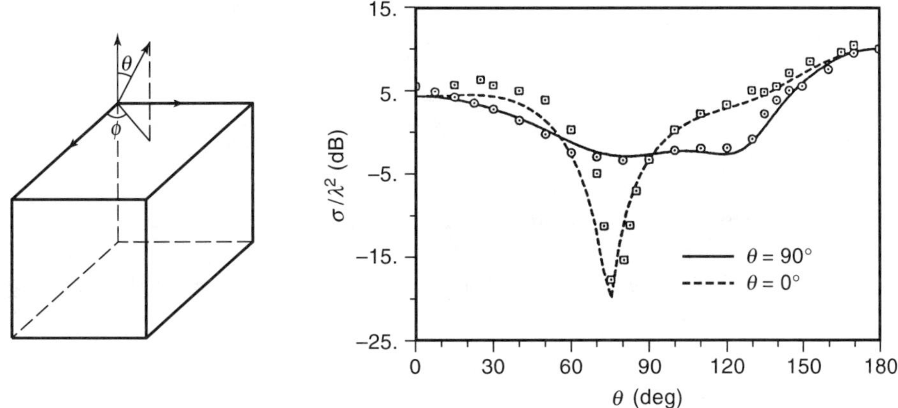

Figure 11.5 Bistatic scattering cross section of a conducting cube of side dimension 0.755λ. Lines represent the result from the CT/LN discretization of the vector Helmholtz equation; circles and squares denote measured data for comparison. After [28]. ©1993 IEEE.

Although local RBCs are approximate, they offer the possibility of terminating the computational domain with little or no fill-in beyond that associated with the vector Helmholtz equation. In many applications, they appear to be sufficiently accurate. The restriction to a spherical boundary places an unacceptable burden on the approach, however, and the following section considers the extension of local RBCs to more general boundary shapes.

11.8 LOCAL RBCS FOR GENERAL THREE-DIMENSIONAL BOUNDARY SHAPES

Since the task of padding the computational domain to a spherical boundary might introduce an unacceptably large number of unknowns, RBCs applicable to more general boundary shapes are desired. Development of RBCs is a recent endeavor, and the testing and evaluation of three-dimensional RBCs to date is far from extensive. Consequently, at the time of this writing it is impossible to survey the local RBCs that have been suggested for general boundary shapes and assess their relative accuracy with any confidence. In this section, several RBCs proposed for general boundaries are briefly surveyed.

As a first choice, the local RBC in (11.77) can be extended to piecewise-planar boundaries by the simple expedient of replacing the spherical normal vector $\hat{r}$ with the general normal vector $\hat{n}$ and replacing the factor $1/r$ in the coefficients with the local curvature (zero). This produces the RBC [28]

$$\hat{n} \times \nabla \times \bar{E}^s = jk\bar{E}^s_{\tan} + \frac{1}{2jk}\nabla \times [\hat{n}(\hat{n} \cdot \nabla \times \bar{E}^s)] + \frac{1}{2jk}\nabla^{\tan}(\nabla \cdot \bar{E}^s_{\tan}) \tag{11.87}$$

One might expect a decrease in the accuracy of the RBC after such a substitution, but reports suggest that good results can still be obtained even if the RBC is less than $\lambda/3$ from the scatterer [8, 28].

D'Angelo and Mayergoyz investigated the second-order Engquist–Majda RBC within a three-dimensional formulation employing CT/LN functions and piecewise-planar boundaries [29, 30]. Along a locally planar boundary, the RBC is equivalent to

$$\hat{n} \times \nabla \times \bar{E}^s = jk\bar{E}^s_{\tan} + \frac{j}{2k}\left(\frac{\partial^2 \bar{E}^s_{\tan}}{\partial t^2} + \frac{\partial^2 \bar{E}^s_{\tan}}{\partial u^2}\right) + \nabla^{\tan}(\hat{n} \cdot \bar{E}^s) \tag{11.88}$$

where (t, u) denote the tangential variables. D'Angelo and Mayergoyz suggested locating the radiation boundary approximately 1λ from the scatterer surface.

Sun and Balanis proposed an alternative family of vector RBCs that were derived from one-way wave equations [31]. Their second-order RBC, applicable to general piecewise-planar boundaries, has the form

$$\begin{aligned} \hat{n} \times \nabla \times \bar{E}^s &= jkp_0\bar{E}^s_{\tan} + q_0\nabla^{\tan}(\hat{n} \cdot \bar{E}^s) \\ &\quad - \frac{jp_1}{k}\nabla \times [\hat{n}(\hat{n} \cdot \nabla \times \bar{E}^s)] + \frac{jp_2}{k}\nabla^{\tan}(\nabla \cdot \bar{E}^s_{\tan}) \end{aligned} \tag{11.89}$$

where the parameters q_0, p_0, p_1, and p_2 are to be obtained by approximating the pseudo-differential operator according to

$$\begin{aligned} &\sqrt{k^2\bar{\bar{I}} - \nabla^{\tan} \times \nabla^{\tan} \times + \nabla^{\tan}\nabla^{\tan}\cdot} \\ &\approx k(p_0\bar{\bar{I}} - p_1\nabla^{\tan} \times \nabla^{\tan} \times + p_2\nabla^{\tan}\nabla^{\tan}\cdot\ ) \end{aligned} \tag{11.90}$$

for the specific boundary of interest. The details of pseudo-differential operator theory are beyond the scope of this text. However, for planar boundaries, a Padé approximation of this operator produces $q_0 = 0.75$, $p_0 = 1$, $p_1 = 0.5$, and $p_2 = 0.25$ for use with (11.89) [31]. These values are reported to produce minimal reflections from planar boundaries over a wide range of incident angles.

It is noteworthy that the three RBCs in (11.87), (11.88), and (11.89) differ only slightly from each other when applied to piecewise-planar boundaries. One might expect that an RBC applicable to continuously curved boundaries would have a similar form, such as

$$\hat{n} \times \nabla \times \bar{E}^s = \bar{\bar{\alpha}} \cdot \bar{E}^s_{\tan} + \bar{\bar{\delta}}\nabla \times [\hat{n}(\hat{n} \cdot \nabla \times \bar{E}^s)] + \bar{\bar{\gamma}}\nabla^{\tan}(\nabla \cdot \bar{E}^s_{\tan}) \tag{11.91}$$

where $\bar{\bar{\alpha}}$, $\bar{\bar{\delta}}$, and $\bar{\bar{\gamma}}$ might be tensor or scalar quantities that are functions of the boundary curvature. The RBCs of this form have been proposed by Stupfel [32] and Chatterjee and

Volakis [33, 34]. Stupfel suggests the parameters [32]

$$\bar{\bar{\alpha}} = \frac{1}{2(jk + 2\kappa_m)} \left[-2k^2 + (3\kappa_2 + \kappa_1) \left(jk + \frac{\kappa_2 - \kappa_1}{4} \right) \hat{t}_1 \hat{t}_1 + (3\kappa_1 + \kappa_2) \left(jk - \frac{\kappa_2 - \kappa_1}{4} \right) \hat{t}_2 \hat{t}_2 \right] \tag{11.92}$$

and

$$\delta = \gamma = \frac{1 - j\kappa_m / k}{2(jk + 2\kappa_m)} \tag{11.93}$$

while Chatterjee and Volakis suggest [33, 34]

$$\bar{\bar{\alpha}} = \delta \, \{4\kappa_m^2 - \kappa_g + jkD + [\kappa_1^2 + \kappa_m(\kappa_1 - \kappa_2) - \kappa_1 D]\hat{t}_1\hat{t}_1 + [\kappa_2^2 + \kappa_m(\kappa_2 - \kappa_1) - \kappa_2 D]\hat{t}_2\hat{t}_2\} \tag{11.94}$$

$$\delta = \frac{1}{D - 2\kappa_m} \tag{11.95}$$

and

$$\bar{\bar{\gamma}} = \frac{\delta}{jk} \left(jk + 3\kappa_m - \frac{\kappa_g}{\kappa_m} - 2(\kappa_1 \hat{t}_1 \hat{t}_1 + \kappa_2 \hat{t}_2 \hat{t}_2) \right) \tag{11.96}$$

In these expressions, $\hat{n}$ denotes the normal vector to the boundary, κ_1 and κ_2 denote the principal curvatures of the boundary, $\kappa_g = \kappa_1 \kappa_2$ is the Gaussian curvature, $\kappa_m = \frac{1}{2}(\kappa_1 + \kappa_2)$ is the mean curvature, and

$$D = 2jk + 5\kappa_m - \frac{\kappa_g}{\kappa_m} \tag{11.97}$$

The expressions for $\bar{\bar{\alpha}}$ in (11.92) and (11.94) and $\bar{\bar{\gamma}}$ in (11.96) are tensor quantities. The parameters in (11.92) and (11.93) apparently neglect derivatives of the curvature while (11.94)–(11.96) take curvature derivatives into account [32–34]. If these parameters are specialized to a spherical boundary shape, both choices simplify to produce the spherical RBC in (11.77).

11.9 RBCS BASED ON FICTITIOUS ABSORBERS

An alternate way of truncating the computational domain in an open-region problem is to surround it with a high-loss medium. Beyond a region of lossy material sufficient to absorb most of the energy, the region can be terminated in a perfect electric or magnetic wall or in a local RBC appropriate for that material. While physically realizable absorbing materials would work in principle, they generally require the introduction of a relatively large number of additional unknowns. Better performance can be realized through the use of fictitious materials that are nonphysical and possibly active [8, 35]. Recently, this idea was extended to the use of a nonphysical anisotropic material that in principle (prior to truncation and discretization) would absorb uniform plane waves from any angle without reflection [36]. Such a material is known as a *perfectly matched layer* (PML).

For instance, consider a medium with relative permittivity and permeability tensors of the form

$$\bar{\bar{\varepsilon}}_r = \bar{\bar{\mu}}_r = \begin{bmatrix} 1 - j\dfrac{\sigma}{\omega\varepsilon} & 0 & 0 \\ 0 & 1 - j\dfrac{\sigma}{\omega\varepsilon} & 0 \\ 0 & 0 & \left(1 - j\dfrac{\sigma}{\omega\varepsilon}\right)^{-1} \end{bmatrix} \tag{11.98}$$

Assuming that the interface between free space and an infinite half-space with the medium parameters given in (11.98) is located at $z = 0$, a wave incident at an arbitrary angle is perfectly absorbed (Prob. P11.20) for both perpendicular polarization,

$$\bar{E}^{\text{inc}} = \hat{y} E_0 e^{-jk(x \sin\theta + z \cos\theta)} \tag{11.99}$$

and for parallel polarization,

$$\bar{H}^{\text{inc}} = \hat{y} H_0 e^{-jk(x \sin\theta + z \cos\theta)} \tag{11.100}$$

or any linear combination thereof. Therefore, if the loss tangent in (11.98) is adjusted appropriately, a relatively small region of such a material (typically much less than a free-space wavelength) will sufficiently absorb an incident wave with minimal reflections. For a numerical implementation, the fields in the absorbing region may be represented using the same basis functions that are employed throughout the rest of the computational domain. Beyond the absorber, the region can be terminated with a perfectly conducting wall or some type of local absorbing boundary condition.

Fictitious absorbing materials appear to provide much lower reflections than the local RBCs discussed in preceding sections without substantial computational overhead [36–39]. They have been used with the finite-difference time-domain method [36, 37] and with frequency-domain discretizations of the vector Helmholtz equation [38–39]. To date, a detailed comparison of their accuracy and relative computational costs with those of the RBCs discussed in previous sections has not been performed.

11.10 VECTOR FORMULATION FOR AXISYMMETRIC HETEROGENEOUS SCATTERERS

The preceding formulations are capable of treating fairly arbitrary isotropic geometries. A heterogeneous scatterer with rotational symmetry, that is, μ_r and ε_r, which do not vary with ϕ, can be treated by a more efficient process that only involves unknowns along a generating sector (Figure 11.6). If the excitation is also rotationally symmetric, as considered in Section 8.8, the formulation is particularly simple since the primary unknown is a single ϕ-component of one of the fields. For a general excitation, a formulation in terms of two ϕ-components (or in terms of the closely related *coupled azimuthal potentials*) is possible [40–42]. It is also possible to pose the problem in terms of a formulation involving three components of a single field [43, 44]. In this section, we review a formulation similar to [44] involving the electric field as the primary unknown.

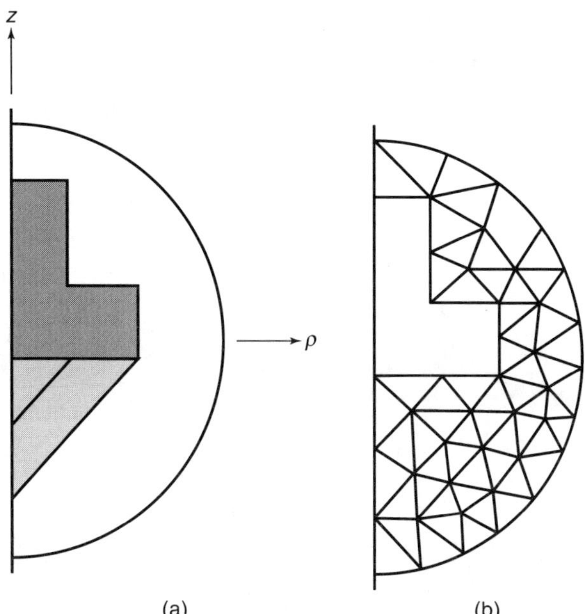

Figure 11.6 (a) Generating sector for an axisymmetric object. (b) Triangular cell mesh for the generating sector.

If the geometry is rotationally symmetric about the z-axis in a cylindrical coordinate system, the electric field can be written in the form

$$\bar{E}(\rho, \phi, z) = \sum_{m=-\infty}^{\infty} \left[\bar{E}_t(\rho, z) + \hat{\phi} E_\phi(\rho, z) \right] e^{jm\phi} \tag{11.101}$$

where the ρ–z plane will be considered the "transverse" plane. The harmonics in (11.101) are coupled only by the excitation. Applied to a single harmonic, the vector Helmholtz equation (11.1) has ϕ-component

$$-\tilde{\nabla} \cdot \left(\frac{1}{\rho \mu_r} \tilde{\nabla}(\rho E_\phi) \right) + jm\tilde{\nabla} \cdot \left(\frac{1}{\rho \mu_r} \bar{E}_t \right) - k^2 \varepsilon_r E_\phi = 0 \tag{11.102}$$

and transverse component

$$\nabla \times \left(\frac{1}{\mu_r} \nabla \times \bar{E}_t \right) + \frac{jm}{\rho^2 \mu_r} \tilde{\nabla}(\rho E_\phi) - \left(k^2 \varepsilon_r - \frac{m^2}{\rho^2 \mu_r} \right) \bar{E}_t = 0 \tag{11.103}$$

where we have suppressed the ϕ-dependence and introduced the pseudo-∇-operator defined by

$$\tilde{\nabla}\Psi = \hat{\rho} \frac{\partial \Psi}{\partial \rho} + \hat{z} \frac{\partial \psi}{\partial z} \tag{11.104}$$

and

$$\tilde{\nabla} \cdot \bar{A} = \frac{\partial A_\rho}{\partial \rho} + \frac{\partial A_z}{\partial z} \tag{11.105}$$

Note that the ordinary curl operator $\nabla\times$ is used in (11.103).

Multiplying (11.102) by a scalar testing function T and integrating over the volume of the three-dimensional region produce the weak equation

$$\int_\rho \int_z \left(\frac{1}{\rho\mu_r} \tilde{\nabla} T \cdot \tilde{\nabla}(\rho E_\phi) - \frac{jm}{\rho\mu_r} \tilde{\nabla} T \cdot \bar{E}_t - \frac{k^2 \varepsilon_r}{\rho} T(\rho E_\phi) \right) \rho \, d\rho \, dz$$
$$= \int_{\partial\Gamma} \left(\frac{1}{\rho\mu_r} T \frac{\partial(\rho E_\phi)}{\partial n} - \frac{jm}{\rho\mu_r} T \hat{n} \cdot \bar{E}_t \right) \rho \, dt \tag{11.106}$$

where t is a local variable tangential to the boundary of the generating sector and the ϕ-integration has been evaluated and omitted (as if the testing function had the appropriate ϕ-dependence). Similarly, multiplying (11.103) with a vector testing function oriented in the transverse (ρ–z) plane and integrating over the volume lead to the weak equation

$$\int_\rho \int_z \left[\frac{1}{\mu_r} \nabla \times \bar{T} \cdot \nabla \times \bar{E}_t + \frac{jm}{\rho^2 \mu_r} \bar{T} \cdot \tilde{\nabla}(\rho E_\phi) - \left(k^2 \varepsilon_r - \frac{m^2}{\rho^2 \mu_r} \right) \bar{T} \cdot \bar{E}_t \right] \rho \, d\rho \, dz$$
$$= - \int_{\partial\Gamma} \frac{1}{\mu_r} (\bar{T} \cdot \hat{n} \times \nabla \times \bar{E}_t) \rho \, dt \tag{11.107}$$

The form of the preceding equations suggests that (ρE_ϕ) and $\bar{E}_t$ be represented by independent expansions, such as

$$\rho E_\phi(\rho, z) \cong \sum_{n=1}^{N_s} e_{\phi n} B_n(\rho, z) \tag{11.108}$$

and

$$\bar{E}_t(\rho, z) \cong \sum_{n=1}^{N_v} e_{tn} \bar{B}_n(\rho, z) \tag{11.109}$$

where $\{B_n\}$ denote scalar Lagrangian basis functions while $\{\bar{B}_n\}$ denote curl-conforming vector basis functions in the transverse (ρ–z) plane. For instance, the generating sector can be modeled by triangular cells (Figure 11.6) and the usual triangular-cell basis functions can be incorporated. The use of $\rho E\phi$ as the primary scalar unknown is not only convenient, it appears to be essential to achieve a proper representation of the nullspace of the curl–curl operator [44].

Boundary conditions must be imposed along the $\rho = 0$ axis, the outer radiation boundary, and the surface of any imbedded perfect conductors. Along the axis, the boundary conditions are $\rho E_\phi = 0$ and $\hat{\rho} \times (\nabla \times \bar{E}_t) = 0$. At the surface of perfect electric conductors, the appropriate conditions are $\rho E_\phi = 0$ and $\hat{n} \times \bar{E}_t = 0$. Along the radiation boundary, some type of RBC must be combined with the weak equations. Observe that the ϕ-component of $\hat{n} \times (\nabla \times \bar{E}^s)$, for the mth harmonic, can be written as

$$\hat{\phi} \cdot \hat{n} \times \nabla \times \bar{E}^s = -\frac{1}{\rho} \frac{\partial(\rho E_\phi^s)}{\partial n} + \frac{jm}{\rho} \hat{n} \cdot \bar{E}_t^s \tag{11.110}$$

which is exactly the expression appearing in the boundary integral in (11.106). Thus, local or global RBCs can be implemented in the usual manner (after separating the transverse and ϕ-components) by substitution into the boundary integrals in (11.106) and (11.107).

Several axisymmetric formulations have been reported. Morgan et al. [40] developed an inward-looking formulation in terms of coupled azimuthal potentials, where an eigenfunction series was used to represent fields in the exterior region. Gordon and Mittra [41] used coupled azimuthal potentials and developed local RBCs based on the series in (11.66). Hoppe et al. [42] discuss several ways of incorporating integral equation RBCs into an axisymmetric formulation employing the ϕ-components of the fields as primary unknowns, which are represented by the usual scalar Lagrangian functions. Khebir et al. [43] used a node-based Lagrangian expansion that treated all three components of the field as independent unknowns but employed the local RBC in (11.77) on a spherical boundary. Their results suggest that the local RBC works well for a variety of scatterers. Reference [44] appears to be the first to use a formulation similar to that outlined above, where scalar and vector basis functions are used for the ϕ-component and transverse components, respectively, of a single field. However, [44] only considered closed-region applications.

11.11 ALTERNATIVE FORMULATIONS FOR THREE-DIMENSIONAL SCATTERING

Previous sections of this chapter have considered formulations based on the vector Helmholtz equations in (11.1) and (11.2), with discretizations obtained from the curl-conforming vector basis functions developed in Chapter 9. Because scalar representations are widely used in other engineering disciplines and are compatible with a substantial amount of general-purpose finite-element software, including field display tools, much interest remains in three-dimensional electromagnetic formulations that permit the use of node-based Lagrangian basis functions. Because of the nature of electromagnetic fields, however (such as the jump discontinuities at material interfaces), such formulations must involve auxiliary potential functions as the primary unknowns instead of the vector fields. Approaches based on potential functions have been discussed by several authors [45–48]; one based on [46] is briefly summarized below.

Suppose we wish to solve Maxwell's equations (1.1)–(1.4) within a source-free three-dimensional region Γ, subject to boundary conditions that specify tangential $\bar{E}$ on a part of the boundary ($\partial\Gamma_1$) and tangential $\bar{H}$ on the rest ($\partial\Gamma_2$). By introducing a vector potential function $\bar{A}$ and a scalar potential function Φ that satisfy

$$\bar{H} = \frac{1}{\mu}\nabla \times \bar{A} \tag{11.111}$$

$$\bar{E} = -j\omega\bar{A} - \nabla\Phi \tag{11.112}$$

one readily obtains the differential equations

$$\nabla \times \left(\frac{1}{\mu}\nabla \times \bar{A}\right) - \omega^2\varepsilon\left(\bar{A} + \frac{1}{j\omega}\nabla\Phi\right) = 0 \tag{11.113}$$

and

$$\nabla \cdot \varepsilon\left(\bar{A} + \frac{1}{j\omega}\nabla\Phi\right) = 0 \tag{11.114}$$

to be satisfied by the potential functions. There is a remaining degree of freedom not specified in (11.111) and (11.112); the choice

$$\nabla \cdot \bar{A} = -j\omega\varepsilon\mu\Phi \tag{11.115}$$

ensures the continuity of both the vector and scalar potential functions at material interfaces [46]. Substitution of (11.115) into the previous equations yields

$$\nabla \times \left(\frac{1}{\mu}\nabla \times \bar{A}\right) - \nabla\left(\frac{1}{\mu}\nabla \cdot \bar{A}\right) - \omega^2\varepsilon\bar{A} - j\omega\Phi\nabla\varepsilon = 0 \tag{11.116}$$

and

$$-\nabla \cdot (\varepsilon\nabla\Phi) - \omega^2\varepsilon^2\mu\Phi - j\omega\bar{A} \cdot \nabla\varepsilon = 0 \tag{11.117}$$

as equations that must be solved to produce the potential functions. Since $\bar{A}$ and Φ are continuous quantities, these equations are appropriate for finite-element discretization using Lagrangian basis functions. In general, there are four unknowns per node associated with the representation. Once the potential functions are determined, the fields can be constructed using (11.111) and (11.112).

Boundary conditions on the fields at conducting surfaces and radiation boundaries must be translated into boundary conditions on the potential functions. On the part of the boundary where the tangential electric field is specified ($\partial\Gamma_1$), appropriate conditions on the potential functions are

$$\hat{n} \times \bar{A} = -\frac{1}{j\omega}\hat{n} \times \bar{E} \tag{11.118}$$

$$\nabla \cdot \bar{A} = 0 \tag{11.119}$$

$$\Phi = 0 \tag{11.120}$$

while on the part of the boundary where the tangential magnetic field is specified ($\partial\Gamma_2$)

$$\hat{n} \times \nabla \times \bar{A} = \mu\hat{n} \times \bar{H} \tag{11.121}$$

$$\hat{n} \cdot \varepsilon\bar{A} = -\frac{1}{\omega^2}\nabla^{\tan} \cdot (\hat{n} \times \bar{H}) \tag{11.122}$$

$$\frac{\partial\Phi}{\partial n} = 0 \tag{11.123}$$

where $\nabla^{\tan} \cdot \bar{V}$ denotes the surface divergence.

The preceding development constitutes one in which the primary unknowns are continuous quantities and thus should be applicable to the use of traditional scalar Lagrangian basis functions. The principal drawback to this approach appears to be the need to use four unknowns per node and the need to differentiate the potential functions in order to construct the fields. The literature provides more details about this formulation, as well as others in which vector and scalar potentials are the primary unknowns [45–48].

11.12 SUMMARY

This chapter has explored three-dimensional formulations for electromagnetic scattering. Different ways of truncating the computational domain for open regions have been emphasized, and a number of RBCs were described. Some are exact prior to discretization, while others are approximate. If high accuracy is desired, it appears that a reasonably efficient formulation can be obtained by using an integral equation RBC in conjunction with an axisymmetric radiation boundary. This chapter has also illustrated the use of curl-conforming vector basis functions to discretize the curl–curl form of the three-dimensional vector Helmholtz equation. An alternative formulation permitting the use of traditional scalar Lagrangian basis functions was briefly summarized. Three-dimensional formulations require substantially more computer resources than two-dimensional approaches and are far from mature at the present time. It is likely that the next decade will see significant advances in these techniques.

REFERENCES

[1] J. S. Savage and A. F. Peterson, "Higher-order vector finite elements for tetrahedral cells," *IEEE Trans. Microwave Theory Tech.*, vol. 44, pp. 874–879, June 1996.

[2] J. C. Nedelec, "Mixed finite elements in R3," *Numer. Math.*, vol. 35, pp. 315–341, 1980.

[3] O. C. Zienkiewicz and R. L. Taylor, *The Finite Element Method*. London: McGraw-Hill, 1988.

[4] Y. Li, S. Zhu, and F. A. Fernandez, "The efficient solution of large sparse nonsymmetric and complex eigensystems by subspace iteration," *IEEE Trans. Magnetics*, vol. MAG-30, pp. 3582–3585, Sept. 1994.

[5] R. F. Harrington, *Time-Harmonic Electromagnetic Fields*, New York: McGraw-Hill, 1961.

[6] J. A. Stratton, *Electromagnetic Theory*, New York: McGraw-Hill, 1941.

[7] W. V. T. Rusch, A. C. Ludwig, and W. C. Wong, "Analytical techniques for quasi-optical antennas," in *The Handbook of Antenna Design*, eds. A. W. Rudge, K. Milne, A. D. Olver, and P. Knight, London, Peregrinus, 1986, pp. 101–109.

[8] J. Jin, *The Finite Element Method in Electromagnetics*, New York: Wiley, 1993.

[9] X. Yuan, "Three-dimensional electromagnetic scattering from inhomogeneous objects by the hybrid moment and finite element method," *IEEE Trans. Antennas Propagat.*, vol. 38, pp. 1053–1058, Aug., 1990.

[10] J.-M. Jin and J. L. Volakis, "Electromagnetic scattering by and transmission through a three-dimensional slot in a thick conducting plane," *IEEE Trans. Antennas Propagat.*, vol. 39, pp. 543–550, Apr. 1991.

[11] J.-M. Jin and J. L. Volakis, "A hybrid finite element method for scattering and radiation by microstrip patch antennas and arrays residing in a cavity," *IEEE Trans. Antennas Propagat.*, vol. 39, pp. 1598–1604, Nov. 1991.

[12] J. L. Volakis, A. Chatterjee, and L. C. Kempel, "Review of the finite element method for three-dimensional electromagnetic scattering," *J. Opt. Soc. Am. A*, vol. 11, pp. 1422–1433, Apr. 1994.

[13] G. E. Antilla and N. G. Alexopoulos, "Scattering from complex three-dimensional geometries by a curvilinear hybrid finite-element–integral-equation approach," *J. Opt. Soc. Am. A*, vol. 11, pp. 1445–1457, Apr. 1994.

[14] E. W. Lucas and T. P. Fontana, "A 3-D hybrid finite element/boundary element method for the unified radiation and scattering analysis of general infinite periodic arrays," *IEEE Trans. Antennas Propagat.*, vol. 43, pp. 145–153, Feb. 1995.

[15] W. E. Boyse and A. A. Seidl, "A hybrid finite element method for near bodies of revolution," *IEEE Trans. Magnet.*, vol. 27, pp. 3833–3836, Sept. 1991.

[16] W. E. Boyse and A. A. Seidl, "A hybrid finite element method for 3-D scattering using nodal and edge elements," *IEEE Trans. Antennas Propagat.*, vol. 42, pp. 1436–1442, Oct. 1994.

[17] T. Cwik, V. Jamnejad, and C. Zuffada, "Coupling finite element and integral equation representations to efficiently model large three-dimensional scattering objects," *Digest of the 1993 IEEE Antennas and Propagation Society International Symposium*, Ann Arbor, MI, pp. 1756–1759, June 1993.

[18] T. Cwik, V. Jamnejad, and C. Zuffada, "A comparison of partitioned and non-partitioned matrix solutions to coupled finite element-integral equations," *Abstracts of the 1993 URSI Radio Science Meeting*, Ann Arbor, MI, p. 253, June 1993.

[19] T. Cwik, C. Zuffada, and V. Jamnejad, "The coupling of finite element and integral equation representations for efficient three-dimensional modeling of electromagnetic scattering and radiation," in *Finite Element Software for Microwave Engineering*, eds. T. Itoh, G. Pelosi, and P. P. Silvester, New York: Wiley, 1996, pp. 147–167.

[20] C. H. Wilcox, "An expansion theorem for electromagnetic fields," *Comm. Pure Appl. Math.*, vol. 9, pp. 115–134, 1956.

[21] A. F. Peterson, "Absorbing boundary conditions for the vector wave equation," *Microwave Opt. Technol. Lett.*, vol. 1, pp. 62–64, Apr. 1988.

[22] A. F. Peterson, "Accuracy of 3D radiation boundary conditions for use with the vector Helmholtz equation," *IEEE Trans. Antennas Propagat.*, vol. 40, pp. 351–355, Mar. 1992.

[23] J. P. Webb and V. N. Kanellopoulos, "Absorbing boundary conditions for the finite element solution of the vector wave equation," *Microwave Opt. Technol. Lett.*, vol. 2, pp. 370–372, Oct. 1989.

[24] V. N. Kanellopoulos and J. P. Webb, "The importance of the surface divergence term in the finite element-vector absorbing boundary condition method," *IEEE Trans. Microwave Theory Tech.*, vol. 43, pp. 2168–2170, Sept. 1995.

[25] J. S. Savage, "Vector finite elements for the solution of Maxwell's equations." Ph.D. dissertation, Georgia Institute of Technology, Atlanta, 1997.

[26] S. P. Castillo and A. F. Peterson, "Solution of the 3-D vector wave equation in an open region using the finite element method," *Abstracts of the 1989 URSI Radio Science Meeting*, San Jose, CA, p. 328, June 1989.

[27] J. P. Webb and V. N. Kanellopoulos, "A numerical study of vector absorbing boundary conditions for the finite element solution of Maxwell's equations," *IEEE Microwave Guided Wave Lett.*, vol. 1, pp. 325–327, Nov. 1991.

[28] A. Chatterjee, J. M. Jin, and J. L. Volakis, "Edge-based finite elements and vector ABC's applied to 3-D scattering," *IEEE Trans. Antennas Propagat.*, vol. 41, pp. 221–226, Feb. 1993.

[29] J. D'Angelo and I. D. Mayergoyz, "Finite element methods for the solution of RF radiation and scattering problems," *Electromagnetics*, vol. 10, pp. 177–199, 1990.

[30] J. D'Angelo and I. D. Mayergoyz, "Three-dimensional RF scattering by the finite element method," *IEEE Trans. Magnet.*, vol. 27, pp. 3827–3832, Sept. 1991.

[31] W. Sun and C. A. Balanis, "Vector one-way wave absorbing boundary conditions for FEM applications," *IEEE Trans. Antennas Propagat.*, vol. 42, pp. 872–878, June 1994.

[32] B. Stupfel, "Absorbing boundary conditions on arbitrary boundaries for the scalar and vector wave equations," *IEEE Trans. Antennas Propagat.*, vol. 42, pp. 773–780, June 1994.

[33] A. Chatterjee and J. L. Volakis, "Conformal absorbing boundary conditions for the vector wave equation," *Microwave Opt. Technol. Lett.*, vol. 6, pp. 886–889, Dec. 1993.

[34] A. Chatterjee and J. L. Volakis, "Conformal absorbing boundary conditions for 3-D problems: Derivation and applications," *IEEE Trans. Antennas Propagat.*, vol. 43, pp. 860–866, Aug. 1995.

[35] J. M. Jin, J. L. Volakis, and V. V. Liepa, "Fictitious absorber for truncating finite element meshes in scattering," *IEE Proc. H*, vol. 139, pp. 472–476, Oct. 1992.

[36] J.-P. Berenger, "A perfectly matched layer for the absorption of electromagnetic waves," *J. Computat. Phys.*, vol. 114, pp. 185–200, 1994.

[37] D. S. Katz, E. T. Thiele, and A. Taflove, "Validation and extension to three dimensions of the Berenger PML absorbing boundary condition for FD-TD meshes," *IEEE Microwave Guided Wave Lett.*, vol. 4, pp. 268–271, 1994.

[38] Z. S. Sacks, D. M. Kingsland, R. Lee, and J.-F. Lee, "A perfectly matched anisotropic absorber for use as an absorbing boundary condition," *IEEE Trans. Antennas Propagat.*, vol. 43, pp. 1460–1463, Dec. 1995.

[39] D. M. Kingsland, J. Gong, J. L. Volakis, and J.-F. Lee, "Performance of an anisotropic artificial absorber for truncating finite element meshes," *IEEE Trans. Antennas Propagat.*, vol. 44, pp. 975–981, July 1996.

[40] M. A. Morgan, S. -K. Chang, and K. K. Mei, "Coupled azimuthal potentials for electromagnetic field problems in inhomogeneous axially symmetric media," *IEEE Trans. Antennas Propagat.*, vol. 25, pp. 413–417, May 1977.

[41] R. K. Gordon and R. Mittra, "PDE techniques for solving the problem of radar scattering by a body of revolution," *Proc. IEEE*, vol. 79, pp. 1449–1458, Oct. 1991.

[42] D. J. Hoppe, L. W. Epp, and J.-F. Lee, "A hybrid symmetric FEM/MOM formulation applied to scattering by inhomogeneous bodies of revolution," *IEEE Trans. Antennas Propagat.*, vol. 42, pp. 798–805, June 1994.

[43] A. Khebir, J. D'Angelo, and J. Joseph, "A new finite element formulation for RF scattering by complex bodies of revolution," *IEEE Trans. Antennas Propagat.*, vol. 41, pp. 534–541, May 1993.

[44] J.-F. Lee, G. M. Wilkins, and R. Mittra, "Finite element analysis of axisymmetric cavity resonator using a hybrid edge element technique," *IEEE Trans. Microwave Theory Tech.*, vol. 41, pp. 1981–1987, Nov. 1993.

[45] J. R. Brauer, ed., *What Every Engineer Should Know about Finite Element Analysis*, New York: Marcel Dekker, 1993.

[46] W. E. Boyse, D. R. Lynch, K. D. Paulsen, and G. N. Minerbo, "Continuous potential Maxwell's equation solution on nodal-based finite elements," *IEEE Trans. Antennas Propagat.*, vol. 40, pp. 1192–1200, 1992.

[47] K. D. Paulsen, W. E. Boyse, and D. R. Lynch, "Nodal based finite element modeling of Maxwell's equations," *IEEE Trans. Antennas Propagat.*, vol. 40, pp. 642–651, June 1992.

[48] K. D. Paulsen, "Finite element solution of Maxwell's equations with Helmholtz forms," *J. Opt. Soc. Am. A*, vol. 11, pp. 1434–1444, Apr. 1994.

PROBLEMS

P11.1 Describe the appropriate boundary conditions to be imposed on the surface of a perfect magnetic conductor, and discuss their implementation in connection with the weak differential equations in (11.6) and (11.7).

P11.2 Evaluate (11.35) and (11.36) for the possible combinations of coordinates associated with element matrix entries for the LT/QN basis and testing functions.

P11.3 Consider an infinite structured mesh obtained by dividing a regular cubical-cell mesh into tetrahedrons, with five tetrahedrons per cube. Determine the number of CT/LN basis functions per cell and the number of LT/QN basis functions per cell associated with this type of mesh. In addition, determine the average number of nonzero matrix entries per unknown for both basis function types. Using the average number of nodes per cell, compare these numbers with similar estimates for linear and quadratic Lagrangian basis functions, assuming that three Lagrangian functions per node are used to represent the same three-dimensional field.

P11.4 By expanding a uniform plane-wave incident field in terms of spherical harmonics, rewrite the RBC in (11.47) in terms of the total electric field.

P11.5 The RBC in (11.47) involves a summation over the spherical eigenfunctions. Discuss the convergence of this summation as a function of the index n.

P11.6 Using the integral formulas

$$\int_0^{\pi} \left[\frac{\partial P_n^m(\cos\theta)}{\partial\theta} \frac{\partial P_q^m(\cos\theta)}{\partial\theta} + \frac{m^2 P_n^m(\cos\theta) P_q^m(\cos\theta)}{\sin^2\theta} \right] \sin\theta \, d\theta = 0 \qquad q \neq n$$

$$\int_0^{\pi} \left[\left(\frac{\partial P_n^m(\cos\theta)}{\partial\theta} \right)^2 + \left(\frac{m P_n^m(\cos\theta)}{\sin\theta} \right)^2 \right] \sin\theta \, d\theta = \frac{2n(n+1)}{2n+1} \frac{(n+|m|)!}{(n-|m|)!}$$

verify the following orthogonality conditions involving the spherical vector wave functions in (11.38) and (11.39):

$$\bar{M}_{mn} \cdot \bar{N}_{mn} = 0$$

$$\int_0^{2\pi} \int_0^{\pi} \bar{M}_{mn} \cdot \bar{M}_{pq} \sin\theta \, d\theta \, d\phi = \begin{cases} \beta_{mn} & p = -m, q = n \\ 0 & \text{otherwise} \end{cases}$$

$$\int_0^{2\pi} \int_0^{\pi} \bar{N}_{mn}^{\tan} \cdot \bar{N}_{pq}^{\tan} \sin\theta \, d\theta \, d\phi = \begin{cases} \gamma_{mn} & p = -m, q = n \\ 0 & \text{otherwise} \end{cases}$$

where "tan" refers to the theta and phi components only,

$$\beta_{mn} = \left(\frac{\hat{H}_n^{(2)}(kr)}{r} \right)^2 \frac{4\pi n(n+1)}{2n+1} \frac{(n+|m|)!}{(n-|m|)!}$$

$$\gamma_{mn} = \left(\frac{\hat{H}_n^{(2)\prime}(kr)}{r} \right)^2 \frac{4\pi n(n+1)}{2n+1} \frac{(n+|m|)!}{(n-|m|)!}$$

Using these orthogonality conditions, derive an alternate form of the spherical eigenfunction RBC, leaving the result in terms of the scattered field.

P11.7 Derive an expression for the scattering cross section as a function of the coefficients of the spherical vector wave functions in (11.38) and (11.39).

P11.8 Using the vector basis function definitions (Chapter 9), demonstrate the equivalence of a CT/LN representation of the magnetic field $\bar{H}$ throughout tetrahedral cells and a CN/LT representation of the equivalent surface current $\bar{J}_s$ on the triangular faces of the same cells.

P11.9 Develop a discretization of (11.52) for the MFIE part ($\alpha = 0$) of the equation using CN/LT basis functions defined on triangular patches and razor-blade testing functions (Section 10.2). Provide explicit entries for the system in (11.53). *Hint:* These entries are similar to those given in Chapter 10.

P11.10 Generalize the outward-looking formulation embodied in Equation (11.62) for the situation when the basis functions used within the interior region Γ are not the same as the basis functions used to represent the equivalent surface current densities $\bar{J}_s$ and $\bar{K}_s$ within the integral equation RBC. In this case, the coefficients of the interior expansion are not directly related to either of those on the surface, and one cannot assume that $\mathbf{e}^{\text{bound}} = \mathbf{k}$. Develop a way of defining one set of coefficients in terms of the other; that is, describe the linear operator f associated with the relation $\mathbf{e}^{\text{bound}} = f(\mathbf{k})$.

P11.11 For body-of-revolution integral equation formulations (Chapter 8), the number of important cylindrical harmonics associated with a normally incident wave is sometimes estimated as $2(6 + k\rho)$, where ρ is the largest radius of the boundary. How many harmonics must be included for a right-circular cylindrical region of radius (a) 1λ, (b) 10λ, and (c) 50λ?

P11.12 Consider a heterogeneous region contained within a right- circular cylindrical boundary having outside length 10λ and radius 3λ, where λ is the free-space wavelength.

(a) Compare the volume of the cylindrical region to that of a sphere of radius 5.9λ (roughly the smallest sphere that encloses the cylinder).

(b) Estimate the number of exterior unknowns needed to pad the tetrahedral-cell computational domain from the cylinder out to a sphere of radius 5.9λ, assuming that CT/LN basis functions are used and that the longest cell edge may not exceed 0.1λ. If the number of unknowns arising within the cylindrical region is 1.5 million, what percentage of the total number of unknowns is used to pad to the spherical boundary?

(c) Repeat part (b) under the assumption that a conformable RBC is imposed on a slightly larger right-circular cylinder whose boundary is a distance of 0.3λ outside the original cylinder.

P11.13 The local RBC in (11.74) can be obtained by equating the left-hand side of (11.73) with zero and using the Helmholtz equation to eliminate the term involving the $\hat{r} \times \nabla \times (\hat{r} \times \nabla \times \bar{E}^s)$ operator. Verify this derivation to obtain (11.74).

P11.14 The spherical vector wave functions $\bar{M}_{mn}$ in (11.38) and $\bar{N}_{mn}$ in (11.39) are associated with the TE-to-r and TM-to-r parts of the scattered electric field, respectively. By applying the differential operators in (11.79)–(11.81) to these functions, verify that the operator in (11.79) only recovers the TE-to-r part of the field while those in (11.80) and (11.81) only recover the TM-to-r part.

P11.15 Using the spherical vector wave functions $\bar{M}_{mn}$ and $\bar{N}_{mn}$ in (11.38) and (11.39), obtain an expression for the error in the second-order local RBCs in (11.74) and (11.77) when applied to the mnth harmonic. (Develop separate expressions for the error in the TE-to-r part of the field and the TM-to-r part of the field.)

P11.16 Study the various differential operators comprising the RBCs in (11.87), (11.88), and (11.89), and try to assess the differences between them. Under what conditions do they become equivalent? (For example, what happens as $r \to \infty$? What happens if $\nabla \cdot \bar{E}^s_{\tan} = 0$?)

P11.17 Show that the conformable RBCs defined by the coefficients in (11.92) and (11.93) and (11.94)–(11.97) reduce to the spherical-boundary RBC in (11.77) if $\kappa_1 = \kappa_2 = 1/r$.

P11.18 Discuss the computation of the scattering cross section, assuming that the vector Helmholtz equation coupled with a local RBC is used to determine the electric field associated with a three-dimensional scatterer. What different approaches are possible? What approximations are involved in each case? Recommend an approach.

P11.19 Suppose a lossy isotropic homogeneous region bounded by a perfect conductor is used to terminate the computational domain in a scattering problem. (In other words, a scatterer surrounded by a region of free space is enclosed within a lossy layer bounded by a perfect conductor.) Assuming that the material has a loss tangent of 1% and has a locally planar surface, how thick must the region be to reduce the reflections from the perfect conductor backing to a level less than that from the interface between the free-space region adjacent to the scatterer and the lossy absorber? Estimate the level of reflection likely from this interface, and comment on the number of additional unknowns required to represent the field throughout this region.

P11.20 (a) Show that a uniform plane wave of the form

$$\bar{E} = \hat{y} E_0 e^{-j\beta_x x} e^{-j\beta_z z}$$

is a valid solution to Maxwell's equations in a medium with relative permittivity and permeability tensors of the form

$$\bar{\bar{\varepsilon}}_r = \begin{bmatrix} a & 0 & 0 \\ 0 & a & 0 \\ 0 & 0 & b \end{bmatrix} \qquad \bar{\bar{\mu}}_r = \begin{bmatrix} c & 0 & 0 \\ 0 & c & 0 \\ 0 & 0 & d \end{bmatrix}$$

provided that

$$\frac{\beta_x^2}{d} + \frac{\beta_z^2}{c} = \omega^2 \mu_0 \varepsilon_0 a$$

Repeat your calculations for a wave of the form

$$\bar{H} = \hat{y} H_0 e^{-j\beta_x x} e^{-j\beta_z z}$$

to show that the phase constants must satisfy

$$\frac{\beta_x^2}{b} + \frac{\beta_z^2}{a} = \omega^2 \mu_0 \varepsilon_0 c$$

(b) Consider a wave incident at an arbitrary angle from free space into a medium with the above permittivity and permeability tensors. Show that the reflection coefficient associated with the perpendicular polarized wave in (11.99) is

$$\Gamma_\perp = \frac{\cos\theta/\eta - \beta_z/\omega\mu_0 c}{\cos\theta/\eta + \beta_z/\omega\mu_0 c}$$

while the reflection coefficient for the parallel polarized wave in (11.100) is

$$\Gamma_\parallel = \frac{\eta\cos\theta - \beta_z/\omega\varepsilon_0 a}{\eta\cos\theta + \beta_z/\omega\varepsilon_0 a}$$

Find conditions on a, b, c, and d that simultaneously eliminate reflections for both polarizations.

P11.21 Develop weak differential equations for the potential function formulation embodied in (11.116) and (11.117), and discuss the implementation of the associated boundary conditions in (11.118)–(11.123).

12

Finite-Difference Time-Domain Methods on Orthogonal Meshes

Up to this point, we have focused on frequency-domain techniques and applications. In this chapter we move to time-domain methods with the goal of simulating the temporal behavior of electromagnetic fields. While integral equation based methods have been employed occasionally in time-domain applications [1, 2], partial differential equation (PDE) based methods have proven to be more practical and have thus seen broader use. Here we discuss in some detail the Cartesian mesh finite-difference time-domain (FDTD) method, the oldest and most widely used of these techniques. This method was first described by K. Yee in 1966 [3]. As computing hardware became faster and more widely available, the method found numerous applications in electromagnetic radiation, scattering, and coupling. Stimulated by these successes, various researchers developed enhancements to the original technique that greatly extended its range of applicability [4–7].

Several key attributes combine to make the FDTD method a useful and powerful tool. First is the method's simplicity; Maxwell's equations in differential form are discretized in space and time in a straightforward manner. Second, since the method tracks the time-varying fields throughout a volume of space, FDTD results lend themselves well to scientific visualization methods. These, in turn, provide the user with excellent physical insights on the behavior of electromagnetic fields. Finally, the geometric flexibility of the method permits the solution of a wide variety of radiation, scattering, and coupling problems.

We will generally limit our discussion in this chapter to the two-dimensional case. This greatly simplifies the exposition of the method while retaining most of its key features. The three-dimensional extension is both notationally and conceptually straightforward. Readers interested in additional information are referred to texts on the subject [8, 9] and to review articles and their extensive bibliographies [10, 11].

12.1 MAXWELL'S EQUATIONS IN THE TIME DOMAIN

The frequency-domain form of Maxwell's equations was presented in Chapter 1. In the time domain, Ampere's and Faraday's laws (the curl equations) take on the forms

$$\frac{\partial \mathbf{D}}{\partial t} = \nabla \times \mathbf{H} - \mathbf{J} \tag{12.1}$$

$$\frac{\partial \mathbf{B}}{\partial t} = -\nabla \times \mathbf{E} \tag{12.2}$$

The divergence equations are unchanged:

$$\nabla \cdot \mathbf{D} = \rho \tag{12.3}$$

$$\nabla \cdot \mathbf{B} = 0 \tag{12.4}$$

In linear, isotropic, time-invariant, nondispersive media, the fluxes (**D** and **B**) and field strengths (**E** and **H**) are related via the material parameters

$$\mathbf{D} = \epsilon \mathbf{E} \tag{12.5}$$

$$\mathbf{B} = \mu \mathbf{H} \tag{12.6}$$

Under these simplifying assumptions, (12.1) and (12.2) can be written as

$$\epsilon \frac{\partial \mathbf{E}}{\partial t} = \nabla \times \mathbf{H} - \mathbf{J} \tag{12.7}$$

$$\mu \frac{\partial \mathbf{H}}{\partial t} = -\nabla \times \mathbf{E} \tag{12.8}$$

These equations form a hyperbolic system of coupled PDEs that are first order in space and time. They can readily be combined to form a single PDE that is second order in both space and time. For example, in the simple case of source-free uniform lossless media, (12.7) and (12.8) can be combined to form

$$\frac{1}{c^2}\frac{\partial^2 \mathbf{E}}{\partial t^2} = \nabla^2 \mathbf{E} \tag{12.9}$$

a second-order wave equation with wave propagation speed $c = 1/\sqrt{\mu\epsilon}$.

12.2 CENTERED FINITE-DIFFERENCE APPROXIMATIONS

Let $f(z)$ be a smooth function. For a sufficiently small interval Δz, df/dz is approximated by a simple two-point finite difference. The highest accuracy occurs when the finite difference is centered as

$$\left.\frac{df}{dz}\right|_{z=z_0} = \frac{f(z_0 + \Delta z/2) - f(z_0 - \Delta z/2)}{\Delta z} + O(\Delta z^2) \tag{12.10}$$

resulting in second-order accuracy. Thus, discretization errors are reduced by a factor of 4 when the mesh size is halved. The FDTD method, like most time-domain PDE methods, is based on the direct solution of (12.7) and (12.8). In the core FDTD algorithm, the continuous derivatives in both space and time are approximated by centered two-point finite differences.

This permits reasonable accuracy at minimal computational expense. The use of higher order methods has been explored [12] but has not yet seen widespread practical use, perhaps due to computational and programming complexity or to the lack of suitable supporting algorithms such as absorbing boundary conditions.

12.3 FDTD SPATIAL DISCRETIZATION

Rather than develop the method in all its detail, consider two-dimensional electromagnetic wave propagation in free space. For TE polarization, Maxwell's curl equations then reduce to

$$\epsilon_0 \frac{\partial E_x}{\partial t} = \frac{\partial H_z}{\partial y} \tag{12.11}$$

$$\epsilon_0 \frac{\partial E_y}{\partial t} = -\frac{\partial H_z}{\partial x} \tag{12.12}$$

$$\mu_0 \frac{\partial H_z}{\partial t} = \frac{\partial E_x}{\partial y} - \frac{\partial E_y}{\partial x} \tag{12.13}$$

The FDTD spatial discretization is developed by positioning the field variables such that all the spatial derivatives can be approximated by two-point, centered differences. The spatial layout is shown in Figure 12.1 with the computational stencil (or molecule) shown in bold. The stencil shows which field variables are computationally connected to which other fields on each time step. The definition of a unit cell within this spatially staggered system is somewhat arbitrary. However, the convention is to define the cells such that the grid lines pass through the electric field variables and coincide with their vector directions. This choice is motivated by the observation that boundary conditions imposed on the electric field are more common in practice than those imposed on the magnetic field. The FDTD programming and mesh generation are thus made somewhat easier.

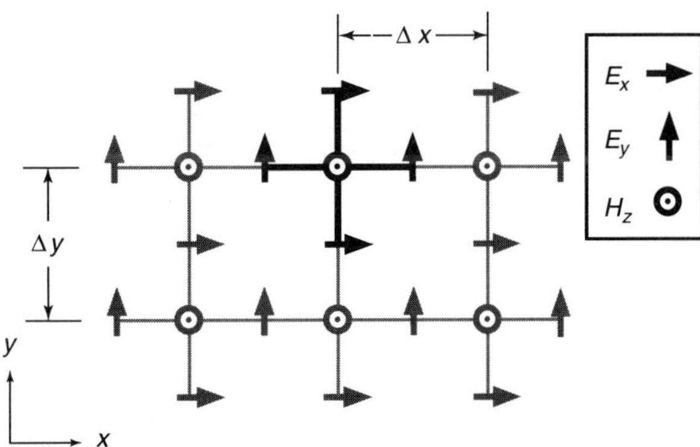

Figure 12.1 Spatial arrangement of field variables in the FDTD method for the two-dimensional TE case.

A word on notation is now in order. A unit cell in the two-dimensional TE case is depicted in Figure 12.2. Let the cell size be $\Delta x \times \Delta y$ and let i and j index the x and y discretizations, respectively. The spatial position of the electric field variables is then

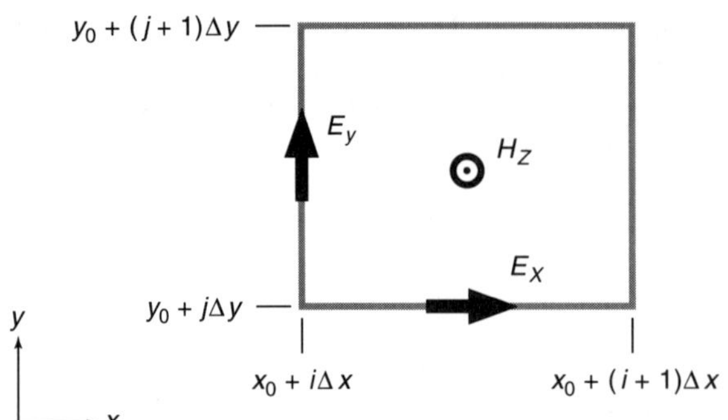

Figure 12.2 The FDTD unit cell for the two-dimensional TE case.

denoted by

$$E_x(i+\tfrac{1}{2}, j) \equiv E_x[x = x_0 + (i+\tfrac{1}{2})\Delta x, y = y_0 + j\Delta y] \tag{12.14}$$

$$E_y(i, j+\tfrac{1}{2}) \equiv E_y[x = x_0 + i\Delta x, y = y_0 + (j+\tfrac{1}{2})\Delta y] \tag{12.15}$$

where (x_0, y_0) is the grid origin. The magnetic field at the cell center is thus $H_z(i+\frac{1}{2}, j+\frac{1}{2})$. This simplified index notation explicitly exposes the half-cell positions of each field variable. Unfortunately, this clear notation cannot be carried over directly into computer software as commonly used programming languages only support arrays with integer indices. Most practitioners adopt the convention of rounding down. For example, the field variable $H_z(i+\frac{1}{2}, j+\frac{1}{2})$ might be associated in code with an array location `HZ(i, j)`. Some authors use this same "rounded-down" notation in publications, thus associating field variables with a particular cell while leaving the precise location within a cell implied. We elect to make the within-cell location explicit.

The FDTD temporal discretization will be discussed in the following section. Superscripts will be used to index time as in

$$E_x^n \equiv E_x(t = t_0 + n\Delta t) \tag{12.16}$$

Using this notation, the spatial FDTD approximation to Equations (12.11)–(12.13) is then

$$\epsilon_0 \dot{E}_x(i+\tfrac{1}{2}, j) = \frac{H_z(i+\frac{1}{2}, j+\frac{1}{2}) - H_z(i+\frac{1}{2}, j-\frac{1}{2})}{\Delta y} \tag{12.17}$$

$$\epsilon_0 \dot{E}_y(i, j+\tfrac{1}{2}) = -\frac{H_z(i+\frac{1}{2}, j+\frac{1}{2}) - H_z(i-\frac{1}{2}, j+\frac{1}{2})}{\Delta x} \tag{12.18}$$

$$\mu_0 \dot{H}_z(i+\tfrac{1}{2}, j+\tfrac{1}{2}) = \frac{E_x(i+\frac{1}{2}, j+1) - E_x(i+\frac{1}{2}, j)}{\Delta y} - \frac{E_y(i+1, j+\frac{1}{2}) - E_y(i, j+\frac{1}{2})}{\Delta x} \tag{12.19}$$

where the dot notation is used to indicate a derivative with respect to time.

The two-dimensional TM case can be derived in a similar fashion. The resultant equations are essentially the dual of (12.17)–(12.19). However, our convention that the discrete electric field variables live on and parallel to the cell edges dictates a half-cell shift in both x and y as shown in Figure 12.3.

An alternate derivation of the FDTD method can be obtained by approximating the integral form of Maxwell's curl equations (Ampere's and Faraday's laws):

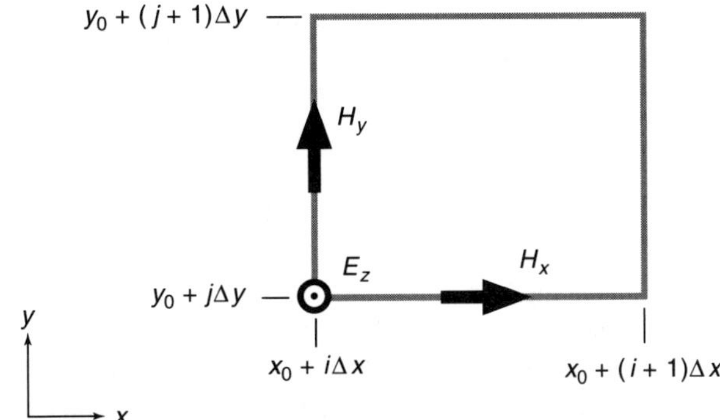

Figure 12.3 The FDTD unit cell for the two-dimensional TM case.

$$\iint_A \dot{\mathbf{D}} \cdot \hat{n}\, dA = \oint_{\partial A} \mathbf{H} \cdot \mathbf{dl} \tag{12.20}$$

$$\iint_A \dot{\mathbf{B}} \cdot \hat{n}\, dA = -\oint_{\partial A} \mathbf{E} \cdot \mathbf{dl} \tag{12.21}$$

where A is some area, ∂A denotes the boundary of that area, and $\hat{n}$ is a unit vector normal to A. Local spatial approximations to these equations form the theoretical basis for a family of related techniques known as finite-volume, finite-integration, or control region methods. These types of formulations have proven useful in developing extensions and generalizations of the basic FDTD method, for example, thin-wire approximations and nonorthogonal grid techniques.

As an example of this integral approach, refer to Figure 12.2 and consider approximating Faraday's law over an FDTD unit cell. Assume that the magnetic field is constant over the cell and that the electric fields are constant along each edge. Then, Faraday's law becomes

$$\begin{aligned} \mu_0 \dot{H}_z(i+\tfrac{1}{2}, j+\tfrac{1}{2})\, \Delta x\, \Delta y &\approx -\sum_{\text{sides}} E_{\text{side}}\, \text{length}_{\text{side}} \\ &= -E_x(i+\tfrac{1}{2}, j)\, \Delta x + E_y(i, j+\tfrac{1}{2})\, \Delta y \\ &\quad + E_x(i+\tfrac{1}{2}, j+1)\, \Delta x - E_y(i+1, j+\tfrac{1}{2})\, \Delta y \end{aligned} \tag{12.22}$$

By dividing the left- and right-hand sides of this equation by $\Delta x\ \Delta y$, this result is shown to be identical to the differential formulation. A similar procedure can also be used for discretizing Ampere's law.

12.4 FDTD TIME DISCRETIZATION

The same centered-difference concepts are used to advance the fields in time. This requires that $\mathbf{E}$ and $\mathbf{H}$ be staggered in time by one half of a time step. Let $\mathbf{E}$ be defined at integer time steps with $\mathbf{H}$ at the half-integer time points. The temporal discretization is then

$$\epsilon_0 \dot{\mathbf{E}}^{n-1/2} \approx \epsilon_0 \frac{\mathbf{E}^n - \mathbf{E}^{n-1}}{\Delta t} = \nabla \times \mathbf{H}^{n-1/2} \tag{12.23}$$

$$\mu_0 \dot{\mathbf{H}}^n \approx \mu_0 \frac{\mathbf{H}^{n+1/2} - \mathbf{H}^{n-1/2}}{\Delta t} = -\nabla \times \mathbf{E}^n \tag{12.24}$$

The spatial discretization is as in the previous section. These expressions can be rearranged to generate a time advancement procedure

$$\mathbf{E}^n = \mathbf{E}^{n-1} + \frac{\Delta t}{\epsilon_0} \nabla \times \mathbf{H}^{n-1/2} \tag{12.25}$$

$$\mathbf{H}^{n+1/2} = \mathbf{H}^{n-1/2} - \frac{\Delta t}{\mu_0} \nabla \times \mathbf{E}^n \tag{12.26}$$

This method, known as leap-frog time integration, is depicted graphically in Figure 12.4. Fields are advanced in time by using information from the opposite field type at an intermediate time point. Note that the method is explicit; calculations depend only on results at earlier times, again contributing to the computational efficiency of the FDTD method.

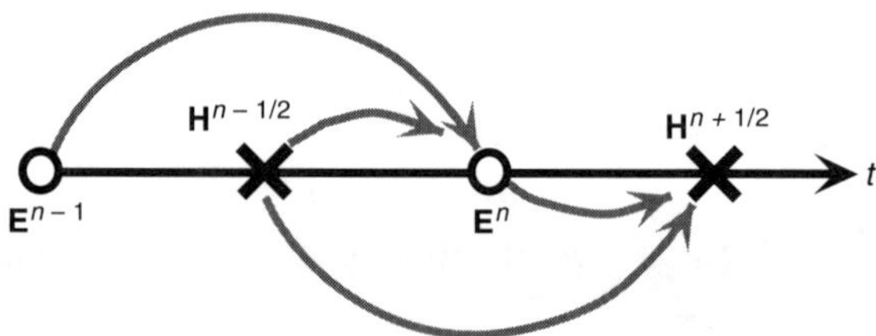

Figure 12.4 Leap frog time integration.

12.5 DIVERGENCE CONSERVATION IN THE FDTD

Up to this point we have worked with Maxwell's curl equations (12.1) and (12.2) while ignoring the divergence equations (12.3) and (12.4). The reasons for this are clear in the continuous analytic case, as divergence conditions are not independent and can be derived from the curl equations. To verify that this remains true under FDTD discretization, we investigate a discrete version of Gauss' law. In free space, Gauss' law states

$$\nabla \cdot \mathbf{D} = 0 \Longrightarrow \oint_{\partial A} \mathbf{D} \cdot \hat{n}\, dl = 0 \tag{12.27}$$

where ∂A is the boundary of some region A with unit normal $\hat{n}$. In the TE polarization, the staggered FDTD mesh lends itself to a discretized version of (12.27) evaluated over the region depicted in Figure 12.5. This region is shifted a half cell from the usual FDTD grid so that electric field unknowns are perpendicular to the region. To simplify the analysis, we have reindexed the cells locally. Equation (12.27) then becomes

$$\oint_{\partial A} \mathbf{D} \cdot \hat{n}\, dl \approx D_x^+ \Delta y + D_y^+ \Delta x - D_x^- \Delta y - D_y^- \Delta x \tag{12.28}$$

Differentiating (12.28) with respect to time and using (12.17) and (12.18) yields

$$\left(\frac{H_1 - H_4}{\Delta y}\right) \Delta y + \left(\frac{H_2 - H_1}{\Delta x}\right) \Delta x - \left(\frac{H_2 - H_3}{\Delta y}\right) \Delta y - \left(\frac{H_3 - H_4}{\Delta x}\right) \Delta x = 0 \tag{12.29}$$

This discrete measure of divergence is conserved locally by the FDTD algorithm. Thus, if the initial conditions are divergence free, then Gauss' law is satisfied throughout an FDTD calculation.

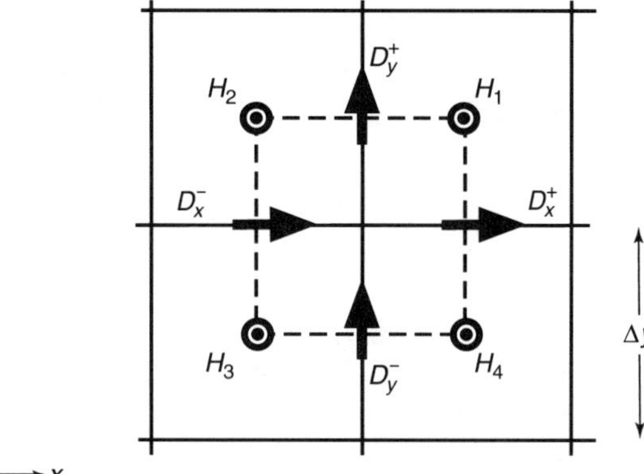

Figure 12.5 Integration region and field variables for an FDTD implementation of Gauss' law.

12.6 EXTENSION TO THREE DIMENSIONS

The three-dimensional case can be derived readily using the same techniques (differential or integral) used previously for the two-dimensional case. Alternatively, one can infer the three-dimensional case by "gluing" together two-dimensional cells from each of the three orthogonal planes, $x-y$, $y-z$, and $z-x$. A full three-dimensional FDTD cell (Yee lattice) is shown in Figure 12.6. As in the two-dimensional case, the custom in programming is to round down half-integer indices. The leap frog integration method is used to advance the time step.

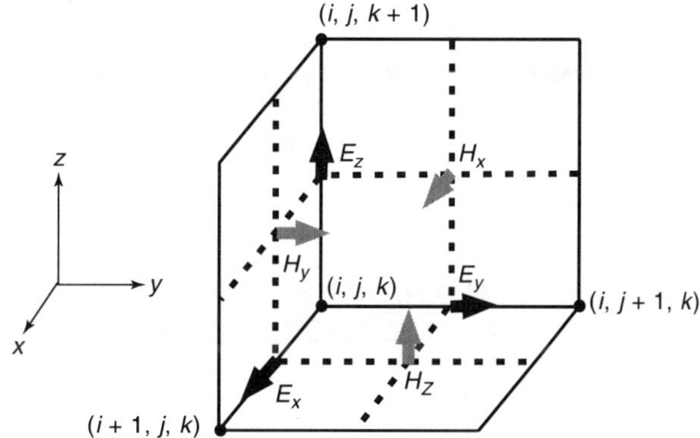

Figure 12.6 Three-dimensional staggered mesh FDTD cell (Yee lattice).

12.7 OTHER COORDINATE SYSTEMS

The FDTD method can also be used in other orthogonal coordinate systems. For instance, consider the two-dimensional TE case in polar coordinates. Maxwell's curl equations in free space are then

$$\epsilon_0 \dot{E}_r = \frac{1}{r}\frac{\partial H_z}{\partial \phi} \tag{12.30}$$

$$\epsilon_0 \dot{E}_\phi = -\frac{\partial H_z}{\partial r} \tag{12.31}$$

$$\mu_0 \dot{H}_z = \frac{1}{r}\frac{\partial E_r}{\partial \phi} - \frac{1}{r}\frac{\partial (r E_\phi)}{\partial r} \tag{12.32}$$

The spatial stencil/FDTD unit cell shown in Figure 12.7 permits centered spatial differencing. Some care is required to handle the $1/r$ factors. For instance, the spatial discretization of (12.32) is

$$\begin{aligned}\mu_0 \dot{H}_z(i+\tfrac{1}{2}, j+\tfrac{1}{2}) = {} & \frac{E_r(i+\frac{1}{2}, j+1) - E_r(i+\frac{1}{2}, j)}{r_{i+1/2}\, \Delta\phi} \\ & - \frac{r_{i+1} E_\phi(i+1, j+\frac{1}{2}) - r_i E_\phi(i, j+\frac{1}{2})}{r_{i+1/2}\, \Delta r}\end{aligned} \tag{12.33}$$

where $r_i = i\,\Delta r$, $r_{i+1/2} = (i+\frac{1}{2})\,\Delta r$, and so on.

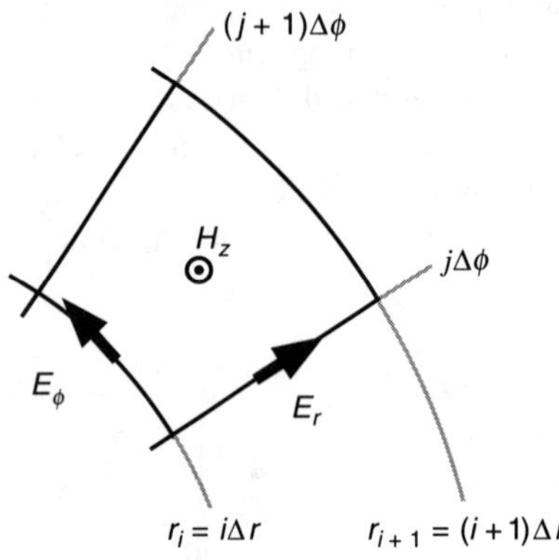

Figure 12.7 Two-dimensional FDTD cell in polar coordinates.

12.8 NUMERICAL ANALYSIS OF THE FDTD ALGORITHM: STABILITY, DISPERSION, AND ANISOTROPY

In this section we investigate the accuracy and stability of the FDTD algorithm. That is, how well do solutions to the finite-difference approximations introduced in the previous sections approximate the true behavior of the continuous PDEs? And, what conditions are required for an FDTD calculation to remain bounded?

For simplicity, consider the case of one-dimensional propagation in the x direction. Maxwell's curl equations can be combined to form a one-dimensional, second-order wave equation

$$\frac{1}{c^2}\frac{\partial^2 E}{\partial t^2} = \frac{\partial^2 E}{\partial x^2} \tag{12.34}$$

It is straightforward to show that the FDTD discretization (staggered spatial mesh with leap frog time integration) of the first-order Maxwell equations is numerically equivalent to discretizing (12.34) with the standard three-point centered finite-difference approximation to the second derivative

$$\frac{E^{n+1}(i) - 2E^n(i) + E^{n-1}(i)}{(c\ \Delta t)^2} = \frac{E^n(i+1) - 2E^n(i) + E^n(i-1)}{\Delta x^2} \tag{12.35}$$

This second-order difference equation is sometimes easier to work with as the half-integer quantities (space and time indices on H) are eliminated.

Stability analysis starts by looking for solutions of Equation (12.35) that are wave like in space,

$$E^n(i) = \xi^n e^{-jkx}\big|_{x=i\ \Delta x} \tag{12.36}$$

where $j = \sqrt{-1}$ and $k = 2\pi/\lambda$ is the wavenumber. The solution will then be growing, shrinking, or oscillating in time depending on the magnitude of ξ. Plugging this expression into the second-order difference equation and eliminating the common factors give

$$\xi^2 - 2A\xi + 1 = 0 \tag{12.37}$$

where

$$A = 1 - 2\left(\frac{c\ \Delta t}{\Delta x}\right)^2 \sin^2\left(\frac{k\ \Delta x}{2}\right) \tag{12.38}$$

hence

$$\xi = A \pm (A^2 - 1)^{1/2} \tag{12.39}$$

A growing (unstable) solution will occur if $|\xi| > 1$. This can only occur if $|A| > 1$. Thus,

$$c\ \Delta t > \Delta x \Longrightarrow \text{instability} \tag{12.40}$$

Alternatively, if $|A| \leq 1$, then $c\ \Delta t \leq \Delta x$ and

$$|\xi| = \left|A + j(1 - A^2)^{1/2}\right| = 1 \tag{12.41}$$

implying an oscillating (wavelike) solution in time.

In a one-dimensional FDTD code, $c\ \Delta t \leq \Delta x$ is required for stability. This constraint is known as the Courant stability condition and is easy to interpret physically. When the Courant stability condition is satisfied, the FDTD grid is causally connected; the speed of light bounds the rate at which information can be transmitted across the mesh.

In higher dimensions, say three, the stability criterion becomes

$$c\ \Delta t \leq \frac{1}{\left(1/\Delta x^2 + 1/\Delta y^2 + 1/\Delta z^2\right)^{1/2}} \tag{12.42}$$

Thus, for the cubical grid case ($\Delta x = \Delta y = \Delta z$), Δt must be chosen such that $c\ \Delta t \leq \Delta x/\sqrt{3} \approx 0.577\ \Delta x$. Similarly, in the two-dimensional square mesh case, stable solutions require that $c\ \Delta t \leq \Delta x/\sqrt{2} \approx 0.707\ \Delta x$. The reason for smaller stability limits in higher dimensions can be in Figure 12.8. Here, a plane wave is propagating on a two-dimensional square-cell FDTD mesh along the mesh diagonals. Projecting lines of constant phase onto an equivalent one-dimensional mesh results in an effective mesh spacing of $\Delta x/\sqrt{2}$. The time step must be reduced in size accordingly to ensure stability on this smaller effective mesh.

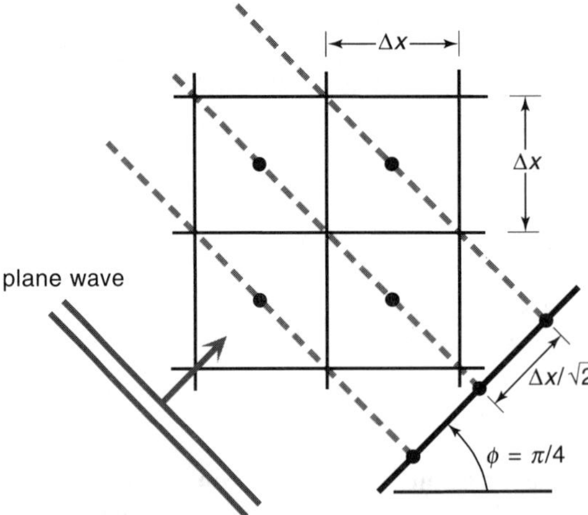

Figure 12.8 Plane-wave propagation along the diagonal of a two-dimensional square-cell ($\Delta x \times \Delta x$) mesh is equivalent to one-dimensional propagation along a mesh with cell size $\Delta x/\sqrt{2}$, thus accounting for the two-dimensional stability limit of $c\ \Delta t \leq \Delta x/\sqrt{2}$.

The stability analysis presented here applies to the uniform media (constant c) case. When a simulation geometry contains several materials with differing values of c, this same Courant condition can be applied, provided the maximum value of c is used.

Satisfaction of the Courant stability limit only ensures that the numerical solution remains bounded. The accuracy of the numerical solution is the next topic to address. In free space all plane waves propagate at the same speed, independent of frequency or wavenumber. This is not the case on an FDTD grid. This phenomenon is known as numerical dispersion, and its analysis is similar to the stability analysis on the previous few pages. First consider the one-dimensional, uniform media case. Assume wavelike solutions:

$$E^n(i) = e^{j(\omega n\ \Delta t - ki\ \Delta x)} \tag{12.43}$$

where ω is the temporal frequency of the wave in radians/sec. We then seek information about the relationship between ω and the wavenumber k.

Substituting (12.43) into the discretized one-dimensional wave equation (12.35) and doing some algebra yields the dispersion relation for the FDTD method:

$$\frac{1}{(c\ \Delta t)^2} \sin^2\left(\frac{\omega}{c}\frac{c\ \Delta t}{2}\right) = \frac{1}{\Delta x^2} \sin^2\left(\frac{k\ \Delta x}{2}\right) \tag{12.44}$$

In the limit, as $\omega\ \Delta t$ and $k\ \Delta x$ go to zero, this equation recovers the dispersion relation for the continuous case:

$$\left(\frac{\omega}{c}\right)^2 = k^2 \tag{12.45}$$

Here the phase velocity $v_p \equiv \omega/k$ is a constant. The FDTD dispersion relation also recovers the exact solution when $c\ \Delta t = \Delta x$ (right at the Courant limit). This is rarely a practical result as it only occurs in the one-dimensional, homogeneous media case.

For finite Δt and Δx, the numerical dispersion relation differs from the ideal, continuous solution case. Define an ideal solution wavelength as $\lambda_0 \equiv cT$, where $T = 2\pi/\omega$ is the temporal period. In the discrete case, the spatial wavelength realized on the grid is $\lambda_g = 2\pi/k$. The phase velocity and the grid wavelength are two measures of numerical dispersion and are related by

$$\frac{v_p}{c} = \frac{\lambda_g}{\lambda_0} \tag{12.46}$$

Define a Courant or stabilty ratio as $\alpha = c\ \Delta t/\Delta x$. Solutions of the dispersion relation (12.44) in terms of these new variables are shown in Figure 12.9. The normalized phase velocity v_p/c is plotted as a function of the normalized cell size for various values of the Courant ratio. The FDTD phase velocity errors are small for well-resolved waves but increase rapidly as the cell size increases. The use of small relative time steps (small Courant ratios) also increases dispersion errors.

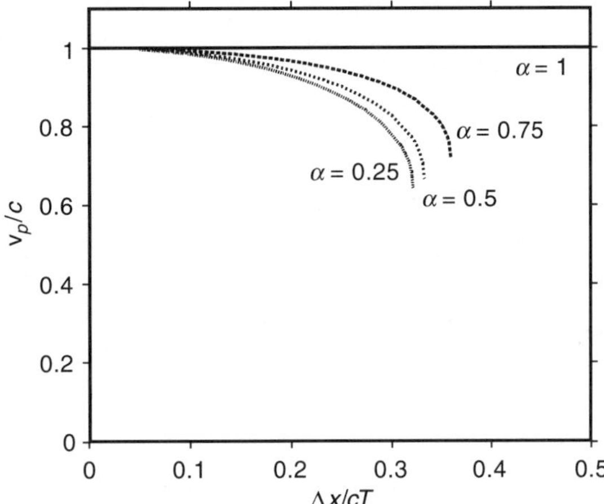

Figure 12.9 Numerical dispersion in the one-dimensional method: the normalized phase velocity v_p/c is plotted as a function of the normalized cell size for various values of the Courant stability ratio α.

A time-limited signal or pulse contains a spectrum of frequencies. As different frequencies propagate at different speeds on the FDTD grid, the impact of numerical dispersion is to distort pulses as they propagate through the mesh. As an example, the left-hand boundary of a one-dimensional FDTD grid is driven with a signal

$$f(t) = \begin{cases} 0.5\left[1 - \cos\left(\dfrac{2\pi t}{T}\right)\right] & 0 \le t \le T \\ 0 & \text{otherwise} \end{cases} \tag{12.47}$$

This pulse has an initial spatial width of cT, and it propagates through the FDTD grid from left to right. Cell size was chosen to resolve the pulse fairly well, $\Delta x/cT = \frac{1}{10}$. The Courant ratio was set to the value commonly used in higher dimensional applications, $c\ \Delta t/\Delta x = \frac{1}{2}$. In Figure 12.10, the initial pulse shape is shown along with its result after propagating down the mesh for four pulse widths. Note the change in pulse shape and the trailing-edge distortion. These effects can, of course, be reduced by using a finer mesh.

In two dimensions, the numerical dispersion relation is

$$\frac{1}{(c\ \Delta t)^2}\sin^2\left(\frac{\omega}{c}\frac{c\ \Delta t}{2}\right) = \frac{1}{\Delta x^2}\sin^2\left(\frac{k_x\ \Delta x}{2}\right) + \frac{1}{\Delta y^2}\sin^2\left(\frac{k_y\ \Delta y}{2}\right) \tag{12.48}$$

where k_x and k_y are the wavenumbers in the x and y directions, respectively. In addition to numerical dispersion effects as observed in one dimension, phase velocity in higher dimensions is a function of the angle of propagation through the mesh. This effect is known as grid anisotropy. In Figure 12.11, the normalized phase velocity is plotted as a function

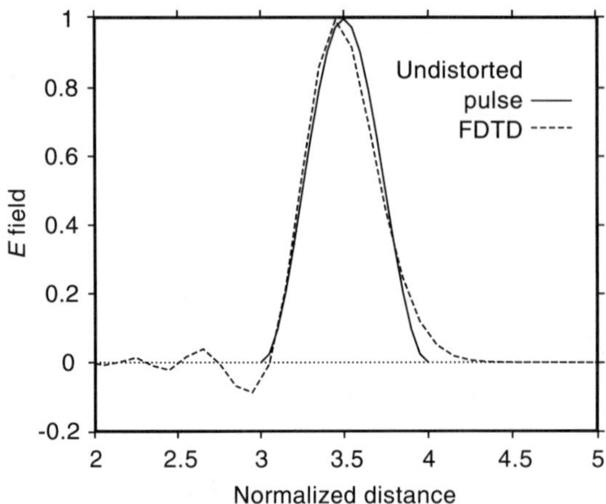

Figure 12.10 Pulse distortion due to numerical dispersion.

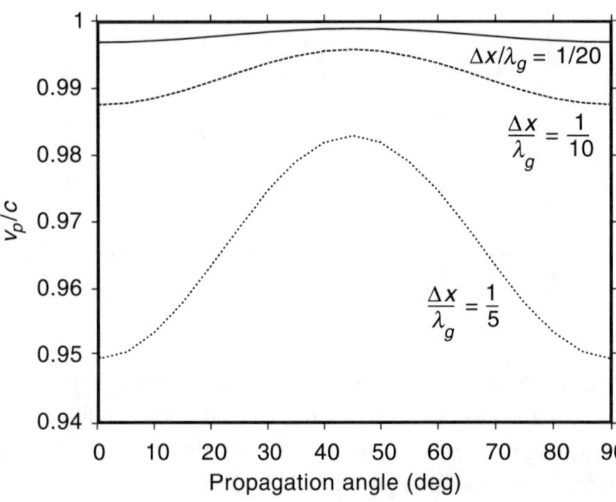

Figure 12.11 Grid anisotropy in the two-dimensional FDTD method. Normalized phase velocity v_p/c is plotted as a function of propagation angle for various cell sizes. Cells are square and $c\ \Delta t/\Delta x = \frac{1}{2}$; 0° and 90° angles correspond to propagation along the grid lines.

of wave angle for various values of the stability ratio. Errors due to grid anisotropy are generally smaller than those due to simple numerical dispersion.

12.9 TREATING LOSSY/CONDUCTIVE MEDIA

Up to this point we have focused on the use of the FDTD algorithm in uniform lossless material. To use this method to solve practical problems, we need to develop mechanisms for handling various boundary conditions, sources, and realistic materials. We first examine a simple modification to the FDTD algorithm that permits the use of lossy, conductive material. A side effect of this algorithm modification is a simple and efficient way to handle p.e.c. boundary conditions.

In materials with finite conductivity, an ohmic loss current is induced with $\mathbf{J} = \sigma \mathbf{E}$. Substituting this into Ampere's law and discretizing in time,

$$\epsilon \dot{\mathbf{E}}^{n-1/2} = \nabla \times \mathbf{H}^{n-1/2} - \sigma \mathbf{E}^{n-1/2} \tag{12.49}$$

Since $\mathbf{E}$ is only known at integer times, the conductive loss term needs to be implemented using a time average

$$\epsilon \frac{\mathbf{E}^n - \mathbf{E}^{n-1}}{\Delta t} = \nabla \times \mathbf{H}^{n-1/2} - \sigma \frac{\mathbf{E}^n + \mathbf{E}^{n-1}}{2} \tag{12.50}$$

Fortunately, $\mathbf{E}^n$ can be isolated algebraically and an implicit calculation avoided:

$$\mathbf{E}^n = \frac{1 - \sigma\ \Delta t/2\epsilon}{1 + \sigma\ \Delta t/2\epsilon} \mathbf{E}^{n-1} + \frac{1}{1 + \sigma\ \Delta t/2\epsilon} \frac{\Delta t}{\epsilon} \nabla \times \mathbf{H}^{n-\frac{1}{2}} \tag{12.51}$$

In addition to handling lossy material, this expression presents us with one way of handling a p.e.c. boundary condition. When $\sigma >> 1$, (12.51) reduces to $\mathbf{E}^n \approx -\mathbf{E}^{n-1}$. Thus, if the components of $\mathbf{E}$ tangential to a p.e.c. boundary are initialized to zero, they will remain nearly zero throughout the calculation. This method for imposing p.e.c. boundary conditions is particularly efficient in general-purpose FDTD codes as it directly employs the ordinary FDTD equations without the need for additional tests or loops.

12.10 FREQUENCY-DEPENDENT MEDIA

The FDTD equations developed in Sections 12.3 and 12.4 are based on the presumption that the instantaneous fluxes and field strengths are linearly related through constant material parameters ϵ and μ. This simple relationship is exact in free-space calculations, and it suffices as a description of many materials over a narrow to moderate frequency range. However, more involved flux–field relationships are required for modeling some common materials (e.g., water), especially when wide-band results are desired. The original FDTD field update equations will need to be modified to accommodate these materials.

Material parameters are typically reported by experimentalists and vendors as functions of frequency; the linear relationship between fields and fluxes holds at each frequency

$$\tilde{\mathbf{D}}(\omega) = \epsilon_r^*(\omega)\epsilon_0\tilde{\mathbf{E}}(\omega) \tag{12.52}$$

where $\epsilon_r^*(\omega)$ is a complex relative dielectric function that may or may not incorporate the bulk conductivity. The time-domain equivalent of (12.52) is the convolution integral:

$$\mathbf{D}(t) = \epsilon_0 \int_0^t \mathbf{E}(t-\tau)\epsilon_r(\tau)\, d\tau \tag{12.53}$$

There are two significant practical problems with a naive direct FDTD implementation of (12.53). First, the convolution integral needs to be reevaluated every time step for each electric field variable within the frequency-dependent media. Second, and more importantly, calculation of the convolution requires storage of the electric field time histories. While we should expect some increase in operation count and memory requirements to be able to handle these materials, the costs articulated here would overwhelm most computational resources when used in three-dimensional simulations with dispersive materials of significant spatial extent.

Luebbers et al. [13] developed a computationally efficient way of implementing (12.53) within the FDTD context by first recognizing that Ampere's law actually deals with $\partial\mathbf{D}/\partial t$ and not $\mathbf{D}$ directly and then by choosing representations of $\epsilon_r(t) \leftrightarrow \epsilon_r^*(\omega)$ that are analytically suitable. In [8], FDTD modifications for handling several different classes of frequency-dependent materials are discussed. We limit our discussion here to the simplest of these, Debye materials.

Debye materials are characterized by a frequency-domain permittivity with a single complex pole

$$\epsilon_r^*(\omega) = \epsilon_\infty + \frac{\epsilon_s - \epsilon_\infty}{1 + j\omega t_0} \tag{12.54}$$

where ϵ_∞ is the relative permittivity at high frequency, ϵ_s is the relative permittivity in the static limit, and t_0 is a characteristic relaxation time for the material. Let the second term on the right-hand side of (12.54) be $\chi(\omega)$, a frequency-dependent electric susceptibility.[1] In the time domain,

$$\chi(t) = \frac{\epsilon_s - \epsilon_\infty}{t_0} e^{-t/t_0} U(t) \tag{12.55}$$

where $U(t)$ is the unit step function.

Discretizing (12.53) results in

$$\mathbf{D}^n = \epsilon_0 \left(\epsilon_\infty \mathbf{E}^n + \sum_{m=0}^{n-1} \mathbf{E}^{n-m} \int_{m\Delta t}^{(m+1)\,Dt} \chi(\tau)\, d\tau \right) \tag{12.56}$$

Developing an analogous expression for $\mathbf{D}^{n-1}$ and defining

$$\bar{\chi}^0 \equiv \int_0^{\Delta t} \chi(\tau)\, d\tau \tag{12.57}$$

and

$$\Delta\chi^m \equiv \int_{m\,\Delta t}^{(m+1)\,\Delta t} \chi(\tau)\, d\tau - \int_{(m+1)\,\Delta t}^{(m+2)\,\Delta t} \chi(\tau)\, d\tau \tag{12.58}$$

lead to the modified FDTD electric field temporal update equation

$$\mathbf{E}^n = \frac{\epsilon_\infty}{\epsilon_\infty + \bar{\chi}^0} \left(\mathbf{E}^{n-1} + \frac{1}{\epsilon_\infty} \mathbf{\Psi}^{n-1} + \frac{\Delta t}{\epsilon_\infty \epsilon_0} \nabla \times \mathbf{H}^{n-1/2} \right) \tag{12.59}$$

where we have introduced a new variable $\mathbf{\Psi}$ to absorb the convolution summation

$$\mathbf{\Psi}^{n-1} = \sum_{m=0}^{n-2} \mathbf{E}^{n-1-m}\, \Delta\chi^m \tag{12.60}$$

Evaluation of (12.60) still appears to require saving the electric field time histories. Fortunately, the exponential form of $\chi(t)$ allows us to greatly simplify this calculation. Evaluating (12.58) yields

$$\Delta\chi^m = (\epsilon_s - \epsilon_\infty)(1 - e^{-\Delta t/t_0})^2 e^{-m\,\Delta t/t_0} \tag{12.61}$$

from which we note that

$$\Delta\chi^{m+1} = e^{-\Delta t/t_0}\, \Delta\chi^m \tag{12.62}$$

and

$$\mathbf{\Psi}^{n-1} = \mathbf{E}^{n-1}\, \Delta\chi^0 + e^{-\Delta t/t_0} \mathbf{\Psi}^{n-2} \tag{12.63}$$

[1] Note that this is a different definition of susceptibility than is found in most introductory electromagnetics textbooks where $\epsilon_r \equiv 1 + \chi$.

The convolution sum is thus reduced to a running sum, eliminating the need to store electric field time histories. Frequency-dependent materials amenable to the Debye model can thus be handled efficiently. When compared with the standard, simple media FDTD equations, the modified equations presented here require several more operations and the addition of an additional auxiliary field variable, Ψ. These costs seem reasonable given the added modeling capability.

Similar modified FDTD schemes have been developed for several other (non-Debye) classes of materials. The basic approach follows that presented here, combining a judicious choice of the analytic form of $\epsilon_r^*(\omega)$ with the use of auxiliary variables to eliminate the time-domain convolution. An alternate ordinary differential equation based approach was developed by Joseph et al. [14]. Following the general linear media constitutive relation given in Harrington [15],

$$\mathbf{D} = \epsilon \mathbf{E} + \epsilon_1 \frac{\partial \mathbf{E}}{\partial t} + \epsilon_2 \frac{\partial^2 \mathbf{E}}{\partial t^2} + \cdots \tag{12.64}$$

Or, even more generally, replace the left-hand side of (12.64) with an expansion in **D** and its derivatives with respect to time. In the frequency domain, the resultant complex dielectric constant is expressed as a simple ratio of polynomials with coefficients and order that can be fit to a wide range of materials. Equation (12.64) (or its generalization) is then approximated using finite differences and used to develop modified FDTD update equations. This technique requires that the flux **D** be updated and stored, thus incurring a significant but reasonable computational cost.

12.11 SIMPLE BOUNDARY AND INTERFACE CONDITIONS

Next we address the FDTD implementation of simple boundary conditions such as perfect electric conductors, perfect magnetic conductors (p.m.c.), and dielectric material interfaces. In Section 12.9 we used the ordinary FDTD update equations and high values of σ to indirectly but efficiently implement a perfect electric conductor. An alternate and perhaps more direct method is to construct the FDTD grid such that the p.e.c. boundaries coincide with the edges of the unit cells and thus with the appropriate tangential components of the electric field. In a total-field formulation of the FDTD (see Section 12.13 for further discussion of total- vs. scattered-field FDTD formulations), the p.e.c. boundary condition is imposed by setting $E_{\tan} = 0$. This can be accomplished either by maintaining the initial conditions throughout a calculation or by resetting the appropriate electric field components to zero on each time step. In a scattered-field formulation of the FDTD, p.e.c. boundaries must be driven with the negative of the instantaneous incident electric field.

It is sometimes useful to set $H_{\tan} = 0$, for instance to model a p.m.c. or, more likely, to invoke a particular symmetry condition. By analogy with the p.e.c. boundary, this is readily accomplished by passing the perfect magnetic conducting boundary through the center of the cells and thus through the tangential magnetic fields. While simple and direct, this approach has the disadvantage that the mesh generation system must be able to deal with boundaries at both unit and half-cell locations. It is thus desirable to develop a method for positioning a p.m.c. at an FDTD cell edge. Consider the FDTD cells and perfect magnetic conducting boundary shown in Figure 12.12. The ordinary FDTD field advance for E_x is

$$\epsilon_0 \dot{E}_x(\text{on grid line}) = \frac{H_z(\text{above}) - H_z(\text{below})}{\Delta y} \tag{12.65}$$

A perfect magnetic conducting boundary is equivalent to an antisymmetry condition on H_z, that is, $H_z(\text{below}) = -H_z(\text{above})$. Hence, at a p.m.c.

$$\epsilon_0 \dot{E}_x(\text{at boundary}) = \frac{2H_z(\text{above})}{\Delta y} \tag{12.66}$$

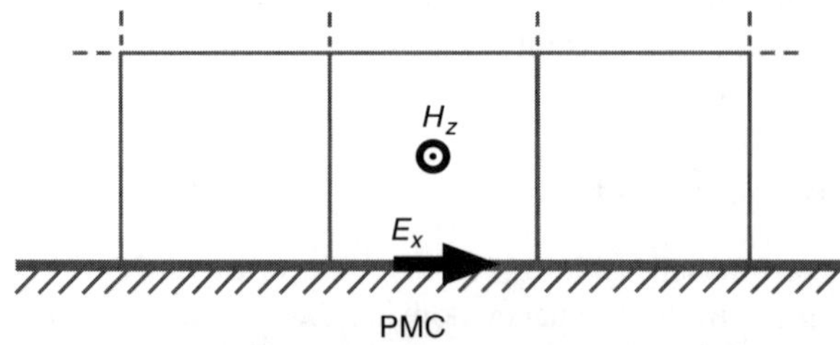

Figure 12.12 Field locations for the PMC boundary condition implementation given by Equation (12.66).

Material parameters such as ϵ, μ, and σ can be set for each field component in each cell. Material interface conditions are generally imposed by simply setting these parameters appropriately. The spatially staggered FDTD mesh makes it difficult to change both the electric and magnetic properties of a material at exactly the same location while simultaneously maintaining continuity of tangential fields and normal fluxes, although this is rarely a significant practical concern.

12.12 ABSORBING BOUNDARY CONDITIONS

To use the FDTD method to solve open-region problems, we require some mechanism for truncating the mesh at a finite distance from the computational region of interest. Ideally, this mechanism [known alternately as an absorbing or radiation boundary condition (ABC or RBC)] should permit electromagnetic energy to pass out of the problem space without distorting the fields or reflecting energy back into computational domain. If possible, the ABC should be computationally efficient, not adding substantial overhead to the core FDTD calculation.

The development of suitable ABCs has been an active area of research for the last two decades. A variety of approaches have been employed, including:

- *Field extrapolation.* In this approach, the fields are extrapolated one cell out from the FDTD computational grid using a cylindrical (two-dimensional) or spherical (three-dimensional) extrapolation or by using Huygens' principle. The extrapolated fields are then used in the ordinary FDTD formulas to update the fields on the boundary. This approach has been reasonably successful, although it forces the user to pick an origin for the expansion. The method may then fail for objects with multiple scattering centers.

- *Impedance conditions*. Electric and magnetic fields perpendicular to the boundary are assumed to be normally propagating plane waves, thus related by the wave impedance. While simple and intuitive, this method only works well when the outgoing waves actually impinge on the boundary at near-normal angles.
- *Surface integrals*. Equivalence principles are used to create virtual surface currents over the exterior of the computational grid; rigorous surface integral formulations are used to extrapolate the fields. While this method can be highly effective, the global nature of the calculations makes the method computationally intractable for most problems.
- *Absorbing material*. This technique wraps the core computational domain with a layer of absorbing material, creating a numerical anechoic chamber. Early versions of this technique were reasonably successful but computationally expensive as thick absorbing layers were required. Recent developments [16] appear to have overcome this difficulty and will be discussed below.
- *One-way wave equations*. This approach has proven to be a productive area for the development of practical and effective ABCs. We will discuss this approach in detail here.

The one-way wave equation approach to the development of ABCs has been employed by many researchers. Seminal works include those by Lindman [17], Engquist and Majda [18], and Mur [19]. Alternative and/or higher order methods are described in [20, 21]. We will focus on the commonly used low-order schemes developed by Mur.

To clarify the discussion, let u represent any generic field component and write the second-order wave equation as

$$\frac{1}{c^2}\partial_t^2 u = \partial_x^2 u + \partial_y^2 u \tag{12.67}$$

where the operator notation ∂_t^2 is used to compactly represent $\partial^2/\partial t^2$. Equation (12.67) supports omnidirectional wave propagation with a dispersion relation

$$\left(\frac{\omega}{c}\right)^2 = k_x^2 + k_y^2 \tag{12.68}$$

The basic concept of a one-way wave equation is a separate PDE that supports wave propagation in some directions but not in others. Ideally, its dispersion relation approximates (12.68) for the desired outgoing directions. Figure 12.13 illustrates the use of this PDE as an ABC. Here, we consider the computational space to be the $+x$ half plane. The standard FDTD equations are used throughout the region $x > 0$. A discrete approximation to a one-way wave equation that supports $-x$ propagating waves only is used as the boundary condition at $x = 0$.

To develop one-way wave equations, we first drop to one-dimension where the second-order wave equation

$$\partial_x^2 u - \frac{1}{c^2}\partial_t^2 u = 0 \tag{12.69}$$

supports bidirectional wave propagation, $u = u(x \pm ct)$. The wave equation can be factored by treating the differential operators ∂_t and ∂_x algebraically:

$$(\partial_x - \frac{1}{c}\partial_t)(\partial_x + \frac{1}{c}\partial_t)u = 0 \tag{12.70}$$

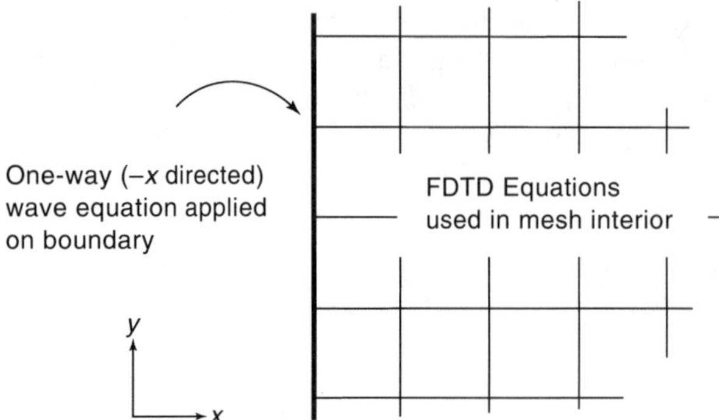

Figure 12.13 One-way wave equation used as an absorbing boundary condition.

Applying each factor to u individually and equating the two results to zero give the one-way wave equations

$$(\partial_x - \frac{1}{c}\partial_t)u = 0 \tag{12.71}$$

$$(\partial_x + \frac{1}{c}\partial_t)u = 0 \tag{12.72}$$

Solutions to (12.71) are $-x$ traveling waves, $u(x+ct)$, while solutions to (12.72) are $+x$ propagating waves of the form $u(x-ct)$. Thus, to simulate an open-region problem, Equations (12.71) and (12.72) are used as boundary conditions on the left and right boundaries, respectively. The use of these one-way wave equations as ABCs is equivalent to the use of an impedance condition.

While the simple impedance condition developed above is occasionally used in higher dimensional computations, it is beneficial to seek one-way wave equations that can better handle nonnormally incident waves. In two dimensions, the second-order wave equation

$$\partial_x^2 u + \partial_y^2 u - \frac{1}{c^2}\partial_t^2 u = 0 \tag{12.73}$$

supports omnidirectional wave propagation. To develop one-way wave equations that are suitable for use as ABCs on the left ($-x$-directed waves) or right ($+x$-directed waves) edges of a computational region, first factor the wave equation as before:

$$\left(\partial_x - \sqrt{\frac{1}{c^2}\partial_t^2 - \partial_y^2}\right)\left(\partial_x + \sqrt{\frac{1}{c^2}\partial_t^2 - \partial_y^2}\right)u = 0 \tag{12.74}$$

Applying each of the factors to u individually gives a one-way wave equation. For instance, on the left wall,

$$\left(\partial_x - \sqrt{\frac{1}{c^2}\partial_t^2 - \partial_y^2}\right)u = 0 \tag{12.75}$$

is the appropriate ABC.

We are then challenged with the interpretation and computation of (12.75). The cavalier algebraic use of partial-derivative operators has led to a pseudo-differential operator that mixes a square-root function with the partial-derivative operators. It is perhaps obvious but worth emphasizing that

$$\left(\sqrt{\frac{1}{c^2}\partial_t^2 - \partial_y^2}\right) u \neq \sqrt{\frac{1}{c^2}\partial_t^2 u - \partial_y^2 u} \; ! \tag{12.76}$$

To interpret this pseudo-differential operator and work toward a computable approximation, consider Fourier transforming u in y and t, with $y \to k_y$ and $t \to \omega$. Thus,

$$u(x, y, t) = F^{-1}\{\tilde{u}(x, k_y, \omega)\} \tag{12.77}$$

and

$$\partial_t u = F^{-1}\{j\omega u\} \qquad \partial_y u = F^{-1}\{jk_y u\} \tag{12.78}$$

The questionable square-root operator is then *defined* by

$$\left(\sqrt{\frac{1}{c^2}\partial_t^2 - \partial_y^2}\right) u \equiv F^{-1}\left\{\sqrt{\left(\frac{j\omega}{c}\right)^2 - (jk_y)^2}\; \tilde{u}(x, k_y, \omega)\right\} \tag{12.79}$$

Although sufficient for placing the pseudo-differential operator on some solid theoretical ground, Equation (12.79) is not directly usable in computations as the inverse Fourier transform is a global operation. Practical one-way wave equations are developed by seeking readily computable approximations to (12.79). To accomplish this, we take the approach followed in [18] and [19] and approximate the square root using a Taylor series expansion. First factor out a $(1/c)\partial_t$,

$$\left(\partial_x - \frac{1}{c}\partial_t \sqrt{1 - \frac{\partial_y^2}{\frac{1}{c^2}\partial_t^2}}\right) = 0 \tag{12.80}$$

We then note that waves incident on the boundary near normal incidence have $|k_x| \approx \omega/c >> |k_y|$. The operators are thus similarly related, $|\partial_x| \approx (1/c)\partial_t >> |\partial_y|$, and the square root can be approximated with a Taylor series. After multiplying through by the denominator, this results in

$$\left(\frac{1}{c}\partial_t \partial_x - \frac{1}{c^2}\partial_t^2 + \frac{1}{2}\partial_y^2\right) u = 0 \tag{12.81}$$

an approximate but computable one-way equation suitable for use on the left boundary.

Discretizing (12.71), the first-order or impedance condition for $-x$ propagating waves, results in the update equation

$$u_{0m}^{n+1} = u_{1m}^{n} - \frac{\Delta x - c\,\Delta t}{\Delta x + c\,\Delta t}(u_{1m}^{n+1} - u_{0m}^{n}) \tag{12.82}$$

where two-point difference approximations centered at $x = \Delta x/2, t = (n + 1/2)\,\Delta t$ have been used. Discretizing (12.81) and centering at $x = \Delta x/2, t = n\,\Delta t$ yield

$$\begin{aligned} u_{0m}^{n+1} &= -u_{1m}^{n-1} - \frac{\Delta x - c\,\Delta t}{\Delta x + c\,\Delta t}(u_{0m}^{n-1} + u_{1m}^{n+1}) + \frac{2\,\Delta x}{\Delta x + c\,\Delta t}(u_{0m}^{n} + u_{1m}^{n}) \\ &+ \frac{\Delta x (c\,\Delta t)^2}{\Delta x + c\,\Delta t}\partial_y^2 u|_{(x,y)=(\Delta x/2, m\,\Delta y)}^{t=n\,\Delta t} \end{aligned} \tag{12.83}$$

Two marginally different discretizations of the second-order y derivative are used in [18] and [19]. Neither has a clear-cut accuracy or efficiency benefit. In Mur's implementation, proper centering of this term is achieved by a spatial average in the x direction,

$$\partial_y^2 u|_{(x,y)=(\Delta x/2, m\,\Delta y)}^{t=n\,\Delta t} \approx \frac{1}{2}\left(\partial_y^2 u|_{(x,y)=(0, m\,\Delta y)}^{t=n\,\Delta t} + \partial_y^2 u|_{(x,y)=(\Delta x, m\,\Delta y)}^{t=n\,\Delta t}\right) \tag{12.84}$$

In contrast, Engquist and Majda average in both space and time:

$$\partial_y^2 u|_{(x,y)=(\Delta x/2, m\,\Delta y)}^{t=n\,\Delta t} \approx \frac{1}{2}\left(\partial_y^2 u|_{(x,y)=(0, m\,\Delta y)}^{t=(n-1)\,\Delta t} + \partial_y^2 u|_{(x,y)=(\Delta x, m\,\Delta y)}^{t=(n+1)\,\Delta t}\right) \tag{12.85}$$

In both cases, the second derivative is discretized using the standard three-point centered-difference approximation. Equation (12.83) is commonly referred to as a second-order Mur condition (or, more simply, Mur-2) by FDTD practitioners.

These discretized ABCs can be analyzed for their accuracy using a procedure similar to the dispersion analysis presented earlier. The finite-difference solution of two-dimensional second-order wave equations supports solutions of the form

$$u(t = n\,\Delta t, x = l\,\Delta x, y = m\,\Delta y) \equiv u_{lm}^n = e^{j\omega n\,\Delta t} e^{jk_x l\,\Delta x} e^{jk_y m\,\Delta y} \tag{12.86}$$

Return to the semi-infinite half-space FDTD grid, as shown in Figure 12.13, with an ABC imposed at $x = 0$. A steady-state, monochromatic plane wave with radian frequency ω impinges on the boundary at an angle θ measured from the $-x$ axis. Since any practical ABC is imperfect, a reflected wave is generated at $x = 0$. As the ABC must match the y variation of the incident field exactly and assuming no frequency conversion on reflection, the total solution is

$$u_{lm}^n = e^{j\omega n\,\Delta t}(e^{jk_x l\,\Delta x} + Re^{-jk_x l\,\Delta x})e^{-jk_y m\,\Delta y} \tag{12.87}$$

where k_x is assumed to be positive and the reflection coefficient R is to be determined. Equation (12.87) will then satisfy the discretized wave equation for all $l > 0$ for any value of R provided that the two-dimensional dispersion relation is satisfied. Note that the incidence angle θ is also related to the dispersion relation. For the positive-sign choice on k_x given above, $\theta = \tan^{-1}(k_y/k_x)$.

Substituting (12.87) into (12.83) gives a means of finding R,

$$R = -e^{jk_x\,\Delta x}\frac{\Delta x \sin(\omega\,\Delta t/2)\cos(k_x\,\Delta x/2) - c\,\Delta t\cos(\omega\,\Delta t/2)\sin(k_x\,\Delta x/2)}{\Delta x \sin(\omega\,\Delta t/2)\cos(k_x\,\Delta x/2) + c\,\Delta t\cos(\omega\,\Delta t/2)\sin(k_x\,\Delta x/2)} \tag{12.88}$$

where $\omega\,\Delta t$ and $k_x\,\Delta x$ are constrained by the dispersion relation (12.48). Applying the same procedure to the second-order Mur condition leads to

$$R = -e^{jk_x\,\Delta x} \times \frac{\Delta x[\sin^2(\omega\,\Delta t/2) - \frac{1}{2}(c\,\Delta t/\Delta y)^2\sin^2(k_y\,\Delta y/2)]\cos(k_x\,\Delta x/2) - c\,\Delta t\cos(\omega\,\Delta t/2)\sin(\omega\,\Delta t/2)\sin(k_x\,\Delta x/2)}{\Delta x[\sin^2(\omega\,\Delta t/2) - \frac{1}{2}(c\,\Delta t/\Delta y)^2\sin^2(k_y\,\Delta y/2)]\cos(k_x\,\Delta x/2) + c\,\Delta t\cos(\omega\,\Delta t/2)\sin(\omega\,\Delta t/2)\sin(k_x\,\Delta x/2)} \tag{12.89}$$

In Figure 12.14, reflection coefficients based on both the first- and second-order ABCs described above are shown for a typical case. The solid line is the first-order ABC (12.82) and the dashed line is the second-order ABC (12.83). Numerical parameters have been chosen to be similar to those commonly used in FDTD codes: $\Delta x = \Delta y$ (square cells), $c\,\Delta t = \Delta x/2$ (time step slightly smaller than the Courant stability limit), and $\omega/c = 2\pi/(10\,\Delta x)$ (cell width is one-tenth of a continuous free-space wavelength).

In the continuous case, both the first- and second-order ABCs are exact at normal incidence. In contrast, the discretized forms of these ABCs are, in general, not exact at normal incidence; $|R(\theta = 0)| \approx 0.02$ for the case shown in Figure 12.14. If the time

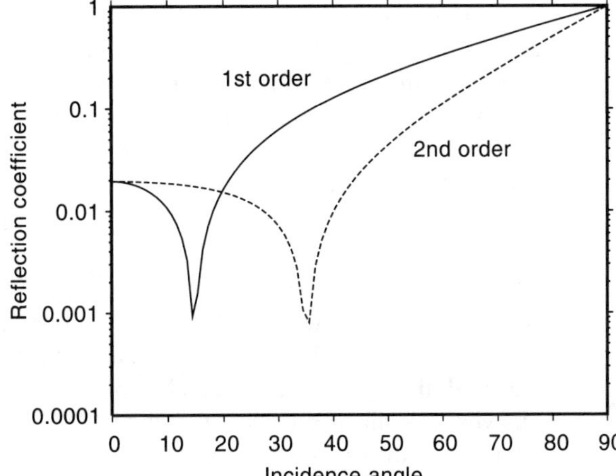

Figure 12.14 Magnitude of the reflection coefficient vs. incidence angle θ for two discretized ABCs. Solid line is the first-order Mur or impedance ABC and the dashed line is the second order Mur condition. Numerical parameters are $\Delta x = \Delta y$, $c\ \Delta t = \Delta x/2$, and $\omega/c = 2\pi/(10\ \Delta x)$.

step were chosen such that $c\ \Delta t = \Delta x$, the discretized ABCs would be exact at normal incidence. Unfortunately, this choice violates the stability criteria for the two-dimensional wave equation and thus cannot be used.

As the angle of incidence increases, the first-order ABC quickly fails while the second-order condition remains useful for a broader range of angles. Arbitrarily defining an ABC's useful range to be that range of angles where the magnitude of the reflection coefficient remains below 0.05, the first-order condition is seen to be useful up to $\theta = 28°$ and the second-order ABC can be used up to about $51°$.

Figure 12.15 demonstrates the effect of the relative cell size on the ABC error. Here the reflection coefficient for the second-order Mur ABC [Equation (12.83)] is plotted as function of incidence angle for three different cell sizes: 5, 10, and 20 zones per continuous free-space wavelength. In each case, $\Delta x = \Delta y$ and $c\ \Delta t = \Delta x/2$. Large reflection coefficients are obtained on the coarse grid, even at near-normal incidence angles. As the grid resolution increases, the quality of the ABC increases and more nearly approximates the ideal case.

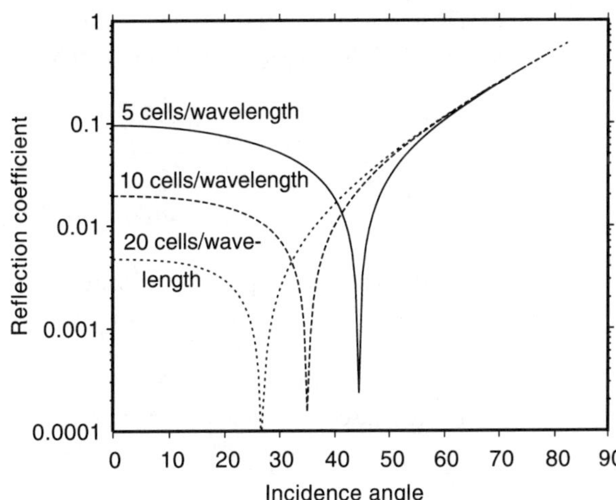

Figure 12.15 Magnitude of the reflection coefficient vs. incidence angle for the second-order Mur ABC with different grid resolutions. Numerical parameters are $\Delta x = \Delta y$ and $c\ \Delta t = \Delta x/2$.

Recent work by Bérenger [16] has revitalized interest in the use of absorbing materials as ABCs. Bérenger's particular approach, termed a perfectly matched layer (PML) in the literature, reduces unwanted reflections from the computational boundary to levels well below those offered by the Mur conditions. The PML is developed by first recognizing that a plane wave is perfectly matched (zero reflection) when normally incident on a half space with material properties

$$\frac{\sigma}{\epsilon_0} = \frac{\sigma^*}{\mu_0} \tag{12.90}$$

where σ^* is a magnetic conductivity. An open-region calculation can then be simulated in a finite-size FDTD grid by surrounding the region of interest with a layer of this lossy material; the layer serves as the absorbing material in a numerical anechoic chamber. The drawback of this approach is that the perfect match fails for waves that are not normally incident on the absorbing layer.

For the two-dimensional TE case, Bérenger overcame this drawback by artificially splitting the transverse magnetic fields into two subcomponents and associating independent magnetic conductivities with each subcomponent. Component-wise electrical conductivities are also introduced. The modified curl equations in the layer then become

$$\epsilon_0 \frac{\partial E_x}{\partial t} = \frac{\partial H_z}{\partial y} - \sigma_y E_x \tag{12.91}$$

$$\epsilon_0 \frac{\partial E_y}{\partial t} = -\frac{\partial H_z}{\partial x} - \sigma_x E_y \tag{12.92}$$

$$\mu_0 \frac{\partial H_{zx}}{\partial t} = -\frac{\partial E_y}{\partial x} - \sigma_x^* H_{zx} \tag{12.93}$$

$$\mu_0 \frac{\partial H_{zy}}{\partial t} = \frac{\partial E_x}{\partial y} - \sigma_y^* H_{zy} \tag{12.94}$$

where $H_z = H_{zx} + H_{zy}$ and σ_x^* and σ_y^* are the component-wise magnetic conductivities. The impact of this particular field/conductivity splitting is that the perfect match is maintained at the free space–PML boundary regardless of incidence angle. For example, in truncating an FDTD grid on the right (outgoing normal direction $= \hat{x}$) boundary, set $\sigma_y = \sigma_y^* = 0$ and let $\sigma_x = \sigma_x^* \epsilon_0/\mu_0$ be a constant. Time-harmonic $+x$-directed solutions in the PML region are

$$H_z = H_0 e^{j\omega t} e^{-j(\omega/c)(x\cos\phi + y\sin\phi)} e^{-(\sigma_x \cos\phi/\epsilon_0 c)x} \tag{12.95}$$

$$\mathbf{E} = (-\hat{x}\sin\phi + \hat{y}\cos\phi)\sqrt{\frac{\mu_0}{\epsilon_0}} H_z \tag{12.96}$$

The form of this solution permits reflectionless transmission into the PML. Energy is dissipated through the exponential decay factor.

When the conductivities in the PML are large, fields decay rapidly and the linear differencing used in the core FDTD algorithm is no longer appropriate. Instead exponential differencing is used. The update equation corresponding to (12.92) is

$$\begin{aligned} E_y^{n+1}(i, j+\tfrac{1}{2}) &= E_y^n(i, j+\tfrac{1}{2}) \\ &\quad - \frac{1}{\sigma_x}(1 - e^{-\sigma_x \Delta t/\epsilon_0}) \frac{H_z^{n+1/2}(i+\frac{1}{2}, j+\frac{1}{2}) - H_z^{n+1/2}(i-\frac{1}{2}, j+\frac{1}{2})}{\Delta x} \end{aligned} \tag{12.97}$$

An infinitely thick absorbing layer is, of course, impractical in a real FDTD application. The simplest solution is to terminate the PML after some distance sufficient to absorb most of the outgoing energy with a simple boundary condition such as a p.e.c. This has the impact of returning a small portion of the outgoing electromagnetic energy into the computational space, thus introducing some error. A second practical issue concerns the thickness of the PML itself. As these cells are added on the exterior of the FDTD region of interest, their number grows rapidly as the thickness increases, especially in three-dimensional applications. Considerable research effort has thus been spent in determining material profiles in the PML that minimize its thickness while maintaining its accuracy. Readers are referred to [22] for details.

12.13 INTERNAL AND EXTERNAL SOURCES

Having developed the core algorithm and a way of getting the waves out of the FDTD box, we next examine methods for driving the FDTD problem space via incident fields and internal sources. One simple source model to include is that of a driven current. Let the current density $\mathbf{J}(\bar{r}, t)$ be specified. Then, maintaining consistency in the time discretization,

$$\epsilon_0 \dot{\mathbf{E}}^{n-1/2} = \nabla \times \mathbf{H}^{n-1/2} - \mathbf{J}^{n-1/2} \tag{12.98}$$

where the standard spatial discretization is used. Note that the current source is located at the same spatial point as the electric field but at the same time point as the magnetic field. Another simple source is a driven voltage gap, such as might occur at an antenna feed. Assume a voltage source applied over a one-cell wide gap in the y direction. Then,

$$E_y^n(i, j) = \frac{V(t = n\,\Delta t)}{\Delta y} \tag{12.99}$$

These methods can be extended to include lumped circuit elements along with the driven sources [8].

A source of external plane waves is required for scattering and coupling applications. As mentioned briefly in Section 12.11, the FDTD can be formulated in terms of either total or scattered fields. While there are advantages to either approach, we have focused on the total-field formulation in this chapter, largely because the handling of p.e.c. boundaries and penetrable materials is straightforward. The use of an ABC, which is designed to remove waves from the FDTD problem space, would appear to be in conflict with the excitation of the incident field in a total-field code. The solution is to divide the problem space as shown in Figure 12.16 into an internal total-field region containing the scattering/coupling object and an external layer of cells containing the scattered field. Since the scattered-field region contains outgoing waves only, an ABC can successfully be placed on the external boundary. To apply the plane-wave source, let the fields on the total–scattered field interface be total fields, as shown in Figure 12.17. The curl equations used to update these fields must then also use total fields. Since only scattered fields are stored in the exterior domain, the FDTD update equations must be modified to add in the incident field to these components. A similar modification is also made to the update equations for the scattered-field components immediately outside the interface. Although the incident field can be calculated analytically, a better solution is obtained by calculating the incident field on a separate (usually one-dimensional) FDTD grid that approximately matches the dispersion characteristics of the main grid.

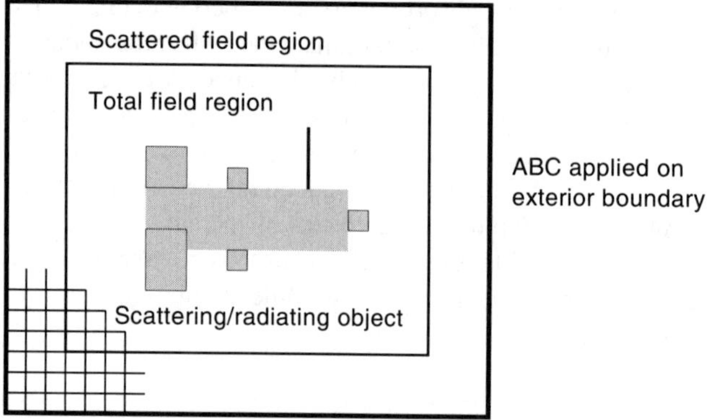

Figure 12.16 Division of the FDTD problem space into an interior total-field region that contains the interacting object(s) and an exterior scattered-field region that permits straightforward implementation of the ABC.

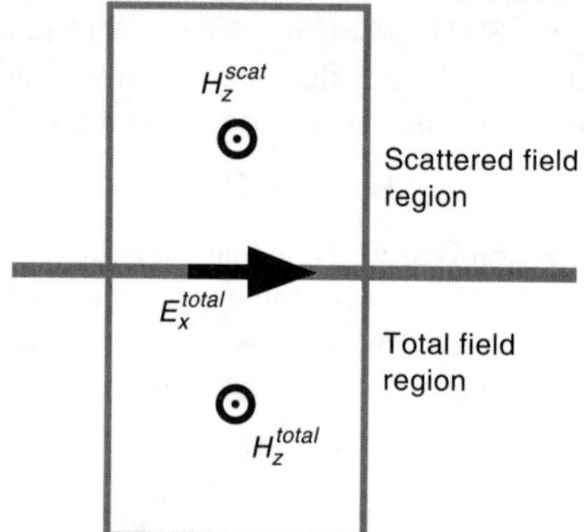

Figure 12.17 Field variables on and adjacent to the total field–scattered field interface.

12.14 FAR-FIELD PROJECTIONS

The FDTD method as developed so far in this chapter suffices for simulating the fields within the computational volume. However, for radiation and scattering calculations we must develop a method for calculating far fields, well outside the computational domain. The approach is conceptually straightforward. Referring to Figure 12.18, define a computational surface inside the FDTD grid that encloses the scattering or radiating object. For simplicity in the subsequent far-field projection equations and in programming the method, this volume is usually rectangular. Tangential electric and magnetic fields are captured on this surface and converted to equivalent surface currents via

$$\mathbf{J}_s = \hat{n} \times \mathbf{H} \tag{12.100}$$

$$\mathbf{M}_s = \mathbf{E} \times \hat{n} \tag{12.101}$$

where $\hat{n}$ is the surface unit outward normal. Fields at any arbitrary point outside this surface can then be computed using vector potentials [23] or the Stratton–Chu representation theorem [24] or by treating the equivalent currents as an array of Hertzian dipoles [25]. The

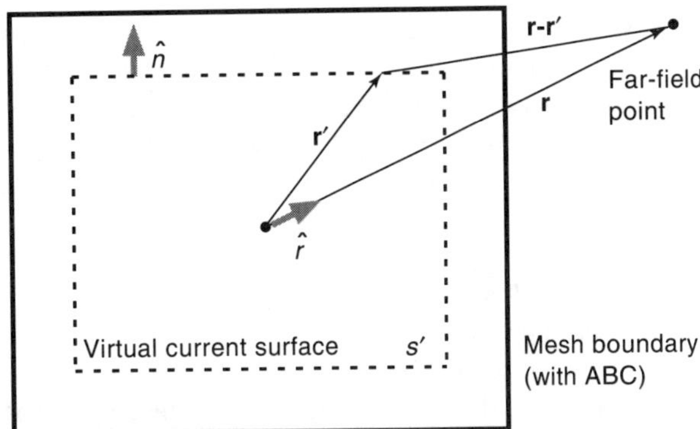

Figure 12.18 Geometry for FDTD far-field projection.

standard far-field approximations are employed to simplify the calculations.

The simplest form of the far-field projection equations occurs in the "time-harmonic" style of FDTD simulations proposed by Taflove [4]. In this method, the excitation is driven with a sine wave at the appropriate frequency and the FDTD simulation is run to steady state. Magnitude and phase information on the equivalent current surface is determined by sampling the last cycle of the sine wave at each measurement point. The frequency-domain form of the vector potentials can then be used to compute far fields.

Other authors use broadband excitations, either because they are interested in the time-varying nature of the solution itself or because broadband frequency-domain results can readily be obtained through a Fourier transform. Time-domain far-field projection algorithms are, however, somewhat more complicated than their frequency-domain counterparts; the equations contain a temporal derivative of the equivalent surface currents. In addition, a proper accounting must be made of the angle-dependent time delays involved in integrating over a spatially distributed source.

To examine these issues, refer to Figure 12.18 for geometry and notation. In the far field, the transverse electric field can be written in terms of the vector potentials as

$$E_\theta = -\mu \frac{\partial A_\theta}{\partial t} - \frac{1}{c}\frac{\partial F_\phi}{\partial t} \tag{12.102}$$

$$E_\phi = -\mu \frac{\partial A_\phi}{\partial t} + \frac{1}{c}\frac{\partial F_\theta}{\partial t} \tag{12.103}$$

where the θ and ϕ subscripts denote vector components in three-dimensional spherical coordinates. Under the standard the far-field approximations, the vector potentials take on the form

$$\mathbf{A} = \frac{1}{4\pi r}\int_{s'} \mathbf{J}_s \left[\mathbf{r}', t - (r - \hat{r}\cdot\mathbf{r}')/c\right] ds' \tag{12.104}$$

$$\mathbf{F} = \frac{1}{4\pi r}\int_{s'} \mathbf{M}_s \left[\mathbf{r}', t - (r - \hat{r}\cdot\mathbf{r}')/c\right] ds' \tag{12.105}$$

where s' denotes the equivalent current surface and the $(r - \hat{r}\cdot\mathbf{r}')/c$ term accounts for the propagation time difference between the origin and the integration point. Note that the use of the equivalence principle requires that $\mathbf{J}_s$ and $\mathbf{M}_s$ be defined on the same surface. Simple averaging is thus used to colocate the spatially staggered FDTD fields.

The far-field projection problem then reduces to one of developing efficient methods for computing (12.104) and (12.105) and/or their temporal derivatives. The key complicating feature is the presence of the integration point $\mathbf{r}'$ in the time delay. The result is that the instantaneous far-field values need to be computed via accumulation over a series of time steps. The length of the accumulation period corresponds to the projected length of the equivalent current source.

A practical challenge is encountered when determining whether to perform the far-field calculations during the FDTD simulation or to save the equivalent currents for subsequent postprocessing. Storing the currents provides the most flexibility but incurs significant storage costs. If the number of desired far-field projection angles and/or frequencies is reasonable, then it generally makes sense to perform the calculations on the fly. Kunz and Luebbers [8] provide a good discussion of the trade-offs between these factors.

12.15 EXTENSIONS TO THE ORTHOGONAL MESH FDTD METHOD

The FDTD method has been used successfully to solve a large number of electromagnetic problems. However, the technique has certain limitations, stemming largely from the use of the uniform orthogonal mesh. This restriction forces curved and diagonal boundaries to be represented by staircased approximations. Various groups have developed FDTD-like algorithms that can be used on nonorthogonal boundary conforming meshes. Approaches include curvilinear coordinate variants of the FDTD [26–28], finite-volume methods [29], and a discrete-surface integral method [30]. These methods allow greater geometric flexibility but incur a higher computational cost. Subgridding approaches have also been explored [31, 32]. These methods allow the orthogonal mesh to be retained throughout most of the computational domain but permit local refinement where needed.

Another successful area of research has been in the development of subcell methods [8, Chapter 10; 33, 34]. These methods modify the difference equations locally, adding algorithms that permit accurate modeling of structures such as wires or slots that are, in at least one dimension, too small to be handled efficiently by the unmodified FDTD method.

REFERENCES

[1] C. L. Bennett and W. L. Weeks, "Transient scattering from conducting cylinders," *IEEE Trans. Antennas Propagat.*, vol. AP-18, pp. 627–633, Sept. 1970.

[2] S. M. Rao and D. R. Wilton, "Transient scattering by conducting surfaces of arbitrary shape," *IEEE Trans. Antennas Propagat.*, vol. AP-39, pp. 56–61, Jan. 1991.

[3] K. S. Yee, "Numerical solution of initial boundary value problems involving Maxwell's equations in isotropic media," *IEEE Trans. Antennas Propagat.*, vol. AP-14, pp. 302–307, May 1966.

[4] A. Taflove and M. E. Brodwin, "Numerical solution of steady-state electromagnetic scattering problems using the time-dependent Maxwell's equations," *IEEE Trans. Microwave Theory Tech.*, vol. MTT-23, pp. 623–630, Aug. 1975.

[5] R. Holland, "Threde: A free-field EMP coupling and scattering code," *IEEE Trans. Nucl. Sci.*, vol. NS-24, pp. 2416–2421, Dec. 1977.

[6] K. Kunz and K. M. Lee, "A three-dimensional finite-difference solution of the external response of an aircraft to a complex transient EM environment: Part I—the method and its implementation," *IEEE Trans. Electromagnet. Compat.*, vol. EMC-20, pp. 328–333, May 1978.

[7] D. E. Merewether and R. Fisher, "Finite difference solution of Maxwell's equation for EMP applications," Electro Magnetic Applications Inc. report EMA-79-R-4, Defense Nuclear Agency, 1980.

[8] K. S. Kunz and R. J. Luebbers, *The Finite Difference Time Domain Method for Electromagnetics*, CRC Press, Boca Raton, FL, 1993.

[9] A. Taflove, *Computational Electromagnetics: The Finite-Difference Time-Domain Method*, Artech House, Norwood, MA, 1995.

[10] A. Taflove and K. R. Umashankar, "The finite-difference time-domain method for numerical modeling of electromagnetic wave interactions with arbitrary structures," in *Pier 2: Progress in Electromagnetics Research*, Elsevier, New York, 1990.

[11] K. L. Shlager and J. B. Schneider, "A selective survey of the finite-difference time-domain literature," *IEEE Antennas Propagat. Mag.*, vol. 37, pp. 39–57, Aug. 1995.

[12] J. Fang, "Time domain finite difference computation for Maxwell's equations," Ph.D. dissertation, University of California at Berkeley, 1989.

[13] R. Luebbers, F. Hunsberger, K. Kunz, R. Standler, and M. Schneider, "A frequency-dependent finite-difference time-domain formulation for dispersive media," *IEEE Trans. Electromagnet. Compat.*, vol. EMC-32, pp. 222–227, Mar. 1990.

[14] R. M. Joseph, S. C. Hagness, and A. Taflove, "Direct time integration of Maxwell's equations in linear dispersive media with absorption for scattering and propagation of femtosecond electromagnetic pulses," *Opt. Lett.*, vol. 16, pp. 1412–1414, Sept. 1991.

[15] R. F. Harrington, *Time-Harmonic Electromagnetic Fields*, McGraw-Hill, New York, 1961.

[16] J. P. Bérenger, "A perfectly matched layer for the absorption of electromagnetics waves," *J. Computat. Phys.*, vol. 114, pp. 185–200, Oct. 1994.

[17] E. L. Lindman, "'Free space' boundary conditions for the time dependent wave equation," *J. Computat. Phys.*, vol. 18, pp. 67–78, 1975.

[18] B. Engquist and A. Majda, "Absorbing boundary conditions for the numerical simulation of waves," *Math. Comp.*, vol. 31, pp. 629–651, July 1977.

[19] G. Mur, "Absorbing boundary conditions for the finite-difference approximation of the time-domain electromagnetic field equations," *IEEE Trans. Electromagnet. Compat.*, vol. EMC-23, pp. 377–382, Nov. 1981.

[20] R. L. Higdon, "Absorbing boundary conditions for difference approximations to the multi-dimensional wave equations," *Math. Computat.*, vol. 47, pp. 437–459, 1986.

[21] Z. P. Liao, H. L. Wong, B.-P. Yang, and Y.-F. Yuan, "A transmitting boundary for transient wave analysis," *Sci. Sinica, Ser. A*, vol. 27, pp. 1063–1076, 1984.

[22] J. P. Bérenger, "Improved PML for the FDTD solution of wave-structure interaction problems," *IEEE Trans. Antennas Propagat.*, vol. 45, pp. 466–473, Mar. 1997.

[23] R. J. Luebbers, K. S. Kunz, M. Schneider, and F. Hunsberger, "A finite-difference time-domain near zone to far zone transformation," *IEEE Trans. Antennas Propagat.*, vol. 39, pp. 429–433, Apr. 1991.

[24] K. S. Yee, D. Ingham, and K. Shlager, "Time-domain extrapolation of the far field based on FDTD calculations," *IEEE Trans. Antennas Propagat.*, vol. 39, pp. 410–413, Mar. 1991.

[25] M. J. Barth, R. R. McLeod, and R. W. Ziolkowski, "A near and far-field projection algorithm for finite-difference time-domain codes," *J. Electromagnet. Waves Applicat.*, vol. 6, pp. 5–18, 1992.

[26] R. Holland, "Finite difference solutions of Maxwell's equations in generalized nonorthogonal coordinates," *IEEE Trans. Nucl. Sci.*, vol. NS-30, pp. 4589–4591, Dec. 1983.

[27] M. Jones, "Electromagnetic PIC codes with body-fitted coordinates," *Proceedings of the 12th Conference on the Numerical Simulation of Plasmas*, American Physical Society, Topical Group on Computational Physics, San Francisco, CA, Sept. 20–23, 1987.

[28] M. Fusco, "FDTD algorithm in curvilinear coordinates," *IEEE Trans. Antennas Propagat.*, vol. AP-38, pp. 76–89, Jan. 1990.

[29] N. Madsen and R. W. Ziolkowski, "A three-dimensional modified finite volume technique for Maxwell's equations," *Electromagnetics*, vol. 10, pp. 147–161, 1990.

[30] N. Madsen, "Divergence preserving discrete surface integral methods for Maxwell's curl equations using non-orthogonal unstructured grids," *J. Computat. Phys.*, vol. 119, pp. 34–45, 1995.

[31] K. S. Kunz and L. Simpson, "A technique for increasing the resolution of finite-difference solutions of the Maxwell equations," *IEEE Trans. Electromagnet. Compat.*, vol. EMC-23, pp. 419–422, Apr. 1981.

[32] S. S. Zivanovic, K. S. Yee, and K. K. Mei, "A subgridding algorithm for the time-domain finite-difference method to solve Maxwell's equations," *IEEE Trans. Microwave Theory Tech.*, vol. MTT-38, pp. 471–479, Mar. 1991.

[33] K. R. Umashankar, A. Taflove, and B. Beker, "Calculation and experimental validation of induced currents on coupled wires in an arbitrarily shaped cavity," *IEEE Trans. Antennas Propagat.*, vol. AP-35, pp. 1248–1257, Nov. 1987.

[34] D. J. Riley and C. D. Turner, "Hybrid thin-slot algorithm for the analysis on narrow apertures in finite-difference time-domain calculations," *IEEE Trans. Antennas Propagat.*, vol. AP-38, pp. 1943–1950, Dec. 1990.

PROBLEMS

P12.1 Discretize the one-dimensional second-order, single field, wave equation (see Equation 12.9) using three-point centered difference approximations to the spatial and temporal second-order derivatives. Show that this discretization is equivalent to the one-dimensional FDTD equations.

P12.2 Assuming staggered field locations as used in the FDTD method, use rooftop basis functions and razor blade test functions to derive the FDTD spatial discretization. Outline how this approach might be used to extend the FDTD method to nonorthogonal meshes.

P12.3 Use the integral form of Maxwell's equations to derive an FDTD-discretization on a uniform mesh of equilaterial triangles.

P12.4 Develop the FDTD equations for the two-dimensional TM case.

P12.5 Derive the one-dimensional dispersion relation and stability conditions for the FDTD equations in conductive media. Derive an approximate dispersion relation in the low-loss case ($\sigma \Delta t / 2\varepsilon << 1$). Some authors use a non-centered approximation to the conductive loss term in Equation (12.49), replacing $\sigma \mathbf{E}^{n-1/2}$ with $\sigma \mathbf{E}^{n-1}$. What impact does this have on dispersion and stability?

P12.6 Show that the time-harmonic PML solutions (12.95)–(12.96) satisfy the PML partial differential Equations [(12.91)–(12.94), with $\sigma_y = \sigma_y^* = 0$ and $\sigma_x = \sigma_x^* \varepsilon_0/\mu_0$]. Consider a time-harmonic plane wave traveling in free space and obliquely incident on a PML half space. Derive the reflection and transmission coefficients as a function of incidence angle; ignore spatial and temporal discretization.

Quadrature

The numerical evaluation of integrals, often called *quadrature*, is a central feature of many computational schemes. This appendix briefly reviews some of the ideas behind single and multiple dimensional quadrature rules.

A.1 ROMBERG INTEGRATION [1–3]

One of the simplest classical quadrature algorithms is the trapezoidal rule. Suppose the integral

$$I = \int_a^b f(x)\,dx \tag{A.1}$$

is to be evaluated based on integrand samples at the endpoints of $N = 2^{n-1}$ equal-sized intervals, treating the function as a trapezoid over each. The resulting approximation can be written as

$$I_{n,1} = \frac{h}{2}[f(a) + f(b)] + h\sum_{i=1}^{N-1} f(a + ih) \tag{A.2}$$

where

$$h = \frac{b-a}{2^{n-1}} \tag{A.3}$$

is the uniform interval size. In the event that this approximation of the integral is not accurate enough, the intervals can be halved, leading to a formula involving 2^n intervals

$$I_{n+1,1} = \frac{1}{2}I_{n,1} + h'\sum_{i=1}^{2^{n-1}} f(a + [2i-1]h') \tag{A.4}$$

where h' denotes the new interval size

$$h' = \frac{b-a}{2^n} \tag{A.5}$$

Observe that the integrand samples used to compute $I_{n,1}$ are reused in $I_{n+1,1}$, saving half the required function evaluations necessary if $I_{n,1}$ was not available. Since the integrands of interest often contain special functions, it is usually desirable to minimize the number of evaluations. Because of sample point reuse, a series of approximations $\{I_{1,1}, I_{2,1}, \ldots, I_{n,1}\}$ can be generated for the same number of function evaluations as $I_{n,1}$. The availability of successive approximations facilitates error estimation.

The error produced by successive applications of the trapezoidal rule has an interesting property, namely that it consists entirely of even powers of $1/N$ [1]. In other words,

$$I - I_{n,1} = \frac{\alpha}{N^2} + \frac{\beta}{N^4} + \frac{\gamma}{N^6} + \cdots \tag{A.6}$$

where $N = 2^{n-1}$. This property can be exploited by a form of Richardson extrapolation in order to obtain better estimates of I. Given two successive approximations for I, the fact that the dominant-error term behaves as $O(N^{-2})$ means that the error drops by a factor of 4, or in equation form

$$I \cong I_{n,1} + \frac{K}{N^2} \tag{A.7}$$

$$I \cong I_{n+1,1} + \frac{K}{4N^2} \tag{A.8}$$

An improved approximation should be possible by solving these two equations for the coefficient K and extrapolating to a better estimate by removing the $O(N^{-2})$ error, to obtain

$$I \cong I_{n+1,2} = \tfrac{4}{3} I_{n+1,1} - \tfrac{1}{3} I_{n,1} \tag{A.9}$$

In general, $I_{n+1,2}$ will be a better approximation to the integral than $I_{n+1,1}$, since its leading-order error term is reduced to $O(N^{-4})$. By successively increasing n and extrapolating, this process can be continued. For instance, the two estimates

$$I \cong I_{n,2} + \frac{L}{N^4} \tag{A.10}$$

$$I \cong I_{n+1,2} + \frac{L}{16N^4} \tag{A.11}$$

can be used to extrapolate to

$$I \cong I_{n+1,3} = \tfrac{16}{15} I_{n+1,2} - \tfrac{1}{15} I_{n,2} \tag{A.12}$$

whose leading-order error is $O(N^{-6})$. By continuing this procedure, one obtains the estimate

$$I_{n+1,m+1} = \frac{4^m}{4^m - 1} I_{n+1,m} - \frac{1}{4^m - 1} I_{n,m} \tag{A.13}$$

for the same number of integrand samples necessary to compute $I_{n+1,1}$. This process of combining trapezoidal rule with Richardson extrapolation is known as *Romberg integration*.

To summarize, Romberg integration requires the use of trapezoidal rule to compute the initial estimate

$$I_{1,1} = (b-a)\left[\frac{f(a)+f(b)}{2}\right] \tag{A.14}$$

and successive estimates $I_{n+1,1}$, according to (A.4). For each $n > 1$, (A.13) is used to extrapolate to $I_{n+1,n+1}$. These estimates can be monitored until the approximation for I appears to converge to desired accuracy.

Computer programs for Romberg integration are widely available [1–3].

A.2 GAUSSIAN QUADRATURE

Romberg integration is based on a uniform subdivision of the interval (a, b). A more efficient type of numerical integration, known as Gaussian quadrature, involves nonuniform intervals. Referring to Equation (A.1), the integral estimate

$$I \cong \sum_{i=1}^{N} w_i f(x_i) \tag{A.15}$$

can be developed so that both the weights $\{w_i\}$ and the abscissas $\{x_i\}$ are selected to optimize the procedure. A common approach is to select these parameters so that the estimate is exact for polynomials up to degree p. For instance, consider a polynomial of degree 3:

$$f(x) = \alpha + \beta x + \gamma x^2 + \delta x^3 \tag{A.16}$$

For simplicity, we restrict our attention to the interval $-1 < x < 1$ and impose left-to-right symmetry on the weights and abscissas.

A rule with the form of (A.15) can be found involving only two sample points, $x = \pm x_1$. These can be determined by equating the rule (A.15) with the exact evaluation of (A.1) for each term in (A.16). This general constraint

$$\int_{-1}^{1} f(x)\,dx = w_{-1} f(x_{-1}) + w_1 f(x_1) = w_1 f(-x_1) + w_1 f(x_1) \tag{A.17}$$

yields the equations

$$2\alpha = w_1\alpha + w_1\alpha \tag{A.18}$$

$$\tfrac{2}{3}\gamma = w_1\gamma x_1^2 + w_1\gamma x_1^2 \tag{A.19}$$

Equations (A.18) and (A.19) can be solved to produce

$$w_{-1} = 1 \qquad x_{-1} = -\sqrt{\tfrac{1}{3}} \tag{A.20}$$

$$w_1 = 1 \qquad x_1 = \sqrt{\tfrac{1}{3}} \tag{A.21}$$

as the weights and samples. If it is desired to integrate a polynomial of degree 4, a third sample point at $x = x_0 = 0$ can be added, producing a set of constraints

$$2\alpha = w_1\alpha + w_0\alpha + w_1\alpha \tag{A.22}$$

$$\tfrac{2}{3}\gamma = w_1\gamma x_1^2 + w_1\gamma x_1^2 \tag{A.23}$$

$$\tfrac{2}{5}\varepsilon = w_1\varepsilon x_1^4 + w_1\varepsilon x_1^4 \tag{A.24}$$

leading to the set of weights and samples

$$w_{-1} = \tfrac{5}{9} \qquad x_{-1} = -\sqrt{\tfrac{3}{5}} \tag{A.25}$$

$$w_0 = \tfrac{8}{9} \qquad x_0 = 0 \tag{A.26}$$

$$w_1 = \tfrac{5}{9} \qquad x_1 = \sqrt{\tfrac{3}{5}} \tag{A.27}$$

Thus, with only three integrand samples, this quadrature rule can exactly integrate fourth-order polynomial functions.

The weights and samples derived above are known as Gauss–Legendre rules. Tables of weights and samples for Gauss–Legendre rules of various orders can be found in most numerical methods textbooks. In practice, higher order rules are developed based on the theory of orthogonal polynomials rather than the cumbersome simultaneous solution of non-linear equations, used here for illustration. Additional information on Gaussian quadrature is available in the literature [1, 3–5].

A.3 GAUSS–KRONROD RULES

When integrating analytic (smooth) functions, Gaussian quadrature is more efficient than Romberg integration in the sense that it generally produces more accurate estimates for a given number of integrand samples. However, Romberg integration permits complete reuse of integrand samples when doubling the number of sample points. Thus, in a situation where no a priori knowledge of the necessary number of samples is available, Romberg integration provides a sequence of integral approximations, at no extra cost, from which convergence can be estimated.

Although a similar sequence cannot be obtained as efficiently with Gaussian quadrature, a partial remedy was provided in 1964 when Kronrod developed $(2n+1)$-point Gaussian quadrature rules that reused the sample points from an n-point Gauss–Legendre formula [6]. This provides two estimates of the integral, one more accurate than the other, for $2n+1$ samples. Since then, similar ideas have been applied to other forms of Gaussian quadrature [7] and incorporated into software libraries such as QUADPACK [8]. Because of the improved efficiency of Gaussian quadrature, the Gauss–Kronrod rules have usually superseded Romberg integration in practice.

A.4 INCORPORATION OF LOGARITHMIC SINGULARITIES

Although standard Gauss–Legendre rules yield relatively poor accuracy if the integrand contains a singularity, generalized Gaussian quadrature rules can be developed for integrals of the form

$$I = \int_a^b w(x) f(x)\, dx \tag{A.28}$$

where $w(x)$ might be singular. For instance, quadrature rules have been developed for a logarithmic singularity [9, 10]. In this case, a quadrature rule using n samples permits the exact integration of functions containing the $2n$ terms $\{1, \ln x, x, x \ln x, \ldots, x^{n-1}, x^{n-1} \ln x\}$.

When high accuracy is required, rules of this type are more efficient than the "singularity subtraction" approach introduced in Chapter 2. For illustration, a FORTRAN subroutine capable of treating a logarithmic singularity is provided in Appendix C.

A.5 GAUSSIAN QUADRATURE FOR TRIANGLES [11]

For evaluating multidimensional integrals, one possible approach is to create "product rules" by nesting single-dimensional quadrature rules. However, these algorithms are far from optimal. Gaussian quadrature rules can be developed specifically for multidimensional domains using ideas similar to those outlined in Section A.2. For triangular regions, consider the quadratic function

$$f(L_1, L_2) = \alpha L_1^2 + \beta L_2^2 + \gamma L_1 L_2 + \delta L_1 + \varepsilon L_2 + \zeta \tag{A.29}$$

where L_1 and L_2 denote two linearly independent simplex coordinates (L_1, L_2, L_3) within the triangle. By imposing triangular symmetry on the sample points and weights, a rule is sought involving the three sample points (η, ξ, ξ), (ξ, η, ξ), and (ξ, ξ, η), each of which is associated with the same weight. After imposing symmetry conditions, the quadrature rule must satisfy the four constraints

$$\iint dA = A \tag{A.30}$$

$$\iint L_1 \, dA = \tfrac{1}{3}A \tag{A.31}$$

$$\iint L_1^2 \, dA = \tfrac{1}{6}A \tag{A.32}$$

$$\iint L_1 L_2 \, dA = \tfrac{1}{12}A \tag{A.33}$$

where A denotes the triangle area. By equating the rule

$$I \cong A \sum_i w_i f(L_{1i}, L_{2i}, L_{3i}) \tag{A.34}$$

with the exact evaluations in (A.30)–(A.33), one obtains a system of four nonlinear equations, from which it is possible to extract the three sample points

$$w = \tfrac{1}{3} \qquad (L_1, L_2, L_3) = \left(\tfrac{2}{3}, \tfrac{1}{6}, \tfrac{1}{6}\right) \tag{A.35}$$

$$w = \tfrac{1}{3} \qquad (L_1, L_2, L_3) = \left(\tfrac{1}{6}, \tfrac{2}{3}, \tfrac{1}{6}\right) \tag{A.36}$$

$$w = \tfrac{1}{3} \qquad (L_1, L_2, L_3) = \left(\tfrac{1}{6}, \tfrac{1}{6}, \tfrac{2}{3}\right) \tag{A.37}$$

Similar rules can be developed for higher order representations, and the literature contains Gaussian quadrature rules sufficient for integrating polynomials up to degree 20 on triangles [12–14].

In the single-dimensional case, orthogonal polynomials can be used to simplify the development of higher order quadrature rules. The lack of an analogous theory for multidimensional orthogonal polynomials has hampered the systematic development of multidimensional Gaussian quadrature rules, and most of the existing rules [13, 14] appear to be the result of numerical solutions of the nonlinear equations.

A.6 GAUSSIAN QUADRATURE FOR TETRAHEDRONS

A few quadrature formulas have been developed for tetrahedral domains using ideas similar to those outlined above. For instance, Keast presents rules up to polynomial degree 9 [14, 15]. The table of coefficients in [15] appears to contain an error for the eighth-degree rule for the weight associated with the sample point in the centroid of the tetrahedron. The correct weight should be negative $(-0.039327\ldots)$.

REFERENCES

[1] W. Cheney and D. Kincaid, *Numerical Mathematics and Computing*. Monterey: Brooks/Cole Publishing, 1980.

[2] C. deBoor, "CADRE: An algorithm for numerical quadrature," in *Mathematical Software*, ed. J. Rice. New York: Academic, 1971, pp. 417–449.

[3] W. H. Press, B. P. Flannery, S. A. Teukolsky, and W. T. Vetterling, *Numerical Recipes*. Cambridge: Cambridge University Press, 1986.

[4] H. Engels, *Numerical Quadrature and Cubature*. London: Academic, 1980.

[5] A. W. Mathis and A. F. Peterson, "A primer on Gaussian quadrature rules," *ACES Newslett.*, vol. 11, pp. 50–59, July 1996.

[6] A. S. Kronrod, *Nodes and Weights of Quadrature Formulas*, New York: Consultants Bureau, 1965.

[7] W. Gautschi and S. Notaris, "Gauss Kronrod quadrature formulas for weight functions of the Bernstein-Szego type," *J. Computat. Appl. Math.*, vol. 29, pp. 199–224, 1989.

[8] R. Piessens, E. deDoncker-Kapenga, C. W. Uberhuber, and D. K. Kahaner, *QUADPACK: A Subroutine Package for Automatic Integration*, Berlin: Springer-Verlag, 1983.

[9] J. A. Crow, "Quadrature of integrands with a logarithmic singularity," *Mathematics of Computation*, vol. 60, pp. 297–302, Jan. 1993.

[10] J. H. Ma, V. Rokhlin, and S. Wandzura, "Generalized Gaussian quadrature rules for systems of arbitrary functions," *SIAM J. Numer. Anal.*, vol. 33, pp. 971–996, June 1996.

[11] J. S. Savage and A. F. Peterson, "Quadrature rules for numerical integration over triangles and tetrahedra," *IEEE Antennas Propagat. Mag.*, vol. 38, pp. 100–102, June 1996.

[12] G. R. Cowper, "Gaussian quadrature formulas for triangles," *Int. J. Numer. Methods Eng.*, vol. 7, pp. 405–408, 1973.

[13] D. A. Dunavant, "High degree efficient symmetrical Gaussian quadrature rules for the triangle," *Int. J. Numer. Methods Eng.*, vol. 21, pp. 1129–1148, 1985.

[14] R. Cools and P. Rabinowitz, "Monomial cubature rules since 'Stroud': A compilation," *J. Computat. Appl. Math.*, vol. 48, pp. 309–326, 1993.

[15] P. Keast, "Moderate degree tetrahedral quadrature formulas," *Comp. Methods Appl. Mechanics Eng.*, vol. 55, pp. 339–348, 1986.

B

Source-Field Relationships for Cylinders Illuminated by an Obliquely Incident Field

The following is a compilation of formulas for the various field components produced by a single strip cell of constant-current density radiating in space. The expressions are used when calculating the moment-method matrix elements for cylindrical scattering problems under the condition that the z-dependence of the excitation is

$$e^{j\gamma z} \tag{B.1}$$

Thus, the incident field may be a plane wave impinging on the scatterer from an oblique angle (not perpendicular to the cylinder axis). This particular z-dependence also arises if a Fourier transformation in z is used to replace a three-dimensional problem involving an infinite cylinder by the superposition of two-dimensional problems.

Figure B.1 illustrates the geometry under consideration. The strip cell of unit current density is centered at the origin, is of cross-sectional length W, and is oriented so that its outward normal vector makes a polar angle ϕ with $\hat{x}$-axis (outward must be defined in the context of a closed cylinder with an inside and outside; our strip is considered to be one of a number modeling such a cylinder). The field components of interest are the $\hat{z}$- and $\hat{T}$-components of the electric ($\bar{E}$) and magnetic ($\bar{H}$) field at some observation point (x, y), where $\hat{T}$ is the tangent vector to a similar strip, with outward normal vector given by the polar angle ψ. The source may be the $\hat{z}$- or $\hat{t}$-component of magnetic or electric current density. We consider only the case where the current density is constant on the strip.

The notation employed will identify both the source component and the field component. For example, $H_z \cdot J_t$ denotes the $\hat{z}$-component of the $\bar{H}$-field produced by the $\hat{t}$-component of electric current density. Appropriate expressions for the fields produced by

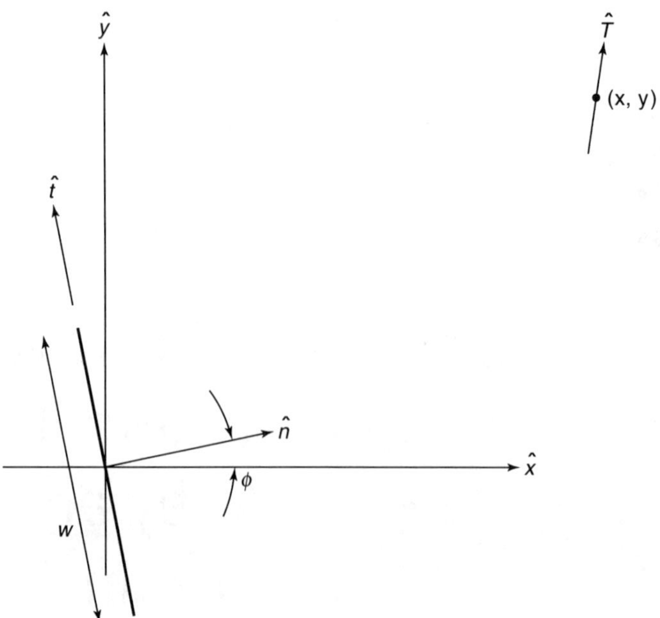

Figure B.1 Geometry of source strip.

sources are found in terms of the vector potentials $\bar{A}$ and $\bar{F}$:

$$\bar{H} = \text{curl } \bar{A} + \frac{\text{grad div} + k^2}{jk\eta} \bar{F} \tag{B.2}$$

$$\bar{E} = \eta \frac{\text{grad div} + k^2}{jk} \bar{A} - \text{curl } \bar{F} \tag{B.3}$$

where the vector potentials are defined as

$$\bar{A}(x, y) = \int_{s=-W/2}^{W/2} [\hat{z}J_z(s) + \hat{t}J_t(s)]\tilde{G}(R; k, \gamma)\, ds \tag{B.4}$$

$$\bar{F}(x, y) = \int_{s=-W/2}^{W/2} [\hat{z}K_z(s) + \hat{t}K_t(s)]\tilde{G}(R; k, \gamma)\, ds \tag{B.5}$$

and

$$\tilde{G}(R; k, \gamma) = \begin{cases} \dfrac{1}{4j} H_0^{(2)}(R\sqrt{k^2 - \gamma^2}) & k^2 > \gamma^2 \\[2ex] \dfrac{1}{2\pi} K_0(R\sqrt{\gamma^2 - k^2}) & \gamma^2 > k^2 \end{cases} \tag{B.6}$$

$$R = \sqrt{(x + s\sin\phi)^2 + (y - s\cos\phi)^2} \tag{B.7}$$

In Equation (B.6), H refers to the Hankel function and K the modified Bessel function of the second kind.

Because of the assumed $e^{j\gamma z}$-dependence, any derivatives with respect to z in the curl, grad, and div operations are replaced by multiplications with $j\gamma$. In some cases, derivatives will be transferred to the Green's function in Equation (B.6), and we note that

$$\tilde{G}'(R; k, \gamma) = \begin{cases} -\dfrac{\sqrt{k^2-\gamma^2}}{4j} H_1^{(2)}(R\sqrt{k^2-\gamma^2}) & k^2 > \gamma^2 \\ -\dfrac{\sqrt{\gamma^2-k^2}}{2\pi} K_1(R\sqrt{\gamma^2-k^2}) & \gamma^2 > k^2 \end{cases} \tag{B.8}$$

For explicit calculations, the vectors $\hat{t}$ and $\hat{T}$ are defined as

$$\hat{t} = -\hat{x}\sin\phi + \hat{y}\cos\phi \tag{B.9}$$

$$\hat{T} = -\hat{x}\sin\psi + \hat{y}\cos\psi \tag{B.10}$$

The $\hat{z}$-component of the $\bar{H}$-field produced at (x, y) by a $\hat{z}$-component of magnetic current density on the strip of Figure B.1 is given by

$$H_z \cdot K_z(x, y) = \frac{k^2-\gamma^2}{jk\eta} \int_{s=-W/2}^{W/2} \tilde{G}(R; k, \gamma)\, ds \tag{B.11}$$

where $\tilde{G}$ is defined in Equation (B.6). A closed-form expression for Equation (B.11) is not available, and in general it must be evaluated numerically. However, in many cases it can be approximated by

$$H_z \cdot K_z(x, y) \simeq \frac{k^2-\gamma^2}{jk\eta} W\tilde{G}(\rho; k, \gamma) \qquad \rho \neq 0 \tag{B.12}$$

where

$$\rho = \sqrt{x^2+y^2} \tag{B.13}$$

and

$$H_z \cdot K_z(0, 0) \simeq -\frac{k^2-\gamma^2}{4k\eta} W \begin{cases} 1 - j\dfrac{2}{\pi}\ln\left(\dfrac{W\sqrt{k^2-\gamma^2}}{6.10482}\right) & k^2 > \gamma^2 \\ -j\dfrac{2}{\pi}\ln\left(\dfrac{W\sqrt{\gamma^2-k^2}}{6.10482}\right) & \gamma^2 > k^2 \end{cases} \tag{B.14}$$

Equation (B.14) is obtained by integrating a small-argument form of the Hankel or modified Bessel function of Equation (B.6). The type of approximation employed here is accurate within a few percent as long as the strip size does not exceed about a tenth of a wavelength.

The $\hat{z}$-component of the $\bar{H}$-field produced by a $\hat{t}$-component of unit magnetic current density on the strip of Figure B.1 may be obtained from the expression

$$H_z \cdot K_t(x, y) = \frac{1}{jk\eta}\hat{z} \cdot \text{grad div}\bar{F}_t \tag{B.15}$$

which reduces to

$$H_z \cdot K_t(x, y) = \frac{\gamma}{k\eta}\frac{\partial}{\partial t} \int_{s=-W/2}^{W/2} \tilde{G}(R; k, \gamma)\, ds \tag{B.16}$$

and finally to the closed-form expression

$$H_z \cdot K_t(x, y) = \frac{\gamma}{k\eta}[\tilde{G}(R_1; k, \gamma) - \tilde{G}(R_2; k, \gamma)] \tag{B.17}$$

where

$$R_1 = \sqrt{(x - \tfrac{1}{2}W \sin\phi)^2 + (y + \tfrac{1}{2}W \cos\phi)^2} \tag{B.18}$$

$$R_2 = \sqrt{(x + \tfrac{1}{2}W \sin\phi)^2 + (y - \tfrac{1}{2}W \cos\phi)^2} \tag{B.19}$$

No $\hat{z}$-component of an $\bar{H}$-field is generated by a $\hat{z}$-component of electric current density; thus

$$H_z \cdot J_z(x, y) = 0 \tag{B.20}$$

A $\hat{t}$-component of $\bar{J}$ does produce a $\hat{z}$-component of $\bar{H}$, according to

$$H_z \cdot J_t(x, y) = \left(\cos\phi \frac{\partial}{\partial x} + \sin\phi \frac{\partial}{\partial y}\right) \int_{s=-W/2}^{W/2} \tilde{G}(R; k, \gamma)\, ds \tag{B.21}$$

For a point (x, y) away from the strip, this expression becomes

$$H_z \cdot J_t(x, y) = \int_{s=-W/2}^{W/2} \left(\cos\phi \frac{\Delta x}{R} + \sin\phi \frac{\Delta y}{R}\right) \tilde{G}'(R; k, \gamma)\, ds \tag{B.22}$$

where

$$\Delta x = x + s \sin\phi \tag{B.23}$$

$$\Delta y = y - s \cos\phi \tag{B.24}$$

and

$$R = \sqrt{\Delta x^2 + \Delta y^2} \tag{B.25}$$

The G' function is defined in Equation (B.8). In general, Equation (B.22) must be evaluated numerically.

As the observation point (x, y) approaches the strip from the outward side (as defined by ϕ), a limiting procedure can be used to compute

$$H_z \cdot J_t(0, 0)\ |_{\text{outside}} = -\tfrac{1}{2} \tag{B.26}$$

If (x, y) approached the strip from the inside, a similar procedure produces

$$H_z \cdot J_t(0, 0)\ |_{\text{inside}} = \tfrac{1}{2} \tag{B.27}$$

The transverse component of the $\bar{H}$-field produced by the $\hat{z}$-component of a magnetic current density may be obtained from the expression

$$H_t \cdot K_z(x, y) = \frac{\gamma}{k\eta} \hat{T} \cdot \text{grad } F_z \tag{B.28}$$

This reduces to

$$H_t \cdot K_z(x, y) = \frac{\gamma}{k\eta} \int_{s=-W/2}^{W/2} \left(\frac{-\Delta x}{R} \sin\psi + \frac{\Delta y}{R} \cos\psi\right) \tilde{G}'(R; k, \gamma)\, ds \tag{B.29}$$

where Δx, Δy, and R are defined in Equations (B.23)–(B.25), and $\tilde{G}'$ is defined in Equation (B.8). In general, Equation (B.29) must be evaluated numerically. For the special case when the observation point (x, y) happens to lie on the source strip, the field vanishes and

$$H_t \cdot K_z(0, 0) = 0 \tag{B.30}$$

The transverse H-field produced by a transverse magnetic current is given by the expression

$$H_t \cdot K_t(x, y) = \frac{1}{jk\eta} \hat{T} \cdot [\text{grad div} + k^2] \bar{F}_t \tag{B.31}$$

which can be expanded to produce

$$\begin{aligned} H_t \cdot K_t(x, y) = \frac{1}{jk\eta} \Bigg[& \left(-\sin\psi \frac{\Delta x_1}{R_1} + \cos\psi \frac{\Delta y_1}{R_1} \right) \tilde{G}'(R_1; k, \gamma) \\ & - \left(-\sin\psi \frac{\Delta x_2}{R_2} + \cos\psi \frac{\Delta y_2}{R_2} \right) \tilde{G}'(R_2; k, \gamma) \Bigg] \\ & + \frac{k}{j\eta} \cos(\psi - \phi) \int_{s=-W/2}^{W/2} \tilde{G}(R; k, \gamma)\, ds \end{aligned} \tag{B.32}$$

where

$$\Delta x_1 = x - \tfrac{1}{2} W \sin\phi \tag{B.33}$$

$$\Delta y_1 = y + \tfrac{1}{2} W \cos\phi \tag{B.34}$$

$$R_1 = \sqrt{\Delta x_1^2 + \Delta y_1^2} \tag{B.35}$$

$$\Delta x_2 = x + \tfrac{1}{2} W \sin\phi \tag{B.36}$$

$$\Delta y_2 = y - \tfrac{1}{2} W \cos\phi \tag{B.37}$$

$$R_2 = \sqrt{\Delta x_2^2 + \Delta y_2^2} \tag{B.38}$$

Although the remaining integral in Equation (B.32) cannot be reduced to a closed-form expression, the approximation employed previously to convert Equation (B.11) to Equations (B.12) and (B.13) may be used for computational purposes.

The transverse $\bar{H}$-field produced by a $\hat{z}$-component of electric current density is given by

$$H_t \cdot J_z(x, y) = \hat{T} \cdot \left(\hat{x} \frac{\partial A_z}{\partial y} - \hat{y} \frac{\partial A_z}{\partial x} \right) \tag{B.39}$$

which reduces to

$$H_t \cdot J_z(x, y) = -\int_{s=-W/2}^{W/2} \left(\sin\psi \frac{\Delta y}{R} + \cos\psi \frac{\Delta x}{R} \right) \tilde{G}'(R; k, \gamma)\, ds \tag{B.40}$$

where Δx, Δy, and R are defined in Equations (B.23) and (B.25). Again, numerical integration must be used to accurately evaluate Equation (B.40). When the observation point lies on the strip, a limiting argument similar to that employed in Equations (B.26) and (B.27) can be used to show that the transverse $\bar{H}$-field an infinitesimal distance outside the strip is given by

$$H_t \cdot J_z(0, 0) \,|_{\text{outside}} = \tfrac{1}{2} \tag{B.41}$$

The transverse $\bar{H}$-field an infinitesimal distance inside the strip is

$$H_t \cdot J_z(0, 0) \,|_{\text{inside}} = -\tfrac{1}{2} \tag{B.42}$$

The transverse $\bar{H}$-field produced by $\hat{t}$-component of electric current density can be found from the expression

$$H_t \cdot J_t(x, y) = -j\gamma \hat{T} \cdot (\hat{x} \cos\phi + \hat{y} \sin\phi) A_t \tag{B.43}$$

which reduces to

$$H_t \cdot J_t(x, y) = j\gamma \sin(\psi - \phi) \int_{s=-W/2}^{W/2} \tilde{G}(R; k, \gamma)\, ds \tag{B.44}$$

The integral can also be approximated according to the procedure outlined in Equations (B.12) and (B.13), if desired. When the observation point approaches the strip, the expression vanishes. Therefore,

$$H_t \cdot J_t(0, 0) = 0 \tag{B.45}$$

The above equations describe the magnetic field produced by a constant electric or magnetic current density. Expressions for the electric field produced by the same sources can be found directly from the above expressions using the principle of duality. These formulas are given as follows:

$$E_z \cdot J_z(x, y) = \eta^2 H_z \cdot K_z(x, y) \tag{B.46}$$

$$E_z \cdot J_t(x, y) = \eta^2 H_z \cdot K_t(x, y) \tag{B.47}$$

$$E_z \cdot K_z(x, y) = 0 \tag{B.48}$$

$$E_z \cdot K_t(x, y) = -H_z \cdot J_t(x, y) \tag{B.49}$$

$$E_t \cdot J_z(x, y) = \eta^2 H_t \cdot K_z(x, y) \tag{B.50}$$

$$E_t \cdot J_t(x, y) = \eta^2 H_t \cdot K_t(x, y) \tag{B.51}$$

$$E_t \cdot K_z(x, y) = -H_t \cdot J_z(x, y) \tag{B.52}$$

$$E_t \cdot K_t(x, y) = -H_t \cdot J_t(x, y) \tag{B.53}$$

Fortran Codes for TM Scattering From Perfect Electric Conducting Cylinders

This appendix illustrates three distinct approaches for the implementation of the pulse basis–delta testing discretization of the EFIE for the TM scattering problem of Section 2.1. Section C.1 provides the specific implementation discussed in Section 2.1, which involved a "single-point" approximation for the off-diagonal matrix entries. Section C.2 illustrates the use of Romberg quadrature with the "singularity subtraction" approach for treating the diagonal matrix elements, as described at the end of Section 2.1. Finally, Section C.3 illustrates the use of a generalized Gaussian quadrature procedure (Section A.4) that can evaluate the singular diagonal matrix entries without special treatment.

The following FORTRAN subroutines are intended for illustration and generally are far from optimal. For instance, it should be more efficient to separately evaluate the real and imaginary parts of the matrix entries and reserve the use of the specialized Gaussian quadrature rule for the singular integrands only.

C.1 IMPLEMENTATION 1: SINGLE-POINT APPROXIMATION

The following FORTRAN computer program illustrates the specific implementation discussed in Section 2.1. The geometry is described by the input data file (CYLFIL), and some additional information is provided through the keyboard. A sample input file and a sample output file are provided after the program listing.

```
c      ptmege      moment-method analysis of pec cylinders for tm-pol
c                  plane-wave excitation -- program finds the current
c                  density and rcs
```

```
c
c               the E-field equation is discretized with pulse basis
c               functions and point-matching (approximate formulas
c               are taken from Harrington, Field Computation by
c               Moment Methods, Krieger reprint, 1982, pp 42 - 46)
c
c     a. f. peterson   feb 11, 1986
c
c     description of input file:
c
c     the model of the cylinder cross-section is read from file
c     'cylfil,' which should be free-formatted and contain the
c     following:
c
c       N
c       x(1) y(1) w(1)
c       x(2) y(2) w(2)
c       .
c       .
c       .
c       x(N) y(N) w(N)
c
c     where
c
c       N is the number of strips in the model
c       (x,y) is the location of the phase center of each strip in the
c             x-y plane
c       w is the width of each strip
c       (x, y, and w are in units of free-space wavelength)
c
c
c     parameter 'na' is the max order of the system matrix and thus
c     the largest size model that can be treated -- adjust accordingly
c
      parameter(na=300)
      complex z(na,na),v(na),j(na),sig,ezpjz
      real x(na),y(na),w(na),theta,arg,mag,rcs,phi,psi
      integer p(na)
c
      open(unit=1,file='cylfil')
      open(unit=4,file='outfil',status='new')
c
c     load model from file #1
c
      read(1,*) n
      do 10 i=1,n
   10 read(1,*) x(i),y(i),w(i)
c
```

```
c     fill impedance matrix
c
      do 20 k=1,n
      do 20 i=1,n
      z(k,i)=ezpjz(x(i),y(i),x(k),y(k),w(i))
   20 continue
      ijob = 0
c
c     generate incident field vector v
c
   30 write(*,*) 'give direction of incident plane wave in degrees'
      write(*,*) '(theta=0. means a wave propagating in the +x'
      write(*,*) 'direction; theta=90. means +y direction, relative'
      write(*,*) 'to the model)'
      read(*,*) theta
      write(4,*)
      write(4,*) 'angle of incidence is ',theta,' degrees'
      theta = .01745329*theta
      do 40 i=1,n
      arg=-6.2831853*(y(i)*sin(theta)+x(i)*cos(theta))
   40 v(i) = cmplx(cos(arg),sin(arg))
c
c     solve system by gaussian elimination
c
      call cgauss(z,na,n,v,j,ijob,p)
      if(ijob.eq.-1) stop
      ijob = 1
c
c     write data to file #4
c
      write(4,*)
      write(4,*) 'current density in magnitude - angle format'
      write(4,*)
      do 60 i=1,n
      mag=cabs(j(i))
      arg=atan2(aimag(j(i)),real(j(i)))*57.2957795
   60 write(4,*) i,mag,arg
      write(4,*)
c
c     compute bistatic scattering cross section
c
      write(*,*) 'give incremental angle for rsc scan in degrees'
      read(*,*) theta
      write(4,*) 'rcs in dB-wavelength'
      write(4,*) '     angle                 rcs'
      write(4,*)
      psi=0.
   70 phi=psi*.01745329
```

```
      sig = (0.,0.)
      do 80 i=1,n
   80 sig = sig + w(i)*cexp(cmplx(0.,6.2831853*(x(i)*cos(phi)+
     +                 y(i)*sin(phi))))*j(i)
      rcs = 10.*alog10(cabs(sig)**2 * 222936.04)
      write(4,*) psi,rcs
      psi = psi + theta
      if (psi.lt.360.) go to 70
c
c     check for additional incident angles
c
      write(*,*) 'want additional incident angles (1=y)'
      read(*,*) i
      if(i.eq.1) go to 30
      write(*,*)
      write(*,*) 'data from this run placed in file: outfil'
      stop
      end
c
c     --------------------------------------------------
c
      complex function ezpjz(xs,ys,xo,yo,ws)
c
c     negative scattered electric field (Ez) at location
c     (xo,yo) due to a unit source (Jz) of width ws at (xs,ys)
c
      complex h02
      wsot=ws*0.1
      arg=sqrt((xo-xs)**2+(yo-ys)**2)
         if(arg.lt.wsot) then
c
c        assume observer at center of source cell
c
         ezpjz=591.7661*ws*cmplx(1.,-.636619772*alog(1.029217*ws))
c
         else
c
c        observer not on source cell; use single-point approximation
c
         arg=arg*6.283185308
         ezpjz=591.7661*ws*h02(arg)
         endif
      return
      end
c
c     --------------------------------------------------
c
c     ZERO-ORDER HANKEL FUNCTION (2ND KIND) FROM ABR & STEGUN PP 369
```

```
c
      COMPLEX FUNCTION H02(A)
      IF(A.GT.3.000) GOTO 5
      BJ=(A/3.00)**2
      BJ=1.0+BJ*(-2.2499997+BJ*(1.2656208+BJ*(-.3163866+BJ*
     &    (.0444479+BJ*(-.0039444+BJ*.00021)))))
      BY=(A/3.00)**2
      BY=2.0/3.1415926*ALOG(A/2.)*BJ+.36746691+BY*(.60559366+BY
     &   *(-.74350384+BY*(.25300117+BY*(-.04261214+BY*(.00427916-
     &   BY*.00024846)))))
      GOTO 10
    5 BJ=3.00/A
      F0=.79788456+BJ*(-.00000077+BJ*(-.00552740+BJ*(-.00009512
     &   +BJ*(.00137237+BJ*(-.00072805+BJ*.00014476)))))
      T0=A-.78539816+BJ*(-.04166397+BJ*(-.00003954+BJ*(.00262573
     &   +BJ*(-.00054125+BJ*(-.00029333+BJ*.00013558)))))
      BY=SQRT(A)
      BJ=F0*COS(T0)/BY
      BY=F0*SIN(T0)/BY
   10 H02=CMPLX(BJ,-BY)
      RETURN
      END
c
c     -------------------------------------------------
c
      subroutine cgauss(a,ia,n,b,x,ijob,p)
c
c     gaussian elimination to solve the complex system ax=b
c
c     (perform forward elimination with scaled partial pivoting on the
c     matrix 'a,' then use back substitution to solve)
c
c     the right-hand side 'b' is destroyed in the process
c
c     ia = row dimension of 'a' as specified in the calling program
c     n = order of the system
c     ijob = 0  means to perform elimination and solve the system
c     ijob = 1  means that forward elimination has already been
c               performed on 'a' by a previous call, so just solve
c     ijob = -1  on return means that 'a' is singular
c     p = array to hold pivot indices
c
c     A. F. Peterson    Feb 11 1986
c     (adapted from the text 'numerical mathematics and computing'
c     by cheney and kincaid,  brooks / cole publishing, 1980)
c
      complex a(ia,n),b(n),x(n),sum
      real c,cmax
```

```
      integer p(n),pk,ia,n,ijob,i,j,k
c
      if(ijob.eq.1) go to 5
c
c     initialize pivot indices and scale factors used when pivoting
c
      do 1 i = 1,n
      p(i) = i
      x(i) = (0.,0.)
      do 1 j = 1,n
    1 x(i) = cmplx(amax1(cabs(x(i)),cabs(a(i,j))),0.)
c
c     loop through one column at a time, sweeping out
c
      do 4 k = 1,n-1
c
c     find the maximum entry of column 'k,' (each normalized to row
c     maximum), ignoring those rows that have already been assigned
c     pivots
c
      cmax = 0.
      do 2 i = k,n
      c=cabs(a(p(i),k))/cabs(x(p(i)))
      if(c.le.cmax) go to 2
      j = i
      cmax = c
    2 continue
c
      if(cmax) 9,9,3
c
c     set pivot index for column 'k'
c
    3 pk = p(j)
      p(j) = p(k)
      p(k) = pk
c
c     perform forward elimination, storing multipliers in 'a'
c
      do 4 i = k+1,n
      sum = a(p(i),k)/a(pk,k)
      a(p(i),k) = sum
      do 4 j = k+1,n
    4 a(p(i),j) = a(p(i),j) - sum*a(pk,j)
c
c     perform forward elimination on the right-hand side 'b'
c
    5 do 6 j = 1,n-1
      do 6 i = j+1,n
```

```
    6 b(p(i)) = b(p(i)) -a(p(i),j)*b(p(j))
c
c     solve by back substitution
c
      x(n) = b(p(n))/a(p(n),n)
      do 8 i = 1,n-1
      sum = b(p(n-i))
      do 7 j = n-i+1,n
    7 sum = sum - a(p(n-i),j)*x(j)
    8 x(n-i) = sum/a(p(n-i),n-i)
      return
    9 write(*,*) 'matrix is singular'
      ijob = -1
      return
      end
```

```
--------------------------------------------------------------
Sample input data (CYLFIL) for circular cylinder,
one wavelength circumference:


          10
   0.1591549       0.0000000       0.1000000
   0.1287590       9.3548924E-02   0.1000000
   4.9181573E-02   0.1513653       0.1000000
  -4.9181588E-02   0.1513653       0.1000000
  -0.1287591       9.3548916E-02   0.1000000
  -0.1591549      -1.3913767E-08   0.1000000
  -0.1287590      -9.3548939E-02   0.1000000
  -4.9181599E-02  -0.1513653       0.1000000
   4.9181603E-02  -0.1513653       0.1000000
   0.1287591      -9.3548872E-02   0.1000000

--------------------------------------------------------------
Corresponding output data:  (OUTFIL)


 angle of incidence is   0.0000000     degrees

 current density in magnitude - angle format

 1   8.1690575E-04   152.8625
 2   9.1447576E-04  -143.4977
 3   2.1823230E-03  -69.01913
 4   3.8987177E-03  -12.58591
 5   5.6035798E-03   27.17590
 6   6.3566919E-03   41.39040
 7   5.6035747E-03   27.17587
```

```
 8   3.8987170E-03   -12.58589
 9   2.1823235E-03   -69.01912
10   9.1447704E-04   -143.4976

 rcs in dB-wavelength
      angle              rcs

   0.0000000          2.831829
    30.00000          2.015279
    60.00000         -0.033761
    90.00000         -1.874606
    120.0000         -2.312734
    150.0000         -2.139701
    180.0000         -2.048646
    210.0000         -2.139700
    240.0000         -2.312733
    270.0000         -1.874606
    300.0000         -0.033763
    330.0000          2.015276
-------------------------------------------------------------
```

C.2 IMPLEMENTATION 2: ROMBERG QUADRATURE

In this example, subroutine EZPJZ is replaced by a routine that employs Romberg integration (Section A.1) to evaluate the matrix entries. An additional parameter Ω_n is required to describe the orientation of each cell, as defined in Section 2.2. In addition, the desired integration accuracy *rerr* is provided from the main routine.

```
      complex function ezpjz(xs,ys,xo,yo,ws,omegs,rerr)
c
c     negative scattered electric field (Ez) at location
c     (xo,yo) due to a unit source (Jz) of width ws and
c     orientation omegs at (xs,ys)
c
c     computed using Romberg quadrature to accuracy rerr
c
      complex zrom1d,zint1,zint2,r(15)
      common xc,yc,com,som
      external zint1,zint2
c
      aerr=0.000001
      wso2=0.5*ws
      wsot=0.1*ws
      arg=sqrt((xo-xs)**2+(yo-ys)**2)
      if(arg.lt.wsot) then
```

```
c
c        assume observer at center of source cell; add singular
c        part of integral (missing from zint1) back in
c
         ezpjz=zrom1d(zint1,0.,wso2,rerr,aerr,15,1,r,ier)+
     +   cmplx(0.,-.636619772*alog(2.058434105*wso2))*wso2
         ezpjz=ezpjz*(2.,0.)
c
         else
c
c        observer not on source cell
c
         xc=xo-xs
         yc=yo-ys
         com=cos(omegs)
         som=sin(omegs)
         ezpjz=zrom1d(zint2,-wso2,wso2,rerr,aerr,15,1,r,ier)
         endif
      ezpjz=591.7661*ezpjz
      return
      end
c
c     ------------------------------------------------------
c
      complex function zint1(u)
c
c     the complex-valued integrand for the electric field
c     (Ez) produced by a unit source (Jz) when the observer
c     IS located on the source cell
c
c     (the singular part of the integrand is removed)
c
      complex h02
c
      arg=6.283185308*u
      if(arg.lt.0.00001) then
      zint1=(1.,0.)
      else
      zint1=h02(arg)+(0.,.6366197723)*alog(5.595404025*u)
      endif
      return
      end
c
c     ------------------------------------------------------
c
      complex function zint2(u)
c
c     the complex-valued integrand for the electric field
```

```
c      (Ez) produced by a unit source (Jz) when the observer
c      is NOT located on the source cell
c
       complex h02
       common xc,yc,com,som
c
       delx=xc-u*com
       dely=yc-u*som
       arg=6.283185308*sqrt((delx)**2+(dely)**2)
       zint2=h02(arg)
       return
      end
c
c      ----------------------------------------------------------
c
       complex function zrom1d(f,a,b,rerr,aerr,maxp,minp,r,ier)
c
c   returns estimate of the integral of a complex-valued function 'f'
c   over the interval (a,b) on the real axis -- Romberg integration
c
c      'f' is an external complex-valued function of the form 'f(x)'
c          where 'x' is real-valued
c          ('f' must be declared 'external' in the calling program)
c   'rerr' is the desired relative accuracy (rerr must be between
c          0.00001 and 1.0)
c   'aerr' is the absolute error desired -- if the estimate falls
c          below this amount, the algorithm will terminate
c   'maxp' specifies the maximum number of function evaluations
c          to be (1+2**(maxp-1))
c   'minp' specifies the minimum number of function evaluations
c          to be (1+2**(minp-1)) -- this may be necessary if the
c          integrand oscillates several times within the interval.
c      'r' is a complex array of length 'maxp' (workspace provided by
c          main routine). On return, contains the last row of the
c          Romberg extrapolation array.
c    'ier' on return is set to zero if accuracy met
c          on return is 1 if rerr is out of range
c          on return is 2 if maxp is less than 1 or minp .gt. maxp
c          on return is 3 if accuracy is not met
c
c   author: A. F. Peterson
c
c   last date revised: Dec 18, 1986
c
       complex r(maxp),f,zsum,rlast
       if((maxp.lt.1) .or. (minp.gt.maxp)) then
          ier=2
          zrom1d=(0.,0.)
```

```
            go to 30
            endif
         if((rerr.lt.0.00001).or.(rerr.gt.1.)) then
            ier=1
            zromld=(0.,0.)
            go to 30
            endif
         ier=0
         zsum=0.5*(f(a)+f(b))
         r(1)=zsum*(b-a)
         n=1
c
c  refine estimate of integral using trapezoid rule
c
         do 20 k=2,maxp
         rlast=r(1)
         del=(b-a)/float(n)
         x=a-0.5*del
         do 10 i=1,n
   10    zsum=zsum+f(x+del*float(i))
         n=n*2
         r(k)=0.5*del*zsum
c
c  update Romberg extrapolation array
c
         w=4.
         do 15 j=2,k
         r(k+1-j)=(w*r(k+2-j)-r(k+1-j))/(w-1)
         w=4.*w
   15    continue
c
         zromld=r(1)
         if(k.lt.minp) go to 20
c
c  check integral estimate:
c
c    if change in estimate is within desired range, or if the
c    estimate falls below the absolute error desired, return
c
         if(cabs(zromld).lt.aerr) return
         if(cabs((zromld-rlast)/zromld).lt.rerr) return
   20    continue
         ier=3
c
c  error
c
   30    write(*,*) 'error in zromld -- ier = ',ier
         return
```

```
      end
```

```
------------------------------------------------------------
Output (OUTFIL) based on Romberg quadrature with minimum
requested accuracy of rerr=0.001:


 angle of incidence is   0.0000000     degrees

 current density in magnitude - angle format

 1  8.6470426E-04   155.8690
 2  9.6364931E-04  -143.5081
 3  2.2039667E-03  -69.17345
 4  3.9260159E-03  -12.26043
 5  5.6881281E-03   27.53558
 6  6.4794831E-03   41.65346
 7  5.6881239E-03   27.53557
 8  3.9260169E-03  -12.26044
 9  2.2039663E-03  -69.17346
10  9.6364942E-04  -143.5081

 rcs in dB-wavelength
      angle              rcs

   0.0000000        2.951957
    30.00000        2.129026
    60.00000        0.055388
    90.00000       -1.814937
    120.0000       -2.238418
    150.0000       -2.033477
    180.0000       -1.929355
    210.0000       -2.033474
    240.0000       -2.238417
    270.0000       -1.814938
    300.0000        0.055385
    330.0000        2.129024
------------------------------------------------------------
```

C.3 IMPLEMENTATION 3: GENERALIZED GAUSSIAN QUADRATURE

In this example, subroutine EZPJZ is replaced by a routine that employs a generalized Gaussian quadrature rule incorporating logarithmic singularities (Section A.4).

```
      complex function ezpjz(xs,ys,xo,yo,ws,omegs)
c
c     negative scattered electric field (Ez) at location
c     (xo,yo) due to a unit source (Jz) of width ws and
c     orientation omegs at (xs,ys)
c
c     computed using 15-point lin/log Gaussian quadrature,
c     so no need to separately evaluate the logarithmic
c     singularity
c
      complex QGLL15,zint1,zint2
      common xc,yc,com,som
      external zint1,zint2
c
      wso2=0.5*ws
      wsot=0.1*ws
      arg=sqrt((xo-xs)**2+(yo-ys)**2)
      if(arg.lt.wsot) then
c
c        assume observer at center of source cell
c
         ezpjz=QGLL15(zint1,0.,wso2)
         ezpjz=ezpjz*(2.,0.)
c
         else
c
c        observer not on source cell
c
         xc=xo-xs
         yc=yo-ys
         com=cos(omegs)
         som=sin(omegs)
         ezpjz=QGLL15(zint2,-wso2,wso2)
         endif
      ezpjz=(591.7661,0.)*ezpjz
      return
      end
c
c     ------------------------------------------------------
c
      complex function zint1(u)
c
c     the complex-valued integrand for the electric field
c     (Ez) produced by a unit source (Jz) when the observer
c     IS located on the source cell
c
      complex h02
c
```

```
      arg=6.283185308*u
      zint1=h02(arg)
      return
      end
c
c     --------------------------------------------------------
c
      complex function zint2(u)
c
c     the complex-valued integrand for the electric field
c     (Ez) produced by a unit source (Jz) when the observer
c     is NOT located on the source cell
c
      complex h02
      common xc,yc,com,som
c
      delx=xc-u*com
      dely=yc-u*som
      arg=6.283185308*sqrt((delx)**2+(dely)**2)
      zint2=h02(arg)
      return
      end
C
C-----------------------------------------------------------------
C
        complex function QGLL15(zfun,a,b)
C
C       This function computes the integral of the complex function
C       'zfun' from a to b with 15-point Gaussian quadrature using
C       a mixture of polynomials and natural logarithms as basis
C       functions.
C       reference: Ma, Rokhlin, and Wandzura, "Generalized Gaussian
C       Quadrature Rules for Systems of Arbitrary Functions,"
C       Research Report, Yale University.
C
C       A. W. Mathis and A. F. Peterson, Feb 17, 1997
C
        complex asum,zfun
        real x(15), w(15)
        integer n
        external zfun
C
        data x/.105784548458629e-3,.156624383616782e-2,
     +          .759521890320709e-2,.228310673939862e-1,
     +          .523886301568200e-1,.100758685201213e0,
     +          .170740768849943e0,.262591206118993e0,
     +          .373536505184558e0,.497746358414533e0,
     +          .626789031392373e0,.750516103461408e0,
```

```
     +         .858255335207861e0,.940141291212346e0,
     +         .988401595986342e0/
      data w/.403217724648460e-3,.3062978434787e-2,
     +         .978421211876615e-2,.215587522255813e-1,
     +         .383230673708892e-1,.588981990263004e-1,
     +         .811170299392595e-1,.102122101972069e0,
     +         .118789059030401e0,.128210316446694e0,
     +         .128163327417093e0,.117489465888492e0,
     +         .963230185695904e-1,.661345398318934e-1,
     +         .296207140035355e-1/
c
      xlength = b-a
      asum = (0.,0.)
c
      do 10 n = 1,15
          asum = w(n)*zfun(xlength*x(n)+a)+asum
   10 continue
      QGLL15 = xlength*asum
      return
      end
```

```
------------------------------------------------------------
Output (OUTFIL) based on 15-point lin/log Gaussian quadrature:


 angle of incidence is    0.0000000     degrees

 current density in magnitude - angle format

 1  8.6470722E-04   155.8694
 2  9.6365029E-04  -143.5080
 3  2.2039628E-03  -69.17333
 4  3.9260103E-03  -12.26035
 5  5.6881257E-03   27.53558
 6  6.4794868E-03   41.65342
 7  5.6881206E-03   27.53556
 8  3.9260117E-03  -12.26037
 9  2.2039623E-03  -69.17331
10  9.6365076E-04  -143.5080

 rcs in dB-wavelength
      angle              rcs

   0.0000000        2.951955
    30.00000        2.129026
    60.00000        0.055388
```

```
90.00000       -1.814939
120.0000       -2.238419
150.0000       -2.033474
180.0000       -1.929351
210.0000       -2.033471
240.0000       -2.238417
270.0000       -1.814939
300.0000        0.055386
330.0000        2.129022
```

Additional Software Available via the Internet

As an additional aid to instruction, a number of FORTRAN routines are available via the Internet. At the time of this writing the codes include the following:

PTME1.F	Code from Appendix C, Section C.1
PTME2.F	Complete code from Appendix C, Section C.2
PTME3.F	Complete code from Appendix C, Section C.3
ETPJT.F	Subroutine implementing the source–field relation for the triangle basis function–pulse testing function approach associated with the TE formulation from Section 2.4
PDSUBS.F	Subroutines implementing the entire set of source–field relations described in Appendix B
TMTEBT.F	Source code for the first-order triangular-cell finite-element treatment of the scalar Helmholtz equation for a two-dimensional region terminated with the second-order Bayliss–Turkel RBC

A README file contains detailed information. Additional codes may be added to this set in the future.

The files are available on EMLIB, and can be accessed via internet browsers at the address http://emlib.jpl.nasa.gov/EMLIB/education.html.

Index

D

E

F

J

K

L

M

N

O

P

Q

R

S

About the Authors

Andrew F. Peterson received the B.S., M.S., and Ph.D. degrees in electrical engineering from the University of Illinois, Urbana–Champaign. Since 1989, he has been a member of the faculty of the School of Electrical and Computer Engineering at Georgia Institute of Technology, where he is an associate professor. He teaches electromagnetic field theory and computational electromagnetics, and conducts research in the development of computational techniques for electromagnetic scattering, microwave devices, and electronic packaging applications.

Dr. Peterson has served as an associate editor of the *IEEE Transactions on Antennas and Propagation*, as chairman of the Atlanta joint IEEE AP-S/MTT chapter, and as the general chair of the 1998 IEEE AP-S International Symposium and URSI/USNC Radio Science Meeting (Atlanta). He has also served for six years as a director of ACES and is a member of URSI Commission B, ASEE, and AAUP.

Scott L. Ray received his B.S. degree in electrical engineering from Michigan State University in 1979. He continued his education at the University of Illinois at Urbana–Champaign, where he received an M.S. degree in 1981 and a Ph.D. in 1984, both in electrical engineering. His academic research focused on electromagnetic applications. Initially, he worked on designing dielectric waveguide antennas for use with millimeter wave integrated circuits. Later, he became interested in numerical techniques and worked on efficient iterative methods for solving integral equation formulations of electromagnetic scattering problems.

In 1984, Dr. Ray joined the Lawrence Livermore National Laboratory, where he first worked in the Fields, Materials, and Plasma Modeling Group within the Engineering Directorate. There he developed an expertise in partial differential equation based time-domain techniques and worked on a variety of finite element/finite difference methods and applications. In 1989, he became project leader for the development of TSAR, a general-purpose FDTD code. In 1991, he moved to the Laboratory's Weapons Program as a Code Team Leader. In 1992, he joined DowElanco (now called Dow AgroSciences), with an initial assignment in the Global Ag Math Modeling and Analysis Group. He is currently the technical leader for R & D Applied Statistics and works in such diverse areas as remote sensing, insect population dynamics, and environmental modeling.

Raj Mittra is professor in the electrical engineering department and a Senior Research Scientist at the Applied Research Laboratory of the Pennsylvania State University. He is also the director of the Electromagnetic Communication Research Laboratory and is affiliated with the Communication and Space Sciences Laboratory of the Electrical Engineering Department. Prior to joining Penn State he was a professor in Electrical and Computer Engineering at the University of Illinois in Urbana–Champaign.

Dr. Mittra is a Life Fellow of the IEEE, a Past President of AP-S, and he has served as the editor of the *IEEE Transactions on Antennas and Propagation*. He won the Guggenheim Fellowship Award in 1965 and the IEEE Centennial Medal in 1984. He has been a Visiting Professor at Oxford University, England and at the Technical University on Denmark, Lyngby, Denmark. Currently, he serves as the North American editor of the journal AE. He is president of RM Associates, which is a consulting organization that provides services to industrial and governmental organizations, both in the U. S. and abroad.

Dr. Mittra's professional interests include the areas of computational electromagnetics, electromagnetic modeling and simulation of electronic packages, communication antenna design including GPS, broadband antennas, EMC analysis, radar scattering, frequency selective surfaces, microwave and millimeter wave integrated circuits, and satellite antennas. He has published over 450 journal papers and 25 books or book chapters on various topics related to electromagetics, antennas, microwaves and electronic packaging. For the last 15 years, he has directed as well as lectured in numerous short courses on Electronic Packaging and Computational Electromagnetics, both nationally and internationally.